THE DIATOMS

THE DIATOMS

BIOLOGY & MORPHOLOGY OF THE GENERA

F.E. ROUND
Department of Botany, University of Bristol

R.M. CRAWFORD
Department of Botany, University of Bristol

D.G. MANN
Department of Botany, University of Edinburgh

CAMBRIDGE
UNIVERSITY PRESS

CAMBRIDGE UNIVERSITY PRESS
Cambridge, New York, Melbourne, Madrid, Cape Town, Singapore, São Paulo

Cambridge University Press
The Edinburgh Building, Cambridge CB2 8RU, UK

Published in the United States of America by Cambridge University Press, New York

www.cambridge.org
Information on this title: www.cambridge.org/9780521363181

First published 1990
Reprinted 1992, 1996, 2000
This digitally printed version 2007

A catalogue record for this publication is available from the British Library

Library of Congress Cataloguing in Publication data
Round, F.E. (Frank Eric), 1927–
Diatoms.
Bibliography: p.
Includes index.
1. Diatoms. I. Crawford, R.M. II. Mann, D.G.
III. Title.
QK569.D54R68   1990   589.4′81   88-35276

ISBN 978-0-521-36318-1 hardback
ISBN 978-0-521-71469-3 paperback

For our families

Contents

Preface

This book was conceived in 1978, some twelve years after the purchase of the first commercial model of a scanning electron microscope had been made for Bristol University by a famous entomologist – the late Professor Howard Hinton – who kindly invited one of us (F.E.R.) to use the machine, soon to be followed by R.M.C. in 1969 and then D.G.M. in 1974. The early years were spent exploring numerous genera, discovering what later became known as rimoportulae and fultoportulae and generally revelling in the extraordinary variety of form and structure. By 1978 we considered we had sufficient information and photographs to begin a survey of the genera. We soon found that many genera were totally unrepresented in our collections and for some common diatoms we were lacking vital detail. In the intervening years we have still not been able to complete all the genera, but we hope that all of those commonly encountered are included.

The introduction will, we trust, provide a fairly comprehensive account of those aspects of diatom biology that must be familiar to anyone who seeks to understand or investigate the evolution and systematics of the group. We have deliberately made little reference to such subjects as the photosynthetic characteristics and physiology of diatoms, not because they are uninteresting or unimportant, but because they have little relevance to our main theme – the circumscription of genera and higher taxa. It may appear that we have devoted too much space to the cell cycle, sexual reproduction and ontogeny, but, aside from their intrinsic interest, these are aspects that must be considered by taxonomists if a natural classification is ever to be evolved. The bias towards wall structure in our illustrations and generic accounts should be regarded as a temporary expedient, not as a model for all future work, which must instead pay more attention to cytological, genetic and reproductive characters.

We have provided a system of classification with many new taxa described at the family level together with 17 new genera, but we have not attempted to discuss the problems of classification in great detail. In developing our classification and making the generic descriptions we have been greatly helped by having available the section of the Index Nominum Genericorum devoted to the diatoms and the compilers (Ross & Sims) have kindly allowed us to reproduce this as an appendix – it helps to define the problems and put our efforts into the context of the total complement of diatom genera. Many of these are undoubtedly synonyms, but it would be wise to check others by SEM. An even larger problem is that many genera are still rather disparate aggregates of clusters, each warranting generic status. We have begun to sort out some of these especially in the raphid and araphid sections but much more remains to be achieved, for instance, in *Amphora* and *'Caloneis'*. The recently established journal *Diatom Research* will, we hope, continue to be a major outlet for taxonomic revisions at all levels, besides dealing with other aspects of diatom biology and ecology.

It has been a deliberate policy not to put magnifications on the illustrations, partly because there are considerable problems in obtaining absolute magnifications unless the electron microscope is calibrated each time it is used but more importantly because diatoms vary enormously in size during the life cycle and it is not wise to place too much emphasis on size as a criterion in diatom systematics.

We have had immense help from our friends at the British Museum (Natural History) (P.A. Sims, T.B.B. Paddock and D.M. Williams), and from others too numerous to mention by name. Every draft and re-draft has been carefully typed by Gillian Lockett who has worked tirelessly on our often inadequate crossed-out, corrected and annotated versions, while in the latter stages of preparation we have had considerable assistance from Linda Medlin. We are also most grateful to R. Ross for checking some of our Latin descriptions.

We wish to acknowledge assistance from an NERC Grant (no. GR3/ 4877) to F.E.R. and R.M.C. and to SERC for a Grant (no. C/68484) to D.G.M. A Royal Society grant also enabled D.G.M. to acquire a photomicroscope.

Biology of diatoms

Preamble

In 1703 an English country gentleman looked with his simple microscope at roots of the pond-weed *Lemna* and '. . . saw adhering to them (and sometimes separate in the water) many pretty branches. compos'd of rectangular oblongs and exact squares'. His descriptions and diagrams. which probably refer to what we now call *Tabellaria flocculosa* (Fig.1a), are the first certain records of a diatom. We do not know the discoverer's name or where he worked; only that his paper was communicated to the Royal Society of London by a Mr C. (which could refer to any of six Fellows) and published in its Philosophical Transactions. He came to the conclusion that the rectangles and squares (which he saw to be 'made up of two parallelograms joyn'd longwise') were plants. which was a remarkable judgement considering that all he had for comparison were macroscopic bryophytes. vascular plants and seaweeds. none of which have anything like the precise geometry of diatoms. Although we have little doubt that the diatom reported by the gentleman was *Tabellaria*, this is not a diatom commonly associated with *Lemna* roots. Leeuwenhoek too probably saw diatoms (e.g. Leeuwenhoek, 1703), but he was not so good an illustrator as his anonymous contemporary and his descriptions of diatom-like organisms are too vague for us to be sure of their identity. Perhaps the next certain record of the group dates from 1753 in Baker's *Employment for the Microscope*. Here Baker describes amongst other things the 'Hair-like insect' – surely an *Oscillatoria* – and also the 'Oat-animal'. which grew in the same mud as the 'Hair-like insect'. and was probably *Craticula cuspidata* (Fig. 1b). By exercising some faith in this identification. it is possible to see in Baker's illustrations the two plastids, the central bridge of cytoplasm containing the nucleus. the two volutin granules. and the raphe-sternum. Some aspects of Baker's description remain enigmatic – the cylindrical fleshy parts supposedly thrust out by the animal. and the jerks and leaps by

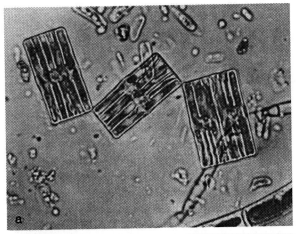

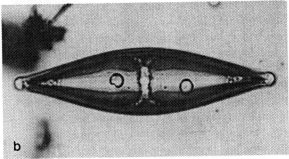

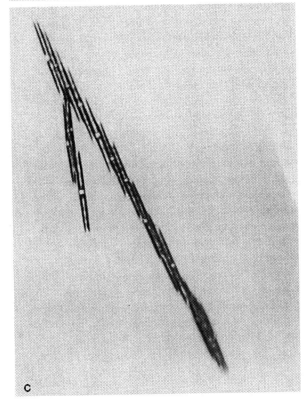

Fig. 1. Light micrographs of (*a*) *Tabellaria flocculosa*; (*b*) *Craticula cuspidata*; (*c*) *Bacillaria paxillifer*.

which it moved – but the 'Oat-animal' was almost certainly a diatom.

Various diatoms were described in the latter half of the 18th century and given Latin binomials. The work of O. F. Müller (e.g. 1783a, b, 1786) is worth a special mention because it was one of his species, *Vibrio paxillifer*, that served as the type of the first diatom genus – *Bacillaria* Gmelin (1791) (Fig. 1c). Müller included two other diatoms in *Vibrio*, one of which (*V. tripunctatus*) is said to be the type of *Navicula*. Müller's illustrations are difficult to interpret, however, and *Vibrio tripunctatus* might well have been a *Nitzschia*. To Müller, *V. paxillifer* and the other *Vibrio* species were animals: he called them 'animalcula infusoria', along with ciliates, amoebae, *Volvox*, dinoflagellates and others.

Indeed, the nature of diatoms was a matter of debate until nearly half-way through the 19th century. Motile, unicellular forms appeared to be animal, and were classified in the animal kingdom by such authors as Bory (1822), and Ehrenberg (e.g. 1838). The protoplast, with its plastids and granules, was interpreted as representing the internal organs of an animal, complete with digestive system. The macroscopic growths of tube-dwelling diatoms (Fig. 22a & p. 518a), on the other hand, fitted well with early naturalists' concepts of plants, as did various colonial forms with their sedentary habit. Thus among the *Confervae* described by Dillwyn (1809) and Smith & Sowerby (1790–1814) are several diatoms – some *Melosira* species, *Tabellaria flocculosa*, *Achnanthes*, *Rhabdonema*, and others. Kützing's monograph of 1844 can be said to mark the end of this early period of uncertainty; in it he treated all the diatoms as plants, whether they were unicellular or colonial, motile or non-motile. From then on almost everyone, though Ehrenberg was a notable exception, classified the diatoms as algae.

Between 1844 and 1900 diatom classification (and indeed many other aspects of diatom biology) progressed only as fast as the development of the microscope. To some extent these were complementary, since for many years diatoms were the preferred test objects for microscope lenses. This was also the golden age of expeditions and collecting, thus making it possible for workers like

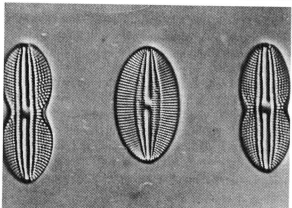

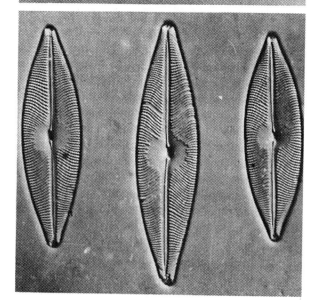

Fig. 2. (*a*) A Victorian slide of diatoms arranged in an intricate pattern. (*b*), (*c*) Modern arrangements of species.

Grunow and Cleve to monograph many diatom genera on a world-wide basis. Microscopy was very popular among better educated gentlemen of several European and North American nations (there seem to have been few lady diatomists). Microscopists vied with each other to resolve the finest details of the diatom valve and exchanged recent acquisitions; interest in diatoms soared to a level never attained before or since. With their almost indestructible shells, conveniently small size, great variety and beauty, diatoms were perfectly 'pre-adapted' to their role as the objects of scientific fashion (Fig. 2b, c).

The latter half of the 19th century has left us with superb collections of types (Fig. 3a) that must be the envy of anyone trying to unravel the nomenclature of less durable plants like the smaller Chlorophyta or many Chrysophyta. This period also bequeathed to us a soundly based classification that has remained stable for around 80 years and within which we and others are only now making relatively minor adjustments. But the fascination with the silica shell, which has continued in the 20th century, has diverted too much attention away from other aspects of diatom biology. True, workers have been interested from the very earliest times in the occurrence of diatoms in various habitats (e.g. Ehrenberg, 1838: Fig. 3b); how diatoms move; how they can be used to study past changes in sea level; or whether they can live at great depths in the oceans; and many papers have been published on diatom physiology, biochemistry, static and dynamic aspects of spindle structure, diatoms as stratigraphic markers, the role of diatoms in the geochemical cycling of various elements, especially silicon, and so on. But basic information on the structure and behaviour of the protoplast during the cell cycle, or on the occurrence and nature of sexuality, is scarce: the last review of plastid shape and arrangement, for instance, was published in 1908, by Heinzerling. Diatom genetics and genecology are only now being developed (e.g. Wood *et al.*, 1987).

In spite of the extensive data obtained from collections such as that from the round-the-world voyage of the *Challenger* (1873–6), the numerous Arctic/Antarctic expeditions and more recently the

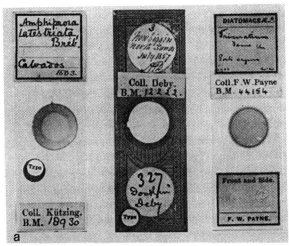

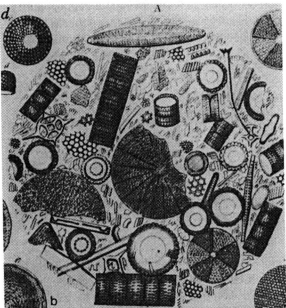

Fig. 3. (*a*) Examples of type slides. (*b*) Diatoms, etc. from marine sediment: an illustration from Ehrenberg's Mikrogeologie. (*c*) Diatomite mine at Lompoc, California. The white deposit is almost pure diatom.

co-operative Indian Ocean survey, we still have much to learn about the distribution of diatoms in the oceans. The role of diatoms in freshwaters has received great attention as the science of limnology has developed, yet only now are we beginning to appreciate the full impact of diatom growth on the complete physical, chemical and biological background of a body of freshwater.

Diatom remains in oceanic and freshwater sediments are valuable indicators of past environments, etc., and have recently been used to detect pH changes attributed to 'acid rain' (see Battarbee, 1986 and references quoted there). From the early days of Kolkwitz & Marsson (1908) diatoms have proved their worth in pollution studies, especially of river systems. Economic usage has developed because deposits of sedimentary diatoms (diatomites, Fig. 3c) are important in many industries owing to the fine structure and inert nature of the siliceous material.

The diatom cell

Diatoms are unicellular, eukaryotic, micro-organisms (Fig. 4a). They are pigmented and photosynthetic, although some at least can live heterotrophically in the dark if supplied with a suitable source of organic carbon. Less than ten species are obligately heterotrophic; all are colourless (apochlorotic) and belong to the genera *Nitzschia* or *Hantzschia* (e.g. Li & Volcani, 1987).

There is nothing particularly unusual about the diatom protoplast; it contains the same organelles as other eukaryotic algae (nucleus, dictyosomes, mitochondria, plastids, etc.). The hallmark of the diatom is its cell wall (Fig. 4b,c). This wall is highly differentiated and almost always heavily impregnated with silica ($SiO_2 \cdot nH_2O$); e.g. in *Aulacoseira italica* subsp. *subarctica*, 60% of the dry weight of the cell is silica (Lund, 1965). Si uptake and deposition involves less energy expenditure than formation of equivalent organic walls (Raven, 1983). The wall is multipartite, always consisting of two large, intricately sculptured units called valves, together with several thinner, linking structures termed girdle elements or cincture (Fig. 5a). The valves lie at each end of the cell and the girdle elements surround the region in between (Fig. 5b). The wall components, often loosely called the frustule (but see below) fit together very closely, so

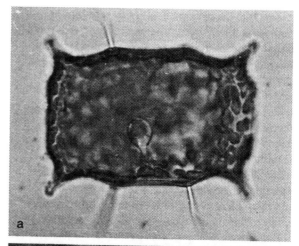

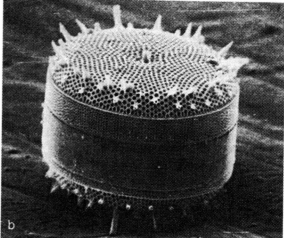

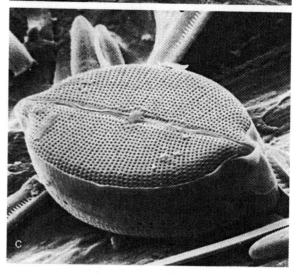

Fig. 4. (*a*) Live cell of *Odontella*. (*b*) *Thalassiosira* frustule, SEM; (*c*) *Mastogloia* frustule, SEM.

that flux of material across the wall must take place mainly via pores or slits in the components themselves; the cytoplasm is totally protected, in spite of the multiplicity of wall components (as many as 50 in some diatoms). Besides silica the wall also contains organic material, which forms a thin coating around the valves and girdle elements and often also a discrete layer beneath them. Probably all diatoms secrete polysaccharides, some of which may diffuse into the surrounding medium; others may remain as a gelled capsule around the cell or as threads, pads, stalks, etc. (Daniel *et al.*, 1987). The outer coating of organic matter often obscures detail when untreated (uncleaned) cells are observed in the scanning electron microscope. When divested of its organic components the frustule falls into its separate parts (Fig. 5a) and it is on this cleaned, fragmentary material that the majority of early observations of diatoms were made.

It is not surprising, perhaps, that the high degree of order shown in the structure of the cell wall is accompanied by great precision and order in the formation of the wall. Each frustule has one valve formed just after the last cell division and an older valve, which may have existed for several or many cell cycles since it was itself formed just after a cell division. The girdle elements also differ in age. One set of elements is associated with the older valve, and one with the newer. Each cell wall, then, consists of two halves. The older valve, together with the girdle elements (cingulum) associated with it, is called the epitheca, while the newer valve and its associated elements is the hypotheca (Figs 5b, 6). The two valves are termed epivalve and hypovalve, while the two sets of girdle elements are referred to as the epicingulum and hypocingulum (collectively the cincture or girdle). It is convenient to refer to the *siliceous* parts of the cell wall collectively as the frustule (literally a 'little bit'!).

New parts of the wall are formed within the protoplasts and are then added to the wall by a form of exocytosis. This means that new valves and new girdle elements are necessarily smaller than the

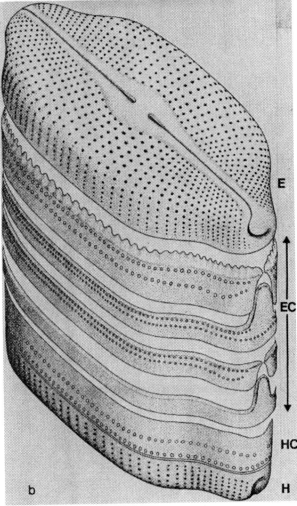

Fig. 5. (*a*) *Cyclotella meneghiniana*. Cleaned, showing frustule, thecae, valves (inside and outside views) and copulae. (*b*) Exploded view of the frustule of a naviculoid diatom, showing the epivalve (E), epicingulum containing 4 bands (EC), the incomplete hypocingulum (HC) and the hypovalve (H).

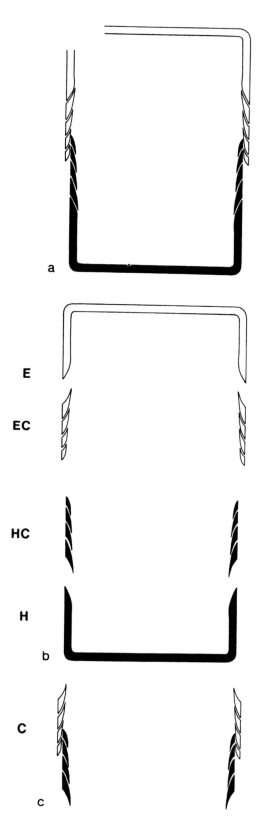

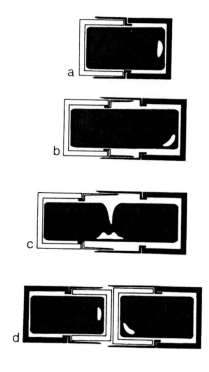

Fig. 6. The components of a frustule. (*a*) A frustule in section, consisting of 2 thecae: an epitheca (white) and a hypotheca (black). The upper surface of the valve is termed the valve face and the downturned part, the valve mantle. (*b*) Expanded frustule: the epitheca consists of the epivalve (E) and several copulae, which together constitute the epicingulum (EC); the hypotheca likewise consists of hypovalve (H) and hypocingulum (HC). (*c*) The 2 cingula together are collectively referred to as the cincture (C) or girdle.

Fig. 7. Growth of the cell during the cell cycle. (*a*) Newly released cell with complete epitheca and hypovalve (in many diatoms some components of the hypocingulum are already present at this stage). (*b*) Uniaxial growth of the cell by sliding apart of epi- and hypotheca, accompanied by addition of elements to the hypocingulum. (*c*) Completion of the thecae: mitosis and cytokinesis. (*d*) Daughter cells beginning to separate: new hypovalves formed beneath the girdle of the parent cell, i.e. beneath the epicingula of the daughter cells.

parts of the cell wall immediately outside them during their formation (Fig. 7). Thus, because the girdle is usually parallel-sided and cylindrical, the hypovalve of a cell is usually smaller than the epivalve. The hypotheca always underlaps the edge of the epitheca, like the two halves of a petri dish, and only in this sense are the petri dish-like illustrations in many texts correct representations; they usually omit the girdle elements. The valves, and to a lesser extent the girdle elements, are fairly rigid structures while surrounding the living cell, and so cell growth can only occur in *one* direction, as the epitheca and hypotheca move apart. As this occurs, more elements are added to the hypocingulum, to accommodate the further expansion of the cell. The hypocingulum is often incomplete until just before cell division, when the last girdle elements are laid down. When the cell divides, the hypocingulum of the parent cell becomes the epicingulum of one daughter cell, and the parental epicingulum becomes the epicingulum of the other daughter cell. Once incorporated in an epitheca, a cingulum will never be added to. It may conceivably lose elements (although we believe this is unusual) but new ones will not be formed to replace them. This is because the girdle elements are themselves formed in sequence, from the valve outward (i.e. towards the middle of the cell). Thus, since the edge of the epicingulum is underlain by the hypocingulum, no girdle element can be formed, exocytosed, and then transported between epicingulum and hypocingulum to take up a position at the edge of the epicingulum. The production of new frustule components within the confines of the parental cell wall usually leads to a decline in mean cell size, one of the best known features of diatom biology. Size is restored via an *auxospore* – a special cell which develops and expands in a highly controlled way before producing a new frustule. Auxospore formation is usually associated with sexual reproduction (see p. 86).

The mode of cell division has other implications

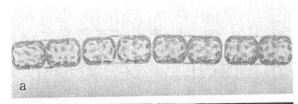

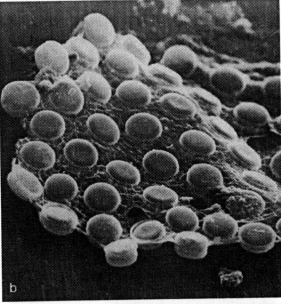

Fig. 8. (*a*) A filament of *Melosira* cells. (*b*) A colony of *Planktoniella*, in which the cells are united laterally by extracellular polysaccharide. (*c*) Valve view of a *Stauroneis* cell containing two large plate-like plastids. (*d*) *Odontella* showing discoid plastids. (*e*) *Melosira* showing dividing plastids.

for diatom biology. After cytokinesis, the new valves are formed back-to-back within the parental cell wall (Fig. 8a). Hence uniseriate filaments of cells are easily built up, especially if siliceous or organic linking structures are produced. Only in a few species do the cells move and re-arrange with their girdles adjacent, e.g. *Planktoniella muriformis* (Fig. 8b) and *Pinnularia cardinaliculus* (Round, 1988). Multiseriate filaments or more complicated thalli, on the other hand, cannot be formed since there is no means by which the diatom can alter its plane of division. In the diatoms, therefore, there are no levels of organisation corresponding to those exhibited by the larger Chlorophyta, Rhodophyta and Phaeophyta. In few other organisms (but compare the Foraminifera, Radiolaria) has the 'Bauplan' of the cell exerted such a profound restriction on the evolution of the group to a higher level of multicellular differentiation.

Except during sexual reproduction, and apparently in those diatoms that live endosymbiotically, e.g. in Foraminifera (Lee & Reimer, 1984), the protoplast is completely contained within the silicified cell wall. The most conspicuous components of the protoplast are the plastids (chromatophores), which are usually brown because the carotenoid pigments (principally β-carotene, diatoxanthin, diadinoxanthin and fucoxanthin; Goodwin, 1974) mask the colour of the chlorophylls (a, c_2, and c_1 or c_3; Stauber & Jeffrey, 1988). Damaged cells, or cells treated with acid, turn green as the carotenoids are destroyed and the chlorophyll is unmasked. In their structure, the plastids are like those of the Phaeophyta, Chrysophyta, Raphidophyta and Xanthophyta, but they are often larger and more elaborate (lobed, dissected, etc.).

Plastids, mitochondria, dictyosomes and other organelles are usually packed into a fairly thin peripheral layer of cytoplasm sandwiched between the cell wall and one or two large central vacuoles (Fig. 8c–e). The nucleus is quite easy to see and either is suspended in the centre of the cell by a bridge or strands of protoplasm or lies to one side of the cell, by the girdle or near one valve. It always changes position prior to cell division unless it already lies alongside the girdle during interphase (see p. 62). Oil bodies, reserve polysaccharides, e.g. chrysolaminarin (see Craigie, 1974) and other inclusions are often visible in the cytoplasm.

Collecting and studying diatoms

There are two principal habitats for diatoms: moist or submerged surfaces (benthic) and open water (planktonic). The most easily sampled micro-habitats are aquatic angiosperms or macro-algae (Fig. 9a,b), although some large algae, e.g. *Fucus* and *Laminaria*, rarely have a rich epiphytic flora, owing to perpetual secretion of extracellular products or to sloughing off of the outer cell layer. The shells, carapaces and skins of freshwater and marine animals also often support a diatom flora and recently diatoms have been discovered attached to the feathers of various seabirds (Croll & Holmes, 1982). The sediments of streams, lakes, salt marshes, sandy beaches, etc., all have floras of their own, which are often extremely rich in motile pennate species. These live on and in the surface layers of sediments and can easily be collected by removing the top 5 to 10 mm of mud and sand. Submerged sediments may be sampled by drawing a glass or plastic tube across the surface of the sediment and allowing it to fill with a mixture of sediment and water (Round, 1953). Deeply submerged sediments that cannot be reached using the simple tube method require the use of some sort of coring device. Direct observation of the silt or sand rarely reveals a rich flora. If, however, the sediment is allowed to settle in the dark and excess water poured off, the sediment mixed and put into dishes, covered with tissue paper (unglazed lens-cleaning tissue is suitable) and illuminated, diatoms and other epipelic algae will move up towards the light, out of the sample and into the tissue. They can then be harvested by removing the tissue or by letting them attach to cover slips placed on the tissue (Fig. 9c is an example). Cover slips have the advantage that the diatoms are easily examined but they are less efficient at removing all the cells; lens tissue has to be torn apart in water on the slide, or the algae removed from it by agitation, before the diatom flora can be examined properly. If two layers of lens tissue are used and only the upper layer removed, a very clean sample can be obtained. Similar techniques can be used for terrestrial soils. With care, quantitative estimates of algal biomass, either as cell densities or as chlorophyll concentrations, can be obtained using appropriate modifications of the above methods (e.g. Eaton & Moss, 1966; Round

& Hickman, 1971; Admiraal, 1984). Sediment communities (Fig. 10a) have recently been observed intact by freezing slivers of sediment and using low temperature SEM techniques (Paterson, 1986; Paterson *et al.*, 1986).

Sand grains (Fig. 10b) taken from fresh or salt waters sometimes have an attached diatom flora, which cannot be sampled by the techniques described so far because the cells have only limited or no power of movement. They can sometimes be removed from the sand by sonication after washing to remove the motile, epipelic algal flora. Neither this, nor the sampling of the epipelic flora, is ever likely to allow counting of more than a bare majority of the algae present. Furthermore, the sample is always likely to be biased, since every species will react differently to the techniques, e.g. some forms are extremely difficult to dislodge, such as the raphid valves of *Cocconeis*, while some motile species are sensitive to high illumination.

For plankton (Fig. 10c, d), water samples can be collected by means of bottles, tubes (Lund, 1949), or by more elaborate samplers where information is required concerning the microstratification of the water column and the planktonic communities it contains (e.g. Croome & Tyler, 1984). Some method of concentration of plankton is usually necessary before study, since even in quite nutrient rich lakes cell densities are often only around 10^4/ml. In early studies, fine plankton nets (mesh size 25–35 μm) were drawn slowly through sufficient lake or sea water to give a suitable concentration of cells. This method is very useful for collecting material for taxonomic purposes or for teaching, but cannot easily be made quantitative. Plankton counts are usually made with the inverted microscope or using counting chambers, after concentration of the original water sample by sedimentation (e.g. Lund, Kipling & LeCren, 1958; Utermöhl, 1958).

Concentrated samples, especially of plankton, often deteriorate quickly and so if the protoplast is to be studied, the samples must be examined as soon after collection as possible. Material should not be crowded into bottles. Observation of fresh material also reveals such features as plastid

Fig. 9. (*a*) Epiphytic diatoms attached to an aquatic moss. (*b*) Diatoms attached to a *Cladophora* cell. (*c*) Epipelic diatoms harvested from brackish sediment onto a coverslip.

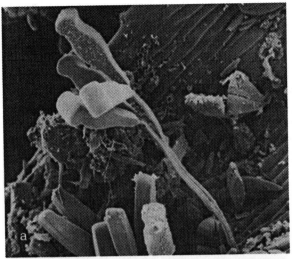

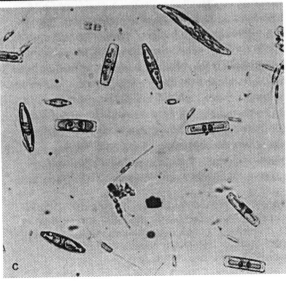

morphology, colony form, overall cell symmetry, mucilage sheaths (dilute Indian ink run under the cover slip often reveals surprising amounts of hitherto undetected mucilage), the presence of fine radiating fibrils of chitin or mucilage, and attachment structures. One can also study the movement of raphid forms. These are all of interest for their own sake, but there is also little doubt that such characters could be used more than they are now in systematics.

To investigate the siliceous exoskeleton in any detail, it is usually necessary to remove the protoplasm, but we have found it extremely valuable to study complete cells in the scanning electron microscope since from such samples the arrangement of girdle elements can be determined. Critical point drying is a valuable technique for delicate diatoms. Removal of the cytoplasm can be done in many ways, the choice depending on whether one wants complete frustules or separate valves and girdle bands. Bacteria and decay alone often remove most of the organic matter, and this leaves the frustules intact. Complete frustules can also be obtained by extracting the cells with 3:1 acetone:water (to remove pigments and lipids), followed by washing and then digestion with 2.5–5% pancreatin solution, phosphate-buffered to pH 7.6, for 3 days at 40°C (von Stosch, 1955; Reimann, 1960; Reimann & Lewin, 1964; von Stosch, Theil & Kowallik, 1973). Gentle treatment with hydrogen peroxide is also effective and can be made more rigorous if necessary by UV-treatment. Methods involving the use of concentrated oxidising acids (e.g. chromic acid, mixtures of concentrated nitric and sulphuric acids, sulphuric acid with addition of KNO$_3$ crystals) usually lead to partial or complete dissociation of the frustule. A common method is to boil the sample for a few minutes in a 1:1:1 mixture of sample, nitric and sulphuric acids. Another is to leave the sample in 70% nitric acid in an oven at 60°C for 24 hours. Recently we have been adding potassium permanganate to material, leaving overnight, then adding concentrated HCl and heating until colourless (Simonsen, 1974). Samples

Fig. 10. (*a*) Marine epipelic diatoms photographed *in situ* using cryo-techniques in the SEM (supplied by D. Paterson). (*b*) Marine epipsammic diatoms on a sand grain. (*c*) Freshwater planktonic diatoms. (*d*) Marine planktonic diatoms.

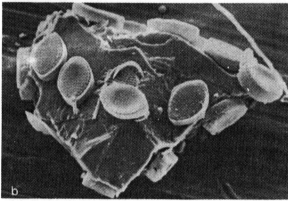

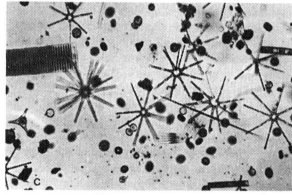

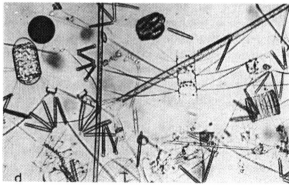

containing calcium carbonate must have this removed first (if sulphuric acid is to be used subsequently) by treatment with hydrochloric or nitric acid followed by washing with distilled water. Samples should not contain an excess of organic matter (e.g. leaves or macro-algae) since this may be converted to carbon particles rather than being completely oxidised – it is preferable to wash or scrape the diatoms from the substrata first. Following any acid treatment the sample must be washed thoroughly to remove all traces of acid, which would otherwise react with the mountant (see below). This can be done by repeated centrifugation and re-suspension in distilled water, or by repeated sedimentation. With sedimentation, however, it is most important to allow enough time for the smaller forms to fall to the bottom: even after several hours, some small forms may be left in suspension (Andrews, 1971). There is no universally best method and every diatomist has some preferred recipe; there is much scope for experimentation. Von Stosch (1970) has described methods for dealing with small quantities of delicate wall elements.

The cleaned sample can be used for both light and electron microscopy. For the former, a drop of suspension is usually dried onto a cover slip and mounted in a high refractive index medium (e.g. Pleurax, von Stosch, 1974; Hyrax; Naphrax, etc.), or, for certain special purposes (e.g. very thin valves, girdle bands, auxospore scales and bands), in air. The high or low refractive index is required to provide contrast between specimen and mountant. Since diatoms have a refractive index close to that of glass, conventional mountants, such as Canada balsam or Euparal, designed to give as perfect an optical match as possible between slide, mountant and cover slip, are almost useless. On occasion, however, with heavily silicified diatoms it may be desirable to reduce the contrast between mountant and specimen by drying the sample onto a microscope slide, covering with immersion oil and observing with immersion objectives (Crawford, 1979); it may sometimes be useful to shadow the specimens, e.g. with gold/palladium, before the immersion oil is added (Crawford, 1975). When material is prepared for light microscopy, the sample should not be dried onto the slide since it may prove impossible to focus upon the specimens

when using high aperture dry or immersion lenses. We have found No. 0 cover slips best.

For scanning electron microscopy (SEM), thoroughly washed samples can be dried directly onto the aluminium or brass specimen stubs, or onto cover slips, which have been attached to the metal stubs. If cover slips are used, care must be taken to ensure good electrical contact between the stub itself and the thin metal coating, usually of gold or gold/palladium, which is applied to cover slip and specimen by sputtering or by evaporation under vacuum. Useful fractures through valves or frustules can be made manually by applying pressure to mounted specimens, or by ultrasonication of suspensions of cleaned material. We usually photograph specimens at 45° tilt since this gives the optimum signal and has the added advantage of providing '3-dimensional' imaging, but to study the structure of the vela it is sometimes better to avoid tilt. The difference in image is shown in Fig. 11a, b.

Samples can also be prepared for both light and scanning electron microscopy by incineration, with or without prior chemical treatment to remove some of the organic material. Heating material on either a slide or cover glass in a muffle furnace to 500° to 600°C for 10 to 60 minutes is usually adequate. Washing with water or HCl to remove salts is often necessary, and perchloric acid has been used as a pre-treatment where some separation of the frustule elements is required (von Stosch, 1982). We strongly advise personal experimentation in the preparation of samples: there is always room for improvement.

For transmission electron microscopy (TEM) of cleaned valves or girdle bands, drops of a suspension of the material in distilled water are placed on carbon-stabilised, plastic-coated grids and the supernatant removed after the diatoms have had time to sediment onto the support films. Valve or girdle elements can be viewed directly, although all that can be seen is a 'silhouette' that usually reveals pore structure but little else. The indirect technique using carbon replicas is valuable where fine detail of surface structures beyond the resolution of the SEM is required (Fig. 11c, d).

Ultrathin sections can be prepared using standard methods (e.g. see Crawford, 1973; Edgar & Pickett-Heaps, 1982, 1983, 1984a; Pickett-Heaps, 1983), the main problem being the rigidity and hardness of the siliceous exoskeleton, which makes diamond knives essential. Differential staining and cytochemistry are also possible using standard techniques (e.g. Lauterborn, 1896; Geitler, 1927a, b; Edgar & Zavortink, 1983, Daniel *et al.*, 1987, with the light microscope; Edgar & Pickett-Heaps, 1982, using TEM).

Preservation of cleaned material poses few problems: it is sufficient to add alcohol at a concentration high enough to prevent the growth of micro-organisms. In our experience formalin is less satisfactory and appears to accelerate the erosion of fine pore occlusions. For this reason it should perhaps be avoided too as a preservative for uncleaned material.

A few remarks upon light microscopy may not be amiss. In the 19th century diatoms were test objects *par excellence* for microscope lenses. Microscopists developed many tricks – increasing the contrast between specimen and mountant using Realgar, or varieties of oblique illumination – to help them resolve fine detail. Such methods are not wholly superfluous today, even though electron microscopy is available, since many type specimens are irreversibly embalmed in Canada balsam. Furthermore, ecologists generally have to make their counts using the light microscope and are likely to have to do so for the forseeable future. But recent optical developments can make life considerably easier for anyone studying diatoms with the light microscope. Differential interference contrast optics are particularly useful for rendering fine striations and details of the valve visible, even in live cells. Phase contrast is preferred by some, but has the disadvantage that the thicker parts of the frustule create bright haloes which can obscure nearby detail. Drawing diatoms using a camera lucida of some sort is still an excellent way to record information and determine important characters such as valve striation density. Photomicrography of diatoms is not easy because of the depth of many specimens compared with the depth of focus of high resolution lenses. Many ecological and palaeontological surveys are virtually useless or have

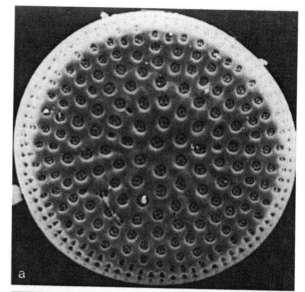

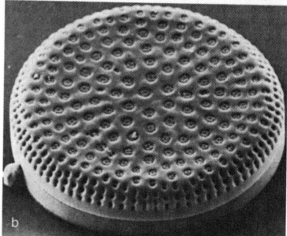

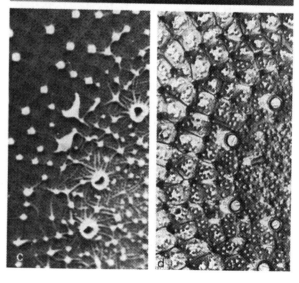

Fig. 11. (*a*) *Psammodiscus*, SEM, taken at 0° tilt. (*b*) The same valve taken at 45° tilt. (*c*), (*d*) Outer valve surface of *Melosira* taken in the SEM (*c*) and a carbon replica (*d*) of the same.

been severely handicapped by the poor quality of their photomicrographs – photomicrography should only be used if the worker has first-class instruments and photographic technique.

Culturing

Diatoms can be grown both on agar gels or in liquid. Many species are easy to grow and often contaminate 'rough' cultures of other algae. On the other hand, some species seem to have very specific requirements and are difficult to culture, whilst others reduce in size (see p. 82) but fail to form auxospores; thus no new large cells are produced and the cultures die out. This is probably the reason for the rarity of diatom species in general algal culture collections. There is much that is unknown concerning the growth cycles of diatoms in cultures, e.g. how can certain species continue through many generations, sometimes reducing in size and then enlarging again, apparently without auxospore formation. Most species in culture are 'weeds' and some are very strange, e.g. *Phaeodactylum*, which is widely cultured but hardly ever recorded in nature. In a taxonomic context, cultures are very valuable for monitoring morphological changes accompanying size reduction and for determining the upper size limit of cells formed after auxosporulation, though it should be remembered that races may exist within many species, with different size ranges (Geitler, 1968a). Keeping cultures until the cell size has reduced to the lower limit often results in small 'round' cells (even of normally linear species) which may be cultural artefacts since they are rarely seen in natural collections (e.g. Jaworski *et al.*, 1988). This may be misleading, however, since the small cells may form in nature but be diluted out of the population when most of it 'rejuvenates' by auxospore formation (see p. 86). If details of the small forms are indiscriminately incorporated into species descriptions, a false impression of the range of variation can easily be given.

The techniques of isolating and culturing diatoms are similar to those applied to other algae (see Stein, 1973; Guillard, 1975; Starr & Seikus, 1987, for many valuable tips, also Eppley, 1977), and will vary according to the objectives of the investigator. Before isolation, it may be necessary to grow up a mixed culture from material collected from nature

in order to obtain a large number of cells, or it may be possible to pick out a small number of cells from a natural population and grow them in a defined culture medium. From either source the initial inoculum can be streaked or spread thinly on agar plates (1–2% made up in water or culture medium), so that individual colonies grow up. The colonies can then be removed with a wire loop or micropipette, or by cutting out a block of agar, and transferred to new plates or liquid. Alternatively, single cells can be picked out by micropipette (mouth operated pipettes give the greatest control). These often fail to yield cultures since the medium often seems to need conditioning; this can be overcome by using a large inoculum, containing many cells. Thus it can be difficult to obtain clonal, as opposed to unialgal cultures, by the pipette method. The cultures can be maintained in various light/dark cycles and at various temperatures depending on individual requirements (e.g. diatoms isolated from sea ice may require low temperature and those from very acid waters, a low pH). Aeration will be required if large volumes of culture are needed. Bacterial contaminants can be eliminated by the addition of antibiotics to the media (e.g. mixtures of penicillin, streptomycin and chloramphenicol are often used in a 10:5:1 ratio, at concentrations between 20–500 mg/l penicillin). However, some diatoms may not survive in axenic culture since they may require substances produced by bacterial metabolism. Media may be defined (i.e. made up only from purified chemicals) or be based on natural freshwater or seawater and enriched with mixtures of chemicals or soil extract. Biphasic media (i.e. soil/water media) are not usually necessary for diatoms. A great variety of media have been used successfully to culture one or more diatom species. We list some of these in Table 1, since the formulae are widely scattered in the literature.

Silicon; occurrence, uptake and deposition

Silicon is, after oxygen, the most abundant element in the earth's crust, but most is unavailable to organisms, being bound in relatively insoluble forms in rocks and underwater deposits. In most of the moist or aquatic habitats where diatoms live, the principal form of silicon in solution is orthosilicic acid, $Si(OH)_4$ (Paasche, 1980). The uptake of this by

Table 1. Suggested media for culturing diatoms

Grundgloeodinium II solution (for freshwater forms in acid oligotrophic habitats)

KNO_3	50.55 mg l^{-1}
Na_2HPO_4	1.42 mg l^{-1}
$MgSO_4$	2.47 mg l^{-1}
$CaCl_2.2H_2O$	0.147 mg l^{-1}
$FeSO_4.7H_2O$	0.278 mg l^{-1}
SiO_2-Sol	12.01 mg l^{-1}
Na_2EDTA	0.744 mg l^{-1}

Freshwater 'WC' medium (Guillard & Lorenzen, 1972)

$CaCl_2.2H_2O$	36.76 mg l^{-1}
$MgSO_4.7H_2O$	36.97 mg l^{-1}
$NaHCO_3$	12.60 mg l^{-1}
K_2HPO_4	8.71 mg l^{-1}
$NaNO_3$	85.01 mg l^{-1}
$Na_2SiO_3.9H_2O$	28.42 mg l^{-1}

Traces

Na_2EDTA	4.36 mg l^{-1}
$FeCl_3.6H_2O$	3.15 mg l^{-1}
$CuSO_4.5H_2O$	0.01 mg l^{-1}
$ZnSO_4.7H_2O$	0.022 mg l^{-1}
$CoCl_2.6H_2O$	0.01 mg l^{-1}
$MnCl_2.4H_2O$	0.18 mg l^{-1}
$Na_2MoO_4.2H_2O$	0.006 mg l^{-1}
H_3BO_3	1.0 mg l^{-1}

Vitamins

Thiamin.HCl	0.1mg l^{-1}
Biotin	0.5µg l^{-1}
B_{12}	0.5µg l^{-1}

Marine enrichment medium 'f/2' (Guillard, 1975)

Major nutrients

$NaNO_3$	75 mg (883 µM)
$NaH_2PO_4.H_2O$	5 mg (36.3µM)
$Na_2SiO_3.9H_2O$	15–30 mg (1.5–3 mg Si or 54–107 µM)

Trace metals

$Na_2.EDTA$	4.36 mg *(ca* 11.7 µM)
$FeCl_3.6H_2O$	3.15 mg (0.65 mg Fe or *ca* 11.7 µM)
$CUSO_4.5H_2O$	0.01 mg (2.5 µg Cu or *ca* 0.04 µM)
$ZnSO_4.7H_2O$	0.022 mg (5 µg Zn or *ca* 0.08 µM)
$CoCl_2.6H_2O$	0.01 mg (2.5 µg Co or *ca* 0.05 µM)
$MnCl_2.4H_2O$	0.18 mg (0.05 mg Mn or *ca* 0.9 µM)
$Na_2MoO_4.2H_2O$	0.006 mg (2.5 µg Mo or *ca* 0.03 µM)

'f/2' (*contd*)

Vitamins

Thiamin.HCl	0.1 mg
Biotin	0.5 µg
B_{12}	0.5 µg

Seawater	to one litre

Chu's freshwater medium (as modified in Bold & Wynne, 1978)

To 1000 ml of autoclaved glass-distilled water, add aseptically 1 ml of each of the following autoclaved stock solutions:

	g/100 ml
$CaCl_2.2H_2O$	3.67
$MgSO_4.7H_2O$	3.69
$NaHCO_3$	1.26
K_2HPO_4	0.87
$NaNO_3$	8.50
$Na_2SiO_3.9H_2O$	2.84
ferric citrate solution	
micronutrient solution	

Ferric citrate solution. To 100 ml of glass-distilled water, add 3.35 g of citric acid and dissolve completely. Add 3.35 g ferric citrate and dissolve by autoclaving; keep sterile.

Micronutrient solution. To 1000 ml of glass-distilled water, add:

$CuSO_4.5H_2O$	19.6 mg
$ZnSO_4.7H_2O$	44.0 mg
$CoCl_2.6H_2O$	20.0 mg
$MnCl_2.4H_2O$	36.0 mg
$Na_2MoO_4.2H_2O$	12.6 mg
H_3BO_3	618.0 mg
Na_2EDTA	50.0 mg

The final solution of this medium is not autoclaved, because to do so causes a precipitate to form. All stock solutions and culture vessels are autoclaved beforehand.

diatoms may at times be so efficient that dissolved silicon may become undetectable in the water using standard techniques of water analysis. The populations built up during the time of plenty frequently decline rapidly, and other algae, not limited by silicate, become dominant. In some situations, e.g. in waters enriched by run-off from silica-rich sites or on sediments, there is a continual supply of silicic acid and the succession of species must be caused by factors other than limitation by silicate (see also Admiraal, 1984).

A number of plants take up and deposit silica in or around their cells and the reader is referred to Raven (1980, 1983) for a detailed discussion. In diatoms, silicic acid is not taken up at a constant rate throughout the cell cycle, being absorbed much more rapidly during the formation of new valves and girdle elements. Uptake across the plasmalemma is an active, carrier-mediated process obeying Michaelis–Menten kinetics (see Paasche, 1980; Sullivan & Volcani, 1981), and takes place against a 30 to 250-fold concentration gradient creating a pool of soluble silicate or organo-silicon compounds (Sullivan, 1986; Blank *et al.*, 1986). Silicon is found not only in the cytoplasm but also in the organelles, especially the mitochondria (Volcani, 1978, 1981). This is interesting in view of the close association of the mitochondria with the silica depositing apparatus (Pickett-Heaps, 1983; Edgar, 1980). Each wall element is formed inside its own vesicle, which is called a silica deposition vesicle (SDV) (see p. 72 and Fig. 12a, b). The membrane of this vesicle is termed the silicalemma.

Silicon is clearly necessary for diatoms because silica forms a major part of the cell wall (a range of 10–72% is quoted by Schmid, Borowitzka & Volcani, 1981). It is also involved in some way with a variety of metabolic processes. In the absence of silicate, cell division in *Cyclotella cryptica* ceases; protein, DNA, chlorophyll and carotenoid synthesis is inhibited; photosynthesis and glycolysis are reduced (Werner, 1978); and lipid synthesis is enhanced and altered (Taguchi *et al.*, 1987; Roessler, 1988). The inhibition of protein synthesis can apparently be traced to a deficiency of the amino acids synthesised from glutamate. This in turn can be related to a great decrease in the pool of α-ketoglutarate which, for example, occurs soon (1 to 2 hours) after the transfer of *Cyclotella cryptica* cells to silicate-deficient medium, and is the earliest (so far reported) metabolic reaction to silicate starvation in this species (Werner, 1978). Werner speculates that the activity of isocitrate dehydrogenase may be regulated indirectly or directly by silicate compounds. The effects of silicon starvation on the synthesis of various polypeptides have been studied in some detail by Volcani and his co-workers (e.g. Okita & Volcani, 1978, 1980; Sullivan & Volcani, 1973, 1976). There is evidence that silicon regulates gene expression (Okita & Volcani, 1980) and two of the four DNA polymerases in *Cylindrotheca fusiformis* are only synthesised if silicate is supplied (Okita & Volcani, 1978).

The soluble silica species within the cell are still not adequately known (Sullivan, 1986), but these are transported into the SDV, where polymerisation occurs to produce solid deposits of hydrated amorphous silica [$SiO_2 . nH_2O$]. Volcani (1981) lists three possible causes of polymerisation: (1) changes in pH or in $Si(OH)_4$ concentration; (2) binding to specific sites in the silicalemma by hydrogen bonding or ionic interaction; or (3) condensation on hydroxyl groups (see also Robinson & Sullivan, 1987). A model involving condensation by the hydroxyl groups of serine and threonine (in a glycine–threonine–serine rich template protein) has been developed by Hecky *et al.* (1973) but remains non-proven though feasible (Volcani, 1981). Pickett-Heaps, Tippit & Andreozzi (1979b) have suggested that fine strands of polysaccharide secreted into the lumen of the SDV by the silicalemma may act as the template for silicification, but again there is no proof. Interestingly, the silica appears to be deposited in different forms (Fig. 12c) and several workers (e.g. Schmid, 1976, 1979a; Schmid & Schulz, 1979; Li & Volcani, 1984; Edgar & Pickett-Heaps, 1984a) have shown that the silica can vary in appearance and structure during the formation of a particular frustule component. Schmid (1976) distinguishes between a growing zone of rough silica and a compacting zone of smooth completed surface.

When polymerised silica first appears within the SDV, the latter is small and occupies only a fraction of the area beneath the cleavage furrow; it may even appear before cleavage is complete (Schnepf, Deichgraber & Drebes, 1980; Li & Volcani, 1985a, b, c). As more silica is polymerised, the SDV grows.

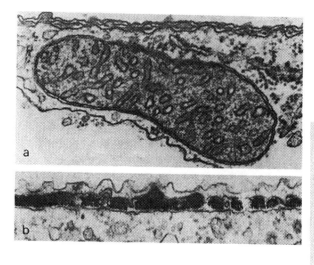

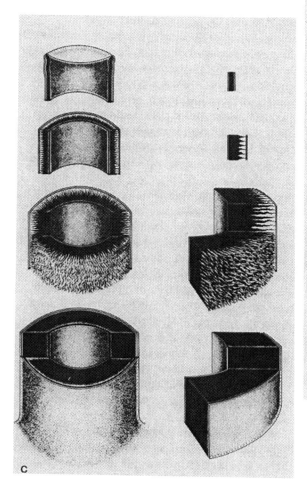

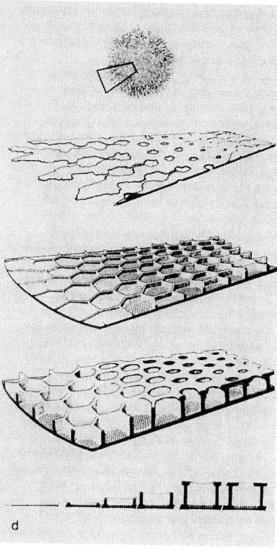

The radial or transapical ribs are initially formed within the SDV as fingers within a glove (Edgar & Pickett-Heaps, 1984a; Li & Volcani, 1985a) though the finger-like projections of the SDV soon become interconnected as the SDV expands. Everywhere, however, the silicalemma is intimately attached to the developing wall elements, suggesting strongly that it is in some way involved in the generation of pattern. In centric diatoms the SDV appears to begin as a small pancake-shaped structure, and from the central boss laid down within this, ribs radiate out as the SDV grows (Fig. 12d). Hence the basic construction of a centric valve is usually a ring of silica subtending a system of radial, branching ribs. This ring (Fig. 24), usually at the centre of the valve, has been termed the annulus (von Stosch, 1977) and is the primary silicification site or pattern centre for the developing valve in centric diatoms. Superimposed on the system of radial ribs, there may be extra valve layers, as in the chambered valves of *Coscinodiscus*, *Thalassiosira*, etc., or thickenings such as the angle 'irons' of *Ethmodiscus*. Schmid (1986) has suggested that vesicles and mitochondria might be involved in the moulding of chambers and alveoli. In pennate diatoms the SDV is at first a long, thin tube, and the pattern laid down as the SDV expands is pinnate. In araphid pennate diatoms, the development of the transapical ribs on either side of the sternum (Chiappino & Volcani, 1977; Mann, 1984b) takes place more or less symmetrically and synchronously. Raphid diatoms are more complicated since the first-formed longitudinal rib or sternum is flanked on one side by the raphe. The pattern of transapical ribs and

striae cannot develop on this side until the sternum has extended around the raphe slits, from both poles and the centre simultaneously (Fig. 13a). Where centrifugal and centripetal trains of deposition meet to complete the raphe-sternum a slight fault is produced in the stria pattern (Mann, 1981a), which was first noted by Voigt (1943) and is now termed a Voigt discontinuity. Further aspects of pattern formation are reviewed by Crawford & Schmid (1986).

The pores, or areolae, through the diatom valve are delimited by the formation of connecting bridges between the primary ribs (e.g. Chiappino & Volcani, 1977; Schmid, 1976, 1979a). This can sometimes be a very complicated process, as shown by Schmid & Volcani (1983) in *Coscinodiscus wailesii*. It is not yet clear how silica deposition is prevented in particular, well-defined areas of the SDV. In the case of the raphe slits (Fig. 13b–d), deposition is apparently prevented by the presence of 'raphe fibres' (Pickett-Heaps *et al.*, 1979b; Edgar & Pickett-Heaps, 1984a; Boyle *et al.*, 1984), while a similar function has been suggested for the rimoportula apparatus (Li & Volcani, 1984; Crawford & Schmid, 1986; see also p. 77).

Cell symmetry

No diatom cell is isodiametric, although the initial cells of some centric species approach this closely. As a result of the differentiation of frustule into valves and girdle, each diatom cell can present itself in two principal orientations or sets of orientations relative to the observer. These are termed valve view, when the valve is seen in 'face view' and the valvar plane is perpendicular to the line of sight; and girdle view, when the valves are seen in profile and the girdle in face view, the valvar plane being parallel to the line of sight (Fig. 14a, b). The proportions can vary greatly, e.g. in some *Coscinodiscus* species the cell is so flat that when seen in girdle view it looks like a coin edgeways-on whilst in other species it more closely resembles a drum.

Cells must usually be seen in valve view for confident identification, but in some genera, e.g. *Chaetoceros* and *Rhizosolenia*, it is the girdle view that is more useful. The valve generally offers more features, or at least more relatively easily determinable features than the girdle. Furthermore,

Fig. 12. (*a*) Early stage in valve deposition: the SDV is closely associated with the plasmalemma along the top of the photograph. (*b*) Later stage with silica deposits more obvious in the SDV. (*c*) Thickening of the central process of *Ditylum* as silica is deposited within the SDV. Different morphological varieties of silica are found during this process (from Li & Volcani, 1984). (*d*) Formation of the chambered valve of *Thalassiosira*. At first an irregular system of ribs is produced centrifugally within the SDV as it grows out beneath the plasmalemma from the centre of the cleavage furrow; the ribs become cross-linked to delimit pores within which fine cribra develop. Then an array of hexagonal chambers is formed towards the outside of the new valve on top of the basic rib framework (from Schmid *et al.*, 1981).

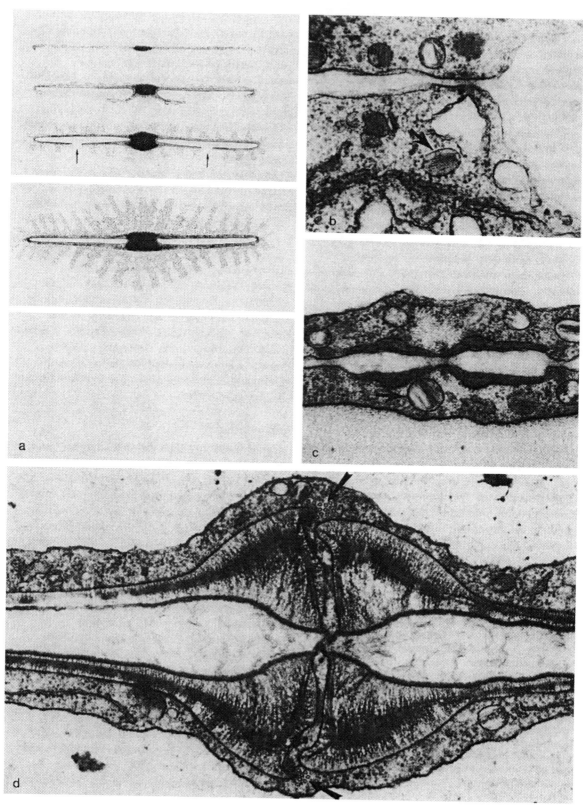

the vigorous cleaning procedures used by diatomists results in partial or complete separation of the frustular components, so that it is very difficult to establish which bands belong to which valve.

Valve outline varies greatly. Many centric diatoms are circular (Fig. 14c), while others, though retaining rotational symmetry, have two, three or more poles (Fig. 15d, e). Semi-circular forms also occur, e.g. *Hemidiscus* (Fig. 15c), *Triceratium* (p. 15e). Pennate diatoms, on the other hand, are almost always bipolar (Fig. 15f), although there are a few exceptions, mainly among the araphid diatoms, e.g. the tetrapolar *Perissonoë* (Fig. 15b), *Centronella* and some forms of *Staurosira*. There is also a curious raphid diatom *Phaeodactylum*, which has a triradiate form amongst its several morphotypes. Triradiate forms of normally bipolar pennate diatoms do occur but are rare, arising when environmental stress is present during auxosporulation. Schmid (1985) has induced triradiate *Achnanthes* by drug treatment. Small 'errors' in valve outline sometimes arise in nature and are maintained for many generations in the population. Sigmoid and arcuate valves also occur as abnormalities in some taxa, but are normal for some pennate genera (see the generic section).

The valve may have a uniform elevation (Fig. 17a) or a gradually sloping profile (Fig. 17b) but in many

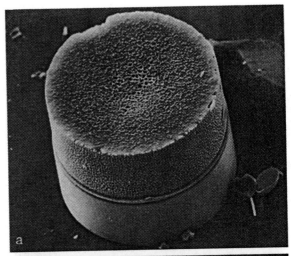

Fig. 13. (*a*) Valve formation in a raphid diatom. At first a narrow rib is produced, extending from one pole of the cell to the other. Then the primary rib recurves at the poles and fuses with 2 secondary ribs, which have extended out from the centre, to enclose the 2 raphe slits. When the ribs fuse, the pattern centre, in this case a raphe-sternum, is complete. Transapical elements develop laterally from the raphe-sternum, which themselves become cross-linked to delimit the valve pores (areolae) (from Chiappino & Volcani, 1977). (*b*)–(*d*) Stages in formation of the valve of *Craticula cuspidata*: (*b*) the early stage of development of the primary ribs (top right), sectioned across the forming raphes. Striated vesicles containing polysaccharide can be seen in (*b*) and (*c*) (arrows). (*d*) Late stage of development: silica apparently being deposited to either side of an initial central layer. Microtubules (arrowed) can be seen associated with dense fibrillar material at the inner openings of the raphe slits. (Micrographs kindly supplied by L. A. Edgar.)

Fig. 14. (*a*), (*b*) A cylindrical frustule with flat valve face and deep vertical mantles (*Melosira*). (*c*) A circular but concentrically and tangentially moulded valve (*Cyclotella*).

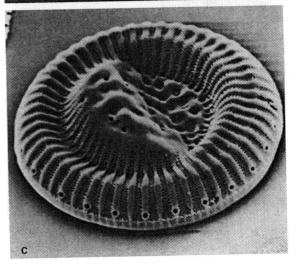

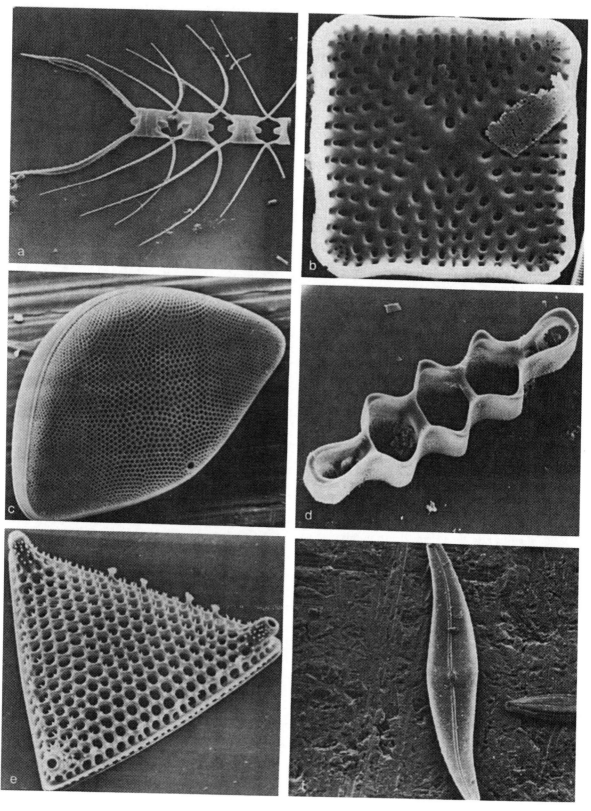

taxa, especially among the more complicated centric species, valve shape is considerably more complex than this, as a result of the presence of horns, setae and other projections (Fig. 16). In a few taxa the shape is complicated further by a torsion of the whole frustule – as though its ends had been held and twisted in opposite directions. Torsion occurs in various unrelated groups, e.g. *Cerataulus*, *Entomoneis* (Fig. 17g, h), some *Surirella* and *Cymatopleura* species, *Scolioneis* and *Scoliotropis*. In other genera, the frustule is curved in girdle view, with one valve being concave and the other convex, as in *Gephyria*, *Achnanthes*, *Rhoicosphenia* or *Campylodiscus* (Fig. 17i, j). *Campylodiscus* exhibits an extra complication in that the principal areas of the two valves (as shown for instance by the positions of the raphe endings) lie at right angles to each other (Fig. 17i, k).

Most diatoms have a relatively simple shape, the valves being radially or at least bilaterally symmetrical, parallel to each other and connected by a cylindrical girdle. But among attached diatoms, and in some free-living ones too, there are often more complex morphologies. In the simplest case, e.g. in many *Eunotia* species, all that is involved is an asymmetry of the valve about the longitudinal plane; the girdle remains cylindrical, i.e. it retains the shape of the valve outline, and the valves remain parallel to each other. Elsewhere, however, the cell exhibits a more pronounced dorsiventrality, in which the girdle is bent, with one side wider than the other, so that the valves are not parallel. In extreme cases, as in some *Amphora* and *Rhopalodia* species (Fig. 17c, d), the valves can lie at almost 180° to each other, the girdle being extremely reduced on the flattened ventral side of the cell. Where the girdle is unequally developed on the two sides of the cell, the valves too are usually asymmetrical, with a deep mantle on one side and a shallow one on the other. Diatoms with such complex shapes are difficult to study or photograph with the light microscope, and can look so different depending on their orientation that they can be very confusing to the inexperienced diatomist.

Fig. 15. (*a*) *Chaetoceros*; (*b*) *Perissonoë*; (*c*) *Hemidiscus*; (*d*) *Terpsinoë*; (*e*) *Triceratium*; (*f*) *Gyrosigma*.

Fig. 16. (*a*) *Eupodiscus*, showing elevations on the valve. (*b*) *Gonioceros*, showing setae on the apices of the valves and hairs on the valve faces. (*c*) *Cerataulus*, showing short spines on the valve face.

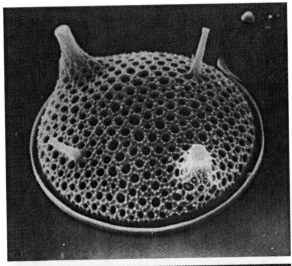

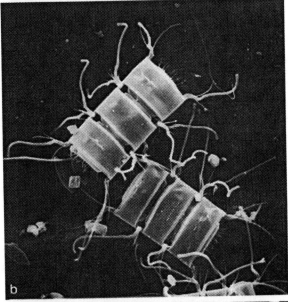

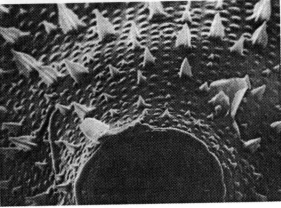

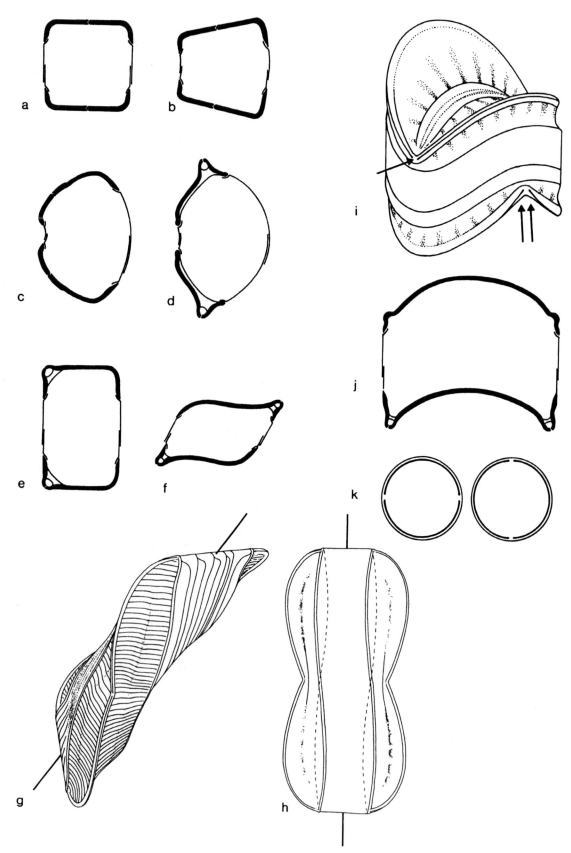

A few diatoms regularly produce cells of different symmetries during normal vegetative growth. The best examples are probably in the genus *Nitzschia*, though similar phenomena occur elsewhere, e.g. in *Rhizosolenia*. In many *Nitzschia* species the valves are asymmetrically developed, with an eccentrically placed raphe system on a high ridge or keel. Some species exist in only one form, in which the raphes of the two valves and their supporting ridges are diagonally opposite; in section, therefore, every cell forms a parallelogram (Fig.17f). Other species, however, have two types of cell – one as above ('nitzschioid' symmetry), and one in which the raphes are on the same side of the cell (Fig. 17e), so that the cell is trapezoidal ('hantzschioid' symmetry). These forms interconvert during vegetative division, according to patterns characteristic of the species, producing 1:2 or 1:1 ratios of nitzschioid: hantzschioid symmetry (Mann, 1980). It must be noted that one type of division, in which a hantzschioid cell gives rise to *two* hantzschioid daughters, never occurs in *Nitzschia*. All hantzschioid cells of *Nitzschia* give rise to one hantzschioid and on nitzschioid daughter. In the related genus *Hantzschia* (Fig. 17e), however, the type of division forbidden in *Nitzschia* ($H - H + H$) is the only one that occurs. In these genera it is unwise for the inexperienced to attempt identification on the basis of isolated valves alone. The only safe way to separate *Hantzschia* and *Nitzschia* is to look at the kinds of cell division and symmetry interconversions present.

Diatom shapes may be described in terms of a series of principal axes and planes (Fig.18). It should be noted that to be of any use in the *quantitative* description of valve shape (which will undoubtedly be an important component in future investigations of morphogenesis and shape change during size reduction; e.g. see Stoermer & Ladewski,1983) these planes and axes must be defined rigorously. It has been usual to equate the apical axis with the midline of the cell, so that in taxa such as *Cymbella* the apical axis is said to be curved. Strictly, however, an axis is 1-dimensional and a plane 2-dimensional, and so in the diagrams (Fig. 18) we have attempted to show how the principal axes and planes are to be understood for a variety of diatom shapes. There seems no reason to perpetuate the inexact system of 'axes' and 'planes' introduced by Müller (1895), attractively illustrated as these were.

Even though cell shape has long been a major criterion in classification at the highest levels and hence has been studied and documented intensively, detailed analysis of symmetry can still be rewarding. The recognition that the valves of raphid diatoms are basically not bilaterally symmetrical is leading to a better understanding of relationships in this group (Mann, 1981a, 1982a, 1983, 1984a, b, Mann & Stickle, 1988), while theoretical arguments have been used by Semina & Beklemishev (1981) in a discussion of the evolution of cell shape.

Fig. 17. Transapical sections through various raphid genera and drawings, showing different conformations of valves and girdle. (*a*) *Navicula*: bilateral symmetry – almost equal development of the two sides of valves and girdle; raphe central. (*b*) *Cymbella*: slight dorsiventrality of girdle and valve; valve faces not parallel; raphe slightly nearer the ventral margin of the valve. (*c*) *Amphora*: marked dorsiventrality of girdle and valve; raphe displaced towards the ventral margin. (*d*) *Rhopalodia*: very marked dorsiventrality of girdle and valves; raphe keeled, subtended by siliceous bridges (fibulae) and displaced towards the dorsal margin of the valve. (*e*) *Hantzschia*: slight or virtually no dorsiventrality of girdle but valves strongly asymmetrical; both raphes on the same side of the cell; both raphes subtended by fibulae, and displaced towards one margin (defined as proximal). (*f*) *Nitzschia*, nitzschioid cell: no dorsiventrality of girdle, but valves strongly asymmetrical; raphes on opposite sides of the cell; raphe subtended by fibulae, displaced towards one margin. (*g*),(*h*) *Entomoneis*, a diatom in which the frustule is twisted about its apical axis: (*g*) 3-dimensional drawing, (*h*) girdle view. (*i*)–(*k*) *Campylodiscus*: each valve has a circular outline (*k*) and is saddle-shaped (*i*), (*j*): the valves are orientated at right angles to each other (*k*, which shows the relative position of the raphe endings in the 2 valves, (arrows), the girdle being non-planar, as in *Entomoneis*).

Life form

Diatoms are basically unicellular. Although some species form colonies, there is little or no differentiation within these, the maximum extent of vegetative differentiation being the production of special separation valves in some chain-forming diatoms, e.g. *Aulacoseira granulata*. Formation of resting stages and auxospores occurs and produces different morphologies, but these should not be confused with the vegetative form. Indeed the wall

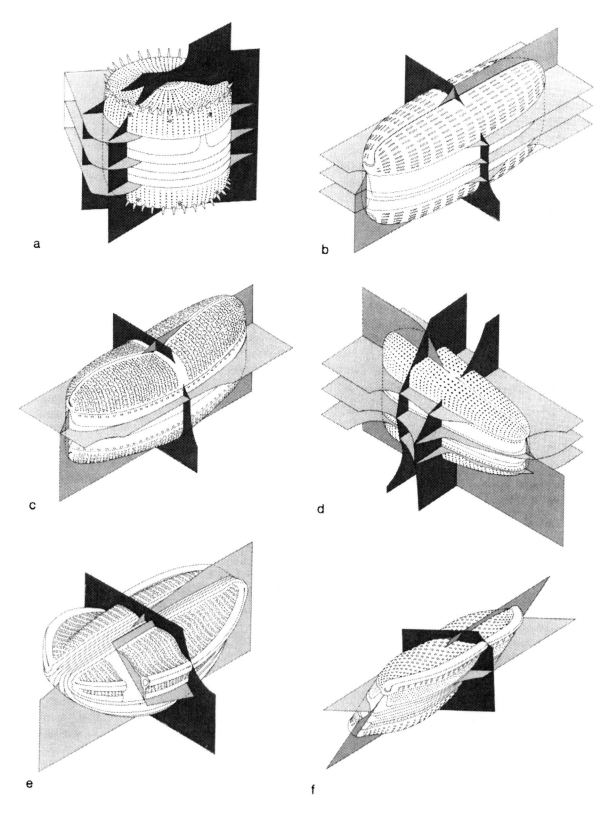

structure and mode of division shown by diatoms makes any organisation other than simple colonies or unicells very difficult to imagine. True multicellular development in plants requires walls that are sufficiently strong to withstand the deformations to which they are subject and yet remain sufficiently plastic to allow the cells to remain in close association with each other during growth. The diatom frustule does not have these properties. It does, however, have great compressional strength, which has allowed the evolution of giant yet rigid forms such as the pillbox-like cells of *Ethmodiscus rex* (up to 2 mm diameter) or the needle-like cells of *Thalassiothrix* (up to 4 mm long). Such forms tend to fragment when sedimenting to the bottom of the ocean, making the remains difficult to identify, e.g. in the belts of '*Ethmodiscus* ooze' found in some tropical regions (Fig. 19a). It appears that in only one group of modern diatoms – the Chaetocerotaceae – is there an actual fusion (Fig. 19b) between siliceous components of the different cells (Fryxell, 1978; Li & Volcani, 1985b; Stockwell & Hargraves, 1986). Many other diatoms form colonies, however, and these are

Fig. 18. Drawings of various genera, showing the positions and meaning of radial, apical, transapical and valvar planes. In each diagram the valvar planes are indicated by pale tone, apical or perapical planes by medium tone, and transapical (or radial) planes by dark tone. (*a*) *Stephanodiscus* (radial symmetry around the pervalvar axis, bilateral symmetry about the median valvar plane): 2 centric radial planes and 3 parallel valvar planes. (*b*) *Navicula* (bilateral symmetry about the apical, median transapical and median valvar planes): apical, median transapical and 3 valvar planes. (*c*) *Achnanthes* (bilateral symmetry about the apical and median transapical planes): apical, median transapical and median valvar planes. (*d*) *Gomphonema* (bilateral symmetry about the apical and median valvar planes): apical, 2 transapical (the median and one other), the median valvar and 2 valve-margin planes. (*e*) *Rhopalodia* (bilateral symmetry about the median transapical and median valvar planes): median transapical and median valvar planes, together with part of a valvar plane at one valve margin (note the vertical orientation of the girdle in this diagram). (*f*) *Psammodictyon* (bilateral symmetry about the transapical plane, 180° rotational symmetry about the apical axis): median transapical and median valvar planes, and an 'apical' plane about which the frustule is diagonally symmetrical.

Fig. 19. (*a*) Fragments of *Ethmodiscus* from deep ocean sediment. (*b*) *Chaetoceros* – fusion of setae. (*c*) *Aulacoseira* – linking spines.

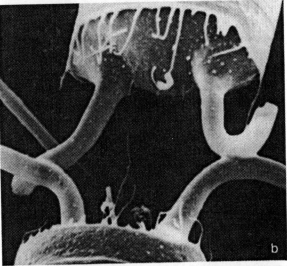

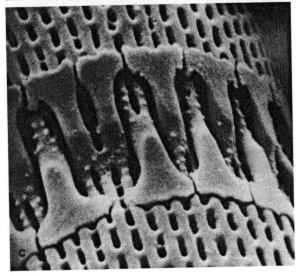

held together by interlocking siliceous spines,
processes or ridges; by pads or stalks of mucilage; or
by threads of polysaccharide (chitinous material?).
These are considered below.

(a) *Linkage by siliceous structures* (Fig. 19c). Many
diatoms form chains in which the cells are
juxtaposed by their valves and in some of these the
cells are linked by spines or other siliceous
structures. Since sibling valves are formed
simultaneously and back-to-back within the parent
frustule, it is possible for interlocking or
interdigitating links to be produced that cannot be
undone once fully formed. Such inextricable
connections are formed by some *Aulacoseira*,
Fragilaria and *Cymatosira* species. Impossibly long
chains do not occur, however, either because special
separation valves are formed (Fig. 20a), which lack
interlocking structures (as in *Aulacoseira granulata*,
Skeletonema costatum) or because cells within the
chain break or die and their girdles fragment (as in
Ellerbeckia arenaria).

The connecting structures may be solid spines
(*Fragilaria*, *Aulacoseira granulata* group) or may be
hollow extensions of tube processes, rimoportulae
(*Stephanopyxis*) or fultoportulae (*Skeletonema*) (see
p. 35ff). It has been suggested that the hollow
structures contain cytoplasm, so that the chains of
cells could be symplastic, but no evidence of this
has yet been produced. In other groups the links are
made via high 'horns' on the valve, e.g. in *Hemiaulus*
species (Ross, Sims & Hasle, 1977).

A few other diatoms can be included in this first
category, although there is probably more to their
mode of colony formation than a geometrical fit
between parts of the silica exoskeleton. In
Rhizosolenia (Fig. 20b), the tube-like extension of the
rimoportula and part of the valve fit into a groove
on the adjacent (sibling) valve, although it is quite
possible that mucilage is also involved in the
linkage. The motile chains of *Bacillaria paxillifer*
(p. 608) are held together by a combination of a
lock-and-key system of grooves formed by the raphe
and raphe-sternum, and raphe secretions. This
remarkable diatom and the few other species of the
genus have intrigued biologists ever since they were
first seen in the 18th century (see section on
motility). From the above it is clear that siliceous
links between cells have been produced

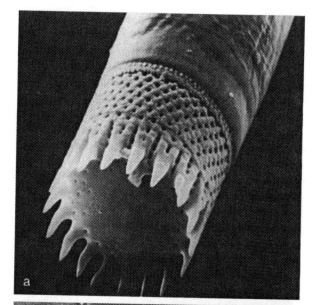

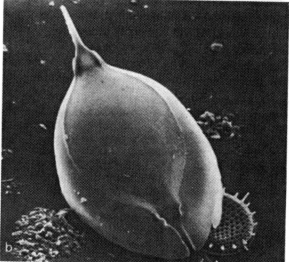

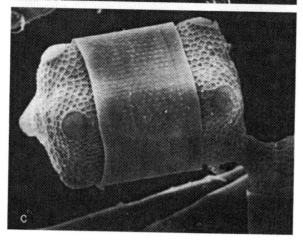

Fig. 20. (*a*) *Aulacoseira* – separation valve (cf. Fig. 19(*c*)).
(*b*) *Rhizosolenia*. (*c*) *Amphitetras*.

independently in a number of lineages which, presumably because of archetypal constraints on their evolution, have used a number of different valve structures to produce very similar linking elements.

(b) *Linkage by mucilage pads or stalks* (Figs 20c, 21). Chains of cells are also produced by diatoms without silica spines or processes. Here adhesion is presumably mediated by polysaccharide material, although the nature of this substance has rarely been established. Cells can be connected over the whole of the valve face, as in *Fragilariopsis*, or over only a part of it, as in the stepped chains of the *Pseudonitzschia* group of *Nitzschia*.

Many araphid pennate and some centric genera form colonies in which the cells are held together by small mucilage pads produced at the poles of the valves. Depending upon whether the two pads made by each cell are usually produced at the same pole or at opposite poles, the colonies formed will be predominantly star-like or predominantly zig-zag. Star-like colonies are formed by *Asterionella*, *Asterionellopsis* and *Nitzschia* amongst others, while zig-zags can be found in *Grammatophora* (Fig. 21a), *Striatella*, *Amphitetras* and several other genera. Some genera, e.g. *Diatoma*, *Tabellaria*, *Thalassiothrix* and *Thalassionema*, can form colonies of either type.

Elsewhere, the pads produced by the diatom cells lead to the formation of colonies, not because the diatoms are attached to one another, but because they form a common attachment to a substratum. Thus, some *Synedra* species form tufted colonies attached to larger plants or inorganic substratum by a small cushion of mucilage (Fig. 21b). It is a small step from the pads of *Synedra* to the stalks of diatoms such as *Licmophora*. In both cases, mucilage is produced from special groups of pores at one or either valve pole, but in *Licmophora* (and sometimes in *Synedra*, e.g. *S. pulchella*, W. Smith, 1853–6) the mucilage secretion is so copious that tall stalks are produced, so that the cells are bound together in groups that resemble fans (Fig. 21c).

In *Gomphonema, Cymbella, Rhoicosphenia, Brebissonia, Didymosphenia* and their allies, the mucilage stalks are again produced from areas of special pores at one or either pole. Here, however,

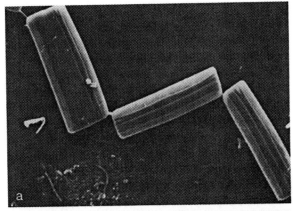

Fig. 21. (*a*) Formation of zig-zag colony: *Grammatophora*. (*b*) Cells radiating from a mucilage cushion: *Synedra*. (*c*) Fans of cells on branched mucilage stalks: *Licmophora*.

the stalks are not coherent but diverge, so that
branching colonies are produced, each branch point
marking a cell division. The lengths of the branches
are frequently unequal so that the branching is not
strictly dichotomous. This may reflect a delay in the
division of one daughter cell like that described by
Müller (1883, 1884) in *Ellerbeckia arenaria*, so that
given more-or-less equal rates of mucilage secretion,
one stalk will branch at a higher point than the
other. Or it could be that one daughter cell might
expand more rapidly than the other daughter cell, so
that its new stalk-secreting machinery, associated
with the hypovalve, is able to start functioning
earlier. Perhaps the observations may have no
significance except as a demonstration of the
principle that no two physical systems are ever quite
identical. In *Rhoicosphenia curvata* the mucilage
contribution of epivalve and hypovalve remain
separate, so that a faint line is visible along the stalk
when viewed from one side. The stalks of
Gomphonema and *Cymbella*, on the other hand, are
single structures with a certain amount of internal
differentiation (Daniel *et al.*, 1987).

 (c) *Inclusion of cells within mucilage tubes,
envelopes or sheaths* (Fig. 22). Tube-dwelling forms
(Fig. 22a) were some of the first diatoms to be noted
by naturalists, since they form macroscopic growths
that superficially resemble simple brown algae such
as *Ectocarpus*. A number of species in several genera,
some only distantly related, form tubes, e.g.
*Berkeleya, Amphipleura, Navicula, Gyrosigma,
Cymbella* and *Nitzschia*. Tube morphology varies. In
Berkeleya there can be tubes within tubes, whereas
elsewhere, e.g. in *Gyrosigma eximium*, the tube
usually contains only a single file of cells. Based on
the structure of the tubes and the cells they contain,
Cox (1977a) has constructed a key to the
identification of British tube-dwelling diatoms. The
tubes are sometimes invaded by 'foreign' diatoms,
e.g. *Nitzschia* or *Cylindrotheca* species, and an
endotubular green alga, *Chlorochytrium cohnii*,
sometimes occurs in the tubes of marine species.

 The tubes of *Berkeleya* (formerly *Amphipleura*)
rutilans consist mainly of xylans and mannans, with
a little mannose and possibly some protein (R.A.
Lewin, 1958). Cox (1981a) has speculated that the

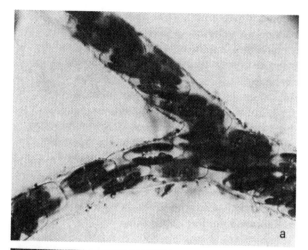

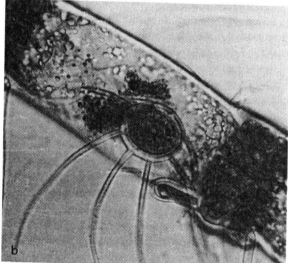

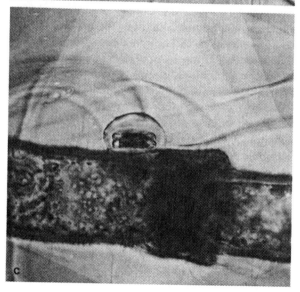

Fig. 22. (*a*) Tube-dwelling diatom: *Parlibellus*. (*b*),(*c*)
Cell in mucilage bubble with long mucilage extensions,
Mastogloia.

outer tubes of some species may be partially silicified, since they sometimes survive acid cleaning. This would be interesting if confirmed because it would presumably indicate extracellular silicification in a group where up to now silicification is known to take place only within intracellular vesicles (SDVs). How diatoms produce tubes is unknown. Cox (1981a) favours release of tube material between the girdle bands, but in view of Edgar & Pickett-Heaps (1982, 1983), observations of polysaccharide secretion through the raphe (and all tube-dwelling diatoms are raphid), this route cannot be ruled out for tube material. The diatoms move within the tubes, and so whatever the mechanism of mucilage secretion, it must exist alongside the raphe-associated motility mechanism.

Aggregation of cells in more-or-less amorphous masses of mucilage is found in some *Cyclotella*, *Thalassiosira* and *Berkeleya* species, whilst some *Mastogloia* species secrete bubbles of mucilage which attach the cells to sand grains (Round, unpublished) or plants (Stephens & Gibson, 1979b). In *Mastogloia*, thick strands of mucilage are also produced, secreted through the pores of the valvocopulae (Fig. 22b, c). Finally, in *Planktoniella muriformis* the cells are united laterally in a sheath of mucilage (Fig. 8b).

(d) *Colonies held together by threads of polysaccharide.* In the fourth group the cells are held together in loose chains by threads or filaments of polysaccharide. This occurs only in a few centric genera, e.g. *Coscinosira* and *Thalassiosira*. The filaments may well be composed of chitin, since similar fibres secreted by *Thalassiosira fluviatilis* and *Cyclotella cryptica* have been found to be a very pure, crystalline form of chitin (McLachlan, McInnes & Falk, 1965; Falk *et al.*, 1966; Blackwell, Parker & Rudall, 1967); they are both secreted through the fultoportulae.

Colonial organisation has been used to a limited extent in systematics. Several genera have traditionally been defined by the form of the colonies they produce, e.g. *Fragilaria*, and Round (1984b) has argued that this can be justified, in spite of recent statements to the contrary (e.g. Lange-Bertalot, 1980a). Furthermore, Hasle (1973a) has used colonial organisation as a diagnostic character for *Bacteriosira*. There is no reason why systematists should not use colony form, and the means diatoms employ to achieve this, as sources of taxonomic information. There is also no doubt that colonial organisation has been subject to strong selection in particular habitats, in relation to attachment, light and nutrient capture (e.g. by raising diatoms out of the still boundary layer around submerged plants and rocks, where they can grow only as fast as molecular diffusion replenishes limiting nutrients), the control of sinking rate, etc. However, it is obvious that the same types of colonial organisation have arisen quite independently in several lineages, e.g. the zig-zag chains in *Tabellaria* and *Grammatophora*. This is not surprising, since diatom colonies are, on the whole, very simple in their form, but we suspect that close scrutiny will reveal differences in, for example, polysaccharide composition or the exact way in which stalks and pads are secreted, between diatoms with the same colonial organisation but different phylogenies.

Valve structure

Valve structure has been studied more than any other aspect of the diatom cell. Diatom classification depends to a great extent upon the intricacies of pore structure, and the arrangement of the wall organelles (ocelli, portulae, raphes, horns, etc.). The anatomy of the valve, imperfectly visible using light microscopy, can now be studied in all its detail using transmission electron microscopy (of sections, whole mounts or replicas) or scanning electron microscopy. Thus it is not surprising that research in this field is still very active. In accordance with our belief that classification should be based on overall similarity, we recommend no slackening in the tempo of research into valve structure; but we would point out that other aspects of diatom structure and behaviour must also be incorporated into the classification. The outline of the valve has historically been a very important character in diatom classification, but whilst it is often a valuable aid to identification and a real indication of affinity, it is not invariably reliable and the totality of the valve structure must be considered. To give an extreme example, *Asteromphalus heptactis* is large and almost circular in outline whereas *A. hustedtii* and *A. petersenii* are small and almost 'naviculoid' and at one time were allocated to a genus classified far from *Asteromphalus*, yet the same complex system of valve chambers is present in both.

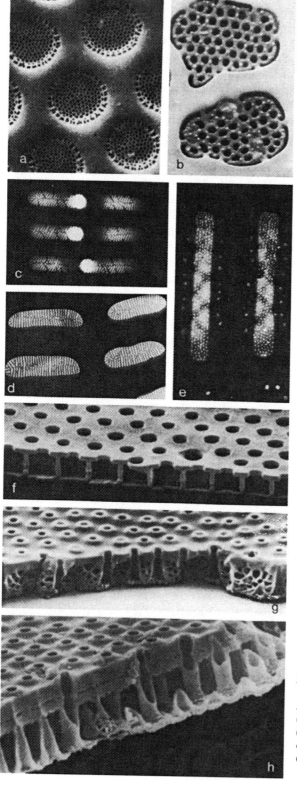

Valves are essentially systems of silica ribs, which grow out from a circular or elongate pattern centre during valve formation. The ribs thus formed are cross-linked by short lateral ribs to form a primary system or network of silica (see Cox & Ross (1980) for further terminology). Some valves are no more elaborate than this basic system of ribs (see p. 310). In other cases the valve structure is complicated by the superimposition of more silica during valve deposition, in the form of an extra framework parallel to or at right angles to the primary network (Fig. 23f). As a result, the valve can become chambered (loculate). The roofs, or in other taxa, the floors of the chambers represent the primary network, while the floors (or roofs), together with the walls of the chambers represent the superimposed, secondary system. The spaces between the ribs and struts of either system are usually occluded by more delicately structured flaps or porous plates of silica (Fig. 23a, b). These occlusions are termed vela (Ross & Sims, 1972) and have been classified as cribra (sing. cribrum), i.e. plates with pores; rotae (sing. rota), i.e. solid discs attached by spokes to the surrounding silica framework; and volae (sing. vola) i.e. flap-like outgrowths from the sides of the pores or from bars crossing the areolae. In addition, the pores of many raphid diatoms are occluded by a very delicate silica membrane called a hymen (pl. hymenes), perforated by round or elongate pores *c.* 5–10 nm in their shortest diameter (Fig. 23c–e; see also Mann, 1981c). The hymenes occlude pores in the primary network. In *Coscinodiscus*, on the other hand, the relatively coarse, sieve-like cribrum across the outer aperture of each chamber is occupying a pore in the secondary network.

Besides these occluded pores through the valve, usually termed areolae, there are often unoccluded pores. These may be single or grouped (as in ocelli, q.v.), or may be accompanied by tube-like or more complicated elaborations of the surrounding silica

Fig. 23. (*a*) Cribra of *Coscinodiscus*. (*b*) Cribra of *Isthmia*. (*c*),(*d*) Hymenate pore occlusions: *Cocconeis*. (*e*) Hymenate occlusion beneath a coarser (outer) cribrum: *Diploneis*. (*f*) Loculate valve: *Pleurosigma*.
(*g*) Loculate areolae with bullulate walls: *Actinocyclus*. (*h*) Tube-like areolae separated by a hypocaust in the outer half of the valve framework (below): *Coscinodiscus*.

(as in portules, q.v.). The presence and arrangements
of these structures are useful in taxonomy.

The silica deposited in either the primary or the
secondary system is usually solid (Fig. 23f).
Fractures through valves sometimes reveal, however,
that the framework of silica is foam-like or bubbly
(Fig. 23g: bullulae, cf. Ross & Sims, 1972), or even
contains large spaces (Fig. 23h). Sections of
Actinocyclus subtilis indicate that organic material is
absent from these spaces (Andersen *et al.*, 1980).
Very fine granules or a rugosity sometimes occur on
valves (especially valve mantles; rarely also on girdle
bands), this being particularly conspicuous on
diatoms living on or attached to coastal sand; we do
not know its significance.

The primary rib system or network is usually
fairly easily distinguished in fully formed valves,
even where massive secondary developments of
silica are present. It can be organised in two main
ways. In some diatoms the ribs radiate out from a
ring (Fig. 24a, d), or annulus; in others they extend
out from an often thicker, principal rib (bar), or
sternum (Fig. 24b, c). There are many modifications
of these two types of organisation and a few diatoms
do not seem to fit into this dichotomy at all. It is
this difference in the organisation of the primary rib
system that underlies the distinction, first made by
Schütt (1896), between centric and pennate diatoms
(see also Mann, 1984b). Centric diatoms commonly
have round or polygonal valves, while pennate
diatoms commonly have bipolar, elongate valves.
But round pennate diatoms can be found, and
bipolar centric diatoms too. It must be emphasised
that the basic distinction depends on the nature of
the pattern centre and not on the outline, although
the latter has become the loose interpretation.
Superimposed on either type of primary network, or
on any of the modifications of the basic types, there
may be a secondary network of some kind. Thus
chambered centric diatoms are found (e.g.

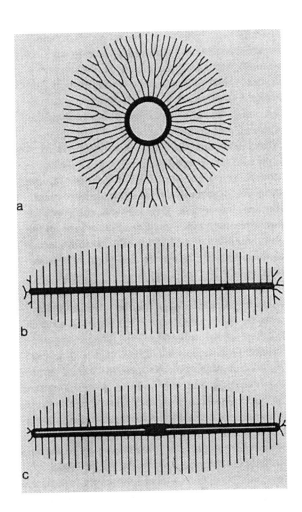

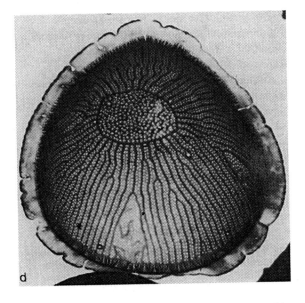

Fig. 24. (*a*)–(*c*) The three principal types of pattern in
diatoms: (*a*) centric, with ribs radiating from a ring
(*annulus*); (*b*) simple pennate, with ribs extending out
from both sides of a longitudinal element (*sternum*);
(*c*) raphid pennate, with ribs extending out from both
sides of a longitudinal element that contains one or two
raphe slits (*raphe-sternum*). (*d*) Centric pattern showing
the annulus and branching rib-system: *Melosira*,
incompletely formed valve.

Coscinodiscus, Thalassiosira) as well as chambered pennate diatoms (e.g. *Ardissonea, Pleurosigma, Psammodictyon*). In some genera the secondary network is incomplete, so that chambering is restricted to particular areas of the valve, e.g. adjacent to the raphe or along the junction of the valve face and mantle.

The centric annulus seems to vary only in its shape and size. It is often circular, but may be elliptical, and it may be rather ill-defined. The area within the annulus is always structurally distinct from the area outside. If it is porous, the pores are more irregularly distributed than elsewhere on the valve, and often of a different size. In some taxa, however, the area within the annulus contains no pores at all.

Most of the modifications of the two basic structural plans (*Baupläne*) concern the pennate group. In the simplest type, a simple sternum runs the length of the valve (Fig. 24b). This arrangement is found in many araphid forms, such as *Diatoma, Tabellaria* or *Asterionella*. The marine littoral genus *Rhaphoneis*, together with its close relative *Delphineis*, also has a simple axial sternum, but this group appears to have diversified giving rise to a number of genera with slightly different organisations. *Perissonoë*, separated from *Rhaphoneis* by Andrews & Stoelzel (1984) consists of tripolar or tetrapolar forms and these have a correspondingly triradiate or tetraradiate sternum. In *Psammodiscus* the sternum seems to have become reduced, so that the ribs it subtends, instead of being arranged like pinnae on a rachis, come to be radial. The pseudocentric organisation of *Psammodiscus* led earlier workers to classify it as a centric diatom in the genus *Coscinodiscus* (Round & Mann, 1980).

Another group of araphid pennate diatoms have a more complicated pattern centre. Here the primary network is subtended by two longitudinal sterna, which appear to interconnect around poles of the valves, forming an elongate ring (*Climacosphenia*) (Fig. 25a). Whether or not the sterna interconnect at both poles, there is a clear difference from the annulus of centric diatoms in that the sterna subtend ribs on both sides. The two centripetally

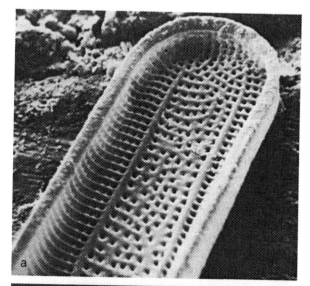

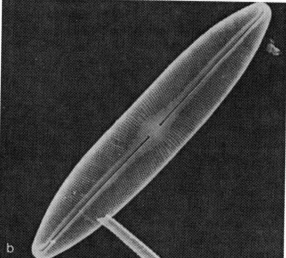

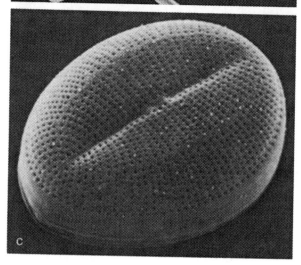

Fig. 25. (*a*) Lateral sterna in *Climacosphenia*. (*b*) Raphe-sternum in a naviculoid diatom. (*c*) Filled-in raphe-sternum in *Cocconeis* (in the pseudoraphe valve).

directed rib systems subtended by the sterna meet along the midline of the valve, where they form a fault line that mimics a simple sternum.

Most pennate diatoms have a more complicated pattern centre than any described so far. In these, an apparently simple sternum is accompanied by one or two raphe slits. The association of the raphe slits with the sternum is usually so close that the two form a compound structure, the raphe-sternum (Figs 24c, 25b). In one small group, however, (the Eunotiophycidae), the raphe slits are not so intimately associated with the sternum, and here the raphe slits often interrupt the ribs extending out from the sternum and subtend a rib system of their own. In most raphid diatoms, the raphe-sternum contains two raphe slits, which occupy almost its whole length. This type of pattern centre may be termed the 'naviculoid' raphe-sternum and it has given rise to three variants. In one, the raphe slits, though initiated, are filled in during valve formation in one of the two new valves (*Cocconeis, Achnanthes*) (Mann, 1982a; Boyle, Pickett-Heaps & Czarnecki, 1984; Mayama & Kobayasi, 1989). The mature (araphid) valve therefore appears to have a simple sternum like that present in many araphid pennate genera (Fig. 25c). Reduction rather than loss of the raphe slits occurs in a second group, which includes *Berkeleya* and *Amphipleura*, while in several 'advanced' raphid diatoms, e.g. some members of the Bacillariales, there is only one raphe slit occupying the whole length of the raphe-sternum. These variants are considered in greater detail by Mann (1984b) and Boyle, Pickett-Heaps & Czarnecki (1984).

A glance at the section on genera will show just how variable the valve shape is and we will not elaborate further on this here. Some valves are hemispherical or smoothly curved, with no distinction between a marginal area and the remainder of the valve. Round & Crawford (1981) have suggested that such shapes are primitive, the natural consequence of the evolution of the diatoms from a spherical or ellipsoidal scaly ancestor. Other

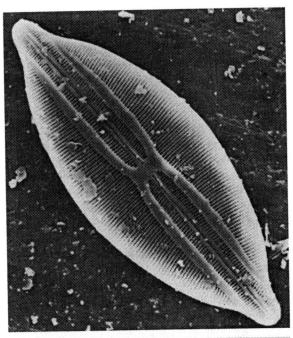

a

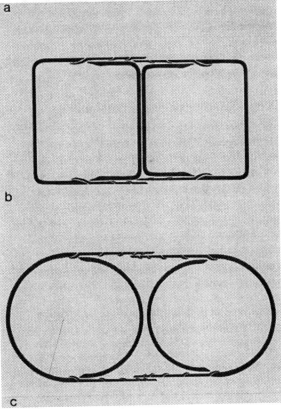

b

c

Fig. 26. (*a*) Lateral sterna: *Lyrella.* (*b*) Interactive division: the protoplasts of the daughter cells remain tightly appressed after cytokinesis, so that sibling valves form against each other. (*c*) Non-interactive division: the protoplasts of the daughter cells retract leaving a space between the sibling valves.

33

diatoms have valves that are clearly differentiated into a flattish top or valve face and downturned side or mantle (Fig. 14a, b). The junction between valve face and mantle is sometimes abrupt, sometimes curved (Fig. 25c), and in colonial species often bears rows of spines which serve to link adjacent cells (Figs, 19c, 20a).

The edge of the valve is usually formed by a continuous strip of silica. This, like the plain or hyaline areas present elsewhere on the valve in some forms, e.g. the lyre-shaped hyaline area in *Lyrella* (Fig. 26a) seems to be formed as a result of controlled in-filling between the primary rib system. The edge of the valve mantle may be modified in a subtle manner to enable the valvocopula to attach. In a few pennate genera, the edge is extended as a short flap, which in girdle view in the light microscope appears as a small ingrowth of the margin, usually termed a pseudoseptum (see p. 471i). In others this development of the margin extends all around the valve, forming a distinct diaphragm as in *Isthmia* (see p. 252). Apart from these internal developments the only notable modification of the valve edge known to us is in *Thalassiosira scotica*, where a projection occurs adjacent to a break in the first girdle band or valvocopula.

Complementarity and heterovalvy

After vegetative cytoplasmic division (see p. 70) the daughter protoplasts may be tightly appressed or may retract from each other. These two types of behaviour have been termed interactive and non-interactive division respectively (Mann, 1984b). Since the new valves are formed in vesicles lying immediately beneath the plasmalemma on the faces of the recently completed cleavage furrow, valves formed during interactive division must have complementary shapes, whereas in non-interactive division, virtually any valve shape is possible (Fig.

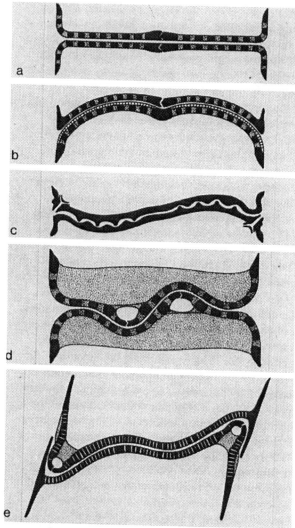

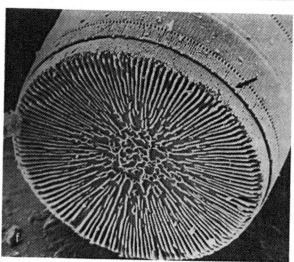

Fig. 27. (*a*)–(*e*) Demonstration of the principle of complementarity; sections through sibling valves in diatoms showing interactive division: (*a*) *Navicula*: planar valve faces; (*b*) *Cocconeis*: heterovalvy, with one convex and one concave valve; (*c*) *Actinoptychus*: radial sectoring of the valves allows concave portions of one valve to complement convex portions of the other, without heterovalvy; (*d*),(*e*) *Denticula* and *Tryblionella*: bilateral asymmetry and opposite orientation of sibling valves allows the development of complex ridge-and-furrow systems. (*f*) Stepped mantle (arrows): *Ellerbeckia*.

26b, c). Most of the diatoms exhibiting interactive division have simple, planar valve faces and hence will fulfil the requirement for complementarity (Fig. 27a). In other genera the valve faces are non-planar but each elevation on one valve is complemented by a hollow in the other (Fig. 27b-e). In the centric diatom *Actinoptychus* elevated radial sectors alternate with depressed sectors, so that the radially symmetrical valves are all alike (Fig. 27c). Similar valves (isovalvy) are also produced in the pennate genera *Plagiotropis*, *Denticula* (Fig. 27d), *Nitzschia* and *Tryblionella* (Fig. 27e). Here the shape of the valve face is rotationally symmetrical about the apical plane. The valve shape is not controlled by shape and orientation of the cleavage furrow: any elevations, projections, etc. on the valve develop after cleavage, while the new valves are being deposited.

Sometimes, however, interactive division involves the formation of dissimilar valves. Thus, in *Achnanthes*, *Cocconeis*, *Gephyria* and others, one valve is concave while its sibling is correspondingly convex. In the genera mentioned, the morphological heterovalvy is accompanied by a structural heterovalvy. In *Cocconeis*, for instance, the concave valve bears a raphe-system, while the convex valve does not (Fig. 27b). *Stephanodiscus* also exhibits morphological heterovalvy, with the production of convex and concave valves, but here there is no structural difference between the two types. Where interactive division occurs with the production of non-planar valves, it is simplest to assume that each pair of complementary features (each hump on one valve and its corresponding hollow on the sibling valve) is produced by a single influence present during valve formation. There is evidence, however, that this application of Occam's razor is not always justified, e.g. in *Stephanodiscus*, Round (1982b) has shown that sibling valves occasionally have the same rather than complementary shapes.

During interactive division the plasmalemma and the associated silica deposition vesicle of each daughter cell are tightly pressed not only against the sibling cell, but also against the frustule of the parent cell, more particularly against its girdle. Consequently, in heavily silicified diatoms, it is common to find that the valve mantle of one daughter cell is 'stepped' (Fig. 27f), the step marking where the valve has been deformed against the edge of the parental hypocingulum (Crawford, 1981).

It is difficult to imagine a function for heterovalvy in solitary planktonic diatoms such as *Stephanodiscus*, unless it is a device to increase surface area. In *Cocconeis* the heterovalvy is clearly associated in some way with the attached existence of the cells. *Cocconeis* cells are always glued to rocks or plants by their concave, raphid valves. Even so, the selection pressures that have resulted in the loss of one raphe-system remain obscure.

There is a more obvious function for the heterovalvy that occurs in some filamentous diatoms, e.g. *Aulacoseira granulata* (see Davey & Crawford, 1986). Here the filaments are held together by linking structures which, once formed, cannot disengage. Filament length can be restricted, however, by the formation of special separation valves, which lack the linking spines (Fig. 20a – spines straight and therefore not linking – contrast (Fig. 19c). Since separation valves are not formed at every division only some of the cells are heterovalvar and clearly these will come to lie at the ends of the filaments.

Portules

There are two major types of portule (Ross & Sims, 1972; Hasle, 1972c): fultoportulae, which are confined to the centric order Thalassiosirales, and rimoportulae, which are widespread. The fultoportula consists of a tube, which penetrates the silica framework (Fig. 28a, c) and is supported internally by two or more buttresses (hence the term fultoportula and its alternative, strutted process). The tube, which may or may not project above the outer surface of the valve, almost always has a simple external opening. Inside, the structure is more complex, since not only is there the internal aperture of the tube but also two to five satellite holes or slits (Fig. 28b), which open into passages connecting the inside of the cell with the central tube (Fig. 28c). It is possible to recognise at least three distinct types of fultoportulae. One has simple buttresses and apertures (Fig. 29a); in another, the satellite passages are extended into tubes, which run up alongside the central tube (Fig. 29b); while in a third group, lobed strips of silica descend into the apertures (Fig. 29c). The functioning of this organelle has not yet been fully elucidated, but in several species of *Cyclotella* and *Thalassiosira*, the fultoportula is clearly involved in the secretion of β-chitin fibrils (Herth, 1978, 1979; Herth & Barthlott, 1979). Beneath the portule is a conical invagination of the plasmalemma at the bottom of which the membrane is specialised, bearing a dense, structured coat on its cytoplasmic surface (Fig. 28d). The β-chitin, which is almost pure and extremely crystalline (Falk *et al.*, 1966; Blackwell, 1969; Blackwell *et al.*, 1967; Dweltz *et al.*, 1968; Herth & Zugenmaier, 1977), is apparently synthesised in the conical invaginations. Herth (1979) suggests either that the chitin synthetases are themselves transmembrane proteins or that they are fixed in the external part of the plasmalemma by transmembrane proteins from the structured coat. Of course, the fultoportula may be the point of exit for other compounds besides chitin, and taxa may vary in what they extrude. The satellite pores are as yet unexplained. Schmid (1984a) has suggested that the fultoportulae may also act as anchorage points for the protoplast especially during morphogenesis. As might be expected for fibril-secreting organelles, the fultoportulae are usually positioned around the valve mantle, but there may be single, scattered or clustered fultoportulae on the valve face as well or instead. Those on the valve face can be important in colony formation (see p. 132).

The rimoportula is morphologically simpler than the fultoportula since there are no satellite pores. It consists of a tube which opens to the inside of the

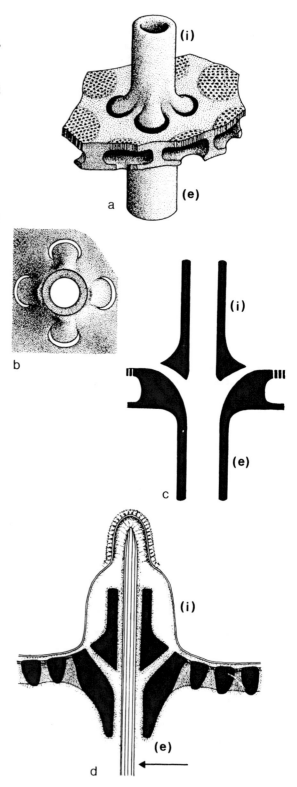

Fig. 28. The fultoportula. (*a*)–(*c*) Siliceous elements: (*a*) 3-dimensional reconstruction, viewed from the inside. Note the central tube (here extending out from the valve as well as in) and the satellite pores; (*b*) view along the axis of the fultoportula from the inside, showing the central tube and 4 satellite pores with slit-like apertures; (*c*) longitudinal section through fultoportula: the satellite pores connect with the lumen of the central tube. (*d*) Longitudinal section of fultoportula and underlying specialised parts of the plasmalemma. Beneath the inner tube of the fultoportula, the plasmalemma is invaginated and forms a pit lined by densely staining particles, which appear to represent the chitin-synthesising apparatus. A chitin thread (arrow) extends from here through the central tube to the exterior. (i) denotes interior of cell and (e) exterior.

Fig. 29. (*a*)–(*c*) Fultoportulae in (*a*) *Thalassiosira*, (*b*) *Stephanodiscus*, (*c*) *Thalassiosira*. (*d*),-(*f*) Rimoportulae in (*d*) *Actinocyclus*, (*e*) *Pleurosira*, (*f*) *Rhabdonema*.

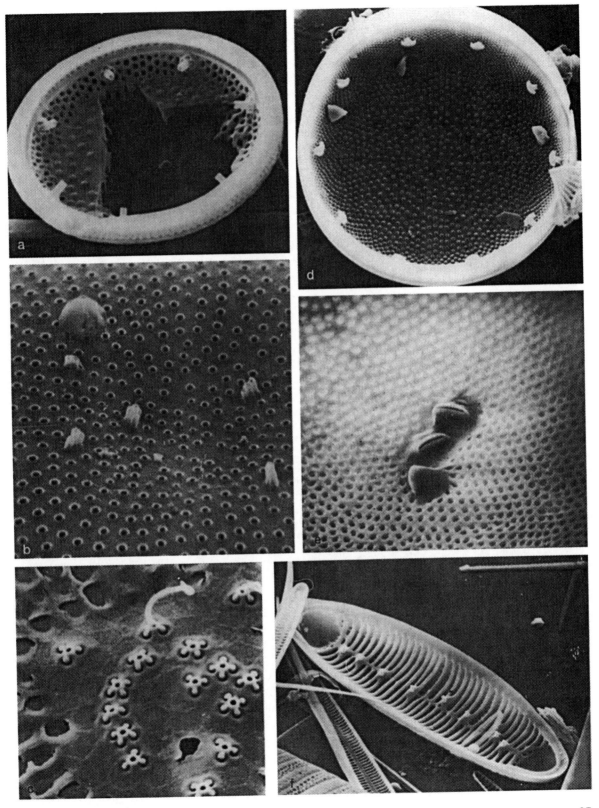

cell by one (rarely two) slits (the term, labiate process, has been used) and to the outside by a simple aperture or a tubular structure open at the apex (Figs 29d–f, 30). As with fultoportulae, there is some variation in structure. In some *Coscinodiscus* and *Palmeria* species the internal opening is not lipped but tubular (see p. 178). Sometimes the inner slit lies more-or-less flush with the valve surface (p. 459k): elsewhere, the rimoportula is raised or even stalked internally, as in *Coscinodiscus*, *Roperia* or *Actinocyclus* (Figs 29d, 30g). The external aperture may be flush with the valve surface, as in *Stellarima* (see p. 180), or raised above it, on a conspicuous tube, e.g. *Lauderia* (see p. 150). Occasionally, the valve mantle itself is formed into folds adjacent to the aperture, as in *Roperia* (see (p. 198). Possibly the most complex rimoportulae are found in *Aulacodiscus*: here the elaboration of the internal component is accompanied by curvature of the slit, while the external tube is capped by a complex asymmetrical dome, perforated by a single aperture that is sometimes extended into three grooves (Venkateswarlu & Round, 1973). The rimoportulae of *Hydrosera* on the other hand have extreme internal development but inconspicuous external apertures (see p. 250).

The arrangement of the rimoportulae tends to be species-specific. In some centric genera there is often a single ring of rimoportulae on the valve mantle (Fig. 29d), sometimes with one or more larger rimoportulae (termed 'macro-rimoportulae' by Brooks, 1975a, b, c). Clustering of rimoportulae sometimes occurs in well-defined areas of the valve, as in *Stictocyclus* (Round, 1978a) or *Isthmia*, where there are dense groups of rimoportulae at the centre of the valve (Round, 1984a). The clusters can be off-centre (e.g. *Pleurosira*, Fig. 29e). Elsewhere, the number of rimoportulae may be reduced to one or two. Thus in the end cells of *Chaetoceros* filaments there is a single central rimoportula, while in several araphid pennate taxa there are one or two, which occupy well-defined positions at one or both poles.

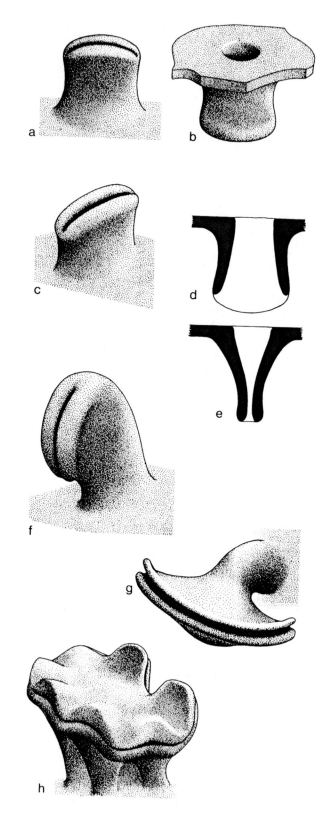

Fig. 30. The rimoportula. (*a*)–(*e*) Simple rimoportula. Internal (*a*), (*c*) views showing its lipped inner aperture and (*b*) its external pore. (*d*),(*e*) Longitudinal and transverse sections through the same rimoportula as in (*a*)–(*c*). (*f*) Curved and shortly stalked rimoportula, e.g. in *Stellarima*. (*g*) Complex, stalked rimoportula, e.g. in *Actinocyclus*. (*h*) Crimped and fluted rimoportula, e.g. in some *Coscinodiscus* spp.

or at the centre. An exceptional arrangement is found in *Rhabdonema* where a row occurs down the midline of the valve (Fig. 29f).

The function of the rimoportulae is less clear than that of the fultoportulae. They were called mucilage pores (*Gallertporen*) by workers such as Hustedt (1927–66) using light microscopy, and it was suggested that they were involved in secretion of mucilage for cell-to-cell or cell-to-substratum attachment. However, in most taxa producing stalks or pads, the principal secreting areas are the fields of small pores present at the poles or on projections. Crawford (1974), however, suggested that the distribution of rimoportulae in *Melosira* and the site of mucilage pad formation between cells is strong circumstantial evidence that the sulphated mucopolysaccharide (Daniel *et al.*, 1987) is passed through these processes. On the other hand, some genera, e.g. *Climacosphenia*, do not possess portules or any other obviously differentiated areas of pores, and yet can attach to rocks or plants perfectly well. It may be, however, that rimoportulae are involved in attachment, but only in the initial phase, when rapid secretion of an unstructured 'adhesive' may be advantageous, the more elaborate and highly organised stalks and pads being produced more slowly from the apical pore fields as the cell establishes itself and the stalk grows. Analyses of the polysaccharides in a variety of haptobenthic taxa indicate that there are indeed several types of secretion (Daniel *et al.*, 1987) but unfortunately they do not allow us to say where the less highly organised adhesive is produced. Crawford (1973a) has also suggested that the rimoportulae are 'utility organelles – passages allowing the flux of various substances from the cell. Recently, rimoportulae have been implicated in the movement of *Actinocyclus* (Andersen, Medlin & Crawford, 1986) and *Odontella* (Pickett-Heaps, Hill & Wetherbee, 1986) – these are the only centric diatoms that have been proved to have some power of movement. Plasmolysis of cells often results in the withdrawal of the cytoplasm from the valves except in the region of the rimoportulae (Schmid 1984a, b) although this was not the case in *Synedra*, studied by Geitler (1949).

A tubular type of process occurs in genera of the

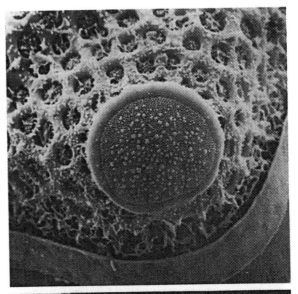

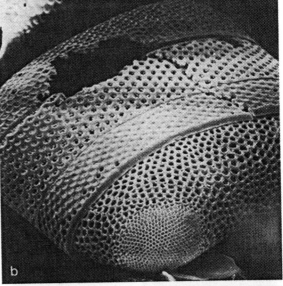

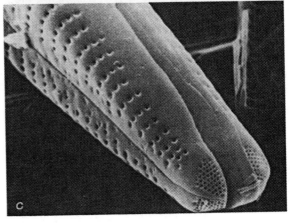

Fig. 31. (*a*) Ocellus of *Amphitetras*. (*b*) Pseudocellus of *Isthmia*. (*c*) Apical pore field of *Gomphonema*.

Cymatosiraceae (see Hasle, von Stosch & Syvertsen, 1983); it occupies a position on the valve face similar to that occupied by rimoportulae in some genera and may in fact be a reduced form of that process.

Two other kinds of process have been reported – the endochiastic areola (von Stosch, 1980b) and the carinoportula (Crawford, 1981b) – neither are widespread; the former occurs in some species of *Coscinodiscus* and the latter in *Orthoseira*.

Ocelli, pseudocelli and pseudonoduli

At the apices of bi-or multipolar cells, there are often areas in which the areolae are smaller and either unoccluded, or occluded by simpler vela than those present elsewhere. A group of such pores surrounded by a discrete rim of silica, as in *Pleurosira*, is called an ocellus (Fig. 31a), the small pores being 'porelli' (Ross & Sims, 1972). The term 'ocellus' is unfortunate since it means 'small eye', but it has become so generally used that it is best retained. Where the areas of finer pores are not so obviously demarcated, but grade into the areolae of the valve face, as in *Isthmia*, the structure is called a 'pseudocellus' (Fig. 31b).

The pseudocelli or ocelli are sometimes raised above the general surface of the valve. In *Eucampia* for instance, they are located on the tips of the elevations (see p. 263). These areas of specialised pores are found in all the major groups of diatoms, although true ocelli are apparently absent in the raphid genera. In all cases the differentiated areas are clearly involved in secretion of mucilage, as in *Gomphonema*, *Cymbella* and *Rhoicosphenia* among the raphid diatoms, *Diatoma* and *Grammatophora* among the araphid pennate, or *Odontella* among the centric diatoms. The area of small pores at the apices of many araphid and raphid pennate diatoms are termed apical pore fields (Fig. 31c).

In the recently described Cymatosiraceae a small area of pores on a slightly raised terminal projection has been termed an ocellulus (see p. 296) (Hasle, von Stosch & Syvertsen, 1983).

In a few centric genera there are structures which in the light microscope can look superficially similar to ocelli, but are really quite different. These 'pseudonoduli' take a variety of forms (e.g. see Simonsen, 1975). Sometimes they are little more than modified areolae (ibid.). In *Stictocyclus*,

however, the pseudonodulus is a flat, plain plate of silica set into the valve (see p. 208). Other pseudonoduli are apparently unoccluded holes through the valve face. It is difficult to believe that all these different forms are homologous. However, until we know something of the ontogeny and function of the pseudonodulus, it is impossible to be sure.

Raphe

A characteristic feature of many pennate genera is the presence of one or two longitudinal slits through the valve, together constituting the raphe system. In most cases there are two raphe slits, arranged end-to-end along or near the midline of the valve. These are separated by a bridge of silica, which is often somewhat thicker than the rest of the valve and has for this reason sometimes been termed the 'central nodule'. The raphe is rarely a simple slit, but often <-shaped in section (Fig. 71c, p. 557h) and it has been suggested (Pickett-Heaps, Tippit & Andreozzi, 1979a) that this structure is a device to prevent the valve splitting longitudinally under turgor and other stresses.

The ontogeny of the raphe (Fig. 32) shows that it, and indeed the raphid valve as a whole, is a fundamentally asymmetrical structure. It was first shown by Geitler (1932) and confirmed and studied in detail by Chiappino & Volcani (1977) that one

Fig. 32. Ontogenetic asymmetry of the valve in naviculoid diatoms. One side of the raphe-sternum and hence of the valve can be considered *primary*, in that it is formed first (*a*). The *secondary* side is produced later by a different process, involving the fusion of arms extending out from both centre and poles (*b*),(*c*). Where the arms fuse, faults often occur in the stria pattern (compare (*c*) with (*d*)): these are the Voigt discontinuities. Where both polar raphe endings are turned towards the same side, as in most raphid genera, they turn towards the secondary side ((*d*), see also Fig. 33). (*e*)–(*j*) The evolution of the raphe: (*e*) the araphid ancestor with rimoportulae at each pole; (*f*)–(*h*) differentiation of the raphe from one rimoportula at each pole. Note the finer striae associated with the raphe because of its essentially transapical orientation, the rotation of the raphe towards the centre, and the parallel movement of the undifferentiated rimoportula to an apical position. (*i*), (*j*) Abutting of the raphe onto the sternum and final integration into the primary pattern-centre. In (*e*)–(*j*) the pattern-centre is shown stippled and the raphe slits are outlined heavily in black (from Mann, 1984b); primary side at right.

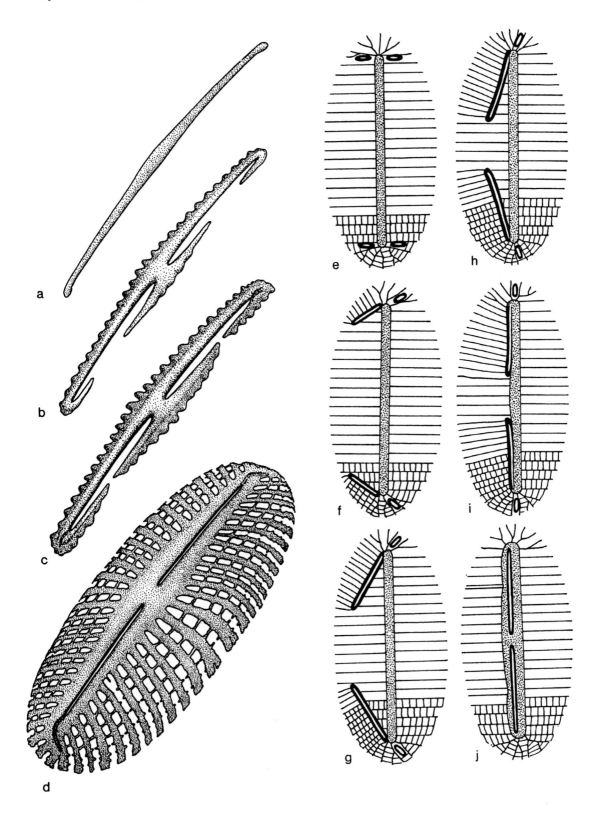

side of the raphe system, the primary side (Fig. 32a), is formed first, and then the secondary side (Fig. 32b, c). In view of this it seems clear that the sternum ('pseudoraphe') of the araphid diatoms (Fig. 24b) must be equated with only one side of the siliceous framework enclosing the raphe (the raphe-sternum). Mann (1984b) has suggested that the raphe-sternum is a composite structure, which has arisen through the ontogenetic and phylogenetic 'fusion' of the araphid sternum with two secondary, originally lateral, pattern centres associated with the raphe slits, which may have developed from two rimoportulae as suggested by Hasle (1974) (Fig. 32e–j). Clearly, however, there are other possible interpretations of the raphe-sternum and how it may have arisen.

The extent, position and structure of the raphe are important taxonomic characters. The raphe may be central or eccentric (Figs 33, 34a), eccentric raphe systems being particularly common in genera with dorsiventral frustules, such as *Amphora*, *Epithemia*, *Rhopalodia*, *Cymbella* or *Hantzschia*. In some cases the raphe extends around most (*Auricula*) or all (*Surirella*, *Campylodiscus*, *Cymatopleura*) of the valve circumference (Fig. 36).

The raphe endings yield much useful information about the relationships between taxa. At the distal end of each raphe slit, near the valve pole, the raphe ends internally in a small structure, which resembles a rolled tongue (Fig. 34b) – hence its name, helictoglossa (Mann, 1977; Cox, 1977b). The helictoglossa varies little in shape, except that in

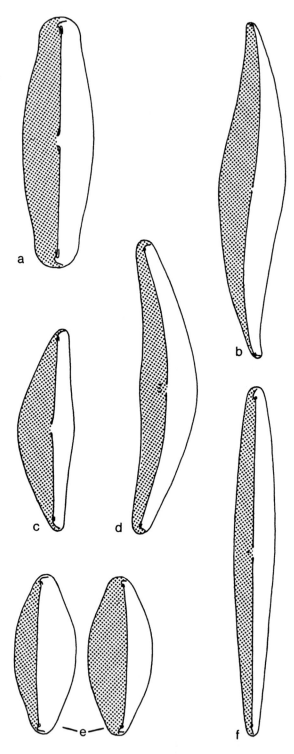

Fig. 33. The extent of the primary (shaded) and secondary sides of the valve in various raphid diatoms; the curvature of the terminal fissures of the raphe is also shown. (*a*) *Sellaphora*; (*b*) *Pleurosigma*; (*c*) *Encyonema*; (*d*) *Cymbella*; (*e*) *Denticula*; (*f*) *Gomphonema*. (In *Denticula* there is a subtle dimorphism: one valve of each sibling pair forms with a narrow primary side, the other with a broad one.)

Fig. 34. The raphe system. (*a*) Simple slits of *Anorthoneis*. (*b*) Internal polar ending, showing ridges on either side of the raphe fissure and the polar helictoglossa in *Pleurosigma*. (*c*) Massive internal development of the raphe-sternum in *Frustulia*. (*d*) External raphe ending with terminal fissure extending beyond the region of the helictoglossa (arrow). (*e*) Fibulae subtending the raphe of a *Nitzschia*. (*f*) Central raphe endings of *Gyrosigma*. (*g*) Internal central raphe endings with 'helictoglossa-like' structures in *Parlibellus*.

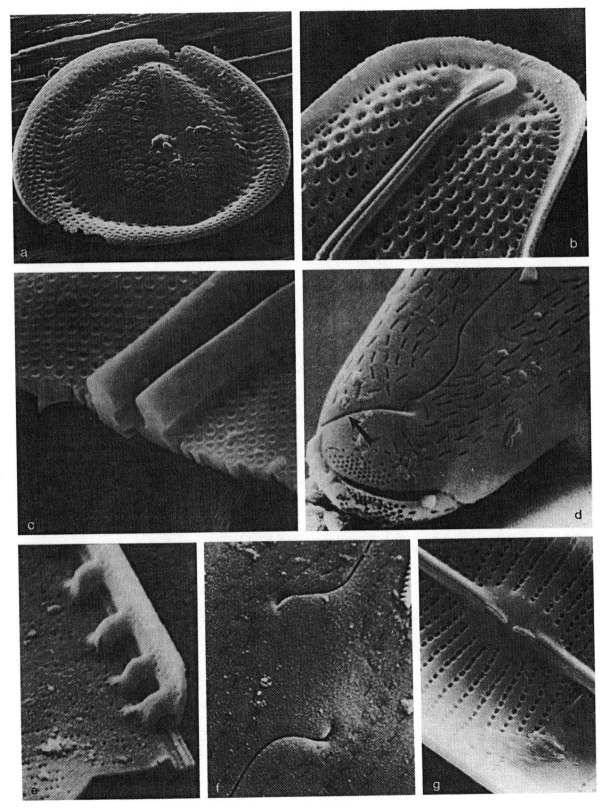

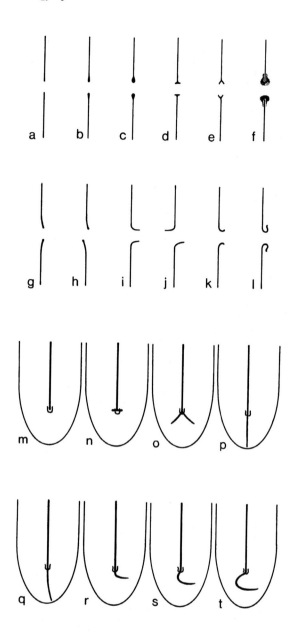

some *Cymbella* species it is turned over to one side (see p. 487j), but it can vary in size from a tiny bump in many of the smaller *Nitzschia* species to conspicuous compressed structures in some of the marine *Pinnularia* species.

The raphe slits may continue for some distance on the outside of the valve beyond the internal position of the helictoglossa. This section of the raphe is thus a blind groove, usually called the terminal fissure (Fig. 35n–t). The terminal fissure is often hooked towards the ontogenetically secondary side of the valve, and indeed, together with the Voigt discontinuities (see p. 40), is a useful marker for this (Figs 33, 34d). Elsewhere, however, the terminal fissure may be double, as in the T-or Y-shaped endings of *Frustulia* (Fig. 35n) or *Neidium* (Fig. 35o), or absent, as in *Cocconeis* or *Anorthoneis* (Fig. 34m). In some of the sigmoid diatoms, e.g. *Pleurosigma* and *Gyrosigma*, one terminal fissure is hooked towards the primary side of the valve, and the other towards the secondary side.

Occasionally, there are helictoglossa-like structures at the internal central end of the raphe slit (Fig. 34g) and in some of the Bacillariales these

Fig. 36. Evolutionary trends in advanced raphid diatoms. The raphe system in naviculoid diatoms runs from pole to pole along the centre of the valve (*a*). The evolution of fibulae seems to have occurred in diatoms with eccentric raphe systems [e.g. (*b*)], and this in turn seems to have allowed further changes in the extent and position of the raphe slits [(*c*)–(*g*)]. In (*b*)–(*g*) fibulae are present internally but are not shown; the central raphe endings or their ontogenetic equivalents are indicated by 'x'. (*a*) *Navicula*: two central raphe slits. (*b*) *Nitzschia*: two eccentric raphe slits. (*c*) *Nitzschia*: central raphe endings lost, raphe continuous from pole to pole. (*d*) *Auricula*: extension of raphe around valve perimiter. (*i*) *Campylodiscus*: polar raphe endings adjacent to each other, creating a raphe that runs around the whole perimeter of the valve. (*f*) *Cymatopleura*: compression of the valve to create a new apical plane, now running from one set of raphe endings to the other. (*g*) *Surirella*: heteropolarity of the new apical axis.

Fig. 37. Keeled raphe systems in genera possessing fibulae. (*a*) Simple shallow (e.g. many *Nitzschia*). (*b*) Deeper keel with fibulae at one level (e.g. *Psammodictyon*). (*c*) Deep keel with fibulae at several levels (e.g. *Entomoneis alata*). (*d*) Fusion of the two sides of the keel to create narrow tubes connecting subraphe canal to cell lumen (e.g. some *Entomoneis* spp.). (*e*) Creation of spaces between the connecting tubes by further fusion and simplification of the keel walls (e.g. some *Surirella, Stenopterobia* and *Campylodiscus* spp.).

Fig. 35. Examples of central (*a*)–(*l*) and external polar (*m*)–(*t*) raphe endings. *Central endings*: (*a*) straight, simple; (*b*) straight, slightly expanded; (*c*) straight, expanded and pore-like; (*d*) T-shaped; (*e*) forked; (*f*) straight, opening into a spathulate groove; (*g*),(*h*) deflected in the same or opposite direction; (*i*),(*j*) bent in the same or opposite directions; (*k*) hooked; (*l*) strongly hooked. *Terminal fissures* (the position of the helictoglossa being indicated by a 'u'): (*m*) absent; (*n*) transverse (polar endings T-shaped); (*o*) double (polar endings forked); (*p*) straight; (*q*) slightly deflected; (*r*) bent; (*s*) hooked; (*t*) strongly hooked.

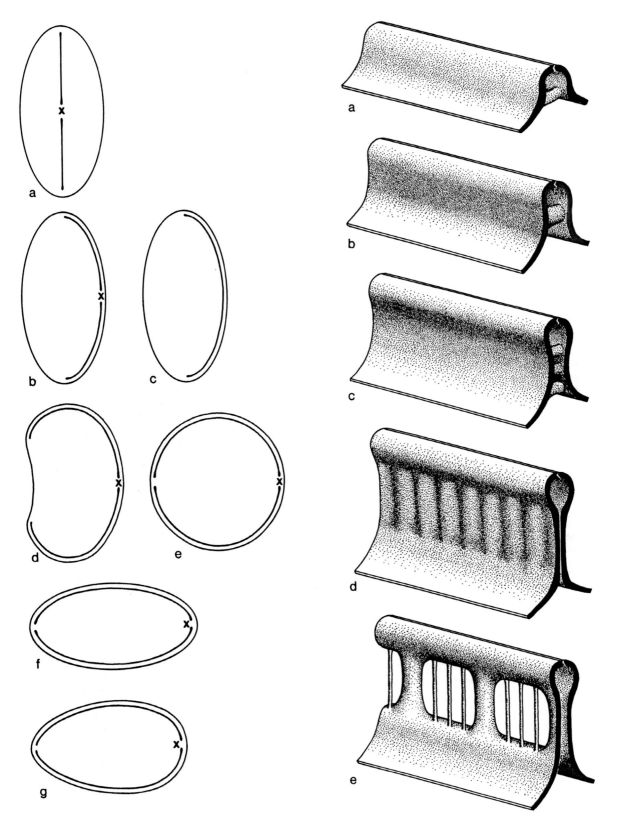

are fused back-to-back to form a prominent, beak-like projection into the cell. Otherwise, the central raphe endings are simple (e.g. *Brachysira*), curved or hooked towards the primary side of the valve (e.g. *Gomphonema, Sellaphora, Pinnularia*), or T-shaped (e.g. *Scolioneis*). In some *Cymbella* and *Pinnularia* species the central internal endings are hidden by a large nodular growth of silica on the primary side of the raphe-sternum, making it appear from the inside as if the raphe is continuous from pole to pole.

The external central raphe endings are very varied – simple or expanded; continued into blind grooves like the terminal fissures; hooked or T-shaped (Fig. 35a–l). Where present, the central fissures may turn towards the same side of the valve, as in *Luticola mutica*, or opposite sides, as in *Neidium* and some sigmoid genera (Fig. 34f).

A good proportion of species in the Bacillariales and Epithemiales have only one raphe slit, which runs unbroken from pole to pole. This appears to be a derived state, which has arisen independently in a number of lineages (Mann, 1982d). In this case, silicification of the secondary side of the raphe proceeds from both poles simultaneously but not from the centre (Mann, 1984b).

In three groups of raphid diatoms – the Epithemiales, Bacillariales and Surirellales – the raphe is subtended internally by bridges of silica, termed fibulae (Fig. 34e). These vary greatly in shape, from narrow ribs to flat sheets or more complex structures, but all cross-link the valve beneath the raphe, like tie-beams and collars in a roof. The effect of this is often to create a canal beneath the raphe, which is partially separated from the rest of the cell lumen: hence the term 'canal raphe', often applied to this type of raphe system. In fibulate diatoms, and in a few other groups, e.g. *Plagiotropis*, there is a tendency for the raphe system to be raised above the general level of the valve on a keel, which can be shallow or deep (Fig. 37a–c). In extreme cases the two sides of the valve may fuse beneath the raphe, or even partly disappear (Fig. 37d, e), leaving only a series of tubes connecting the raphe canal to the rest of the cell. Such fenestrated keels occur in *Entomoneis, Surirella* and

Fig. 38. (*a*) Segmented girdle: *Urosolenia*. (*b*) Open valvocopula with a fimbriate edge to the pars interior: *Pleurosira*. (*c*) Closed copula with lateral extensions in one half: *Rhabdonema*.

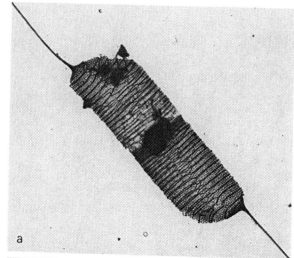

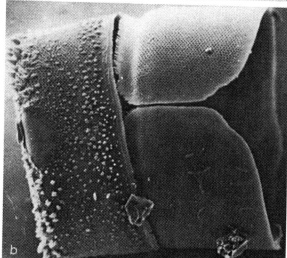

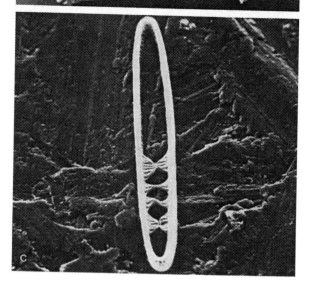

Campylodiscus (see p. 647h). Surveys of keel and fibula morphology have been made by Paddock & Sims (1977, 1981).

Girdle bands (copulae)

The elements of the silica exoskeleton lying between the two valves have been neglected by most light microscopists, partly because of their small size and delicate structure, and partly because after acid cleaning, the frustule usually falls apart; in a mixed sample it is very difficult to decide which valves and girdle bands belong together. (Fig. 4b, c illustrates cells with bands in position and Fig. 5a isolated bands and diaphragms).

The major function of the girdle bands is clearly to enclose and protect the cell and yet at the same time accommodate the increase in cell volume during the cell cycle. But the variety of girdle band structure suggests strongly that this is not the only function. There are three main types of band – scales, split rings and closed rings (see von Stosch, 1975, who gives an excellent summary of girdle structure). The scale, or segmented type, as found for instance in *Stephanopyxis* or *Ditylum*, may be the primitive condition (Round & Crawford, 1981). Here each girdle element is short and some occupy only a fraction of the circumference of the girdle. Partly segmented girdles are also found in 'rhizosolenioid' genera (Fig. 38a), and the pennate genera *Subsilicea* and *Denticula*. In *Denticula* the half bands are probably not primitive: this genus is fairly closely related to various *Nitzschia* species that have girdles of split rings (e.g. *N. denticula*, *N. sinuata*), but is unlikely to have been ancestral to them. Furthermore, even in the centric genera, which are often regarded as having preceded the pennate (although we would not accept this until more evidence is forthcoming), the arrangement of the scales (in whorls) suggests that extant segmented girdles, although they may be primitive in being segmented, nevertheless differ considerably in organisation and development from the ancestral condition. In an algal group very similar to the

Fig. 39. (*a*) Ligula (arrowed) and antiligula (arrowhead): *Corethron*. (*b*) Fimbriate valvocopula lying in position inside valve mantle: *Pleurosira*. (*c*) Grouped fimbriae on valvocopula: *Odontella*. (*d*) Scalloped pars interior, each projection corresponding to a valve rib: *Synedra*.

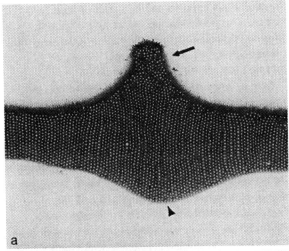

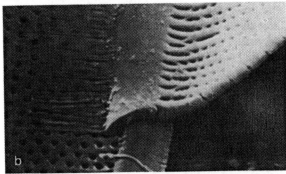

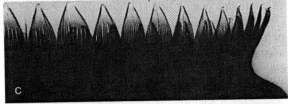

diatoms, the Chrysophyta, the scales of such forms as *Synura* and *Mallomonas* are arranged spirally (Round, 1981a), while elsewhere (and apparently in the auxospores of centric diatoms) the scale arrangement shows no particular order.

Split or open rings (Fig. 38b) are found in a wide range of genera from all the major diatom groups, and are by far the commonest type of girdle element. The splits of adjacent bands are nearly or, in most pennate species, exactly 180° apart, and each split is closed by a tongue-like section (*ligula*) on the adjacent (younger) abvalvar band. Sometimes this does not completely fill the gap and the remainder is closed by a small projection (*antiligula*) on the adjacent (older) advalvar band (Fig. 39a). Hence the girdle is almost always complete, with no obvious line of weakness. In the Chaetoceraceae the final girdle bands in some species are modified to allow the newly forming setae to extend between the epitheca and hypotheca (see p. 337g).

Closed bands – complete hoops around the whole cell – are less common and possibly the most advanced type. They occur in the Biddulphiales, e.g. in *Hydrosera* and *Isthmia*; in *Climacosphenia* and *Rhabdonema* (Fig. 38c) among the araphid forms; and in some raphid diatoms, e.g. some populations of *Hantzschia marina* and *H. amphioxys*.

As far as is known, within a theca the advalvar part of each girdle element always underlaps the edge of the adjacent girdle element (Fig. 6) or, in the case of the first band (*valvocopula*), the edge of the valve (Figs 38b, 39b, d). Each girdle band, therefore, consists of a pars interior (the underlapping part) and a pars exterior. These two parts usually differ from each other in size and structure. The pars interior is often modified to fit closely around the element it underlaps (Fig. 39b). This is especially marked in the bands next to the valve, since the valve is usually much thicker than the girdle bands and may indeed bear ridges or pseudosepta near its margin, around which the first band must fit. The fit is often so exact that valve or band must be broken, or powerful force applied (e.g. by ultrasonication), before the two will come part (e.g. *Isthmia*: Round, 1984a; *Rhoicosphenia*: Mann, 1982a). The edge of the pars interior is often elaborated into thin projections or fimbriae (Latin: fringe) but these can be classified into two, probably quite distinct types. In the first,

exemplified by many centric diatoms, the fimbriae appear to bear no relation to the structural elements of valve or pars exterior, either spatially or in terms of the linear densities of individual elements (i.e. of fimbriae and striae: Fig. 39b, c). In the other type, instead of the fimbriae there are flat extensions (flaps) which correspond exactly in their linear density and often their structure to elements (ribs/costae, etc.) of the valve (Fig. 39d). Thus in *Hantzschia virgata*, the pars interior of the most advalvar band resembles a sliver taken off the inside of the marginal area of the valve: *in vivo* it fits against the valve in such a way that each valve pore is underlain by a pore in the band (Mann, 1981b). Such exact correspondence between band and valve, or band and band, raises interesting questions about how the pattern of striae and pores is generated. Either one element acts as a blueprint for the other, or the patterns of both are determined by the same morphogenetic agent. In either case the control must operate across membranes and any intervening cytoplasm.

There may be a determinate or indeterminate number of bands in the girdle, and bands may be added to the hypotheca throughout the cell cycle (e.g. *Striatella*, *Phaeodactylum*: Roth & De Francisco, 1977; Borowitzka & Volcani, 1978) or only at particular stages. In *Navicula pelliculosa*, for instance, two bands are produced soon after cell division, but the third is not added until just before the next division (Chiappino & Volcani, 1977). Similarly, all the girdle bands may be alike, or differentiated into two or more types. Unfortunately the relationship between girdle development and girdle morphology is not yet understood. Von Stosch (1975) has suggested a terminology based on the morphology and position of the bands. It can be applied only if the whole theca is known (it is essential, therefore, to study epithecae that have not lost any elements from their abvalvar ends), and proceeds as follows. Start from the open end of the girdle (i.e. distant from the valve): if all the bands are similar to the band at this end then all can be called pleurae. However, the band next to the valve is always slightly different from the others since it never bears a ligula. This band is called the valvocopula and it may have special features, e.g. the apical elongate pores in *Grammatophora* (see p. 437f). If, on the other hand, one encounters a 'marked and abrupt change of structure or/and morphology' (von Stosch, 1975), when proceeding from the open end of the girdle towards the valve, then only the more abvalvar bands are pleurae, the differentiated bands nearer the valve being copulae. This terminology is logical but it should not be construed that all copulae, for instance, are necessarily homologous structures. The distinctions made are morphological, and the purpose of the classification is purely pragmatic. In diatoms with more than two types of band, the terminology would have to be adapted.

We believe that the terminology of copulae and pleurae is quite impractical and has to be abandoned, for the following reasons:

(i)   it is rarely possible to be sure that a complete cingulum has been observed;

(ii) in many cases there is a gradual, not an abrupt, change in band morphology across the cingulum;

(iii) there is no absolute measurement of difference that can be used to validate distinctions between types of band;

(iv) where there are abrupt changes in girdle band morphology in a cingulum there may be 2, 3, 4 or more discrete kinds of bands making a bipartite classification unsatisfactory;

(v)   over and above all these, we think it is unsound to employ a terminology that can only be applied *a posteriori*, after subjective interpretation of the structure of the whole cingulum. The description should be as simple and flexible as possible.

We therefore advocate using the term 'copula' for any element of the cingulum. The element(s) adjacent to the valve, which are always recognisable by their position and development and which are usually considerably modified to fit under the valve margin, simply require the prefix valvo-. The copulae can be numbered in order of their formation, i.e. in order from the valve outwards. Here it is not necessary to know the whole structure of the girdle before description can proceed. We realise that our proposals conflict with earlier usage but we feel they will provide a more usable and rational system.

Copulae are almost always smooth-surfaced structures and only occasionally is there slight granular ornamentation on the outside. No portulae of any sort occur; any such structure would impede

the sliding apart of the two halves of the cell.

A feature peculiar to a few araphid genera is the formation of thin inwardly directed septa (reaching nearly to the centre of the cell) on some of the copulae (Fig. 40a). The function of these is unclear. The only other marked elaboration of the girdle occurs in *Mastogloia*, where the valvocopula bears a complex system of internal bulbous chambers (partecta) and tubes, which open to the outside by relatively large apertures (Fig. 40b). Otherwise, copulae are usually simply structured (but see also *Campyloneis*). Sometimes, especially towards the abvalvar end of the cingulum there are plain, non-porous bands of silica – these occur, for example, in *Rhabdonema*, where there is in addition a series of scales between the poles of the valves and the valvocopulae (Pocock & Cox, 1982). Sometimes all the copulae are plain but often some at least bear several rows of pores of similar structure to those of the valve, except that the pores tend to be smaller and the pore occlusions simpler.

Internal valves

Under certain circumstances, some diatoms produce 'internal valves', i.e. valves that are not associated with a proper division of the cell in which they are formed, but appear superfluous. A cell can build up considerable numbers of such internal valves: Hustedt (1927–66) illustrated a theca of *Eunotia serpentina* with six valves! Internal valves appear to be formed in many cases as a reaction to an increase in the osmotic potential of the surroundings, and so they are particularly common in taxa living in subaerial and saline habitats. Schmid (1979b) induced their formation experimentally by manipulating the ionic concentration of the medium, but the function of these valves, or at least, the reason for their formation, was not determined. In other cases, however, e.g. in *Eunotia soleirolii* (Fig. 40c: see also von Stosch & Fecher, 1979) and probably also in *Meridion circulare*, the formation of internal valves is clearly associated with dormancy (see below).

Internal valves are formed by one of three pathways (Geitler, 1980), all of which involve DNA replication and mitosis as a first step (Fig. 41a–e). In the first, there is an unequal cytokinesis, with most of the organelles remaining in the larger daughter cell, though parts of the plastids may remain in the

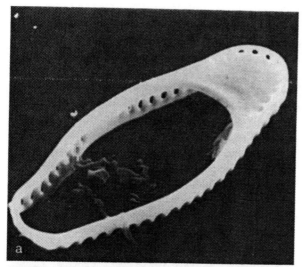

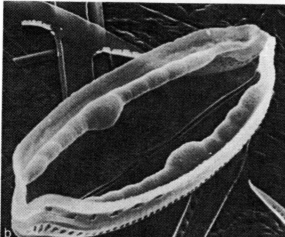

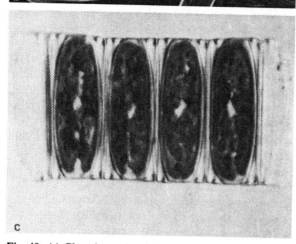

smaller cell (Fig. 41c). The smaller cell produces only a reduced, lightly silicified valve and its nucleus soon disintegrates. The valve produced by the larger daughter cell develops more strongly and becomes the functional hypovalve of the cell. It is often strongly convex as a result of the collapse in turgor of the sibling cell as it degenerates. The second pathway also involves an unequal cytokinesis, but here the smaller cell is so reduced that it produces no valve (Fig. 41d). In the third pathway, both nuclei remain in the same daughter cell (Fig. 41e), and the single internal valve is produced either following an extremely unequal cytokinesis or perhaps after a unilateral plasmolysis of the cell (von Stosch & Kowallik, 1969; Geitler, 1970c, 1980). All three can be found in some taxa, e.g. *Hantzschia amphioxys*. Thus the principal developmental difference between internal and normal valve formation is that only one functional valve is produced per mitosis, not two as in normal division (Fig. 41a, b). Internal valves usually differ little from normal valves in their basic structure. However, because they are formed either against one of the parental valves or more-or-less free within the cell, not against the sibling valve, their shape is more rounded than usual and where the valves bear pseudosepta or fibulae, the internal valves are moulded accordingly.

Fig. 41. (*a*)–(*e*) Formation of internal valves in *Hantzschia amphioxys*: (*a*),(*b*) normal equal cell division (*a*), with the formation of two similar valves; (*c*) unequal division. Fragments of the plastids, a nucleus and a small amount of cytoplasm are segregated into the smaller residual cell, which forms a feebly silicified non-functional valve; (*d*) more unequal division: smaller cell with nucleus but no plastids or hypovalve; (*e*) extremely unequal division (or plasmolysis?), both nuclei remaining in the larger cell though one becoming pyknotic; new internal hypovalve very strongly convex (after Geitler, 1980). (*f*)–(*k*) Formation of resting spores in *Stephanopyxis turris*: (*f*) cell at end of interphase, with nucleus near hypovalve; (*g*) mitosis, accompanied by plasmolytic cleavage; (*h*) formation of what will be the resting spore epithecae; (*i*) contraction of protoplasts away from the spore mother cell epithecae; (*j*) unequal cleavage of the spore mother cells, with the formation of a functional spore cell and a small residual cell; (*k*) completion of the resting spores with the formation of the spore hypothecae. Degeneration of the residual cells (after von Stosch & Drebes, 1964).

Fig. 40. (*a*) Closed copula with polar septum: *Rhabdonema*. (*b*) Valvocopula with bulbous partecta: *Mastogloia*. (*c*) Resting cells in *Eunotia*.

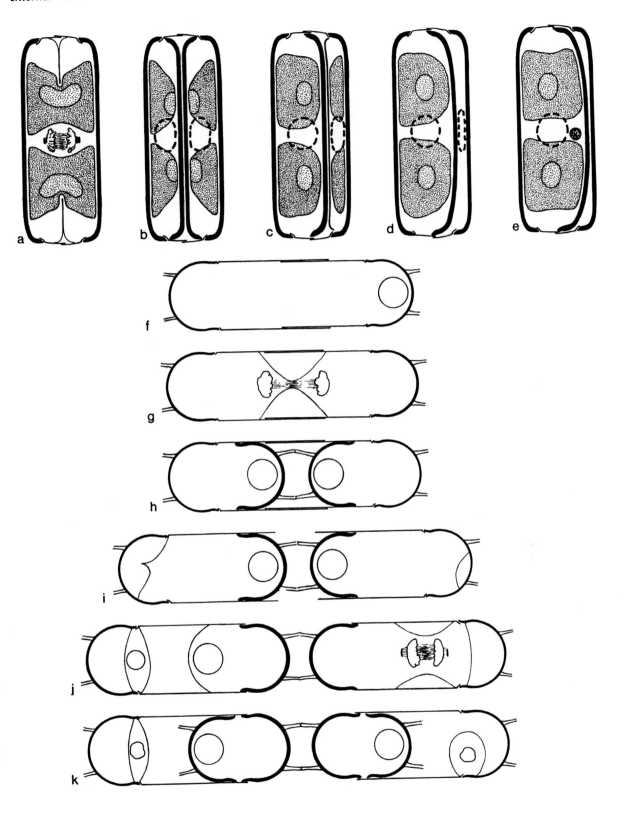

Resting stages and resting spores

Some diatoms, especially those whose natural habitat is soil or rock, can survive desiccation for a long time. Indeed, the ability to withstand periods of drought, or to escape in some way from their effects, is a prerequisite for terrestrial life. Subaerial or aerophytic diatoms do not appear to produce morphologically distinct resting stages in response to environmental stress: the vegetative cells themselves become dormant. The same is true of some freshwater planktonic algae that form resting stages as part of their annual growth cycle. In *Aulacoseira italica* subspecies *subarctica* and *A. granulata* the dormant stages differ in appearance from actively growing vegetative cells only in their thicker cell walls, their greater oil content and their rounded plastids (Lund, 1954). Such cells can survive for several decades if incorporated into anoxic sediments (Stockner & Lund, 1970).

Elsewhere, the induction of dormancy is associated with the formation of morphologically distinct cells, which can be called *resting spores*. Very few pennate diatoms are known to produce resting spores. The best investigated is *Eunotia soleirolii*, in which the whole course of spore formation and germination has been followed *in vivo* by von Stosch & Fecher (1979). Some other *Eunotia* species (*E. pectinalis* and *E. faba*), together with *Diatoma anceps*, *Meridion circulare*, *Nitzschia grunowii* and *Achnanthes taeniata* also seem to form spores (for references, see von Stosch, 1967, von Stosch & Fecher, 1979). In *Craticula cuspidata* a complex resting spore is produced in response to desiccation. In the course of its formation, two characteristic types of internal valve are formed within the cell wall of the vegetative cell, which itself lies within a mucilage envelope (Schmid 1979b). One consists of a scaffold of ribs (the raphe-sternum and a number of transverse bars) which is known as the 'craticula' valve, and the other is a valve of more conventional appearance but with radial instead of orthostichous striae. The latter has been described as the var. 'heribaudii'; as a phenotypic variant it should probably be accorded no formal taxonomic recognition.

Von Stosch & Fecher (1979) found that the resting spores of *Eunotia soleirolii* differed in five respects from the vegetative cells:

(i)   they have a different ontogeny;
(ii) the thecal morphology is different;
(iii) the spores are rich in storage products, especially oil;
(iv) they are unable to germinate without a cold treatment;
(v)   they can survive for 3 years or more.

But the spores do not differ greatly in appearance from the vegetative cells and in some centric diatoms too the spore valves, although recognisable as such, differ relatively little from the vegetative valves (e.g. *Thalassiosira nordenskioeldii* and *Detonula confervacea*: Syvertsen, 1979). In some other centric diatoms, however, the resting spore thecae deviate so much in their shape and structure (e.g. see Hargraves, 1979, 1986) that, were they never seen in association with the vegetative thecae, they would probably be classified in different families or orders (Fig. 42). Indeed, the spores of *Chaetoceros* may have been classified as distinct genera (*Goniothecium*, *Periptera*, *Omphalotheca*, *Xanthiopyxis*: Tappan, 1980, but see the discussion in Hargraves, 1986). The resting spore, like the vegetative cell, has a cell wall composed of two thecae. In addition the resting spore valves are usually very much thicker than the vegetative valves, are far less porous and may bear warts or spines of various kinds. Their shape is usually less elaborate than the valves of the vegetative cells. The frustules are unusual in that they very rarely contain a well-developed girdle.

The cytological events surrounding resting spore formation are known in detail for only a few species (e.g. *Stephanopyxis*: von Stosch & Drebes, 1964; *Bacteriastrum*, Drebes, 1972; *Chaetoceros didymum*: von Stosch, Theil & Kowallik, 1973). The work of von Stosch and his colleagues has shown that the formation of the resting spore is no exception to the rule that every valve is formed as a consequence of a mitosis (Geitler, 1963a; von Stosch & Kowallik, 1969). This mitosis may or may not be associated with cytokinesis. In some cases, as in the formation of the spore epivalves in *Stephanopyxis turris* (von Stosch & Drebes, 1964), not only is the mitosis cytokinetic, but the cell division is equal, as in vegetative cells (Fig. 41f–h). Elsewhere, as for instance in the formation of the resting spore hypovalve in *Stephanopyxis* (ibid.), the cell division is unequal, leading to the production of a large, viable spore cell and a smaller residual cell, which is often depleted in organelles and later aborts (Fig. 41i–k). Elsewhere again, the formation of the resting spore valve takes place beneath an area of the plasmalemma that has contracted away from the parent frustule as a result of a spontaneous plasmolysis (e.g. in *Chaetoceros*, von Stosch, 1967; von Stosch, Theil & Kowallik, 1973). In this case, after mitosis one nucleus aborts during the formation of each spore valve.

A vegetative cell (mother cell) may give rise to just one resting spore, or to two or four. This, together with the kinds of division involved in spore valve formation (see above) allows us to classify resting spores into various types (von Stosch, 1967; Ross *et al.*, 1979; Syvertsen, 1979). Von Stosch's classification (1967) is the most complete and should be consulted for further information. Three types of spore can be distinguished, which (following Ross *et al.*, 1979) are:

(i) exogenous: mature resting spore not enclosed by its parent frustule (Fig. 43b);

(ii) semi-endogenous: one valve of the mature spore enclosed within the parent cell, the other valve free (Fig. 43c);

(iii) endogenous: mature spore completely enclosed within the parent cell (Fig. 43d).

These three types are distinguished purely on the basis of the relationship of the *mature* spore to the parent frustule and so it is possible to use the classification whether or not the ontogeny of the spores has been worked out. For the most part, however, these different end-products arise by characteristic developmental pathways, which are given in Fig. 43a, this being an elaborated version of a scheme put forward by Syvertsen (1979). It must be emphasised, however, that spore ontogeny cannot always be guessed from the characteristics of the mature spore: thus *Chaetoceros didymum* has 'semi-endogenous' mature spores, yet the spore ontogeny is fully endogenous (von Stosch, Theil & Kowallik, 1973).

A number of external factors have been found to induce spore formation or to exert effects upon spore development. These include the availability of various nutrients (N, P, Fe, Si), temperature, light intensity and pH. In almost every species, nitrogen deficiency appears to be an effective and usually the

Fig. 42. Resting spores. (*a*) *Leptocylindrus*; (*b*) & (*c*) *Chaetoceros* (?).

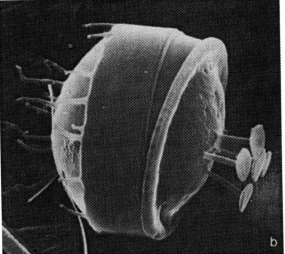

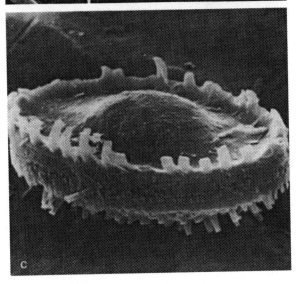

most effective or even the sole inducer (see Hargraves & French, 1983). Not all clones of a spore-forming species seem to be capable of spore formation (Hargraves & French, 1983; Hargraves, 1984), or perhaps it is that different clones are induced to produce spores by different environmental conditions. Furthermore, not all cells of a clone are necessarily able to form spores. Just as it is only cells of a restricted size range that have the potential to become gametangia (see p. 87), so also in some centric diatoms only particular stages in the size reduction cycle can produce spores. Thus in *Leptocylindrus danicus*, resting spores (Fig. 42a) are formed only within the expanded auxospores, while in *Stephanopyxis palmeriana*, only cells in the lower half of the size range produce spores. On the other hand, all cells of *Chaetoceros diadema* appear to have the potential to form them (Hargraves & French, 1983).

As has already been noted, the relatively undifferentiated resting stages of *Aulacoseira italica* subspecies *subarctica* have been found to survive for considerable periods in anaerobic sediments. The function of the spores in this and other freshwater *Aulacoseira* species seems to be to tide the diatom over a period of the year when active growth in the plankton is impossible. Thus, when turbulence in the euphotic zone is reduced, at the onset of summer or inverse stratification, the *Aulacoseira* cells sink rapidly to the sediments and remain there, dormant, until they are resuspended by more turbulent conditions (Lund, 1954, 1955, 1971). The viable spores buried many centimetres down in the sediments must be regarded as relicts, since they have very little chance of ever being returned to the plankton.

Data on the longevity of other resting stages or spores are scanty, and are summarised by Hargraves & French (1983). None of the morphologically differentiated resting spores of marine centric diatoms are known to last longer than two years. In general their survival is best in the dark and at low temperatures, but in these marine forms the spores do not appear to be able to withstand the conditions accompanying anoxia in sediments, e.g. high sulphide and ammonia concentrations. Hence it is unlikely that spores sinking to the ocean bottom could act as a sedimentary inoculum for future planktonic growths, even in coastal regions. Resting

spores survive periods of darkness better than vegetative cells and generally contain large amounts of storage material (Hargraves & French, 1983) but their function remains mysterious. Since the production of a spore involves at least two mitoses and the deposition of two valves, together with the necessary accompanying rearrangements of the protoplast, spore formation cannot be regarded as adaptive in relation to short term environmental change. In *Chaetoceros didymum*, for instance, spore formation takes 6–48 h (von Stosch, Theil & Kowallik, 1973). Hargraves & French (1983) suggest that spores may be important in promoting sinking towards the end of a diatom bloom, when the diatom populations may suddenly become nutrient limited and subject to high light intensities. If so, however, it is not clear why spore formation should be so complex, nor why their morphology should differ so much from that of the vegetative cells. Extra density and hence more rapid sinking could be achieved without significant change of morphology, merely by the production of thicker vegetative valves. Resting spores are usually more rounded than the vegetative cells and this might be regarded as an adaptation to allowing higher sinking rates, were it not that in many cases the spores remain partially or completely enclosed in the parental frustule. Perhaps the spores function in reducing the effects of grazing (acting as a predation-resistant stage; Hargraves & French, 1983). Clearly, far more work needs to be done on the pattern of spore production and the fate of spores and vegetative cells during the growth of natural diatom populations. Quantification of the loss processes affecting the marine phytoplankton, in the same way as has been done for freshwater phytoplankton (e.g. see Reynolds, 1984), may yield the vital clue to spore function, which of course may differ in different taxa.

Just as encystment appears to be induced most often by nutrient (especially N-) deficiency, so

Fig. 43. (*a*) The formation of different types of resting spore. 1. Normal equal vegetative division. 2–4. The formation of exogenous (2), semi-endogenous (3) and endogenous (4) resting spores. Note the formation of thin supernumerary valves in the residual cells produced during pathways 3 and 4. (*b*)–(*d*) *Thalassiosira nordenskioeldii*: sections through exogenous, semi-endogenous and endogenous resting spores. (After Syvertsen, 1979.)

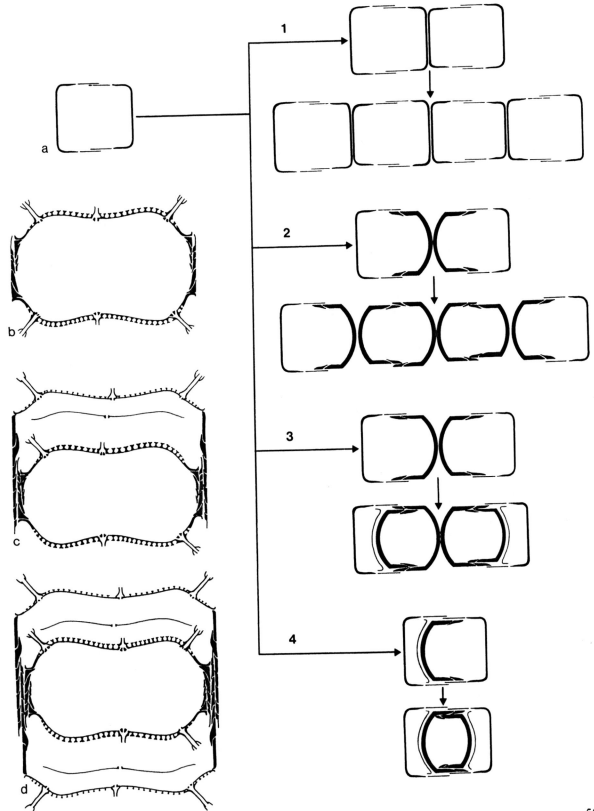

excystment is usually promoted by resupply of nutrients and light (Hargraves & French, 1983). *Eunotia soleirolii* requires a cold period before excystment but this is unusual. Centric diatom resting spores have instead a short period of obligate dormancy during which the spores will not germinate, after which, providing the nutrient and light conditions are suitable, excystment will occur.

Excystment sometimes occurs simply by division of the resting spore and the production of new vegetative valves. Hence a chain of cells may be produced with the resting spore valves retained at the ends of the chain. Elsewhere the spore valves are discarded, as a result of the formation of a new vegetative cell within the spore frustule. This is accomplished via two acytokinetic mitoses or two unequal divisions of the spore (e.g. see von Stosch & Drebes, 1964; von Stosch, Theil & Kowallik, 1973; von Stosch & Fecher, 1979). Germination of *Aulacoseira granulata* resting cells has been studied in some detail, using light and electron microscopy, by Sicko-Goad *et al.* (1986) and Sicko-Goad (1986).

The organic casing

If the wall components are obtained from living diatoms (e.g. by cracking the cells with glass beads and then isolating the walls by differential centrifugation), and treated with hydrofluoric acid, the silica is dissolved but 'phantom' cell walls remain. This is the organic casing, which coats all the siliceous components. It is often so thin that it is not observable in material prepared for sectioning and TEM observation but it can obscure detail in uncleaned cells viewed by SEM.

According to von Stosch (1981), the casing consists of the coating membranes and the diatotepic layer which in some diatom groups is laid down internally between the frustule and the plasmalemma, ('membrane' means here a thin layer, and not a semi-permeable protein-lipid structure as in the plasmalemma, endoplasmic reticulum, etc.). The diatotepic layer has been well studied in some diatoms, most of them centric, and we simply do not know how widespread this layer is within the group.

The valves and copulae are formed within the cell, in silica deposition vesicles and then exocytosed from the cell. Several models of valve release have been discussed by Schmid (1986a, b, 1987) and Crawford & Schmid (1986) and include

loss of membrane profiles, their incorporation into the organic component of the frustule or withdrawal of membranes from outside the new valve to become part of the new plasmalemma. Together with any remnant organic material present within the silica deposition vesicle, they might form an ephemeral primary coating around the frustule. Subsequently, secondary coating membranes may be formed around the frustule, by deposition of intercellular and interparietal (i.e. within the cell wall) solutes. Thus in *Bellerochea yucatanensis* the valve becomes partially separated from the plasmalemma by a spontaneous plasmolysis and acquires a rather heavy secondary coating (von Stosch, 1981). The primary and secondary coatings are distinguished on the basis of their development, but are difficult to separate visually. They stain only lightly with basic dyes but may stain densely with uranyl acetate and lead citrate. In *Bellerochea malleus* and *B. horologicalis* the coatings stain with PAS reagent, which reacts with the vicinal hydroxyl groups of carbohydrates.

The diatotepic layer is secreted against the inner side of the frustule. It consists largely of acidic polysaccharides, and, at least in *Phaeodactylum* (Ford & Percival, 1965; Schmid *et al.*, 1981) and *Navicula pelliculosa* (Volcani, 1978, 1981; Schmid *et al.* 1981) contains a sulphated glucuronomannan. The diatotepic layer was first investigated in detail by Liebisch (1929), who noted its presence (his 'Pektinmembran') in centric (e.g. *Melosira, Odontella, Amphitetras*), araphid (e.g. *Rhabdonema*) and raphid diatoms (e.g. *Achnanthes, Mastogloia, Bacillaria, Surirella*). The diatotepum has been investigated in thin section by von Stosch (1981) who found that it may be homogeneous in appearance, laminate or fibrous. Within a particular diatom it may be differentiated laterally in relation to pores and other structures in the overlying silica (von Stosch, 1981). There is always a sharp boundary between the diatotepum and any other organic components of the wall. Unfortunately the results of ultrastructural studies cannot yet be fully integrated with the biochemical evidence accumulated by Volcani and his co-workers (e.g. see Volcani, 1978, 1981; Schmid *et al.*, 1981). These workers have documented changes in the organic components of the cell wall during the cell cycle but the relatively poor resolution of temporally separated events in their experiments (even though their cultures were synchronised by silicate starvation or manipulation of light–dark cycles), makes it impossible to judge whether certain constituents of the wall form part of the coating membranes or part of the diatotepic layer.

Particularly well-developed diatotepic layers are found in diatoms with partially or poorly silicified frustules. Thus in *Subsilicea fragilarioides* (von Stosch & Reimann, 1970), the diatotepic layer is several times thicker than the silica of the cingulum, while in *Phaeodactylum* the diatotepic layer constitutes virtually the whole wall in some of the morphotypes. In other diatoms, e.g. *Bellerochea, Papiliocellulus*, the wall is split, the discrete diatotepic layer lying for the most part at some distance from the siliceous components and their coating membranes. This may involve the valve (e.g. *Bellerochea*: von Stosch, 1977) or the girdle (e.g. *Striatella*: Roth & de Francisco, 1977). We illustrate (Fig. 44) the diatotepum in *Triceratium* (Fig. 44a) where discrete pores are found in that layer; in *Odontella* (Fig. 44b) where several layers of material are arranged in similar fashion to the helicoidal array of microfibrils reported by Neville (1986,1988) from insect cuticle and other organisms; and in *Melosira* (Fig. 44c, d) where the diatotepum has been shown to be continuous with the organic component of the auxospore perizonium (arrowhead).

The sulphated glucuronomannan mentioned above can be fairly confidently reckoned to be part of the diatotepic layer. It is formed during the later stages of maturation and in *Navicula pelliculosa* occurs in fine fibrils making up more than 70% of the total polysaccharide of the organic casing (diatotepum plus coating membranes). The casing also contains other polysaccharides but the locations of most of these are not known with certainty. In *Navicula pelliculosa* sulphated polysaccharides containing fucose, mannose and

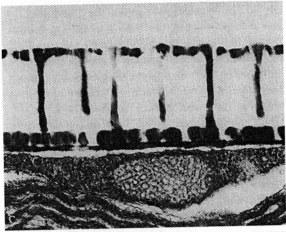

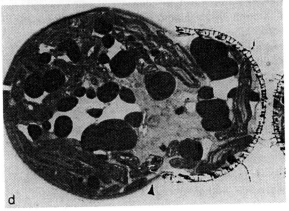

Fig. 44. The diatotepum in some centric diatoms. (*a*) In *Triceratium* a pore in the diatotepum is found beneath the areolae (arrow) but not always in section (arrowhead). (*b*) In *Odontella* the polysaccharide is found in several layers resembling helicoidally arranged microfibrils. (*c*) The thin diatotepum of the vegetative valve of *Melosira nummuloides* is continued into the perizonium of the auxospore as a greatly thickened layer, (*d*). (*a*), (*b*) courtesy R. A. Andersen; (*c*), (*d*), Crawford, 1973b.

galactose, together with smaller amounts of xylose and rhamnose are synthesised during the early stages of valve formation and form part of the primary coating membrane of the wall – probably either as components of the silicalemma or as components of the matrix within which the valve is silicified, inside the SDV (Volcani, 1978,1981). The organic casing has also been shown to contain ribose, arabinose, galacturonic acid, uronic acid and various other sugars (Hecky *et al.*, 1973; Schmid *et al.*, 1981).

Recently, Waterkeyn & Bienfait (1987) have shown that strips of callose, a β1–3 linked glucan, form a gasket at the abvalvar edge of the hypocingulum. This apparently helps to maintain the integrity of the cell wall as the epitheca and hypotheca slide apart during the cell cycle. When the hypotheca becomes an epitheca at the subsequent cell division, the callosic strip is lost, a new one being produced at the edge of the new hypocingulum.

Protein and lipid also occur in the organic casing. The ratio of protein to carbohydrate varies, from only 0.33 in *Phaeodactylum* (where the diatotepic layer is especially well developed) to 6.5 in *Cyclotella cryptica*. In addition to the 20 common amino acids, the proteins of the wall contain some unusual amino acids, including 2,3-cis-4-trans-3,4-dihydroxy-L-proline (Nakajima & Volcani, 1969), ε-trimethyl-L-δ-hydroxylysine and its phosphorylated compound (Nakajima & Volcani, 1970). These amino acids are interesting since they are analogues of amino acids (hydroxyproline and hydroxylamine) found in collagen, and could perhaps form part of a similar fibrous protein in the diatom wall.

The amino acid composition of the casing differs from that of the cytoplasm, with threonine, glycine and especially serine in relatively higher amounts in the casing, and glutamic and aspartic acid, tyrosine, phenylalanine, methionine and cystine in lower proportions (Schmid *et al.*, 1981). In *Cyclotella cryptica* and *C. stelligera*, serine accounts for 23 and 25% of total wall amino acids and Hecky *et al.* (1973) have suggested that silicification itself is mediated by a protein template, in which serine and threonine present a layer of hydroxyl groups on which silicic acid molecules can condense and polymerise. Embryonic walls, however, have much less serine and threonine than mature walls, and so unless the threonine- and serine-rich template

protein is present in very small quantities, perhaps in the silicalemma, the model proposed by Hecky *et al.* seems unlikely.

The lipid content of the cell wall, like the sugar and amino acid content, is unusual, consisting mostly of non-phosphatides – neutral lipids, free fatty acids, mono-and di-galactosyl diglycerides and sulpholipids; of the small phosphatide content, phosphatidyl glycerol is the main component (Kates & Volcani, 1968). Again, like the sugars and amino acids, the lipid composition of the casing differs from that of the cytoplasm. For instance, oleic acid is up to 25 times as abundant in the casing as in the cytoplasm, while the polyunsaturated fatty acid content of the casing is relatively low. The significance of such differences, however, is unclear. In five out of the six species studied by Kates & Volcani, lipid accounted for only 0.8 to 5% of the organic casing; in *Cyclotella cryptica*, however, the content was 13%.

The organic coat around the silica (but presumably not the diatotepum, which is internal to the valve) may help protect the frustule against dissolution. Certainly, acid-cleaned frustules dissolve far more readily than the intact walls of heat-killed diatoms (Lewin, 1961). Removal of cations from intact walls increases dissolution, while addition of Fe^{++} or Al^{+++} ions to acid-cleaned cells reduces dissolution. This suggests that one function of the organic casing may be to complex cations such as iron or aluminium, hence minimising loss of silica from the frustule.

The diatotepic layer presumably has a different function. Von Stosch (1981) suggests that the principal function is to help contain the protoplast. It forms a matrix in which the silica elements are partially embedded or onto which they are stuck, and hence it helps to maintain the integrity of the frustule. The diatotepum may also decrease the effective pore size of areolae and other pores through the frustule and hence modify the permeability of the whole wall.

The protoplast

Plasmalemma

In normal, turgid cells the plasmalemma presses closely against the cell wall and in most cases, therefore, against the silica frustule. Where the valve is chambered, the protoplast may (e.g. *Pinnularia*) or may not (e.g. *Thalassiosira*, *Coscinodiscus*) extend into the chambers, depending on whether the larger apertures of the chambers, i.e. the foramina, are internal or external and whether the diatotepum lines the silica components or lies at some distance from them. Plasmolysis results in the withdrawal of the plasmalemma from the siliceous components except in certain well-defined areas. In some cases attachment is maintained at the rimoportula (e.g. *Thalassiosira*: Schmid, 1984a, b, 1987), elsewhere at the ocellus (e.g. *Lampriscus*: Round, unpublished observations). The plasmalemma also can remain tenaciously attached to the abvalvar end of the hypocingulum (see Geitler, 1949), as might be expected if the cells, especially those with complex shapes, are to be able to divide equally (Mann, 1984b). The possible importance of plasmalemma anchoring sites is also evident from Schmid's (1984b) studies of valve shaping in *Thalassiosira eccentrica*.

Plastids

Plastid shape and number vary greatly within the diatoms as a whole but are relatively constant within many natural groups. As a source of taxonomic information they have been unjustly neglected, except by a few, e.g. Mereschkowsky (1901a, b, c, 1902, 1902/3, 1903a, b) and more recently Cox (e.g. 1981b). Identification of diatoms in ecological studies could often be aided by the use of plastid characters but this information is almost totally absent from floras.

Centric diatoms usually have discoid plastids located in the peripheral cytoplasm (Figs 4a, 8d, e), or in the strands of cytoplasm that radiate out from the nucleus when this is central, or in both. In some genera, e.g. *Chaetoceros* and *Biddulphia*, thin extensions of the valve also contain cytoplasm and chloroplasts. Streaming of the plastids has been observed in *Rhizosolenia* (Round, unpublished observations), and elsewhere slow movement occurs in response, for example, to changing light intensity: in bright light they may congregate at the centre of the cell, while in dimmer conditions they spread out around the cell periphery (see Haupt, 1983).

In pennate diatoms there are usually only a few plastids per cell (two in Figs 8c & 47), or even just one, although in some araphid forms (e.g. *Asterionella*, *Diatoma vulgare*) there are many small plastids as in centric diatoms. Some species of *Nitzschia* (*N. longissima*, *N. ventricosa*) and *Pleurosigma* also have small plastids, but here the resemblance to the centric diatoms is clearly secondary. The one or two plastids found in many raphid diatoms are often elaborately lobed and dissected. Their form, position and behaviour during the cell cycle promise to be of great importance for systematics. We illustrate a range of pennate plastid types in Fig. 45 while some aspects of the changes occurring during the cycle are discussed on p. 82. In raphid diatoms, chloroplast division occurs at a well-defined stage of the cell cycle, although this may be before (e.g. *Sellaphora bacillum*: Mann, 1984a) during (e.g. *Nitzschia*: Geitler, 1975a) or after (e.g. *Donkinia recta*, Cox, 1981c) cytokinesis.

In section (Fig. 46a), each plastid can be seen to be surrounded by four membranes, the outer two representing the plastid endoplasmic reticulum (interpreted as host vacuole membrane and eukaryotic endosymbiont plasmalemma by Whatley & Whatley, 1981) and the others the two membranes of the organelle itself. Inside the plastid are a series of more-or-less parallel lamellae, each composed of three appressed thylakoids, which run the length of the plastid. Surrounding these is a lamella, the 'girdle lamella' which is continuous around the ends of the plastid (Fig. 46a). The plastid DNA is localised in a ring around the edge of the plastid, just beneath the girdle lamella (Coleman, 1985); in TEM sections it often appears as a paler, fibrillar area.

The plastid probably always contains at least one pyrenoid (Fig. 46a), although it is difficult to be sure of this since pyrenoids are often not readily distinguishable with the light microscope. There is evidence for a membrane around the pyrenoid in some taxa (*Melosira varians*: Crawford, 1973a) but not others ('*Caloneis*' *amphisbaena*: Edgar, 1980) while the presence of a lamella traversing the pyrenoid is similarly inconstant (ibid.). The pyrenoid matrix is granular and sometimes paracrystalline or even crystalline (Holdsworth, 1968; Taylor, 1972). Pyrenoid shape varies greatly, from lenticular in most taxa with discoid plastids, to rod-shaped (e.g. in *Navicula sensu stricto*), spherical (e.g. *Cymbella ehrenbergii*) or polyhedral (e.g. the tetrahedron of some *Sellaphora* species). In some cases the pyrenoid is deeply embedded within the plastid (e.g. *Melosira*: Crawford, 1973a), but elsewhere it is more superficial and may project into the cell lumen, when it may be penetrated by simple or branching tubular invaginations of cytoplasm (Tschermak-Woess, 1953; Edgar, 1980; Mann, 1984a). The pyrenoids appear to persist throughout the plastid division cycle and usually divide at the same time as the plastid, although there may then be some change of shape.

The pyrenoids are not associated directly with any intra- or extra-plastidial reserve product: this contrasts with the starch sheaths present in the Chlorophyta. The plastid stroma may contain osmiophilic globules. The predominance of the brown carotenoids usually masks the green of the chlorophylls. The plastid colour is somewhat variable, however, and some taxa are characterised by a paler, or greener, hue (e.g. *Anomoeoneis sphaerophora*, *Pinnularia viridis*). Colouration also varies with habitat. In well-illuminated places, the plastids may be very pale, while populations growing on deep sediments – e.g. at 10 metres in clear lakes and at 50 metres or more along some ocean coasts – tend to be deep brown.

Mitochondria

With care, these may be observed in living cells using light microscopy (e.g. Geitler, 1937, 1949; Fig. 51(E)j). The mitochondria are of the villate type, in which the inner membrane forms tubular projections into the stroma, as in other 'chromophyte' phyla (Fig. 46c). Where the cytoplasm extends into channels or chambers in the cell wall (e.g. in *Pinnularia*, *Nitzschia* and *Surirella*) mitochondrial profiles can often be seen within the channel (e.g. Edgar, 1980). The peripheral position suggests a relationship with transport of substances across the plasmalemma and in raphid genera they are also closely associated with the raphe slits.

Dictyosomes

As with mitochondria, the dictyosomes can often be distinguished in live cells as small plate-like or rod-like structures, depending on whether they are seen in face view or profile. They were illustrated by early light microscopists (Lauterborn, 1896) but their nature was confirmed only recently following TEM studies (Fig. 46b). Static and dynamic aspects of

their distribution in the cell have been studied particularly intensively in several pennate genera, by Gschöpf (1952). Many diatoms have a shell of single dictyosomes parallel and close to the nuclear membrane (Crawford, 1973a), as well as others

Fig. 45. Some examples of plastid arrangements in raphid diatoms. In every case the plastid content of only *one* theca is shown in valve view; the content of the other theca should be assumed to be similar, the mirror image of what is drawn, except in (*a*), (*e*) and (*h*) (see below). Light stipple is used for parts of the plastids seen in face view; dark stipple for parts seen in profile. (*a*) One plastid lying against one valve and one side of the girdle, e.g. *Achnanthidium*, *Diadesmis*. (*b*) One plastid lying against one side of the girdle, with many lobes extending below the valves, e.g. *Epithemia*, *Rhopalodia*. (*c*) One plastid lying against one side of the girdle, where it is deeply constricted, with two lobes extending beneath each valve, e.g. the *Amphora ovalis* group. (*d*) As (*c*), but only one lobe beneath each valve, though this is deeply indented below the raphe slits, e.g. *Encyonema*. (*e*) One H-shaped plastid, consisting of 2 girdle-appressed plates connected by a narrow isthmus which lies against the epivalve (e.g. *Sellaphora*, *Fallacia*) or hypovalve (some *Pinnularia* spp.). (*f*) One plastid, consisting of 2 plates, which are diagonally opposite each other against the girdle and connected by a thin central strip, e.g. *Proschkinia*. (*g*) One plastid, consisting of 2 large valve-appressed plates connected by a narrow isthmus near one pole, e.g. *Cymatopleura*, *Surirella*, *Petrodictyon*. (*h*) 2 plastids, symmetrically placed one on each side of the median transapical plane, each lying against one side of the girdle and one valve, e.g. *Nitzschia*, *Tryblionella*, *Denticula*. (*i*) 2 plastids, symmetrically placed one on each side of the median transapical plane, each consisting of 2 girdle-appressed plates connected by a broad central pyrenoid-containing isthmus, e.g. some *Achnanthes*, *Biremis*, *Hantzschia*. (*j*) As (*i*), but the plastids rotated through 90°, so that the plates lie against the valves, e.g. some *Cavinula*, *Mastogloia*, *Aneumastus*. (*k*) 2 plastids, one lying against each side of the girdle and extending below each valve, e.g. *Navicula*, *Craticula*, *Stauroneis*. (*l*) 2 plastids, one lying against each valve, deeply indented beneath the raphe and transversely, e.g. *Petroneis*. (*m*) 2 plastids, one on each side of the median transapical plane, each consisting of 2 valve-appressed plates, which are indented beneath the raphe and connected across the cell lumen by a narrow isthmus, e.g. some *Lyrella*. (*n*) 4 plastids, lying against the girdle, one in each quadrant of the cell, and extending below the valves, e.g. *Neidium*, *Scoliotropis*. (*o*) 4 plastids, 2 lying against each valve, e.g. some *Lyrella*. (*p*) 2 or 4 contorted, ribbon-like plastids, e.g. *Pleurosigma*. (*q*) 4 or more plastids of the type shown in (*i*), e.g. some *Climaconeis*. (*r*) Many discoid or grain-like plastids, e.g. a few species of *Nitzschia*, *Pleurosigma*, *Amphora*.

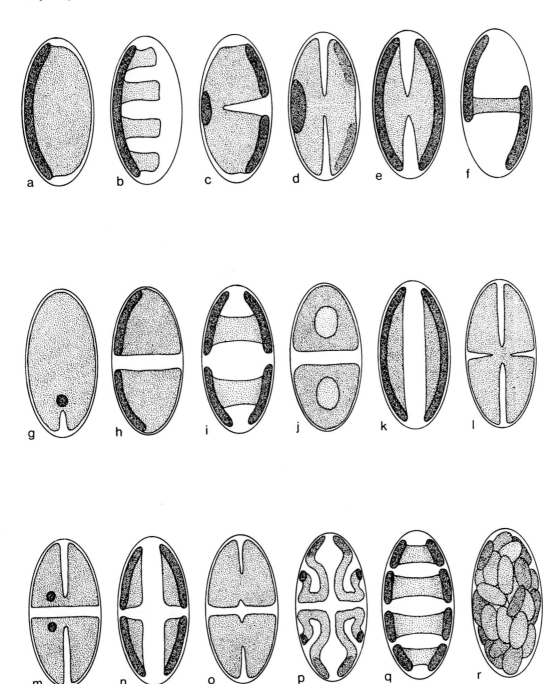

scattered in the cytoplasm around the nucleus or elsewhere in the cell, while in *Synedra* a line of dictyosomes runs from the nucleus towards either pole (Geitler, 1949; Gschöpf, 1952; Mann, unpublished). The dictyosomes of several *Pinnularia* and also *Striatella* (de Francisco & Roth, 1977) species are particularly characteristic since near the nucleus they tend to be paired (Fig. 51(E)i), each pair being oriented at right angles to the nuclear envelope and associated with a centrally placed evagination of both membranes of the latter. This would appear to be a device allowing more dictyosomes to be grouped around the nucleus (Edgar, 1980), and it is interesting that pairing seems to occur only in large diatoms in which the surface area to volume ratio for the nucleus is likely to be less favourable for dictyosome packing.

Nucleus

During interphase, the nucleus usually lies centrally, in a bridge of cytoplasm or near the centre of one valve, and it is usually internal to the plastids (Fig. 47). In a significant minority of taxa, however, the nucleus lies to one side of the cell, adjacent to the girdle. Whatever the interphase position, the mitotic position is apparently always lateral, mitosis taking place just beneath the girdle on one side of the cell (Lauterborn, 1896; Geitler, 1931; von Stosch, 1951b; von Stosch & Drebes, 1964; Mann & Stickle, 1988). This presumably results from the need for the cell to position the dividing nucleus accurately with respect to the plane of cleavage. The positioning is probably mediated by the microtubule organising centre (centrosome), which moves to a central, subgirdle position ahead of the nucleus and there contributes to the formation of the spindle (see p. 66).

Most diatom nuclei are ellipsoidal or spherical, especially where they are suspended in the centre of the cell by a bridge or strands of cytoplasm. There are some interesting exceptions. For example, in *Lauderia* the nucleus consists of two lobes, one by each valve, connected across the centre of the cell by a very thin strand (Holmes, 1977) while in *Surirella* the nucleus can be H-shaped (e.g. Drum &

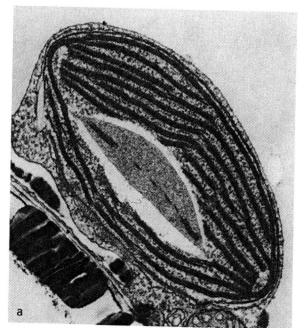

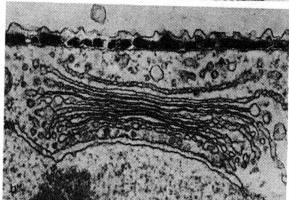

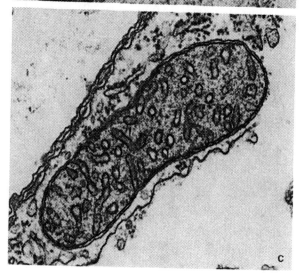

Fig. 46. (*a*) Plastid showing an immersed lenticular pyrenoid, lamellae composed of three thylakoids and a girdle lamella. (*b*) A dictyosome lying between a forming valve and the nucleus. (*c*) A mitochondrion with tubular cristae.

Pankratz, 1964). The significance of this is unknown. The gomphocymbelloid genera frequently have a nucleus that is bean- or kidney-shaped in valve view, presumably because it has to fit around the large central pyrenoid.

The interphase nucleus sometimes 'exhibits a relatively homogeneous fine granular appearance with little indication of chromatin aggregates' (Duke & Reimann, 1977), but as Geitler (1979) has pointed out, heterochromatic structures are quite widespread in diatoms. In some cases this manifests itself in a granular appearance of the interphase nucleus (e.g. *'Caloneis' amphisbaena*; Edgar, 1980), but elsewhere heterochromatic 'plaques' may be present, as at either side of the nucleus in some forms of *Navicula cryptocephala* (Geitler, 1951), or a massive central compound chromocentre, as in other races of *N. cryptocephala*, *N. viridula*, *N. radiosa* and *N. tripunctata* (Geitler, 1951a, 1958b, 1979). Such features can be demonstrated relatively easily using acetocarmine techniques and may provide additional taxonomic data. The nucleus also contains one or more nucleoli. The nuclear envelope is double, as is usual in eukaryotic organisms, with conspicuous pores of a complex structure (Drum & Pankratz, 1964; Crawford, 1973a; Edgar, 1980).

At metaphase the chromatin forms a dense ring in which it is usually impossible to distinguish individual chromosomes. Chromosome counts, where they have been made, have thus been determined principally from meiotic, or occasionally from mitotic prophase (e.g. Geitler, 1927b, 1973). More than 40 counts have been made on various centric andpennate diatoms, the diploid number varying from 8 (*Navicula peregrina*) to c. 130 (*Surirella saxonica*): for details and references, see Cave (in Altman & Dittmer, 1962) and Kociolek & Stoermer (1989). Few generalisations can be drawn from the data available at present.

Vacuole

Diatoms frequently have extensive vacuoles (Fig. 47). Sicko-Goad, Stoermer & Ladewski (1977) found that the vacuolar volumes were considerably greater in diatoms than in other algal phyla, amounting to as much as 61% of total cell volume in *Stephanodiscus binderanus* and 35% in *Fragilaria capucina*. As a consequence of this, the metabolising volumes of their cells were the lowest estimated. Marine planktonic species often have very extensive vacuoles and it has been suggested that regulation of the ionic content of the vacuolar sap may be important in buoyancy control (Gross & Zeuthen, 1948; Anderson & Sweeney, 1978; Beklemishev, Petrikova & Semina, 1961). The vacuolar sap is generally colourless but blue in one species, *Navicula ostrearia*.

In many pennate diatoms there are two large vacuoles, rather than one per cell, these being separated by the central plasma mass containing the nucleus (Fig. 47). It is interesting that in such forms, although the two vacuoles often become tightly appressed as a result of the movement of the nucleus to the side during mitosis, there appears to be no tendency for the vacuoles to fuse (Mann, unpublished).

In many taxa, especially of the centric and araphid pennate groups, there are no obvious features associated with the vacuole. Some raphid diatoms, however, have volutin granules (Fig. 1b). These are usually spherical and apparently structureless (LM) and lie either in the principal vacuoles, (e.g. in *Craticula cuspidata* or *Sellaphora bacillum*) or in smaller, subsidiary vacuoles lying either near the nucleus (*Navicula oblonga*) or the poles (*Cymbella ehrenbergii*). The granules often disperse during cell division and reform afterwards (*Craticula cuspidata*, *Navicula oblonga*), but in *Sellaphora bacillum* one segregates to each daughter cell so that only one new granule is produced by each sibling (Mann, 1985).

The cytoplasm is separated from the vacuoles by a tonoplast but there are no details concerning this membrane.

Storage materials

A β1–3 linked glucan, chrysolaminarin, is commonly found in diatoms along with other carbohydrates (Darley, 1977; Waterkeyn & Bienfait, 1987) and volutin or polyphosphate. The latter is found conspicuously in some raphid pennate genera and the number and position of the accumulation bodies can be used to distinguish taxa. Perhaps the most studied storage products are the various lipids (Lewin & Guillard, 1963; Kates & Volcani, 1966; Darley, 1977). The fatty acid composition differs

somewhat from that of green algae and higher
plants, notably in the absence of linolenic acid from
most species (Darley, 1977). The majority of species
that have been investigated are marine but there
appears to be little significant difference in the fatty
acid complement between freshwater and marine
taxa or between centric and pennate diatoms
(Darley, 1977). However, Opute (1974) showed that
lipid accumulation began only at the end of the
exponential phase in culture, reaching a peak during
the stationary phase. Taguchi *et al.*, (1987) have since
demonstrated that silicate deficiency leads to lipid
synthesis and Roessler (1988) has found that the
lipid is derived partly from newly assimilated
carbon and partly by conversion of previously
assimilated carbon in the form of non-lipid
materials.

Electron microscopists have paid little attention to
the localisation of storage materials in the cell but it
appears that lipids may occur inside and outside the
cell vacuole in *Melosira*, while polyphosphate has
been found only within the vacuole (Crawford,
1973a).

The cell cycle

Like the cells of other organisms during the cell
cycle, diatoms must grow to around twice their
initial volume, double their number of
mitochondria, plastids, dictyosomes and other
organelles, replicate their chromosomes, and then
segregate a full complement of components to each
daughter cell. In addition, each daughter cell must
synthesise a new set of wall elements (hypotheca), to
accompany the set (epitheca) inherited from its
parent.

Because of the structure of their walls, diatom
cells can increase in size only in one direction, as
the epitheca and hypotheca slide apart. This feature,
present in only a few other unicellular organisms,
e.g. the fission yeast *Schizosaccharomyces pombe*
(e.g. see Mitchison, 1957, 1971) makes diatoms
particularly suitable for cell cycle studies, since good
estimates of changes in cell size can be obtained by

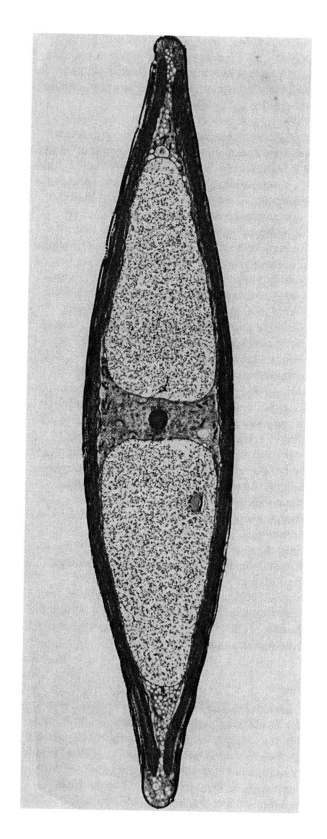

Fig. 47. A longitudinal section in the median valvar
plane through *Craticula cuspidata*. The nucleus (with a
dark nucleolus) lies in the cytoplasmic bridge. The two
vacuoles have fine granular material in them which
may be a fixation artefact. (Micrograph kindly supplied
by L. A. Edgar.)

making one set of measurements, of pervalvar depth. Surprisingly few workers have taken advantage of this and so there is little information about the course of cell expansion, except in the detailed studies of von Stosch (e.g. von Stosch & Drebes, 1964; von Stosch, Theil & Kowallik, 1973) and more recently Olson *et al.* (1986a). These show that expansion is sometimes continuous but elsewhere discontinuous and restricted to particular phases of the cell cycle. In *Chaetoceros didymum*, for instance, expansion is restricted to the second half of interphase and mitotic prophase, and to a short period after cytokinesis, during the formation of the new setae (von Stosch, Theil & Kowallik, 1973); for the first 4 h of interphase, cell length remains more-or-less constant. In *Stephanopyxis turris* too, most of the increase in cell volume takes place towards the end of interphase and during prophase but here, although the period of expansion is longer than in *Chaetoceros*, it occupies a relatively smaller part of the cell cycle (von Stosch & Drebes, 1964). Thus, whereas in *Chaetoceros* cell length remains constant for a third to a half of the 9 h cell cycle, in *Stephanopyxis* the proportion is nearer two-thirds (of a cycle of 24 h). In several raphid diatoms, e.g. *Sellaphora pupula*, the principal phase of expansion takes place just after valve formation is completed, as the cells separate (Mann, unpublished). Olson *et al.* (1986a) analysed the growth of five centric diatoms and found that some (*Thalassiosira weissflogii* and *Lauderia borealis*) increase in size continuously during the cell cycle, while others (*Odontella aurita*, *Coscinodiscus* spp. and also *Stephanopyxis turris*, already studied by von Stosch & Drebes, 1964), expand only at particular times.

Synchronous division has been induced in several species by manipulation of the light–dark regime or by silicon starvation (e.g. von Denffer, 1949; Lewin *et al.*, 1966; Darley & Volcani, 1971). The principal aim of synchronisation has been to produce a single, well-synchronised burst of division, in which the different stages of silicon uptake, metabolism and deposition can be separated and studied. Thus, we have fairly extensive knowledge of the short period during which the new valves are formed (e.g. see Schmid, Borowitzka & Volcani, 1981; Volcani, 1978, 1981; Sullivan & Volcani, 1981), especially for two diatoms, *Navicula pelliculosa* (which will have to be transferred out of *Navicula*, probably to *Sellaphora*!) and *Cylindrotheca fusiformis*. By contrast, although with appropriate light–dark regimes synchrony can sometimes be maintained over several cycles of growth and division, little is known about the other, greater part of the cell cycle.

If *Cylindrotheca fusiformis* cells are starved of silicon, they stop dividing. The cells grow to mature size but are arrested in the G_1 phase of the cell cycle, i.e. before DNA replication occurs (Darley & Volcani, 1969). Addition of silicate to silicon-starved cells releases them from the arrest and DNA synthesis begins soon afterwards. In cells that have been synchronised by light–dark cycles followed by silicon starvation (Darley & Volcani, 1969), DNA content increases between 0.5 and 3.5 h after the addition of silicate, while mitosis, cytokinesis and valve formation must all follow shortly afterwards because the daughter cells separate between 4 and 7 h after the resupply of silicate. From this it seems clear that in this species at least, the G_2 phase (between DNA synthesis in the S phase, and mitosis, M) must be extremely short. In *Thalassiosira weissflogii*, on the other hand, Olson *et al.* (1986b) have shown by flow cytometry that rapidly growing cells spend roughly a third of their 8 h cell cycle in each of G_1, S and G_2 + M (once allowance is made for G_1 cells that have not separated from each other). Nitrogen limitation, however, leads to a great lengthening of G_1, so that it becomes the dominant phase, S and G_2 + M remaining little changed. When *Navicula pelliculosa* is starved of silicon, it arrests at the very beginning of the cell cycle, after DNA replication, mitosis and cytokinesis but before valve formation and cell separation (Darley & Volcani, 1971): the relative lengths of the G_1 and G_2 phases are apparently unknown. Thus, in spite of the considerable amount of research accomplished in the last 30 years, our knowledge of the cell cycle as a whole remains remarkably scanty, even in 'experimental organisms' such as *Navicula pelliculosa*, *Cylindrotheca fusiformis* or *Cyclotella cryptica* (for which, see Werner, 1978). Even so, there have been attempts by phytoplankton ecologists to develop models to explain the control of the cell cycle and the phasing of division in relation to light–dark cycles (e.g. see Heath & Spencer, 1985).

Some aspects of the diatom cell cycle have been studied in detail, however, especially the visible events of nuclear and cell division. This has been done by a combination of electron microscopy of thin sections and careful light microscopy of living cells. As a result we now have an excellent picture of the process and to some extent an understanding of the mechanism of mitosis in a variety of diatom genera, including *Melosira* (Tippit, McDonald & Pickett-Heaps, 1975), *Diatoma* (Pickett-Heaps, McDonald & Tippit, 1975; McDonald *et al.*, 1977; McIntosh *et al.*, 1979; McDonald, Edwards & McIntosh, 1979), *Fragilaria* (Tippit, Schulz & Pickett-Heaps, 1978; Schulz & Jarosch, 1980), *Surirella* (Tippit & Pickett-Heaps, 1977; Pickett-Heaps, Schmid & Tippit, 1984), *Pinnularia* (Pickett-Heaps, Tippit & Andreozzi, 1978a, b, 1979c; Soranno & Pickett-Heaps, 1982; Hinz, Spurck & Pickett-Heaps, 1986), *Nitzschia* and *Hantzschia* (Pickett-Heaps, Tippit & Leslie, 1980a, b; Tippit, Pickett-Heaps & Leslie, 1980). In addition Manton, Kowallik & von Stosch (1969a, b, 1970a, b) have described mitotic and meiotic division in *Lithodesmium*. The structure of the diatom mitotic apparatus, with its remarkably compact and orderly central spindle (Fig. 48c–e), makes it particularly suitable for detailed analysis (e.g. see Pickett-Heaps & Tippit, 1978; Pickett-Heaps, Tippit & Porter 1982) and for experimentation, e.g. using ultraviolet light to destroy specific small parts of the spindle *in vivo* (Leslie & Pickett-Heaps, 1983, 1984). It has also become possible to isolate the spindles from *Stephanopyxis turris* cells and reactivate them *in vitro* (Cande & McDonald, 1985, 1986), offering further opportunities for experimentation.

These recent advances were foreshadowed in an early paper by Lauterborn (1896), which is remarkable for its accuracy and insight (see also Pickett-Heaps, Schmid & Tippit, 1984). Lauterborn investigated mitosis and cell division in *Surirella* and to a lesser extent in several other freshwater diatoms, by light microscopy of living or fixed and stained cells. Other genera were subsequently studied by Cholnoky, Geitler, Karsten and others (see Fritsch, 1935).

The unusual nature of the diatom spindle was evident from these early papers, but it is largely the investigations of Pickett-Heaps and his co-workers that form the basis of new interpretations of the mechanisms of mitosis in other groups of organisms (Pickett-Heaps, 1986) and for the following account.

During interphase the nucleus is usually accompanied by a small body, the centrosome (Figs 48a, 49a), which is rarely obvious in the light microscope, except in large diatoms such as *Surirella capronii*, studied by Lauterborn (1986), and *S. robusta* (Pickett-Heaps, Schmid & Tippit, 1984). In preparation for or during prophase, the centrosome becomes the focus for numerous radiating microtubules (Fig. 49b) – hence its alternative name, the microtubule centre, MC, or microtubule organising centre, MTOC – and moves to the side of the cell, to the midline of the girdle. As it does so, the nucleus is also drawn to the side, apparently through interaction with the system of microtubules radiating from the centrosome. During this movement the nucleus is often distorted, one side being extended into a conical beak pointing towards the centrosome (e.g. see Lauterborn, 1896). It must be noted, however, that the nucleus does not always move before mitosis. There are some diatoms in which the nucleus already lies against the centre of the girdle during interphase; in such forms (which include *Cymbella*, *Gomphonema*, *Amphora ovalis*, *Epithemia* and others) there is no nuclear migration.

At some stage a small, disc-like, laminated structure appears near the centrosome (Figs 48a, b, 49c, d). The association between them is so close that it is hard to avoid the conclusion that it is the centrosome that gives rise to the layered structure. In *Surirella ovalis* the layered structure, which is the precursor of the spindle, appears before the migration of centrosome and nucleus to the broad end of the cell, and before the nucleus enters prophase (Tippit & Pickett-Heaps, 1977). In *Lithodesmium undulatum*, however, the spindle precursor is apparently present throughout interphase (Manton, Kowallik & von Stosch, 1969a), and it has been shown too that in *Hantzschia*

Fig. 48. Formation of the mitotic spindle. (*a*) Early stage: spindle precursor (arrow) adjacent to densely staining spherical microtubule organising centre or centrosome (see also Fig. 49(*c*)). (*b*) Spindle precursor enlarging. (*c*) A longitudinal section of an expanding spindle: the spindle precursor has split into two polar plates separated by two interdigitating sets of microtubules. (*d*) Transverse section of the hollow spindle. (*e*) Expanded spindle with chromosomes stretched over it (at metaphase): the dark masses on either side of the spindle are the chromosomes (see also Fig. 49(*h*)). (Photographs kindly supplied by Professor J. D. Pickett-Heaps.)

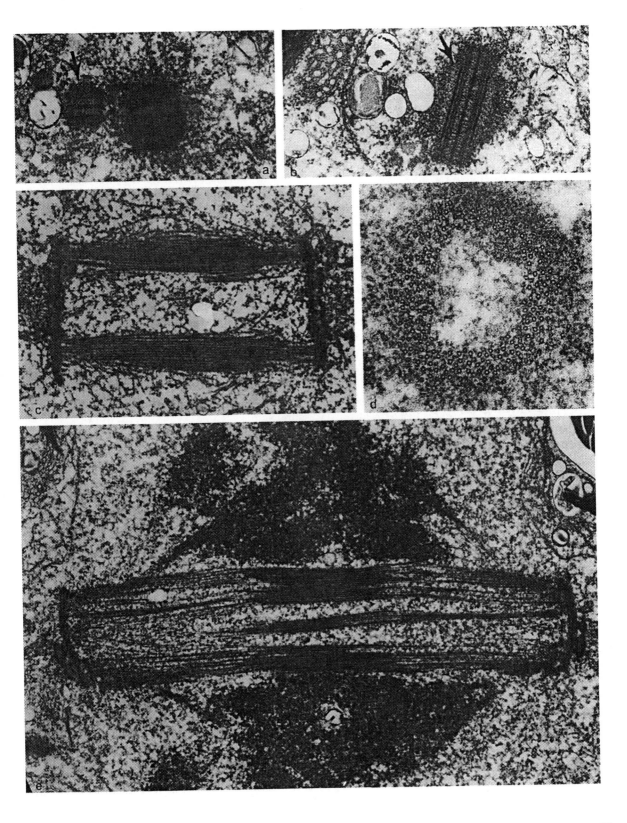

amphioxys the spindle precursor develops close to the centrosome soon after the preceding cell division, towards the end of valve formation (Pickett-Heaps & Kowalski, 1981).

During prophase the spindle precursor, now positioned just beneath the centre of the girdle on one side of the cell, splits into two equal halves separated by a band of densely-packed, parallel microtubules (Fig. 49d–f). The halves of the precursor form the polar plates of the spindle (Fig. 48c) and persist throughout mitosis, until they detach from the spindle at telophase and disappear. Each polar plate is a thin, layered structure. It is often associated with a cap of finely fibrillar material (e.g. Pickett-Heaps, Tippit & Leslie, 1980b), which is sometimes apparently enclosed by membranes (Pickett-Heaps, Schmid & Tippit, 1984), and there may also be a prominent 'polar vesicle' (Fig. 49d, e). The functions of these structures are unknown. As mitosis proceeds the cap and the vesicle disappear (Fig. 49f–i).

The polar plates are often concave during the early development of the spindle (Fig. 49e–g) and lie at an angle to each other, the spindle being broader towards the cytoplasm than towards the cell wall. At first, all the microtubules found between the polar plates are continuous from pole to pole (Fig. 49f). They lie in orderly arrays, part or all of the spindle exhibiting tetragonal or hexagonal packing. In addition, some microtubules can be seen radiating into the cytoplasm from around the edges of the polar plates.

In some species, e.g. *Melosira varians* (Tippit, McDonald & Pickett-Heaps, 1975), *Surirella ovalis* (Tippit & Pickett-Heaps, 1977) and *Fragilaria capucina* (Tippit, Schulz & Pickett-Heaps, 1978), the central band of microtubules (the central spindle) is a solid cylinder, with microtubules inserted over the whole inner surface of each polar plate. In other diatoms, by contrast, the central spindle is hollow (Figs 48c, d, 49f–i: Pickett-Heaps, Tippit & Andreozzi, 1978a; Pickett-Heaps, Schmid & Tippit, 1984). It is probably no coincidence that the diatoms with hollow spindles (*Pinnularia* species and *Surirella robusta*) are also the largest that have been investigated, with the largest nuclei. Since any interactions between chromosomes and spindle must take place around the periphery of the central spindle, we can predict that an increase in the

volume of chromatin to be segregated during mitosis will have to be accompanied by an even greater increase in the surface area of the spindle. It is unlikely, however, that the force required to separate the chromosomes needs to increase more than linearly with the increase in chromatin mass. Thus as chromatin mass increases by an amount x, the surface will need to increase by an amount ax^2, but the number of microtubules probably only by an amount bx, where a and b are constants. Hence the change from a solid to a hollow spindle.

During prophase the nuclear envelope begins to break down. The spindle, which has hitherto been entirely extranuclear, sinks down into the nucleus during prometaphase and concomitantly enlarges markedly (Fig. 49g, h). The centrosome disappears and is not seen again until telophase. The elongation of the spindle is caused in part by growth of the central spindle microtubules, but it also involves sliding between microtubules of opposite polarities (e.g. Pickett-Heaps & Tippit, 1978; Tippit, Schulz & Pickett-Heaps, 1978; McIntosh *et al.*, 1979; McDonald, Edwards & McIntosh, 1979). After the late prophase or prometaphase elongation, the central spindle can be seen to consist of two interdigitated half-spindles, which at this stage overlap to a considerable extent (Figs 48c, 49g, h). The chromosomes now become arranged around the spindle and attach to it in some way, still not fully understood, involving interactions between kinetochores and microtubules attached distally to the polar plates, and also involving a collar of material that encircles the central spindle between the chromatin and each pole (Tippit & Pickett-

Fig. 49. Diagrammatic sections of mitosis in *Surirella*. (*a*) Centrosome during interphase. (*b*) Microtubules forming at the centrosome. (*c*) Appearance of a tiny spindle precursor. (*d*), (*e*) Growth of the spindle precursor accompanied by electron dense material and vesicles. (*f*) Degeneration of the centrosome and growth of the spindle microtubules. (*g*) Two sets of microtubules becoming obvious. (*h*) Maturing spindle at metaphase with chromosomes stretched over the two interdigitating half-spindles. (*i*) Telophase: interdigitating spindles only overlapping slightly following sliding apart of the microtubules; chromosomes now positioned at or beyond the poles. (*j*) Separation of the polar plate from the spindle microtubules (at or following telophase). (*k*), (*l*) Budding off and separation of a new centrosome from the polar plate.

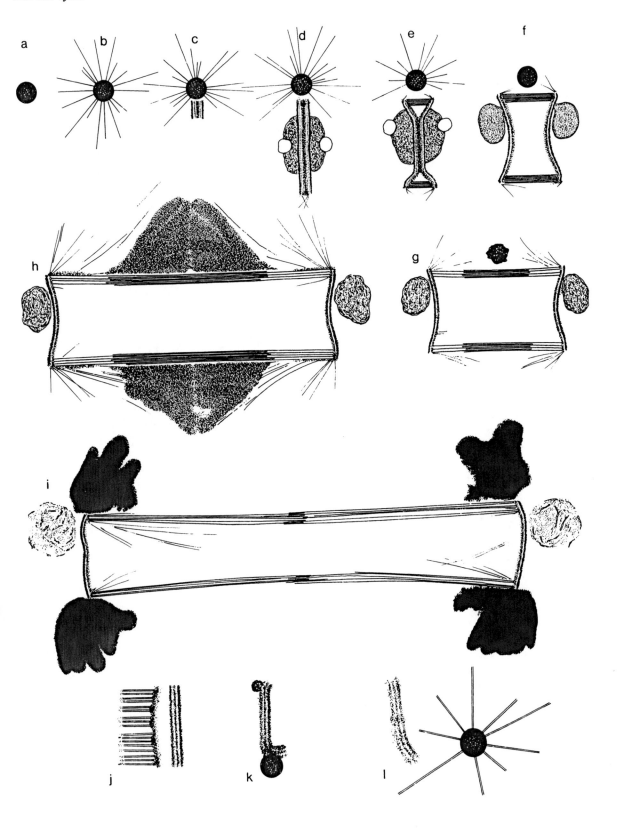

Heaps, 1977; Pickett-Heaps & Tippit, 1978; Pickett-Heaps, Tippit & Andreozzi, 1978a, b; Pickett-Heaps, Tippit & Leslie, 1980a, b; Pickett-Heaps, Tippit & Porter, 1982; Pickett-Heaps, 1986). Two processes occur simultaneously at anaphase: movement of the chromosomes to the spindle poles, and elongation of the central spindle by further sliding apart of the two half spindles (Figs 48e, 49h, i). This latter process has been studied intensively *in vivo* (Tippit & Pickett-Heaps, 1977; McDonald *et al.*, 1977; Tippit, Schulz & Pickett-Heaps, 1978; McIntosh *et al.*, 1979; McDonald, Edwards & McIntosh, 1979; Pickett-Heaps, Tippit & Leslie, 1980a, b; Leslie & Pickett-Heaps, 1983, 1984) and has also been reproduced *in vitro* (Cande & McDonald, 1985, 1986). The elongation is caused by interactions between microtubules, making them slide apart, not by microtubule polymerisation, nor by forces applied at the spindle poles (Cande & McDonald, 1986). At telophase the two halves of the spindle separate and disassemble (Soranno & Pickett-Heaps, 1982), the polar plates separate from the spindle (Fig.49j and see Tippit & Pickett-Heaps, 1977) and the nuclear envelope reforms. A new centrosome is formed close to the daughter nucleus (Fig. 49k, l), apparently from the remains of the polar plate (Tippit & Pickett-Heaps, 1977; Pickett-Heaps & Kowalski, 1981).

While the nucleus is dividing, a cleavage furrow appears and cuts the cell in two. In diatoms with circular valves the outline of the furrow (i.e. the appearance of the furrow in valve view) is also circular (Fig. 50a and see Schmid & Volcani, 1983), whereas in bipolar diatoms the furrow becomes visible at the pole long before it begins to cut through the central part of the cell (Fig. 50b; Mann, 1985). But even though the furrow itself may not at first encircle the cell in a bipolar diatom, the effector of cleavage – a narrow contractile band of microfilaments running around the whole perimeter of the cell (Pickett-Heaps, McDonald & Tippit, 1975; Tippit & Pickett-Heaps, 1977) – is apparently complete from a very early stage (Pickett-Heaps, McDonald & Tippit, 1975; Mann, 1985). It has been suggested that the bundle of microfilaments is initiated close to the dividing nucleus and grows out from here around the cell (Schmid & Volcani, 1983) but this requires confirmation. The profile of the cleavage furrow (i.e. its appearance in girdle view)

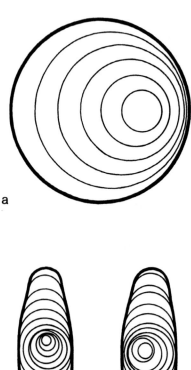

a

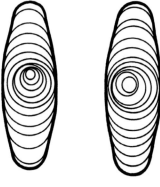

b

Fig. 50. Successive stages of cleavage in (*a*) *Coscinodiscus wailesii* (after Schmid & Volcani, 1983); (*b*) 2 cells of *Sellaphora pupula*. The outline of the cleavage furrow is shown moving into the cell as seen in the median valvar plane.

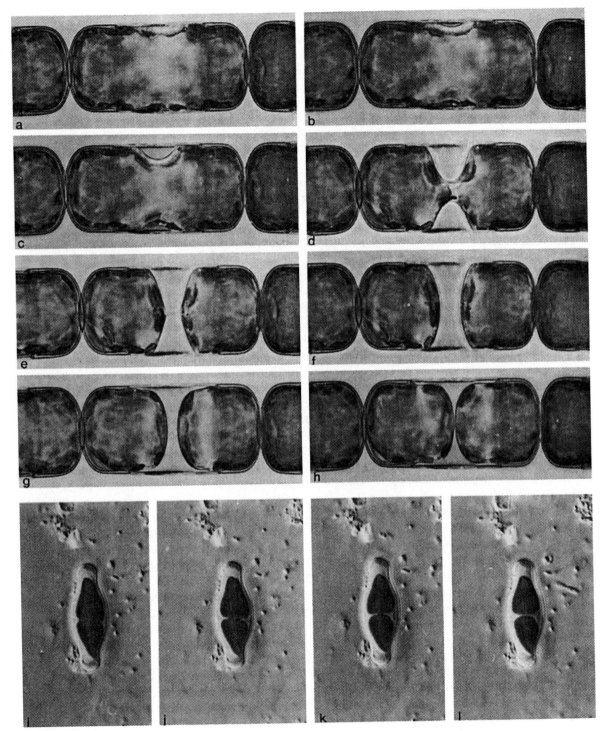

Fig. 51(A). a–h, Cell division in *Melosira moniliformis* in girdle view. Following nuclear division and cell elongation (a), the protoplast cleaves (b–e), at first unequally. The daughter protoplasts expand before the new hypovalves are formed (h). The sequence spans two minutes. i–l, Plastid division in *Navicula capitata*. The sequence spans 20 minutes.

and the time taken for cleavage vary from species to species (Fig. 51). Slight plasmolysis accompanies cleavage in many marine centric diatoms, e.g. *Stephanopyxis turris* (von Stosch & Drebes, 1964) and *Melosira moniliformis* (Crawford, 1981a), so that the daughter cells are partly or wholly separated from each other within the parent frustule. Indeed, it is this withdrawal of the protoplasts from each other that allows the formation of elaborate, non-complementary valve shapes such as those in *Odontella* or *Chaetoceros* (von Stosch, Theil & Kowallik, 1973). The time course of cleavage has not been studied in enough diatoms for many generalisations to be possible. The whole process takes a few minutes in fairly small marine diatoms such as *Melosira moniliformis* (Crawford, 1981a), *Stephanopyxis turris* (von Stosch & Drebes, 1964) or *Chaetoceros didymum* (von Stosch, Theil & Kowallik, 1973), but is much longer in *Surirella robusta* (over 20 min.) and *Coscinodiscus wailesii* (40 min.), two large species studied by Pickett-Heaps, Schmid & Tippit (1984) and Schmid & Volcani (1983). Pickett-Heaps, Tippit & Leslie (1980a) noted an abrupt stimulation of the rate of cleavage at one stage in the division of *Hantzschia amphioxys*, but in *Sellaphora* (= *Navicula*) *pupula* there is apparently only a gradual acceleration of the shortening of the contractile ring (Mann, 1985 and unpublished).

The two half-spindles have often disengaged and begun to disassemble before cytokinesis is complete (e.g. Soranno & Pickett-Heaps, 1982), but sometimes the spindle persists, at least in part, slowing down (e.g. in meiosis I of *Lithodesmium undulatum*: Manton, Kowallik & von Stosch, 1970a) or even preventing the final separation of the cells. In *Chaetoceros rostratum* the daughter cells remain connected by a cytoplasmic bridge until valve formation is well under way, when a siliceous septum is formed across the tube between the cells (Li & Volcani, 1985b).

It is not known how the plane of cleavage is determined, but it is not unlikely that the cell uses its wall as a 'map' and positions the cleavage bundle in relation to a particular wall component. The edge of the hypocingulum or somewhere near it is an obvious candidate for this role. Where a cell divides unequally, either after a mitosis or after meiosis I, the plane of division usually bears a fixed relation to the polarity of the cell. Thus, for instance, in the

formation of the first resting spore valve of *Eunotia soleirolii* (von Stosch & Fecher, 1979), or during the formation of the single gamete in a *Sellaphora pupula* gametangium (Mann, unpublished), the plane of cleavage is always displaced towards the epivalve.

Following cleavage the new valves are formed, each within a single silicon deposition vesicle (SDV). Various cytoplasmic structures are closely associated with the SDV and are presumably involved in morphogenesis. These include the nucleus; various systems of microtubules, some of which are connected to the centrosome; the centrosome itself; microfilaments; endoplasmic reticulum; and the 'raphe fibre' and 'labiate process apparatus', in diatoms that have raphes or rimoportulae. Li & Volcani (1985a, c) have shown that the SDVs of *Ditylum brightwellii* (see also Pickett-Heaps, Wetherbee & Hill, 1988), *Stephanopyxis turris*, *Odontella aurita* and *O. sinensis* appear during cleavage, and this is also true in *Attheya decora* (Schnepf, Deichgräber & Drebes, 1980), but in other species investigated so far, the SDV has been detected only after cytokinesis is complete (e.g. Chiappino & Volcani, 1977; Pickett-Heaps, Tippit & Andreozzi, 1979b; Pickett-Heaps & Kowalski, 1981; Edgar & Pickett-Heaps, 1984a). The centrosome, having re-formed at or just after telophase, now usually migrates to lie just below the young SDV. In raphid diatoms, in which the SDV is at first a long thin tube, the centrosome often comes to lie centrally, where the 'central nodule' is formed between the two raphe slits (e.g. Pickett-Heaps, Tippit & Andreozzi, 1979b; Pickett-Heaps & Kowalski, 1981). There is no obligate relationship between centrosome and SDV, however, since in some diatoms, e.g. *Diatoma* (Pickett-Heaps, McDonald & Tippit, 1975) and *Ditylum* (Li & Volcani, 1985a), the centrosome or other microtubule centre remains in a depression in the surface of the nucleus, or lies free in the cytoplasm during valve formation. Likewise there is usually a close relationship between the nucleus and the forming valve, but not always. When, after

Fig. 51 (B). a, Plastids of *Nitzschia* sp. almost fully cleaved by cleavage furrow. b–e, Cell cleavage of *Craticula cuspidata* in valve view. f–i, Cell cleavage of *C. cuspidata* in girdle view. The sequences span 4 and 18 minutes.

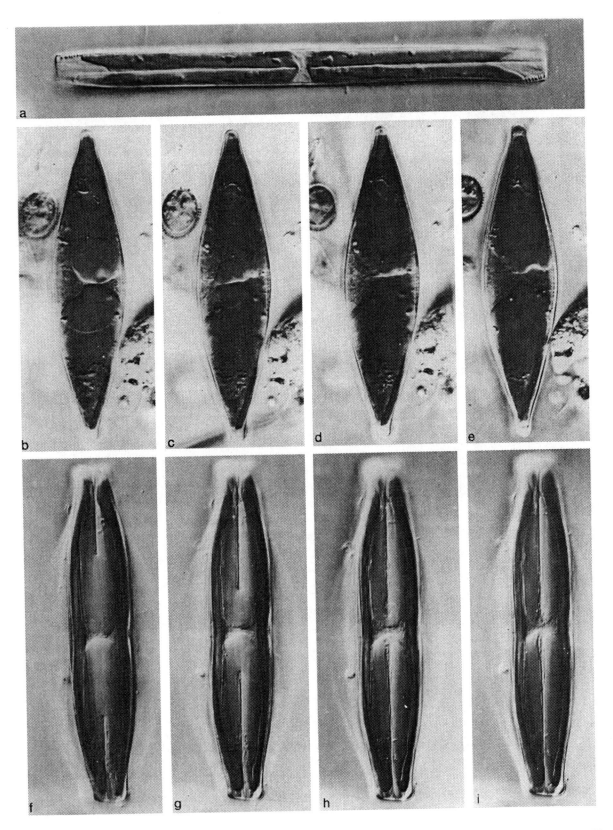

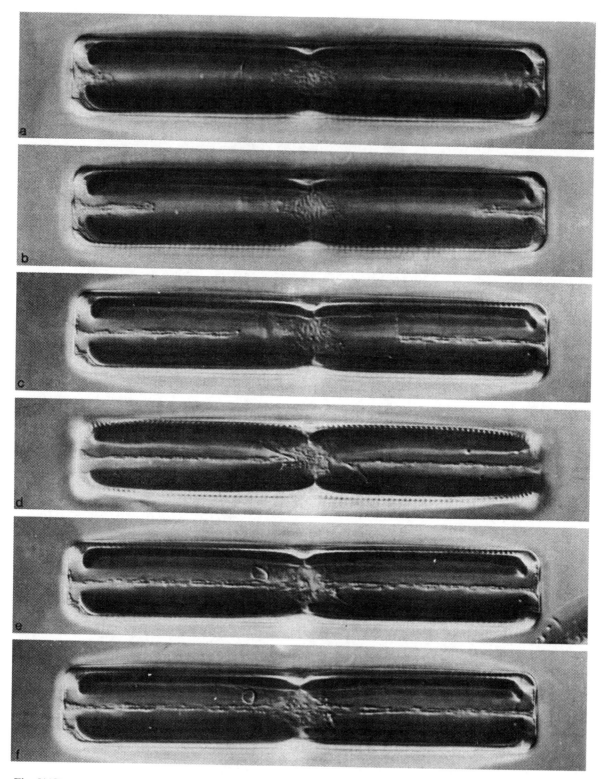

Fig. 51(C). Cell division of *Navicula oblonga* in girdle view. The protoplast is fully cleaved in f. The sequence spans 8 minutes.

cytokinesis, the centrosome moves to the surface of the cleavage furrow, the nucleus is often drawn along behind it, coming to lie below the centre of the SDV as it develops. Thus, in *Craticula cuspidata* the nucleus, central during interphase, moves to the girdle before mitosis but moves back to the centre during the first few minutes after cytokinesis (Edgar & Pickett-Heaps, 1984a; Mann unpublished). In many other diatoms too the nucleus remains close to the new valve as it forms, even if during interphase the nucleus lies at the opposite end of the cell (e.g. in *Melosira varians*: Tippit, McDonald & Pickett-Heaps, 1975). But in *Sellaphora pupula*, *Cymbella*, *Placoneis* and a number of other raphid diatoms (Mann, unpublished), the nucleus is never central in the cell, nor is it centred with respect to the forming raphe system.

The SDV now expands laterally. As with the cleavage bundle, there must be some kind of cortical differentiation within the cell or outside it, perhaps involving the cell wall via transmembrane proteins, that 'tells' the expanding SDV where to stop. Control is precise, since it is rare to find valves with abnormally deep mantles: parent valves and daughter valves usually correspond almost exactly in their length, breadth *and* depth. Bundles of microfilaments can sometimes be seen flanking the SDV on either side (in raphid diatoms the microfilaments are initially on one side only, reflecting the asymmetry of the raphe-sternum and its formation), and these may perhaps be involved in the expansion of the SDV (Pickett-Heaps, Tippit & Andreozzi, 1979b; Edgar & Pickett-Heaps, 1984a). Microtubules can also be found close to forming valves.

In raphid diatoms the developing raphe fissures are occupied by organelles called 'raphe fibres' by Pickett-Heaps, Tippit & Andreozzi (1979b), who discovered them in *Pinnularia* and suggested that they might be involved in the generation of the shape of the raphe slit, by slow lateral movements during valve formation. Some kind of interaction seems likely between the raphe fibre and microtubules because treatment with microtubule inhibitors leads to a fragmentation and disruption of the raphe system (Schmid, 1980; Blank &

Fig. 51(D). Cell division sequence in *Amphora* sp. in girdle view lasting 6 minutes.

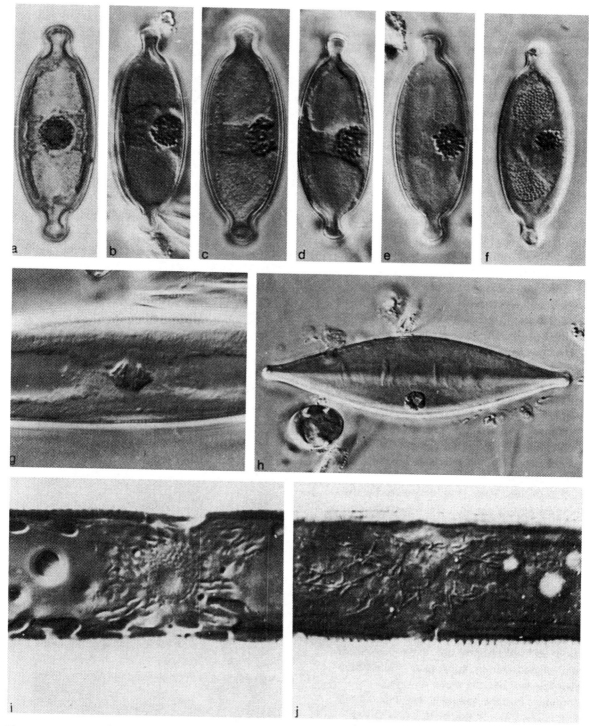

Fig. 51(E). a–f, Nuclear division stages in *Pinnularia amphisbaena*: a, interphase; b, early prophase; c, mid-prophase; d, late prophase; e, metaphase; f, telophase. g, *Craticula cuspidata*, spindle and chromosomes, prophase. h, *C. cuspidata*, metaphase nucleus. i, j, double dictyosomes and mitochondria respectively in *Pinnularia*.

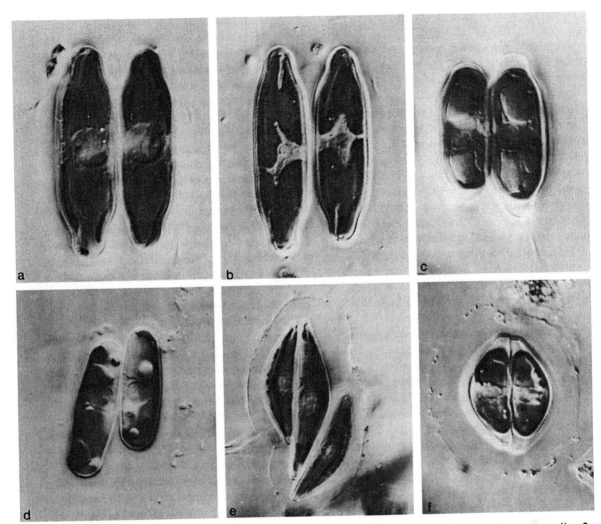

Fig. 51(F). Sexual reproduction in pennate diatoms. a, b, Paired cells of *Neidium*; c, zygotes of *Neidium*; d, Paired cells of *Sellaphora*, e, Rare triplet of *Cymbella* in meiotic prophase; f, Mucilage envelope surrounding 2 zygotes of *Cymbella cistula*.

Sullivan, 1983); the stria pattern is relatively little affected.

The formation of the rimoportulae has been studied in several genera by Li & Volcani (1985a, b, c). Their claims that rimoportulae are the centres of silicification in centric diatoms (1985c) cannot be upheld in the face of evidence from comparative morphology (e.g. this volume) and studies of taxa such as *Coscinodiscus wailesii* (Schmid & Volcani, 1983). Li & Volcani's studies do show, however, that each forming rimoportula is subtended by a layered structure, which is attached to the inner profile of the SDV. This structure, which has been found in *Ditylum, Stephanopyxis, Odontella* and *Chaetoceros*, appears just before the formation of the rimoportula and disappears when this is complete. It presumably functions in some way to prevent silica deposition in the lumen of the rimoportula and to control the development of the internal lips of the portule. The raphe fibre apparently operates in a similar way and it has been suggested that these two organelles may be homologous (Li & Volcani, 1985c; Pickett-Heaps, Wetherbee & Hill, 1988).

We have discussed elsewhere the sequence of

silica deposition in diatoms and the pattern centres characteristic of different groups (p. 31). The cytoplasmic basis of pattern formation remains unknown. Diatoms are able to produce intricate patterns and to reproduce them faithfully division after division. The spacings of ribs and pores are remarkably constant within species or populations, but how they are generated is a mystery (see, however, Lacalli, 1981 for some ideas). The basic pattern of ribs and pores is generated as the SDV expands; vertical differentiation, with the addition of additional layers above or below the basic framework, may take place later, when the SDV occupies the whole area of the presumptive valve. Following the completion of the valve it is released to the cell exterior by a process that is not yet fully understood (see Crawford & Schmid, 1986; Schmid, 1986a, b). Later in the cell cycle, new girdle bands are added to the hypotheca, each being formed in its own, separate SDV.

Before leaving valve formation we should mention some changes in protoplast structure and arrangement, occurring either during valve formation or between cell divisions, which affect the overall symmetry or topography of the valve, rather than the basic structure of sternum or annulus, ribs and areolae. Immediately after the cleavage furrow has cut through the cell, each daughter cell is bounded on one side, not by a cell wall but by a naked membrane. This surface is therefore more-or-less evenly curved, partially planar, or wholly planar, depending upon the extent to which the parent cell plasmolyses during cleavage. If there is no plasmolysis, the naked part of one protoplast will be pressed against the naked part of the other protoplast and the boundary between them will therefore usually be planar. If on the other hand the daughter protoplasts withdraw from each other completely, as a result of plasmolysis, then their naked surfaces will usually be evenly curved. This difference in cleavage underlies the distinction we have made elsewhere between interactive and noninteractive division (p. 34). Following cleavage, however, the cells sometimes modify their shape, so that the new valves are not planar or evenly curved. This is especially common in cells exhibiting non-interactive division and is elegantly demonstrated by von Stosch, Theil & Kowallik's (1973) *in vivo* observations of *Chaetoceros* or the recent EM studies

of *Surirella robusta* by Pickett-Heaps *et al.* (1988), but interactive types may also change the shape of the cleavage furrow as valve formation proceeds, producing complementary humps and hollows in the sibling valves. The modification of the initial shape of the cleavage furrow takes place progressively as the valve matures. Thus, in *Cymatopleura*, the circumferential raphe system, together with the fibulae and the transapical ribs are all present, though incompletely thickened, before the valve face becomes thrown into the series of undulations so characteristic of the genus (Mann, unpublished observations). Again, in *Tryblionella* the initially planar cleavage furrow is gradually bent during valve formation to produce the curved profile of the mature valve (Pickett-Heaps, 1983, figs 13, 15 of '*Nitzschia*' *tryblionella*). It is not known how these changes are brought about, although by analogy with other organisms we may suspect that microtubules are involved.

Another change that occurs in a few diatoms is a movement of the whole silica-depositing apparatus – SDV, centrosome, and associated mitochondria, microfilaments, microtubules, etc. – across the cell during valve formation. Thus, in *Hantzschia amphioxys*, the SDVs of both daughter cells arise more-or-less centrally, as in naviculoid diatoms, but then migrate across to the side, near the girdle (Pickett-Heaps & Kowalski, 1981). A similar phenomenon occurs in *Nitzschia*, except that here the SDVs of the daughter cells move in opposite directions, so that they come to lie near opposite sides of the girdle (Pickett-Heaps, 1983). The effect of the movement in *Hantzschia* is to ensure that every cell has both raphe systems on the same side of the frustule (constant hantzschioid symmetry). In *Nitzschia*, however, various types of behaviour are shown, depending on the species. In some cases all

Fig. 52. Nuclear movements during the cell cycle. (*a*) In a centric diatom (*Melosira*) spanning interphase (I), through metaphase (II), valve formation (III), back to interphase (IV), (V) movement of the nucleus in anticipation of division and (VI) return movement, in girdle view. (*b*)–(*g*) Similar movements in, for example, (*b*) *Cymbella, Eunotia, Rhoicosphenia*; (*c*) *Petroneis, Lyrella*; (*d*) *Navicula, Pinnularia*; (*e*) *Sellaphora, Fallacia*; (*f*) *Surirella, Petrodictyon*; (*g*) *Cymatopleura*. (I–V valve view, VI girdle view). I. Interphase, II metaphase, III valve formation, IV interphase, V next metaphase, VI interphase cell in girdle view.

The cell cycle

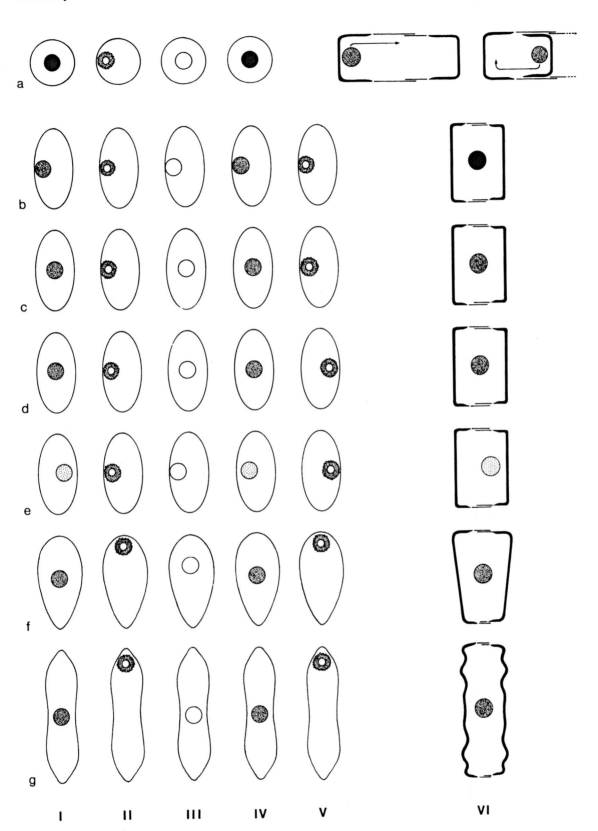

a   b   c   d   e   f   g

I   II   III   IV   V   VI

79

cells have their raphe systems on diagonally opposite sides of the frustule (constant nitzschioid symmetry), but elsewhere nitzschioid and hantzschioid cells may be produced in a 1:1 or 1:2 ratio (Mann, 1980a).

Achnanthes exhibits another interesting variation. The frustule of an *Achnanthes* cell is heterovalvar: one valve has a raphe, the other does not. After mitosis and cytokinesis in *A. coarctata*, the centrosome, accompanied by the nucleus, moves to the centre of the cleavage furrow and becomes associated with the newly formed SDV. Initially the development of both new valves is identical: both begin to form raphe fissures. But in one daughter cell, before silica deposition has progressed very far, the centrosome (microtubule centre) loses its association with the SDV and moves away from it, together with the microtubules and other associated organelles. At the same time the SDV moves across to the side of the cell, near the girdle, and the incipient raphe fissure is filled in. Thus, instead of having a central raphe-sternum, one valve comes to possess a plain, bar-like sternum or 'pseudoraphe', at the junction of valve face and one mantle (Boyle, Pickett-Heaps & Czarnecki, 1984). In other 'monoraphid' diatoms too the pattern-centre of the raphe-less valve lacks a raphe, not because a raphe is never present but because it is filled in during development (e.g. *Cocconeis*: Mann, 1982a; *Achnanthidium*: Mayama & Kobayasi, 1989).

We have mentioned previously that the nucleus moves to one side of the girdle for mitosis (Fig. 52a) unless it is there already, but we did not specify which side. The position of the mitotic nucleus is often constant in relation to the structure and symmetry of cell wall and protoplast. Furthermore, there is also often a strict relationship between the position of the dividing nucleus and subtle aspects of symmetry in the daughter cells. Such matters are difficult to study in radially organised centric diatoms, since although their valves and girdle are usually polar – e.g. as expressed in the eccentric pseudonodulus of *Actinocyclus*, or the single eccentric rimoportula of some *Thalassiosira* species – the polarity is cryptic and difficult to observe in living, dividing cells. In bipolar centric diatoms such as *Chaetoceros*, however, the nucleus can be seen to divide in the middle of the girdle, on one of the broader sides of the cell (von Stosch, Theil &

Kowallik, 1973). The raphid diatoms are better known. Here the asymmetry of the valves, shown in the deposition of one side of the raphe-sternum first, then the other (p. 40), bears an apparently constant relation to the position of the mitotic nucleus. The nucleus always divides below the middle of one of the broader sides of the cell, and it is on this side that the first-formed, primary sides of the new valves appear (Mann, 1981a, 1983, 1984a). But this side need not be the same side of the cell on which the nucleus last divided. Thus two types of frustule can be produced – *cis* frustules, in which the primary halves of epivalve and hypovalve both lie on the same side of the cell, and *trans* frustules, in which they lie on opposite sides. Where nuclear division always takes place on the same side of the cell (Fig. 52b, c), all the frustules have *cis* symmetry; this occurs in *Cymbella*, *Gomphonema*, *Lyrella*, *Petroneis*, *Placoneis*, *Anomoeoneis* and others. If division can take place on either side with more-or-less equal frequency, the result is a 1:1 ratio of *cis:trans*; this is found in *Neidium* and (with some modification) in *Brachysira*. In a third group, however, there is a strict alternation between one side of the cell and the other (Fig. 52d, e), producing a 1:2 ratio of *cis:trans* (e.g. see Mann, 1984a, b; Mann & Stickle, 1988). Many of the naviculoid genera, e.g. *Navicula*, *Pinnularia*, *Sellaphora*, *Gyrosigma* and *Frustulia*, exhibit this 'oscillation'. These cryptic symmetry relationships are intriguing. Thus, although *Navicula* and *Petroneis*, for example, have central interphase nuclei and protoplasts that are apparently quite symmetrical about the apical plane, there must be some invisible polarity that enables a *Petroneis* cell

Fig. 53. Plastid movements during the cell cycles of (*a*) *Nitzschia*, (*b*) *Sellaphora*, (*c*) *Neidium*. Each sequence begins with mid-interphase (I) and goes through mitotic prophase (II), cytokinesis (III), early valve formation (IV), late valve formation (V) and interphase (VI), except in (*a*), where V is interphase. Each stage is presented in valve and girdle view. Special points: in (*a*) III the cleavage furrow constricts each plastid in two. In (*b*), the single H-shaped plastid moves around onto the girdle and divides (II) prior to cleavage (III); then the daughter plastids invaginate transversely and rotate to restore the interphase state (IV–VI). In (*c*) the plastids move beneath the valves (I, II) before cleavage (III), and then divide longitudinally (IV); the daughter plastids move back to the girdle (V) and invaginate in preparation for the next division (VI).

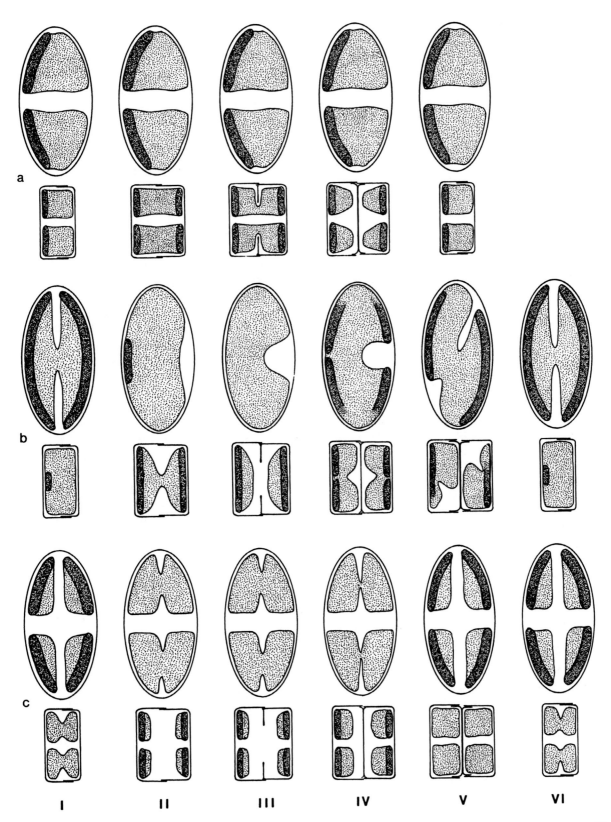

to place its mitotic apparatus unerringly on the same side of the cell (Fig. 52c), and a *Navicula* cell to alternate the apparatus unerringly from one side to the other (Fig. 52d)! The basis of this is unknown.

Where the raphe system extends around the whole circumference of the valve, as in *Surirella* and its allies, the nucleus always migrates to the same pole for mitosis (Fig. 52f), and it is presumed that the break in the raphe system formed here in each daughter cell corresponds to the central raphe endings of naviculoid diatoms (Lauterborn, 1896; Tippit & Pickett-Heaps, 1977; Schmid, 1979a, b; Pickett-Heaps *et al.*, 1988). Even in the isopolar diatom *Cymatopleura*, the nucleus always moves to the same pole for mitosis (Fig. 52g and see Mann, 1987).

There are, then, cycles of cell growth and division, cycles of nuclear movement, growth and division, and cycles of centrosome movement and activity. Alongside these, however, there is also a plastid cycle, which is poorly known in centric and araphid diatoms but increasingly well understood (and taxonomically useful) in raphid diatoms. In many centric diatoms there does not seem to be a particularly well organised method of segregating the plastids to the two daughter cells at division, but with large numbers of plastids this probably does not matter. At division in *Melosira*, approximately half the plastid complement is often concentrated near the hypovalve, and half near the epivalve, leaving a relatively 'empty' zone beneath the girdle where the cleavage furrow forms. The time of plastid division is unknown in most species but in *Stephanopyxis* it occurs around the middle of interphase (von Stosch & Drebes, 1964). A particularly detailed sudy of plastid division has been made in the araphid pennate diatom *Diatoma* by Ettl & Brezina (1975) and Ettl (1978).

Raphid diatoms usually have only one, two or four plastids and so it is essential that there is an accurate method of division and segregation if both daughter cells are to be viable. The simplest way of doing this is for the plastids to be centred with respect to the prospective plane of division, so that cells can be cut into more-or-less equal halves by the cleavage furrow (Fig. 53a). This kind of division occurs in some *Hantzschia* species, *Nitzschia*, *Cymbella*, *Gomphonema*, *Placoneis*, *Achnanthes* and others. Following valve formation each plastid grows and/or moves onto the hypovalve, re-establishing its symmetrical position with respect to the median valvar plane. Elsewhere, however, the plastids divide independently of the cell and may do so just before, e.g. *Sellaphora* (Fig. 53b), *Amphora ovalis* or a short while after cytokinesis, e.g. *Neidium* (Fig. 53c). Furthermore, division may be transverse (*Navicula*, *Diploneis*, *Gyrosigma*) or longitudinal (*Neidium*). During interphase the plastids of raphid diatoms usually change little in shape or position. Movement, plastic deformation and division of the plastids are usually restricted to premitotic (late interphase or prophase) and postmitotic (during valve formation and cell separation) phases. As they take up their mitotic positions or regain their interphase positions, the plastids often perform complex and highly organised rotational or translational movements (e.g. Ott, 1900; Mereschkowsky, 1903a; Cox, 1981c; Mann, 1985, 1989b). These are the subject of a detailed review (Mann & Stickle, in preparation) and will not be considered further here, although we have used much evidence from cell cycle studies during our revision of the raphid genera.

In summary, the diatom cell cycle is poorly known, except with respect to a few, relatively short phases (e.g. mitosis and valve formation). In few species do we have a complete account, even of the visible changes in cell size and structure. From the patchy knowledge available, however, it is clear that the cycle is highly organised, especially in raphid diatoms, with many processes taking place in parallel, in a highly co-ordinated progression from division to division. Investigation of the cell cycle is essential not only to satisfy the curiosity of cytologists and morphogeneticists, but also to enable the taxonomist to make proper use of data derived from cell wall or protoplast.

Vegetative multiplication and cell size reduction

Because valves and copulae are formed within the confines of the parent frustule, diatoms have the peculiar property that within a population, mean cell size usually decreases with each cell division (Fig. 54). There are a few records (e.g. Geitler, 1932, 1980; Wiedling, 1948; Locker, 1950), mainly from experimental studies, of taxa in which no size reduction occurs but it is not known how often this

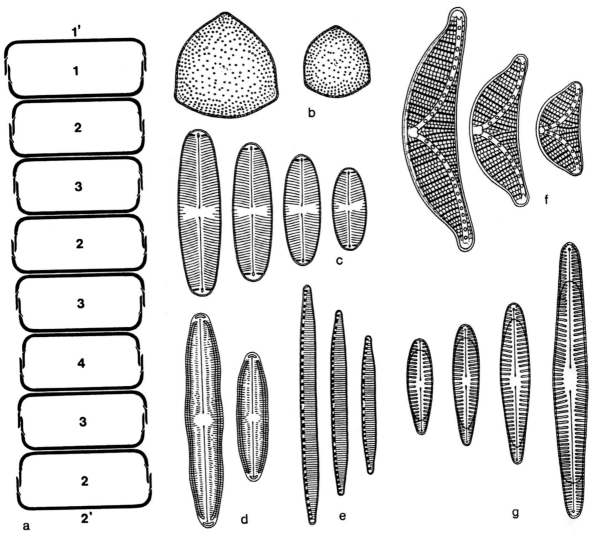

Fig. 54. (*a*) Division of one cell to form a chain of eight in which cell 1 is the largest, i.e. the same size as the original cell, and 2, 3 and 4 are the sequentially reduced sizes of cell. Valves 1′ and 2′ are the epivalve and hypovalve of the original cell. (*b*)–(*g*) Examples of size and shape change during sequences of normal cell divisions (very many divisions are necessary for the change from the largest to the smallest cells): (*b*) *Stictodiscus* (after Hustedt), (*c*) *Sellaphora*, (*d*) *Brachysira*, (*e*) *Nitzschia*, (*f*) *Epithemia*, (*g*) *Rhoicosphenia*.

is the case in nature. Indeed it is theoretically possible, and occurred in cultures of *Eunotia pectinalis* (Geitler, 1932), that the daughter valves could be larger than the parental valves, providing the girdle is sufficiently arched (see discussion in Round, 1972a and Crawford, 1981). Even in diatoms that have never been observed to form auxospores, however (and in taxa that in culture do not reduce in size) the same kind of size and shape variation can usually be found in nature that has been demonstrated elsewhere to be associated with a typical size reduction cycle, involving size restitution via an auxospore (but see Round, 1972a; Mann, 1988b). Measurements of valve diameter or length over a period of time in natural populations often reveal periods of little change followed by slight decrease in size (Wesenberg-Lund, 1908; Round, 1982b). This variation is not understood. In the largest and most heavily silicified diatoms, the process of size reduction can be demonstrated easily even in a single cell, since the hypovalve is visibly smaller than the epivalve. Occasionally,

83

disturbances during cell division can cause abrupt and drastic changes in dimensions, e.g. to less than half the length of the parent cell (Locker, 1950). For the most part, however, each size reduction is very slight, but over many generations, of course, the effect can be considerable. In large *Coscinodiscus* species, for instance, valve diameter can reduce from 600 to 250 μm (*C. wailesii* in culture: A.-M. Schmid, personal communication) or from 200 to 50 μm (*C. asteromphalus*: Werner, 1971a, 1978), and this will of course involve an even greater decrease in cell volume even if, as often occurs, the cell decreases proportionately less in its pervalvar dimension (von Stosch & Drebes, 1964). Such great changes lead to considerable problems in the classification and identification of diatoms, especially since the full range of size from initial cell to gametangium is unknown or poorly documented in most species.

During size reduction in centric diatoms the main features of the valve remain little altered (Fig. 54b). In pennate diatoms, the valve structure and pattern remain almost the same but the valve outline can change from linear or linear-lanceolate to oval or even nearly circular (e.g. in some *Rhaphoneis* and *Delphineis* species). This is because the width of pennate diatom valves usually decreases proportionately far less than the length (Fig. 54c–g): in some cases, there may even be an increase (Geissler, 1970a, b; Mann, 1981b). Shape change during size reduction tends to follow certain rules, which were formulated by Geitler (1932), whose paper on the life cycles of pennate diatoms is essential reading for anyone interested in diatom form and growth. Besides causing absolute and relative changes in valve dimensions, the process of vegetative division in diatoms often produces changes in the shape of particular parts of the valve. Diatoms with rostrate or capitate poles often lose these during size reduction, assuming an altogether simpler outline, while in *Gomphonema*, *Rhoicosphenia*, *Meridion* and other heteropolar diatoms, heteropolarity is much more marked in smaller cells (Fig. 54g); possible reasons for this are explored by Mann (1984d). Cells kept in culture without undergoing auxospore formation can become so small that division is rare or impossible: they often form abnormally large numbers of copulae and their pervalvar axes elongate enormously. Cultural studies of diatoms are extremely valuable but care should be taken to exclude the cultural anomalies (such as increase in the pervalvar axis or production of round or mis-shapen valves at the lower size limit) from type descriptions; there are many such anomalies but they are not often observed in natural populations.

Although most diatoms divide to produce valves that are virtually identical to the parent valves, except in size, a few species are pleomorphic: the new valves differ from the parents in the degree of silicification and even to some extent in the arrangement of the areolae. These variants have led to some taxonomic problems, e.g. in *Thalassiosira* (Hasle, Heimdal & Fryxell, 1971) and *Aulacoseira* (Müller, 1903, 1906), but they are rare.

The way diatoms divide inevitably results in certain properties of their cells and colonies. Newly formed valves are always hypovalves and the hypotheca of the mother cell forms the epitheca of one of the two daughter cells. The end valves of a complete filamentous colony (as in *Melosira*, *Fragilaria*, etc.) are always epithecae and valves lying back-to-back in such a colony must have been formed during the same cell division.

It might be thought that as a result of division from a single cell, the total number of cells after a certain time (t) would be approximately $2^{t/t_c}$ where t_c is the duration of the cell cycle; and that the cells within this total would have sizes distributed according to the binomial rule. However, this will happen only if the length of the cell cycle is the same for each cell, whatever its size and therefore 'age'. There is evidence from Müller's work (1883, 1884), however, that at least in *Ellerbeckia* (*Melosira*) *arenaria* the smaller of the two daughter cells divides half as quickly as the larger: i.e. it 'misses' a division. In this case the population reduces in size less per division, and if the population is examined

Fig. 55. Life cycle of a diatom; based on *Stephanodiscus* except that the position of the auxospore (c) and the formation of motile gametes (a) has not been seen in this particular genus. (d) Auxospore wall breaking open to reveal initial cell. (e) First division of initial cell to form two new normal hypovalves back to back. (f) One of the cells from stage (e) with a normal valve and an initial cell valve. (g) A cell formed following several divisions of (f) (the other cell is not illustrated but will be a replica of (f). (h), (i) Vegetative size reduction. (i) Small cell which will give rise to either the male or female (or + or −) gametes.

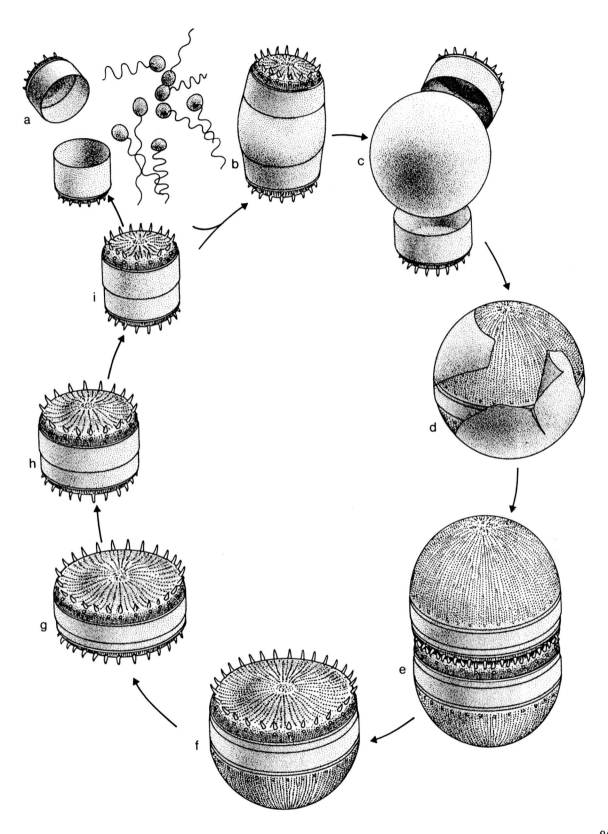

there will appear to be a disproportionate number of smaller cells, i.e. the mode is displaced towards the lower end of the size range.

The course of size reduction has rarely been followed in nature. Chain-forming diatoms offer one opportunity to make such studies, and this is the basis for Müller's work on *Ellerbeckia* (*Melosira*) *arenaria*. Another possibility is to study populations that have undergone a fairly well-synchronised burst of auxosporulation. Here one can not only watch how the mean size of a population decreases, but also calculate the average residence time of cells in the water column or habitat, using the initial valves (which usually differ in shape and/or pattern from normal valves) as markers. This has been done for *Stephanodiscus* by Round (1982b); the initial valves could be detected in the population for only a little over a month and this at a time (November) when turbulence in the water column was likely to be high. Early this century several workers studied size changes in populations of freshwater plankton (summarised in Hutchinson, 1967; Mann, 1988b). In many cases the sample size was too small or the time scale of the investigation too short for confident interpretation of the changes observed. The best studies were those of Nipkow (1927), who examined a long 28-year sequence of laminated sediments from Lake Zürich, as well as shorter sequences from elsewhere. He was able to demonstrate that planktonic species exhibit sexual cycles (from auxospore to auxospore) of up to 20 or more years in length, although the pattern of change in some species was extremely complex. More recently Nipkow's data have been reanalysed and augmented by Mann (1988b) in a review of the diatom life cycle in nature. Lewis (1984) has suggested that long life cycles may have evolved because the ability to reproduce sexually is retained but the costs of sex minimised.

Sexual reproduction

Following a phase of vegetative multiplication, during which the cells have become smaller, size must be restored by auxospore formation (vegetative enlargement is known in a few diatoms, e.g. *Skeletonema costatum*: Gallagher, 1980). This is preceded by meiosis and sexual reproduction, or by some modification of these (see Fig. 55 for a summary of the life cycle).

The study of meiosis, gamete formation, plasmogamy, karyogamy and other aspects of sexual reproduction in diatoms has lagged behind work on the vegetative cell. The pennate genera have been more extensively studied and it was not until 1950 that von Stosch finally established that oogamy occurs in centric diatoms, although this had been suggested as a possibility by Geitler (1932). Since 1950 a number of centric species and one araphid pennate genus (*Rhabdonema*) have been studied and found to exhibit oogamy (von Stosch, 1950, 1951a, b, 1954, 1956, 1958a; Geitler, 1952a; von Stosch & Drebes, 1964; Drebes, 1966, 1969, 1974; von Stosch, Theil & Kowallik, 1973). Autogamy or apomixis occurs in some centric species (Iyengar & Subrahmanyan, 1944; Drebes, 1977a) but these are apparently in every case modifications of basically oogamous processes (Drebes, 1977a). The reproduction of many pennate genera has been documented and a great variety of behaviour shown (summaries in Geitler, 1932, 1973, 1984) yet even here the number of species in which sexual reproduction has been observed is very small relative to the total number of species recognised. It may be that sexuality has been lost in many species, but it is equally possible that the appearance of asexuality is an artefact of the methods used to study diatoms. Few diatomists ever see sexual reproduction and even fewer observe it in natural populations; for reasons why this may be so, see Mann (1988b). Our experience indicates, however, that in spite of the great length of the sexual cycle in some diatoms (apparently lasting well over a decade, for instance, in *Aulacoseira* (*Melosira*) *islandica* var. *helvetica*: Nipkow, 1927; Mann, 1988b), relative to the time taken to complete sexual reproduction and auxospore development (often approximately a week), auxospore formation can often be detected in a few species in any sample, at least in epiphytic and epilithic communities, although the proportion of cells involved may be small. But for this to be done, samples must be examined while fresh, and the observer must be familiar with the kinds of behaviour exhibited by diatom cells during sexual reproduction. It is partly in the hope of stimulating diatomists to look more carefully for sexual stages that we have illustrated auxospore formation in so many different pennate genera here.

Sexuality, where it occurs, it always linked to the

restoration of maximal cell size, which is achieved by the swelling of a specialised zygotic cell, the auxospore. Diatom cells are apparently not always sexually mature (Geitler, 1932). Werner (1971c) has shown, for instance, that *Coscinodiscus asteromphalus* cannot be induced to produce gametes at cell diameters above 100 μm or below 70 μm. Any cells that become too small for sexual reproduction divide on until they die. The physiological basis of this size effect, which is not restricted to *C. asteromphalus* (e.g. Geitler, 1932; Drebes, 1966) is not known, although means of manipulating cell size artificially are available for some genera (von Stosch, 1965).

Oogamy

In this process, large non-motile female gametes are fertilised by small, motile flagellate male sperm (Fig. 49 l), or rarely (*Rhabdonema*) by small, non-motile naked spermatia. Chemotaxis and recognition phenomena are presumably involved here as in other groups of algae, but the basis of these remains uninvestigated.

Drebes (1977a) has reviewed aspects of sexual reproduction in centric diatoms. Sperm formation takes place in a number of ways. In Fig. 56 we illustrate *Chaetoceros* in detail (based on von Stosch, Theil & Kowallik, 1973). In the simplest case, apparently undifferentiated vegetative cells produce the spermatocytes directly, without intervening mitoses: this has been documented in *Cyclotella* by Geitler (1952a) and Schultz & Trainor (1968). Elsewhere the spermatocytes are produced by differentiated, mitotically active cells, the spermatogonia. In *Melosira* (von Stosch, 1951a, b), the spermatogonia divide several times, producing chains of 2 to 32 cells. Each cell is rather short, thin-walled and poor in plastids relative to the vegetative cells. A series can be traced from *Melosira* to such forms as *Guinardia flaccida* and *Arachnoidiscus argus*, in which the spermatogonia become progressively simpler and more reduced. In *Stephanopyxis* (von Stosch & Drebes, 1964) and *Biddulphia* (von Stosch, 1954, 1956), the spermatogonia form rudimentary siliceous thecae, although these all remain enclosed within the mother cell, while in *Odontella sinensis*, *O. regia* and *O. mobiliensis*, thecae are produced only after the first spermatogonial division. Thecae are absent in

some other genera, e.g. *Coscinodiscus* (von Stosch & Drebes, 1964; Drebes, 1974), *Rhizosolenia* (Drebes, 1974), some *Chaetoceros* (von Stosch, Theil & Kowallik, 1973), *Lithodesmium*, *Streptotheca* and *Bellerochea* (von Stosch, 1954), while in *Guinardia* and *Arachnoidiscus* (von Stosch & Drebes, 1964), not only are thecae not produced, but the spermatogonia fail to separate after each mitosis, so that the spermatocytes are multi-nucleate plasmodia.

The spermatocytes swell during prophase so that the spermatogonial thecae and the mother cell thecae usually split apart (Fig. 56f, g). Hence the spermatocytes are often liberated into the surrounding medium and may complete their development there (Fig. 56h–l). Sperm production may occur by one of two methods. In the hologenous type found in *Skeletonema* (Migita, 1967), *Lithodesmium* (von Stosch, 1954), *Bacteriastrum* (Drebes, 1972), *Chaetoceros* (von Stosch, Theil & Kowallik, 1973), *Odontella sinensis*, *O. regia* (von Stosch, 1954, 1956), *Bellerochea malleus* and *Streptotheca tamensis* (von Stosch, 1954), the whole of the spermatocyte protoplast is distributed equally among the four sperms, equal cytokineses occurring after both meiotic divisions (Fig. 57a–c). The merogenous type of formation is found in *Melosira*, *Stephanopyxis*, *Actinocyclus*, *Actinoptychus* (von Stosch & Drebes 1964), *Biddulphia rhombus* and *Pleurosira* (= *Biddulphia*) *laevis* (von Stosch, 1956). Here meiosis is completed before any division of the spermatocyte occurs (Fig. 57a, d, e). Then the four uniflagellate sperm detach themselves from the spermatocyte, leaving behind a residual body containing all the plastids (Fig. 57f). A variant is found in *Odontella* (= *Biddulphia*) *granulata*, in which cytokinesis occurs after meiosis I (von Stosch, 1956), so that when the sperms detach themselves, two residual bodies are produced. In *Coscinodiscus*, on the other hand, a mass of cytoplasm containing the plastids is cut off from the spermatocyte before meiosis I is complete, formation of the sperm thereafter taking place hologenously.

With either method of sperm production, the flagella grow out in pairs near the nuclei shortly after completion of meiosis I (Figs 56i, j, 57b–d). In *Lithodesmium undulatum*, the only diatom in which meiosis has been investigated fully using TEM techniques (Manton, Kowallik & von Stosch, 1969a, b,1970a, b), the flagellar bases are closely associated

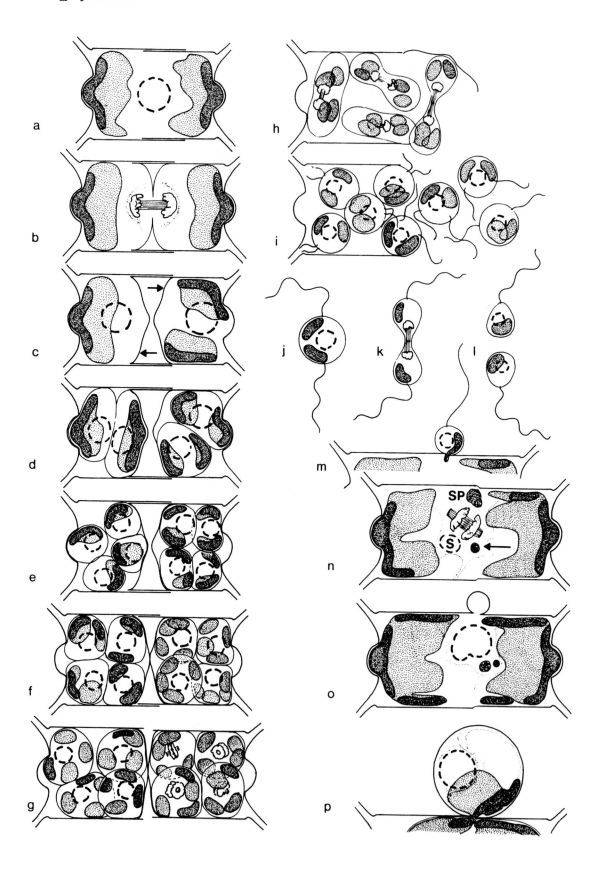

with the poles of the spindle and hence separate from one another as the spindle elongates during meiosis II. The single flagellum is anterior and pleuronematic (von Stosch, 1954, 1958a; Drebes, 1966, 1974; Manton & von Stosch, 1966; Schultz & Trainor, 1970; Heath & Darley, 1972; von Stosch, Theil & Kowallik, 1973), but is unusual in lacking a central pair of microtubules in its axoneme (Manton & von Stosch, 1966; Heath & Darley, 1972).

Recently (see Hasle, von Stosch & Syvertsen, 1983), flagellate male gametes have been discovered in taxa (*Plagiogrammopsis, Cymatosira, Leyanella* and *Arcocellulus*) which were formerly regarded as araphid pennate diatoms. The ultrastructure of the valves, however, coupled with the production of motile gametes, places this group in the centric series.

The production of egg cells is rather simpler than that of sperm, since no mitoses intervene between vegetative cell and oocyte (except in *Melosira nummuloides*: Rieth, 1953), the oocytes being distinguishable from vegetative cells by their greater length, denser cytoplasm and numerous plastids. Three types of egg formation can be distinguished, however, according to the number of eggs produced and whether or not cytokinesis occurs after meiosis I (Drebes, 1977b). In *Lithodesmium undulatum*,

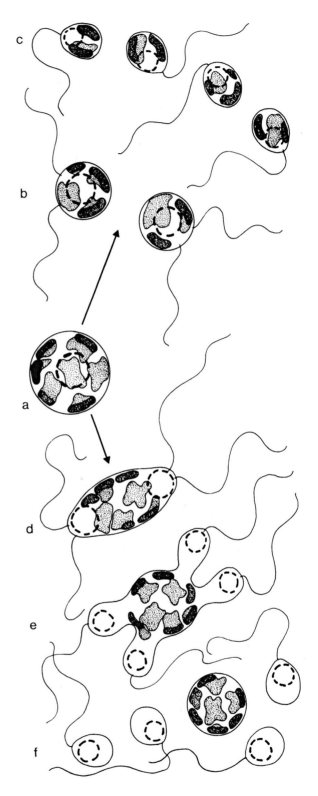

Fig. 56. Life cycle of *Chaetoceros didymum*. (*a*) Small cell, containing two plastids and a central nucleus. (*b*) First spermatogonial division. (*c*) Formation of two lightly silicified spermatogonial thecae (arrows). (*d*) Following the second spermatogonial division: protoplasts rounded up and contracting. (*e*) Following the last spermatogonial division: each spermatogonium contains four spermatocytes. (*f*),(*g*) Swelling of the spermatocytes, causing separation of the parent cell thecae. (*h*) Release of the spermatocytes, which undergo the first meiotic division. (*i*) Outgrowth of the flagella. (*j*)–(*l*) Second meiotic division and separation of mature sperm. (*m*) Fertilisation of an oogonium by penetration of the sperm between the oogonial thecae. (*n*) Sperm nucleus (s) within the oogonium; second meiotic division of the oogonial nucleus in progress. A pyknotic nucleus from the first meiotic division is visible (at the arrow), as is the sperm's plastid (sp). (*o*),(*p*) Following karyogamy between sperm and egg nuclei: the zygote begins to expand and migrate out from the oogonial frustule, forming a lateral auxospore.

Fig. 57. Spermatogenesis in centric diatoms. (*a*)–(*c*) Hologenous formation of sperm. (*d*)–(*f*) Merogenous formation, involving the separation of the four sperm from a residual body containing the plastids.

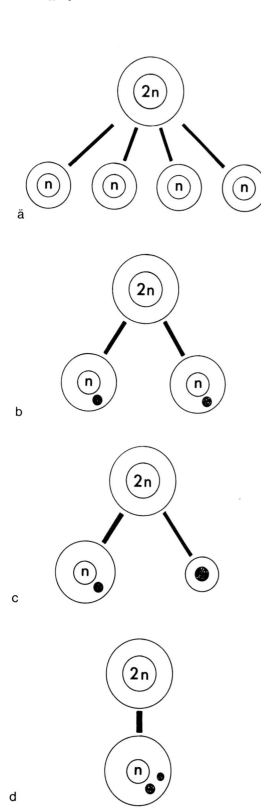

Odontella (= *Biddulphia*) *mobiliensis* (von Stosch, 1954), *Odontella granulata* (von Stosch, 1956), *Odontella regia*, *O. sinensis* (Drebes, 1974) and *Attheya* (Drebes, 1977), two eggs are produced, an equal cytokinesis occurring after meiosis I but not after meiosis II; one haploid nucleus becomes pycnotic in each cell. In a second type, known only in *Biddulphia rhombus* and *Cerataulus smithii* (von Stosch, 1956), there is an unequal cytokinesis after meiosis I. The smaller cell degenerates; the other completes meiosis, but one nucleus becomes pycnotic so that only one egg cell is produced. Finally, in other centric diatoms investigated, e.g. *Melosira moniliformis* (von Stosch, 1951a, 1958a), *Stephanopyxis turris* (von Stosch & Drebes, 1964), *Skeletonema costatum* (Migita, 1967), meiosis proceeds without cytokinesis, one nucleus becoming pycnotic at each division so that again only one egg cell is produced (Fig. 58d).

Fertilisation of the egg cannot occur until part or all of the egg is exposed, by partial or complete separation of the oocyte thecae, although even in *Biddulphia rhombus* where the egg is extruded, the egg remains enclosed by a weakly silicified membrane which the sperm must penetrate (von Stosch, 1956). Thecal separation, and indeed sperm penetration, accompanied by the loss of the latter's flagellum, may occur even while the oocyte is still in meiotic prophase (Fig. 49 I; von Stosch, Theil & Kowallik, 1973), but elsewhere the eggs become exposed only after meiosis is complete. Karyogamy follows plasmogamy but not necessarily immediately (Drebes, 1977a).

Rhabdonema exhibits an essentially oogamous method of reproduction, but the male gametes are non-flagellate. Spermatogonia divide mitotically a number of times, as in some centric species, resulting in the formation of a chain of short cells,

Fig. 58. Nuclear fates during gametogenesis in diatoms. (*a*) All haploid products of meiosis survive, as in the spermatogenesis of centric diatoms. (*b*) Equal cytokinesis after meiosis I, one haploid nucleus degenerating in each cell after meiosis II, e.g. oogenesis in *Lithodesmium* and *Odontella*, gametogenesis in *Craticula*, *Neidium*. (*c*) Unequal cytokinesis after meiosis I, the larger cell behaving as in (*b*), the smaller one degenerating, e.g. oogenesis in *Cerataulus*, gametogenesis in *Sellaphora*. (*d*) Degeneration of one nucleus after each meiotic division, e.g. oogenesis in *Chaetoceros didymum*, *Stephanopyxis turris*.

poor in plastids, which function as spermatocytes. In each spermatocyte, a cytokinesis occurs between the two meiotic divisions; subsequently one of the two haploid nuclei degenerates in each cell, so that only two gametes are produced per spermatocyte. There is apparently some evidence for a merogenous development of the male gametes but the details are unpublished. The single egg cell of *Rhabdonema adriaticum* is produced in the same way as described above for *Biddulphia rhombus* and *Cerataulus smithii*, while *Rhabdonema minutum* and *R. arcuatum* produce two cells per oocyte. Apposition of the gametes takes place through passive movement of the colonies containing them, the amoeboid movement of the male gamete to the girdle region of the oocyte wall, and injection of its nucleus into the egg. This genus, in terms of sexual reproduction, seems to be intermediate between centric and pennate diatoms (von Stosch, 1958b, 1962a).

Physiological anisogamy and isogamy

Within the pennate diatoms, excluding *Rhabdonema*, sexual reproduction involves morphologically indistinguishable gametes (Fig. 64b). In this sense, therefore, they are isogamous. However, although not morphologically differentiated, the gametes in many taxa behave differently, one remaining in the gametangium while the other moves to effect plasmogamy (Fig. 59b–d, e–j, l, m); these gametes, then, are physiologically anisogamous. In other taxa, all the gametes move and are therefore apparently truly isogamous (Figs 60g–l, 61n–r, 62a–d). All allogamous taxa are alike, however, in that vegetative cells function directly as gametangia, meiosis taking place in each with the production of one or two non-flagellate gametes (Figs 64a, 61e, j). Often the paired cells become invested in a distinct mucilaginous envelope (Figs 59a, k, 61a–j, e). Apparently in no case, however, does plasmogamy occur more than a few microns away from the gametangial thecae and, in the anisogamous forms, the zygotes are formed within the thecae (Figs 59d, j, m, 62i, j), although of course these must have separated to some extent to allow gamete movement. The gametes are apparently capable of moving only very limited distances and copulating cells are always closely associated. Motile raphid diatoms may, when sexually active, produce hormones that attract other receptive cells; in the araphid pennate

species, on the other hand, apposition of gametangia must come about largely passively, although some can make sluggish movements (e.g. *Synedra tabulata*: Hopkins, 1969).

There are seven main types among the allogamous pennate diatoms, although variations in the arrangement of the gametes, the means by which the gametes gain access to each other, and the pairing configurations, have allowed Geitler (1973) to split these into 18 subgroups. Using Geitler's notation, the seven main types are:

I. Two functional gametes produced in each gametangium (and hence two zygotes per copulation).
   (a) Gametes anisogamous, differentiated into active and passive types (based on motility).
      1. One active and one passive gamete produced by each gametangium (Figs 59a–d, e–j, 62e–l).
      2. Both active gametes produced by the same gametangium (Fig. 60a–f).
   (b) Gametes strictly isogamous. Apical axes of the gametangia and auxospores bearing a fixed relation to each other.
      1. Apical axes of gametangia and auxospores at right angles to eachother (Fig. 61n–r).
      2. Apical axes of gametes and auxospores parallel. No rearrangement of the gametes, so that they remain on either side of the median valvar plane.
   (c) Gametes more or less isogamous, though behaving somewhat unpredictably (Fig. 61a–f). Gametes undergoing re-arrangement to lie either side of the median trans-apical plane, and moving fairly freely in a soft mucilaginous envelope (copulation jelly). In consequence the auxospores are either somewhat irregularly orientated or more or less parallel to each other and to the gametangia (Fig. 61g).
II. One functional gamete produced in each gametangium.
   (a) Gametes isogamous (Figs 60g–m, 61h–k, 62a–d)
   (b) Gametes anisogamous (Fig. 59 l–n)
Full details of the variants can be found in Geitler (1973).

In Type I diatoms, cytokinesis occurs after

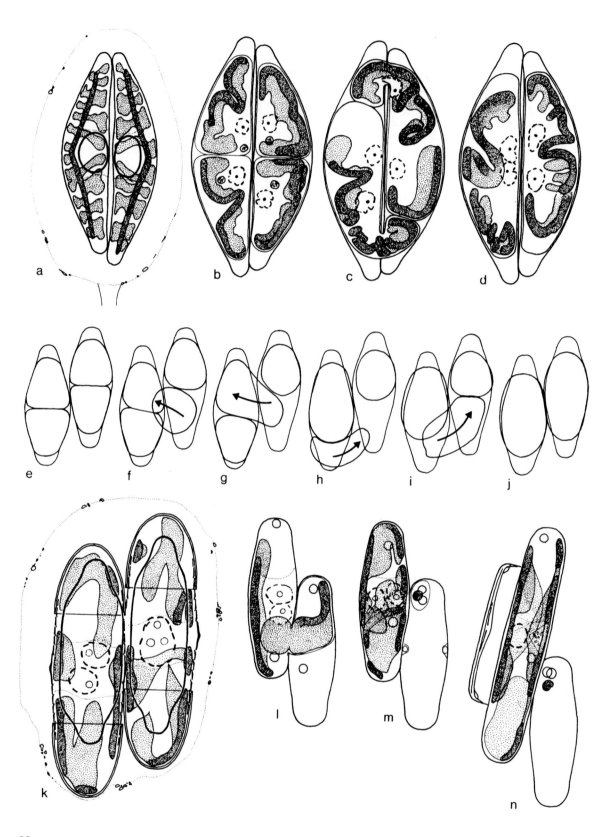

meiosis I, with abortion of one haploid nucleus in each cell after meiosis II. Where only one gamete is produced, one of the two cells produced after meiosis I may degenerate, or the two meiotic divisions may proceed without cytokinesis, one nucleus aborting after each division (Fig. 58).

Variation in reproductive behaviour among pennate species to some extent mirrors variation in other characters. Most *Cymbella* and *Gomphonema* species, for instance, have a variant of type Ia1, in which the gametes become re-arranged in the gametangium before migrating to effect plasmogamy, and in which migration does not take place through morphologically distinct passages in the copulation envelope (Fig. 59c–j). Several *Nitzschia* species, on the other hand, exhibit a different kind of type Ia1 behaviour, involving formation of a single copulation tube between the centres of the gametangia (Mann, 1986a). Furthermore, apart from *Navicula halophila* (Subrahmanyan, 1947), which needs to be rechecked, type Ia2 (Fig. 60a–f) is known only from the araphid pennate group where it has been documented in *Synedra ulna* (Geitler, 1939a, b), *Synedra rumpens* (Geitler, 1952b), *Synedra amphicephala* (Geitler, 1958b), *Diatoma elongatum* (Tschermak-Woess, 1973), *Licmophora gracilis* (Mann, 1982c), and perhaps in *Meridion circulare* (see Geitler, 1973). Elsewhere, however, variation may occur even within a species or species aggregate. Some races of *Cocconeis placentula*, for instance, are anisogamous, while others are isogamous (Fig. 61h, i); still others are parthenogenetic (Geitler, 1982).

In anisogamous type Ia1 cells, fusion (which has been followed *in vivo* in only a few species) sometimes takes place between adjacent gametes, as

in *Cymbella cistula* (Geitler, 1954), and *C. lanceolata* (Fig. 59b–d), sometimes between diagonally opposite gametes, as in *Gomphonema parvulum* (Fig. 59c–j: Geitler, 1932); the latter is certainly not universal, as implied by Drebes (1977a).

Automixis and parthenogenesis

The majority of centric and pennate taxa investigated are at least facultatively anisogamous. A number, however, are automictic – and here a separation can be made (following Hartmann, see Geitler, 1973) into paedogamy, where two gametes from the same gametangium fuse (Fig. 62m–r), and autogamy, where fusion of gametic nuclei occurs without cleavage of the gametangial protoplast (Fig. 62s–u) – or parthenogenetic. In the centric series, autogamy has been recorded in *Cyclotella meneghiniana* (Iyengar & Subrahmanyan, 1944) and *Melosira nummuloides* (Erben, 1959); it is possible that allogamy also occurs, however (although this has never been observed), since flagellate male cells are formed (Schultz & Trainor, 1968; von Stosch & Drebes in Drebes, 1977a). *Detonula* (= *Schroederella*) *schroederi* is apparently autogamous or parthenogenetic (Drebes, 1974), while parthenogenetic development of the auxospore occurs in *Melosira moniliformis* var. *octogona* and *Actinoptychus undulatus* (Broer, Behre & von Stosch in Drebes, 1977a).

Among pennate diatoms, paedogamy is known in *Synedra ulna*, a few populations of *Cymbella* and *Gomphonema* species, in *Nitzschia frustulum* var. *perpusilla* (Fig. 62m–r and see Geitler, 1973 for

Fig. 59. (a)–(d) *Cymbella lanceolata*: (a) paired gametangia in meiotic prophase, invested in a mucilage sheath; (b) gametangia following rearrangement of the gametes to lie on either side of the median transapical plane; (c) gamete migration following plasmogamy; (d) zygotes within the gametangia. (e)–(j) *Gomphonema parvulum*: fertilisation and migration of gametes, showing physiological anisogamy. (k) *Neidium*: expanding zygotes (auxospores), in one of which the nuclei have fused. (l)–(n) *Sellaphora pupula*: plasmogamy and gamete migration (l),(m) and expansion of the auxospore (n). *Sellaphora* gametangia produce only one gamete apiece; contrast *Cymbella* ((b)–(d)) and *Gomphonema* ((e)–(j)). [(e)–(j) After Geitler, 1932.]

Fig. 60. (*overleaf*) (a)–(c) *Synedra ulna*: formation of two gametes per gametangium and fusion to produce two zygotes. (d), (e) *Synedra rumpens*: physiological anisogamy of type Ia2, producing both auxospores within only one of the gametangial frustules. (f) *Licmophora anglica*: similar in behaviour to *Synedra*: expanding auxospores. (g)–(j) *Eunotia flexuosa*: (g) paired cells, (h) formation of papillae from the ends of cells, (i) isogamous fusion, (j) expansion of the auxospore. (k)–(m) *Eunotia arcus*: (k) unequal division of the gametangial cell produces a functional gamete and a relatively large residual cell containing a plastid; (l) expanding auxospore; degenerating remnants of plastids are visible within the gametangial frustules, derived from the residual cells; (m) initial epivalve within the expanded auxospore. (After Geitler, 1939b, 1951b, c, 1952c, and original).

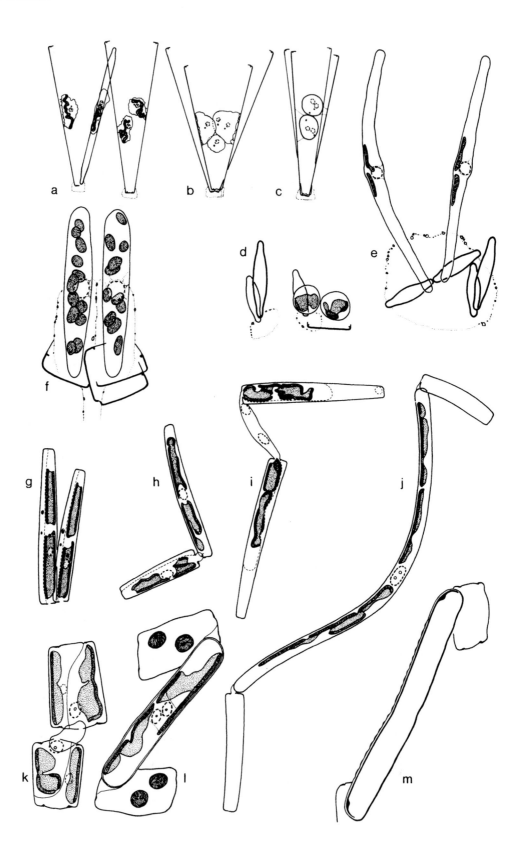

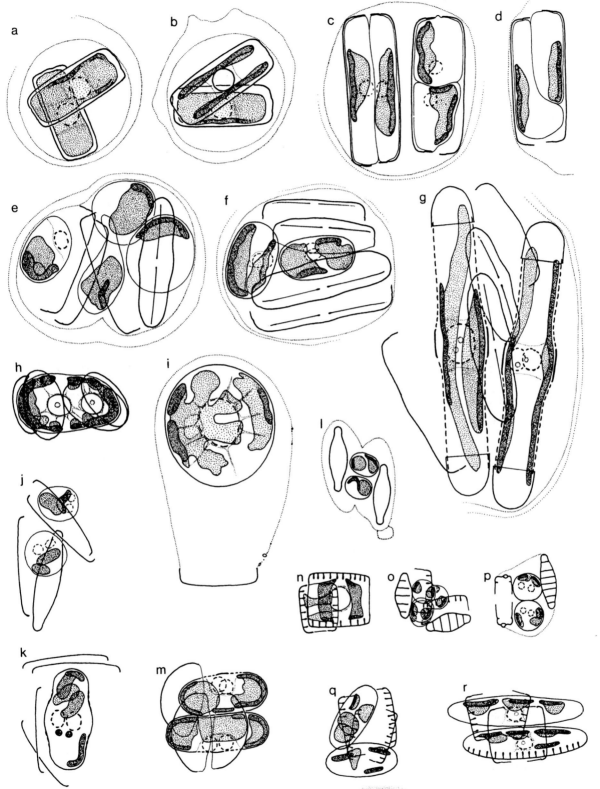

references), *Amphora* cf. *veneta* (Thaler, 1972) and *Achnanthes subsessilis* (Karsten, 1897); autogamy in *Denticula tenuis* (Fig. 62s–u) and *Encyonema minutum* (Geitler, 1958b); and automictic development of some kind in *Synedra vaucheriae* (Geitler, 1958b) and *Navicula minima* (Granetti, 1968). Early reports of parthenogenesis in pennate species must be treated with some caution (see Geitler, 1932, for references); *Rhabdonema* and *Meridion*, for instance, were originally thought to be parthenogenetic, whereas they are now known to be allogamous. Parthenogenesis is well documented, however, in *Cocconeis placentula* vars *klinoraphis* and *lineata* (Geitler, 1982).

In every case the restriction or circumvention of recombination and sexuality is clearly secondary. The automictic and parthenogenetic taxa are closely related to diatoms with a normal sexual cycle, and are scattered among the different diatom groups. Their distribution suggests that, here as elsewhere (Maynard Smith 1978), whatever the short-term advantages of abandoning sex, the apomicts and automicts have sacrificed their evolutionary prospects.

Auxospore development

There is some confusion in the use of the terms zygote, auxospore and initial cell. Plasmogamy produces a binucleate cell which on fusion of the nuclei becomes a zygote. The binucleate phase may be transient or last for many hours while the cell is expanding (Figs 59k, 66a–d). We consider this cell to be an auxospore from the moment of plasmogamy (or from the completion of meiosis in an autogamous species).

The auxospore, which may be free in the medium or associated more or less closely with the gametangial thecae (Figs 63a, 64a), first forms an organic wall of polysaccharide material. In some centric genera, and in *Rhabdonema*, this wall also contains numerous siliceous scales. Some centric species form nothing more complicated in their auxospore walls than simple round, radially constructed scales (Fig. 64b–d). Examples of such taxa are *Melosira varians*, *M. nummuloides*, *Ellerbeckia arenaria* (Crawford, 1974a), *M. lineata* (Crawford, 1978), *M. moniliformis* (Crawford, 1974a); *M. moniliformis* var. *octogona* (von Stosch, 1982). *Orthoseira* (*Aulacoseira*) *dendrophila* (Crawford, 1981), *O. epidendron* (Roemer & Rosowski, 1980 as *Melosira roeseana*), *Stephanopyxis turris* and *Actinoptychus* (see von Stosch, 1982). In these forms the expansion of the auxospore is more-or-less isometric, except where it is constrained in its expansion by the oocyte thecae (Figs 63a, 64a) as, for example, in *Melosira nummuloides* (Crawford, 1974a). The expanded auxospore is thus nearly spherical, which is perhaps not surprising given the simple radially

Fig. 61. (*previous page*) (*a*)–(*g*) '*Caloneis*' *ventricosa*: (*a*) paired cells within mucilage capsule: (*b*) gametangia in meiotic prophase; (*c*) after meiosis II, gametes before (left) and after rearrangement; (*d*) gametes undergoing rearrangement in one gametangium; (*e*) gametes (4) released within the mucilage capsule by separation of the gametangial thecae; (*f*) two zygotes, before expansion; (*g*) nearly fully expanded auxospores. (*h*),(*i*) *Cocconeis*: (*h*) isogamy, (*i*) expanding spherical auxospore raised in mucilage sheath. (*j*),(*k*) *Navicula cryptocephala* var. *veneta*: (*j*) a single gamete is produced by each gametangium (the supernumerary nuclei from meiosis II survive so that the gametes are binucleate); (*k*) the single auxospore during expansion. (*l*) *Brachysira exilis*: paired zygotes within a shortly stalked mucilage sheath. (*m*) *Amphora* (*ovalis* group): expansion of auxospores at right angles to the apical axes of the gametangia (contrast Fig. 59(*n*)). (*n*)–(*r*) *Denticula tenuis*: (*n*) paired cells in meiotic prophase; (*o*) gametes; (*p*) zygotes formed isogamously between the gametangial frustules; (*q*) expanding auxospores; (*r*) fully expanded auxospores, one with initial epivalve. (After Geitler, 1952a, 1953, 1954b, and original.)

Fig. 62. (*a*)–(*d*) *Nitzschia amphibia*. (*a*) production of papillae from the cell apices; (*b*)–(*d*) expansion of the auxospore following isogamous fusion. (*e*)–(*g*) *Nitzschia palea*: (*e*) gametangia containing two gametes apiece, connected by a central copulation tube; (*f*) physiological anisogamy: a pair of gametes has fused and migrated into one gametangium; the other pair has yet to move; (*g*) zygotes lying within the gametangial frustules. (*h*)–(*j*) *Surirella peisonis*: (*h*) paired gametangia after meiosis I; (*i*) young auxospores, following physiologically anisogamous fusion: one zygote lies within each gametangium; (*j*) expanding zygotes. (*k*),(*l*) *Cymatopleura solea*: (*k*) paired gametes in each gametangium; (*l*) migration of gametes through copulation canal. (*m*)–(*r*) *Nitzschia frustulum* var. *perpusilla*: (*m*) meiotic prophase in unpaired gametangium; (*n*)–(*p*) gametes before, during and after rearrangement; (*q*) paedogamy results in the formation of a single zygote lying within the gametangium; (*r*) expanded auxospore. (*s*)–(*u*) Autogamous development of the auxospore in a race of *Denticula tenuis*. (After Geitler, 1928, 1953, 1969b, 1970a; Thaler, 1972, and original.)

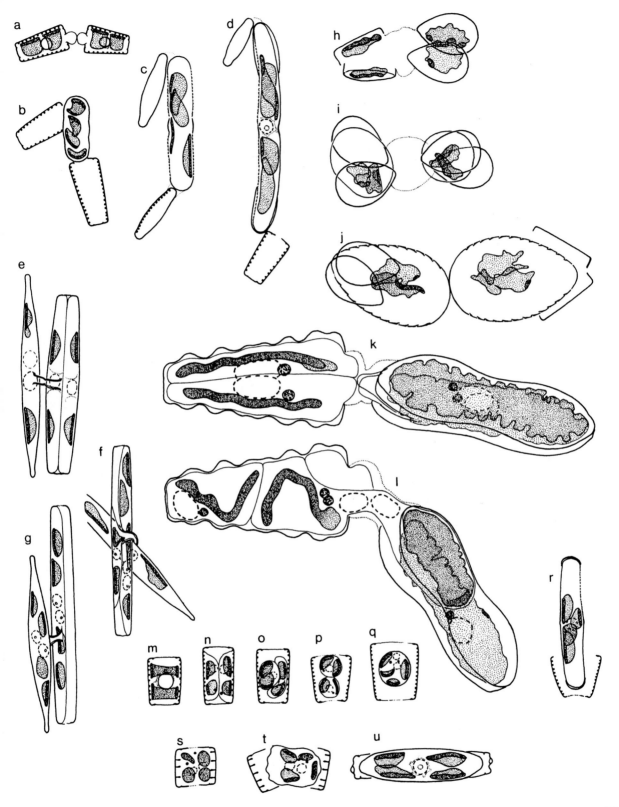

symmetrical valve shape in these taxa. In *Stephanodiscus* the auxospore wall appears to lack scales but further studies are required (Round, 1982b).

Of course, not all centric diatoms are radially symmetrical since bi- or multipolar forms occur. Here, although the initial auxospore expansion may be isometric, the primary auxospore envelope being a scale invested organic wall, as in *Melosira*, etc., the auxopore later develops anisometrically, this being accompanied by the formation of a more complicated system of silica bands and hoops (Fig. 65) which has been termed a 'properizonium' (von Stosch, Theil & Kowallik, 1973; von Stosch, 1982). Unlike the perizonium of the pennate diatoms, however (see below), the properizonium is not separated spatially from the primary auxospore wall but is physically and developmentally continuous with it (von Stosch, 1982). Properizonia have been found in *Chaetoceros didymum* (von Stosch, Theil & Kowallik, 1973), *C. protuberans, Attheya decora, Bellerochea malleus, Lithodesmium undulatum* (Fig. 58), *Odontella aurita, O. regia, Biddulphia pulchella, Triceratium antediluvianum* (von Stosch, 1982) and *Bacteriastrum hyalinum* (Drebes, 1972), and take a variety of forms. There are a number of similarities to the pennate perizonium (see below): in both, there are closed and open hoops of silica, and in both they overlap each other unilaterally, as in the girdle of vegetative cells. Just as there may be a differentiation into scaly and properizonial layers, so there may also be differentiation within the scaly layer itself. In *Odontella*, for instance, the outer scales bear dichotomously branching siliceous spines (Fig. 56c), while the inner scales are circular plates as, for instance, in *Melosira* (von Stosch, 1982). Spiny scales are also present in some forms related to *Cymatosira* (ibid.). Finally, between the scaly and properizonial layers, there may be an irregular layer of coarsely structured elements – the epizonium. This has not yet been fully investigated.

In pennate diatoms, the expansion of the auxospore is largely bipolar. The primary organic wall is often ruptured more or less equatorially at an

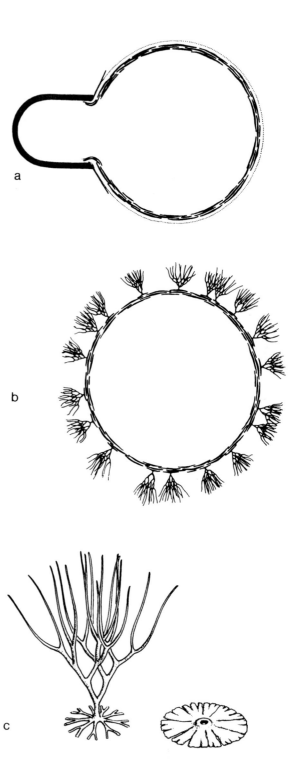

a

b

c

Fig. 63. *(a) Melosira nummuloides*: auxospore attached to theca of oogonium; auxospore wall with silica scales within an organic matrix. *(b), (c) Odontella*: *(b)* free auxospore with spiny and plain scales; *(c)* two types of scale, one with dichotomously branched projections and the other flat, with a central annulus.

early stage and persists during auxospore expansion as two caps, one over each pole of the auxospore (Fig. 66a–e; and see Mann & Stickle, 1989). In other cases, the primary wall appears to disperse. Only in *Rhabdonema* are scales found in the primary wall of the auxospore (von Stosch, 1958b, 1962a), although in *Neidium* the primary wall of the zygote is silicified and may be homologous to the scale layers of centric forms (Mann, 1984c). As the auxospore develops, transverse silica bands are deposited, adjacent to and just within the polar caps; hence expansion occurs only at the poles (Fig. 66f–j). In some ways then the development of this type of auxospore is the reverse of thecal growth in the vegetative cell, i.e. the new elements are being added not at the centre of the cell but at both ends. The balance between tip growth and wall solidification (cf. the fungal hypha) is apparently delicate, and if for any reason the auxospore is checked in its growth, symmetrically placed bulges or constrictions may be produced. The transverse silica bands of the perizonium are thin and often have dissected, fimbriate margins (Figs 67a–c, 68d). The central band is a complete ring in *Rhoicosphenia* (Fig. 67a) and *Pinnularia*, and probably some other raphid diatoms (Mann, 1982b), but incomplete in *Rhabdonema* (von Stosch, 1962a, 1982). In all these, however, the central or primary transverse band is bilaterally symmetrical, unlike the other transverse bands, which are all split rings (Fig. 67b) and differentiated into 'pars interior' and 'pars exterior' (Figs 67b, c, 68d), just as girdle bands are. All the open ends of the transverse bands lie on the same side of the auxospore, and curve inwards towards the primary band (Fig. 67c). The bands overlap each other from the centre outwards, so that the transverse series may be likened to a bipolar cingulum (Figs 67c, 68d). The presence of these bands and the overlaps between them leads to the corrugated appearance (Fig. 66j) (*Wellblechstruktur*) described from many naviculoid taxa by Karsten (e.g. 1899), and which was noted as early as 1855 by Griffith (1855, 1856).

In *Rhabdonema*, *Rhoicosphenia* and *Pinnularia* there is a longitudinal series of bands in addition to

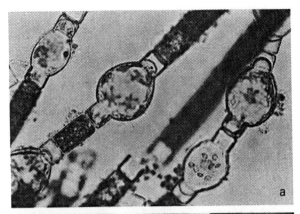

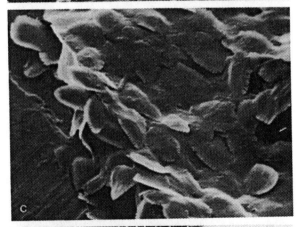

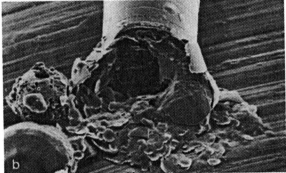

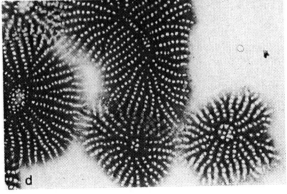

Fig. 64. (*a*) Auxospores of *Melosira varians*. (*b*), (*c*) Collapsed auxospore of *Ellerbeckia* showing scaly covering. (*d*) Individual scales of *Melosira nummuloides*: note the annulus at the centre of each scale.

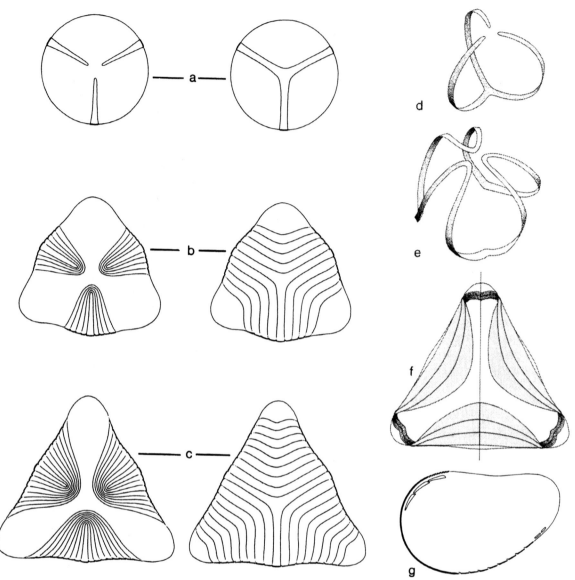

Fig. 65. Development and structure of the auxospore casing of *Lithodesmium undulatum*. (a)–(c) Three stages in the formation of the properizonium, in ventral (left) and dorsal view. (a) Young spherical stage with triradiate primary band (see (d)). (b), (c) Tripolar expansion of auxospore and sequential formation of the initial series of properizonial bands (see (e)), which is complete in (c). (d) Three-dimensional drawing of the triradiate primary band. (e) Drawing of one of the other closed properizonial bands of the initial series. (f) Properizonial bands formed towards the end of auxospore expansion beneath the initial series (c). (g) Section through the auxospore along the line indicated in (f), showing the positions of the late properizonial bands in relation to the initial series. (After von Stosch, 1982.)

Fig. 66. (a)–(e) *Craticula cuspidata*. Bipolar expansion of the auxospore. Note that the ruptured organic wall of the zygote persists as two caps, one over each end of the auxospore. In (e) the initial epivalve is present (striae drawn only at the centre). (f)–(j) '*Caloneis*' *ventricosa*: (f) ventral view of auxospore at early stage of expansion. The broad primary transverse perizonial band is a complete hoop; the other bands are split rings; (g)–(j) sequential formation of the transverse perizonium (in section) during bipolar expansion.

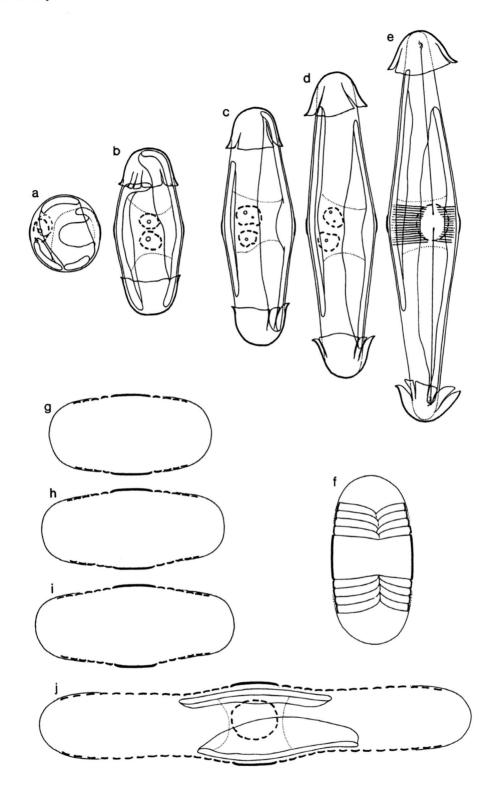

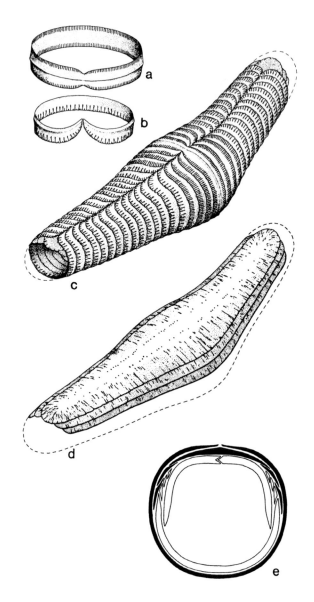

the transverse series (Fig. 67d, e). In *Rhoicosphenia*, this consists of a central, primary longitudinal band which overlaps two secondary bands on either side (Mann, 1982b). These bands lie beneath the 'suture' line on the ventral side of the auxospore, formed by the open ends of the transverse series. Von Stosch (1962a, 1982) found only longitudinal bands in the perizonium of *Achnanthes longipes*, although Karsten (1899) illustrated a ringed perizonium, suggesting the presence of transverse rings of some sort. There is clearly some variation in perizonium structure within the pennate diatoms, as evidenced by *Achnanthes longipes* and by light microscope observations of *Nitzschia fonticola* and *Epithemia adnata* (Geitler, 1932). The majority of taxa, however, appear to have a well-developed transverse series, of which the central, primary band is often somewhat wider and differently structured (e.g. Carter 1865; Karsten, 1899; Geitler, 1968a, 1969b; Mann, 1989a, unpublished). In *Surirella peisonis* and *Cymatopleura solea*, growth of the auxospore is unipolar, with the primary band at the broader pole (Thaler, 1972; Schmid in von Stosch, 1982; Mann, 1987): this correlates with the morphology of the vegetative cell, since the broader pole of at least some *Surirella* species corresponds to the centre of other bi-raphid diatoms (Schmid, 1979a; Mann, unpublished).

Once expansion of the auxospore is complete, the first thecae of the new generation are laid down (Figs 67e, 68, 69). The formation of the initial epivalve and subsequently the formation of the initial hypovalve are both preceded by an acytokinetic mitosis, the supernumerary nuclei becoming pycnotic. The initial thecae usually have a modified morphology because, formed as they are beneath the auxospore wall, they are not subject to the same constraints as the valves and bands of vegetative cells (Fig. 68). Sometimes the auxospore plasmolyses spontaneously before the formation of one or both initial valves, as in *Melosira*, *Odontella*

Fig. 67. The perizonium of *Rhoicosphenia curvata*. (*a*) The primary transverse perizonial band. (*b*) One of the split perizonial bands. (*c*) The complete perizonial casing in ventral view, showing the suture formed by the ends of the transverse perizonial bands. (*d*) As (*c*), but the transverse perizonium removed to reveal the longitudinal perizonial bands. (*e*) Transverse section of the initial cell, showing the 5 longitudinal perizonial bands (in black) beneath the ventral suture, and the initial epitheca and hypovalve.

Fig. 68. (*a*)–(*c*) *Aulacoseira ambigua*: (*a*) initial cell; (*b*) initial cell beginning to expand; (*c*) new large-celled filament alongside original-sized ones. (*d*)–(*f*) *Rhoicosphenia*: (*d*) initial epivalve with transverse perizonial bands of the auxospore; (*e*) initial epivalve; (*f*) large cell bearing the initial hypovalve (the raphid valve) and to the left a small original cell.

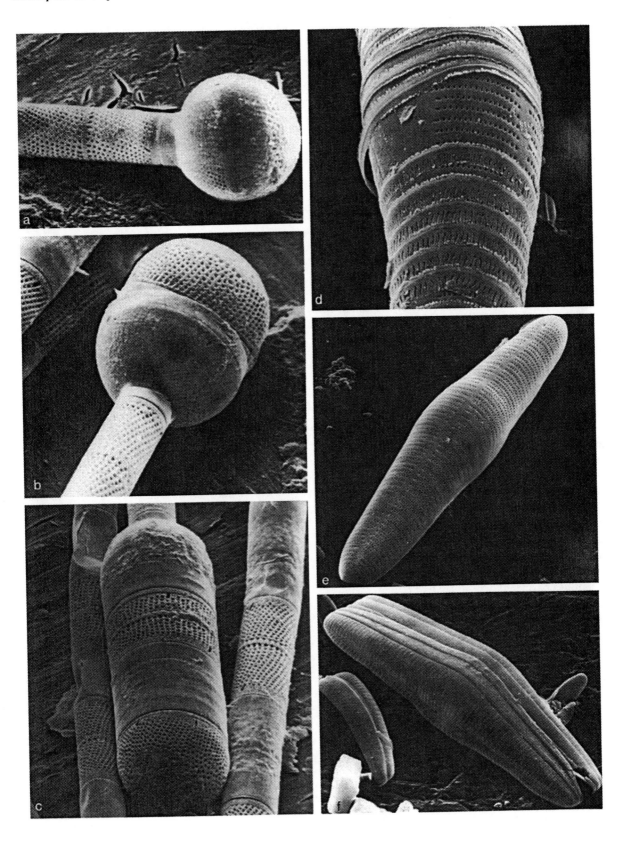

(Fig. 69) and *Navicula cryptocephala* (Geitler, 1968a), in which case they depart to a lesser or greater extent from the auxospore shape. Elsewhere, however, as for example in *Pleurosira* (Ehrlich, Crawford & Round, 1982b, as *Cerataulus*), *Rhoicosphenia* (Mann, 1982b), there are no such plasmolyses and the initial valves are moulded directly against the inner surface of the auxospore casing. In centric and pennate diatoms, therefore, the initial valves tend to have a much more rounded morphology than the valves of vegetative cells (Fig. 68 and for *Stephanodiscus* see Round, 1982b). Besides their modified shape, the initial valves often have a rather simpler structure. In *Melosira* and *Stephanodiscus*, for instance, the initial valves may lack spines (Crawford, 1975; Round, 1982b), while in some raphid diatoms, pseudosepta are present in all valves except the initial epivalve (Geitler, 1932; Mann, 1984d). The girdle is often rather simpler in initial cells, both in centric (Crawford, 1975; Round, 1982b) and pennate taxa (Mann, 1984d).

Motility

Some araphid pennate diatoms seem to have limited powers of movement (e.g. *Synedra tabulata*: Hopkins, 1969) and rotational movement has been observed in the centric diatom *Actinocyclus subtilis* (Medlin, Crawford & Andersen, 1986) and 'shuffling' in *Odontella* (Pickett-Heaps, Hill & Wetherbee, 1986). Otherwise motility is restricted to the uniflagellate sperm of centric diatoms and to those pennate diatoms that possess a unique organelle – the raphe system. Flagellum structure and activity are the subjects of much research and many reviews, and will not be considered here since diatom sperm function like other flagellate cells, although the absence of central microtubules from the axoneme presumably has some effect upon the way in which the flagella beat.

The gliding movement of raphid diatoms has been studied by various workers, and many theories have been advanced as to the mechanism. Some have suggested that the raphe is occupied by streaming cytoplasm, others that small flagella protrude through the raphe slits (for reviews of early theories see Drew & Nultsch, 1962 and Edgar & Pickett-Heaps, 1984b). Recent ideas have in common that the movement of the cell relative to

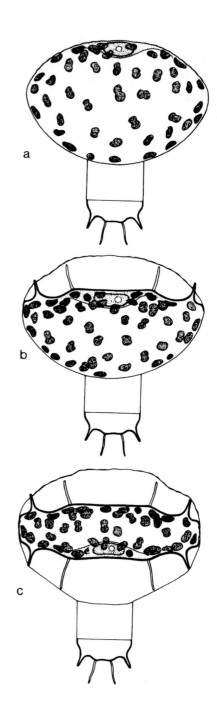

Fig. 69. (a)–(c) *Odontella regia*: (a) expanded auxospore; (b) contraction of the protoplast and formation of the initial epivalve; (c) renewed contraction and formation of hypovalve. Note the position of the nucleus adjacent to the forming valves.

the substratum is considered to be mediated by the secretion of material from the raphe (Jarosch, 1962; Hopkins & Drum, 1966; Harper & Harper, 1967; Gordon & Drum, 1970; Edgar & Pickett-Heaps, 1983, 1984b). Diatom movement appears to be smooth over short periods of time between reversals or stopping, but is in fact jerky, sudden accelerations and decelerations alternating with periods when the diatom is stationary or moving with constant velocity – it is necessary to use high speed cinematography to study these movements (Edgar, 1979). Reversing is frequent. Such behaviour is understandable given that diatom locomotion, involving relatively slow movement of a small cell, is dominated by viscous forces. Edgar (1982) calculates that with a Reynolds number of the order of 10^{-4}, a $10 \times 10 \times 100$ µm diatom moving at 10 µm s^{-1} cannot 'coast' or 'freewheel'. Movement is directional, the path taken corresponding fairly closely to the course of the raphe system (Fig. 70) – curved where the raphe is curved (e.g. some *Nitzschia* species with eccentric raphe systems), straight where the raphe is straight (e.g. *Navicula*, *Pinnularia*), and even sigmoid where the raphe is sigmoid (e.g. *Pleurosigma angulatum*). Other information relevant to an understanding of raphe-associated motility is that movement can only occur on a solid surface (Fig. 71a); that polysaccharide is secreted into the raphe slit (Fig. 71b–d: Edgar & Pickett-Heaps, 1982; Edgar, 1983) and is present within the whole length of the slit; and that a somewhat discontinuous trail of this material is left behind by the diatom as it moves (Edgar & Pickett-Heaps, 1983, 1984b and references therein). Based on these pieces of evidence, and ultrastructural observations of protoplast structure in the vicinity of the raphe, Edgar & Pickett-Heaps (1983, 1984b) have outlined a hypothesis which goes a long way towards explaining the mechanism of motility. It is suggested that the motive force is generated by interaction between actin filaments and transmembrane structures which are free to move within the cell and raphe, but fixed to the substratum at their distal ends (Figs 71c, d, 72). The transmembrane structure presumably includes an ATPase and a protein able to make translational movements within the plasmalemma (as in the fluid mosaic model of membrane structure; Singer & Nicholson, 1972). The transmembrane structure is itself connected to filaments of acid mucopolysaccharide, which can become attached to the substratum at their distal ends. Thus, as the transmembrane structures are moved along the raphe by their interactions with the bundles of actin filaments beneath (Edgar & Zavortink, 1983), the cell moves relative to the substratum (Fig. 72). Edgar & Pickett-Heaps (1983) have proposed that the flow of mucopolysaccharide along the raphe is made easier by a hydrophobic lipid coat over the silica of the raphe-sternum. When the mucopolysaccharide reaches the end of the raphe, it is detached from the plasmalemma at the helictoglossa and continues within the terminal fissure (if present) before being left behind as a sticky but ephemeral trail. Edgar & Pickett-Heaps (1983) speculate that secretion from just one vesicle might be sufficient for cell displacement of half the cell length (or a full cell length where the central raphe endings are absent; see Mann, 1982d), so that movement need not involve significant loss of fixed carbon as would have to occur to satisfy some earlier models of locomotion. Raphe mucopolysaccharide material can be seen in the cell within vesicles, whose contents have a characteristic striated, fibrillar appearance (Edgar & Pickett-Heaps, 1982). Some of these vesicles, called 'crystalloid bodies' by Drum & Hopkins (1966), are also clearly associated with the microfilament bundles (Edgar & Pickett-Heaps, 1983), and appear to move along them, judging by the extremely rapid bi-directional movement of tiny granules visible using the light microscope (Edgar & Pickett-Heaps, 1983). The exact sites where polysaccharide is secreted into the raphe are unknown, but since the two main bundles of microfilaments underlying the raphe system are usually well separated at the poles and centre, but merge and apparently block access to the raphe in between, it seems likely that secretion is limited to areas close to the ends of the raphe slits (Edgar & Pickett-Heaps, 1983; Edgar & Zavortink, 1983). Since some diatoms, e.g. various *Nitzschia* and *Denticula* species lack central raphe endings, these are probably not critical to raphe function.

Cooksey & Cooksey (1980; Cooksey, 1981; Cooksey & Cooksey 1986) have shown that calcium is necessary for adhesion and movement in raphid diatoms, and have recently begun to investigate the mechanism of chemotaxis in *Amphora* (Cooksey &

Cooksey, 1988).

One colonial diatom, *Bacillaria*, possesses cell motility but this is restrained so that the cells slide rhythmically along one another, stop and reverse at an apparently pre-determined point along the valve. If single cells are isolated they continue this rhythmic motility.

Ecology

Diatoms are distributed in all waters except the hottest and most hypersaline. They are abundant in the phytoplankton (Fig. 73a–c) and phytobenthos of marine and fresh waters, whatever the latitude. Hardly a sample can be taken from aquatic habitats that does not contain some cells.

The growth and behaviour of planktonic diatom populations are subject to two predominant special influences. The first is the availability of silicate, which the cells need to make their frustules. The second is the tendency for diatom cells to sink, often quite rapidly, as a result of the high density of their siliceous walls.

Both in lakes and in the sea, diatoms can deplete the silicate to very low concentrations or even until it is analytically undetectable by standard methods. The relationship between diatom growth and silica availability has received considerable attention: see the reviews of Lund (1965), Paasche (1980), Lund & Reynolds (1982) and Reynolds (1984). In the open ocean, the level of silicate is often extremely low and growth of the diatoms is dependent upon transport across the thermocline from the silicate-rich deep water. Indeed, diatom populations are often found

Fig. 70. Tracks of moving raphid diatoms. (a) *Navicula*; (b) *Pleurosigma*; (c) *Gyrosigma*; (d) *Cymatopleura*; (e) *Hantzschia*; (f) *Nitzschia*; (g) *Amphora*.

Fig. 71. *Craticula cuspidata*. (a) Transapical section of cell on, but not touching, the surface of a resin film. (b) Polysaccharide material extending through the raphe slit. (c) Section through the raphe slit showing fibrils extending the raphe slit from the interior to the exterior and attaching to the substratum. (d) As (c) but sectioned along the raphe. (Fig. 71 kindly supplied by L. A. Edgar.)

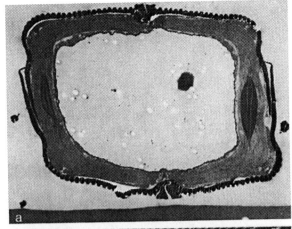
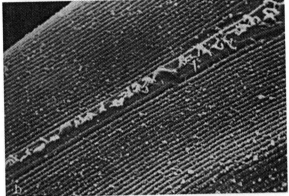
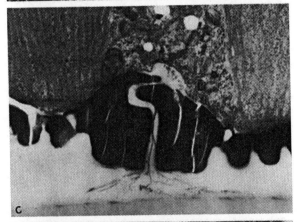

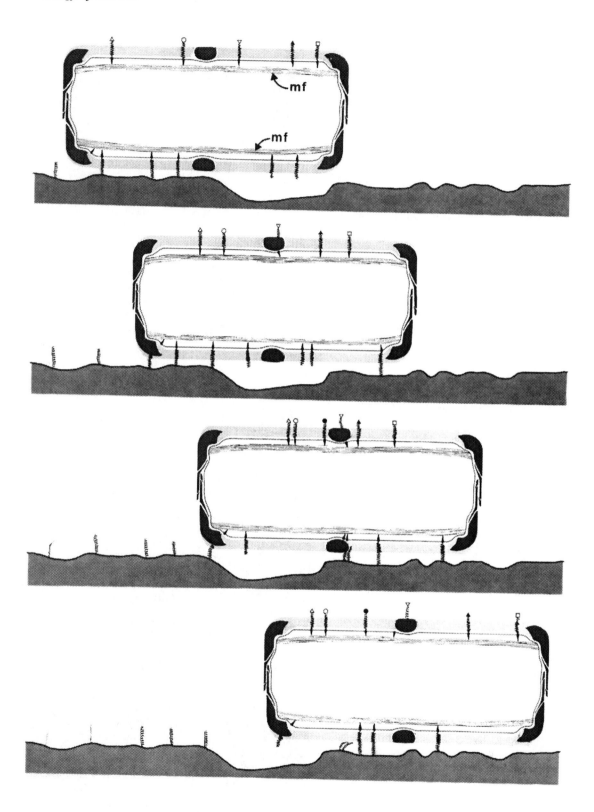

near the thermocline, which may be only 30–40 m below the surface in temperate waters, but 100–120 m in tropical oceans. Little light penetrates to such depths but recently algae (red and green) have been found occasionally growing *in situ* at depths down to 250 m, so clearly growth of diatoms may well be possible at extremely low irradiances.

Planktonic diatoms are invariably as dense as or denser than water; they also lack any form of propulsion. In still water, then, they will usually sink quickly out of the photic zone and be deposited onto the sediment and there preserved (Fig. 75a) to varying degrees depending on the conditions. It has been shown that the density of *Cyclotella meneghiniana* can vary over the complete range quoted for diatoms – so density, and hence sinking rate, varies considerably according to the physiological state of the population (Oliver, Kinnear & Ganf, 1981). No water is ever completely still. Part or all of the photic zone is often turbulent from a combination of winds, currents and convection. In these circumstances, dense algae like diatoms can be kept in suspension. However, the efficiency with which turbulence keeps diatoms in suspension remains critically dependent on the quiescent sinking velocity of the diatoms (Reynolds, 1984). It is therefore not surprising to find that many planktonic diatoms exhibit features that from theory and experiment can be plausibly argued to be devices to slow the quiescent sinking velocity. These adaptations to suspension include the following (see also Walsby & Reynolds, 1980):

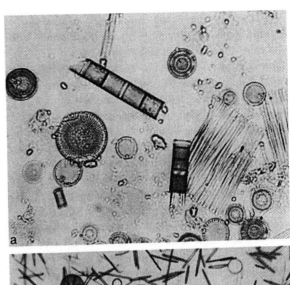

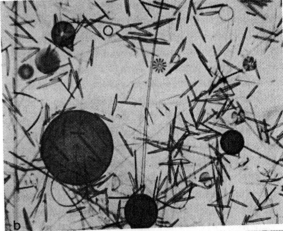

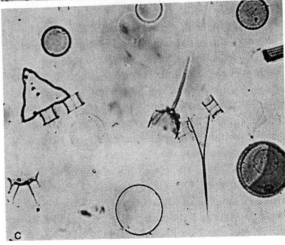

Fig. 72. Schematic sectional diagram of a raphid diatom moving over an irregular surface. Beneath the plasmalemma and underneath the raphe slits are bands of microfilaments (mf). There are probably transmembrane proteins which interact with the microfilaments causing the movement of the polysaccharide fibrils along the raphe. On the upper valve these fibrils collect particles which can be moved relative to the cell (in either direction). The fibrils produced by the valve adjacent to the substratum attach to it and as they are moved along the microfilaments so the cell is moved relative to the substratum. The connection between the polysaccharide fibrils and the transmembrane proteins are presumably broken at the ends of the raphe and the fibrils are left behind as an ephemeral trail.

Fig. 73. Plankton. Cleaned sample from (*a*) a lake, (*b*), (*c*) ocean.

1. All other things being equal, small size can be regarded as an adaptation. In viscous situations (and most planktonic algae live in a world dominated by viscous, not turbulent, forces), the tendency of a particle to sink, expressed by its weight, increases with the cube of the linear dimension while the surface drag resisting sinking increases only as the square. This is presumably part of the explanation why the summer diatom plankton in dimictic or warm monomictic lakes is often composed of relatively small-celled centric species (Hutchinson, 1967). In the spring and autumn, when the whole or most of the water column is turbulently mixed, larger celled forms such as *Aulacoseira*, *Asterionella* or *Tabellaria* are at less of a disadvantage. It must always be remembered, however, that suspension, although a major influence, is only one of the factors affecting the phytoplankton. Grazing, or the need to minimise concentration gradients of nutrients between cell and environment (which is helped by high sinking rates!) may select for shapes and sizes of plankton which in themselves hinder suspension.

2. Presence of a low density mucilage sheath or envelope around the cells will decrease the sinking rate of a cell or colony up to a certain critical sheath thickness. Above this thickness the increase in sinking rate due to the lower surface area to volume ratio of the larger unit will outweigh the decrease in sinking rate produced by the lower mean density (Hutchinson, 1967).

3. In marine diatoms, regulation of the ionic content of the cell sap contained in the vacuole may partially offset the excess density produced by cytoplasm and cell wall. This works by replacement of heavy ions, e.g. calcium and magnesium, by lighter ions, e.g. sodium or potassium (Anderson & Sweeney, 1978). This method of reducing sinking rates is ineffectual in freshwaters because of the low ionic content of the water.

4. The shape of the planktonic unit (cell or colony) influences its sinking velocity. Cylindrical shape of single cells (*Synedra* or *Nitzschia*) or chains (*Aulacoseira*) can be regarded as an adaptation to a planktonic existence, providing the cylinder is more than seven times as long as broad (Hutchinson, 1967). Elaborations of the cell shape that increase the surface area to volume ratio will increase the viscous drag on the cell as it sinks and hence slow its descent, providing the elaborations do not unduly increase the mean density of the cell. Thus, in *Thalassiosira*, artificial removal of the long chitin fibrils extruded through the fultoportulae leads to an increased sinking rate, even though the fibrils themselves are considerably denser than the naked cell (Walsby & Xypolyta, 1977; Walsby, 1988). The effective surface area of the diatom is increased by the fibrils more than enough to offset the higher mean density of the cell.

Another facet of the problem of the suspension should be mentioned and this is the question of entrainment in turbulent water. To some extent the shapes of phytoplankton cells are probably not adaptations minimising the quiescent settling rate. Instead, they serve to improve the efficiency with which cells are entrained within larger scale water movements (Reynolds, 1984).

Silicate availability and the stability of the water column seem to determine the more general features of planktonic diatom ecology. Within the limits they set, other parameters – such as the light climate, the amounts or relative amounts of other nutrients, or the incidence of parasitism and grazing – decide which species will be present and when.

The brief, over-simplified account given above probably underestimates the degree to which the ecology of diatoms, as a group of planktonic algae, differs from the ecology of other groups of planktonic algae. We may predict that planktonic diatoms will be found to have other general ecological characteristics that are unique to the group. For instance, unlike flagellate algae, diatoms do not have to expend energy in regulating their water content via some form of contractile vacuole. Unlike other algae, diatoms take up silicate actively and then polymerise it in a highly organised pattern during a discrete and relatively *short* phase of the cell cycle. Again, diatoms as a group have a particular, characteristic set of accessory photosynthetic pigments, which will 'harvest' the incident light in a particular way (Prézelin & Boczar, 1986): this will differ somewhat from the way chlorophytes, cyanophytes, cryptophytes or dinoflagellates harvest the light. These general biological characteristics are presumably reflected in general features of diatom ecology, but they have not received as much attention as silicate availability and buoyancy.

One feature often commented upon in the literature is that large planktonic species are often present only as occasional cells in samples. This may reflect a rather slow rate of growth (though other equally large forms can form massive populations) or that the 'taxon' is growing actively elsewhere, e.g. in deep layers of the ocean.

Benthic diatoms are much less well understood ecologically than planktonic diatoms. The benthos is more diverse than the plankton, both in terms of the numbers of species and the life forms present. Unfortunately, the benthic diatom communities and their environments are far more difficult to sample and quantify than those of the plankton, and so they have been largely ignored by ecologists. The distinction between plankton and benthos is not absolute. For instance, some lake diatoms spend part of the year dormant on the bottom and part growing actively in the water column (e.g. *Aulacoseira* spp: Lund, 1954, 1955) and along some sea coasts it is possible on occasion to find dense populations growing in the 'surf zone'. These communities, often dominated by small-celled species of *Chaetoceros* and *Asterionellopsis*, are sometimes suspended at the water surface, sometimes deposited on the sandy sediment by the retreating tide. They control their buoyancy in such a way that there is no net tendency for them to be swept out to sea by the ebb tide. Deposits left behind on the beach can be very unsightly and up to 15 cm thick! (Lewin & Norris, 1970; Lewin & Mackas, 1972; Lewin & Hruby 1973; Lewin, 1974; Lewin, Hruby & Mackas, 1975; Lewin & Rao, 1975).

Within the benthos a division can be made between the diatoms that live attached to the substratum, be it rock or plant, and those that live free, on or in sediments (Round, 1981b). In neither case is suspension a significant problem and in many benthic habitats (except for some epilithic assemblages and for epipelic communities living on highly calcareous or highly organic sediments) silicate is unlikely to be reduced to limiting concentrations. We can therefore predict that on the whole benthic diatom populations will behave quite differently from planktonic populations.

Consider first the diatom assemblages that live freely on and in sediments (Fig. 74a–c). These

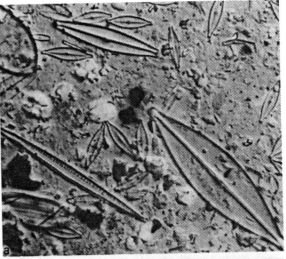
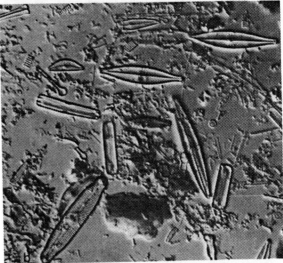
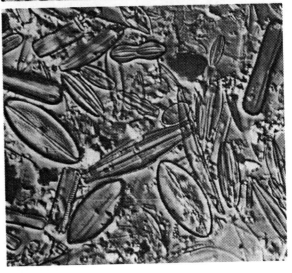

Fig. 74. Epipelon. Examples of cleaned material from (*a*) river, (*b*) lake, (*c*) marine sediments.

epipelic and endopelic diatoms (Round, 1979b) live in an almost two-dimensional world. Below them are the sediments themselves, often highly reduced, with high nutrient concentrations yet where toxic substances, such as sulphide, may also be present in appreciable quantities (see the comprehensive discussion in Admiraal, 1984). Except where the sediments are composed of translucent material such as quartzitic sand, light does not penetrate far below the surface. Above the thin surface layer where the epipelic flora live, there may be an appreciable depth of water containing plankton, or only a water film separating the algae from the air. In either case the concentrated, essentially two-dimensional epipelic community is subject to grazing by a variety of invertebrate and vertebrate animals living above or within the sediments. The whole community is also liable to be buried by the deposition of fresh sediment or disturbance of the sediment. In view of these characteristics, it is not surprising that the epipelon is composed of motile species (Fig. 73a, b, c), which migrate up and down in the topmost sediments in relation to environmental cycles (e.g. light–dark cycles, in some instances modified by tidal disturbance). Several epipelic diatoms have been shown to exhibit endogenous rhythms of migration, moving into and out of sediments, even when kept under constant conditions of light and temperature. *Hantzschia virgata* exhibits a remarkable tidal rhythm (Palmer & Round, 1967). In constant conditions the cells emerge approximately 50 minutes later each day (in phase with the tides in nature) and even re-phase to the morning tide when the low tide they have been 'tracking' reaches the evening (see also Happey-Wood & Jones, 1988). The non- diatom members of the epipelon (e.g. desmids, blue- green algae, chrysophytes, euglenoids, dinoflagellates and cryptomonads) are also motile and some (e.g. *Euglena*: Palmer & Round, 1965; Round & Palmer, 1966) have been shown to make rhythmic migrations like those of the diatoms.

The attached communities are less likely to become buried beneath sediment but they are even more liable to grazing than the epipelic communities, which may be able to minimise losses to surface feeders by migrating to depth during the dark period. There are various modes of attachment to the substrate but these fall into two categories. In one, the cells are adnate, i.e. closely appressed to the substratum, as in *Cocconeis, Amphora, Epithemia* or *Rhopalodia* (Figs 9a, 76e). In the other, the cells are pedunculate, i.e. attached to the substrata by stalks or pads, e.g. *Ardissonia*, some *Cymbella* and *Gomphonema* species (Figs 9b, 76a). The adnate forms are rarely colonial whereas the pedunculate often are. The development of pedunculate forms on the surface transforms the structure of the community from two-dimensional to three-dimensional. Within the microscopic 'forests' produced, some species can be seen to be canopy-formers, (e.g. long-stalked species of *Cymbella*) whereas others form the 'shrub' and 'field' layers (e.g. *Rhoicosphenia* and small *Gomphonema*). These communities are probably the most complex in which diatoms play a major role.

The attached communities may be classified according to their substrata; thus epipsammic, epilithic, epiphytic and epizoic communities may be distinguished, growing on sand grains (Figs 18b, 75b–d), rock (stones) (Fig. 76b–d), plants (Fig. 75e, f) and animals (Fig. 76e). There are overlaps between these communities in the species they contain and, except for the epipsammon, all contain very similar life-forms. The epipsammon is quite distinct, consisting of very small appressed species or species with very short stalks. These occupy depressions or crevices in the surface of the sand grain, where they are buffered from the severe abrasion between grains that occurs when the sand is disturbed. Occasionally larger-celled diatoms occur here and can bridge the gaps between grains, cementing them together (e.g. *Psammodiscus, Raphoneis, Cerataulus*). Sometimes too, species of *Mastogloia* live here in blobs of mucilage attached to the sand. Co-habiting with the epipsammic community, there may also be an epipelic community, moving on and between the sand grains. Some epipsammic species also move, but usually only very slowly from grain to grain (Harper, 1977).

There are differences between epilithic and epiphytic assemblages and between the epiphytic communities growing on different plants. Eminson

Fig. 75. (*a*) Example of cleaned oceanic sediment. (*b*) Epipsammic diatoms from a freshwater lake. (*c*),(*d*) Epipsammic diatoms from marine sand. (*e*) Epiphytes from freshwater *Cladophora*. (*f*) Epiphytes from freshwater *Phragmites*.

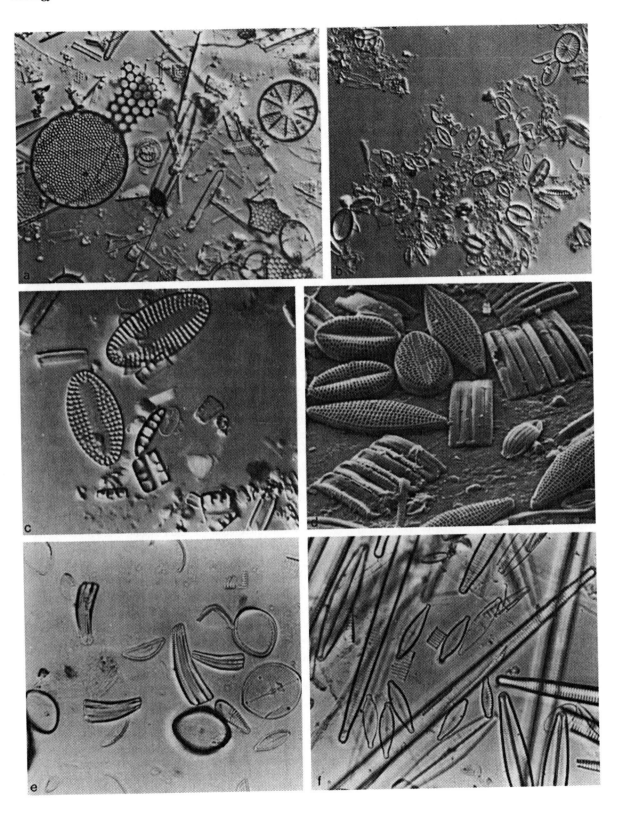

& Moss (1980) have shown that there is very high host specificity in infertile lakes but that this almost disappears in more eutrophic situations, where the high external (limnetic) nutrient concentrations swamp any small differences between the host plants in the rate of nutrient secretion or release. The whole question of host specificity, however, needs much greater study.

The epiphyton and epilithon are best developed in submerged habitats, but a small specialised flora is also to be found colonising the spaces between the leaves of bryophytes, or the leaves of angiosperms in tropical rain forest, or growing on terrestrial soils and damp rock faces. These diatoms are often collectively referred to as sub-aerial forms and include species like *Hantzschia amphioxys*, *Diadesmis* spp., *Pinnularia borealis* and a number of other raphid taxa, together with several large chain-forming *Orthoseira* species.

The epiphytic and epilithic communities are not, of course, pure diatom assemblages but have other algae intermingled. Together they often produce copious amounts of mucilage, in which other, non-attached algae live, including several diatom species: this relatively neglected community was termed the 'metaphyton' by Behre (1956).

Some diatoms grow attached to animals. The episarc of hydroids (Fig. 76e) forms a very favourable habitat and is sometimes brown from the growth of both adnate and pedunculate species. The shells of many molluscs and Crustacea are sometimes colonised selectively by particular taxa, e.g. the genus *Pseudohimantidium* is only known to occur on a few crustaceans, notably *Coryaeus* species, where it forms a monoculture. *Synedra cyclopum* occupies a similar niche on freshwater Crustacea. Organisms with long-lived shells or tests are not the only animals to be colonised by diatoms. The mucilage of the freshwater colonial ciliate *Ophrydium* is colonised by raphid diatoms such as *Cymbella cesatii* (Geitler, 1968b, 1975b). Vertebrates may also bear a diatom flora. Whales often have rich growths of *Bennettella* and *Epipellis* (related to *Cocconeis*) on their skins (Hart, 1935; Holmes, 1985), and recently some diving birds have been shown to bear dense growths of diatoms on their ventral body feathers (Croll & Holmes, 1982; Holmes & Croll, 1984).

Only a few diatoms are themselves hosts for other algae or diatoms. The most commonly encountered hosts are large freshwater species of *Surirella*, *Cymatopleura* and *Nitzschia* (Fig. 76f). These support growths of *Synedra parasitica* and *Amphora ovalis* var. *pediculus* (*sensu* Hustedt, 1930). On *Surirella*, *Synedra* seems to be confined to the girdle and this is a common position for the epiphytes in other cases, e.g. *Cocconeis* species on *Biddulphia* and *Isthmia*. Whether the preference for the girdle is a result of mechanical or physiological factors remains to be determined. Cells and small colonies of the haptophyte *Phaeocystis* are often seen on *Chaetoceros* species. Some freshwater planktonic diatoms, (e.g. *Asterionella, Fragilaria, Stephanodiscus*) bear choanoflagellates. Apart from these examples, algal or other epiphytes are rarely found on diatoms and this may be because of the presence of a mucilage sheath (a similar situation exists in filamentous Zygnemaphyceae). A report (Jolley & Jones, 1977) of a symbiotic relationship between a diatom and *Flavobacterium* may indicate that other associations should be sought and investigated.

A few diatoms act as hosts for endosymbiotic organisms. The cyanobacterium *Richelia intracellularis* occurs as an endophyte of *Rhizosolenia* species and has been shown to fix nitrogen (Mague, Mague & Holm-Hansen, 1977). Endosymbionts, again cyanobacteria, occur in *Epithemia* and *Rhopalodia* species (Drum & Pankratz, 1965a; Geitler, 1977) but these need further study. They occur on the dorsal side of the cell in small numbers, close to the nucleus. The freshwater populations of *Epithemia* and *Rhopalodia* also fix atmospheric nitrogen via their symbiont.

Finally the recent fascinating series of discoveries of diatoms living endosymbiotically in Foraminifera (Lee *et al.* 1979, 1980 a, b, c, 1982; Lee & Xenophontes, 1989) further extend the habitats of diatoms. These diatoms do not form frustules within their hosts, but produce normal siliceous shells when isolated and grown separately. A number of taxa have been isolated, most of them raphid forms (Lee & Reimer, 1984). It has also been suggested that

Fig. 76. (*a*) A dense community of freshwater epiphytic diatoms. (*b*),(*c*) Epilithon from acid oligotrophic rivers. (*d*) Epilithon from a eutrophic river. (*e*) Diatoms (*Cocconeis*) on the episarc of a hydroid. (*f*) A diatom (*Synedra parasitica*) epiphytic on a diatom (*Nitzschia sigmoidea*).

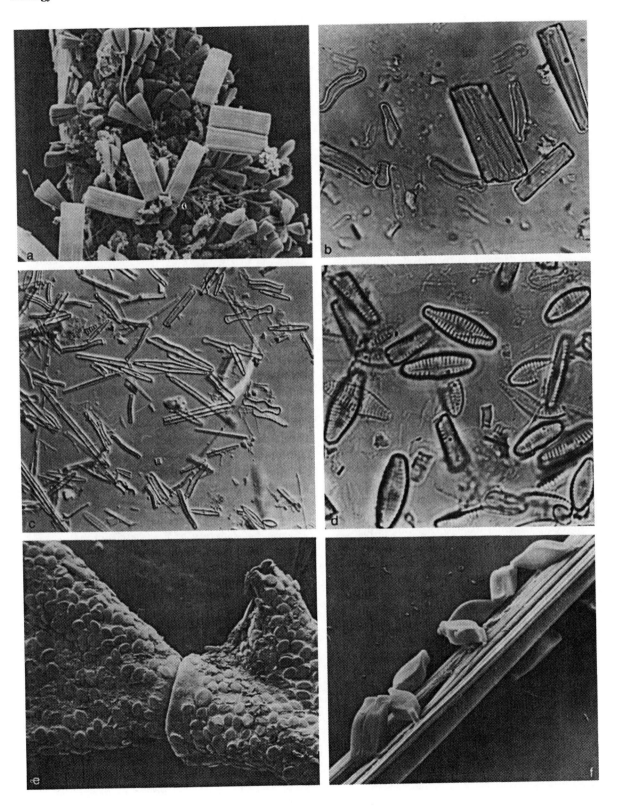

the endosymbiont of the dinoflagellate *Glenodinium foliaceum* is a diatom that can no longer form a frustule (Dodge, 1983).

Palaeoecology

The siliceous exoskeleton, especially the valves and even more so the resting spores, where these are produced, is resistant to decay. It dissolves only slowly, except in alkaline conditions or, surprisingly, in the interstitial water of peat bogs. Thus, once incorporated into sediments, diatom shells often remain there, forming a record of the populations from the benthos or the water column above. Marine floras are rarely as well preserved as freshwater ones. Dissolution seems to be more rapid in the sea, and the longer journey to the sediments gives greater opportunity for fragmentation. Lizitzin (1971) has estimated that in the sea more than 90% of diatom valves dissolve before burial. Nevertheless, the sedimentary record is invaluable in both habitats for reconstructing the history of events both in the water and, to some extent, on the land.

Many lakes lying in temperate latitudes have a relatively short history, since the major glaciations of the Pleistocene scoured away any earlier deposits. Their sediments therefore represent at most around 15 000 years sedimentation. The lowest sediments are often glacial clays, poor in organic remains, as a result of low productivity and high abiogenic sedimentation rates. In Britain, the earliest late glacial assemblages are characterised by an abundance of alkaliphilous species. Later, the diatom record indicates increasing acidity of the waters as the soils and the catchment areas matured and became leached and podsolised (Round, 1957, 1961; Haworth, 1969; Crabtree, 1969; Evans, 1970; Evans & Walker, 1977; Walker, 1978). This sequence, which has been confirmed and extended by chemical analyses (Mackereth, 1965, 1966), is contrary to that predicted by some theories of lake evolution (e.g. Pearsall, 1921). In North America, however, the opposite sequence can also be demonstrated, where lakes originated on acid sediments and became more alkaline as the land flora and drainage changed (Haworth, 1972). Studies of diatoms in recent sediments can be a great help in dating anthropogenic eutrophication, pollution or acidification of lake systems (Round, 1961; Stockner, 1972; Flower & Battarbee, 1983; Battarbee *et al.*, 1985; Engstrom *et al.*, 1985; Jones *et al.*, 1986). The

sedimentary record can be extremely detailed, as has been shown by Haworth's (1980) comparisons of the plankton records kept by the UK Freshwater Biological Association with the diatom content of the topmost sediments, or Simola's studies of annually laminated sediments (e.g. Simola, 1977).

Outside the present temperate and sub-arctic zones, lake deposits may represent much longer periods of time, and these too are now under study (e.g. Stager, 1984; Stager *et al.*, 1986). The larger lakes, formed in tectonic basins, e.g. Lakes Biwa, Baikal and Tanganyika, have been in existence for millions of years and must contain a fascinating diatom record reflecting climatic, geomorphological and biological change.

The diatoms preserved in marine sediments represent an even longer time span – back to the Cretaceous at least (Harwood, 1988). Recently there has been an upsurge in interest in fossil marine diatoms, stimulated by the revolution in geologists' thinking accompanying the acceptance of the dominant role of plate tectonics in moulding continental and oceanic crust form and structure. Much material has been recovered by the Deep Sea Drilling Programme, of which only a fraction has been examined in detail (see refs. in Barron 1985a, Barron *et al.*, 1985). Independent dating of cores can be carried out using radiometric and geomagnetic methods, and the diatoms themselves used for stratigraphical cross correlation. Geomagnetic dating is also useful for freshwater deposits, but here it is the finer details of the wanderings of the magnetic poles that are used (Thompson, 1984) rather than the periodic reversals of the earth's magnetic field employed in the longer time-scale dating of marine sediments (Burckle & Opdyke, 1977).

Marine deposits (Fig. 75a) tend to contain the more robust centric genera, together with the more highly silicified parts of other forms, e.g. the 'spines' of *Rhizosolenia*. The appearances and disappearances and, to some extent, the relative frequencies of taxa, are used for stratigraphical correlation, which therefore depends upon the assumption that such events are (within a reasonable span of geological time) simultaneous throughout the area under study. Stratigraphical correlation can also be carried out using analyses of the $O^{18}:O^{16}$ ratio in diatom silica (Labeyrie, 1974;

Labeyrie & Juillet, 1982; Labeyrie, Juillet & Duplessy, 1984), which at the same time yields valuable information on the characteristics of ocean surface waters during the last few million years (Labeyrie, Juillet & Duplessy, 1984; Labeyrie *et al.*, 1986).

The margins of the continents are subject to inundation or terrestrialisation as sea levels change with the waxing and waning of the ice caps, or as the crust itself is uplifted, depressed or folded. As these changes occur, the diatom floras will shift accordingly, so that cores taken around the continental margins often reveal a complex history of sea level changes in the diatom record (e.g. Buzer, 1981; Robinson, 1982). Such information is of interest to the geomorphologist concerned with isostatic readjustment after an ice age, or to archaeologists investigating the environment of historic or prehistoric man.

Concepts in diatom systematics

Few of the difficulties encountered in diatom systematics are peculiar to diatoms. Thus, different biologists have different ideas about how to define species and other taxonomic categories. Among diatom taxonomists, there have been some who have taken a narrow view of species; others prefer a wider concept, as can be seen from the recent revisions of *Nitzschia* and other genera made by Lange-Bertalot (e.g. 1976; Lange-Bertalot & Simonsen, 1978). The problem is mainly one of defining the range of morphotypes to be included in a single species and distinguishing where the breaks occur between these and their nearest neighbours. At the generic level and above, again some prefer to split, others to lump, e.g. contrast the opinions of Round (1978a, b) and Simonsen (1979) on how to treat *Chrysanthemodiscus* and *Stictocyclus* at the family level. There are, however, one or two special features of diatom variation. The formation of the new valves within the confines of the old cell wall, while in one sense leading to less morphological plasticity (Round, 1981c) since the valves once fully formed are rigid and unexpandable, at the same time introduces extra, sometimes random elements into variation and evolution. Thus, for instance, any chance plasmolysis during valve formation, affecting the margins of the new valves, may produce effects, not only on the shape of these valves, but also on

the shape of all succeeding generations until the next occurrence of auxospore formation. This might be a long time in those taxa in which there is little or no size reduction (see Round, 1972a; Crawford, 1981b; Mann, 1988b). In this case unequal competition could occur in the same habitat between cells of the same genotype but different phenotype.

There is also the interesting question of whether diatoms have evolved means of reducing competition between cells at different stages in the size reduction cycle, just as the males and females of dioecious higher plant or animal species tend to occupy different niches. Because of their different surface to volume relationships, small and large cells of the same diatom clone will almost inevitably have different physiological characteristics so that co-existence on the basis of identical competitive ability is impossible, unless elaborate mechanisms are present to compensate for the size effects. This being so, it may be advantageous for the larger and smaller cells to diverge sufficiently in their physiology in order that they can co-exist stably – limitation by different resources might be a possible 'solution'. Alternatively, if auxospore formation was phased, taking place perhaps at the same time each year, this might avoid the worst effects of intra-clonal competition. In spite of their diplontic life cycle, diatoms are probably subject to the same kinds of evolutionary pressures as algae with more complex life cycles, involving an alternation of generations: in these, niche separation between, for example, the *Conchocelis* and thalloid phases of *Porphyra* is more obvious, but the problem is essentially the same for diatoms. Evidence in diatoms for changes correlated with size reduction over and above direct effects via area–volume relationships is as yet rather limited, but Werner's data for *Coscinodiscus asteromphalus* (1971a, b, c, 1978) and the polymorphism observed in *Mastogloia* and *Coscinodiscus* (Stoermer, 1967; Holmes & Reimann, 1966) suggests that some form of niche separation may be occurring.

Shape and pattern changes during the life cycle have sometimes confused taxonomy, e.g. the classification of the initial valve of *Anomoeoneis sphaerophora* as an independent species or as the variety *costata*, and the separate naming of the internal valve of *Craticula cuspidata*, as the var.

heribaudii (see Schmid, 1979b). Furthermore, some diatoms respond to changes in salinity and silicate availability by changing the degree and pattern of silicification in their valves (Belcher *et al.*, 1966; Hasle *et al.*, 1971; Schultz, 1971; Fryxell & Hasle, 1972; Booth & Harrison, 1979). The problems these phenomena cause for taxonomists are relatively easily overcome, through careful study of natural populations or by culturing. If two or more morphs exist side by side in nature and it is suspected that they may be different products of the same genotype, then one good test (as in desmids: see Brook, 1981) is to search for 'Janus cells', i.e. cells with one valve of one type and one of another. This test is useful in any organism in which different parts of the cell or body are formed at different times, and possibly, therefore, under different conditions. The most extreme example of polymorphism in the literature, apart from *Phaeodactylum* is Wood's (1959) account of two Janus cells found in plankton hauls, each with one valve of '*Coscinodiscus lineatus*' and one of '*Asteromphalus roperianus*'; this report needs to be checked. It is most important that floras and monographs document properly the co-variation of different characters during the life cycle. Descriptions usually give only the extremes of variation in each character (e.g. length, striation density and valve outline), which is scarcely adequate. Good illustrations of the range of variation are invaluable but rarely supplied.

The systematics of diatoms are based almost entirely on phenetic data and even these are virtually confined to the structure and shape of the siliceous parts, especially the valve. Little attention has been paid to the protoplast or to aspects of reproductive behaviour as sources of taxonomic information. This bias has become even more pronounced recently with the introduction of more powerful microscopical techniques. There is virtually no information about the extent of hybridisation between morphotypes: just some isolated observations of reciprocal crosses between forms of *Biddulphia* (von Stosch, 1962b) and of hybridisation between *Cocconeis placentula* populations (Fig. 77b) and between two 'varieties' of *Gomphonema constrictum* (Fig. 77a and see Geitler, 1958a, 1969a). Mann (1984e) has shown that some 'forms' of *Sellaphora pupula* will only cross with like forms

while others can apparently interbreed (Fig. 77c), and so the question of using the biological species concept in the diatoms does appear to be feasible (see also Mann, 1988a, 1989). On the whole, however, taxa from form to family and beyond are defined purely on the basis of comparative morphology and anatomy.

While there is obviously an urgent need for studies of reproductive isolation between diatom populations, it should not be thought that this could ever be used as the sole basis for species level taxonomy. Experience with other groups of organisms shows that the biological species concept is not always useful. In apomictic groups, for instance, every clone is reproductively isolated from every other and thus would qualify as a 'species'. In addition, it is impracticable to use the biological species concept extensively, in view of the huge labour required to investigate even a small group of species. Furthermore, it is not as objective as its more ardent supporters claim, since intrinsic

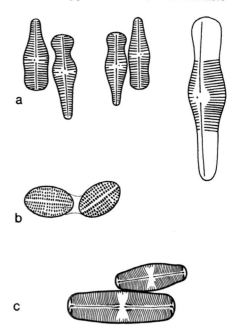

Fig. 77. Hybrid pairings between different morphotypes. (*a*) *Gomphonema constrictum* (? = *truncatum*) vars *constrictum* and *capitatum*: two pairs and a hybrid initial cell. (*b*) *Cocconeis placentula* vars *euglyptoides* and *pseudolineata*: copulating cells. (*c*) *Sellaphora pupula* races (After Geitler, 1958a, 1969a and original).

barriers to hybridisation may be only partly effective, with all variations from almost complete interfertility to almost complete intersterility, while extrinsic barriers (geographical features, etc.) may in nature provide insurmountable barriers to crossing between potentially compatible strains. But in spite of these drawbacks, studies of reproductive isolation between morphologically distinct populations would give some kind of standard against which to judge the range of morphotypes to be included within a single species (Mann, 1984e, 1988a, 1989).

We believe that a classification should be useful for a great variety of purposes, e.g. providing a basis for phylogenetic reconstruction; aiding communication between physiologists, biochemists and others; predicting unknown properties of taxa; enabling ecological surveillance, etc. In order to do this the taxonomist must as far as possible use all available information and employ it dispassionately, assessing relationships on the basis of *overall* similarity. The outcome of this phenetic approach is a natural classification (Davis & Heywood, 1963; Sneath & Sokal, 1973). Weighting of taxonomic evidence is a dangerous procedure, to be avoided wherever possible. It appears to work in certain groups only because a lucky or inspired choice of the characters to be weighted has selected a character-complex consisting of several or many logically independent but empirically correlated character states. In such a case all the taxonomist can usually offer as justification for his action is his intuition, although this may be supported by some appeal to the likelihood that the particular character may be more stable or more critical to the survival of the organisms possessing it.

Of course, diatoms do not vary in morphological characteristics alone. Forms originally classified in *Thalassiosira pseudonana*, for instance, have been found to exhibit distinctive, stable physiological characters. Estuarine and oceanic clones of this species vary in their growth responses to temperature and salinity (Guillard & Ryther, 1962), vitamin B12 requirements (Guillard, 1968), uptake characteristics for nitrate (Carpenter & Guillard, 1971) and silicate (Nelson *et al.*, 1976), sensitivity to polychlorinated biphenyls and other chemicals (Fisher *et al.*, 1973; Fisher, 1977; Murphy & Bellastock, 1980), and cell division periodicity (Nelson & Brand, 1979). The work of Brand *et al.*

(1981) suggests, but does not prove, that there is genetic exchange between clones; the work as a whole indicates the existence of neritic and oceanic races. Indeed, Hasle (1978) has separated parts of *Thalassiosira pseudonana* into *T. guillardii*. Gallagher (1980, 1982) has demonstrated that in morphologically *identical* clones of *Skeletonema costatum* isolated from a single bay at different times of the year there are distinct genotypic variants, recognisable from their electrophoretic banding patterns; this variation is coupled with differences in growth rate, chlorophyll content and carbon uptake characteristics. Medlin *et al.* (1988) examined the small subunit rRNA sequencing of *Skeletonema* and found differences between geographical races.

In these two examples then, there is much physiological variation with only slight or no morphological variation, while in other taxa an equivalent amount of variation may be spread more evenly between these two categories. Thus, in *Fragilaria crotonensis* two morphological variants have been detected, which also differ in their susceptibility to parasitism by chytrids (Canter & Jaworski, 1982, 1983; Crawford *et al.*, 1985). It would probably be better, and it would certainly be more natural, if all genotypic variation was treated similarly by taxonomists, whether morphological, physiological or biochemical, but this, like the biological species concept, is often impractical. Classification and identification in diatoms will inevitably have to be based on cytological and morphological characteristics for the foreseeable future. Electrophoretic analysis of proteins may help clarify relationships within species and species aggregates, or between selected examples of higher taxa, but it will be difficult to integrate the information from this rather crude characterisation of the protein complement with data on complex valve and protoplast structures.

Since diatom species are at present defined phenetically, there is no fundamental difference between their definition and the definitions of genera, families, or whatever. Thus, for example, a genus is merely a cluster of species between which, in the opinion of taxonomists, the differences are nowhere large enough to allow further subdivision. As in other categories, no character need be common to *all* the species included in the genus and no single diagnosis or definition is necessarily possible; taxa must often be defined polythetically, i.e. a genus could be defined by the possession of character states A,B,C,D, *or* B,C,D,E, *or* A,C,D,E, etc. When we seek to delimit taxa at whatever rank, we are looking for discontinuities in the variation pattern. The question thus arises as to how large this discontinuity must be to justify separation at a particular taxonomic level. As far as species are concerned, some authors recommend a minimum number of discontinuities for separation (references in Davis & Heywood, 1963), but laudable though standardisation may be as a general aim, it would not be easy to put into practice especially at higher levels of the taxonomic hierarchy. Our attitude to the number of discontinuities required must be tempered to some extent by the pattern of variation between and within taxa in the character hyperspace (i.e. a multi-dimensional space in which each character state is expressed as a value along an axis orthogonal to each other character's axis).

With no absolute guide to the size of discontinuity for separation at a given rank, and no universally applicable, objective criteria for the delimitation even of species, still less any other taxon, we must admit that our taxonomy is never likely to be free from bias and distortion. However, in many cases the data make it obvious that taxonomic revision is necessary. Thus there can surely be no question about the need for *Placoneis* to be separated from *Navicula* (Cox, 1987), *Anomoeoneis* to be separated from *Brachysira*, or *Psammodiscus* from *Coscinodiscus*. The problems of generic circumscription, familial and ordinal limits have been constantly in our minds as we prepared the generic section. Fortunately, many genera are so obviously natural and so clearly defined by the possession of unique features, e.g. the complete cross bars on the valvocopulae of *Climacosphenia*, that no problems arise.

It is important to realise that no character can be assumed *a priori* to be invariable within taxa of a given rank. Conclusions about the worth of a character in one group cannot be transferred to a different group, e.g. the fact that velum structure is constant in genera like *Stephanodiscus* cannot be used as an argument that velum structure will be constant in every genus and hence that a genus such as *Mastogloia* with its range of velum structure should *necessarily* be split. Thus, whether or not

Simonsen (1979) is right to consider it impossible to separate the Cymbellaceae and Gomphonemataceae from the Naviculaceae on the basis of cell symmetry, he was in our view wrong to attempt to justify this conclusion by appealing for consistency in the weighting of symmetry evidence in different diatom families. If our decisions concerning the limits of taxonomic groups are based on overall similarity, it is likely that differences in symmetry will sometimes coincide with boundaries best recognised at family level but elsewhere only with species boundaries. Thus symmetry may be useful at a higher level in the Gomphonemataceae than in *Climaconeis* and its allies. Or again, judging by overall similarity, the formation of massive costae is a species characteristic in *Podocystis* and *Isthmia*, but useful in the diagnosis of genera elsewhere, e.g. in the Bacillariaceae (*Denticula*). 'Trivial' ornamentation such as small granules, warts, spinules, etc., should be treated with reserve since it rarely appears to have any taxonomic significance; even so, its taxonomic significance should not be prejudged and can only be determined in retrospect.

Another point worth making is that if phenetic analysis reveals the existence of a cluster, then it is really a matter of taste at what level this cluster is recognised, providing that its *relative* position in the taxonomic hierarchy remains the same. Thus the arguments of Hasle (1972b) against keeping *Fragilariopsis* separate from *Nitzschia*, or Simonsen's (1979) arguments against the Gomphonemataceae and Cymbellaceae, are largely invalidated because these authors continue to recognise these groups, but at a lower taxonomic level: *Fragilariopsis* is retained as a section of *Nitzschia*, and the *Gomphonema* and *Cymbella* groups are recommended to be recognised as subfamilies or tribes within the Naviculaceae. If *Fragilariopsis* exists as a natural entity, then there is no particular reason why it should not be accorded generic rank, rather than sectional, and in view of the size of the genus *Nitzschia* and the great diversity within it, it would seem reasonable to retain Hustedt's *Fragilariopsis*, though with an amended description to meet Hasle's criticisms. All we can hope to do in taxonomy is to rank different clusters correctly in a hierarchy. No cluster can have an absolute claim to be called a genus, family, etc.

The taxonomic system set up by light microscopists was based on relatively few characters, though probably more than for most other unicellular eukaryotic groups. The introduction of electron microscope techniques has greatly increased the number of characters available (Mann, 1978, listed over 40 characters useful in classification of *Nitzschia* species; see also Mann, 1982d) and this has led to a much improved understanding of many groups, an understanding that is helped further when protoplast characters, the method of auxospore formation, auxospore structure, ecology and other phenomena are also taken into account. The new characters are just as valid as the old and where they overwhelmingly indicate relationships at variance with those previously accepted, revision is necessary. It should come as no surprise that a few genera have to disappear and an even greater number be erected or resurrected. In the past, workers such as Hustedt clearly recognised that many genera were not homogeneous but on light microscope evidence they were reluctant to split genera. Contrary to some opinion, Hustedt was a rather conservative taxonomist and this resulted in 'lumping' at the generic level, and recognition of many infrageneric groups. This does not do justice to the diversity within the diatoms, nor, in spite of appearances, does it make the classification easier to operate. For example, in Hustedt (1930, 1927–66) and Cleve-Euler (1951–55), the larger genera are split into sections and each section is provided with a separate key to species. Thus the same difficulty is encountered in using the classification for identifying diatoms as if all the sections were treated as genera. Furthermore, many authors describing new species of these larger genera do not specify which section or subgenus they belong to. Thus it becomes very difficult to integrate newly described taxa into the established framework. If the sections were genera instead, this difficulty could not arise, which is not an argument for raising all existing sections to generic rank, but to shun categories between genus and species until such time as the classification has achieved a degree of stability and naturalness much closer to that present in angiosperm taxonomy. We believe it is time to make a fresh start in many groups of diatoms insofar as this is permitted by the International Code of Nomenclature, and that the only way to make significant progress is by careful and thorough monographing of well-defined genera.

The floristic approach, in which authors describe the diatoms of a circumscribed and usually small area, has been useful for taxonomy in the past and even now yields valuable ecological and biogeographical data; it is, however, to be discouraged since it produces piecemeal, inconsistent classifications.

Having established a generic description, which by its nature has to be broad and brief, allowing as much variation as possible but highlighting the absolutely distinctive features, one then has to ask how far it can be stretched. Each taxonomist will set a limit to this and this is then his limit for the genus and any taxa with features outside this have to constitute new genera. There is no *a priori* reason why genera should be large, or, for that matter, that any higher systematic unit should be sizeable; the object is to establish natural boundaries and 'classes' of small membership are no less desirable than large (a commonly held belief among protistologists). If during revision of a genus a species is found which is anomalous in a number of respects, one should not be afraid to separate it into its own monotypic genus. There is no doubt that it is easier to 'lump' species into genera and widen the generic descriptions but this obscures much of interest. We have therefore made many changes in the circumscription of genera and erected many new genera. We realise that this will not be welcomed by all and that enormous nomenclatural difficulties remain, e.g. in genera such as *Navicula*, and in the biddulphioid/hemiauloid/triceratioid clusters. In spite of the difficulties, these splits must be made since the segregate genera, for instance of *Navicula*, are often less closely related to each other than they are to other long established diatom genera. There is still more splitting to be done and we sometimes indicate this in the generic section, where we could not complete the work ourselves.

Relative to many groups of algae the nomenclature of the diatoms is fairly well documented. Thanks to VanLandingham (1967-79), who built on the earlier indices of F. W. Mills (1933-5) and others, there is a catalogue of generic and specific names in the Bacillariophyta, while the Index Nominum Genericorum[1] also gives

[1] The diatom section was compiled by Ross & Sims and later additions made by Sims & Williams. The compilers have very generously allowed us to reproduce this as an appendix – see p. 682.

information about the nomenclatural status and typification of genera; such compilations are invaluable, as are accounts of important collections of type specimens (e.g. Simonsen, 1987).

Evolution and phylogeny

Neither topic has received much attention though a kind of phylogeny grew up out of the early taxonomic treatments based mainly on morphology of the vegetative cell; this classification goes back to Pfitzer (1871) and Schütt (1896). It is interesting that although by 1928 Karsten could provide an excellent account of diatom structure, reproduction, sexuality, etc., he could only comment briefly (a mere half-page) on evolutionary relationships, and that only on Pascher's (1921) view that the diatoms were allied to the Chrysophytes and then he doubted the validity of Pascher's conclusion. By 1964, Hendey was only able to say 'Despite recent researches upon the subject, the ancestry of diatoms is somewhat obscure'. It is still obscure! The similarity to the 'chrysophytes' *sensu lato* has been discussed by Round (1981a) and there the inclusion of the diatoms within a single 'heterokont' taxon was not supported though a general branching phylogeny from a common ancestor was one of the possibilities considered. Gradually the acceptance of a general 'heterokont' or 'chromistan' line of evolution has been gaining ground and we now believe that the scheme in Round (1981a) involving a common ancestor for a whole cluster of 'brown pigmented' forms is in line with modern concepts incorporating biochemical and 'symbiotic' evidence. The latest and most detailed proposals are by Cavalier-Smith (1986) who proposes a kingdom Chromista in which the diatoms form a class (Diatomophyceae) closely related to his new class Synurea, which is distinct from the Chrysophyceae and contains the genera *Synura*, *Mallomonas* and *Mallomonopsis*. Cavalier-Smith proposes that the divergence of these two classes occurred before the loss of chlorophyll c_2 in the Synurea. It is perhaps significant that the most impressive similarity we have noted between a diatom and a member of another group is that between *Corethron* and *Mallomonas* (Crawford & Round, 1989), where the bristles on the valves and scales are virtually indistinguishable, although there are also similarities between diatoms and another chrysophyte group, the Parmales (Mann &

Marchant, 1989). Cavalier-Smith also proposes a superclass Raphidoista to include the above two classes plus the Dictyochea, Pedinellea and Raphidomonadea. All are in a new subclass Ochrista in a phylum Heterokonta. We cannot go into details of the newly proposed system but it does have much to recommend it and is well argued and documented – the kingdom is based on the premise that tubular mastigonemes and chloroplast endoplasmic reticulum evolved only once and that all organisms possessing them evolved from a single common ancestor. It is interesting to note that amidst all the argument two aspects are regular features: (1) the taxonomic status (hierarchy) of the clusters and (2) the grouping of the clusters. The grouping into one kingdom to stand alongside the Biliphyta (Red Algae) and Viridiplantae (all green algae, non-vascular and vascular Cryptogams and Spermatophytes) is a major step, but one which in no way alters the basic concept of the diatoms as a group. It is only the hierarchy – class rather than phylum – which is affected. The clustering of this class with the others of the Chromista likewise has no effect on the discussion of diatoms *per se*.

The origin of the diatoms from spherical uniformly scaly cells as proposed by Round & Crawford (1981) fits perfectly well with Cavalier-Smith's (1986) concept of their splitting off from an ancestor which also gave rise to the early Synurean monad. Cavalier-Smith considers it was a monad with an anterior flagellum and this would of course fit with the retention of a motile male gamete such as survives in the centric series, though now having lost the scales – certainly scales would not be needed in a gamete and indeed amongst other scaly monads, the scales are not retained when gametes are produced. Alternatively, the diatoms may have arisen from cyst-like forms similar to the extant Parmales (Mann & Marchant, 1989). There is still no fossil evidence, and it is not likely to be found easily. Even if an Ur-diatom were found and shown to be a scaly cell, it would be very difficult to establish it if had flagella. So evidence has to be indirect and circumstantial. Further detailed electron microscopic studies on motile male gametes of diatoms are urgently required – a scaly gamete would add further evidence!

Equally, evidence for evolution *from* this Ur-diatom is elusive since the early fossil record seems to have been destroyed by diagenesis. The truncation of the fossil record occurred after the early stages of evolution and the available deposits contain only the record of 'modern' diatoms. Several early genera are considerably more complex than modern genera and there is much evidence for the extinction of these complex forms. Amongst the remainder there has been either slow phyletic change or little change at all. The bursts of 'symbiotic events' (quantum evolution) giving rise to mitochondria, chloroplasts and chloroplast endoplasmic reticulum (Cavalier-Smith, 1985) must have been followed by a further burst to produce the complex scales (or two bursts, the first to form simple scales and the second to evolve the complex diatom frustule – the Synuran scales could have more easily developed from simple scales). The components of the diatom wall are so precise and similar in their basic form in all genera that they could only have evolved once. After this had arisen the diatoms may then have undergone rapid change to produce diverse clusters (orders?) which then underwent little change ('punctuated equilibria'). This 'quantum burst' of evolution occurred before the point in geological time when fossilisation became possible, so it too is unlikely to be easily documented. Hence attempts at the construction of phylogenies have of necessity been based upon phenetic data – at least this has the advantage that the fossil species can be included but it is often difficult to find clues to phylogenetic distance between genera or clusters of genera. Such phylogenies as have been produced are little more than convenient (although sometimes inaccurate) clusterings for easy classification and the common fallacy that classificatory groups reflect (or even are) phylogenies is evident in some writing. Andrews (1981) comments 'the arrangement of a classification usually implies phylogenetic relationships between its units. Such assumed relationships may be dangerously misleading in classifications based only on morphologic considerations'. This approach yielded the centric/pennate series and the araphid/monoraphid/biraphid series. More and more the traditional limits of these are breaking down, especially between the centric and araphid series (cf. the removal of the cymatosiroid group to the centric). The biraphid series may, however, be a more natural clustering and may have existed

parallel with the centric series for a very long time. Determining how one cluster was derived from another is fraught with difficulties but phyletic gradualism is probably everywhere evident in the form of minor phenetic change, e.g. in the gradual simplification of 'hemiauloid' genera or the rise of many minor variants in some 'naviculoid' genera. These changes are of course coupled with both speciation and extinction. Speciation has led to the development of populations which are sometimes only slightly distinct at the phenetic (morphological) level but sexually incompatible (cf. the 'species' of *Sellaphora pupula* and other taxa studied by Mann, 1984e, 1988a, 1989). It is possible to distinguish small variations in morphology in populations of what would normally be recognised as a single species, e.g. in *Cyclotella meneghiniana* (Round, unpublished) or in *Thalassiosira pseudonana* (Brand et al., 1981). In the latter case the variants also exhibit physiological (growth) differences. The study of Brand et al. is extremely interesting, showing that several clones from a single water bottle sample from an oceanic warm core ring varied considerably whereas several coastal (neritic) clones, from widely separated areas of the world, all had similar temperature/growth responses. Such differences have yet to be explained in genetic terms but it does emphasise that species cannot be *fully* described or defined morphologically and even less so by use of type material in which a single valve is designated as the type (though we agree that this must be done to conform with the International Code and also because chaos would result if it were not done). An element of discretion must therefore enter into the concept of a species.

Fossils

A well-documented fossil record of diatoms extends back to the middle Cretaceous. Earlier records, e.g. Jurassic, need careful checking: some old records are due to contamination which is very easy unless great care is taken in sampling and preparation. The vast majority of studies are on Eocene/Miocene deposits (e.g. Barron, 1985b; Kim & Barron, 1986). These deposits are scattered world-wide and usually consist of fine white/grey material which is extremely light, specific gravity around 0.45 (Tappan, 1980). Mining of this powdery material, usually from the surface (Fig. 2c), is a considerable industry since 'diatomite' is used in a range of industries – food/drink processing, paint, cosmetics, insulation, toothpaste etc. The most extensive deposits, however, are those lying at the bottom of the ocean where many are indicated on maps as 'diatomaceous ooze', e.g. see Lizitzin (1967). The major belts occur around Antarctica, along the equator and in the Arctic seas (see Barron, 1987, for a review). A few areas, however, are poor in diatoms, being dominated by red clays, e.g. the Atlantic. As with all fossil assemblages, the diatom content is only a selection of the species from the contributing live communities. Some freshwater deposits seem to have only a single dominant throughout most of a core, e.g. a triangular *Cyclotella* in some Miocene deposits in S. Spain (Servant-Vildary, 1986) and it is hardly conceivable that the parent community was so constituted. Late and post-glacial freshwater deposits are, however, almost 100% representative of the communities since incorporation into sediments is rapid and dissolution slow. In marine environments the long journey to the sea bed results in breakage (via animal guts) and also slow dissolution – nevertheless, many sediments are rich in diatoms. The rich deposits, some of which are almost pure diatom, can contain well over 10^7 diatoms per gram dry weight.

Pre-Cretaceous sediments have been so altered by diagenesis that diatoms are virtually absent, although Radiolaria and sponge spicules are present. But there is no reason to doubt the occurrence of diatoms far back beyond that of the fossil record. The upper Cretaceous sediments have diatoms entirely of marine origin; almost all of these are centric species and most of the genera are extinct. They were noticeably highly silicified, though some of this may be secondary silicification. Also, the genera exhibited are of a somewhat more complex structure than is seen in extant genera. The Tertiary period yields a much greater diversity of diatoms, and there does seem to be a 'flowering' of marine diatoms in the Miocene, by which time the morphological forms were very similar to those of the present day. A late Miocene extinction seems to have taken place. The earliest freshwater diatom fossils seem to be late Eocene and by the Miocene, floras with forms virtually identical to the modern species were abundant.

It is easier to document the disappearance of fossil genera from the geological record than it is to detect the precise point in time of the appearance of new genera and in general it seems that many species have had a long history, often with little morphological change in form (e.g. *Actinoptychus senarius* quoted by Andrews, 1979). However, views have been expressed that the pennate genera are now 'flowering' in the way that centric forms did in the Miocene.

Accounts of genera – preliminary notes

We have found the taxonomic notes on the genera and species of freshwater diatoms in the volumes of Patrick & Reimer (1966, 1975) and of marine forms in Hendey (1964) to be indispensable. Likewise the more general comments on the genera in Hustedt (1927–66, 1930) contain a wealth of valuable information. The VanLandingham *Index* (1967–79) has provided us with a ready entry into the mass of names applied to the diatoms and an easy source of data on the possible number of taxa in each genus.

The generic names are quoted with the authority, date and reference, followed by the citation of the type species. We have checked many of the citations with the literature but almost every generic citation is a small research project and there are still problems to be solved.

The descriptions and notes of interest have been kept to a minimum to allow as much space as possible for the illustrations. More detailed descriptions and illustrations at the light microscope level can be found in the major floras (Schmidt, 1874–1959; Peragallo & Peragallo, 1897–1908; Hustedt, 1930, 1927–66; Cleve-Euler, 1951–55; Hendey, 1964; Patrick & Reimer, 1966, 1975; Germain, 1981; Krammer & Lange-Bertalot, 1986, 1988) and also in other publications referred to in the text. We are aware that within the space we have allocated to the text, we cannot possibly give generic descriptions which will be inclusive of all the variation. We hope we have explained most terms adequately in the introduction. In addition, however, reference should be made to Anon. (1975), Ross *et al.*, (1979), von Stosch (1975) and Barber & Haworth (1981).

The first paragraph of each generic account describes how the cells look in live material, noting any special problems, but excluding detailed comment on valve shape or the appearance of the cell in girdle view, which is frequently quadrate/

oblong. The form of plastids is given, where known, but only in general terms (unfortunately this information is often lacking or based on preserved material which is unreliable), and this is followed by a note on the ecology of the genus. Geographical data are rarely given since so many genera are cosmopolitan. The paragraph ends with a note on the habitats occupied by the genus. Argument often arises over marine/freshwater distribution and when we refer to one or the other we are simply stating the habitat of the overwhelming majority of species; we are well aware of the problems relating to describing species as living in brackish waters. Marine species can occur inland but then they are growing in exceptional sites where saline groundwater emerges, or in evaporation basins, or where they have been carried upstream in rivers, possibly adapting to reduced salinity.

The second paragraph describes the morphology and structure of the valves and copulae, in external and internal detail, and we hope that light microscopists will be able to interpret these descriptions, even though the detail usually refers to what can be observed in the electron microscope.

The third paragraph gives any special points of interest.

We have not indicated size on the illustrations since we do not feel it is necessary or instructive to provide them; there is usually considerable variation within both species and genera. In many publications the scales give only spurious exactitude; the magnification factors, especially on the SEM, are notoriously inaccurate unless the apparatus is calibrated regularly – it has been known for the magnification indicators on SEMs to be up to 50% in error! Furthermore, measurement from the apparently three-dimensional image produced by the SEM is difficult owing to the angle of tilt, beam rotation and barrel distortion.

Table 2. Summary of classes and subclasses

Division BACILLARIOPHYTA

Coscinodiscophyceae Round & Crawford, *class. nov.*
Fragilariophyceae Round, *class. nov.*
Bacillariophyceae Haeckel 1878, *sensu emend.*

Class
Coscinodiscophyceae (Centric diatoms)

Subclasses
Thalassiosirophycidae Round & Crawford, *subclass. nov.*
Coscinodiscophycidae Round & Crawford, *subclass. nov.*
Biddulphiophycidae Round & Crawford, *subclass. nov.*
Lithodesmiophycidae Round & Crawford, *subclass. nov.*
Corethrophycidae Round & Crawford, *subclass. nov.*

Cymatosirophycidae Round & Crawford, *subclass. nov.*
Rhizosoleniophycidae Round & Crawford, *subclass. nov.*
Chaetocerotophycidae Round & Crawford, *subclass. nov.*

Class
Fragilariophyceae (Araphid, pennate diatoms)

Subclass
Fragilariophycidae Round, *subclass.nov.*

Class
Bacillariophyceae (Raphid, pennate diatoms)

Subclasses
Eunotiophycidae D.G. Mann, *subclass. nov.*
Bacillariophycidae D.G. Mann, *subclass. nov.*

Table continues overleaf

Division of subclasses into orders, suborders, families and genera

All the genera dealt with in the atlas are recorded in this table; some we have not been able to investigate are also included, but are marked with an asterisk. It is not a complete listing of genera; for this see Appendix III. Latin diagnoses and transfers are included in Appendix I. The other genera are not included here because we are uncertain of their affinities. The typefaces used for each classification are denoted thus:

SUBCLASS *ORDER* Family *Genus*

Class: COSCINODISCOPHYCEAE
THALASSIOSIROPHYCIDAE
1. *THALASSIOSIRALES* Glezer & Makarova 1986
   Thalassiosiraceae Lebour 1930
   *Thalassiosira, Planktoniella, Porosira, Minidiscus, Bacteriosira**
   Skeletonemataceae Lebour 1930, *sensu emend.*
   Skeletonema, Detonula
   Stephanodiscaceae Glezer & Makarova 1986
   Cyclotella, Cyclostephanos, Stephanodiscus, Mesodictyon, Pleurocyclus*, Stephanocostis**
   Lauderiaceae (Schütt) Lemmermann 1899, *sensu emend.*
   Lauderia

COSCINODISCOPHYCIDAE
1. *CHRYSANTHEMODISCALES* Round, *ord. nov.*
   Chrysanthemodiscaceae Round 1978
   Chrysanthemodiscus
2. *MELOSIRALES* Crawford, *ord. nov.*
   Melosiraceae Kützing 1844, *sensu emend.*
   Melosira, Druridgea
   Stephanopyxidaceae Nikolaev
   Stephanopyxis
   Endictyaceae Crawford, *fam. nov.*
   Endictya
   Hyalodiscaceae Crawford, *fam. nov.*
   Hyalodiscus, Podosira
3. *PARALIALES* Crawford, *ord. nov.*
   Paraliaceae Crawford 1988
   Paralia, Ellerbeckia
4. *AULACOSEIRALES* Crawford, *ord. nov.*
   Aulacoseiraceae Crawford, *fam. nov.*
   Aulacoseira, Strangulonema †
5. *ORTHOSEIRALES* Crawford, *ord. nov.*
   Orthoseiraceae Crawford, *fam. nov.*
   Orthoseira

† We are not able to place this with any confidence – it resembles *Aulacoseira* but may eventually require its own family.

6. *COSCINODISCALES* Round & Crawford, *ord. nov.*
   Coscinodiscaceae Kützing 1844
   Coscinodiscus, Palmeria, Stellarima, Brightwellia, Craspedodiscus
   Rocellaceae Round & Crawford, *fam. nov.*
   Rocella
   Aulacodiscaceae (Schütt) Lemmermann 1903
   Aulacodiscus
   Gossleriellaceae Round, *fam. nov.*
   Gossleriella
   Hemidiscaceae Hendey 1937 *emend* Simonsen 1975
   Hemidiscus, Actinocyclus, Azpeitia, Roperia
   Heliopeltaceae H. L. Smith 1872
   Actinoptychus, Glorioptychus, Lepidodiscus
7. *ETHMODISCALES* Round, *ord. nov.*
   Ethmodiscaceae Round, *fam. nov.*
   Ethmodiscus
8. *STICTOCYCLALES* Round, *ord. nov.*
   Stictocyclaceae Round 1978
   Stictocyclus
9. *ASTEROLAMPRALES* Round & Crawford, *ord. nov.*
   Asterolampraceae H. L. Smith 1872
   Asterolampra, Asteromphalus
10. *ARACHNOIDISCALES* Round, *ord. nov.*
    Arachnoidiscaceae Round, *fam. nov.*
    Arachnoidiscus
11. *STICTODISCALES* Round & Crawford, *ord. nov.*
    Stictodiscaceae (Schütt) Simonsen 1972
    Stictodiscus

BIDDULPHIOPHYCIDAE
1. *TRICERATIALES* Round & Crawford, *ord. nov.*
   Triceratiaceae (Schütt) Lemmermann 1899
   Triceratium, Odontella, Lampriscus, Sheshukovia, Pseudoauliscus, Eupodiscus, Pleurosira, Amphitetras, Cerataulus, Auliscus
   Plagiogrammaceae De Toni 1890
   Plagiogramma, Glyphodesmis, Dimeregramma, Dimeregrammopsis
2. *BIDDULPHIALES* Krieger 1954
   Biddulphiaceae Kützing 1844
   Biddulphia, Biddulphiopsis, Hydrosera, Isthmia, Trigonium, Terpsinoë, Pseudotriceratium
3. *HEMIAULALES* Round & Crawford, *ord. nov.*
   Hemiaulaceae[1] Heiberg 1863
   Hemiaulus, Eucampia, Climacodium, Cerataulina, Trinacria, Abas, Briggera, Pseudorutilaria, Keratophora, Kittonia, Strelnikovia, Riedelia*, Baxteriopsis**

    Sphynctolethus*, Ailuretta*
   Bellerocheaceae[1] Crawford, *fam. nov.*
    Bellerochea, Subsilicea
   Streptothecaceae Crawford, *fam. nov.*
    Streptotheca, Neostreptotheca*
 4. *ANAULALES* Round & Crawford, *ord. nov.*
   Anaulaceae (Schütt) Lemmermann 1899
    Anaulus, Eunotogramma, Porpeia*

LITHODESMIOPHYCIDAE
 1. *LITHODESMIALES* Round & Crawford, *ord. nov.*
   Lithodesmiaceae Round, *fam. nov.*
    Lithodesmium, Lithodesmioides*, Ditylum

CORETHROPHYCIDAE
 1. *CORETHRALES* Round & Crawford, *ord. nov.*
   Corethraceae Lebour 1930
    Corethron

CYMATOSIROPHYCIDAE
 1. *CYMATOSIRALES* Round & Crawford, *ord. nov.*
   Cymatosiraceae Hasle, von Stosch & Syvertsen 1983
    Cymatosira, Campylosira, Plagiogrammopsis, Brockmanniella, Minutocellus, Leyanella, Arcocellulus, Papiliocellulus, Extubocellulus,
   Rutilariaceae De Toni 1894
    Rutilaria, Syndetocystis

RHIZOSOLENIOPHYCIDAE
 1. *RHIZOSOLENIALES* Silva 1962
   Rhizosoleniaceae De Toni 1890
    Rhizosolenia, Proboscia, Pseudosolenia, Urosolenia, Guinardia, Dactyliosolen
   Pyxillaceae (Schütt) Simonsen 1972
    Pyxilla, Gladius*, Gyrodiscus*, Mastogonia*, Pyrgupyxis*

CHAETOCEROTOPHYCIDAE
 1. *CHAETOCEROTALES* Round & Crawford, *ord. nov.*
   Chaetocerotaceae Ralfs in Pritchard 1861
    Chaetoceros, Gonioceros, Bacteriastrum
   Acanthocerataceae Crawford, *fam. nov.*
    Acanthoceros

   Attheyaceae Round & Crawford, *fam. nov.*
    Attheya
 2. *LEPTOCYLINDRALES*[2] Round & Crawford, *ord. nov.*
   Leptocylindraceae Lebour 1930
    Leptocylindrus

Class: FRAGILARIOPHYCEAE
FRAGILARIOPHYCIDAE
 1. *FRAGILARIALES* Silva 1962 *sensu emend.*
   Fragilariaceae Greville 1833
    Fragilaria, Centronella, Asterionella, Staurosirella, Staurosira, Pseudostaurosira, Punctastriata, Fragilariaforma, Martyana, Diatoma, Hannaea, Meridion, Synedra, Ctenophora, Neosynedra, Tabularia, Catacombas, Hyalosynedra, Opephora, Trachysphenia, Thalassioneis, Falcula, Pteroncola, Asterionellopsis, Bleakeleya, Podocystis
 2. *TABELLARIALES* Round, *ord. nov.*
   Tabellariaceae Kützing 1844
    Tabellaria, Tetracyclus, Oxyneis
 3. *LICMOPHORALES* Round, *ord. nov.*
   Licmophoraceae Kützing 1844
    Licmophora, Licmosphenia*
 4. *RHAPHONEIDALES* Round, *ord. nov.*
   Rhaphoneidaceae Forti 1912
    Rhaphoneis, Diplomenora, Delphineis, Neodelphineis, Perissonoë, Sceptroneis
   Psammodiscaceae Round & Mann, *fam. nov.*
    Psammodiscus
 5. *ARDISSONEALES* Round, *ord. nov.*
   Ardissoneaceae Round, *fam. nov.*
    Ardissonea
 6. *TOXARIALES* Round, *ord. nov.*
   Toxariaceae Round, *fam. nov.*
    Toxarium
 7. *THALASSIONEMATALES* Round, *ord. nov.*
   Thalassionemataceae Round, *fam. nov.*
    Thalassionema, Thalassiothrix, Trichotoxon
 8. *RHABDONEMATALES* Round & Crawford, *ord. nov.*
   Rhabdonemataceae Round & Crawford, *fam. nov.*
    Rhabdonema

---

[1] There is considerable confusion in the literature concerning the status of these groups and of the exact allocation of genera (see Ross & Sims, 1985, 1987).

[2] This group is tentatively placed until further studies can be made.

9. *STRIATELLALES* Round, *ord. nov.*
 Striatellaceae Kützing 1844
  Striatella, Microtabella, Grammatophora
10. *CYCLOPHORALES* Round & Crawford, *ord. nov.*
 Cyclophoraceae Round & Crawford, *fam. nov.*
  Cyclophora
 Entopylaceae Grunow 1862
  Entopyla*, Gephyria
11. *CLIMACOSPHENIALES* Round, *ord. nov.*
 Climacospheniaceae Round, *fam. nov.*
  Climacosphenia, Synedrosphenia
12. *PROTORAPHIDALES* Round, *ord. nov.*
 Protoraphidaceae Simonsen 1970
  Protoraphis*, Pseudohimantidium

Class: BACILLARIOPHYCEAE
EUNOTIOPHYCIDAE
1. *EUNOTIALES* Silva 1962
 Eunotiaceae Kützing 1844
  Eunotia, Actinella, Semiorbis, Desmogonium*
 Peroniaceae (Karsten) Topachevs'kyj & Oksiyuk 1960
  Peronia

BACILLARIOPHYCIDAE
1. *LYRELLALES* D. G. Mann, *ord. nov.*
 Lyrellaceae D. G. Mann, *fam. nov.*
  Lyrella, Petroneis
2. *MASTOGLOIALES* D. G. Mann, *ord. nov.*
 Mastogloiaceae Mereschkowsky 1903
  Aneumastus, Mastogloia
3. *DICTYONEIDALES* D. G. Mann, *ord. nov.*
 Dictyoneidaceae D. G. Mann, *fam. nov.*
  Dictyoneis
3. *CYMBELLALES* D. G. Mann, *ord. nov.*
 Rhoicospheniaceae Chen & Zhu 1983
  Rhoicosphenia, Campylopyxis, Cuneolus, Gomphoseptatum, Gomphonemopsis
 Anomoeoneidaceae D. G. Mann, *fam. nov.*
  Anomoeoneis, Staurophora
 Cymbellaceae Greville 1833
  Placoneis, Cymbella, Brebissonia, Encyonema, Gomphocymbella
 Gomphonemataceae Kützing 1844
  Gomphonema, Didymosphenia, Gomphoneis, Reimeria, Gomphopleura*
4. *ACHNANTHALES* Silva 1962
 Achnanthaceae Kützing 1844 *sensu emend.*
  Achnanthes

Cocconeidaceae Kützing 1844
 Cocconeis, Campyloneis, Anorthoneis, Bennettella*, Epipellis*
Achnanthidiaceae D. G. Mann, *fam. nov.*
 Achnanthidium, Eucocconeis
5. *NAVICULALES* Bessey 1907 *sensu emend.*
Neidiineae D. G. Mann, *subord. nov.*
 Berkeleyaceae D. G. Mann, *fam. nov.*
  Parlibellus, Berkeleya, Climaconeis, Stenoneis
 Cavinulaceae D. G. Mann, *fam. nov.*
  Cavinula
 Cosmioneidaceae D. G. Mann, *fam. nov.*
  Cosmioneis
 Scolioneidaceae D. G. Mann, *fam. nov.*
  Scolioneis
 Diadesmidaceae D. G. Mann, *fam. nov.*
  Diadesmis, Luticola
 Amphipleuraceae Grunow 1862
  Frickea, Amphipleura, Frustulia, Cistula*
 Brachysiraceae D. G. Mann, *fam. nov.*
  Brachysira
 Neidiaceae Mereschkowsky 1903
  Neidium
 Scoliotropidaceae Mereschkowsky 1903
  Scoliopleura, Scoliotropis, Biremis, Progonoia, Diadema*
Sellaphorineae D. G. Mann, *subord. nov.*
 Sellaphoraceae Mereschkowsky 1902
  Sellaphora, Fallacia, Rossia*, Caponea*
 Pinnulariaceae D. G. Mann, *fam. nov.*
  Pinnularia, Diatomella, Östrupia*, Dimidiata*
Phaeodactylineae J. Lewin 1958
 Phaeodactylaceae Silva 1962
  Phaeodactylum
Diploneidineae D. G. Mann, *subord. nov.*
 Diploneidaceae D. G. Mann, *fam. nov.*
  Diploneis, ?Raphidodiscus
Naviculineae Hendey 1937
 Naviculaceae Kützing 1844
  Navicula, Trachyneis, Pseudogomphonema, Seminavis, Rhoikoneis, Haslea, Cymatoneis
 Pleurosigmataceae Mereschkowsky 1903
  Pleurosigma, Toxonidea, Donkinia, Gyrosigma, Rhoicosigma*
 Plagiotropidaceae D. G. Mann, *fam. nov.*
  Plagiotropis, ?Stauropsis, Pachyneis*
 Stauroneidaceae D. G. Mann, *fam. nov.*
  Stauroneis, Craticula
 Proschkiniaceae D. G. Mann, *fam. nov.*

Proschkinia

6. *THALASSIOPHYSALES* D. G. Mann, *ord. nov.*
   Catenulaceae Mereschkowsky 1902
   Catenula, Amphora, Undatella
   Thalassiophysaceae D. G. Mann, *fam. nov.*
   Thalassiophysa

7. *BACILLARIALES* Hendey 1937 *sensu emend.*
   Bacillariaceae Ehrenberg 1831
   Bacillaria, Hantzschia, Psammodictyon, Tryblionella, Cymbellonitzschia, Gomphonitzschia*, Gomphotheca, Nitzschia, Denticula, Denticulopsis*, Fragilariopsis, Cylindrotheca, Simonsenia*, Cymatonitzschia*, Perrya*

8. *RHOPALODIALES* D. G. Mann, *ord. nov.*
   Rhopalodiaceae (Karsten) Topachevs'kyj & Oksiyuk 1960
   Epithemia, Rhopalodia, Protokeelia*

9. *SURIRELLALES* D. G. Mann, *ord. nov.*
   Entomoneidaceae Reimer in Patrick & Reimer 1975
   Entomoneis
   Auriculaceae Hendey 1964
   Auricula
   Surirellaceae Kützing 1844
   Hydrosilicon, Petrodictyon, Plagiodiscus, Stenopterobia, Surirella, Campylodiscus, Cymatopleura

Note: A preliminary classification, conceived by two of us (F. E. R. & R. M. C.) is expected to appear after publication of this book in *Protoctista* by L. Margulis, D. Chapman & J. Corliss. It is the result of our early thoughts on the subject and is made redundant by the above.

Generic atlas

Centric group

Centric genera are abundant in the plankton and as fossils, and these are areas that have been worked intensively – unlike the benthic habitats. This has led to a fairly well-defined system of classification and we have not attempted to make many changes – indeed, we do not have the expertise for this; in addition, data on the protoplast are generally lacking or inadequate but when obtained will certainly lead to a better classification. We have, however, made some changes to the families, mainly by splitting off groups of genera with features which do not fit them for inclusion in the older systems. The notable addition to the centric series is the inclusion of the Cymatosiraceae, which form a very distinct group on frustule characters. We have erected only a few new genera and resurrected a few others (see *Ellerbeckia*, *Orthosira*, *Limnosolenia*, *Gonioceros*). Other workers have been active (see *Azpeitia*, *Stellarima*, *Briggera*), though we have not always been able to include their new genera owing to lack of material (e.g. of *Mesodictyon*, *Pleurocyclus*, *Sphynctolethus*, *Ailuretta*, *Dicladiopsis*, *Maluina*, *Bonea*, *Hyperion*, *Strelnikovia* and almost all the Pyxillaceae). Within certain groups there is intense taxonomic activity leading to clearer definitions of the genera (in the splitting of *Coscinodiscus* – see papers by Sims, Fryxell, Hasle in particular; in the Hemiauloid group – see Ross & Sims; in the *Cyclotella/Stephanodiscus/Cyclostephanos* group – see papers by Håkansson, Stoermer & Theriot; and in *Thalassiosira/Porosira/Coscinosira* – see works of Hasle & Fryxell). However, in all these and others there is still much work to be completed, especially in the purely fossil groups. The Biddulphiophycidae stand out as a subclass which requires intensive study – the triangular/multiangular taxa are particularly troublesome and we have simply had to exclude certain taxa from our consideration (e.g. *Solium*). Studies of these groups require a coordination of work on both freshwater and fossil material, coupled with a thorough study of the Russian literature, which in the fossil field is considerable and important.

Many of the genera have highly complex loculate valves and information obtained from fractures is essential for understanding relationships. The simplest example is the confusion which existed in the genus *Thalassiosira* (and still partially exists in *Coscinodiscus*) until workers such as Hasle and Fryxell showed that the cribra were on different sides of the loculus in the two genera.

Thalassiosira P. T. Cleve 1873. Bih. Kongl. Svenska. Vetensk.-Akad. Handl. 1: 6

T: *T. nordenskioeldii* Cleve

Cells discoid to cylindrical,[a] solitary, or joined by threads or valve to valve to form loose chains; or in mucilage masses. Plastids numerous, discoid. Mainly in the marine plankton.

Valve circular,[b-e] with a flat valve face and short downturned mantles or sometimes almost watchglass-like.[d, f] Areolae usually loculate,[j] arranged in radial rows, tangential rows, or arcs; varying in size and prominence. The areolae open to the outside by circular foramina,[b-f] sometimes with finger-like projections (e.g. *T. ferrelineata*); internally they are occluded by slightly raised cribra. Valve mantle plain or with spines of varied form;[h-k] mantle edge often very prominently ribbed and rimmed. Fultoportulae are present in a ring around the valve mantle opening externally by short tubes; in some species scattered or grouped fultoportulae occur also on the valve face.[f, g] The internal fultoportula structure varies; very short[k] to very long tubes[g, h] are surrounded by (3 or) 4 buttresses. Each valve usually has one rimoportula, which opens externally through a larger and more obvious tube, located somewhat internal to the marginal ring of fultoportulae; in some cases, however, the external tube can be thin and located close to fultoportulae (e.g. *T. conferta*); in a few species a rimoportula may occur in the centre of the valve face. The internal rimoportula opening is usually elongate and sessile. The mantle edge may be thickened internally in some species. Copulae numerous, split and ligulate. The valvocopula is more obviously areolate than the other copulae, with a loculation similar to that of the valve and a more conspicuous row of areolae adjacent to the valve mantle. A slight septum occurs on the valvocopula of *T. trifulta* which also has a complex triradiate ligula.

This is a large genus – descriptions of species (many new) are to be found in the numerous papers of Hasle & Fryxell. Most species (*c.* 100) are marine, although up to 12 have been recorded in freshwaters (Hasle, 1978); of the latter *T. weissflogii* (= *T. fluviatilis*) is probably the best known. There is considerable variation within the genus, which is one of the largest in our collection; we cannot do justice to it in our illustrations. Makarova (1988) provides a comprehensive account of the genus in seas around Russia.

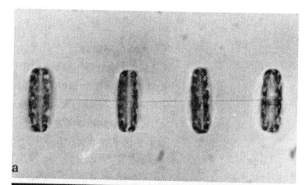

a

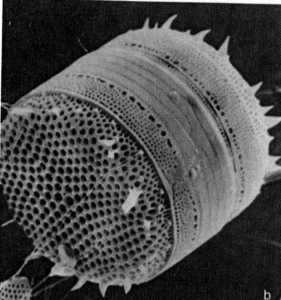

b

c

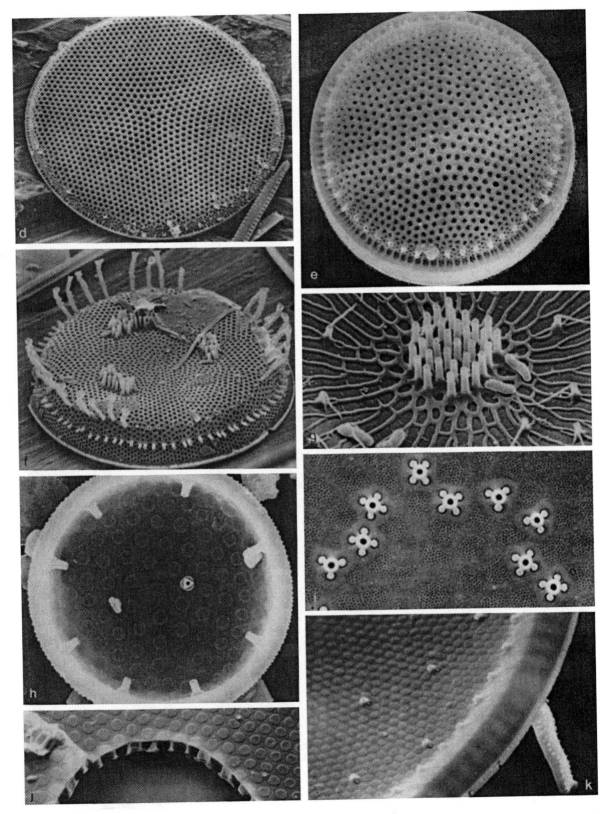

Planktoniella F. Schütt 1893. Das Pflanzenleben des Hochsees: 20

T: *P. sol* (Wallich) Schütt (= *Coscinodiscus sol*)

Cells discoid, solitary[a] (*P. sol*) or in flat colonies[b] (*P. muriformis*). A wing of stiffened mucilage (?) is produced around each cell so that it lies in a flat plate.[c] In *P. sol* the mucilage forms a series of radial chambers and appears to be secreted through the girdle bands. Plastids small and discoid. *P. sol* is a widespread open ocean form, mainly tropical in distribution, while *P. muriformis* has only been recorded in tropical coastal regions.

Valves discoid, with a flat valve face and a shallow mantle,[d-g] the lower part of which has modified areolae; fultoportulae lie at the junction between the mantle and valve face. Areolae chambered,[i, j] opening internally via fine cribra[h] and externally by larger round foramina,[e-g] which can, however, be partially occluded by siliceous ingrowths[j] arranged in arcs. A single fultoportula occurs near the centre of the valve face and a ring of fultoportulae around the mantle.[h] A single large rimoportula in *P. muriformis* occurs slightly internal to the marginal ring of fultoportulae. Internally the rimoportulae and fultoportulae are unexceptional – the fultoportulae tend to be very shallow, almost flush with the surface. Copulae split, with vertical rows of small areolae and a row of larger areolae along the advalvar margin.[e-g] It is not clear how the chambering/sectoring of the mucilage wing is produced in *P. sol*, since there seem to be no breaks in the copulae or their areolation, and the spacing of the fultoportulae bears no relationship to the chambers.

This marine genus is one of several which have fultoportulae and rimoportulae, and the view might be taken that all could be included in *Thalassiosira*. However, that genus is already large and extremely diverse and we believe it is preferable to maintain the two genera, on the basis of the extended wing. The two species mentioned above have been discussed in detail by Loeblich *et al.* (1968), Gerloff (1970) and Round (1972c). Two other species reported in the literature require further study.

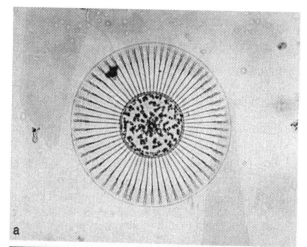

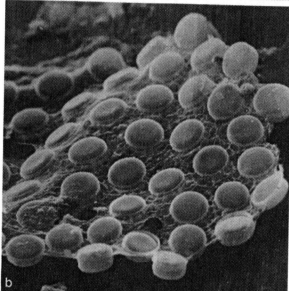

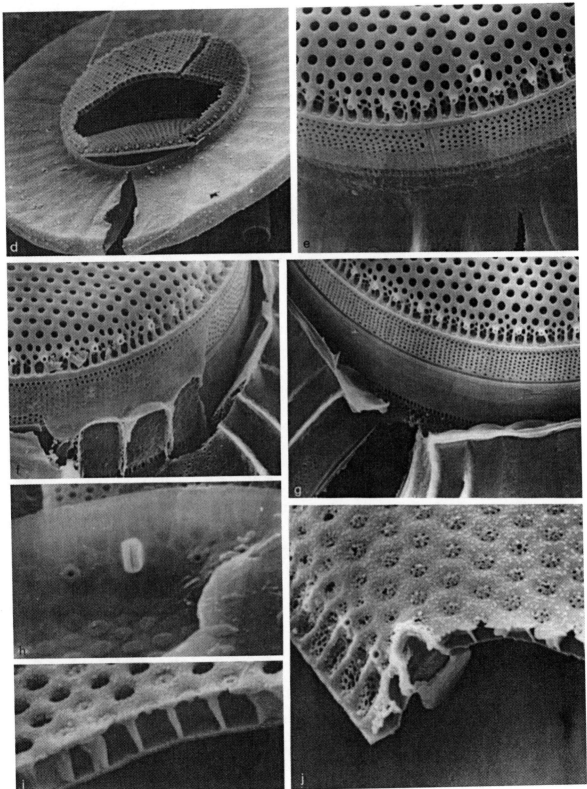

Porosira E. Jørgensen 1905. Bergens Mus. Skr. 7: 97

T: *P. glacialis* (Grunow) Jørgensen (= *Podosira hormoides* var. *glacialis*)

Cells short, cylindrical, attached by mucilage to form filaments of spaced cells. Numerous fine polysaccharide threads also radiate from the valves. Plastids disc-like. Marine, planktonic.

Valves watchglass-like with a central annulus and radiating rows of areolae;[a-d] no real mantle. Areolae loculate, opening to the exterior by elongate foramina.[e, f] Internally the areolae are closed by finely perforate cribra. Fultoportulae are scattered over the whole of the valve[a-d, i] but become more frequent around the margin; they have simple openings or short stalks externally and are unremarkable internally, resembling those of *Thalassiosira*. A single flattened rimoportula,[g, h] with a short stalk is located to the inside of the denser region of fultoportulae;[b, i] the external opening is on a slight elevation.[h] Copulae numerous, split, with areolae similar to those of the valve.[j]

This genus is close to *Thalassiosira* and *Lauderia* but Hasle (1973a) considered it sufficiently distinct to maintain as a separate genus. In *Lauderia*, cell-to-cell contact is maintained via prolongations of the processes, while in *Thalassiosira* it is via the central group of fultoportulae.

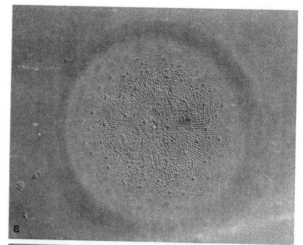

a

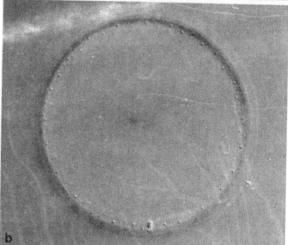

b

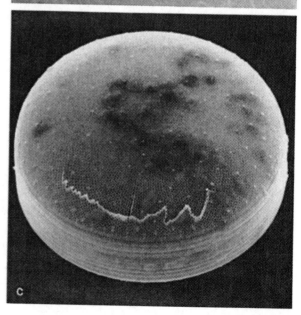

c

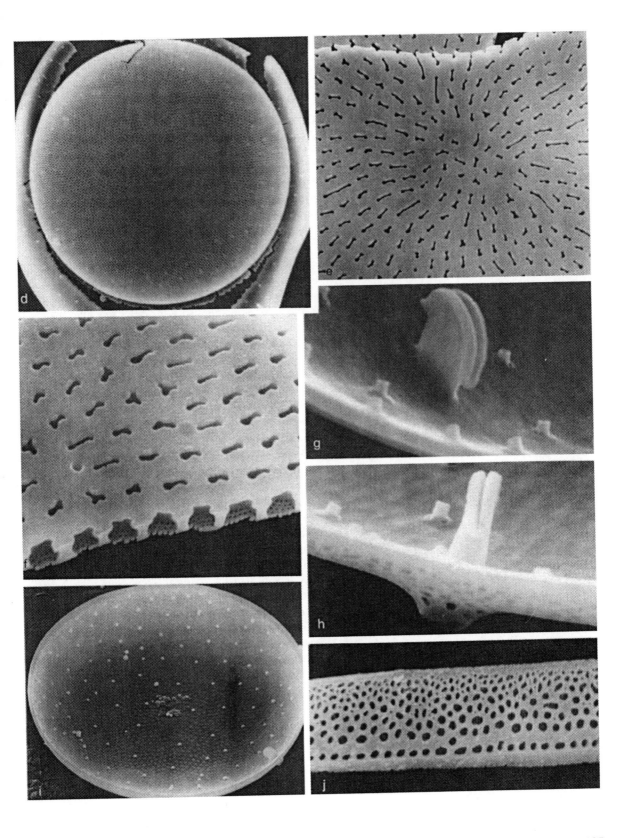

Minidiscus G. R. Hasle 1973. Norweg. J. Bot. 20: 67.

T: *M. trioculatus* (F. J. R. Taylor) Hasle
(= *Coscinodiscus trioculatus* Taylor)

Cells very small,[a, b] usually less than 10 μm in diameter, barrel-shaped, unicellular or in clusters with little detail observable in the light microscope. Widely distributed in coastal phytoplankton. Monotypic until Takano (1981) described a second species *M. comicus* and Rivera & Koch (1984) a third, *M. chilensis*.

Valves slightly or more prominently convex[c] (*M. comicus*) with an outer plain region or with an almost loculate outer ring. Areolae covering the whole valve face surrounded by raised network of silica[c,d] or areolae restricted to the marginal region of the valve[f] sometimes with the outer row thickened only on the inner walls. 2–3 (4) raised pores[c, d] occur, which are the external openings of the fultoportulae; they are rather distinctive with a plain rim of silica around the central tube – a feature which shows up very well in TEM. A single rimoportula is also present in the central region of the valve face between two of the fultoportulae.[e, f] Internally some species appear to have an almost plain surface in SEM but TEM reveals the characteristic fine pores on the inner face of the areolae (see Hasle, 1974a; Takano, 1981). Copulae not studied.

Originally described as a *Coscinodiscus* sp. by Taylor (1967). Its position in the Thalassiosiraceae is dependent upon the presence of fultoportulae, but their disposition is somewhat unusual in that a marginal ring is absent (Hasle, 1973b). In addition, there is a distinct rim of plain silica around the external openings of the fultoportulae. Another characteristic of this genus seems to be the existence of only two apertures on either side of the central tube of the fultoportula, though more rarely three have been found.

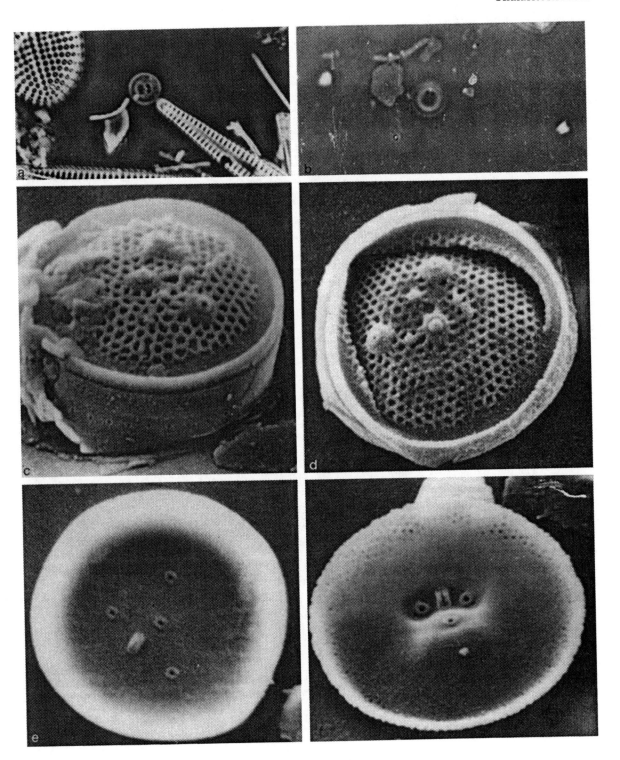

Skeletonema R. K. Greville 1865. Trans. Microsc. Soc. Lond. Ser. 2, **13**: 43

T: *S. barbadense* Greville

Cells joined by long marginal processes to form filaments, which appear in the light microscope like short beads joined by numerous fine threads.[a] Plastids disc-like or cup-shaped. Living in the coastal marine plankton.

Valves circular; valve face convex to flat; mantles deep. Valves with a prominent network of costae externally, becoming pseudoloculate near the margin.[c-e] Internally with distinct cribra on the flat[i] or slightly ribbed[h] surface. A single ring of processes occurs around the top of the mantle.[b-f] These are closely associated with a ring of fultoportulae, the external openings of which are short tubes hidden in the bases of the processes. The processes are semi-circular in cross-section[e] and expand at their apices to form 'knuckles'[f] which interlock with the processes from the adjacent cell; they are sometimes much longer than the cell and can interlock with either one[f] or two[e] processes of the sibling valve. Occasional valves produce flattened spinulose processes[g] and on these there is also a central rimoportula, with a tubular external opening. Occasional rimoportulae also occur around the valve mantle. The internal endings of both the fultoportulae and rimoportulae are very small though typical in form.[h] Girdle composed of a split valvocopula and numerous, finely porous and delicate copulae, each with a distinct pars media.[j] In intact filaments the new sibling valves and processes are covered by the copulae whilst the processes linking old pairs of valves are exposed.

Occurs in coastal waters throughout the world where it can be an extremely common diatom. Many ecological studies have been made on it and, because it also grows easily and rapidly in culture, it has been used for physiological studies. Several species are reported in various oceanic regions but all these need re-investigation. In addition several fossil species have been described.

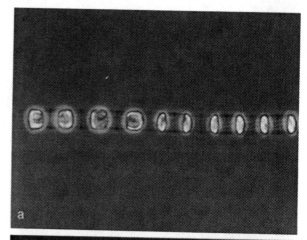

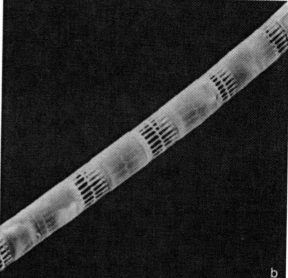

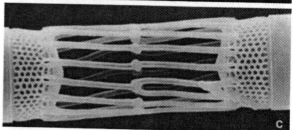

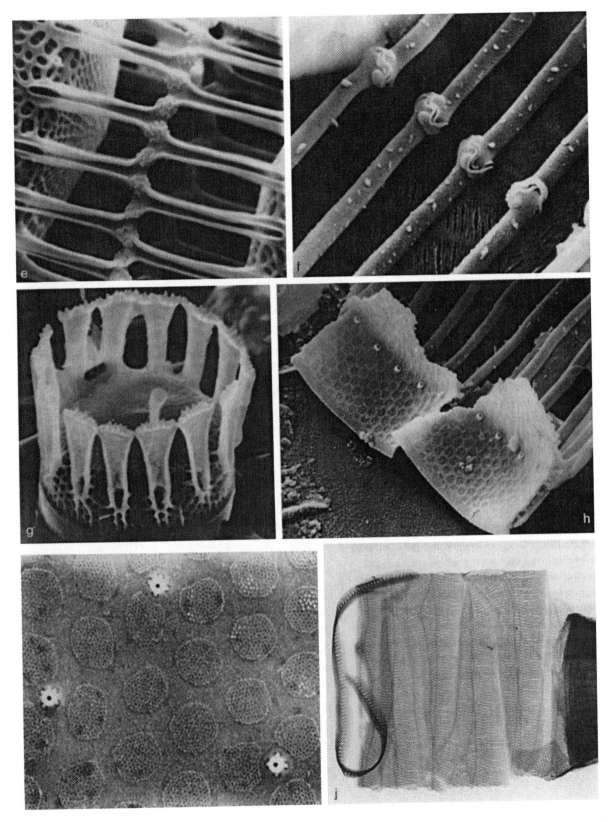

Detonula Schütt 1894. Rep. Sci. Res. Voy.
'Challenger', Bot. 2: 89, 1880

T: *D. pumila* (Castracane) Shütt (= *Lauderia
pumila*)

Cells barrel-shaped, connected by short processes[h]
and mucilage threads into long filaments. Plastids
numerous, discoid. Originally all the species were
placed in *Lauderia* but see the discussion in Hasle
(1974a) who recognised *D. confervacea*, *D. pumila* (we
illustrate only this species) and *D. moseleyana*. A
fairly common marine planktonic genus sometimes
forming blooms in coastal waters.

Valves saucer-like[c, f] with radiating costae between
which small areolae occur; near the centre these are
more-or-less round[i] but they become elongate half-
way to the margin. A central fultoportula[c, f, i]
(occasionally two) occurs with a short simple exit
tube through which a thread of polysaccharide is
produced connecting the valve to its sibling.[h]
Around the edge of the valve face there is a ring of
fultoportulae[c-g], their exit tubes having conspicuous
upright flanged extensions.[d, e] Adjacent cells
interlock by means of these flanges. A single plain
exit tube of a rimoportula is found in the
circumferential ring (arrow).[g] Internally the central
fultoportula may appear raised and has a variable
number of buttresses.[i] The fultoportulae in the
circumferential rings are low structures internally,
usually with 4 buttresses. The internal part of the
rimoportula is much larger and usually lies at right
angles to the valve margin.[g] Copulae numerous,[b, e]
all split with ligulae. The distal one in Hasle's fig. 74
of *D. pumila* is plain, but otherwise they are porous,
each bearing several transverse rows of areolae.[e]
Resting spores occur and are more coarsely silicified
spiny versions of the vegetative cells (see Hasle,
1974a, for illustrations).

The genus is distinguished from *Lauderia* by the
absence of valve face spines. Hasle (1973a)
considers there is no justification for retaining
Pavillard's name *Schroederella*.

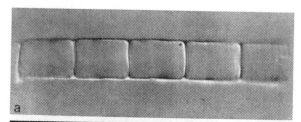

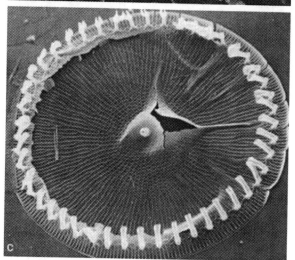

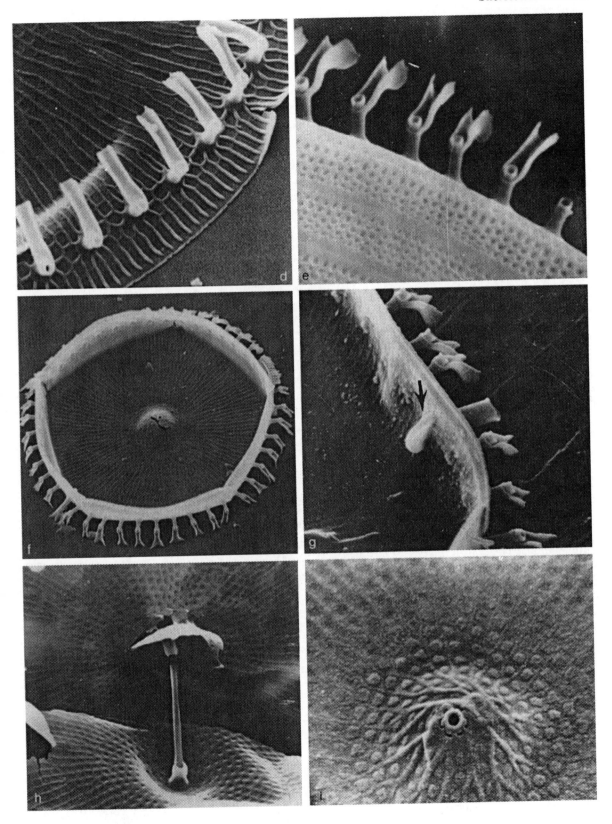

Cyclotella F.T. Kützing ex A. de Brébisson 1838. Consid. Diat. 19

T: *C. operculata* (C. A. Agardh) Brébisson (= *Frustulia operculata*)

Cells short, drum-shaped;[b] free-living or forming filaments, chains or rarely clusters, united by mucilage. Plastids numerous, discoid. Mainly freshwater and planktonic, with two species occurring in shallow coastal waters, probably as an evolutionary invasion from brackish waters.

Valves circular (oval and triangular species have been found in fossil material from the Miocene, and oval ones very rarely in modern samples), with either a tangential[b] or concentric[c-e] undulation of the valve face. Rows of areolae extending from valve centre (or absent from centre), becoming grouped into fascicles on the outer region of valve face,[c-e] and continuing to the valve edge without a break; no distinctive valve mantle in most species; scattered areolae in the central region in some species. Central area often ornamented with warts, granules, etc.;[b-d] small warts or spines often scattered between the areolae on the valve mantle,[b, c] forming patterns. Openings of portules flush with valve surface, often on hyaline ridges or strips and sometimes surrounded by spines, one of which may be prominent. Mantle edge often banded by a grooved thickening.[b] Areolae poroidal, closed internally by cribra.[h, i] Fultoportulae present in a ring near the valve margin[f, g, j] and often also scattered over the centre of the valve.[g-i] Wall infolded in marginal area with fultoportulae opening on ridges;[g, j] sometimes with secondary ridges within the folds which may be expanded into chambers.[f, g] Rimoportulae few, lying on the folds[j] or at the edge of the central area.[g] Copulae numerous, split.[b]

A fairly well circumscribed genus of around 100 spp. Differs from *Stephanodiscus* in the nature of the central area, absence of a distinct single ring of spines, absence or only slight development of the external tubes of the portules, and the distinct folding or chambering of the interior of the valve margin. For details and comparison with other 'cyclostephanoid' genera, see Theriot, Stoermer & Håkansson (1987). There are many problems in the LM identification of species. *C. meneghiniana* is very common and has been discussed and figured in detail by Schoeman & Archibald (1980). Further details of this can be found in Hoops & Floyd (1979).

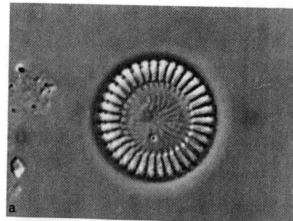

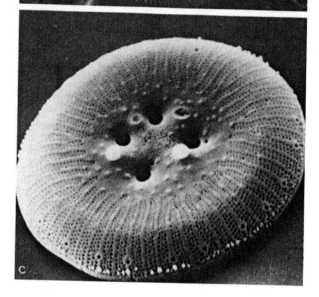

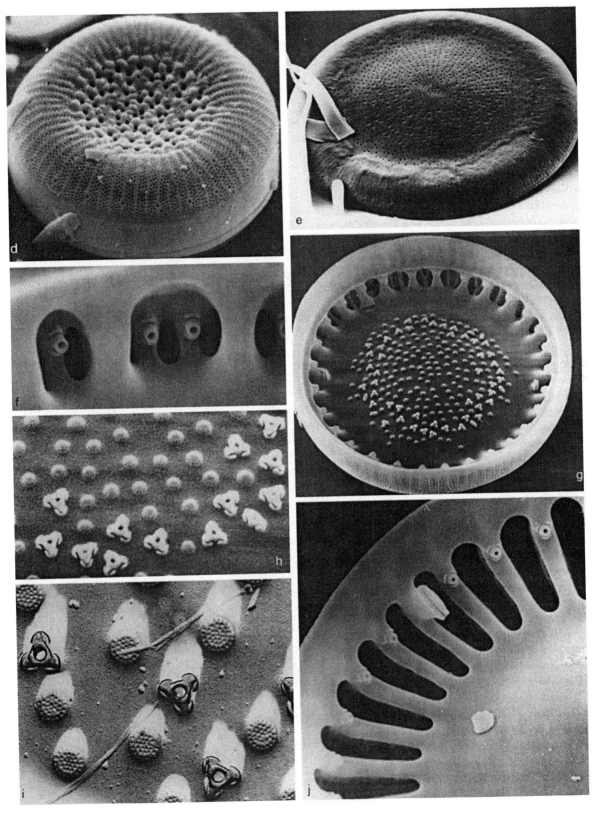

Cyclostephanos F. E. Round, in Theriot *et al.*
1987. Br. phycol. J. **22**: 346

T: *C. novaezeelandiae* (Cleve) Round
(= *Stephanodiscus novaezeelandiae*; lectotype
selected by E. Theriot *et al.* 1987, Br. phycol. J.
22: 346)

Cells short, disc-shaped;[a] solitary or in short[a] chains.
Valves circular, with a concentric undulation and
spines around the margin.[a-f] Plastids discoid.
Freshwater, planktonic. Previously included within
Stephanodiscus.

Valves with radiating lines of areolae,[a-h] grouped
into fascicles around the margin and separated by
raised, rounded ridges. Areolae continuing down
valve mantle in separate fascicles. Spines occur on
ridges at the edge of the 'valve face' and the
fultoportulae open below these, usually close to the
valve edge, which may be well developed. No clear
distinction between valve face and mantle. Central
region of valve face variously nodulose or plain.
Areolae simple, internally with domed cribra.[g, i]
Struts or ribs extend in from the valve edge onto the
valve face internally between the fascicles of
areolae[g-i] but they do not not form distinct
chambers in species investigated to date. There is a
ring of fultoportulae on the mantle, each opening
internally onto one of the struts between the
fascicles;[i] scattered fultoportulae may also occur on
the valve face: all these have only two satellite pores.
A single rimoportula may be present.[i] Copulae
porous, split, with ligulae; areolae much finer than
on the valve.[j]

This genus was erected by Round (1981d) to
include *Stephanodiscus dubius* and *S. novae-zeelandiae*
since these differed from other *Stephanodiscus*
species in having the fascicles of areolae continued
down the valve 'mantle' and internal ribbing in the
'mantle' region; the name was made valid in Theriot
et al. (1987). *Cyclostephanos* differs from *Cylotella* in
the extension of the rows of fascicles to the centre of
the valve and the presence of a single ring of spines
around the valve face. The internal folds/ struts are
reminiscent of *Cyclotella* but are not developed to
the same extent. They do, however, suggest a
relationship, hence the generic name. *C. damasii*
(Hust.) Stoermer & Håkansson (1983) can be added
to the list of species. Further discussion and figures
of S. African forms can be found in Schoeman *et al.*
(1984) and Theriot, Stoermer & Håkansson (1987)
discuss in detail the rimoportulae in this genus and
in *Cyclotella*.

Stephanodiscus C. G. Ehrenberg 1845. Ber. Bekanntm. Verh. Königl. Preuss. Akad. Wiss. Berlin, **1845**: 80

T: *S. niagarae* Ehrenberg

Cells discoid or barrel-shaped, often with delicate organic threads radiating from around the edge of the valve.[a] A ring of spines is present around the valve face. Plastids numerous, discoid. A common freshwater planktonic genus in lakes, rivers and reservoirs.

Valves shallow, saucer-like. Valve face concentrically undulate[b, c] or almost flat;[f, h] in sibling pairs the elevations of one valve fit into depressions of the other. The areolae are simple round pores occluded internally by domed cribra;[j] they are arranged in files usually containing a single row of areolae at the centre, but become double or multiple towards the edge of the valve face. The areolae of the valve mantle are slightly closer together and often occur in diagonal rows without intervening costae; the mantle edge is often vertically grooved. Spines occur in a single row around the edge of the valve face, where they are attached to small raised pads at the ends of costae;[b-h] spines simple, except in *S. binderanus*[i] where they fork. Fultoportulae present, arranged in a ring around the mantle, always occurring beneath certain of the spines;[b-h] also occurring on the valve face,[f,h] singly or in a ring. The fultoportulae may have either subsidiary tubes or buttresses.[f, j] One or a few rimoportulae occur in the mantle and externally have tubular extensions. [arrows, b, d] Copulae numerous, areolate, split, with ligulae.

This genus has proved exceedingly troublesome especially to ecologists and not all the problems are yet solved. *S. binderanus* was transferred from *Melosira* by Round (1972b). Round (1981e) discusses the history of the genus and gives descriptions of three species (*S. niagarae*, *S. rotula*, *S. minutula*) and there is an excellent account of the variability of *S. niagarae* by Theriot & Stoermer (1981). Håkansson & Locker (1981) also discuss some of these species. It is important to note that *S. astraea*, once thought invalid, has been rediscovered (Håkansson & Locker, 1981 – see also Håkansson, 1986).

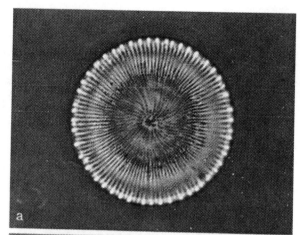

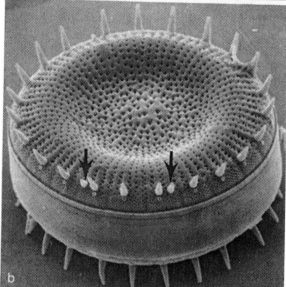

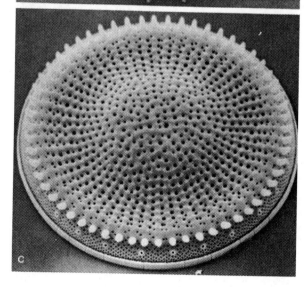

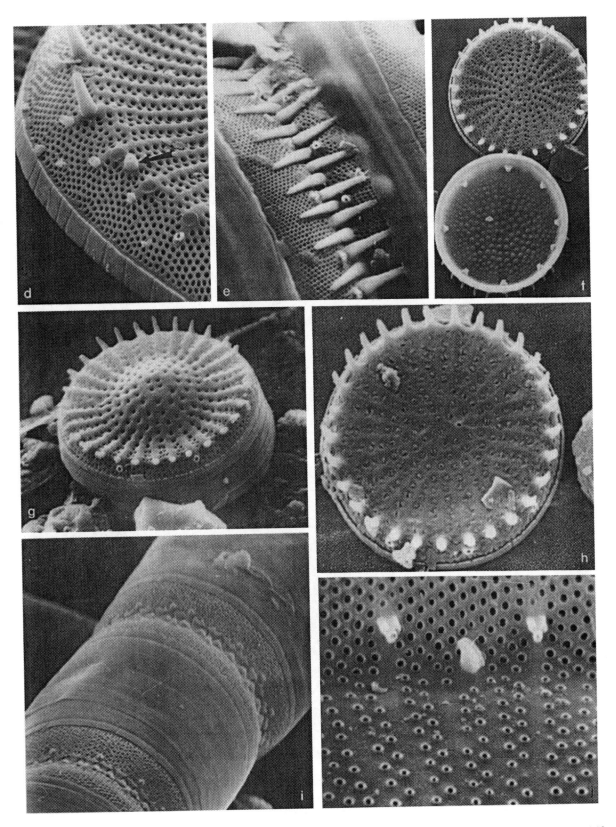

Lauderia P. T. Cleve 1873. Bih. Kongl. Svenska Vetensk.-Akad. Handl. 1(11): 8

T: *L. annulata* Cleve

Cells shortly cylindrical, joined to form straight filaments. In girdle view[a] the well-rounded valves give a beaded appearance to the filament. Plastids numerous, discoid, lobed. Common and widespread in the phytoplankton of the oceans. A small genus of two (possibly five) recorded species but all reduced to a single taxon by Hasle (1974).

Valves very delicate,[b] with fine costae radiating from a central annulus;[b-e] fine pores penetrate the basal layer between the costae. Around the circumference only, anastomosing costae are present externally,[c, d] producing a weakly loculate structure. The central area of the valve may be plain or indistinctly porose. Around the circumference there are numerous fultoportulae in several ranks;[b-d, f, g] a few fultoportulae also occur scattered on the valve face. The fultoportulae have well-developed external tubes, whose rims are slightly serrate and whose bases merge into the framework in the marginal region. Internally the fultoportulae are small and have only feebly developed buttresses.[g, h] A ring of longer, thinner processes is also present at the boundary between the loculate and costate regions but these do not arise from fultoportulae and are probably occluded. These are the spines referred to by Gran (1900) which in TEM[b] can be seen to be solid structures. A single fairly large rimoportula is located near the edge of the valve,[arrows, c, d] but just within the costate area. Copulae loculate[f, i] as in the marginal region of the valve, although they are unusual in that the loculae are much larger than on the valve; split.

This is a very characteristic genus which cannot be closely allied to other genera in the Thalassiosirales; hence our proposal to place it within its own family.

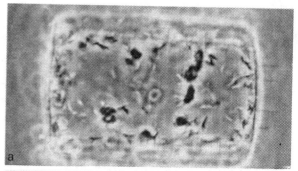

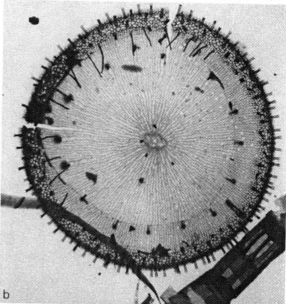

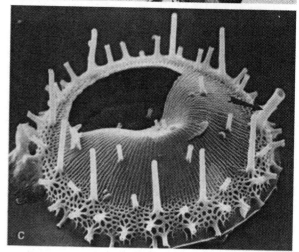

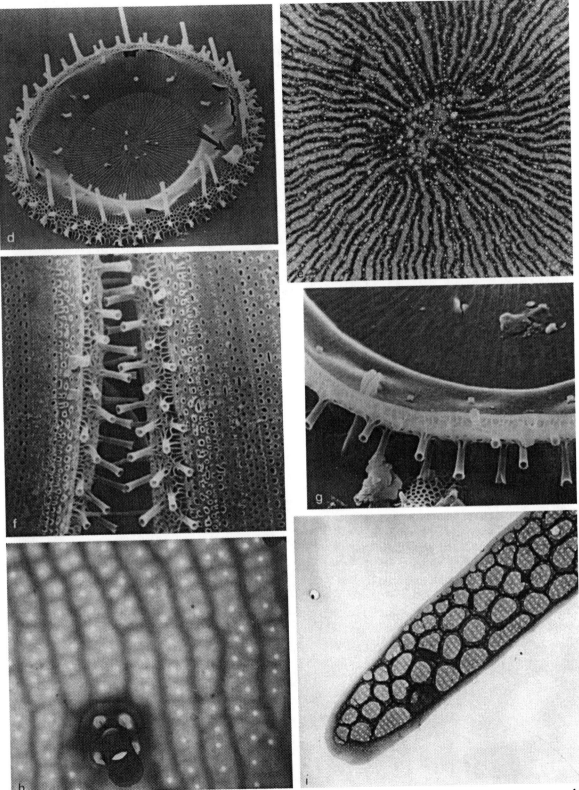

Chrysanthemodiscus A. Mann, 1925. Bull. U.S. Natl. Mus. 100(6): 58

T: *C. floriatus* A. Mann

An epiphytic, marine diatom with delicate cylindrical cells[a] (often 4–6 times as long as wide) and convex valves, attached end-to-end by means of mucilage pads arising from the valve centres. Sometimes, however, cells are attached by pads near the valve margins and appear heterovalvate – one being domed and the other flatter and notched. Plastids numerous, discoid. Often abundant on macroscopic algae along tropical shores but only rarely reported, e.g. from Oahu, Hawaii (Round, 1978b), off Florida (Gibson & Navarro, 1981) and from the coast of Oman (samples from S. Hiscock).

Valves domed[a, b, g] with concentric undulations or rises (see Gibson & Navarro). Centrally with a cluster of 'pegged', round areolae[d] (probably of the rota type) within the annulus, and outside this with radiating rows of areolae whose lengths change as they approach the centre.[c-f] The vela are delicate and in most specimens only 'pegs' remain of what were possibly rotae located in somewhat radially elongate areolae. No distinct valve mantle.[h] Internally no more detail is obvious. There are no portulae. Cingulum composed of numerous complete copulae[g] of similar form, with vertical rows of areolae of similar structure to those of the valve. Proximal edges of bands slightly fimbriate.[i]

Round (1978b) suggested that this was a distinctive diatom which should be placed in its own family, the Chrysanthemodiscaceae. Simonsen (1979) removed it to the subfamily Stictodiscoideae of the Biddulphiaceae, a subfamily containing so many diverse diatoms in his treatment as to be unacceptable: we retain the original family. Whilst preparing this account we found that Takano (1965) had rediscovered *C. floriatus* in waters off Japan and his light micrographs are the best we have seen of this genus. He also points out that *Melchersiela* described by Teixeira (1958) is probably congeneric.

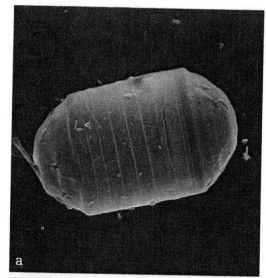

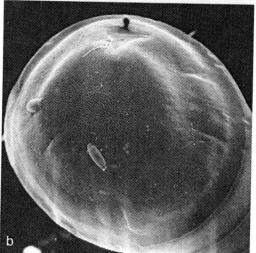

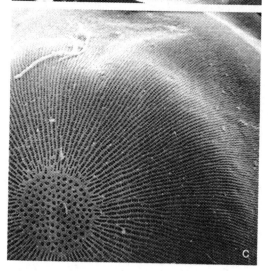

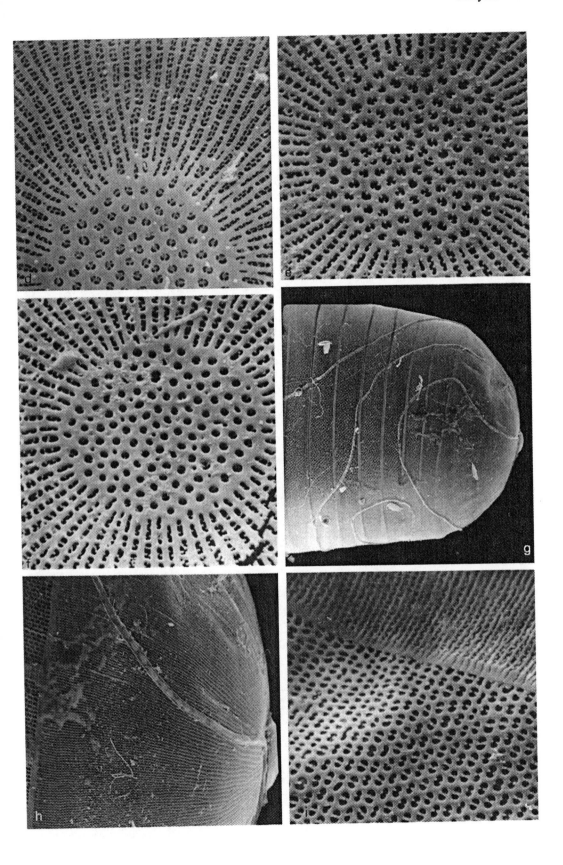

Melosira C. A. Agardh 1824. Syst. Algarum xiv: 8 (nom. cons.)

T: *M. nummuloides* Agardh (typ. cons.)

Cells cylindrical[a-d] to subspherical, united in filaments by mucilage pads secreted onto their valve faces; in addition irregular spines may assist the linkage. Cells united distinctively into pairs or triplets[a] by their cingula. Plastids lobed, small plate-like, lying in the peripheral cytoplasm. A common genus in freshwater (*M. varians*) and marine epibenthic habitats (*M. nummuloides*, *M. moniliformis*).

Valve face flat[d] or domed,[c] covered with small spines or granules; a more or less well-developed corona[arrow. c] consisting of larger irregular spines is sometimes developed. This may be surrounded by a carina (a flat collar-like structure) as in *M. nummuloides*.[arrow. h] Valve mantle not readily distinguishable from valve face in most species. Valve mantle edge having a milled appearance.[d] The valve structure is loculate,[h] the loculus being open to the outside via a number of small simple pores[f] and to the inner surface by somewhat larger pores, which may be partially or completely bridged by silica struts (in the latter instance forming rotae).[g] The pattern of these inner pores is independent of the loculi (cf. *Stephanopyxis*). The loculi may be randomly arranged or lie in rows radiating from the centre of the valve. Rimoportulae occur usually in a ring near the mantle edge and sometimes scattered or grouped on the valve; there is a circular external aperture surrounded by an irregular rim.[c, f] The internal apertures are in the form of elongate slits also surrounded by an irregular, low rim. Copulae split, ligulate, with regular longitudinal rows of small pores;[i] pars media distinct and near advalvar edge. The two valves are closely associated during the greater part of the cell cycle and the cingula of adjacent daughter cells overlap one another considerably.

The genus as conceived by previous workers has included a great range of taxa, some of which have been transferred to other genera (see *Aulacoseira*, *Paralia*, *Orthoseira* and *Ellerbeckia*). For references on *Melosira* and related genera see Crawford, 1988.

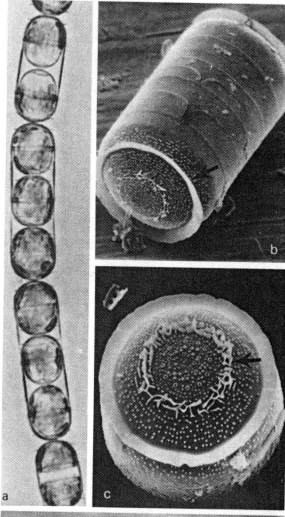

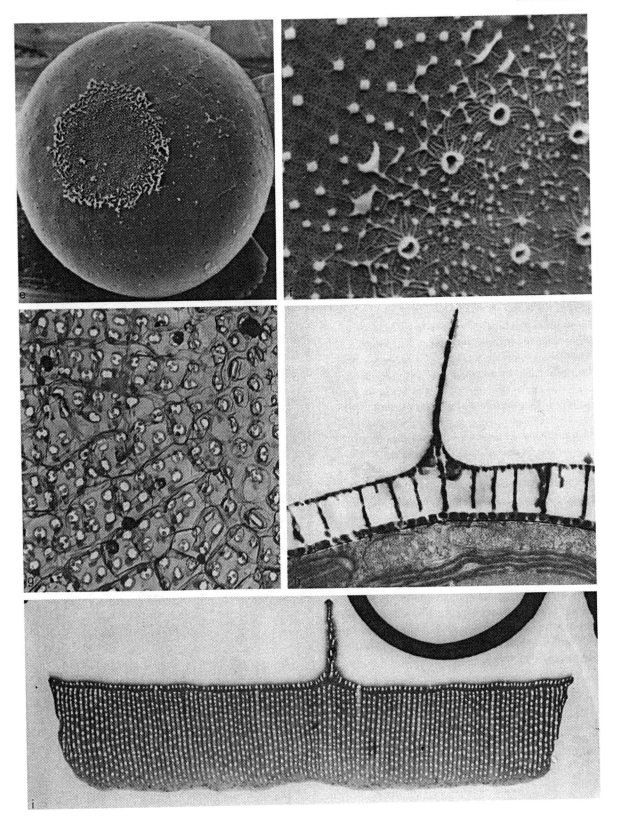

Druridgia A. Donkin 1861. Quart. J. Microsc. Sci. Ser. 2, **1**: 13

T: *D. compressa* (T. West) A. Donkin (= *Podosira compressa* = *D. geminata* Donkin)

Cells usually united in pairs;[a] elliptical in valve view,[b] oval in girdle view.[a, d] Each cell with four plastids. A monotypic genus (*D. geminata* and *D. compressa* are synonymous) occurring along marine beaches. Exact micro-habitat unclear but may occur attached to sand grains or free near sediments (see Hendey & Crawford, 1977).

Valves oval, with a central hyaline area[c] from which the valve surface falls in all directions, steepening to an almost vertical mantle though this is not topologically distinguishable. Radiating rows of domed areolae with 1–5 external pores cover the external valve surface;[f, h] the rows become vertical on the mantle. Beneath each dome is a chamber in the silica wall which opens internally[g, i] through a fine rota (see Hendey & Crawford for further detail). The rimoportulae are grouped in a very unusual manner. On one side of the valve they occur in a row along the margin;[d] this row curves up and over the poles[c] but ceases before reaching the mid-point of the far valve edge, above which there are two clusters of rimoportulae.[c] The rimoportulae themselves are unusual in that they have outer openings with 'peg-like' projections. Internally the rimoportulae are simple sessile structures.[g] The cingulum consists of at least 5 copulae, the valvocopula being a plain split ring. The copulae also have a plain region with fimbriae extending towards the valve and the usual type of ligula.[k]

Druridgia is a rarely recorded but highly distinctive diatom, especially at the EM level. In the past it has been confused with *Podosira* but the very precise loculate form of the wall of *Druridgia* is more similar to that of *Melosira* and accordingly we place *Druridgia* in the Melosiraceae.

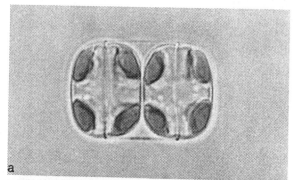

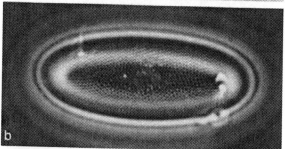

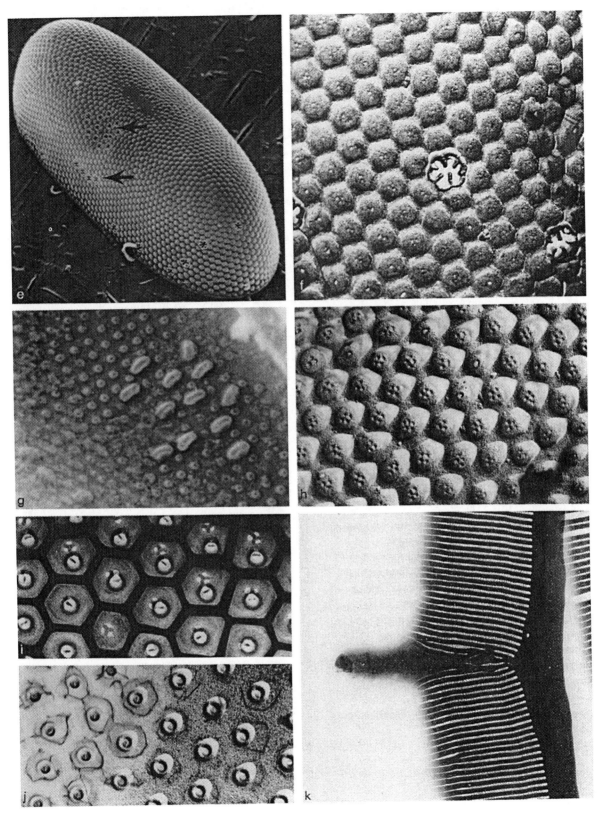

Stephanopyxis (C. G. Ehrenberg) C. G.
Ehrenberg 1845. Ber. Bekanntm. Verh. Königl.
Preuss. Akad. Wiss. Berlin, **1845**: 80
(= *Pyxidicula* Ehrenberg subgen. *Stephanopyxis*)

T: *S. aculeata* (Ehrenberg) Grunow
(= *Pyxidicula aculeata*)

Cells cylindrical to almost spherical,[a, b] joined by
processes into long filaments. Plastids discoid,
lobed. A fairly common marine planktonic genus
tending to be tropical in distribution but carried into
colder waters by currents. Also common in fossil
deposits where it exhibits a larger number of species
– some may be resting spores.

Valves domed,[c, d] hemispherical to discoid;
without obvious distinction into valve face and
mantle. Areolae large, hexagonal, opening outwardly
by large foramina in shallow chambers;[e, g] forming a
base to the areolae is a layer of silica with a
continuous series of poroids containing rotae,[e-g]
arranged in rows radiating from the valve centre.
Areolae more regularly arranged towards the valve
centre. A ring of tubular (sometimes half-tubular)
processes is present; in some fossil forms additional
processes occur in the centre.[d] The processes arise as
extensions of the walls of the areolae and articulate
with those of the sibling valve. Low-lying
rimoportulae occur in a ring beneath the processes,
into which they open. A further ring of rimoportulae
occurs around the valve margin,[i] in the marginal
row of loculate areolae, and these open to the
outside by simple slits.[l] The cingulum is composed
of a complete (?) narrow valvocopula and numerous
segmented bands.[k, l] All have simple pores.

It is reported (e.g. Hendey, 1964) that strands of
cytoplasm pass up the processes and thus connect
cell to cell; this requires confirmation and if this is
so it is the only diatom we know which has such
interconnections. *Stephanopyxis* does not fit easily
into any higher grouping of genera, though
Simonsen (1979) placed it in the Melosiraceae.
Recently Nikolaev (1984) has erected a new family
Stephanopyxaceae and this appears to be a more
logical step but must be emended to
Stephanopyxidaceae (see Silva, 1980). Further EM
detail is provided by Round (1973).

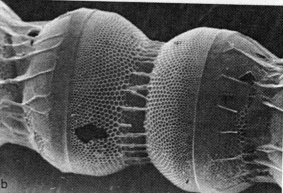

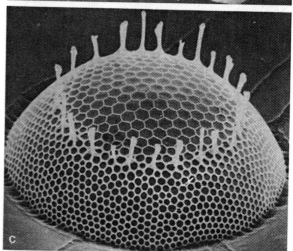

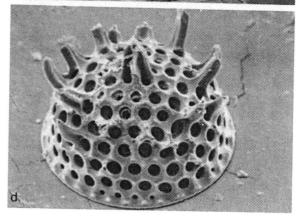

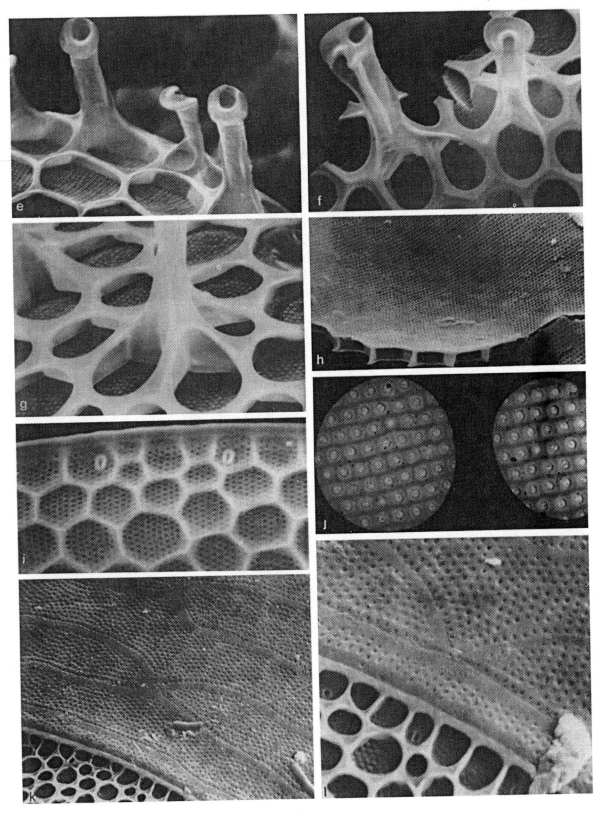

Endictya C. G. Ehrenberg 1845. Ber. Bekanntm. Verh. Königl. Preuss. Akad. Wiss. Berlin, 1845: 71

T: *E. oceanica* Ehrenberg

Cells barrel-shaped, joined into short chains. Plastids not investigated. A modern and fossil genus so far only observed by us as separate valves. Reported as a benthic marine genus.

Valves broadly cylindrical with sharply defined mantle and valve face.[a, b] In well-preserved specimens there is a narrow flange at the valve rim in which there is a line of pores.[a] Towards the centre of the valve face are a number of projections. Mantle width one-third to one-half as deep as the valve face. The areolae are variable in size and lie in a heavily silicified framework. They are loculate, with a foramen on the outside[d, e] and fine pores on the inside.[e, i] These pores are arranged in rows radiating from the valve centre,[g] being independent of the pattern of the foramina on the outside; in this *Endictya* resembles *Melosira* and *Stephanopyxis*. Consequently they may be expected to be poroids with rotae in the living cell. Between mantle and valve face on the inner surface are the internal apertures of rimoportulae.[h, i] These correspond in position to the pores in the flange on the outer surface and run between areolae in fractured specimens.[f] Copulae have not been observed here, but Hustedt (1930) recorded many 'rings' forming the cingulum.

This is a small genus with only 10 species recognised by VanLandingham (1969). Hustedt (1927–66) reports colony formation in modern material but there do not appear to be true spines and the absence of any other linking mechanism on the valve margin suggests that these cells are solitary. As Hustedt commented, there is considerable confusion in the literature and the taxon has been variously placed in *Orthosira*, *Melosira*, *Dictyopyxis*, *Coscinodiscus*, *Stephanopyxis* and *Craspedodiscus*! The genus is clearly in need of re-assessment.

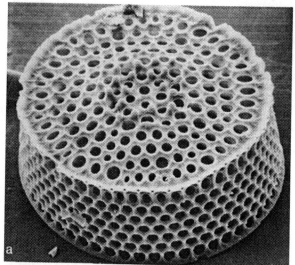

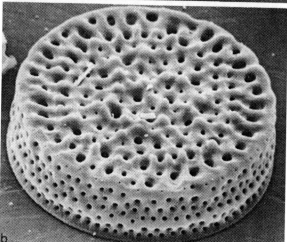

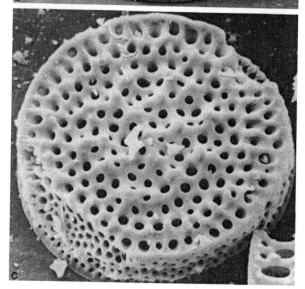

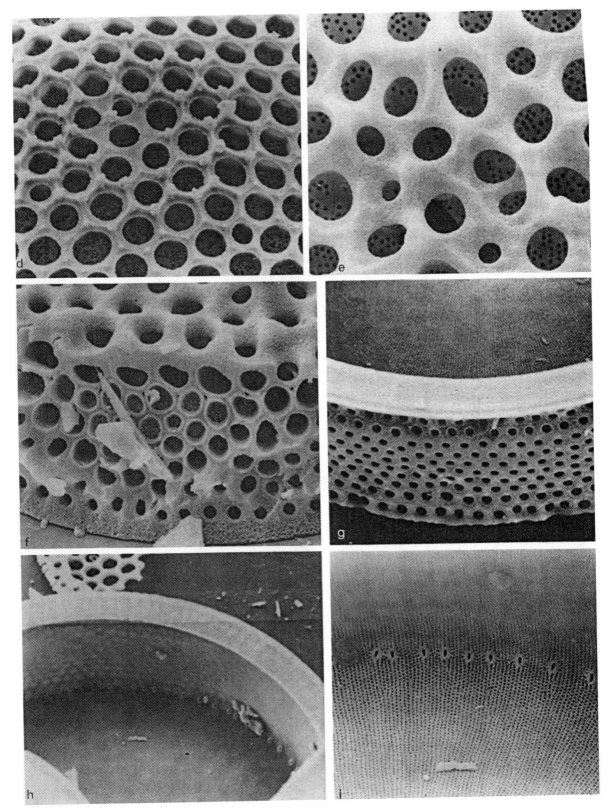

Hyalodiscus C. G. Ehrenberg 1845. Ber.
Bekanntm. Verh. Königl. Preuss. Akad. Wiss.
Berlin, **1845**: 71

T: *H. laevis* Ehrenberg

Cells lens-shaped, occurring singly or, more
commonly, in pairs united by copulae. Plastids
several, irregularly discoid. A common marine genus
often attached by thick mucilage pads to seaweeds,
in short filaments of up to four cells, but also found
in turbulent inshore waters or lying unattached on
the sediments.

Valves deep, watchglass-shaped to hemispherical,[a, b] sometimes with a flat area at the centre;
the mantle edge is turned outwards. Arrangement of
the areolae may be weakly or strongly fasciculate.
Valve structure with narrow bullulae;[g] the areolae
open to the inside by round pores with rotae,[f] but at
the valve centre the areolae are closed on both the
inner and outer surfaces by a plain mass of silica:[c]
the genus owes its name to this central feature. In
transmission electron micrographs the areolar
pattern can be seen continuing into the hyaline
area; it is very variable in extent (see the three light
micrographs).[h-k] Rimoportulae occur in a ring
around the mantle edge[d] and also sparsely scattered
over the rest of the valve. Internally the
rimoportulae are simple and sessile.[l] On the outside
it is difficult to distinguish the outer openings of the
rimoportulae from the areolae, though in some
valves there are slightly wider openings, which
almost certainly represent the rimoportulae.[arrow, c]
The copulae have not been investigated in detail but
appear to be very thin, porous bands.[m]

The relationship of this genus to *Podosira* needs
re-investigation. Only a few species are regularly
found and the others, e.g. in VanLandingham (1971),
are often single records. The cells are attached to
each other by conspicuous blocks of mucilage which
must be secreted from around the central hyaline
area yet this is not a place where there are any
obvious secretory organelles.

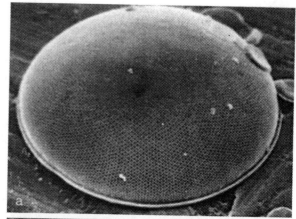

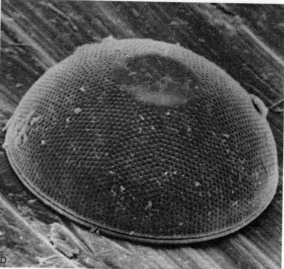

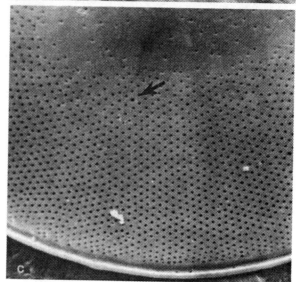

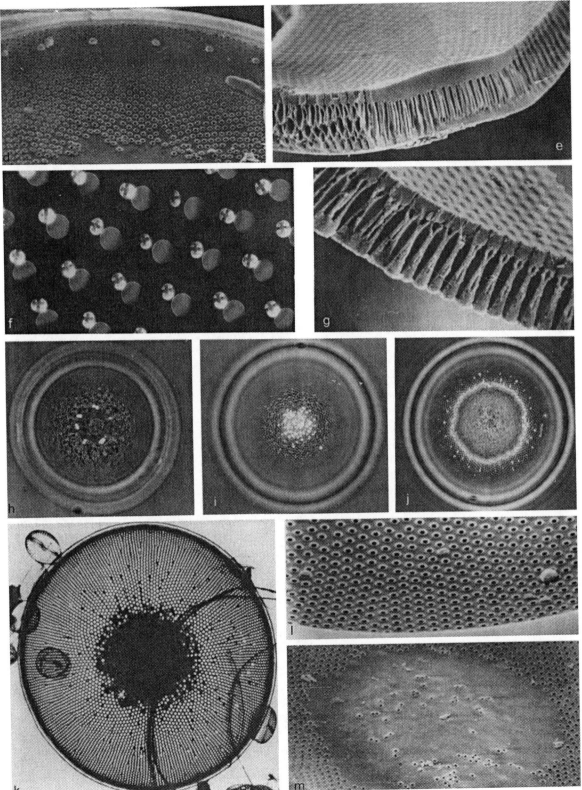

Podosira C. G. Ehrenberg 1840. Ber. Bekanntm. Verh. Königl. Preuss. Akad. Wiss. Berlin **1840**: 161

T: *P. moniliformis* (Montagne) Ehrenberg (= *Trochosira moniliformis*)

Cells spherical to sub-spherical, single or united in pairs or rarely triplets.[a-c] Approximately 12 irregularly discoid plastids. Benthic, attached by several thick polysaccharide stalks to seaweeds.[b]

Valves hemispherical or watch-glass-shaped;[d, e] no distinct valve mantle; some species have a flat central area. Areolae loculate, arranged in decussate rows within radial sectors. Valve thick, the narrow loculi[g] opening to the outside by pores that are often too small to be visible even with the SEM; to the inside the loculi open through what appear to be small open foramina but which in fact contain rotae.[g, h] The edge of the valve is smooth. Rimoportulae are scattered over the whole of the valve.[e] Their simple openings often provide the only markings on the outside;[c, d] internally they are small oval structures without stalks. The rimoportulae replace areolae; their apertures are not aligned in any particular direction. Copulae numerous, narrow, with simple longitudinal slits.[i, j] The valvocopula lacks slits and inserts against a shallow shelf on the valve edge.[f]

Characterisation of this genus and its relationship to *Hyalodiscus* needs further examination. For example, *P. stelliger* was removed from *Hyalodiscus* by A. Mann (1907) but its clear area of the valve face and well-marked areolae arranged in fascicles suggest it should be returned to its original genus. Bullulate (*Hyalodiscus*) and non-bullulate (*Podosira*) loculate systems may be valid differences between the genera.

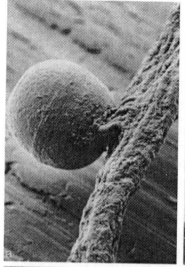

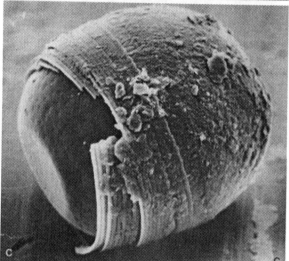

Paralia P. A. C. Heiberg 1863. Consp. Crit. Diat. Dan.: 33

T: *P. marina* (W. Smith) Heiberg (= *Orthosira marina*)

Cells shortly cylindrical, linked to form straight chains.[a, b] Plastids several per cell, small, discoid. Commonly found in marine inshore plankton but probably belonging on sandy sediments. A small genus, containing at least two species – the fossil *P. sulcata*[f, h–j] and the extant *P. marina*.[a– c, g]

Valves robust, circular, with radial markings on the valve face. Valve face and mantles sharply differentiated. Sibling valves within chains linked via well-developed interlocking ridges and grooves, and by marginal spines;[c, d, f, i] cameo (relief) and intaglio valves occur, comparable with those of *Ellerbeckia*. The relief valves have stepped mantles.[j] At the ends of each chain are separation valves, with reduced ridges and no marginal spines.[b, e, h] Valves composed of a thick inner layer penetrated by tubes, which connect the cell lumen with a series of chambers lying beneath a thinner outer layer;[g] the chambers open to the outside by large holes positioned around the edge of the valve face and on the valve mantle (the inner part of the valve face is non-porous).[d–f, h] Copulae numerous, open, bearing slits (*P. marina*[d]) or without perforation (*P. sulcata*[j]). Unlike most other genera, the valvocopula does not underlap the epivalve and is only inserted in a shallow groove between the two mantles of the cell. Consequently the copulae are easily detached.

Many varieties have been described in *P. sulcata* (e.g. in A. Schmidt 1874–1959, T. 175–178) and require re-investigation, especially in view of the heterovalvy occurring in the genus. *P. marina* is unlikely to be confused with any other living diatom. The heterovalvy and the structure of the valves suggest a close relationship to *Ellerbeckia* and *Trochosira*, which can be recognised by placing both in a new family, the Paraliaceae. A detailed account of *Paralia* is given by Crawford (1979b).

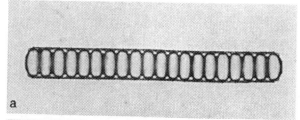

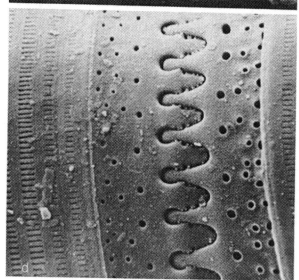

Ellerbeckia R. M. Crawford, 1988. In 'Algae and the aquatic environment' (ed. F. E. Round): 421

T: *E. arenaria* (Ralfs ex Moore) Crawford (= *Melosira arenaria*)

Cells large, shortly cylindrical, linked to form curved chains of up to 30 or more cells.[a, b] Plastids numerous, small discoid. Mainly in freshwater, with both fossil and recent forms known. Sometimes visible to the naked eye!

Valves robust, circular, with radial markings.[c] Valve face and mantles sharply differentiated.[f, g] Sibling valves different and complementary, the 'cameo' valve bearing a system of ridges[g, j, k] and the 'intaglio' valve an equivalent system of grooves to accommodate them.[f, h] Intaglio valves have a plain mantle,[f] cameo valves a stepped mantle,[g, j] as a result of the formation of the cameo valves beneath the edge of the parental hypotheca (Crawford, 1981b). The valves also have different curvatures and when the concavo-convexity is as pronounced as in *E. arenaria* forma *teres*,[a, b] this results in cells of unequal pervalvar dimensions:[b] large cells have two (convex) intaglio valves, small cells have two (concave) cameo valves, and intermediate cells have one of each. Valve face without pores except for a peripheral ring in some species. Valve mantle very thick, consisting of a massive inner layer penetrated by long tubes,[d] which connect the cell lumen with a series of shallow chambers; these run the length of the mantle and open to the outside via a finely porous layer of silica.[i] A unique type of tubular process is present on the mantle; these are easily seen in LM in the forma *teres*. Each process takes the place of a mantle tube and opens to the inside by a small pore at the top of a dome-like projection.[d, e & insert] The valvocopula is closed but the other copulae are open. All have rows of small round pores. Valvocopula not easily detached from its valve, bearing a series of crenulations which interdigitate with projections from the edge of the valve mantle;[i]

Paralia has a similar valve structure and also exhibits heterovalvy, justifying its inclusion in the same family as *Ellerbeckia* (the Paraliaceae). Species of *Ellerbeckia* have been reported from marine as well as freshwater localities, providing a further link with *Paralia*. Moyiseeva & Genkal (1987) have recently combined taxa with *Paralia* that we consider find their proper place within *Ellerbeckia*.

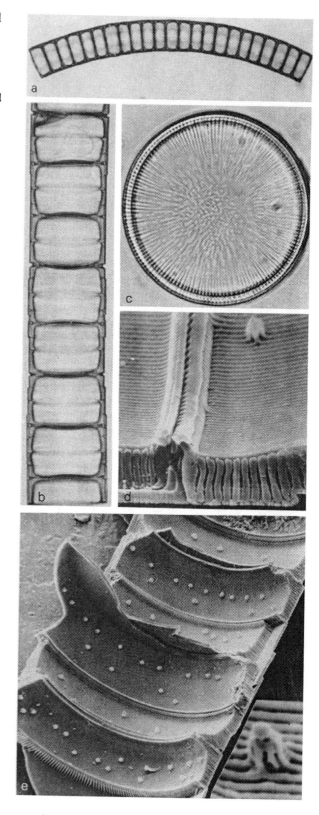

Aulacoseira G. H. K. Thwaites 1848. Ann. Mag. Nat. Hist. Ser. 2,1: 167

T: *A. crenulata* (Ehrenberg) Thwaites
(= *Gaillonella crenulata*)

Cells linked tightly to form long straight, curved or even coiled filaments.[a, b] Plastids discoid. A common freshwater, planktonic genus previously placed in *Melosira*.

Valves circular.[f] Valve face plain or with scattered poroids, which are often restricted to its periphery. Valve mantle deep,[b, c, e] making a right angle to the planar valve face from which it is sharply differentiated; with vertical or curved rows of areolae. Mantle edge plain. Valve face/mantle junction provided with spines, which are expanded at their apices and fit with the spines of adjacent cells to form a linkage breakable only by damaging the spines. In some species the spines have two 'roots' that straddle a row of pores on the mantle,[b, e] others have a single 'root' that runs between pore rows.[c, d, f] Separation cells[e] occur at intervals in *A. granulata*, the spines on these being lanceolate and of variable length; the areolae of separation valves are often in straight rows, those on other valves spiralling away from the sibling junction in opposite directions.[b] Inside the valve mantle, at the junction between the plain mantle edge and the areolate portion, there is a thickening (ringleiste),[g] which is hollow in *A. ambigua*.[h] Mantle areolae simple, round to rectangular, containing volate occlusions.[h, i] Small rimoportulae occur inward of the ringleiste and open through it to the outside. The copulae are split rings with ligulae;[j] they are finely areolate, with an advalvar pars media.

Simonsen (1979) resurrected the long disused name *Aulacosira* (*sic*) and the common species *italica*, *granulata*, *islandica*, *nyassensis* and *distans* of the genus *Melosira* are placed here (see also Haworth, 1988). *M. varians* has a very different wall structure and remains in *Melosira* (see p. 154). Davey & Crawford (1986) showed that the frequency of separation valve formation in *A. granulata* controls filament length, which in turn (Davey, 1986, 1987) may influence the buoyancy of this diatom. Why some species, e.g. *A. italica* subsp. *subarctica*, only appear to have separation valves while others, e.g. *A. ambigua*, only have linking valves is unexplained.

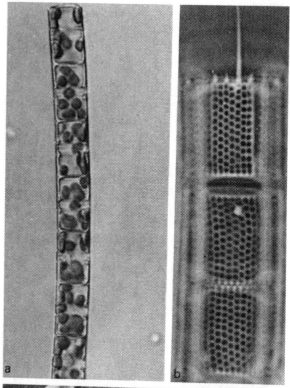

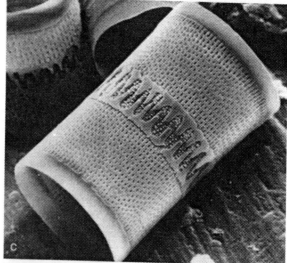

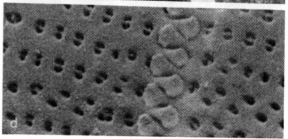

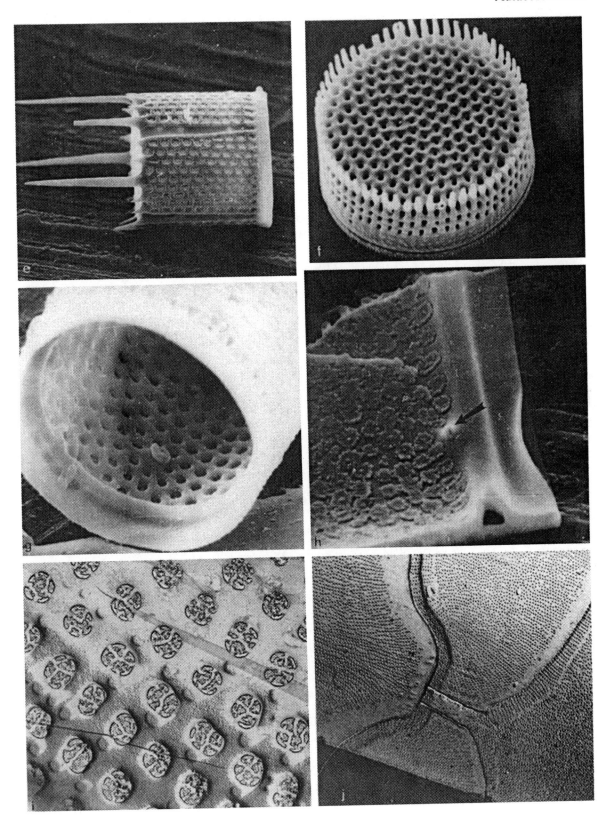

Strangulonema R. K. Greville 1865. Trans. Microsc. Soc. Lond. Ser. 2, **13**: 43

T: *S. barbadense* Greville

Cells cylindrical, forming chains linked by spines arising from a valve face 'collar'.[a-f] A fossil genus found in sediments of Eocene to Oligocene age.

Valves circular, with deep mantles and a smooth raised 'collar' on the valve face, from which flattened furcate spines arise.[b-f] The collar turns outwards around its apex and, besides spines, also bears downwardly directed projections, which may originally have linked with the remnants[c, d] of upwardly directed projections visible on the valve mantle, to form a network over the depression formed around the collar. The interlocking spines are non-separable and in the second form illustrated seem to be completely fused between sibling cells.[e-h] Areolae large, round or elliptical. The valve face areolae radiate from the centre and are slightly smaller than those of the mantle (P. A. Sims, personal communication); on the valve mantle the areolae are in vertical rows and in addition there are small pores through the framework.[d] No vela have been detected in our material, which is eroded. We cannot comment on the occurrence of portulae. Copulae wide, complete, with widely spaced areolae.

According to VanLandingham (1978) this is a monotypic genus but our material suggests that this is not so. Without further information, especially on presence or absence of portulae, it is difficult to place this genus in a family. However, the indented 'collar' may represent a 'pseudosulcus' such as found in *Aulacoseira granulata* and its relatives, and this might indicate a relationship.

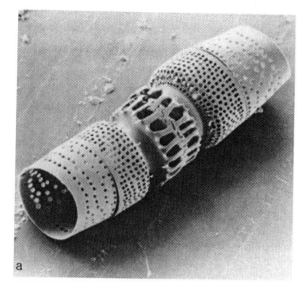

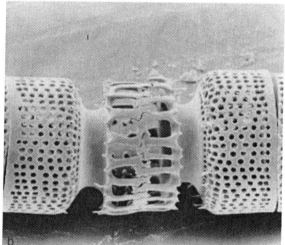

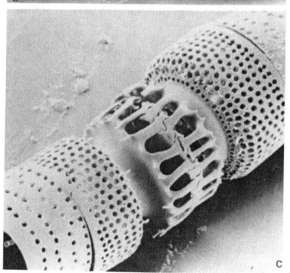

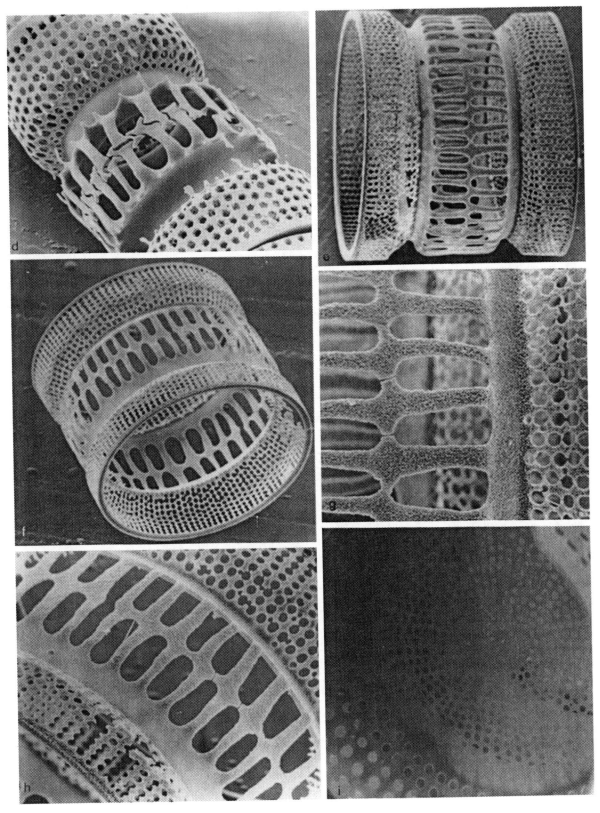

Orthoseira G. H. K. Thwaites 1848. Opusc. Crypt. **74** (18): 7

T: *O. americana* (Kütz.) Thwaites (= *Melosira americana*)

Cells cylindrical, united to form short filaments with valve faces closely appressed.[a, b] Interlocking spines can be seen between cells[a, b] and a characteristic thickening of the girdle bands can be seen in mid-focus.[c] The plastids are numerous, small and discoid. They are peripheral[a] while the nucleus is suspended centrally on cytoplasmic bridges through the vacuole. A small genus of sub-aerial diatoms commonly found among bryophytes especially in alkaline areas.

Valve views of the acid-cleaned valves show 2–5 characteristic tube processes at the centre of the face.[d] Rows of areolae radiate from near the centre and pass over the valve rim and down the mantle.[e] At the rim in most species there are well-defined spines. These are simple and triangular in side view[f-h] or pyramidal with stellate bases[i] and are found between rows of areolae. In *O. dendrophila* the ring of spines is interrupted by clusters of distinctive areolae.[f] In intact sibling valves these clusters of pores are precisely aligned with each other.[g] The valve is a simple laminate layer of silica.[m] While the areolae are covered on the inner surface by a velum or some other form of siliceous layer[m] the pores in the clusters are left unoccluded.[k] Though it is improbable that the marginal spines serve to link neighbouring cells, they would prevent torsion or rotation of siblings *vis-à-vis* one another.[g] Cells are probably bound together by mucilage secretions passed through the tube process. These are unique fluted passages through the valve with simple internal openings[k] and well-defined collars to the outside,[i, l] described as 'carinoportulae' by Crawford (1979a). The cingulum is composed of a number of bands whose pars interior is greatly thickened. In *O. dendroteres* the bands are so angled as to give the effect of a spiral arrangement.[n] The bands are sufficiently robust to leave a step on the mantle of one of the sibling valves[e, g] similar to that observed in *Ellerbeckia*.[j]

Auxospores are covered by a layer of scales[j] similar to those reported from *Melosira* (Crawford, 1974a).

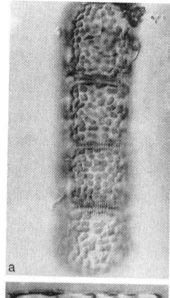
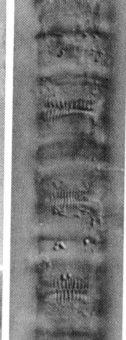
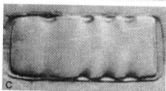
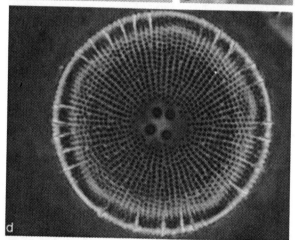
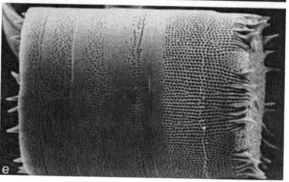

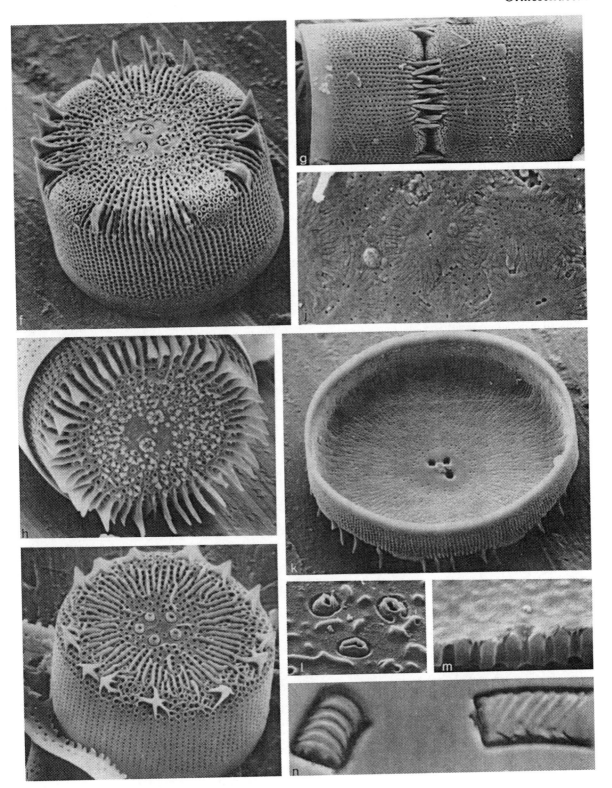

Coscinodiscus C. G. Ehrenberg. Abh. Königl. Acad. Wiss. Berlin, 1838: 128

T: *C. argus* Ehrenberg

Cells discoid,[a, b] sometimes thin (like a coin) or more barrel-shaped, occasionally with valve mantle deeper on one side. Plastids numerous, discoid. Free-living, marine and often abundant in the phytoplankton. A few species reported from lakes but usually only in high conductivity waters and many records need checking. Widely distributed in the fossil record.

Valves saucer- to petri-dish shaped. Valve face flat, sometimes depressed centrally; areolae radiating from central annulus,[c] sometimes sectored. Ovoid and triangular forms occur as fossils. Valves loculate with complex external vela with central small pores and an external ring of larger openings.[d, e] Vela often becoming diamond-shaped on valve mantle. The internal openings are foramina, often rimmed and becoming smaller on the valve mantle.[f-h] Rimoportulae present at intervals around the valve:[f] external openings simple;[i] internally with small cup-shaped expansions.[g, h] Macro-rimoportulae sometimes occur at intervals in the ring;[g, h] these are larger, sometimes with flat apical expansions and curled; the macro-rimoportulae open via larger external apertures. Copulae poroidal, split and ligulate.[i]

This is a large genus and requires extensive study since there is much variation at the ultrastructural level. Fryxell (1978) has documented the typification. Several species have been removed to *Thalassiosira*, others to *Actinocyclus* – many of the changes involve species living in tropical lakes. *C. nitidus* has been transferred to *Psammodiscus* by Round & Mann (1980). Many small centric forms need checking and the position of the vela determined; any forms found to have fultoportulae are almost certainly *Thalassiosira* species. Sims (1989) has reviewed three subgroups of *Coscinodiscus* and discussed their stratigraphical distribution and phylogeny.

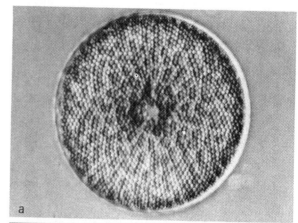

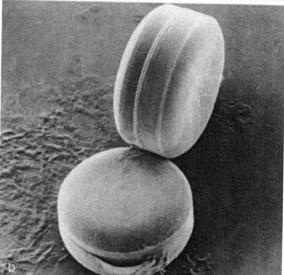

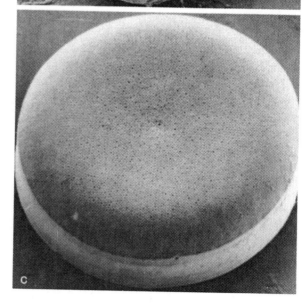

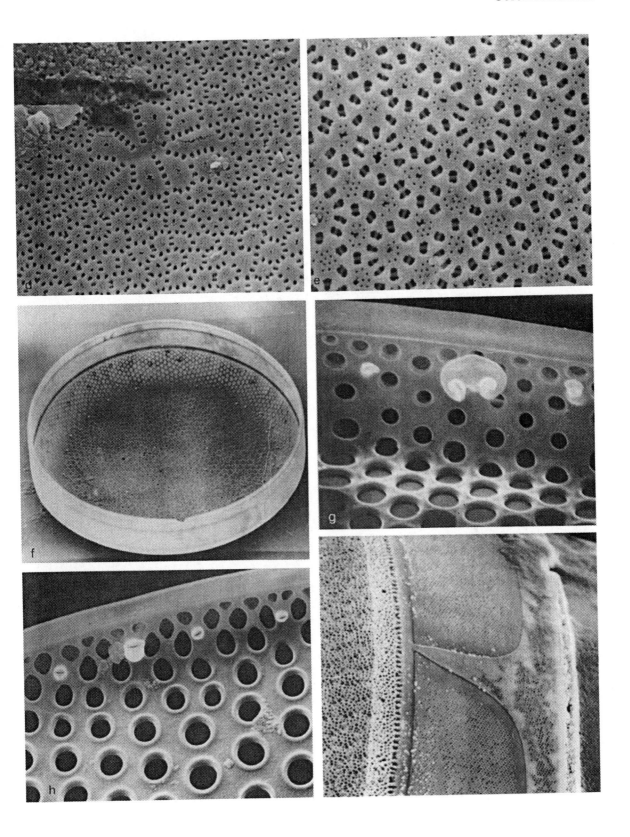

Palmeria R. K. Greville 1865. Ann. Mag. Nat. Hist. Ser. 3, **16**: 1

T: *P. hardmanniana* Greville

Cells in the form of a segment of a sphere. Plastids numerous, discoid, very small.[a] Marine and planktonic, probably limited to the tropics; according to Simonsen (1974) it is neritic.

Valve semi-circular,[b] with a flat valve face and an extensive mantle on the convex side, which rises from the corners to the centre of the semi-circle; there is almost no mantle along the straight edge. Areolae loculate[h, i] (as in *Coscinodiscus*), externally with a cluster of central pores surrounded by a hexagonal outer series;[c] internally with simple round, rimmed foramina.[d-i] The areolae radiate from a plain area in the centre of the valve face.[d, e] Hyaline rays run from near the centre and terminate near the valve margin at rimoportulae.[e-g] Around the arc of the convex side the rimoportulae are located on the curved junction between valve face and mantle;[g] on the straight edge they lie near the margin itself. Macro-rimoportulae occur, replacing the smaller rimoportulae on each valve near the centre of the convex, dorsal side. The other rimoportulae are relatively simple and tube-like internally.[e-g] The rimoportulae are further apart along the straight edge than elsewhere. Copulae not known.

It is important to distinguish this 'hemi-discoid' diatom from the genus *Hemidiscus*. *Palmeria* is clearly associated with *Coscinodiscus* on account of its wall structure and the type of rimoportulae whereas *Hemidiscus* has a pseudonodule and at present is allied to *Actinocyclus*. Simonsen (1972) reinstated *Palmeria* as a separate genus and, as with *Hemidiscus*, this can only be justified at present on the single criterion of cell shape. We would agree with Simonsen that it is more practical to maintain *Palmeria* as a genus and not 'crowd the genus *Coscinodiscus* more than is necessary'. The 'tubular' type of rimoportula occurs in some *Coscinodiscus* spp. whereas the 'slit' type occurs in others, in which the more complex macro-rimoportulae are lacking.

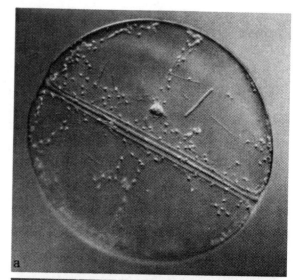

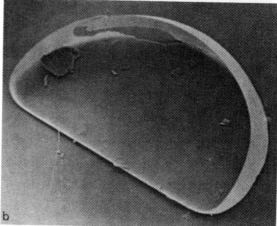

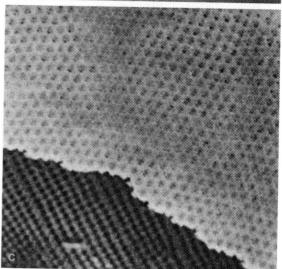

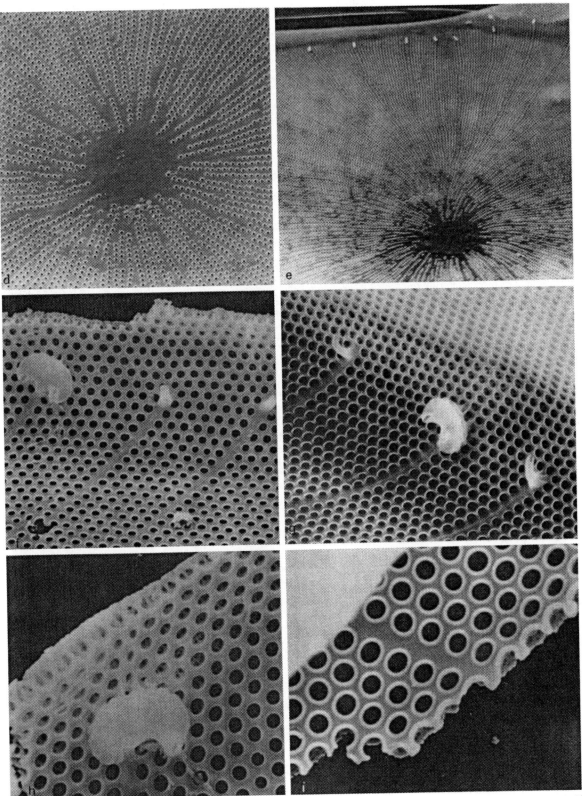

Stellarima G. R. Hasle & P. A. Sims, 1986. Br. phycol. J. **21**: 111

T: *S. microtrias* (Ehrenberg) Hasle & Sims (= *Symbolophora microtrias*)

Cells discoid, solitary. Plastid morphology unknown. *Stellarima* species have usually been placed in *Coscinodiscus*. Nikolaev (1983) recognised that they do not belong there and resurrected *Symbolophora* Ehrenberg, placing in it *S. microtrias* as the type, together with five other species, all previously in *Coscinodiscus*; he also created the new family Symbolophoraceae. However, the type of *Symbolophora* is in fact an *Actinoptychus* (*Cymatogonia*) and therefore the name *Symbolophora* cannot be used for these species; hence Hasle & Sims (1986) have erected the genus *Stellarima*. A marine planktonic genus – our material came from the Antarctic, like most material studied by others, but also some samples were from the coast of South Africa.

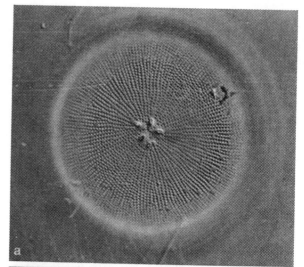

Valves circular,[a, b] convex, without a distinct valve mantle. Striae uniseriate,[c-f] radial, some reaching the centre, others shorter; fasciculate in resting spores (Syvertsen, 1985). Areolae loculate;[h] outer openings simple[c] or with cribra:[j] some scattered cribra [arrows, j] have one central pore and are more regular than the others; internal apertures with domed closing plates.[e, f, h] Rimoportulae central, arranged in a ring[d-f] in the centre of the valve: 1–15 have been observed. Externally the rimoportulae have slit-like openings;[b, c, i] internally they are elongate and raised, their lips being orientated radially. No other portules present. Copulae not observed.

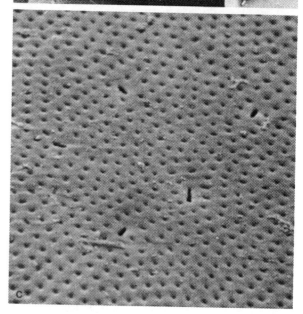

This genus differs from *Coscinodiscus* in having no marginal rimoportulae but only central radiate rimoportulae and in having internal closing plates to the areolae. It is therefore more akin to genera such as *Actinocyclus*. Resting spores and comments on palaeoceanographic and evolutionary problems in this genus are discussed by Sims & Hasle (1987).

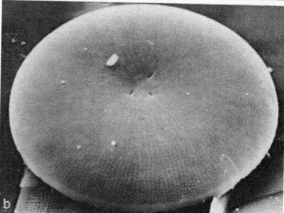

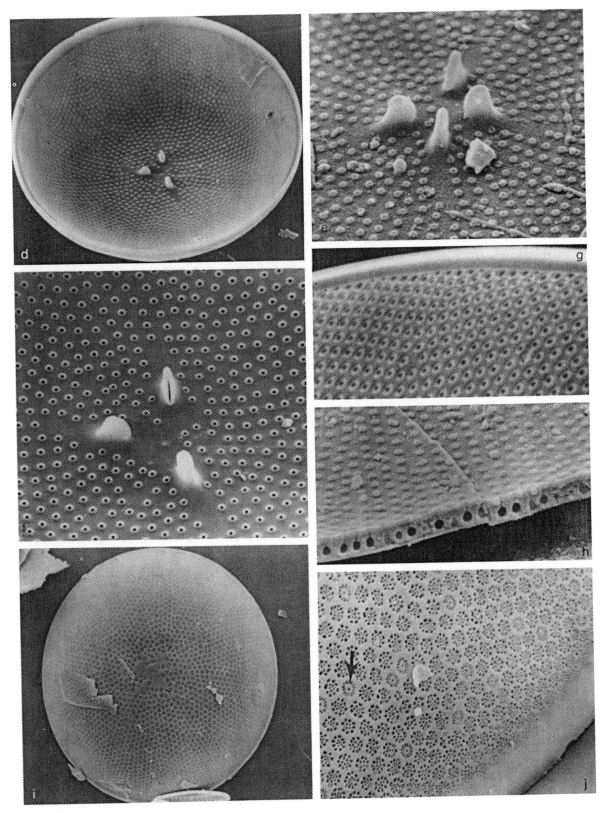

Brightwellia J. Ralfs in A. Pritchard 1861. Hist. Infus., ed. 4: 940

T: *B. coronata* (Brightwell) Ralfs
(= *Craspedodiscus coronatus*)

A fossil genus in which whole cells have not been seen. Present in sediments of Oligocene to Eocene age.

Valves round, shallow and saucer-like, with a ring of radially elongate (oval) openings externally[a,b] separating a central zone of areolae from a marginal zone. Areolae loculate, opening internally by rimmed foramina[f, i] and externally through cribra. The cribra probably have large central pores surrounded by a ring of smaller pores;[b-e] our specimens were eroded but there appear to be bars occluding the pores. The pores form curving rows, in spite of 'belonging to' different areolae. A small plain central area is present. The inner foramina are rather irregularly arranged in the centre but in radial rows in the marginal region. The large oval openings seen on the outside seem to open into chambers with a single small, round internal foramen.[h, i] The ring of chambers forms a more expanded part of the valve which is revealed internally as an annular swelling.[f] Close to the valve edge there are small pores [arrows, g] which are almost certainly eroded rimoportulae. The cingulum is unknown.

The valve structure is 'coscinodiscoid' with the addition of the ring of enlarged chambers. Hustedt's (1927-66) illustration shows hexagonal chambers. The genus also shows distinct similarities to *Craspedodiscus*, which also has a ring of enlarged areolae (see Gombos, 1982), whilst a central cluster of such chambers occurs in *Hyperion* (Gombos, 1983). These developments of the valve areolae into localised enlarged chambers seem to be a characteristic of fossil genera. Andrews (1986) has reviewed and illustrated several *Brightwellia* species, and discusses the relationships and distribution of the genus.

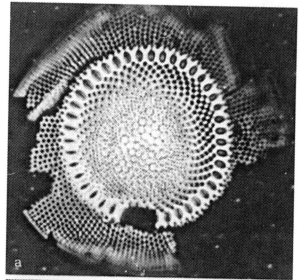

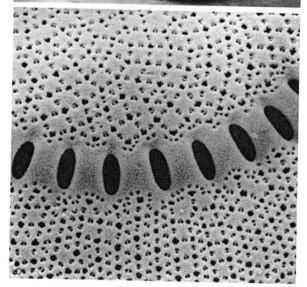

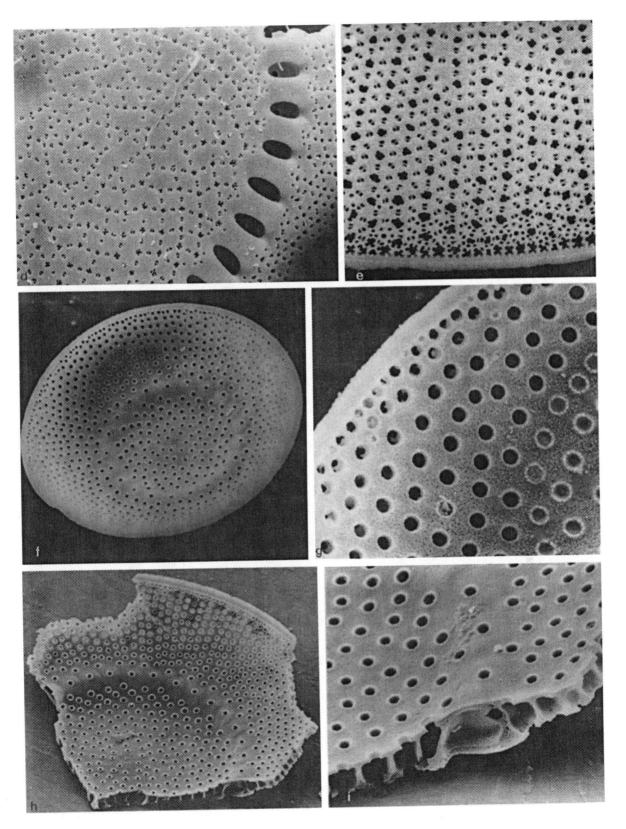

Craspedodiscus C. G. Ehrenberg 1844. Ber. Bekanntm. Verh. Königl. Preuss. Akad. Wiss. Berlin, **1844**: 261

T: *C. elegans* Ehrenberg

A small centric genus of 14 species (VanLandingham, 1969), all fossil. Cells probably discoid (girdle view not seen). Presumably marine in distribution and by analogy with modern genera, planktonic.

Valves circular, concentrically waved. Small areolae at centre becoming larger towards the periphery, arranged in eccentric intersecting whorls. Valve mantle shallow, with smaller areolae and a ring of small pores through the framework,[d] which are probably the external openings of rimoportulae. Framework between the areolae massive with small external warts which are possibly the anchoring points of an outer siliceous meshwork, fragments of which are still present on the valve mantle of some cells. One areola[arrow, c] possibly still has a velum in it – if this is real then it is of the cribrate type; the structure of the areolae would then be very similar to that in *Coscinodiscus*. Internally, repeating the external pattern, there are small central foramina, larger peripheral foramina and smaller mantle ones. A small plain area is present centrally on some valves and not others (therefore slightly heterovalvar?). Small triangular openings can be seen along the mantle internally, probably representing the eroded stalks of a ring of rimoportulae. Copulae not seen.

The genus needs considerable re-investigation. Superficially it resembles *Coscinodiscus*, but this genus is itself in need of revision, and is indeed being split up (cf. *Stellarima*, *Azpeitia*, etc.). Hence for the moment we retain it.

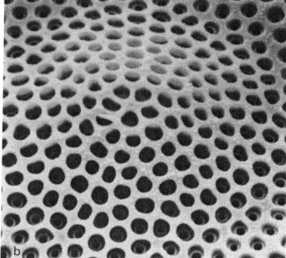

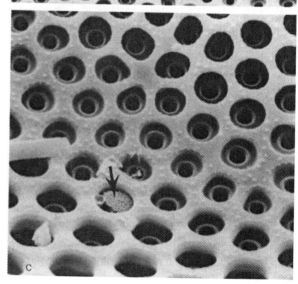

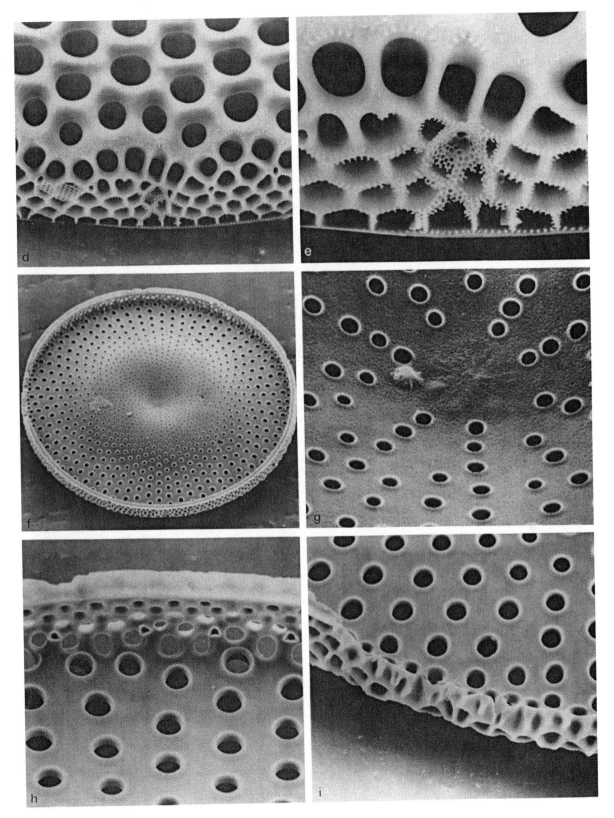

Rocella G. D. Hanna, 1930. J. Paleont., 4: 415–6

T: *R. gemma* Hanna

Cells discoid. Marine fossil, from Oligocene to Miocene. A small genus, probably planktonic.

Valves circular,[a–c] flat, with short downturned mantles. Mantle edge thick, almost septum-like, crimped in some valves.[i] Central areolae large, marginal and mantle areolae decreasing in diameter. Our material is too eroded to have vela but small spines,[d, e] around the edges of the areolae point to their existence. Internally there is a conspicuous central 'plate' of silica[f, h] with thick ribs branching and running out to join the inwardly developed mantle. A central rimoportula[f–h] opens internally on a slight protruberance and externally by means of a simple large pore.[a–c] A further pore is present internally adjacent to the rimoportula in our material. Cingulum incompletely observed but copulae apparently complete, narrow with a single row of areolae[e, f, j] and at least the valvocopula bearing an internal flange.[f]

The systematics of this genus have given us some problems and finally it was decided to allocate it to a new family. The simple central rimoportula, the internal ribbing and the structure of the valvocopula make it difficult to place this genus in any previously described family – further studies are required on this and many other fossil genera before relationships can be clarified.

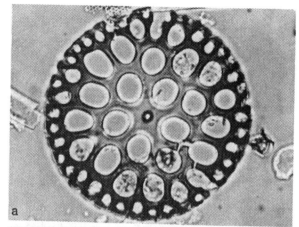

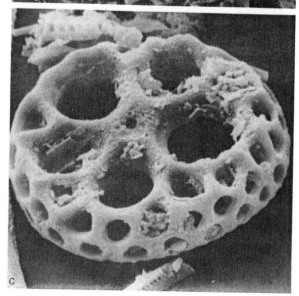

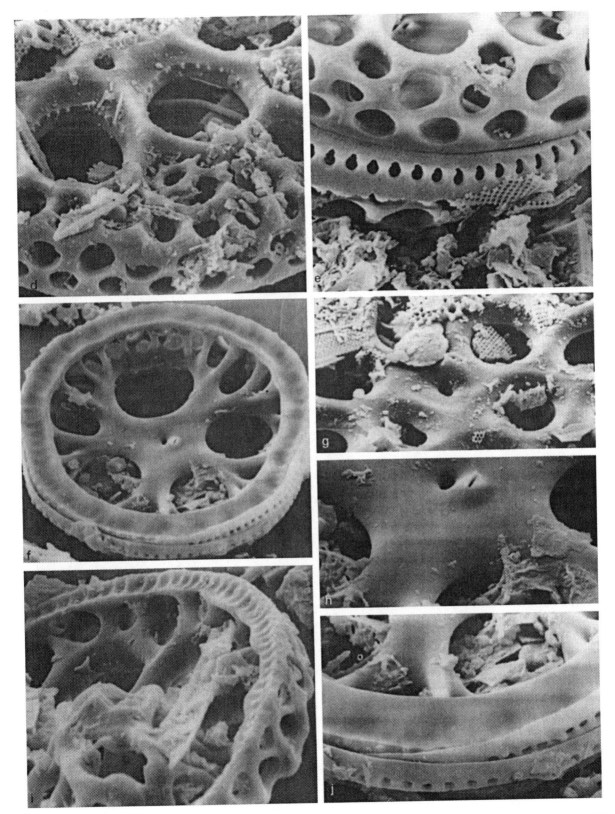

Aulacodiscus C. G. Ehrenberg 1844. Ber.
Bekanntm. Verh. Königl. Preuss. Akad. Wiss.
Berlin, 1844: 73 (nom. cons.)

T: *A. crux* Ehrenberg

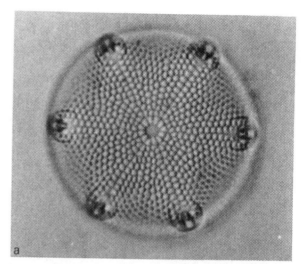

Cells solitary, circular in valve view with
conspicuous marginal processes.[a, c] Truncated
oblong in girdle view with the marginal processes
protruding.[b] Plastids not observed. A marine genus
living associated with inshore sediments and
sometimes attached to sand grains. Can form
massive growths in the surf zone, e.g. along Copalis
Beach, Oregon coast during certain years in the
1920s and 1930s (Lewin, 1974) and along California
beaches recently (Holmes & Mahood, 1980). A very
large, mainly fossil genus going back to the
Cretaceous and exhibiting considerable diversity of
valve and rimoportula form.

Valve with a conspicuous hyaline area at the
centre from which rows of loculate areolae radiate.[c]
The valve surface is often corrugated and the valve
mantle is not differentiated from the rest of the
valve. The areolae are closed by cribra[f, g] externally
and open internally by large foramina;[g, h] we have
some evidence that there is an internal occlusion of
some kind in some species. Rimoportulae marginal,
complex, opening externally either via a tube[c] or
onto a cap extending towards the valve centre.[b, d, e]
In capped forms, the processes often have 'fork-like'
slits. The rimoportula is unique. The tube passes
through the valve and ends internally in a curved
slit, almost flush with the internal layer of the wall.[e, h]
This slit is either E-shaped or C-shaped and may be
oriented with the opening of the E or C pointing to
the centre or tangentially. Sometimes clear hyaline
rays extend from the centre to the rimoportulae,
these being best seen on the inside of valves.[e]
Copulae numerous,[i] split,[j] ligulate and antiligulate,
fimbriate (at least on the valvocopula), with vertical
rows of small areolae.

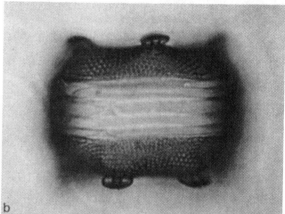

Reviewed by Rattray (1888a) and Ross & Sims
(1970) and studied intensively by Burke &
Woodward (1963-74). The structure of *A. amherstia* is
discussed by Venkateswarlu & Round (1973) and by
Sims & Holmes (1983) (who regard it as *A. johnsonii*
var. *amherstia*) and *A. kittonii* by Holmes & Mahood
(1980). Both Holmes & Mahood and Ross & Sims
(1970) consider that the morphology of the
rimoportula is a useful specific taxonomic character.

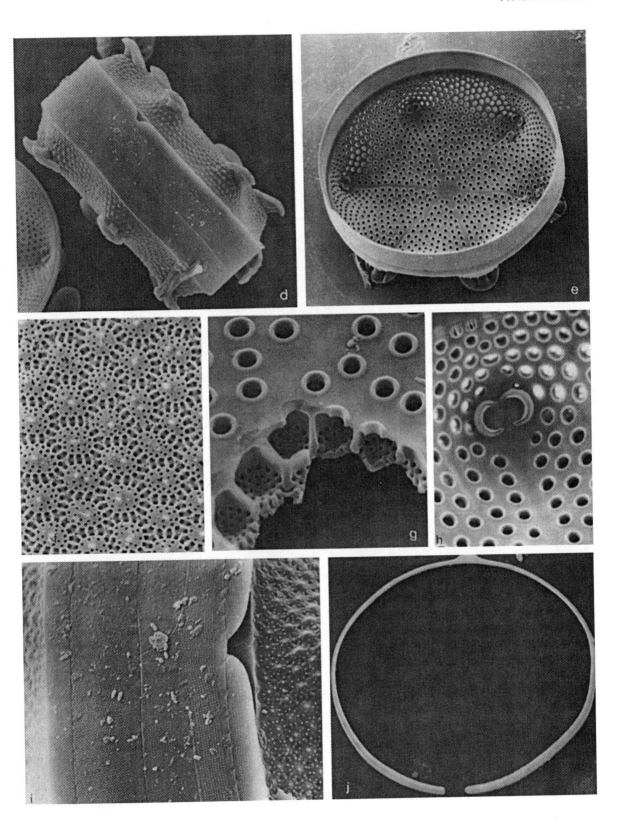

Gossleriella F. Schütt 1893. Pflanzenleben der Hochsee: 20

T: *G. tropica* Schütt

Cells solitary, discoidal with a ring of closely spaced, thick and thin, radiating spines;[a] the spines are all of similar length (approximately half the cell's diameter). Plastids numerous, small platelets. A rather rare tropical marine planktonic genus (see Hargraves, 1976, for some distributional data).

Valves circular. Valve face flat; mantle shallow and plain. Areolae in simple radiating rows, visible only at high (SEM) magnification.[b, c] According to Hargraves (1976) the mantle region is somewhat loculate. Internal vela are present but their form has not yet been determined; internally the areolar openings appear as raised rings[c] but there is probably further detail to be determined. A single rimoportula is present near the centre of the valve.[d] The copulae are more distinctly areolate[f, g, i–k] than the valve; the areolae are in precise quadrate arrangement, each with a central bar (rota).[k] The valvocopula attached to the epivalve bears an external ring[a, b, e–j] which is fimbriate on the abvalvar edge and has deep flask-shaped slits on the advalvar edge. At intervals prominent spines radiate from the abvalvar region of the valvocopula: these are mostly single but at intervals two spines fuse to form a thicker spine.[e–g] According to Hargraves the ring of spines is not fused with the valvocopula but may have cytoplasmic connections (more likely mucilage connections). There is variation in this ring structure in our samples from different oceans and the final details need to be worked out.

In some cells there is also a remarkable internal set of spines.[arrow, b] It is not clear how these are formed and we know of no other such internal elaboration in centric genera. However, it is possible that this is a partially formed ring of spines prior to liberation from the cytoplasm; this would be possibly only if the structure is initially flexible.

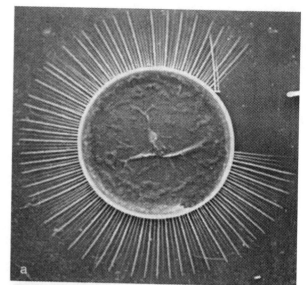

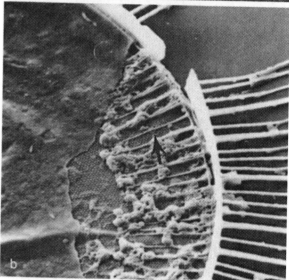

Hemidiscus G. C. Wallich 1860. Trans. Microsc. Soc. Lond. Ser. 2, **8**: 42

T: *H. cuneiformis* Wallich

Cells shaped like orange segments, the valve faces lying at an acute angle to each other.[b, h] Plastids small, discoid. The genus is sometimes recorded in lists of plankton under the incorrect name *Euodia*. A marine planktonic genus of widespread distribution, particularly in the warmer oceans but carried in warm currents to temperate seas.

Valves cuneiform,[a, h] often with a central swelling on the ventral side. Valve face flat, mantle shallow and distinguished by its smaller areolae.[b–f] Valve areolae radiating from an indistinct central annulus; pattern somewhat irregularly sectored (fasciculate). Areolae loculate,[g] opening externally through cribra[c] sunk below the valve surface; internally with round, rimmed foramina.[d–g] Breaks in the valve show the internal expansion of the areolae and a slight bullulate structure within the walls.[g] Rimoportulae confined to a row along the ventral margin. Internally they are shortly stalked auricular structures,[f] the slits being more-or-less parallel to the valve margin. The external openings of the rimoportulae are expanded parallel to the plain valve margin. A single pseudonodulus occurs, slightly inward of the line of rimoportulae.[c, h] The pseudonodulus seems to open by means of a simple aperture.[h, i] Copulae wider on dorsal than ventral side, split, and ligulate,[h, i] the splits occurring a little way from the apices. Substructure of the copulae not known but they appear to be solid.
they appear to be solid.

The structure of *Hemidiscus* is very similar to that of *Actinocyclus* and some workers include it within that genus but we propose to keep it separate until further studies can be made. See also Simonsen (1972), who discusses this and the unrelated genus *Palmeria*. The ventral row of spinulae referred to in the literature are almost certainly the rimoportulae and not external ornamentation of the valve.

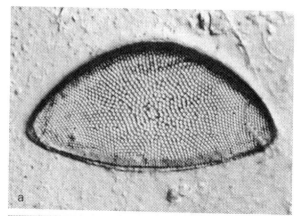

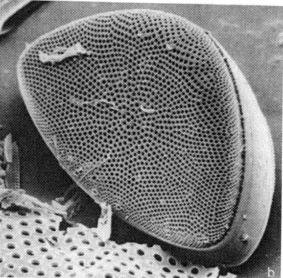

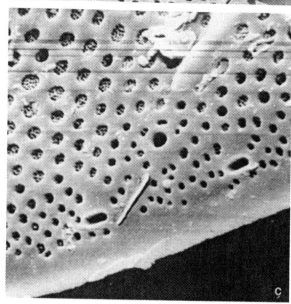

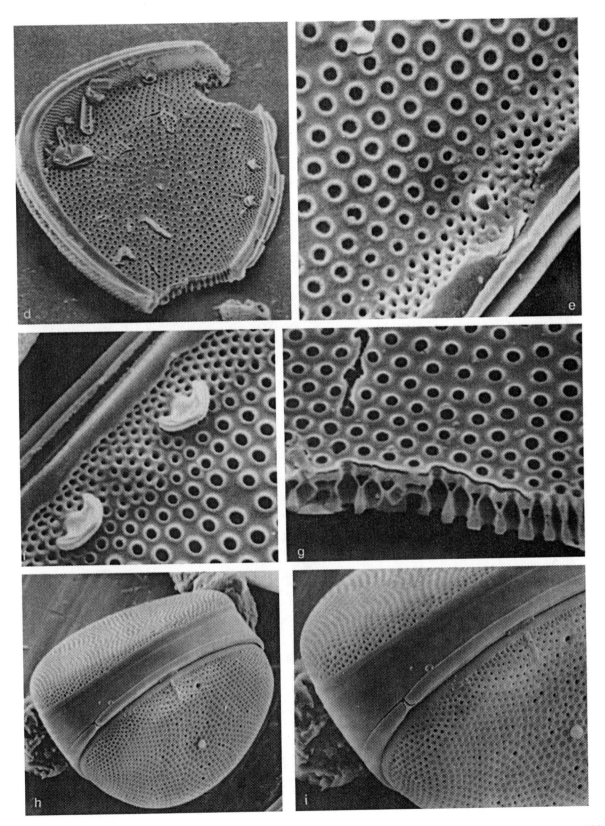

Actinocyclus C. G. Ehrenberg 1837. Ber. Bekanntm. Verh. Königl. Preuss. Akad. Wiss. Berlin, **2**: 61

T: *A. octonarius* Ehrenberg.

Cells barrel-shaped,[b] probably mainly epiphytic on seaweeds but often encountered in nearshore plankton. Plastids numerous, discoid.[a]

Valves circular (rarely elliptical or triangular) with radial sectoring;[b] valve face planar or concentrically waved. Surface corrugate [b, c] with tiny external pores (too small to be seen by SEM) and a single marginal 'pseudonodulus';[c, f] sometimes with a distinct rim at the junction of face and mantle.[c] Mantle often distinct with large, simple rimoportula openings.[c, d, f] Some species with stepped mantles. Valve framework bullulate.[i-k] Internally, areolae seen to be arranged in variously organised fascicles with domed closing vela[e] and conspicuously expanded rimoportulae usually angled to the valve rim.[e, g] Pseudonodulus in a clear area. Copulae open. Valvocopula massive, fimbriate, underlapping a shelf projecting from the valve rim. Other copulae (2 or more) smaller.

Rattray (1890) was the first to discuss the genus in detail and the status of the pseudododulus has been re-examined recently by Simonsen (1975). *f* shows a closing plate with an unperforate outer membrane and *g* the inside view suggesting this is the 'operculate' type of Simonsen. *h* on the other hand shows the inner view of Simonsen's 'areolate' form. Thin sections confirm that the pseudonodulus is not perforate. The function of the pseudonodulus is not known and there is clearly much still to be learnt about the range of structure. Simonsen uses it as a criterion to distinguish the family Hemidiscaceae (*Actinocyclus*, *Hemidiscus* and *Roperia*). A freshwater species *A. normanii* has been recorded under various names during the last century (see Hasle, 1977) and caused much confusion. Hasle has shown clearly that it is truly an *Actinocyclus*. It is one of several basically marine algae which are becoming established in freshwaters, probably due to increasing eutrophication. Several recent publications have dealt with *Actinocyclus* species (e.g. Fryxell & Semina, 1981; Watkins & Fryxell, 1986). The bullulate wall is likely to prove typical of this genus and species without this feature may belong elsewhere (Andersen *et al.*, 1980).

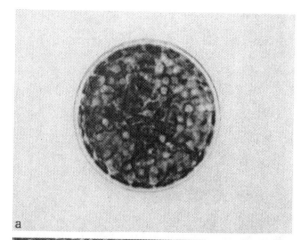

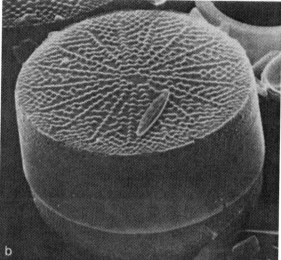

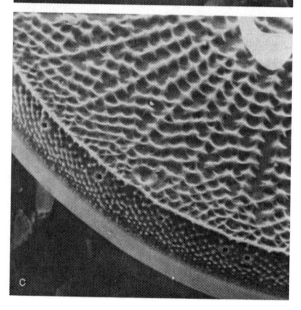

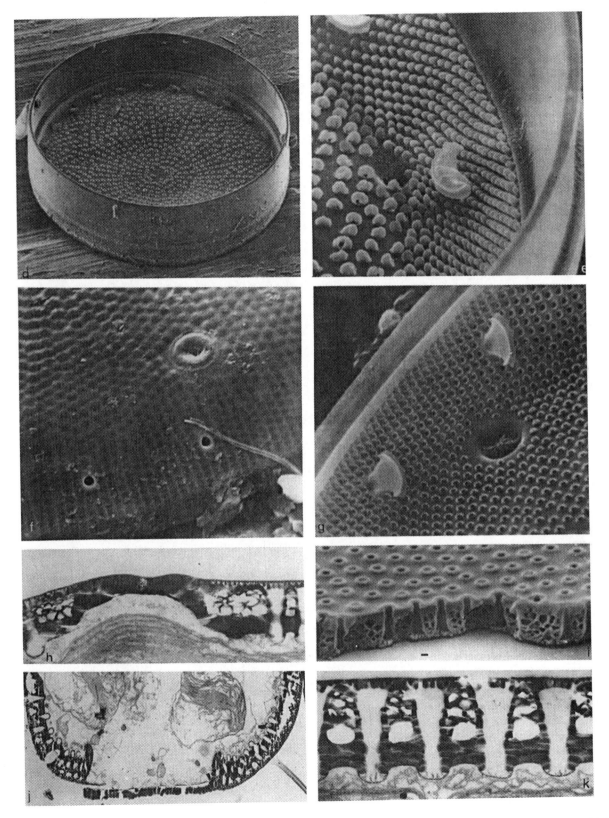

Azpeitia M. Peragallo, 1912. In Tempère & H. Peragallo, Diat. Monde Entier, ed. 2: 326

T: *A. obscurum* (Greville) Sims in Williams (= *Triceratium obscurum*)

Cells discoidal, solitary. A genus of species formerly included in *Coscinodiscus* but separated on the basis of possessing a central 'nodule'.[b, c] Plastids not observed. Marine, planktonic; modern species tend to be tropical/subtropical in distribution and many fossil species are recorded, also from areas believed to have been tropical.

Valves circular (triangular and multiangular in some fossil forms), flat, with a shallow but distinct valve mantle and often with a conspicuous raised central tube;[h] immediately adjacent to this is a depression, into which the external tube of the sibling valve fits.[c, f, h] The tube is the external extension of the central rimoportula. A distinct annulus is present in some fossil species (P.A. Sims, personal communication). Areolae in radiating rows, loculate. The inner openings are round foramina with raised rims, both towards the valve interior and towards the loculus itself;[i] externally there are depressed cribra.[f-h] The edge of the valve face is modified to form a solid ring of silica, within or below which the rimoportula openings are located.[b, d-g] In *A. africanus* each opening extends into a circumferential slit.[b] The areolae continue beneath the ring and tend to change shape, becoming diamond-shaped, often with one point truncated.[f, g] Rimoportulae shortly stalked; one central rimoportula[b, c] and others in a ring around the valve face/mantle junction.[d] Copulae apparently plain.[g]

The structure of the valve clearly places this genus in the Coscinodiscaceae but the central structure and distinct valve rim make a separate genus preferable. When the remainder of the species referred to *Coscinodiscus* are examined it is likely that others will be transferred to *Azpeitia*; we have observed several as yet unidentified species with conspicuous central rimoportulae and complex marginal thickenings. Certainly the common '*Coscinodiscus*' *nodulifer* has to be transferred to *Azpeitia* and these are illustrated in Figs f–i. '*C*'. *furcatus* also belongs here. Many species are important indicator species in the fossil record (Eocene to Recent). Fryxell *et al.* (1986) place this genus in the Hemidiscaceae, since some of the fossil species have a pseudonodulus.

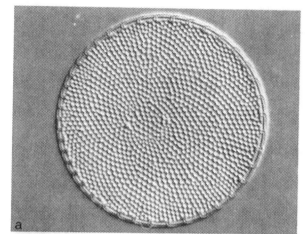

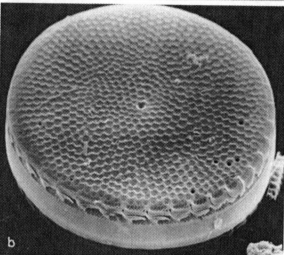

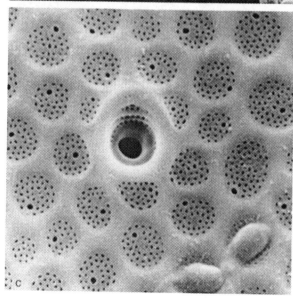

196

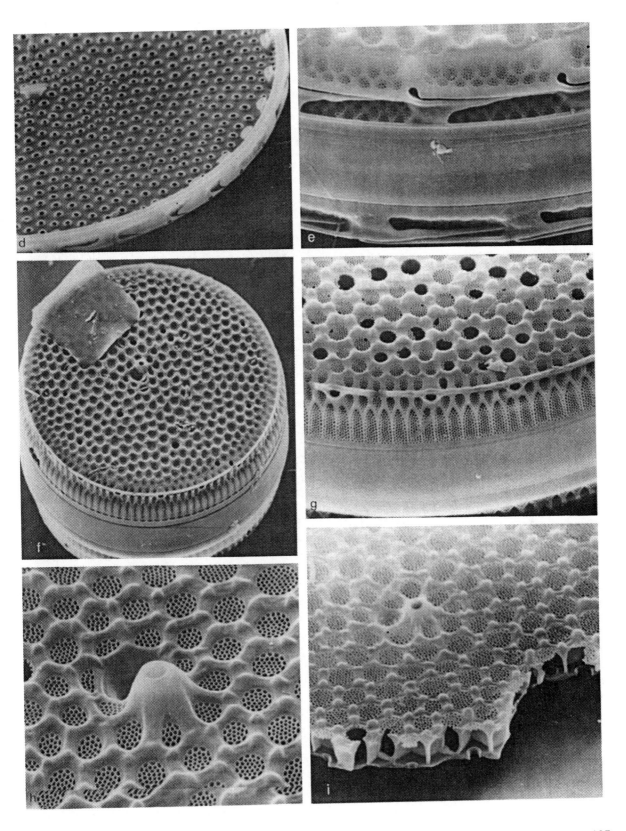

Roperia A. Grunow ex J. Pelletan 1889. Les
Diatomées, **2**: 158

T: *R. tessellata* (Roper) Grunow (= *Eupodiscus
tessellatus*)

Cells discoid, solitary. Plastids numerous, discoid. A
small marine planktonic genus.

Valves circular,[a, e] with a flat valve face and
distinct mantles. A large opening (pseudonodulus)[a-f]
occurs near the valve rim; it appears open in our
preparations for the EM, but it is most likely
occluded by a silica membrane or an organic wall
in the live cell. Areolae loculate, arranged in slightly
curved decussate rows. The loculi open externally
via cribra which are almost flush with the outer
surface of the valve;[b, c, k] internally they have large
round foramina.[e-h] The areolae are circular on the
valve face, becoming diamond-shaped[b, c, h, j] or
merely continuous areas of pores on the mantle.
Rimoportulae occur in a ring around the mantle.[g, h]
Their external openings are slit-like,[b, c] angled[i, k] and
occur in a ring at the junction of the valve face and
mantle. At this point there is either a rim or a series
of flaps above the rimoportula openings that are
often angled in relation to the mantle edge; some
valves seem to combine these two features.
Internally the rimoportulae are stalked and may be
tilted at an angle to the mantle margin. Girdle
bands not investigated in detail but appear wide,
split and ligulate.

The genus is small: only *R. tessellata* is recorded
commonly, most other records being mis-
identifications of *Eupodiscus* (see VanLandingham,
1979 who reports only one other species, *R.
marginata*). The combination of the pseudonodulus
and the silica flaps around the edge of the valve
face enable this taxon to be distinguished from
Coscinodiscus. A pseudonodulus also occurs in
Actinocyclus but that genus has bullulate valves and
simple openings to the rimoportulae. The
arrangement of the rimoportulae around the margin
of the valve is in some ways reminiscent of some
species of *Azpeitia*.

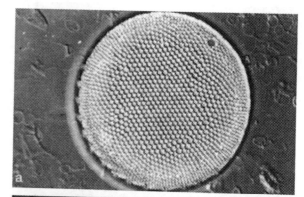

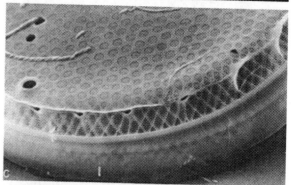

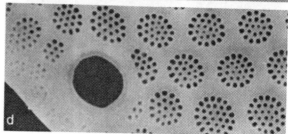

Actinoptychus C. G. Ehrenberg 1843. Abh. Königl. Akad. Wiss. Berlin, **1841** (1): 400, 437

T: *A. senarius* Ehrenberg

Cells discoid, solitary.[a] Very common in neritic collections, probably mainly a component of the assemblage lying loose or attached to other algae on coastal sediments. Plastids several, irregular plates. Many fossil species.

Valves sectored[a-c] (6 in the common *A. senarius* [= *A. undulatus*] but up to 20 in others) so that alternate sectors are elevated or depressed. Central area plain[g, h] or granulate.[f] Areolae in radiate striae, opening by simple pores to the outside but over much of the surface the silica is corrugated or pitted, delimiting groups of areolae.[d, e] Internally, uneroded areolae are closed by domed vela. External tubes of rimoportulae prominent,[d] usually located at distal points on radii of elevated sectors.[c] Depressions (termed pseudopores by Andrews, 1979) occur at the corresponding position on the depressed sectors.[arrow, e] The internal openings of the rimoportulae tend to lie at right angles to the valve margin[i] and are either straight or curved. Valve framework bullulate. Margin of valve face often produced into a thickened rim or marked with special ornamentation below which the distinctively patterned valve mantle extends. The valve mantle often has spines, wart-like outgrowths, siliceous ridges, etc., and the edge is produced into a smooth flange. Copulae plain, split and wide.

Andrews (1979) shows that the arrangement of rimoportulae can vary, with more than one in each raised sector, and that they can also occur on the edge of the depressed sectors, at least in *A. heliopelta*. A genus with numerous species and varieties (VanLandingham records 150 possible valid species) many of which require re-investigation. Its ecology also requires further study. The primary valves were formerly described as the genus *Debya* Pantoscek.

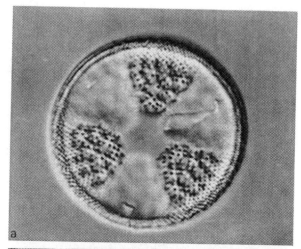

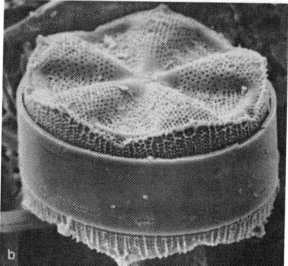

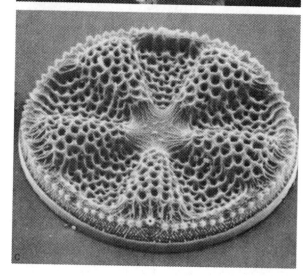

Glorioptychus G. D. Hanna, 1927. Occas. Pap.
Calif. Acad. Sci. 13: 19

T: *G. callidus* Hanna

Cells probably discoidal but only valves seen. A
marine, monotypic fossil genus described from the
Upper Cretaceous of California.

Valves circular[a] in face view with a clear central
area surrounded by an inner series of three raised
and three lowered sectors, and then, around this, an
outer ring of nine raised and nine lowered sectors.
The morphology is thus based on a tri-radiate
system. Valve face areolate,[b, c] with irregular solid
ridges forming a meshwork around groups of
areolae; at the junctions of the ridges the silica is
raised, which may indicate a more 'spiny' surface of
the valves prior to fossilisation. The valve face
terminates in a continuous low ridge[b, c] separating
off a shallow valve mantle in which the areolae run
in vertical rows. The low ridge undulates somewhat,
rising up along the raised sectors of the valve face.
Internally the openings of the areolae are simple
and radiate in rows from the central area.[d-f] Three
of the outer raised sectors occur on the same radii
as the raised sectors of the inner ring and on these
outer sectors the wide slit-like openings of the
rimoportulae are to be found.[g] The areas around the
slits are verrucose. The rimoportulae open internally
by narrow snake-like slits[g] orientated at right angles
to the valve margin.

This genus is clearly allied to *Actinoptychus*, as
Hanna (1927) believed from his LM observations on
the structure and sectoring of the valve. Our
illustrations are almost certainly of the same species
as that studied by Hanna. We thank G. W. Andrews
for material of this and the following genus.

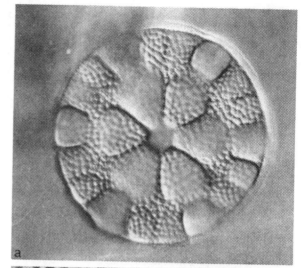

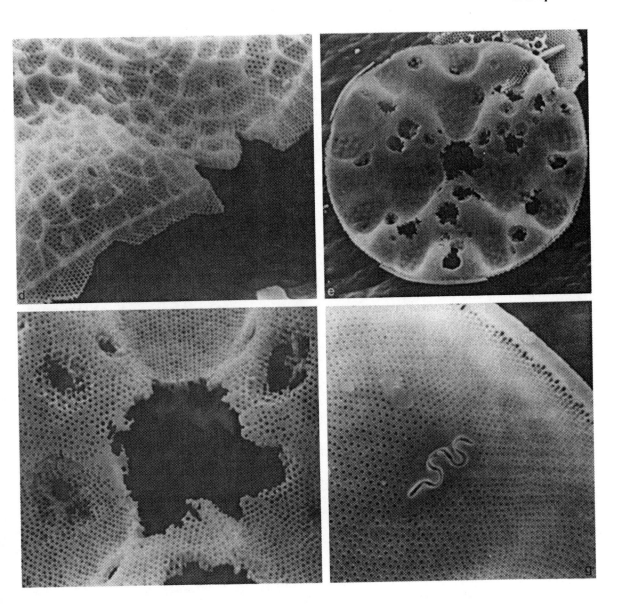

Lepidodiscus O. N. Witt 1886. Verh. Russ.-Kais.
Mineral Ges. St. Petersburg, Ser. 2, **22**: 163

T: *L. elegans* Witt.

Cells probably discoid, but only valves seen.[a] A
small genus of fossils (7 spp. according to
VanLandingham, 1971), most occurring in Eocene
sediments of marine origin.

Valves circular,[a, d] with concentric undulations of
the valve face and in addition a ring of prominent
alternating elevations and hollows around the
circumference. The system of elevations and hollows
is relatively slight in our material; other species have
very pronounced external ridges and furrows (P. A.
Sims, personal communication) and this results in
very dissimilar sibling valves. Areolae in fascicles
(striae) radiating from a central annulus, within
which the areolae are randomly scattered; the
fascicles increase in width towards the margins[e] and
then split up again in the undulate marginal region.[g]
The areolae are circular and appear to be simple
passages through the massive framework of the
valve.[c] Valve mantle short and areolate; on a ridge
between it and the valve face there appears to be a
series of openings, each surrounded by paired flaps
of silica;[b] these are presumably the openings of the
rimoportulae: a similar arrangement has been seen
in several *Actinoptychus* species. Internally the valve
surface is plain; the plain strips between the
fascicles become 'buttresses' around the margin,
connecting the valve face to the edge of the
mantle.[d-g] Within the depressions between the
buttresses the rimoportulae can be seen, one per
depression, their apertures orientated radially.
Copulae unknown.

A little known genus which requires further study,
especially of the valve mantle and of material with
copulae attached. From these admittedly
preliminary observations the genus seems close to
Actinoptychus.

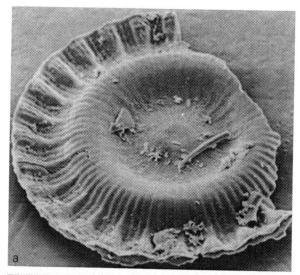

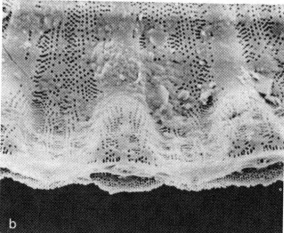

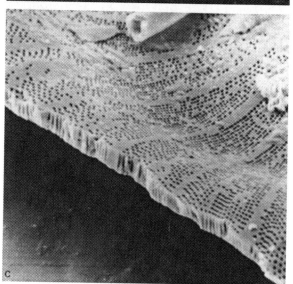

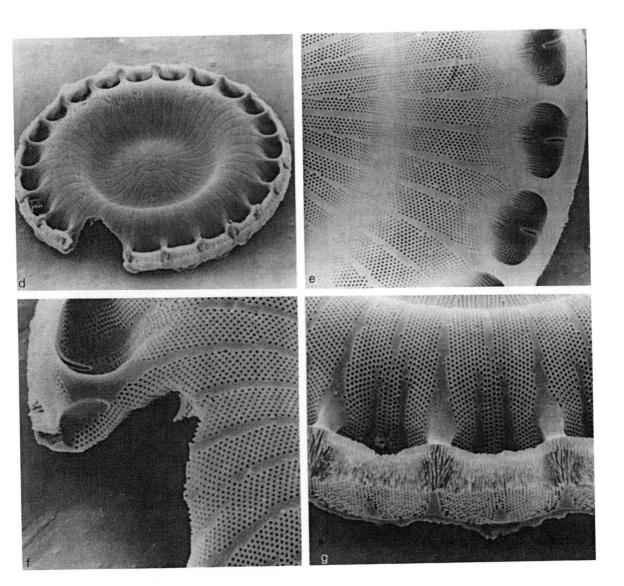

Ethmodiscus F. Castracane 1886. Rep. Sci. Res.
Voy. 'Challenger', Bot. 2: 166

LT: *E. gigas* Castracane, Rep. Sci. Res. Voy.
'Challenger', Bot. **2**, p. 169, pl. XIV, Fig. 5
(lectotype selected by F. E. Round)

Cells drum-shaped with flat[a] or domed[b] valves.
Free-living and the largest celled of any centric
form, attaining 2–3 mm in diameter. Plastids not
observed. A tropical marine planktonic genus, but
whole cells rarely observed. Fragmented material
forms large sedimentary deposits in tropical seas.

Valves flat or domed, with a plain central area[c]
from which radiate rows of simple loculate areolae.[m]
These are interrupted by a plain ridge at the edge of
the valve face[d, e] and then continue down the valve
mantle in vertical rows. The areolae have inner and
outer vela with fine pores.[g] There is considerable
variation in valve structure – some valves have
rows of 'bottle-like' projections,[e, f] some have no
annulus, or marginal plain ridge. Relating these
variants to cells which are also possibly heterovalvar
(at least four slightly different valve types have been
found) and usually fragmented, has proved difficult
(see Round, 1980). Rimoportulae occur on both
valve face and mantle. External slit-like openings
of rimoportulae occur in a ring[l] around the annulus
or scattered in the centre. Other rimoportulae can be
found around the edge of the valve face[j] and, on the
mantle near the valve margin,[i] there are
rimoportulae with small circular pores. Internally
the central rimoportulae are sessile or slightly
stalked and curved;[k] those around the valve face
edge are small, no larger than the areolae, those at
the mantle edge are slightly larger and occur
amongst the rows of areolae. Copulae split,
areolate.[n]

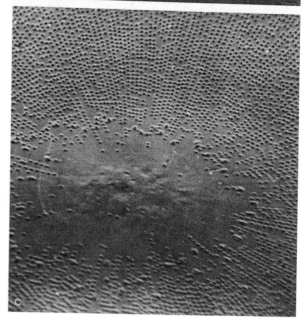

This genus needs to be investigated by
observation of whole cells,[*] preferably from culture.
It is an unusual genus in that three different kinds
of rimoportulae occur on a single valve – a feature
we have not noted in any other genus. Samples
collected in nets are often fragmented owing to
pressure on the fine bubble-like cells. Further
studies are also required on its ecology. The genus
has been confused with *Coscinodiscus* in early
literature and also placed in a genus *Antelminellia*. It
is clearly not close to *Coscinodiscus* and we separate
it into its own order and family.

[*]A paper including such observations is in press (Rivera *et al.*, 1989).

Stictocyclus A. Mann 1925. Bull. U.S. Natl. Mus., **100** (6): 146

T: *S. stictodiscus* (Grunow) Ross (= *Actinocyclus stictodiscus*; = *S. varicus* A. Mann, nom. illeg.)

Cells cylindrical, lying with the multiple copulae visible; attached loosely by the edges of the valve faces to form filaments. Plastids discoid, lobed. Recorded from tropical regions, but only rarely; epiphytic. The one known species has sometimes been put in *Actinocyclus* or *Ethmodiscus*, but was confirmed in *Stictocyclus* (as *S. stictodiscus*) by Ross & Sims (1973).

Valves petri-dish like,[d] with prominent radial ribs, which become indistinct towards the centre. The ribs become thicker and buttress-like at the angle between valve face and mantle and thin out in the centre and along the valve mantle. Valve face flat,[a-c] sharply differentiated from the deep, vertical mantles. Note: Fig. a has a central segment of the sibling valve lying on the complete valve. Areolae arranged in radial rows, several occurring between each pair of ribs. Towards the centre, where the ribs disappear, the areolae are replaced with scattered puncta, which in SEM are seen to be elongate openings of the rimoportulae. A large pseudonodulus occurs on one side[c] and is closed by a fine plate sunken below a slight rim. Areolae somewhat irregular in outline, with small inwardly projecting pegs supporting a velum. The rimoportulae are short and curved internally[g, h] and open by elongate slits externally;[b] they are scattered over the central area and also extend slightly amongst the rows of areolae, where they penetrate the siliceous framework rather than replacing the areolae. Copulae complete, numerous, areolate as in the valves, with long fimbriae on their advalvar edges.[i]

This genus was studied by Ricard (1970) and Ricard & Gasse (1972), from material collected in Tahiti, and its basic structure described. Round (1978) reported on the SEM structure of material collected in Hawaii and discussed its classification, placing it in a new family, the Stictocyclaceae. Its relationship to other families is still not clear since the presence of a pseudonodulus removes it from the Coscinodiscales and points to a position near the Eupodiscales. However, this order does not possess central rimoportulae and has valves that tend to be loculate (see Round, 1978a, for further discussion).

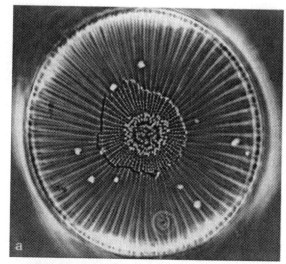

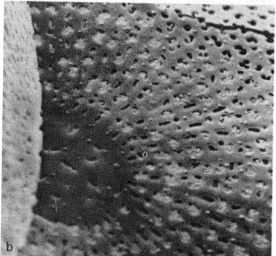

Asterolampra C. G. Ehrenberg 1844. Ber. Bekanntm. Verh. Königl. Preuss. Akad. Wiss. Berlin, 1844: 73

T: *A. marylandica* Ehrenberg

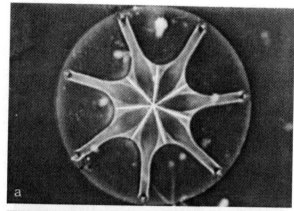

Cells discoid, solitary.[a] Plastids reported as numerous discoid bodies. A marine, planktonic genus of about 30 species; mainly tropical.

Valves watch-glass (saucer-) shaped,[b] without a distinct valve mantle but with a radiating system of raised rays (Gombos, 1980) which expand to touch each other in the centre (modern species) or stop short of the centre, which is then areolate (fossil species). The elevations are unperforate. At the outer ends of the rays there is a fine membrane[d] which is usually removed by the cleaning procedure, leaving an aperture (ray hole).[b, e] Immediately below this on the outer rim is the opening of a rimoportula.[e] Between the rays the valve is areolate,[c] each areola being loculate and opening to the outside by simple pores which form a system of obliquely crossing rows.[c-e] The areolae usually appear to open internally by simple foramina[f-i] but we have seen specimens with an inner 'closing' membrane (cf. *Actinocyclus*). Internally the rays are seen to be the plain bases of open chambers the walls of which rise to the centre and sometimes curve over to form a partial roofing.[f, g] It is the walls of these chambers which form the sets of radiating curved lines seen in the light microscope. A kidney-shaped rimoportula lies almost flat on the valve at the outer end of each internal chamber.[h, i] Adjacent to the chambers the foramina of the areolae are elongate, giving a 'ribbed' appearance.[g, h] The two valves of any one cell are displaced half a sector relative to one another, so that the cell viewed from above has the radiating elevations alternating and not superimposed. Copulae not investigated.

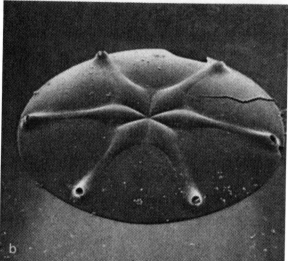

Asterolampra rarely occurs in large populations and could only be confused with *Asteromphalus*; this, however, always has one ray narrower than the others and is thus not symmetrical. The use of the term 'pseudonodulus' for the 'ray hole' (Fryxell & Hasle, 1974) may be acceptable, though Simonsen (1975) disagrees. Its structure is similar to that in *Actinocyclus* (where, however, there is only a single pseudonodule). The fossil genera *Discodiscus*, *Bergenia* and *Rylandsia* are also characterised by rays (see Gombos, 1980, for details).

Asteromphalus C. G. Ehrenberg 1844. Ber. Bekanntm. Verh. Königl. Preuss. Akad. Wiss Berlin, **1844**: 198

T: *A. darwinii* Ehrenberg (lectotype selected by C. S. Boyer 1927, Proc. Acad. Nat. Sci. Philad. 78 Suppl.: 72

Cells discoid to slightly pear-shaped, rarely naviculoid, undulate in girdle view. Plastids reported as numerous discoid bodies. Marine, planktonic in warm waters but never occurring in large quantity: about 35 species.

Valves shaped like watch-glasses,[b] without a distinct valve mantle; some species slightly extended along one radius to give an obovate form. The valve has very distinctive plain rays (see Gombos, 1980, for terminology) alternating with areolate areas. The broad inner parts of the plain sectors abut at the centre or off-centre on the valve face; one ray differs from the others (the median ray of Greville, 1860) having a thinner distal section[a] and forming a focus from which the others radiate. At the marginal (outer) end of each ray is a thin plate which is usually lost leaving a large aperture (ray hole);[b, c] below this is a smaller opening, representing the external aperture of a rimoportula. This is absent on the median ray but here the thin plate is missing, the outer part of the process is morphologically different and the rimoportula must open into the ray itself for a process can be seen on the inside.[d, f, g] The rays are long chambers opening internally by ray slits which taper towards the margin. The valve edge is plain usually with a notch offset from the end of the thin ray.[arrow, d] The areolae are loculate. Externally their apertures are variable, though basically cribrate.[e, h-j] Internally they have simple foramina.[d, f, g] Flattened kidney-shaped rimoportulae occur marginally, opposite the ends of the ray slits;[d, f, g] the rimoportula opposite the narrow ray slit is often larger. In some species small rimoportulae also occur near the central plain area.[f] Copulae not investigated.

This genus and *Asterolampra* are clearly closely related and both are found in tropical oceans. The rare 'naviculoid' forms were described by Kolbe (1955) as *Liriogramma* but are only boat-shaped species of *Asteromphalus* – they occur in the Gulf of California (Round, unpublished observation).

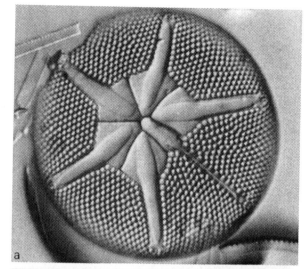

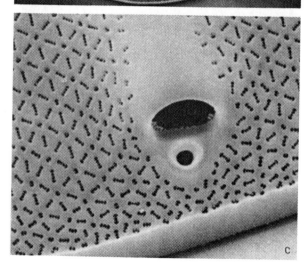

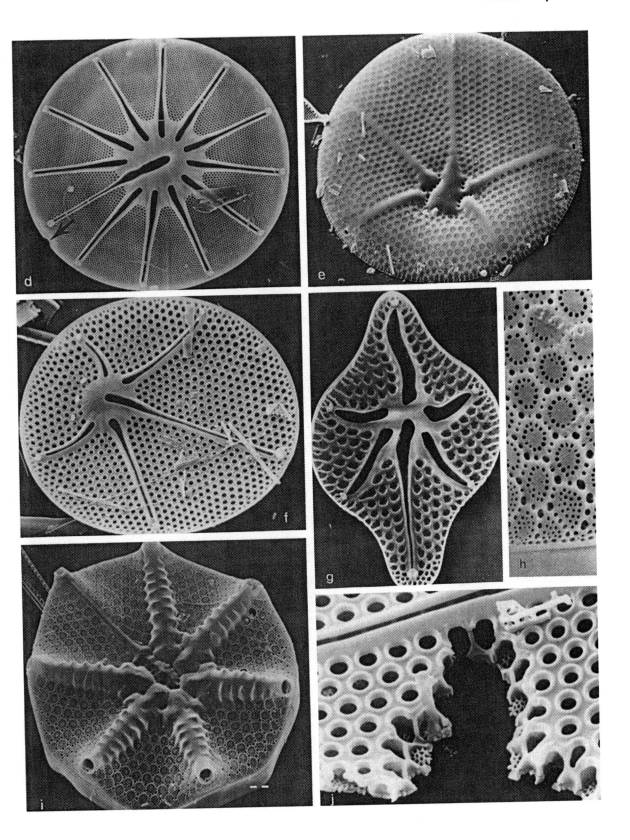

Arachnoidiscus H. Deane ex A. Pritchard 1852.
Hist. Infus., ed. 3: 318 (nom. cons.)

T: *A. japonicus* Shadbolt ex Pritchard

Cells discoid, epiphytic on seaweeds. No
information on plastids or other aspects of the live
cell. Also fossil, extending back to the Palaeocene.

Frustules heterovalvar; valves petri-dish shaped.[a, b]
One valve with a plain central area,[c] the other with
the plain centre ringed by elongated radial slits,[b, d, e]
which are the openings of rimoportulae. Striae
radiating, some short, broken by a plain region at
valve rim. Pattern somewhat sectored[a, b] (especially
obvious in light and transmission electron microscopy). One valve,which may be the plain-
centred or the radial-slit type, has a distinct step in
the mantle.[c] Areolae occluded by volae,[e-g] in
depressions of a massive costate system. Valve
mantle areolae smaller and without surrounding
costae. Fine diagonally crossed rows of 'pin-holes',
which do not penetrate to the inside, are sometimes
observable at the junctions of the external
costae.[arrows, f, g] Internally, massive radial and
tangential ribbing occurs,[h-k] superimposed on which
is a system of costae radiating from a flange around
the central ring, or from intermediate points, to a
flange on the valve mantle edge.[k] The internal and
external flanged portions have been termed the
umbilicus by Helmcke & Krieger (1953–77, pl. 321)
but this does not seem an appropriate term though
their reconstruction is fairly accurate. One valve has
a ring of slits internally in the depressed centre,[h, i]
which may be interpreted as rimoportulae, whilst
the other valve has a series of openings in its
sunken centre,[j] which do not seem to penetrate
through to the outside.[c] The cingulum appears plain
but requires further study.

The genus is not often encountered, occurring
mainly on seaweeds, especially on the tropical
coasts around the Pacific, where it sometimes occurs
in such abundance that the dried seaweed appears
covered with a white powder. Hustedt (1927–66)
doubted its occurrence in European waters and it
may be virtually confined to the Pacific. Several
fossil species are recorded. We assume that the ring
of slits around the centre of one valve are
rimoportulae though they do not have very typical
form. However, P. A. Sims (personal
communication) confirms this with excellent
illustrations of rimoportulae in *A.* cf. *russicus*.
Stictodiscus is possibly the only near relative.

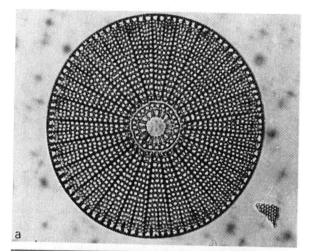

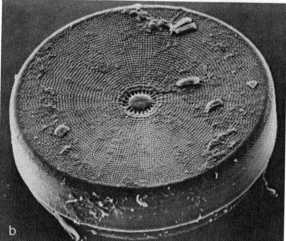

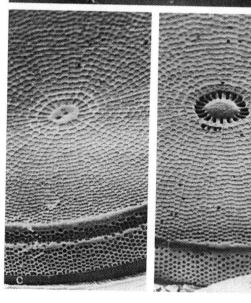

Stictodiscus R. K. Greville 1861. Trans. Microsc. Soc. Lond. Ser. 2, **9**: 39

LT: *S. rota* (Ehrenberg) Greville (= *Discoplea rota*; lectotype selected by C. S. Boyer 1927, Proc. Acad. Nat. Sci. Philad. 78 Suppl.: 69)

Cells discoid.[a-c] Plastids not observed. This is a genus known mainly from fossil marine material. Hustedt (1930) comments on its scarcity and occurrence in deep water habitats.

Valves circular,[a, b] triangular[a, c] to multiangular. Valve face flat or slightly bowed; mantles distinct but shallow. Areolae in rows radiating from the centre, although often the central area itself is almost devoid of areolae.[c] Raised siliceous thickenings often surround the widely spaced inner areolae externally,[b, d] while towards the edge of the valve face there can be conspicuous radial thickenings and less prominent tangential ones separating the outer files;[b, f] a distinct network is visible in the LM. The valve mantle has a similar structure to that of the valve face. No other surface structures present. Internally the valve face can be plain or bear radial and tangential thickenings.[g] No portules or ocelli are present. Copulae not seen.

This is a very distinctive genus easily recognised by the valve markings. The early separation of the multiangular forms into *Stictodiscella* is not recommended. The classification alongside *Arachnoidiscus* in a Stictodiscineae (Hustedt, 1927–66) or Stictodiscoideae, with several other genera (Simonsen, 1979), may be valid for some species referred to *Stictodiscus*; others seem further removed from the *Arachnoidiscus* form. However, it is clear that this and several other circular to multiangular genera need very careful detailed study before final decisions can be made. Most of the nearly 70 spp. listed by VanLandingham (1978) have rarely been recorded and should be checked; there is a possibility that some are resting spores. We thank P. A. Sims for the material photographed here.

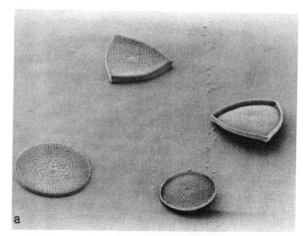

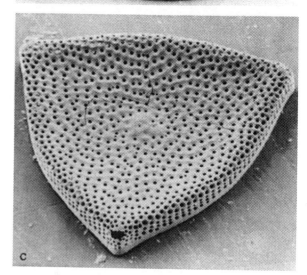

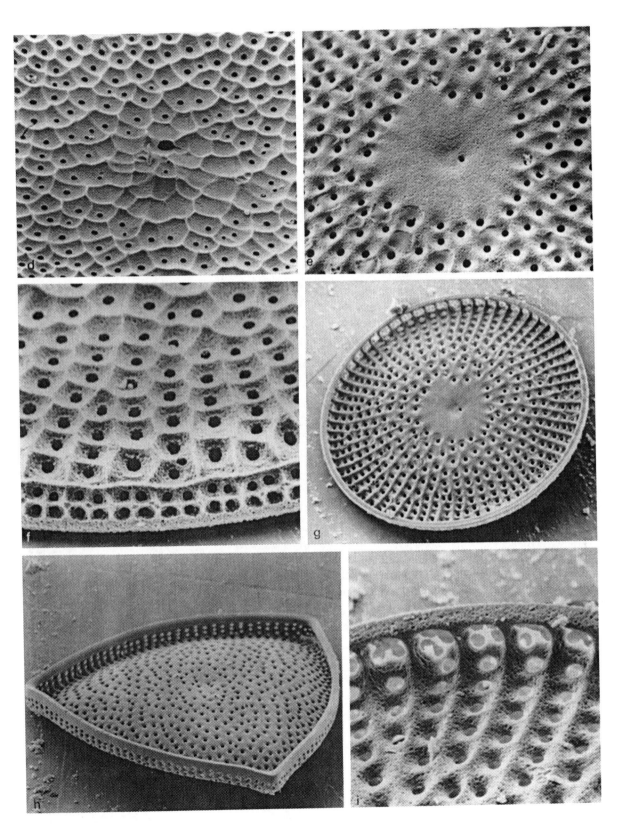

Triceratium C. G. Ehrenberg 1839. Ber.
Bekanntm. Verh. Königl. Preuss. Akad. Wiss.
Berlin, **1839**: 156

LT: *T. favus* C. G. Ehrenberg (*vide* A. Mann,
Contr. U.S. Natl. Herb. 10: 295, 1907)

Cells free-living or attached; usually triangular in
valve view[a] and narrowly oblong in girdle view,
with elevations at the corners and a slight central
convexity.[b] Marine, coastal.

Valves triangular or sometimes square,[d] shallow,
often ornamented with simple[c] or branched spines.[f]
Valve face flat or slightly convex; mantles very
shallow. Areolae loculate,[k] opening externally via
large foramina;[a-i] the bases of the locules are formed
by a continuous sheet of silica with rows of pores
radiating from a central annulus.[e-g, i] Pores often
eroded but *in vivo* with domed coverings; internally
the pores are clustered[j, k] but they still clearly radiate
from the centre of the valve (see also Miller &
Collier, 1978). Margin of valve face raised,[c, d] with a
single row of stalked or spathulate[c] collared tubes,[i]
which are the exits of rimoportulae. Corner
elevations present, ending in ocelli.[g, h] In some
species there is a conspicuous rimoportula opening
adjacent to each ocellus. Valve mantle often with a
ridge below which there is a fine row of pores just
above the recurved mantle edge. Copulae simply
porous; the valvocopula is finely fimbriate (Miller &
Collier, 1978. We are not entirely clear about the
arrangement of the copulae, as their interpretation
does not seem possible).

There are many more triangular diatoms in need
of investigation. Over 400 spp. of *Triceratium* are
recognised by VanLandingham (1978)! How many
actually conform to the features above is not known
but since the above observations are based on
material which agrees with the type we take them to
be representative of *Triceratium sensu stricto*. There
are other triangular taxa which clearly do not
belong to *Triceratium* as we have defined it and
these we recognise as members of *Amphitetras*,
Biddulphia, *Cerataulus*, *Lampriscus*, *Sheshukovia* and
Trigonium. In the detailed review by Glezer (1975),
Triceratium as recognised by us is maintained;
Glezer considers that another 30 spp. also belong in
Triceratium. The genus ranges back to the Eocene.

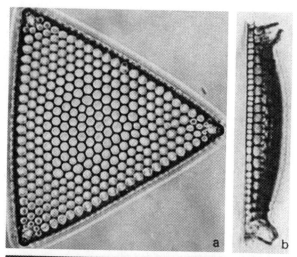

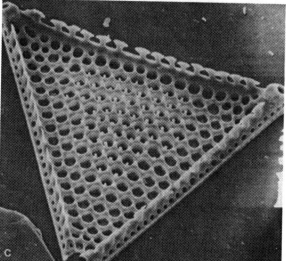

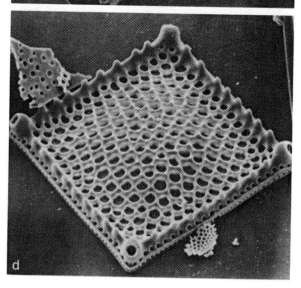

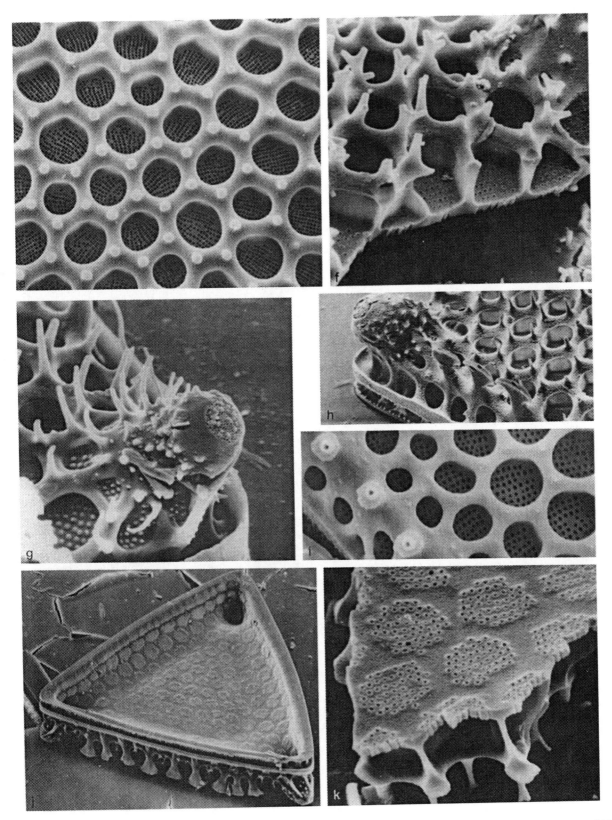

Odontella C. A. Agardh 1832. Consp. Crit. Diat.: **56**

T: *O. aurita* (Lyngbye) Agardh (= *Diatoma auritum*)

Cells oblong in girdle view, with long 'spines' and raised apical elevations.[a–d] Often forming chains linked by the processes, with the tubular spines crossing each other. Plastids many, small, discoid. Marine, planktonic or epiphytic. Very abundant throughout the oceans.

Valves elliptical or lanceolate, with no separation into face and mantle. Valve face plain or with fine granules, spinules or spines;[c, e] sometimes with two ridges[e] (which may be fimbriate) running on either side delimiting an elliptical area in the centre. At each end there is an elevation, sometimes low and blunt, elsewhere horn-like,[a] which bears an ocellus.[h] Wall loculate, with fine external pores[g] and round internal foramina.[i] The edge of the valve mantle is sometimes recurved[e] so that a groove runs around just above the free edge. The spines,[e, f] which are very variable in length, are actually the exit tubes of the rimoportulae, and are placed in the centre of the valves or close to the bases of the elevations, diagonally opposite each other; they can have small apical spinules.[f] Internally the rimoportulae are sessile and lie in slight depressions.[i] Copulae split, with ligulae and clustered fimbriae along the advalvar edge;[j] areolae simple, in vertical rows. The valvocopula is modified to fit the 'sculptured' edge of the valve mantle.

Odontella differs from *Biddulphia* in having distinct ocelli, rimoportulae that often have long external tubular extensions, and a loculate valve structure, with small external pores and simple foramina internally. We include in *Odontella* the following '*Biddulphia*' species: *aurita, obtusa, edwardsii* (possibly = *rhombus*), *mobiliensis, regia* and *sinensis*. The last two species are delicate planktonic forms. The coarser structure of *O. aurita,* and *O. edwardsii* and some others requires further investigation but for the moment we leave these taxa here.

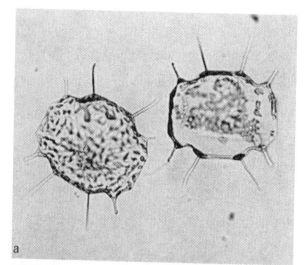

a

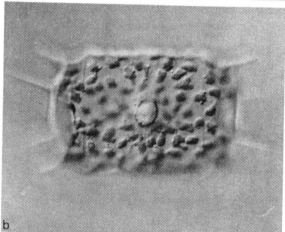

b

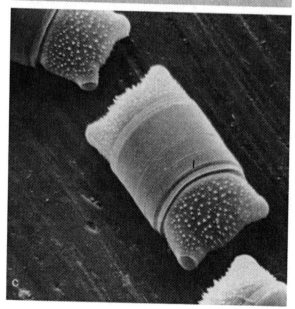

c

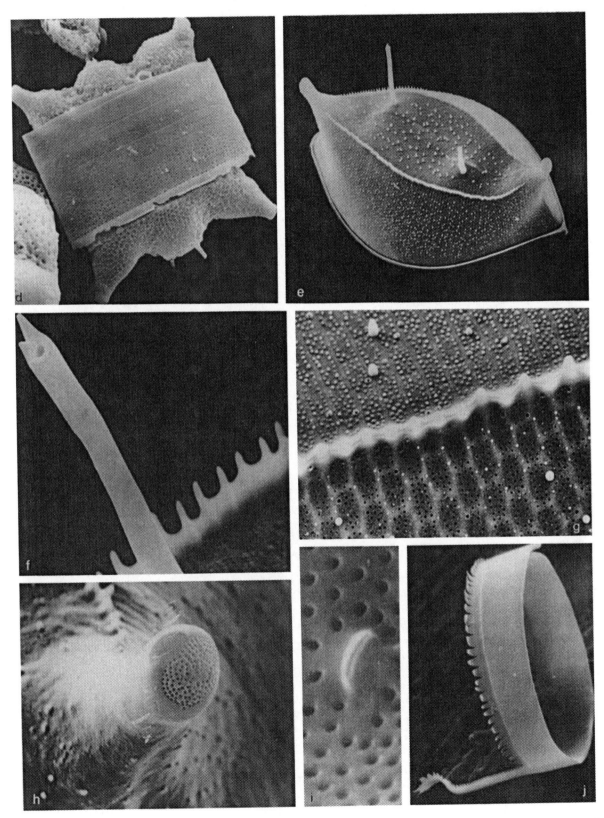

Lampriscus A. Schmidt 1882. Atlas Diat. T. **80**, fig. 11.

T: *L. kittonii* Schmidt

Cells forming long, filamentous colonies. In girdle view rectangular with slight valve elevations, which link the cells. Plastids small, discoid, usually lying in cytoplasmic strands. Mainly tropical; epiphytic on marine algae or attached to other submerged objects.

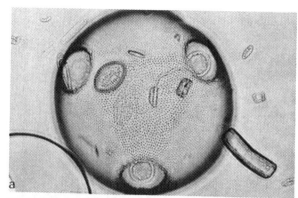

Valves with conspicuous pseudocelli which may superimpose a triangular/quadrangular pattern within the circular outline[a] of the valve, or impose an overall angularity on the valve, making it triangular,[b, d] quadrangular or multiangular. Striae radiate from the centre[a, d, g] and continue without a break down the deep valve mantle.[c] Areolae elliptical, occluded by rotae;[c] occasional simple pores present. Pseudocelli lying on slight elevations; separated into inner and outer portions by a circular hyaline rim located towards the inner margin of the pseudocellus,[a, d, f] which may be plain or bear small spines[c] and an additional large spine.[c-f] This unusual structure can appear in the LM as a circular spot with an outer arc. The pores of the pseudocellus are much smaller than the valve areolae and tend to form linear arrays both inside[h] and outside the rim;[c, e] the orientation of these arrays differs and the external rows have a line of discontinuity at the angle of the valve.[e] Copulae numerous; valvocopula with a slight septum hooking over a slight ridge on the valve. Areolae in straight lines with simple rotae;[i] the pegs of these rotae may be at right angles to those on the adjacent mantle.[i]

The diatoms included here were originally placed in *Triceratium*, removed to *Biddulphia*, and then to *Lampriscus*. In Hustedt (1927–66) and other texts the principal species is '*Triceratium shadboltianum*'; Hustedt discusses some of the problems associated with this form. Species concepts in this genus still require detailed study and some other biddulphioid taxa may have to be transferred into it. We are not certain of its taxonomic position; the interlocking of valve and valvocopula is reminiscent of *Isthmia* but the spines on the pseudocellus suggest a link with the hemiauloid series. All Schmidt's (1874–1959) illustrations are from the Pacific or Caribbean and ours were collected in Hawaii and Queensland. Navarro (1981) also records the genus from Puerto Rico.

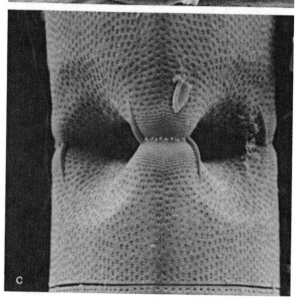

Sheshukovia Glezer 1975. *Bot. Zhr.* **60** (9): 1307

T: *S. kolbei* var. *uralensis* (Jousé) Glezer
(= *Triceratium kolbei* var. *uralensis*)

Cells colonial, forming zig-zag chains; triangular in valve view, square in girdle view. Plastids numerous, irregularly discoid. A common marine epiphytic genus.

Valves triangular[a-d, f] with slightly raised, rounded elevations at the angles,[c, i] which are separated from the rest of the valve by shallow furrows. Valve surface smooth or granular. Areolae large, containing cribra, arranged in rows radiating from the centre, where there may be a plain area; the central areolae are usually larger than those towards the edge of the valve face or on the mantle. Mantles distinct, vertical, areolate, the areolae being smaller beneath the elevations.[c] Internally the valves have a plain central area and marginal ribbing[f-h] – some ribs may be more developed.[b] Ocelli raised on elevations; with a plain rim, surrounding rows of small pores.[i] Rimoportulae central, opposite angles with long external tubes[d, e] and scattered with short external tubes[e, g] – probably variable amongst species. Cingulum of complete (?) copulae,[c, i] with vertical rows of pores.

This genus has previously been placed in *Triceratium*, *Biddulphia* and *Trigonium*. Recently Glezer (1975) has created the new genus *Sheshukovia* and our concept of the genus is from her account. This and other triangular genera require detailed study before any conclusions can be drawn about generic limits and relationships.

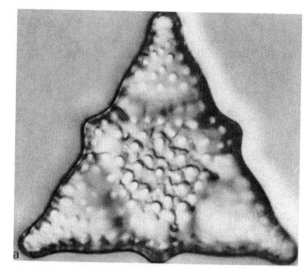

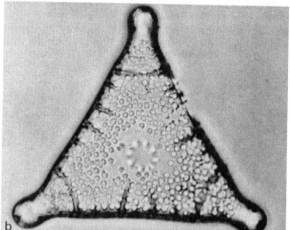

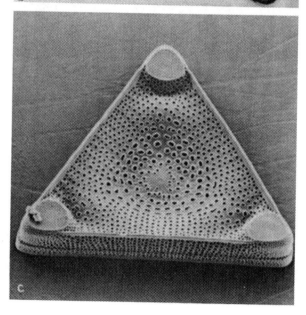

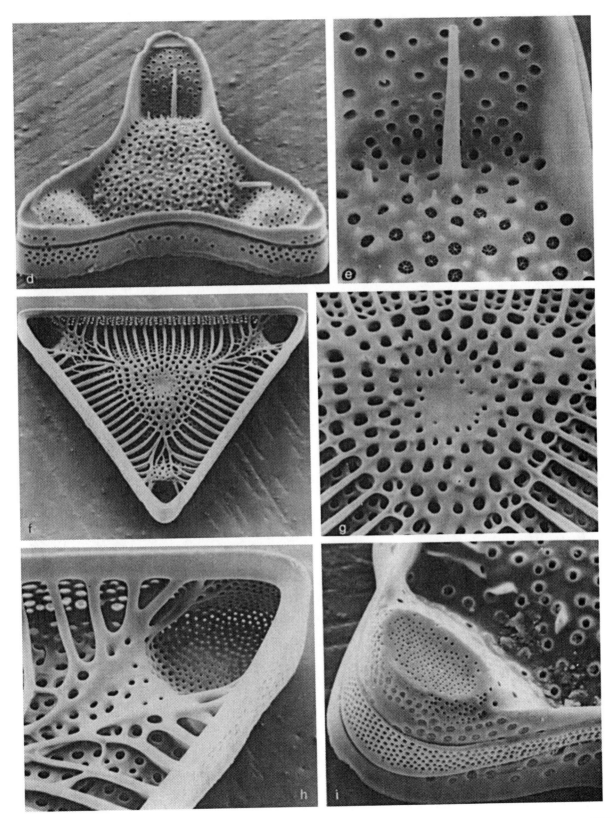

Pseudauliscus A. Schmidt 1875. *Atlas Diat.*
T. 32, fig. 29

T: *P. peruvianus* A. Schmidt

An entirely fossil genus of ocellus-bearing centric
diatoms; cell shape and habit unknown. Distributed
in the Miocene (?).

Valves circular,[a, b] shallow, with a series of
concentric undulations on the valve face and a short
but distinct mantle. The valve face is separated from
the mantle by a prominent rim. Striae uniseriate,[b]
radiating from the centre, which is irregularly
areolate, continuing down the mantle which is
variously ribbed.[b, d, e] Areolae round, apparently
simple; vela missing (eroded?) in samples
investigated, although occasionally there are areolae
that appear domed and perforate.[c] Two conspicuous
rimmed ocelli[a, b, d-f] occur on the valve face, on
opposite sides of the valve; the small pores (porelli)
within the ocellus are in radial rows. Two
rimoportulae present, one[g] on either side of the
ocellar axis, each about halfway between the centre
and the edge of the valve face; the rimoportula axis
is perpendicular to or oblique to the ocellar axis.
Girdle unknown.

VanLandingham (1978) includes this genus in
Auliscus but the organisation of the valve, areolae,
etc. is, we believe, sufficiently distinctive to maintain
a separate genus. Schmidt (1874–1959) also
maintained this distinction, based on the contrast
between the sculptured valve face of *Auliscus* and the
simply porous one of *Pseudauliscus*.

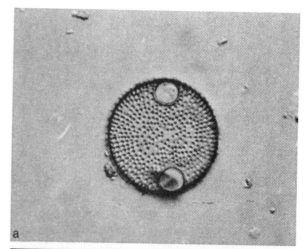

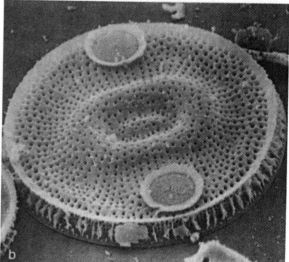

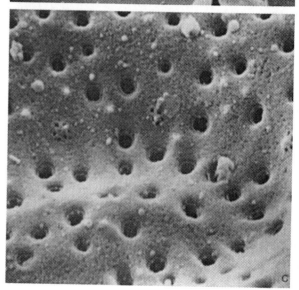

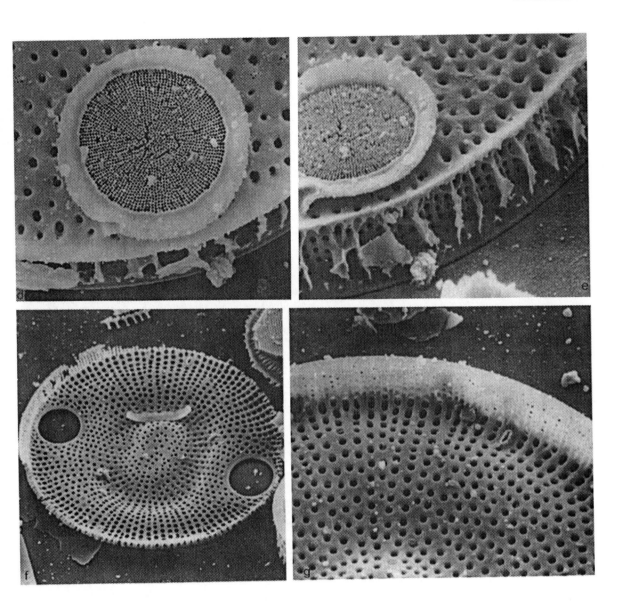

Eupodiscus J. W. Bailey 1851. Smithsonian Contr. Knowl. **2** (8): 39 (nom. cons.)

T: *E. radiatus* Bailey (typ. cons.)

Cells solitary, discoid. Plastids not observed. Occurring in marine habitats; probably epiphytic or even benthic but not often recorded except in loose-lying populations on coastal sediments; such material may be fossil since there seem to be no records of living cells.

Valves circular, with a radiating system[a, b] of raised hexagonal thickenings.[c–g] Within the hexagons rows of pores occur[f, g] and overall these form a radiating system. Each hexagon corresponds to an areola; these are loculate, opening to the outside by several tiny pores and to the inside by a fairly small foramen.[g–i] Within the locules the foramen has a slightly raised rim and is open,[g] whilst on the internal surface of the valve there is a dome[i] that suggests there may be some form of closing membrane; we have not been able to study the detail. The margin of the valve face is 'scalloped' forming a narrow ridge or wing.[b–e] Interrupting the ridge are four equally spaced low ocelli.[b] The valve mantle has continuing rows of fine pores but is also variously ornamented with spines, dendritic outgrowths,[c–e] etc., and often has a circumferential ridge above the outwardly turned mantle edge. The internal openings of the ocelli are sunken.[h, i] The ocellus pores are round and arranged in more-or-less concentric rows. According to Sullivan (1986), there are no portulae present in this genus (there are none in our material) but in material photographed by Sims there are minute rimoportulae on the mantle between the ocelli (P. A. Sims, personal communication). In many species the internal and external development of the rimoportula is very slight. Copulae not studied.

There has been confusion in the taxonomy and *E. radiatus* Bail. has been conserved over *E. (Tripodiscus) germanicus* Ehr. Little seems to be known about the ecology and distribution of this characteristic genus.

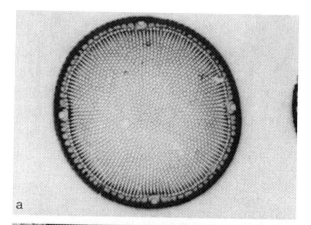

a

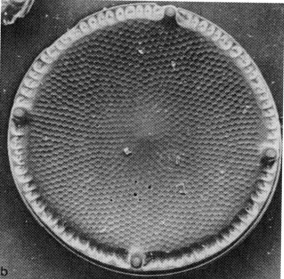

b

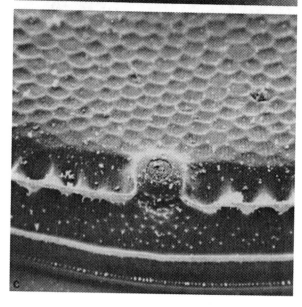

c

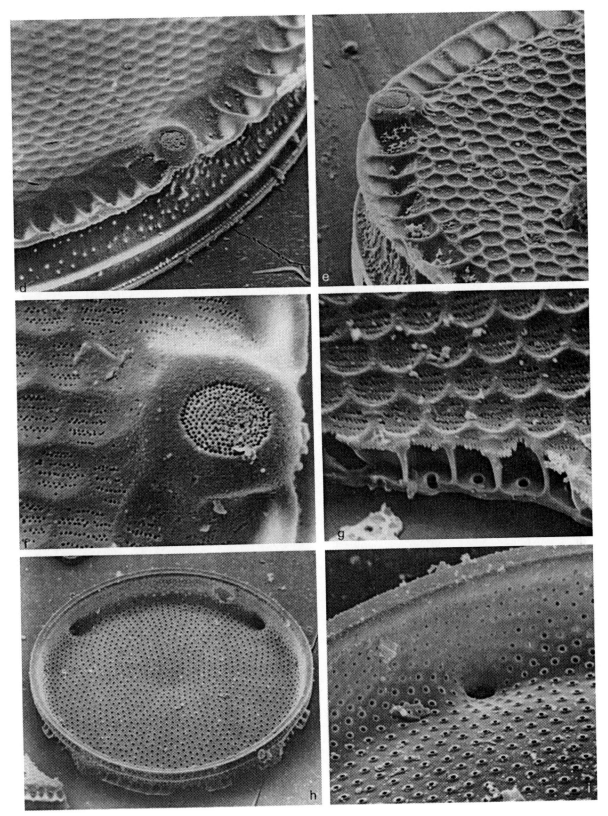

Pleurosira (G. Meneghini 1845. Atti. R. Istit Veneto Sc. Lett. Arti. 1845–6: 95) V. B. A. Trevison 1848. Saggio di una monografia delle alghe coccotalle

T: *Pleurosira laevis* (Ehrenberg) Compère
(= *Biddulphia laevis* = *P. thermalis* Meneghini)

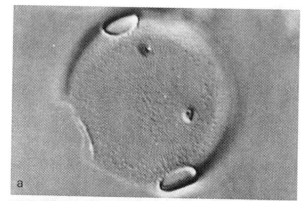

Cells cylindrical, united in straight or zig-zag filaments by means of mucilage pads extruded from ocelli;[b–d] long in girdle view. Plastids discoid. The genus is probably one of the few which finds its ideal niche in brackish waters (estuaries, ditches, etc.), especially in tropical or subtropical regions. It can extend inland into waters of high salinity and will tolerate culture in seawater though it is not usually found in truly marine habitats.

Valves circular[a–e] (subcircular) with a flat valve face and deep mantles. Striae uniseriate, radiating from the centre[e, f] and continuing without a break down the mantle; valve surface sometimes granulose. Areolae simple, penetrating a thick wall of silica;[h] straight or enlarged slightly within the thickness of the valve framework – neither Ehrlich *et al.* (1982a) nor Compère (1982) could confidently detect a velum. Mantle edge recurved,[e] sometimes fimbriate. Two ocelli, each surrounded by a distinct rim and bent over the junction between valve face and mantle,[e, f] sometimes slightly raised but never on elevations. Rimoportulae present, lying in small clear areas, either off-centre in two groups along an axis at right angles to the axis between the ocelli[g, i] (*P. laevis*), or in a circle (*P. socotrensis*); internally sessile, lying in depressions; externally with slit-like openings. Valvocopula split and fimbriate. Copulae split, fimbriate, ligulate, and anti-ligulate.[j] All bands finely areolate.

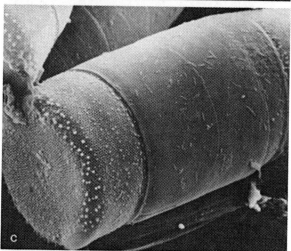

The genus *Proteucylindrus* Li & Chiang is considered synonymous (Compère, 1982). *Cerataulus* encompasses different species in which there is a twist in the axis between the ocelli (see p. 234). As mentioned in Ehrlich *et al.* (1982a), *Pleurosira* is the correct generic name for the diatom which has long been called *Cerataulus laevis* and Compère (1982) validated this – see these papers and Ehrlich *et al.* (1982a, b) for detail.

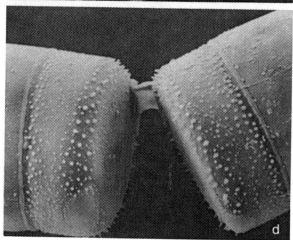

Amphitetras C. G. Ehrenberg 1841. Abh. Königl. Akad. Wiss. Berlin, **1839**: 122

T: *A. antediluviana* Ehrenberg

Cells forming zig-zag colonies[a] attached to seaweeds, often with conspicuous mucilage pads between cells;[c] these pads appear 'collar-like' in some of our material.[d] Plastids numerous, plate-like. Not uncommon and probably of world-wide distribution.

Valves quadrate,[b] often with concave sides. Superimposed upon the quadrate morphology there is a central circular dome with a depressed centre.[b] Prominent ocelli occur on low elevations at the angles.[e, f] Valve framework massive, sometimes granulose or spinulose, and sometimes with branching spinules near the mantle edge. Areolae occluded by cribra located near the outside surface,[g, h] some of which bear fine spinules.[g] The edge of the valve mantle is formed into internal and external flanges; the latter is turned upwards parallel to the valve mantle and is often quite deep and in some valves finely granular. Internally the underlying centric radiating system of costae is obvious at the centre of the valve;[j] here there are a few simple pores through the framework.[h] Rimoportulae absent. Copulae[k] are complete, areolate with cribra[i] of the same form as those of the valve but lying flush with the siliceous framework which is more laminate than that of the valve. The valvocopula has an inwardly directed pars interior, which underlaps the valve.

The siliceous framework of the valve is very similar to that of *Isthmia*. The genus *Amphitetras* has formerly been placed in *Biddulphia, Triceratium* and *Stictodiscus*. With the redefinition of these genera, *Amphitetras* remains as a distinct entity. An extremely detailed and fascinating account of *A. antediluvianum* is given by Liebisch (1928) in which he also figures auxospore formation involving expansion of the cytoplasm in the distal region of the hypocingulum which then splits as further expansion of the auxospore wall occurs. The primary cell is formed within the auxospore while still attached to the hypovalve, as in *Melosira varians*. The classification of genera in the order Biddulphiales is still in a state of flux. *Amphipentas* is possibly to be included within *Amphitetras* but a position in *Stictodiscus* also needs consideration.

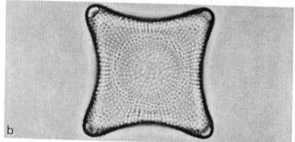

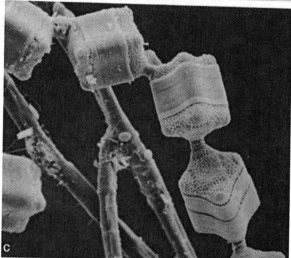

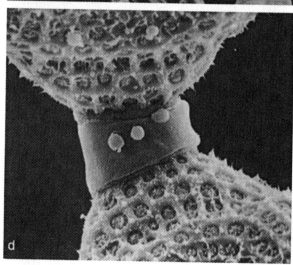

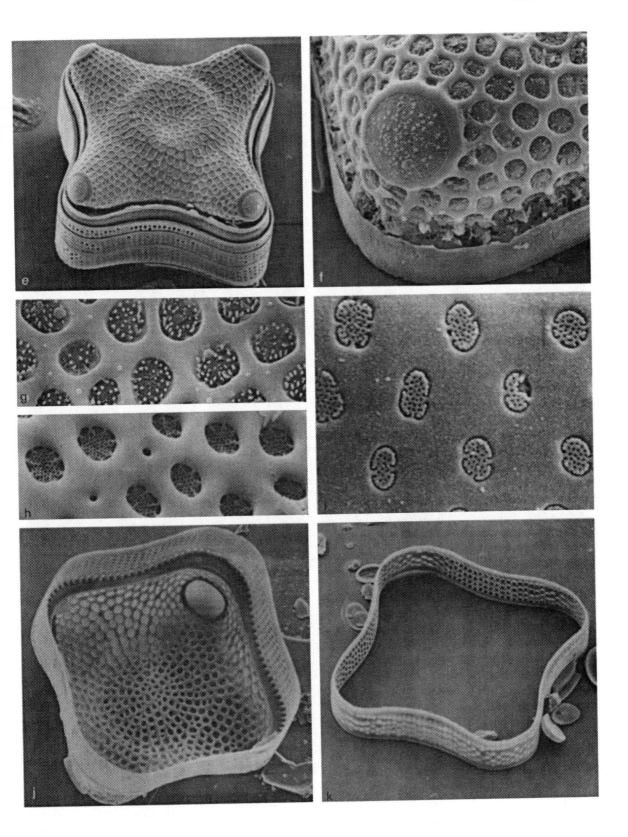

Cerataulus C. G. Ehrenberg 1843. Ber. Bekanntm. Verh. Königl. Preuss. Akad. Wiss. Berlin, **1843**: 270

T: *C. turgidus* (Ehrenberg) Ehrenberg (= *Denticella turgida*)

Cells rectangular in girdle view with conspicuous ocelli twisted out of the pervalvar plane.[b] Elliptical[c] or circular[a] in valve view. Plastids numerous, discoidal. A marine genus often found attached to one or more sand grains by mucilage extruded through the ocelli. Also frequently found coated with fine silt and clay particles, sometimes mixed into the plankton but strictly a benthic genus.

Valves elliptical or circular and twisted, with large ocelli on short elevations pointing in opposite directions.[b-d] The torsion varies between slight and extreme and makes the valve margin non-planar. Surface of valve with spines or granules,[b-h] especially on the mantle. Areolae radiate from the centre.[f] Areolae loculate[h] (see also Ross & Sims, 1971) with fine outer pores and larger round foramina internally. Outer pores clustered into radiating multiseriate rows, except within a central annulus, where they are scattered.[f] The valve mantle may be ornamented with circumferential ridges,[b-d, g] 2 (4) – sometimes forked exit tubes at the apices of the rimoportulae are situated near the margin of the valve face,[c, d, f] midway between or near to the ocelli, one or two on either side of the valve in our specimens. Ocelli with wide plain margins[d, e] and porelli in sectors. Copulae curved to follow the torsion of the valves, split, with ligulae.[b, i]

The systematics of this and other 'biddulphioid' taxa needs considerable attention and other species need detailed study, e.g. a triangular species, *C. subangulatus*, has been described from Oamaru, which Schmidt (1874–1959, taf. 116) considered might be a new genus. Hustedt (1927–66) clearly considered that the degree of torsion of the cell was an important distinction but even this requires further investigation. Only *C. turgidus* is regularly recorded and VanLandingham has only a few other records. We suggest that *C. smithii* belongs with other species in the 'reticulata' group of *Biddulphia* which requires re-investigation since they may form a new genus. *C. laevis* has already been returned to *Pleurosia* (see p. 230).

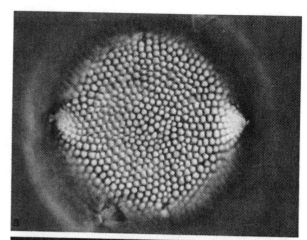

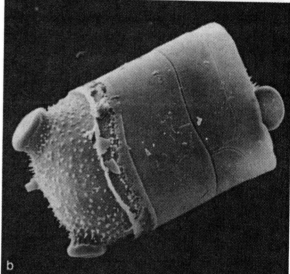

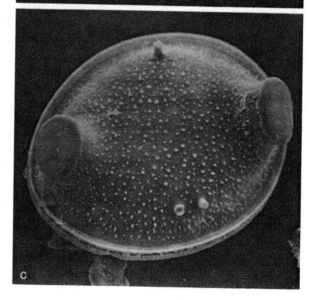

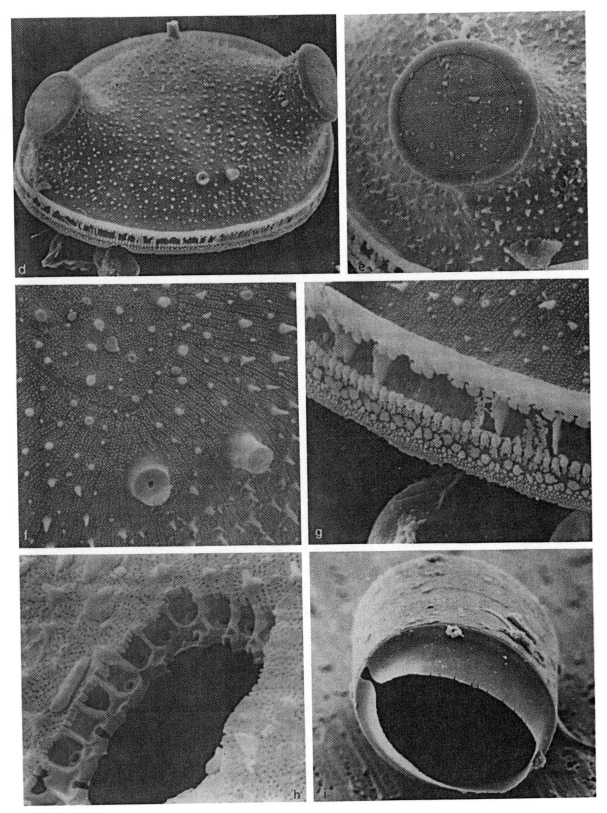

Auliscus C. G. Ehrenberg 1843. Ber. Bekanntm. Verh. Königl. Preuss. Akad. Berlin, **1843**: 270

T: *A. cylindricus* Ehrenberg

Cells solitary, circular or elliptical in valve view, rectangular in girdle view with elevations. Probably attached to sand grains in the marine inshore region but rarely recorded in live material. Many fossil species extending back to the Cretaceous. Hustedt (1927–66) comments upon the variability of the valve markings of this genus.[b, c]

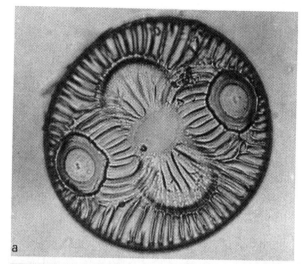

Valves with two or three conspicuous ocelli on elevations and a valve face that is ridged or variously moulded.[a–c] The ocelli have radiating sectored rows of porelli[f, g] and are surrounded by thick ridges of silica; they are often located slightly oblique to the centre line of elliptical forms and a second axis of markings is situated at right angles to the ocellar (elliptical) axis. Fine and relatively inconspicuous rows of areolae radiate in groups between the ridges.[d] The valves are extremely thick.[h] The outer apertures of the areolae appear unoccluded whilst the inner almost certainly have domed vela.[g] The areolae of the valve mantle form distinct rows and a clear circumferential ridge, sometimes developed into a distinct rim, divides these from the valve face areolae.[d] Internally the ocelli are sunken, with no indication of the circumferential thickenings so obvious on the outside.[g] Rimoportulae appear to be absent in our material; P. A. Sims (personal communication) has fossil material with extremely inconspicuous marginal rimoportulae and Simonsen (1979) also reports their presence in fossil but not modern material. The copulae are split and are coarsely or finely poroid.[i]

Rattray (1888b) revised the genus but there has been little systematic work since, though many species have been described, most of them fossil. The species of the genus *Pseudoauliscus* Leudiger-Fortmorel are usually included within *Auliscus*. The genus is unlikely to be confused with any other and is clearly distinguished by the massive ocelli.

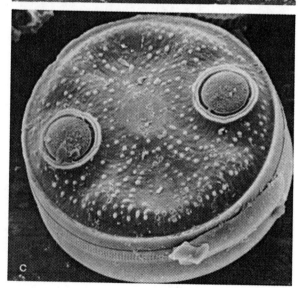

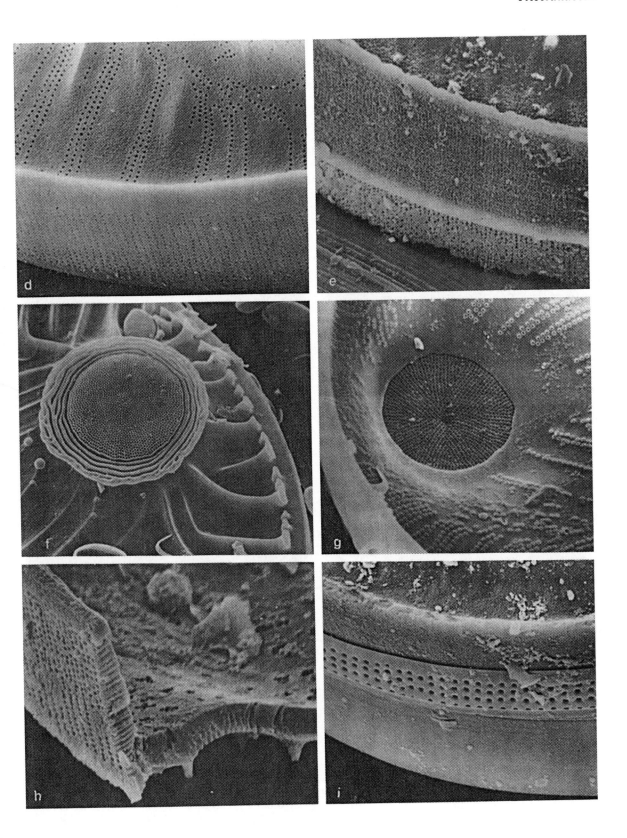

Plagiogramma R. K. Greville 1859. Quart. J. Microsc. Sci. 7: 207

T: *P. staurophorum* (Gregory) Heiberg (= *Denticula staurophora = P. gregorianum* nom. illeg.)

Cells united into short chains, oblong in girdle view, sometimes with expanded apices and with conspicuous inwardly-directed thickenings. Four plastids, lying adjacent to the valves. Marine, benthic.

Valves elliptical[a-c] or lanceolate, sometimes with rostrate apices; usually with a slightly domed valve face; with or without a clear distinction into valve face and mantle. Edge of valve face with short spines;[b, c, h] the striae continue beneath these onto the mantle. The wide valve margin is finely granulose[c] or plain.[h] Striae transverse, uniseriate, apparently uninterrupted by a sternum[c, f] or with a trace of a sternum;[g, h] striae missing centrally, so that a plain transverse area (fascia) is produced, which is sometimes raised. Transapical ribs sometimes thickened internally, so that the areolae open into distinct grooves. Areolae large, elliptical, with cribra lying close to the external surface.[c, d] Cribra, at least in *P. staurophorum*, in the form of delicate porous plates, suspended by pegs.[e] Apices with distinct pore fields[a-c, g, h] which may be papillose, with small plate-like projections around the lower margin – this may not be a characteristic of all the species. Internally conspicuous pseudosepta occur on either side of the central stauros,[f] and in *P. pulchella* also across the apices.[i] No rimoportulae. Copulae split, often narrower and curved towards the valves at the apices.[b] Valvocopula deep, probably lacking areolae;[b] subsequent copulae thinner, each with a single row of small areolae.

This is a fairly distinctive genus now that several species have been transferred to *Plagiogrammopsis* and *Brockmanniella* by Hasle, von Stosch & Syvertsen (1983), but it is still large. Most of the species listed in VanLandingham require further study.

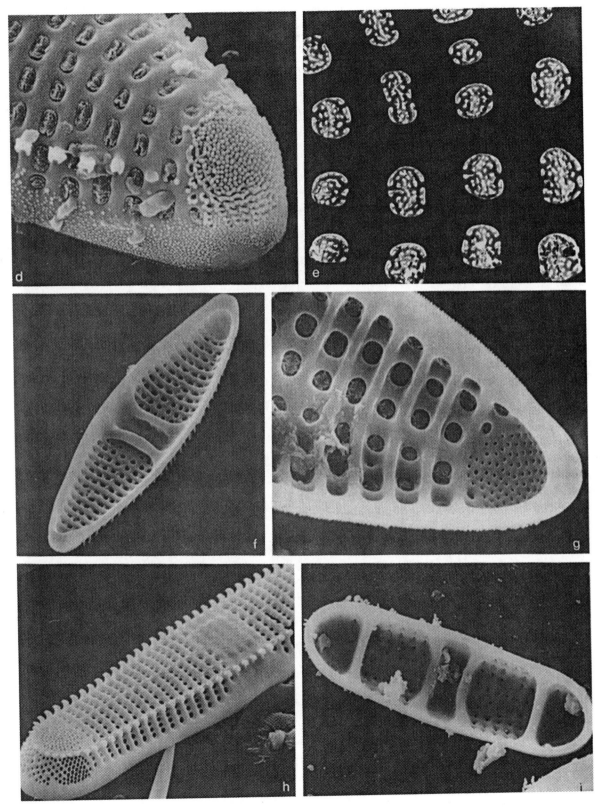

239

Glyphodesmis R. K. Greville 1862. Quart. J. Microsc. Sci. Ser. 2 (**2**): 234

T: *G. eximia* Greville

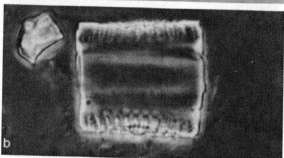

Cells colonial, forming chains; rectangular in girdle view.[b] Plastids not observed. A common marine genus associated with sediments but habitat not well known.

Valves elliptical to lanceolate;[a, c-f] valve face with central hyaline raised area and sternum, from which radiate robust ribs, these being especially prominent internally;[b] areolae large, occluded by cribra lying close to the outer surface.[d] Complex linking spines[c-g] occur on the ribs at the valve face/mantle junction and below these on the mantle is a single row of areolae; below this again is a plain or granulose margin. Apices with prominent ocelli, elevated above the valve surface;[c-g] the plain rims of the ocelli are further elevated. Below each ocellus is an apical pore field, whose porelli tend to be larger than those of the ocellus. Rimoportulae absent. Cingulum of at least two copulae. The valvocopula appears finely granulose,[i, j] while the second copula is even more finely granulose and has a row of advalvar pores.

The taxon *G. acus* discussed by Fryxell & Miller (1978) does not have ocelli and its inclusion in *Glyphodesmis* must therefore be suspect. The possession of ocelli suggests that *Glyphodesmis* is not an araphid genus but possibly belongs in the Triceratiaceae. However, for the present we propose to leave it in the Plagiogrammaceae. The type of the genus has been studied in detail by Sullivan (1988).

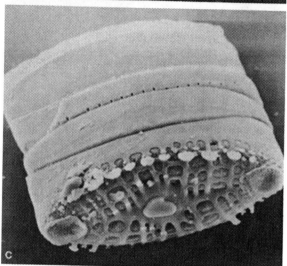

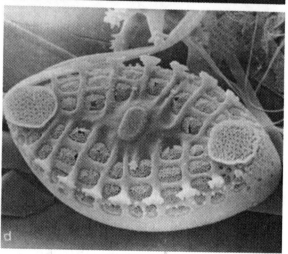

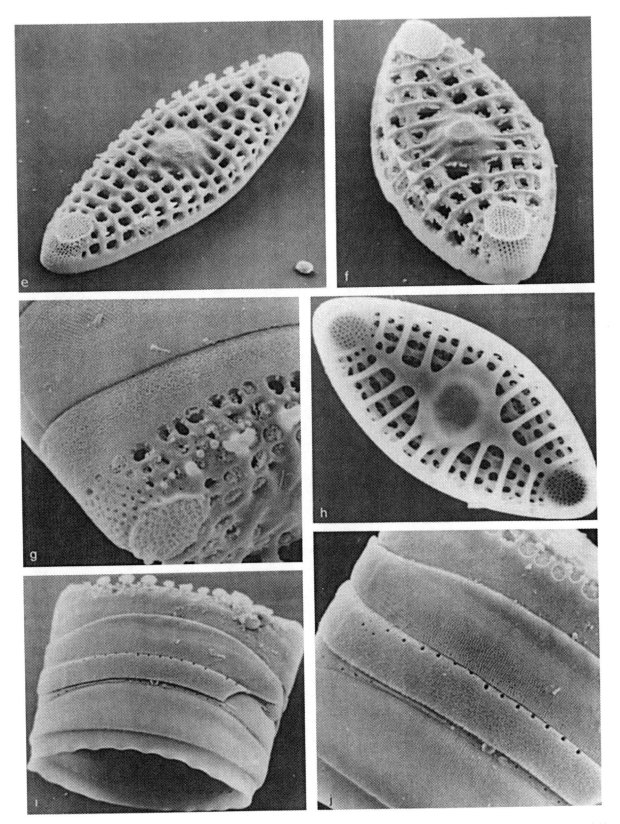

Dimeregramma J. Ralfs in A. Pritchard 1861.
Hist. Infus., ed. 4.: 790

T: *D. minor* (Gregory) Ralfs (= *Denticula minor*)

Cells more-or-less rectangular in girdle view, joined
to form filaments.[a-c] Cells touching along the
complete valve face or only at centre and apices.
Plastids plate-like, probably two lying alongside the
girdle. Lives in the marine littoral but details not
known; probably associated with sediments or sand
grains.

Valves linear to elliptical,[d, f, j] sometimes sub-
rostrate. Valve face almost completely flat or slightly
elevated at centre and apices. Sternum narrow or
centrally expanded. Striae uniseriate, slightly radiate,
containing round areolae that open externally into
grooves sunken between prominent transapical ribs.
Areolae with cribra near outer face.[f, k] Mantles
shallow; smooth or finely granular.[g] Bifurcating
nodulose spines arise from the transapical ribs near
the valve margin,[d-h] apparently involved in linking
the cells into filaments. Small nodules also occur on
the ribs. No rimoportulae. Apical pore plates
present,[f, i] in some samples with minute spinules
externally. Copulae split, curved at apices, and with
single rows of areolae.[h, j] Valvocopulae much wider
than subsequent copulae and narrowed at the
abvalvar edge at both poles; second copula with
broad 'ligula' to fill the wide gap between split end
of the valvocopula. At least four copulae in the
complete cingulum.

A common genus, having affinities with
Plagiogramma (excl. *Plagiogrammopsis*) and
Glyphodesmis. These genera form a distinctive group
from the view both of structure and ecology and we
place them in a family (Plagiogrammaceae).

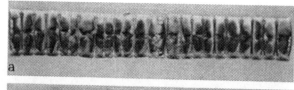

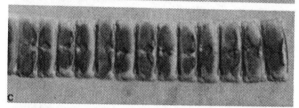

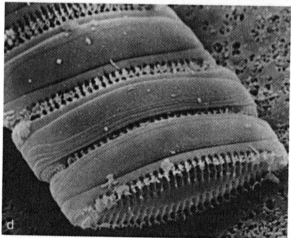

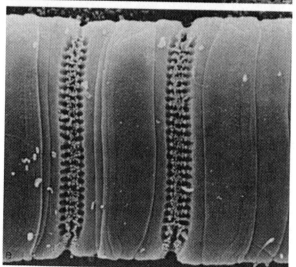

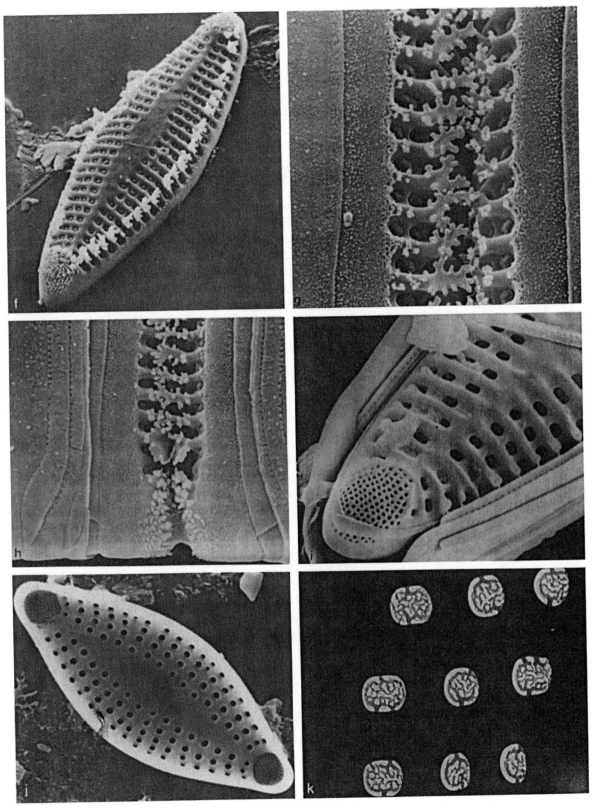

Dimeregrammopsis M. Ricard,1987. Atlas du phytoplancton marin **2**: 85

T: *D. furcigerum* (Grunow) Ricard

Cells lanceolate, apices attenuate. Oblong in girdle view. A marine genus recorded first from the Balearics but present in material from Barbados kindly supplied by George Andrews. The only taxon we could connect with our material is *Dimeregramma furcigerum*, which during the final stages of preparation of our book has been made into a new genus *Dimeregrammopsis*.

Valves lanceolate, rounded in transapical section, without a conspicuous valve mantle.[a, c] Areolar rows transverse with no trace of a sternum, though one is illustrated in Peragallo & Peragallo (1897–1908), Hustedt (1927–66) and in Ricard (1987). A row of simple spines occurs along either edge of the valve. An apical pore field is present at either end and an elongate furrow lies along the apical axis at each end of the valve.[a, c] Internally the valve is costate with no central break in the transverse costae.[b, d–f] Beneath the external apical furrow there is a slightly raised area in which two rows of pores occur. Girdle bands appear plain except for an advalvar row of areolae.

Although placed in *Dimeregramma* by early workers it has none of the major characteristics of that genus (see p. 242). The distinctive apices without rimoportulae preclude a position in other araphid genera. *Plagiogramma* (*P. interruptum*) also has an apical structure which in the light microscope looks similar but *Plagiogramma* has distinct pseudosepta and this new genus does not. The girdle view drawn in Hustedt (1927–66) shows a discrete thickening of the cell at the apices corresponding to the apical structure. LM specimens on BM slide 31004 from Pensacola, Montogmery & Miller (1978) figure the same taxon (as *Dimeregramma*) and show the external apical furrow very distinctly.

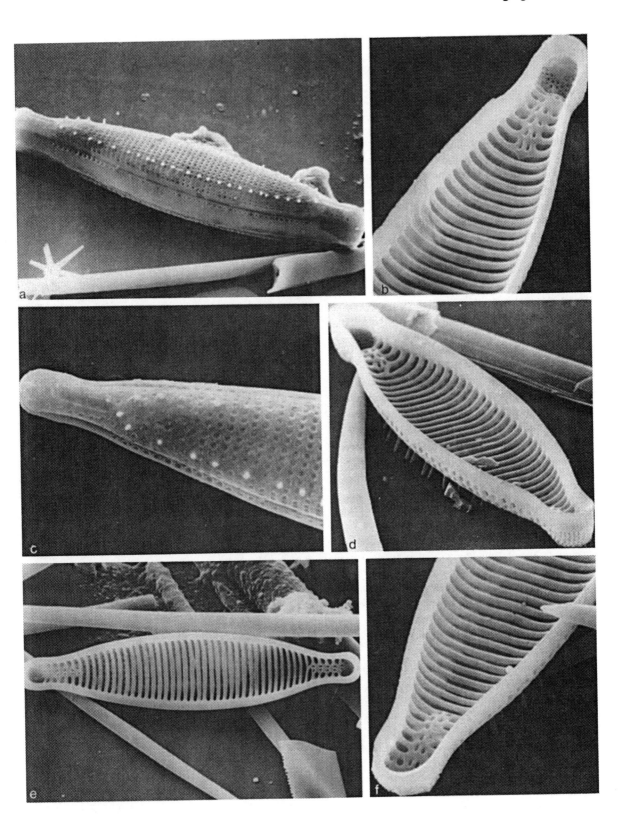

Biddulphia Gray 1821. *A Natural Arrangement of British Plants* 1: 294

T: *B. biddulphiana* J. E. Smith (Boyer)
(= *Conferva biddulphiana*; = *B. pulchella* Gray, nom. illeg.)

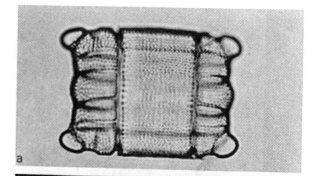

Cells robust, rectangular in girdle view,[a] elliptical in valve view, with prominent elevations at the poles. Normally seen in girdle view, often growing in zig-zag chains attached to filamentous seaweeds,[b] etc. Often found also in inshore plankton samples. Plastids numerous, discoidal. A very common marine genus but taxonomically extremely confused (see below).

Valves bipolar[c] /lanceolate[d] to almost circular, often with wavy margins. Valve surface often furrowed[g] with various thickenings, spines (often conspicuous in the central region)[e] or ridges.[b, c, g] Apices bearing rounded pseudocelli on low or extended elevations.[a–d, f, g] Valve mantle not well-defined but extreme edge of valve often recurved and variously moulded. Areolae large with perforate vela of the cribrum type.[e, f] Simple pores also occur occasionally in the valve framework.[e] Internally with conspicuous plain ridges (pseudosepta)[h, i] beneath the external indentations of the valve. One to several rimoportulae present, sessile, clustered in centre;[i] external tubes often stout and surmounted by two spines.[e] Cingula with a complete, closed valvocopula and 3–4 split copulae; areolae large, in rows; fluting of girdle corresponding to that of valve edge.[c, d]

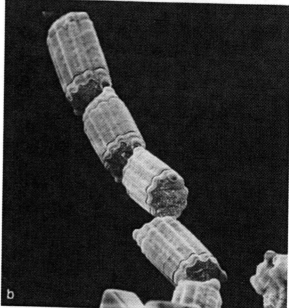

Ross & Sims (1971) concluded that several species of *Triceratium* also belong in *Biddulphia* but these we have not investigated. At the moment the best practice seems to be to distinguish groups of species which clearly fall into genera where the type species is well defined. *Biddulphia* has well-defined pseudocelli, cribra, marginal external ridges, internal pseudosepta and central rimoportulae. The genus includes *B. biddulphiana*, *B. regina* and *B. tuomeyi*. Other species formerly in *Biddulphia* have been transferred to *Odontella*, *Pleurosira*, *Biddulphiopsis*, etc. Hoban (1983) provides further detail of the type species together with a discussion of the family limits within the confused group of 'biddulphioid' forms.

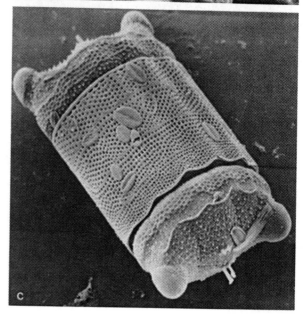

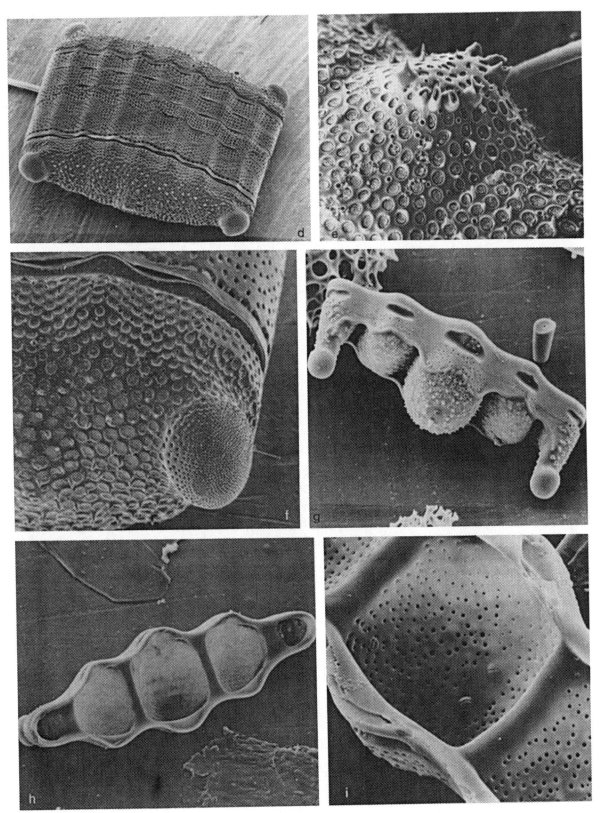

Biddulphiopsis H. A. von Stosch & R. Simonsen 1984. Bacillaria 7: 12

T: *B. titiana* (Grunow) H. A. von Stosch & R. Simonsen (= *Cerataulus titianus*)

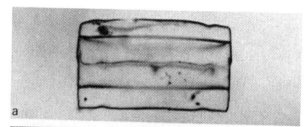

Cells rectangular with rounded corners in girdle view;[a] forming chains[b] which may branch to form small colonies, and which are attached to substrata (usually other algae). Small plate-like plastids occur in the peripheral cytoplasm and in conspicuous strands radiating from a prominent central nucleus. A widespread marine genus of two species, *B. titiana* and *B. membranacea*, which were formerly placed in *Biddulphia* (see the detailed account of von Stosch & Simonsen, 1984).

Valve outline oval[d] to subcircular with deep vertical mantles.[b, c] In girdle view the valve faces may be slightly concave but vary from this through flat to slightly convex; their centres may be slightly vaulted and notched towards the margins.[a] Areolae scattered in a central elongate field,[d, i] becoming organised outside this into uniseriate, radial rows and then even more precisely aligned down the mantles;[c] occluded by rotae.[c, h, i] Scattered pores also occur.[h] Apices differentiated into areas of smaller pores forming rather indistinct pseudocelli.[c] The mantle edge is extended inwards slightly to form a pseudoseptum,[d–f] which is especially prominent at the apices. A row of rimoportulae occurs adjacent to the pseudocelli,[e, f, h] sometimes forming an irregular ring around it. Internally the rimoportulae are shortly stalked, the slits not being aligned in any way. The cingulum consists of four closed copulae. The valvocopula has a narrow septum[f] which fits over the valve pseudoseptum, while its abvalvar edge is extended at the apices forming 'false antiligulae' (von Stosch & Simonsen, 1984); the next copula is moulded to fit against this edge. The advalvar edges of the copulae have short, thin projections[g] (fimbriae) with a few interspersed pores, internal to which is a plain area. The areolae run in rows down the copulae and have single pegged rotae or multiple pegged, elongate rotae.[j]

The genus is distinguished from *Biddulphia sensu stricto* (see p. 246) by the arrangement of the rimoportulae around the ill-defined pseudocelli, the lack of distinct apical elevations and the unusual form of the copulae. Von Stosch & Simonsen discuss these points in more detail and provide excellent illustrations.

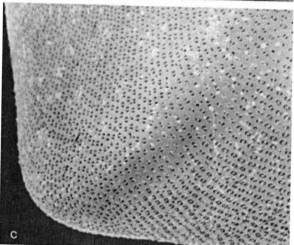

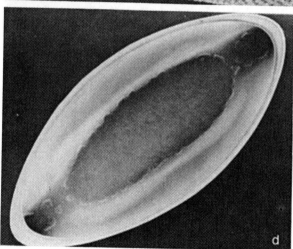

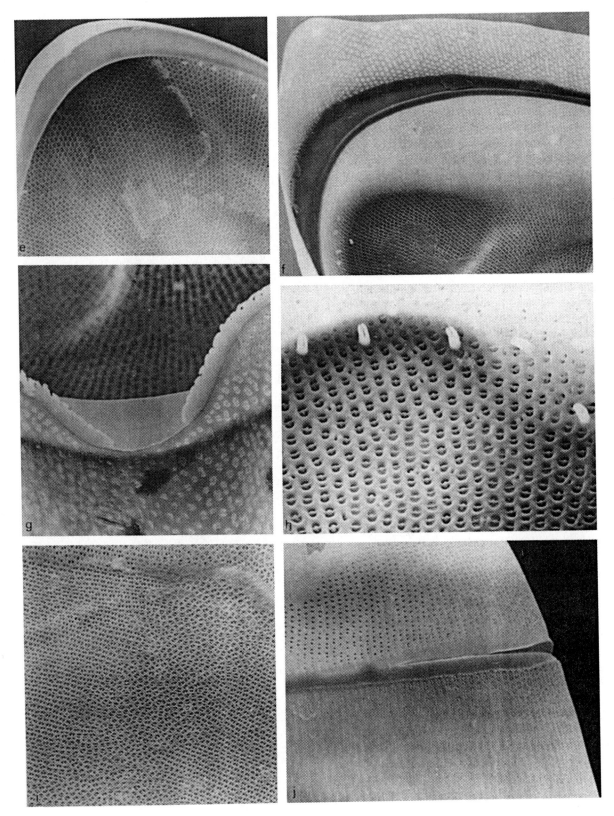

Hydrosera G. C. Wallich 1858. Quart. J. Microsc. Sci. 6: 231

T: *H. triquetra* Wallich

Cells triangular and undulate in valve view, rectangular in girdle view; forming short chains by mucilage pads secreted at all three valve angles; attached to the substratum by secretion at one angle. Plastids: numerous, elliptical platelets (Li & Chiang, 1977). A marine epiphytic genus usually found in estuarine situations; it may penetrate upstream in rivers and even into other freshwater sites (Li & Chiang, 1977). Generally regarded as tropical but now recorded in the R. Thames, U.K. A small distinctive genus with two commonly reported species.

Valves triangular, their sides undulate.[a-c] Valve face flat, curving into a fairly deep mantle; both valve face and mantle coarsely areolate, the areolae being somewhat smaller on the mantle.[b] Pseudocelli on each corner,[c] sometimes with small blunt spines around the outer rim; similar projections may occur on other parts of the valve. Areolae loculate,[d,h,i] with foramina externally and small round pores internally, the inner pores forming rows radiating from the valve centre.[f-i] Small domed structures on the outer valve face[d] in the pits between the outer ornamentation. No vela have been detected; nor did Li & Chiang find any though further investigation is necessary. Valve mantle edge plain. The valve has one conspicuous stalked rimoportula, with an S-shaped slit internally.[e-g] Plain ridges or folds (pseudosepta) extend across the bases of the three corners of the triangular valve.[e,f] Between the valve mantle and the rimoportula, which is located opposite one of the central undulations of the wall, is a region with three cavities and a short ridge[g] (the cavities were termed pseudonoduli by Li & Chiang, 1977). The opening of the rimoportula and any connection with the bases of the cavities are not obvious from the outside. Copulae[j] closed, finely porous, with a fimbriate edge on the valvocopula; four(?) per cell. Li & Chiang report a complex overlapping of the valvocopula onto the pseudosepta.

The 'pseudonoduli' do not appear comparable to those in *Actinocyclus*. The internal folding of the valve, loculate structure and small domed external openings in the primary layer of the wall are reminiscent of other Biddulphiaceae though the relationships of the genera are unclear. Further details are given in Qi (1984) and Qi *et al.*, (1984).

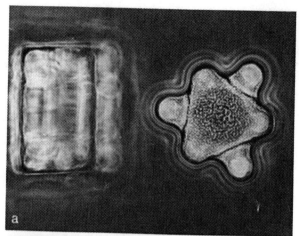

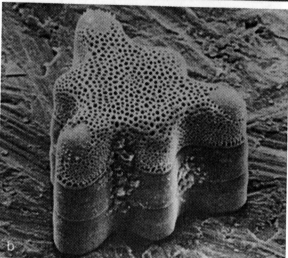

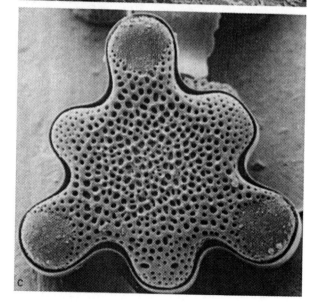

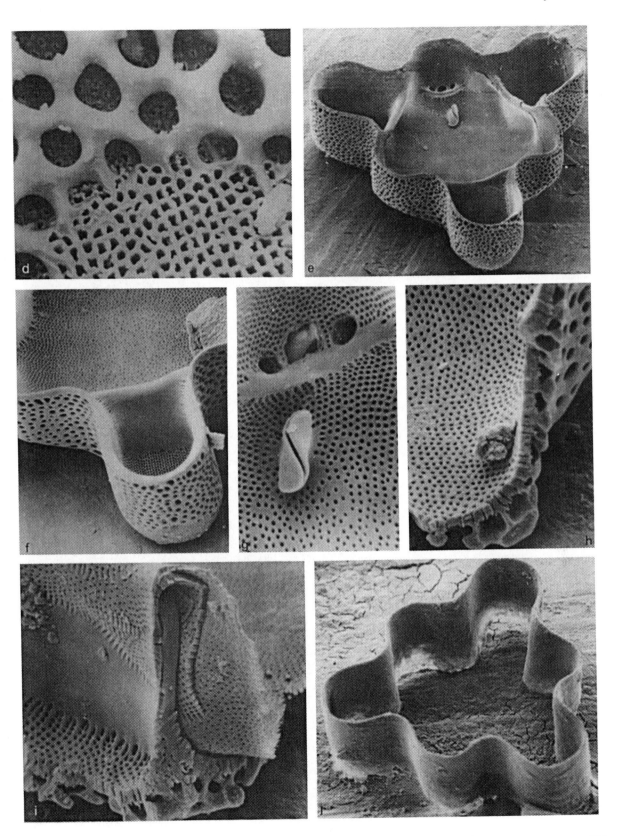

Isthmia C. A. Agardh 1832. Consp. Crit. Diat.: 55

T: *I. obliquata* (J. E. Smith) Agardh (= *Conferva obliquata*; lectotype selected by F.T. Kützing 1833. Linnaea 8: 579, 610)

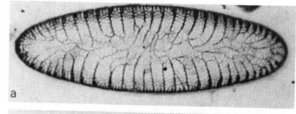

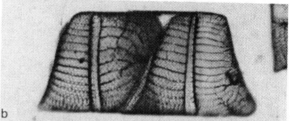

Cells either rhomboidal or trapezoidal in girdle view[b, c] and attached by massive mucilage pads to form complex branching colonies. Plastids discoid. Cells epiphytic on seaweeds, e.g. species of *Polysiphonia, Ceramium,* etc.

Valves oval (bipolar) in valve view and with no distinction into valve face and mantle.[a-c] Valves heteropolar, one pole being much more elevated than the other. Frustules slightly heterovalvate, one valve being produced into a more bulbous extension than the other and having a conspicuous pseudocellus at this pole.[e, h] Valves with a massive siliceous framework more-or-less equally developed on both the outside and inside of the valve (*I. enervis*)[e, f] or the internal framework enlarged in a 'girder-like' manner[g-j] (*I. nervosa*). Areolae elliptical to polygonal, containing complex thick cribra.[f] In addition simple pores penetrate the framework in various places. There are no obvious rimoportula openings externally, but internally bulbous and slightly flattened rimoportulae can be found clustered on the valves;[(arrow. h)] the exact distribution has not been determined. A large flanged pseudoseptum extends inwards from near the edge of the valve mantle;[g] this is supported by struts arising from the valve surface.[j] Copulae complete, with smaller areolae and cribra than on the valve – both valve (below) and copula (above) are figured in Fig. f. The pars interior of the valvocopula forms a strutted 'septum'[i] which hooks over the valve pseudoseptum. The copula associated with the epivalve of the trapezoidal cells of *I. enervis* has a distinct notch.[d, e]

This is a most striking genus with many complicated features (see Round, 1984a, for further detail). It is not easy to fit into the classification of the centric group (see Round & Crawford, 1981) – most authorities have placed *Isthmia* in the Biddulphiaceae.

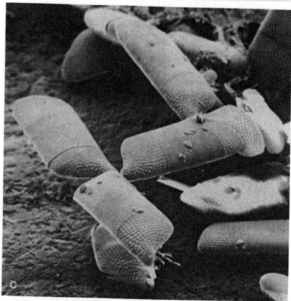

Trigonium P. T. Cleve 1868. Ofvers Kongl.
Vetensk.-Akad. Forhandl, **1867**: 663

T: *T. arcticum* (Brightwell) Cleve (= *Triceratium arcticum*)

Cells tri-[a] to multiangular, rectangular in girdle
view.[b] Plastids discoid. A marine genus growing in
zig-zag chains attached to seaweeds; cosmopolitan
but most widely distributed in warm waters. Also
fossil, from the Late Eocene to the present.

Valves triangular to multiangular, elevated at the
corners;[c] with deep, vertical mantles. Surface
apparently smooth but at high magnification small
granules can often be seen.[d, e] Areolae more-or-less
simple, arranged in rows radiating from the centre,[a]
becoming smaller at the angles to form pseudocelli;[f]
closed externally by a layer of silica in which there
is a ring of usually six pores, each closed by rotae.
A septum occurs at each corner.[h, i] Our figure
suggests that there is a septum on the valvocopula.
Rimoportulae low lying, clustered in the centre of
the valve;[j] their external openings are
inconspicuous. Cingulum deep, containing porous,
complete copulae.

Hendey (1964) clarified Cleve's description and
rejected later modifications. His concept of the
genus seems now to be accepted (e.g. by Glezer,
1975 and Hoban, 1983) although Simonsen (1974)
has a different concept, using the name for
Triceratium alternans. Keeping to the Cleve–Hendey
concept helps to bring some order to the confusion
of tri- and multiangular taxa. Hoban (1983)
discusses and figures the type species of *Trigonium*;
our material of another species (*T. formosum*?) is
clearly in the same genus.

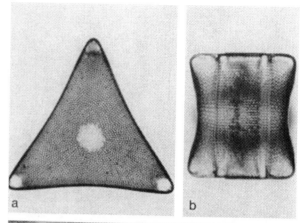

a   b

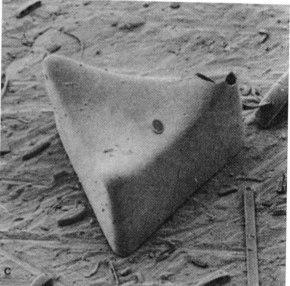

c

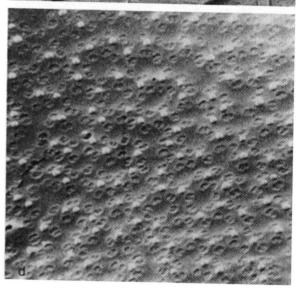

d

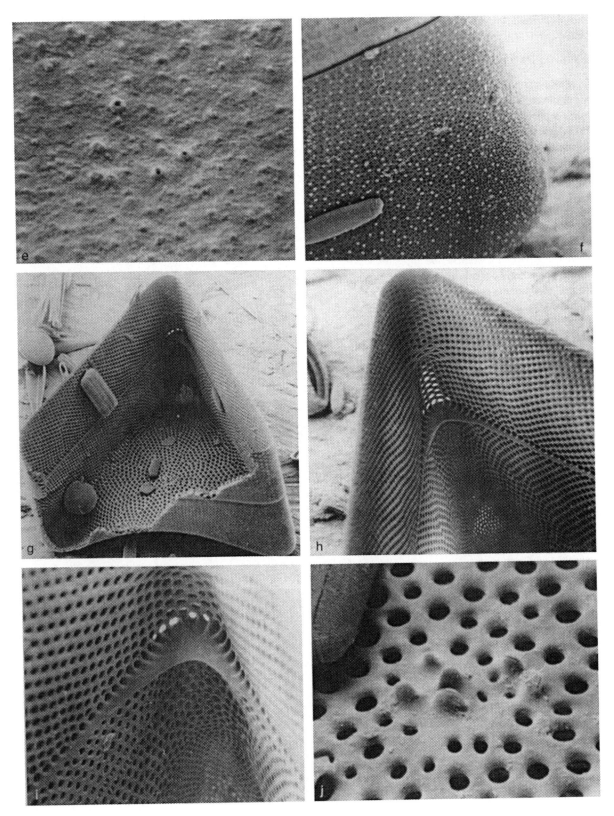

Terpsinoë C. G. Ehrenberg 1841. Abh. Königl. Akad. Wiss. Berlin, 1843: 402

T: *T. musica* Ehrenberg

Cells tabular, notched, with conspicuous pseudosepta in girdle view; markedly elongate, undulate in valve view. Plastids not observed. Forming epiphytic zig-zag colonies in brackish/freshwater; often on wet rock faces in tropical regions.

Valves tri-undulate[a-d] with narrowed slightly capitate apices ending in finely porate pseudocelli. Valve mantle sharply downturned. Valve surface corrugated with irregularly sized and spaced areolae; no obvious annulus. Corrugation somewhat thicker between valve face and mantle and finer on valve mantle. An inconspicuous indentation[arrow, d] occurs where the rimoportula opens. Internally with conspicuous pseudosepta[f] between constrictions; sometimes other lesser developed pseudosepta present especially near apices; irregularly porate with an off-centre rimoportula in the central inflation. Rimoportula with an S-shaped slit.[g, h] Copulae several, plain and complete.[b, i]

Terpsinoë is a very distinctive genus with a very precise ecological distribution. Its structure and ecology are very close to that of its relative *Hydrosera*, cf. the valve surface, pseudocelli, pseudosepta and rimoportula. In girdle view the pseudosepta resemble musical notation – presumably the derivation of the specific epithet of the type. The study of auxospores of *T. musica* by Müller (1889) is one of the earliest and illustrates the intimate relationship between the initial hypovalve and the mother cell.

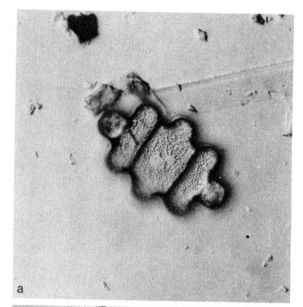

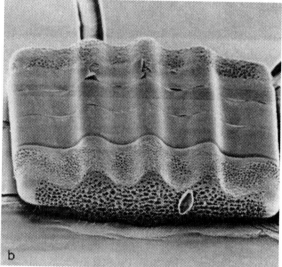

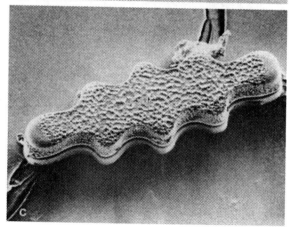

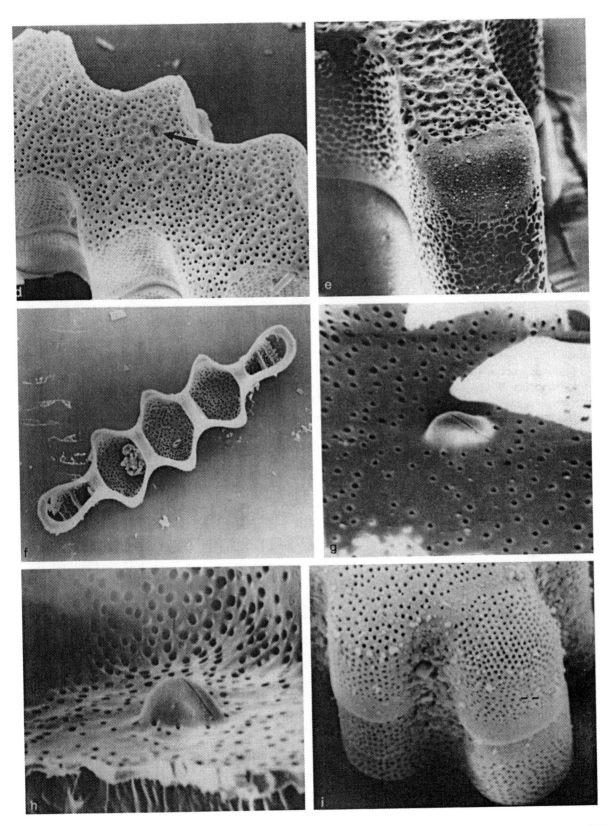

Pseudotriceratium A. Grunow 1884. Denk. math-nat. Kl. Kaiserl. Akad. Wissensk. **48**: 83

LT: *P. fallax* Grunow (lectotype selected by Glezer, 1975)

Cells triangular in valve view. Habit and plastids not known. A marine planktonic genus possibly mainly tropical/subtropical in distribution. Species of the genus have formerly been placed in *Triceratium*.

Valves triangular.[a, b] Valve face flat to slightly bowed; valve mantle indistinct. External areolation of clustered pores in central region merging into isolated pores elsewhere.[b-d] Cribra not observed. Internally with large circular apertures;[i] valve loculate. Rimoportulae occur in the apices only,[g, h] or there and along the lateral margins; in some species an additional central rimoportula occurs. External openings of the rimoportulae not expanded other than into a slight flange around the oval aperture;[d-f] internally short, sessile. Copulae not observed.

One of our specimens seems close to Simonsen's (1974) illustration of *P. punctatum* (see Simonsen for numerous synonyms). The genus differs greatly from *Triceratium*, notably in the absence of ocelli, and from *Trigonium* by the lack of both pseudocelli and the central cluster of rimoportulae. As with other triangular genera, *Pseudotriceratium* also needs a detailed study.

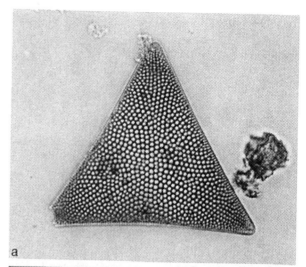

a

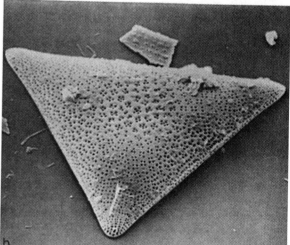

b

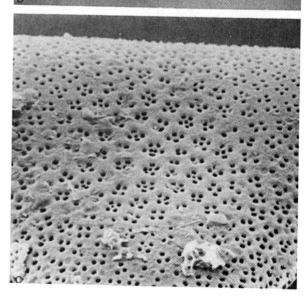

c

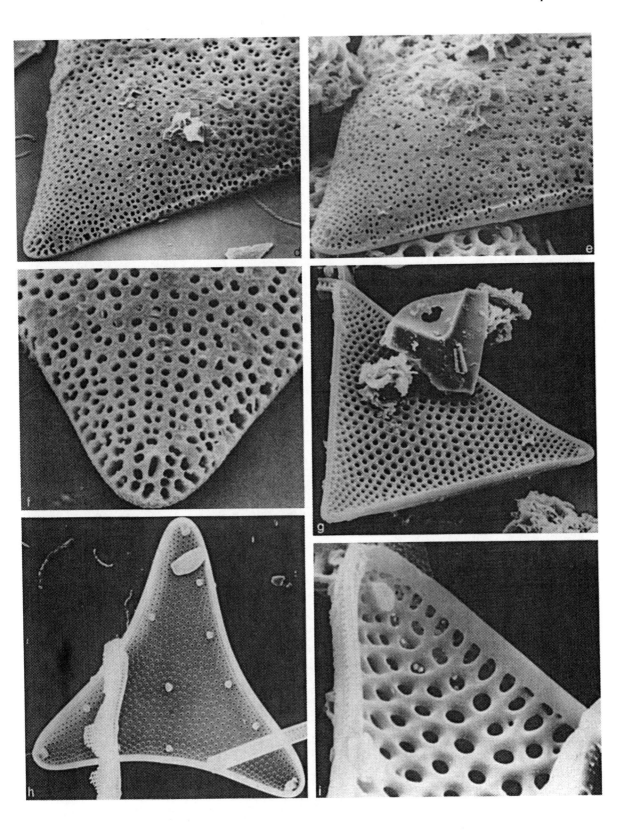

Hemiaulus C. G. Ehrenberg 1844. Ber. Bekanntm. Verh. Königl. Preuss. Akad. Wiss. Berlin, **1844**: 199

T: *H. antarcticus* Ehrenberg

Cells united by short or long processes to form straight[a] or curved chains. Plastids small, discoid. A common marine planktonic genus with many fossil representatives (up to 90 spp. according to VanLandingham, 1971) but many require re-investigation (see Ross & Sims, 1985, for the removal of some to *Briggera* – also p. 272).

Valves elliptical[i] with long thin processes[b–f] linking by apical spines (the species originally in *Hemiaulus* but having bulbous endings to the processes no longer belong here). Valve face curved, merging imperceptibly with the deep mantles. Areolae: simple round pores (*H. hauckii*)[c, h] or large elliptical to rectangular holes closed by complex cribra[b, g, i] (*H. sinensis*: figs 20–33 of Ross, Sims & Hasle, 197). One rimoportula usually present in a central or offset position[g] (cf. *Eucampia*) but sometimes absent. Copulae more finely areolate[i] than the valves, split with pointed ends.

At the LM level *Hemiaulus* is not usually confused with *Eucampia* but in the electron microscope there are similarities (see also Syvertsen & Hasle, 1983). The absence of pseudocelli of the *Eucampia* type we regard as sufficient to maintain the genera separate. The three genera *Hemiaulus*, *Eucampia* and *Climacodium* form a distinct group of genera with two elongated processes on each valve, both processes linking to form chains. However, *Eucampia* has distinct features (e.g. the costate ocellus) allying it to *Cerataulina*. Recently Ross & Sims (1985) have erected the genus *Briggera* to encompass a number of fossil species which have swollen ends to the processes and interlocking spines (see p. 272). Sims (1986), in describing two genera from the Eocene not illustrated here, *Ailuretta* Sims and *Sphynctolethus* Hanna, provides a link with present-day members of the subfamily Hemiauloideae, namely *Cerataulina*, *Eucampia* and *Hemiaulus*. There is little doubt that selection since the Cretaceous has eliminated many species of *Hemiaulus*, leaving only the lightly silicified, rather delicate forms.

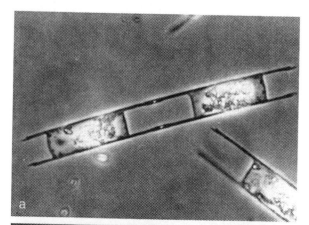

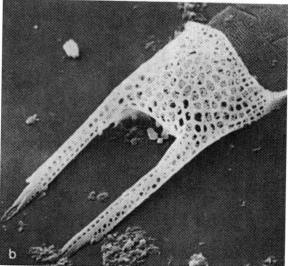

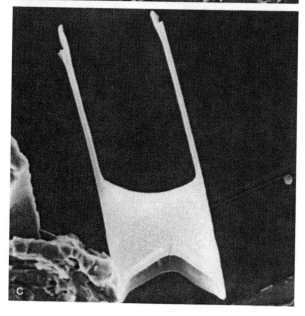

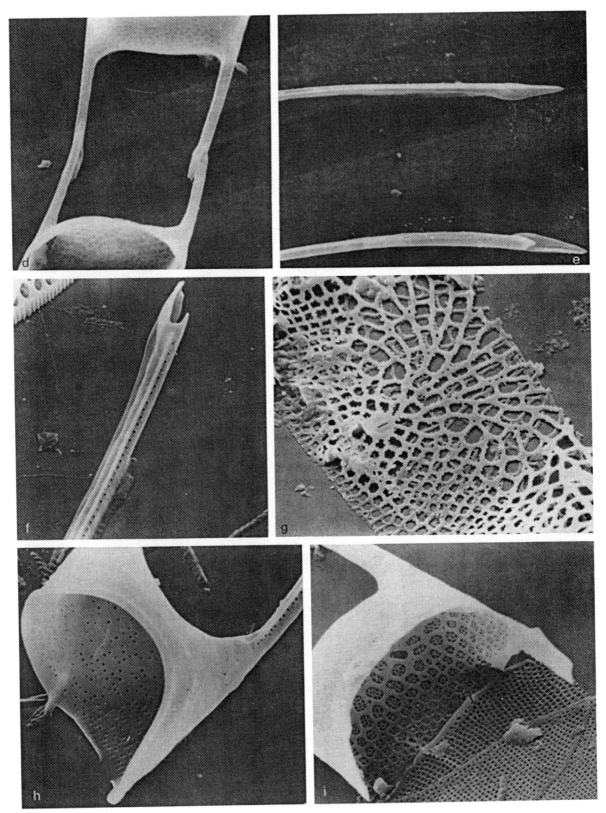

Eucampia C. G. Ehrenberg 1839. Ber.
Bekanntm. Verh. Königl. Preuss. Akad. Wiss.
Berlin, **1839**: 156

T: *E. zodiacus* Ehrenberg

Cells interlocking by two apical elevations to form chains,[a, b] often curved in spirals due to slightly unequal development of the valves and girdle. Plastids many, discoid. A common marine planktonic genus of only 5 species (VanLandingham, 1969).

Valve elliptical to linear.[d] Rows of large areolae radiate from a central annulus[d, f] within which a rimoportula and some areolae are located. The vela are complex cribra,[e–i] resembling those in *Achnanthes* and *Isthmia*. Apical elevations flattened, with rows of small pores separated by ridges[e, i, j] which may[i] or may not[e] cross the central area: this structure is termed an ocellus and certainly functions as such, but its structure is rather unlike that of the ocelli in, for example, the Eupodiscaceae; it does, however, resemble that of *Cerataulina*. Simple pores occur in the valve framework[g, h] particularly where new rows are intercalated. Internally the valve is without any special differentiation except that the single central rimoportula protrudes;[h] according to Ross, Sims & Hasle (1977), there is slight development of internal costae. Copulae numerous,[c] areolate,[k] split and ligulate. The areolae are smaller and with more regular pores than on the valves. Resting spores are formed inside vegetative cells and according to Hoban *et al.* (1980), the first valve has ocelli and coarser valves, while the second has pointed processes but no ocellus. These spores (of *E. balaustium*) have been described in the literature as separate species of *Eucampia*. See also Steyaert & Bailleux (1975a, b) who also report that valves at the ends of filaments do not have ocelli – these may therefore be separation valves.

This genus is very close to *Climacodium* and *Hemiaulus* but there are no spines at the apices of the elevations as in *Hemiaulus* and a ridged ocellus is prominent in *Eucampia* but not in *Hemiaulus*. Ross *et al.* (1977) discuss the relationships in more detail and report that *E. balaustium* is congeneric with the type species *E. zodiacus*. Syvertsen & Hasle (1983) give a detailed account of *Eucampia* with which the genus *Muelleria* is synonymous.

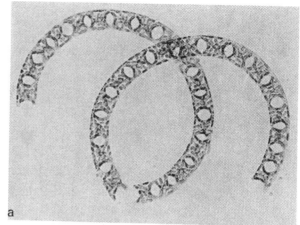

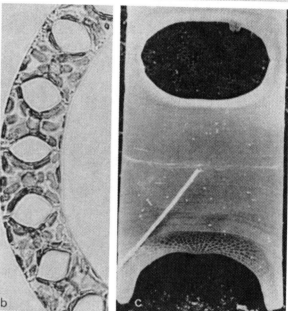

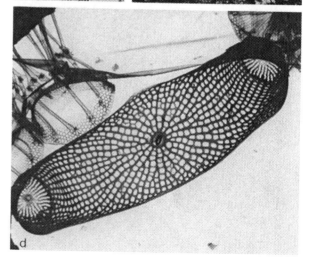

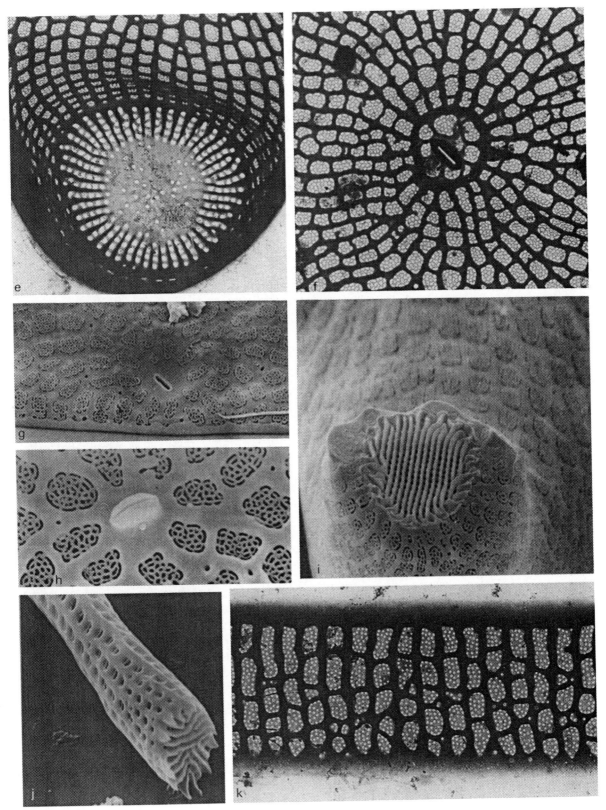

Climacodium A. Grunow 1868. Reise Fregatte Novara, Bot. 1, 102.

T: *C. frauenfeldii* Grunow

Cells H-shaped, joined by the extended valve apices to form straight filaments.[a-c] Plastids numerous, discoid. A marine plankton genus mainly distributed in tropical/sub-tropical oceanic waters.

Valves narrow with apical processes and deep mantles without obvious linking structures at apices of the processes.[c, h] The valve surface appears structureless in the SEM,[b-h] except for a small, rimmed pore midway between the two processes.[g] This pore may not be present on all valves and varies in its appearance.[f, g] Cingulum not observed and may be lacking except during division (Hustedt, 1927–66). Only an indistinct line[arrow, e] indicates the junction between the two thecae of a cell.

Cells are presumably linked into filaments by mucilage extrusion at the apices of the processes. Apart from the type species only one other valid species, *C. biconcavum* Cleve, has been described – originally as *Eucampia*. The genus is in need of study; TEM is necessary to determine the nature of the valve though there may be no areolation and little silicification. Its taxonomic position near *Eucampia* is based merely on general form and a thorough study especially of cultured material (cf. that on *Streptotheca* by von Stosch) should be made.

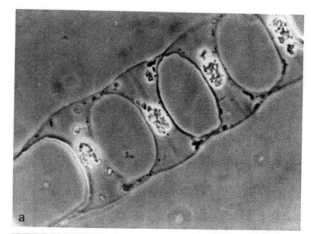

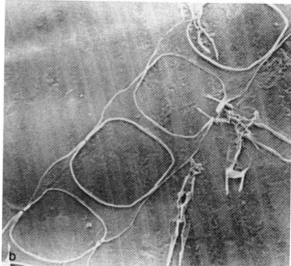

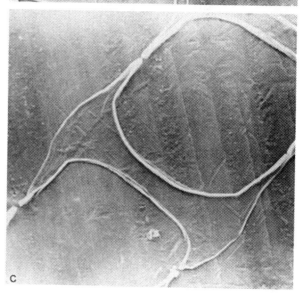

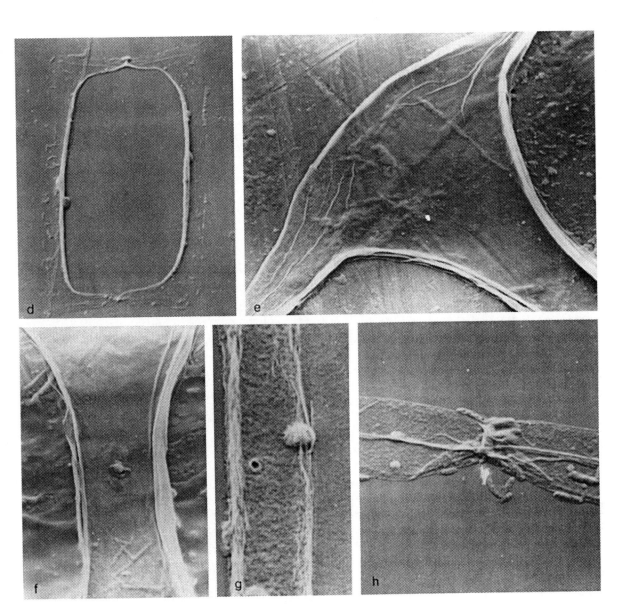

Cerataulina H. Peragallo 1892. Diatomiste 1: 103

T: *C. bergonii* (H. Peragallo) Schütt (= *Cerataulus bergonii*)

Cells narrowly cylindrical, joined in chains.[a-c] The straight or slightly twisted chains of cells appear to have the valve faces fairly closely apposed, but the linkage between them is not strong and in the absence of 'hooked' interlocking devices the filaments easily break up into individual cells (cf. *Lauderia* and *Guinardia*). Two small submarginal elevations are present on each valve. Plastids discoid, lying in the peripheral cytoplasm. A delicate marine planktonic species common near coasts and penetrating into brackish waters.

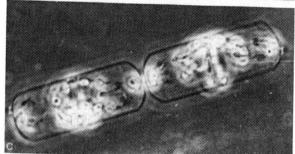

Valves circular, slightly mis-shaped[d, e, h] by indentations to accommodate the spines of sibling valves, and by the formation of two wings[d-h] subtending the two ocelli[d] and extending in opposite directions. The wings have a spine close to the ocellus and then continue, forming a low ridge running round onto the mantle. The ocellus has very distinct tangential bars.[d-f, i] The valve mantle is deep. The areolae are developed in a thin framework of radiating costae[l] and are closed by fine porous cribra of varying complexity.[i-l] Occasional simple pores also occur.[j-l] A single rimoportula with a slit-like outer opening is located marginally (*C. daemen, C. dentata*)[e] or centrally (*C. pelagica*).[h] Internally the low rimoportula is obvious[j] and the bars of the ocellus are continued to the inside. Copulae numerous,[n] split, with ligulae and antiligulae, the splits occurring adjacent to each other or staggered.

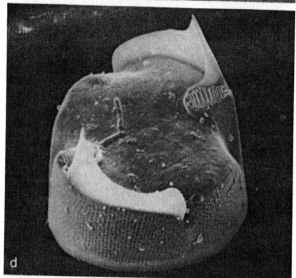

Hasle & Syvertsen (1980) give an excellent detailed account of *Cerataulina* and its resting spore, described earlier as a separate genus, *Syringidium*. The systematic position of this genus, close to *Eucampia*, depends on the valve/areola structure and on the similarity of the ocellus. As others have commented, the *Cerataulina/Eucampia*-type ocellus (termed the 'costate ocellus' by Syvertsen & Hasle, 1983) is somewhat specialised and should not be used as a feature to associate these genera with others having different kinds of ocelli, such as are found, for instance, in the Eupodiscaceae.

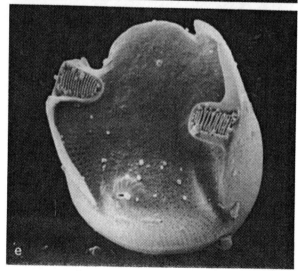

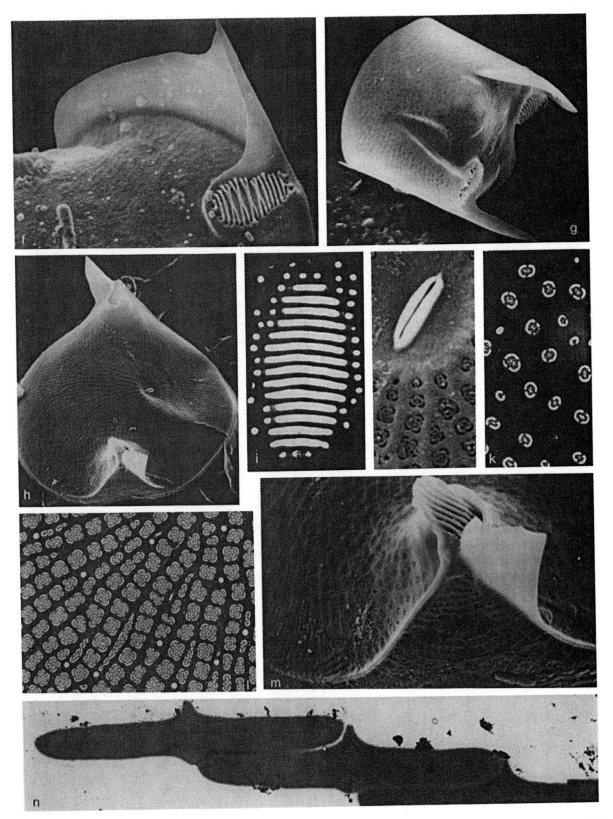

Trinacria P. A. C. Heiberg 1863. Consp. Crit. Diat. Dan. **49**

LT: *T. regina* Heiberg (Vide C. S. Boyer, Proc. Acad. Nat. Sci. Philad. 78, Suppl. 142, 1927)

Cells tri- (quadr-)angular, attached in chains by the extended apices. A marine, fossil genus occurring in the Early Eocene.

Valves tri- (quadr-)angular,[a] elevated at the corners,[b, c] where short and a longer spine(s) occur and interlock neatly with adjacent cells.[d, e] Valve face raised in the centre[b, d] and variously ornamented (e.g. with small spines);[b, h] ridged along the valve face/mantle junction, the ridges running up the elevations; valve mantle shallow (the internal surface has folds in some species: see Ross, Sims & Hasle, 1977). Areolae radiating from the centre,[g, j] cribrate; continuing down the valve mantle and becoming smaller up the elevations. At the apices of the elevations and surrounded by the spines, there are a few pores[c, h] (indistinct in our specimen). Isolated simple pores penetrate the framework of valve face and mantle. Three rimoportulae occur as a central ring[small arrow, i] and three others occur near the corners;[large arrow, i, j] the internal development is slight but the external tubes can be quite long.[b, f] Ross *et al.* (1973) figure a pseudocellus on the outer face of the elevation of *T. exsculpta*. Copulae not known.

The overall morphology of this genus places it in the Hemiaulaceae (see Ross, Sims & Hasle, 1977, for a detailed discussion and their view that this group should be a subfamily of the Biddulphiaceae). Some of the fossil species have been investigated in great detail by Sims & Ross (1988). We thank P. A. Sims for material for this and the subsequent five genera.

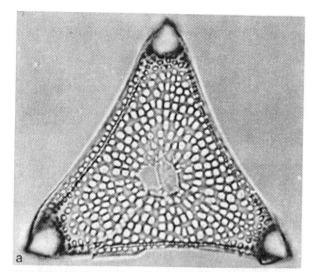

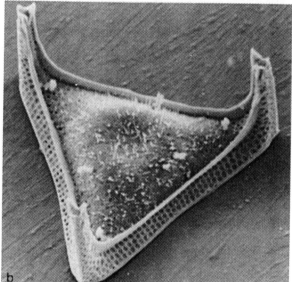

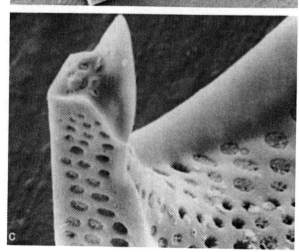

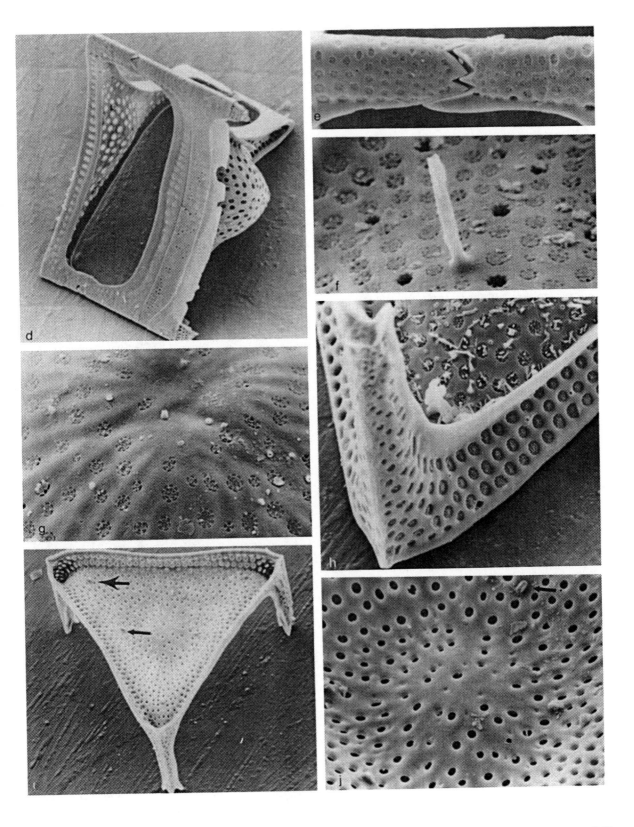

Abas R. Ross & P. A. Sims 1980. Bacillaria 3: 119

T: *A. wittii* (Grunow) Ross & Sims
(= *Syringidium wittii*)

Cells united by long processes to form curving filaments.[a] A fossil genus previously classified amongst the diatom spore forms under the name *Syringidium*, some of which must remain as spores (cf. *Cerataulina*). *Abas* is known from Eocene to Oligocene strata of tropical sites.

Valves circular with radiating areolae[b] closed by cribra. Valve mantle shallow and vertical with scattered areolae[c-f] and a slight spinose rim[d] between the mantle and valve face. The mantle edge is as wide as the porous part and has an internal flange.[g] Two massive cylindrical elevations arise from the valve rim and fuse completely with those from the adjacent frustule.[a] The elevations bend and (almost) touch centrally between the frustules[e] – here there are a few pores in the elevations on the inner (i.e. adjacent) side. Also arising from the valve rim equidistant from the elevations is a spear-like projection[c] with an apical barb near the tip. This is the exit tube of a rimoportula. A group of small spines also occur on the centre of the valve (reported in Ross & Sims but not seen by us). Valvocopula with a curved flange (pars interior) fitting inside the flange on the inside of the valve mantle.[f] The valvocopula is deep with undulating rows of areolae and a plain abvolvar rim.[f] Ross & Sims also record a further copula and one hyaline pleura. All are complete.

Ross & Sims (1980) discuss this genus in detail and comment on its systematic position. They tentatively suggest a place in the Hemiauloideae and reinforce this by comparison of its girdle structure with that of *Dextradonator*. It is a very unusual diatom in that the two elevations are completely fused. We know of only one other group in which a fusion occurs but this fusion is lateral and the setae then separate again distally (i.e. some species of the Chaetoceraceae). A fusion such as in *Abas* raises interesting points from a morphogenetic standpoint since the separate silicalemmas would need to fuse.

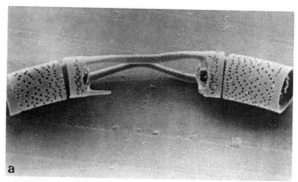

a

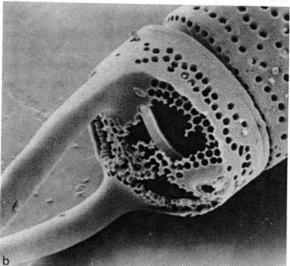

b

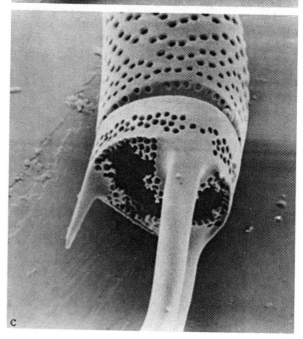

c

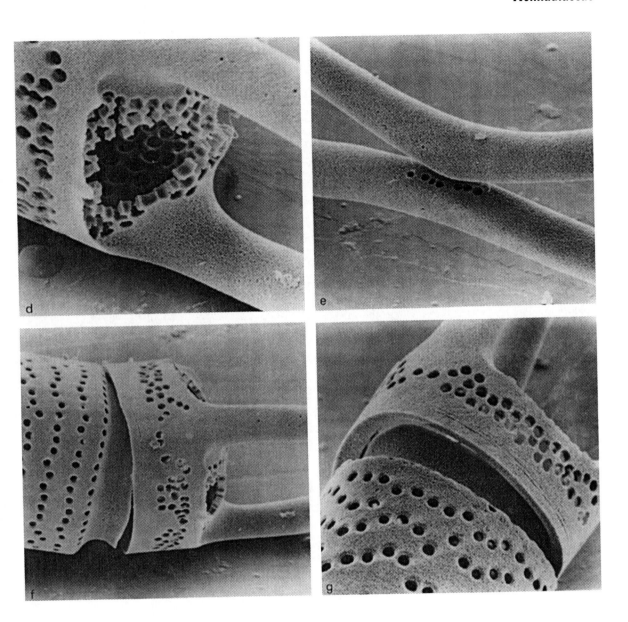

Briggera R. Ross & P. A. Sims 1985. Bull. Br. Mus. Nat. Hist. (Bot.) **13** (3): 291

T: *B. ornithocephala* (Greville) Ross & Sims (= *Hemiaulus ornithocephalus*)

Cells 'hemiauloid' thus forming colonies linked by apical elevations.[b] The genus is fossil and has been separated from *Hemiaulus* by Ross & Sims (1985). Found in marine sediments of upper Cretaceous to Lower Miocene age.

Valves elliptical,[a] undulate with the two apical swellings extended into massive elevations ending in swollen apices.[b, e, g] Valve surface areolate with simple but deep pores[c, d, f] penetrating the silica; often with scattered spines. The central region of the valve face is separated from the ends by clear transverse bands of silica.[c] The valve mantle is plain or slightly areolate and raised into a rim at the valve face/mantle junction.[c] The elevations have a porous area forming a pseudocellus[b, c, e–h] on each apical swelling, facing outwards. The inward face of the swelling is produced into a series of ridges or spines[e–f] (5 or fewer) which interlock with those on the adjacent valve. Internally the only conspicuous features are the two ridges of plain silica traversing the valves.[i] Rimoportulae absent or 1–5 scattered near the centre of the valve; each has a short external tube and an internal slit (rimoportulae were not present on the material we use for illustration). We have not seen the cingulum but it is reported to consist of 3–4 (or more) open copulae (Ross & Sims, 1985).

Ross, Sims & Hasle (1977) considered the genus *Hemiaulus* in detail and at that time species such as *capitatus*, *includens* and *haitensis* were included and figured: these should now be placed in *Briggera*. Ross & Sims (1985) have dealt with the new genus in a most comprehensive manner, together with *Strelnikovia*, *Keratophora*, *Thaumatonema* and *Dicladiopsis*. *Hemiaulus* differs from *Briggera* principally in having much more slender elevations, which have apical spines that do not interlock and much smaller or almost non-existent ocelli. In addition, the valve of *Hemiaulus* does not have such clear transverse plain areas on the valve face and the areolae are more closely spaced.

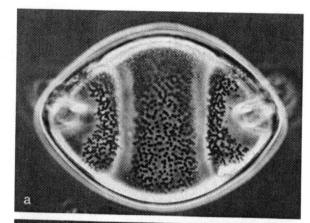

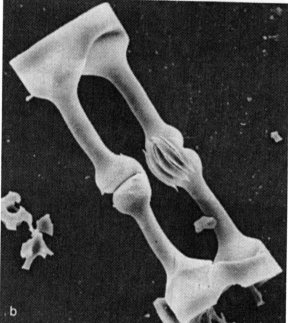

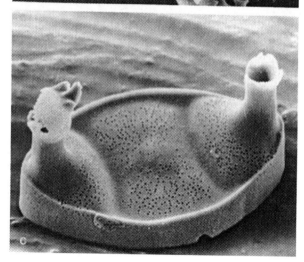

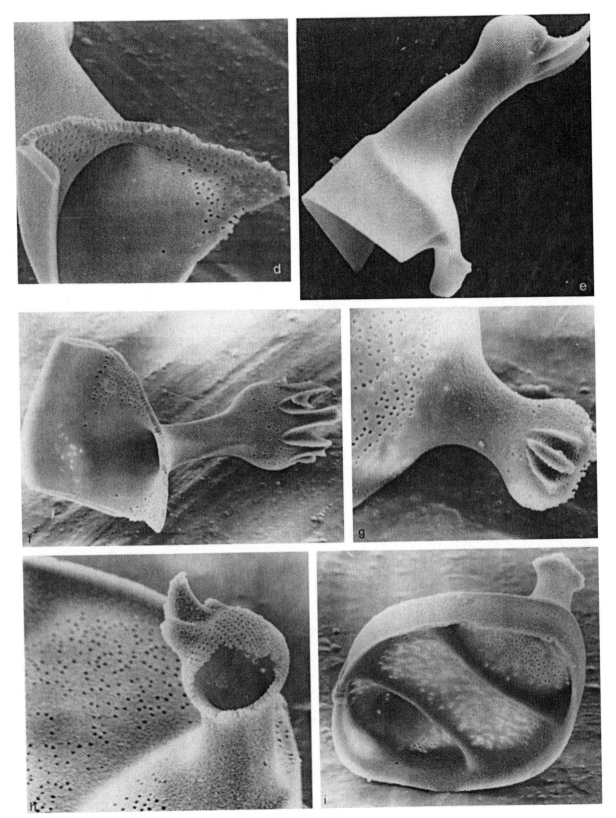

Pseudorutilaria (E. Grove & G. Sturt ex De Toni & Levi) Grove & Sturt ex De Toni 1894. Syll. Alg. **2**: 854

T: *P. monile* Grove & Sturt ex De Toni

Cells bipolar, with apical elevations. A fossil marine genus found mainly in the Eocene but extending into the lower Oligocene; it is monotypic if *Rutilariopsis* is a valid genus (see VanLandingham, 1978).

Valves linear to linear-lanceolate, with crenulate margin and a marked central expansion.[a, b] Valve face undulate, domed at the centre, and with a marginal flange separating it from the shallow vertical mantle which has an upturned edge;[c-f] apices elevated. Areolae distant and scattered on the central swelling, more closely spaced over the remainder of the valve and mantle. On the central swelling there appear to be one or two external tubes, which might possibly be the exit tubes of rimoportulae. The apical elevations probably contain ocelli.

The specimen is eroded and we cannot determine adequate detail for a full description. The affinities probably lie with the genera *Solium* Heiberg and the new genus *Monile* Ross & Sims. We have not been able to illustrate these or the other new genera *Maluina* Ross & Sims and *Bonea* Ross & Sims. For a detailed description of species of *Pseudorutilaria* and the new genera, see Ross & Sims (1987).

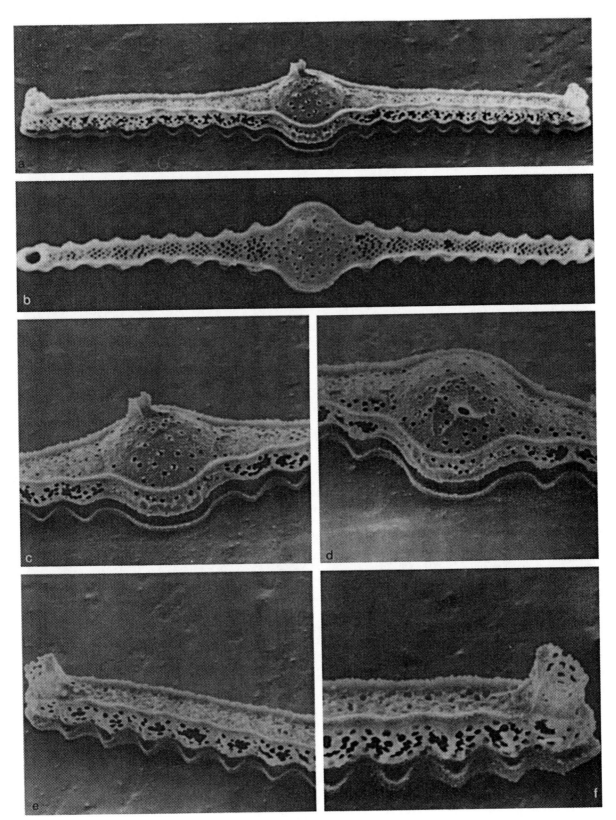

Keratophora J. Pantocsek 1889. Beitr. Kennt. Foss. Bacill. Ungarns 2: 85

T: *K. nitida* Pantocsek

Cells ellipsoidal in valve view with two elevations arising from transapical ridges; the elevations unite the cells into inseparable chains (Ross & Sims, 1985). Rectangular in girdle view. A small fossil genus (3 spp. in VanLandingham, 1971, but monotypic according to Ross & Sims, 1985) known from Hungary and Russia, in deposits of Eocene age; presumably marine.

Valves oval, traversed by two raised, transapical ridges from the centres of which arise two elevations;[a, b] these bend towards the apices but finally bend again so that the flat tops of their slightly swollen apices lie parallel to the plane of the valve.[c, e, f] The elevations have pores radiating from the outermost point, forming a lateral ocellus.[d-f] The flat top is also porous[d, g] but less densely so, and is surrounded by a ring of ridges (these are the remains of eroded linking spines, which are illustrated well by Ross & Sims, 1985); in our material the ridges are higher on the side opposite the ocellus. The valve surface is areolate with an indistinct radiating pattern; the areolae are round. No vela are present in the material investigated, but Ross & Sims illustrate a substantial velum. There is a prominent, vertical valve mantle which at high magnification appears to be finely ridged or possibly finely porous.[e, f] There are no obvious portules either in our material or that of Ross & Sims. The cingulum is illustrated in Ross & Sims and consists of porous bands, which are split and ligulate; further work is necessary to reveal the full detail.

This is a genus close to *Kittonia* though differing in the structure of the processes which are reminiscent of those in 'hemiauloid' genera. According to Ross & Sims the genus is monotypic; their excellent paper also gives details of the history of the genus and compares it with *Dicladiopsis*, *Strelnikovia* and *Thaumatonema*, all fossil genera which we have not been able to study.

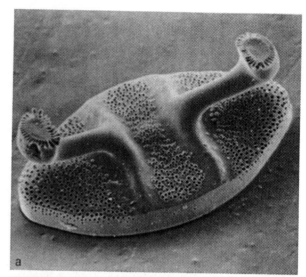

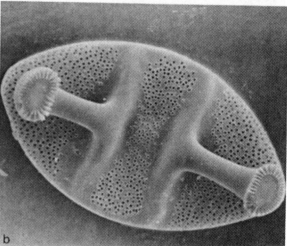

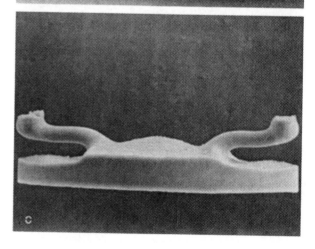

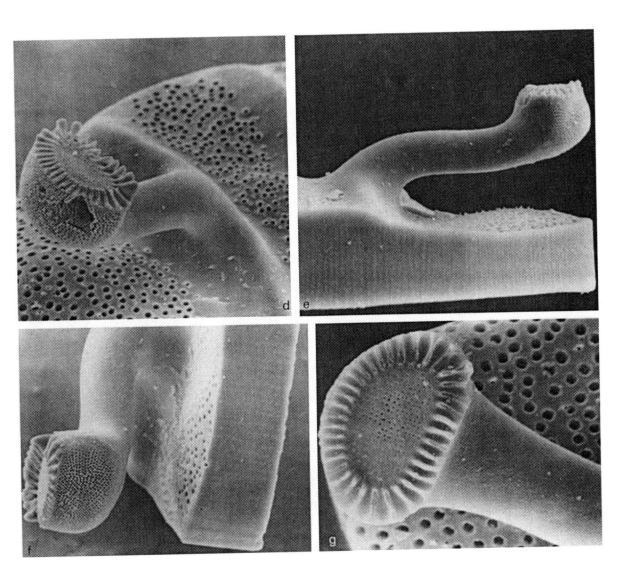

Kittonia E. Grove & G. Sturt 1887. J. Quekett
Microsc. Club, Ser. 2(2): 74

T: *K. elaborata* (Grove & Sturt) Grove & Sturt
(= *Biddulphia elaborata*)

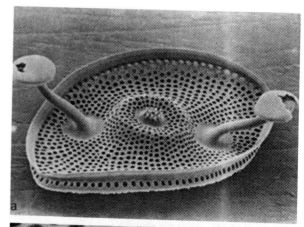

Cells elliptical in valve view with two conspicuous
processes ending in flattened swellings. A fossil
genus of nine species (VanLandingham, 1971) but
requiring re-investigation. Prominent in the Eocene
Oamaru deposit of New Zealand.

Valves elliptical, shallow, with a prominent
marginal flange and two (in some species three)
extended processes.[a, c] These arise from clear areas
of the valve face about half-way between centre and
margin; the axis on which they lie is at
approximately 45° to the apical and transapical axes
of the valve. The processes are at first plain, tubular
and bent outwards, but end in flattened swellings,
lying parallel to the valve, perforated by rows of fine
pores forming a type of 'ocellus'.[c, i] A clear central
area is present bearing a cluster of short tubes
externally[b] (the openings of the rimoportulae).
Surrounding this central area the valve rises and
then falls and rises again to the marginal flange.[a, d]
Rows of large areolae radiate from the centre, the
areolae containing fine cribra.[h] Internally the
framework is conspicuously ribbed between the
uniseriate rows of areolae.[c-h] Fine slits/pores occur
on the ribs both on the tops[arrow, g] and sides;[arrow, h]
these open to the outside through the simple pores.[b]
A narrow valve mantle extends beneath the
marginal flange and is provided with a row of
areolae.[a, d] The mantle extends beneath the row of
areolae and in this region is finely perforate.
Rimoportulae clustered, central, with tubular
extensions externally, but sessile internally and lying
in a slight depression.[g] Small pores occur around
and between the rimoportulae. We have no data on
the cingulum.

Kittonia is one of the most striking of diatoms. Its
classification in the 'biddulphioid' section is based
on the assumption that the pores at the apices of the
processes form ocelli. The structure of the vela is
very much like that in *Biddulphia*.

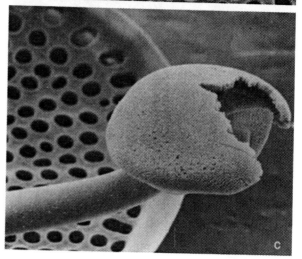

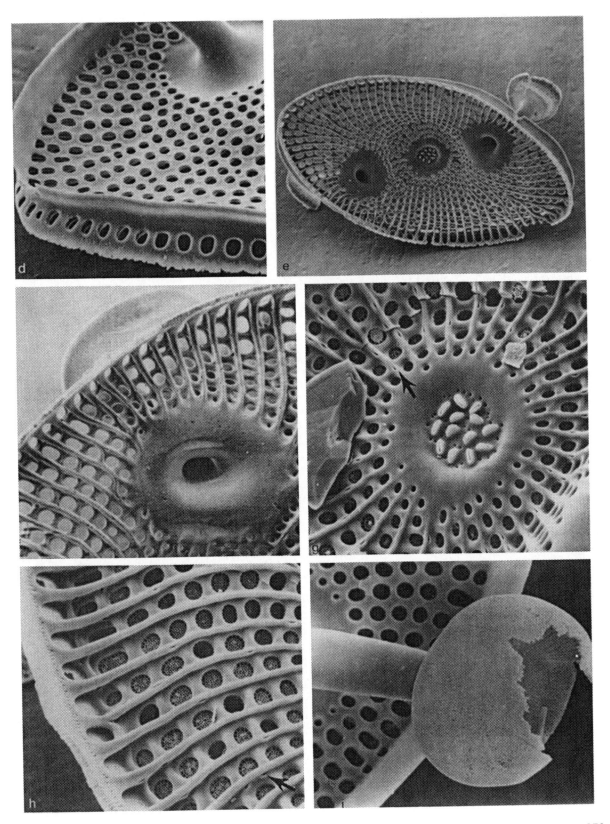

Bellerochea H. van Heurck 1885. Syn. Diat. Belgique: 203

T: *B. malleus* (Brightwell) van Heurck
(= *Triceratium malleus*)

Cells rectangular in girdle view, joined to form long filaments in which small gaps occur between cells due to curvature of the valve faces.[a, b] Cells weakly silicified. Plastids numerous small plates. A widely distributed marine, planktonic genus, containing three species.

Valves triangular (sometimes bi- or quadrangular); valve face depressed slightly internal to the angles. Face and girdle formed by an open meshwork of siliceous costae radiating from a central annulus,[h, i] or not even reaching the centre (*B. malleus*); costae sometimes forming small 'C' shaped swirls.[h] Minor costae link the principal ones to delimit areolae; no vela can be detected. At the junction between valve face/mantle a slight marginal ridge is developed [i] and at the angles (elevations) of the valve there is what appears to be an ocellus, with distinct longitudinal ridges.[d] The marginal ridges interlock or even fuse with those of adjacent cells. Rather delicate rimoportulae occur (arrows) either centrally or marginally; they have long external tubes.[h] Elements of the cingulum numerous [e, j] but even more lightly silicified than the valves and we have not been able to resolve any detail (but see von Stosch, 1977).

The genus is discussed in great detail in the excellent study of von Stosch (1977) who discusses the diatotepic layer forming a 'wall' internal to the siliceous framework. The only point we can add and which is significant is the possible formation of an ocellulus which, if confirmed, will require a re-assessment of the classification of this genus and possibly ally it to genera of the Hemiaulales rather than the Lithodesmiales.

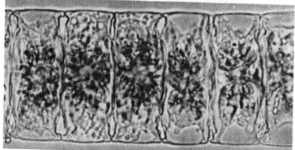

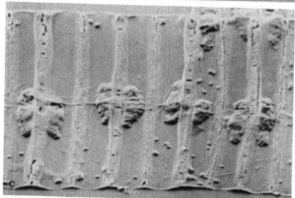

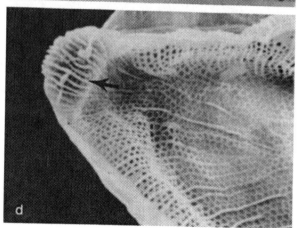

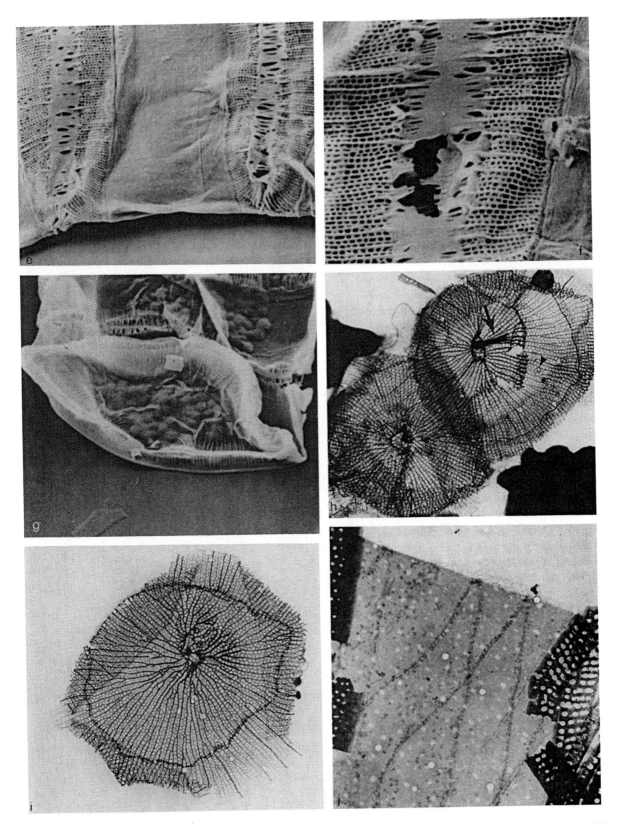

Subsilicea H. A. von Stosch & B. E. F. Reimann 1970. Nova Hedw., Beih. **31**: 13

T: *S. fragilarioides* von Stosch & Reimann

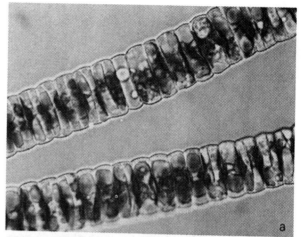

Cells firmly linked together to form long filaments; oblong with rounded apices in girdle view.[a] One plastid (rarely up to 8 according to von Stosch & Reimann, 1970). Found in the marine littoral associated with sediments but carried up into the plankton by turbulence.

Valves linear to elliptical;[c, f] very lightly silicified, the siliceous part comprising a central rib (sternum) from which lateral extensions traverse the valve face[c, d, g] and continue down the valve mantle, branching at their ends. There is some indication of a ridge (upward extension)[g] along the junction of the valve face and mantle but this is not obvious in von Stosch & Reimann's TEM sections. At the apices of the sternum there are small semi-circular areas of pores,[e, h] which appear from their structure and position to be clusters of porelli forming small ocelli (ocelluli). There are no true areolae or cribra but the lightly silicified framework is supported by a relatively well-developed organic component (see von Stosch & Reimann). There are no portules or other valve organelles. The cingulum is most striking, being composed of numerous scales.[j, k] Each scale has a thicker pars media and a slightly thickened abvalvar edge, the remainder being merely a delicate membrane with a fimbriate advalvar margin. The scales vary in size across the cingulum. The scale forming the valvocopula is apparently a complete hoop; the adjacent scales are elongate while the others become shorter and more numerous towards the abvalvar edge of the cingulum.

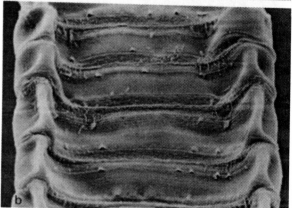

Von Stosch & Reimann (1970) placed this genus in the Fragilariaceae but it seems most unlikely that it can remain there. On valve structure there are similarities to *Papiliocellulus* but these we regard as superficial due simply to the extremely slight silicification. Scale-like cingula are rare in pennate genera (though there are scales at the apices of *Rhabdonema*). From habitat considerations and structure, it is more likely that *Subsilicea* is an unusual centric genus perhaps in, or close to, the Bellerocheaceae.

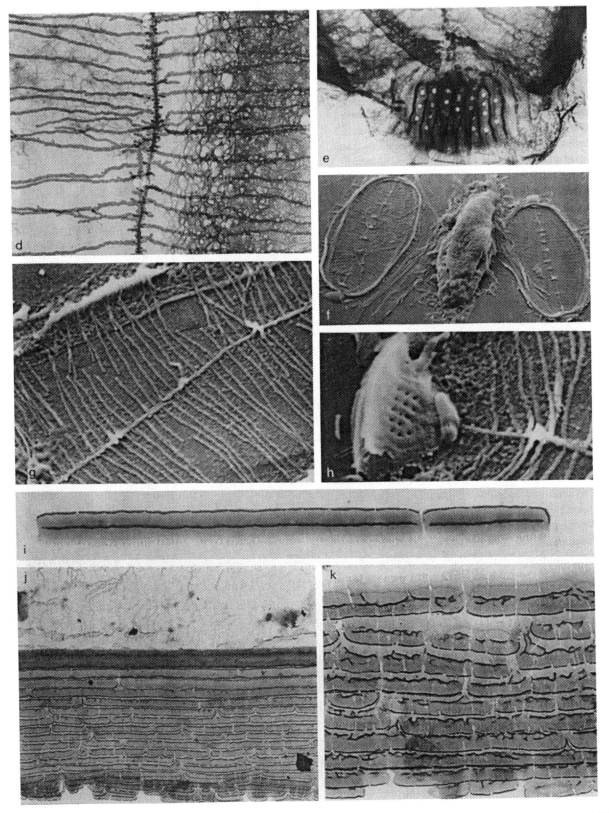

Streptotheca W. H. Shrubsole 1890. J. Quekett
Microsc. Club, Ser. 2 (4): 259

T: *S. thamesis* Shrubsole

Cells joined to form long, ribbon-like filaments,
which are often twisted; in these there are no
intervening apertures between the cells, which are
square in girdle view.[a] Plastids numerous, discoid,
often located in cytoplasmic strands radiating from
the central nucleus. A marine planktonic genus.

Valves linear[b] to elliptical, rounded at the apices
and sometimes slightly swollen in the centre. Valve
face flat; mantle very shallow,[d] slightly more
pronounced at the apices[f] where it bears slit-like
openings. The valve is unusual in that it consists of
a simple flat lamina, which is porous but without
any ribbing, except for slight vein-like thickenings[d, e]
at isolated points on the valve face and along the
valve mantle (see also von Stosch, 1977). The
slight thickenings on the valve surface are
sometimes associated with slightly more pronounced
pores[d-f] than those scattered over the remainder of
the valve. The areolae are apparently without vela.
There is a central rimoportula with a short external
tube[arrow, h] (see also Hasle, 1975). The cingulum
consists of numerous simple porous bands,[b, c, g-i]
which both Hasle and von Stosch thought might be
half-bands: we still cannot resolve this point. In this
diatom the greater part of the cell covering is
composed of copulae.

The taxonomic position of this diatom has been
discussed by both Hasle (1975) and von Stosch
(1977), possible places for it being near the
Chaetoceraceae or Rhizosoleniaceae, or in the
cluster containing *Lithodesmium*, *Bellerochea* and
Ditylum. Von Stosch clearly preferred a position in
the latter cluster, associating it particularly with
Neostreptotheca and *Bellerochea*. The valve structure
of *Bellerochea* is not at all like that of *Streptotheca*
but for the moment we see no better alternative.

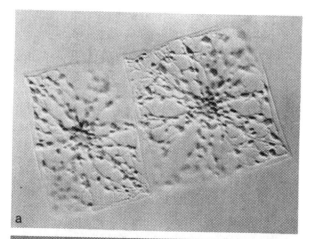

a

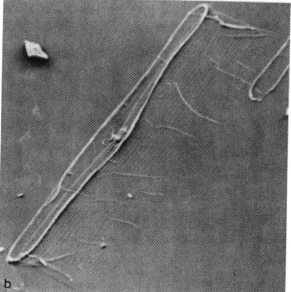

b

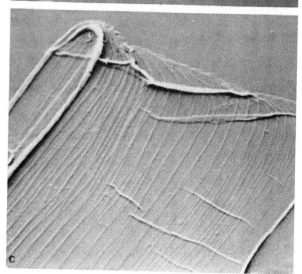

c

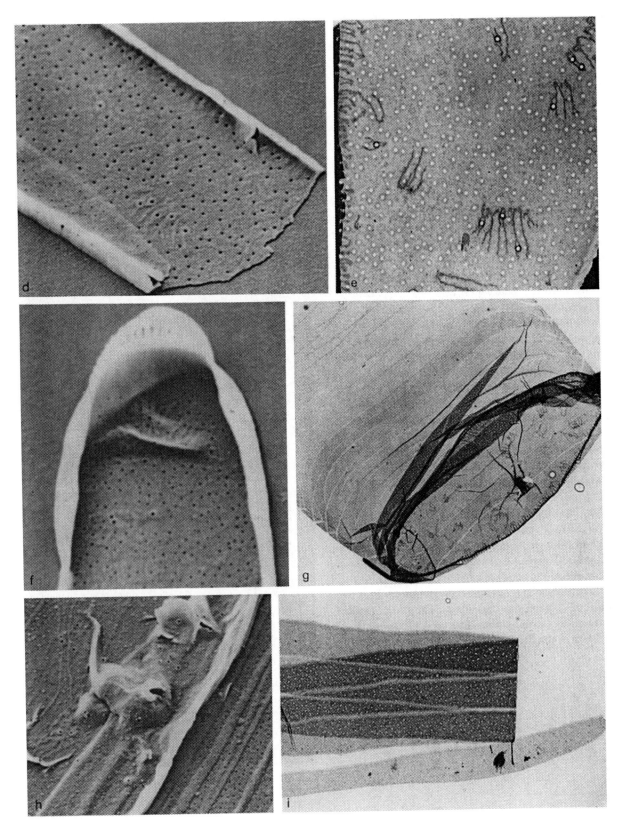

Anaulus C. G. Ehrenberg 1844. Ber. Bekanntm. Verh. Königl. Preuss. Akad. Wiss. Berlin, **1844**: 197

T: *A. scalaris* Ehrenberg

There is doubt about the typification of this genus – *A. scalaris* is a large diatom and requires re-investigation. Drebes & Schulz (1981) cultured and described a new species *A. creticus* and our description is based on material of that species kindly supplied by Dr G. Drebes. The cells occur singly[a,b] or in short chains[c] attached at the poles and centres of the valve and leaving narrow spaces between the adjacent valve faces. Plastids single and lobed. Marine planktonic or perhaps associated with sediments.

Valves narrowly ovoid in valve view (slightly indented at the pseudosepta) with two conspicuous pseudosepta crossing the valve.[c–i] In girdle view cells oblong with short apical processes and a central rimoportula having an external tube of variable length.[c–i] Valve surface areolate except in the regions of the pseudosepta. Areolae with delicate vela. Areolae are scattered on the valve face and in vertical rows on the mantle. The apical processes terminate in ocelli with no obvious porellar plate; Schulz *et al.* (1984) record fine granular material at the base of the ocellus in sections observed in TEM. The rimoportula is tubular both externally and internally (see Drebes & Schulz, 1981, figs 10 & 19). The pseudosepta are internal thickenings of the valve. The cingulum is composed of several (5–7) open areolate copulae.[j]

Until the type species and others are investigated (VanLandingham lists 19 spp.) it is not possible to comment in further detail. Drebes & Schulz consider that it is a heterogeneous group at the moment. *Anaulus* has been placed either in a family Anaulaceae (Hendey, 1964; Round & Crawford, 1988) or in the Biddulphiaceae (Simonsen, 1979).

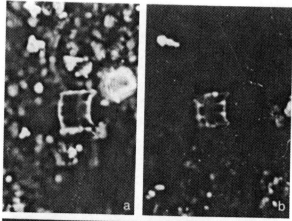

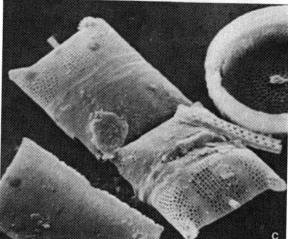

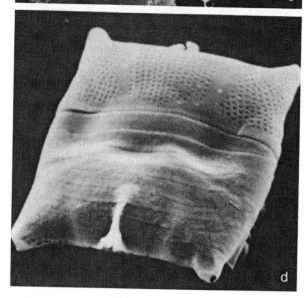

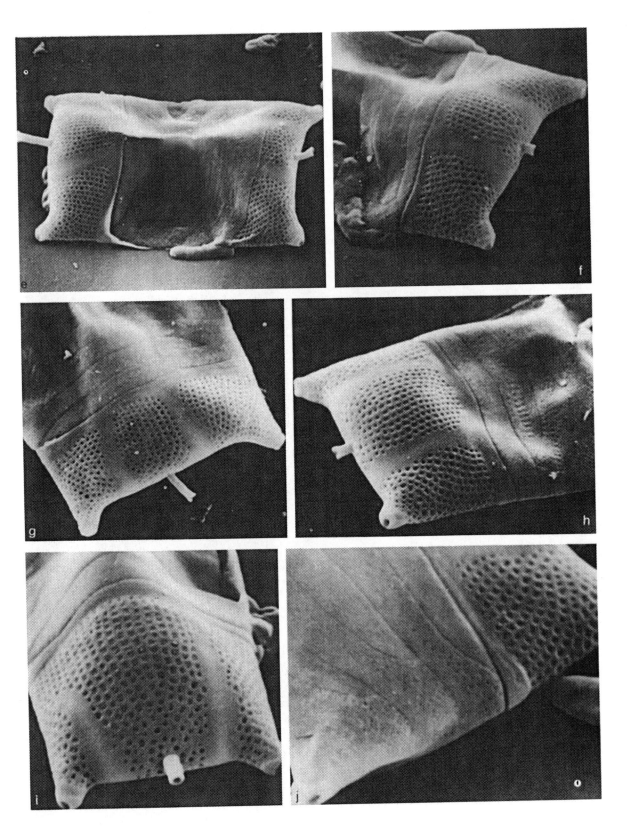

Eunotogramma Weisse, Melange. Ecol. Bull. St. Petersbourg 2: 243, 1855

LT: *E. laevis* Grunow in Cleve et Möller, Slide no. 257 (selected by F. E. Round)

Cells asymmetrical about the apical plane:[b, c] dorsal margin convex and ventral margin more-or-less straight or slightly convex; rectangular in girdle view.[a] Solitary or forming short chains. Plastids not observed. A widespread and common marine genus, but rarely recorded owing to its small size and preponderant occurrence attached to sand grains – also fossil.

Valves sometimes slightly inflated in the centre of the ventral margin, to elongate, slightly arcuate,[b-f] in some with an abrupt downturned mantle and slight ridge at the valve face/mantle junction.[e] Areolae radiating from a central annulus[f] or in transverse rows interrupted by plain cross-bars;[d] valve surface sometimes spinulose[d] or rugose. Vela not studied. A small rimoportula opening occurs near the central ventral margin[d, f] but some valves appear to have a row of rimoportulae along the dorsal margin. Internally with well-developed transverse costae[i, j] curving from mantle edge to mantle edge. Copulae appear to be incomplete bands but more studies are needed.

Only 13 spp. are recognised by VanLandingham (1969) and they certainly need re-investigation. It is usually placed close to *Anaulus* in classificatory systems. Our specimens are not easy to study and there are many points requiring clarification. The number and position of the rimoportulae appears rather variable as does the ornamentation of the valve face.

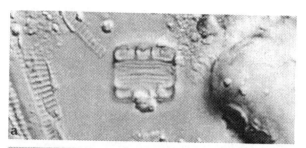

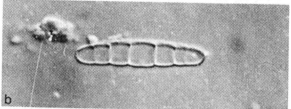

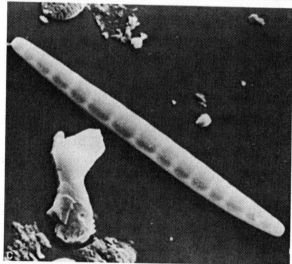

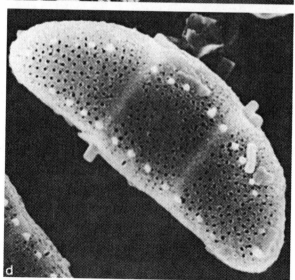

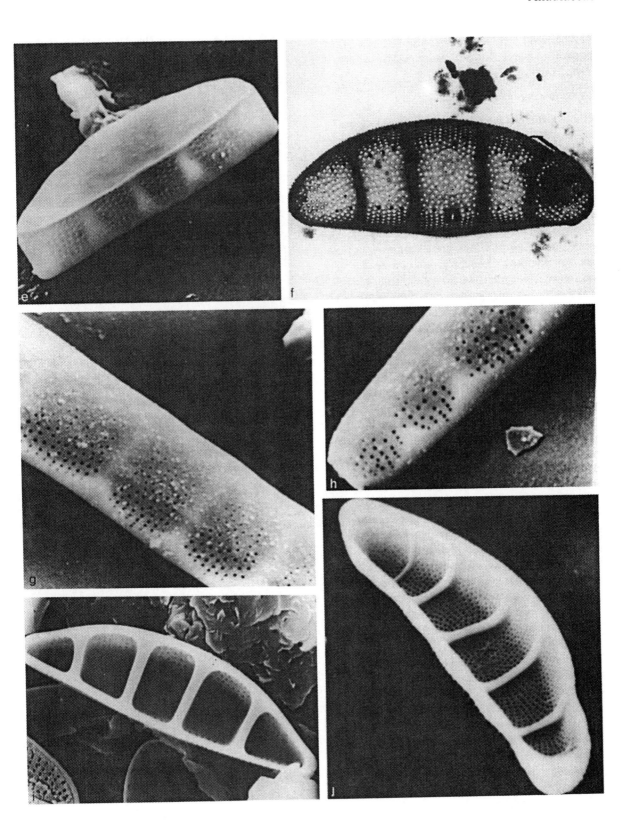

Lithodesmium C. G. Ehrenberg 1839. Ber.
Bekanntm. Verh. Königl. Preuss. Akad. Wiss.
Berlin, **1839**: 156

T: *L. undulatum* Ehrenberg

Cells rectangular in girdle view and joined to form
filaments.[a] Plastids numerous, discoid. A marine
planktonic genus of few species, only the type being
frequently recorded.

Valve triangular (sometimes quadrangular), with a
central tubular spine and a marginal ridge;[b-d] an
elevation at each angle; a fold or undulation across
the angles, which creates a more-or-less circular,
'centric' middle part[e] on which the triangular shape
seems superimposed. Uniseriate rows of areolae
radiate from the centre.[g] The marginal ridge is high
and extends upwards from the edge of the valve
face, even around the elevations; it continues the
line of the mantle. It has previously been known as
the 'lamella' (*Verbindungslamelle* of Hustedt, 1930)
and has pores that are much larger than the areolae
on the valve itself;[b, c, f] they are probably
unoccluded. The marginal ridges of sibling valves
overlap.[d] At the angles of the valves, the ridge is
lower and leaves a large hole opening into the
intercellular space.[d] At this point the ridge of one
valve appears to change from overlapping to
underlapping. Hustedt (1930) carefully pointed out
that the cells were connected by the marginal ridge
and that these had nothing to do with the girdle
bands. The central long tube forms the external
opening of the rimoportula which internally has two
slits in line (Hasle, 1975, fig. 1496) and has been
termed a bilabiate process (von Stosch, 1977). The
cingulum is composed of segments[i] arranged in four
rows. Each row consists of three or more segments
(depending upon the number of sides to the valve),
which are areolate and ligulate[h] (see also von
Stosch, 1980a).

The structure of this genus and of *Ditylum* have
many features in common and the early light
microscopists were correct in grouping the two
together. Von Stosch (1980a) emphasises the
difference between the lamellate marginal ridge of
Lithodesmium and the row of double spines in
Ditylum. Apart from overall shape neither genus has
much in common with the *Triceratium* group, which
is in a confused taxonomic state. They are often
allied with *Triceratium* itself, but we do not accept a
close relationship.

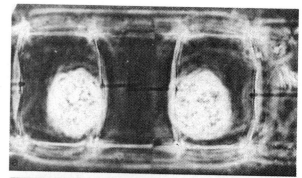

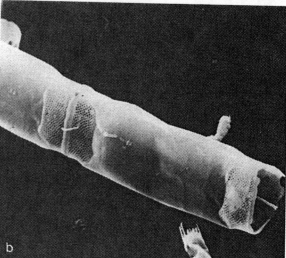

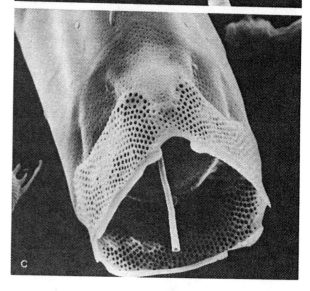

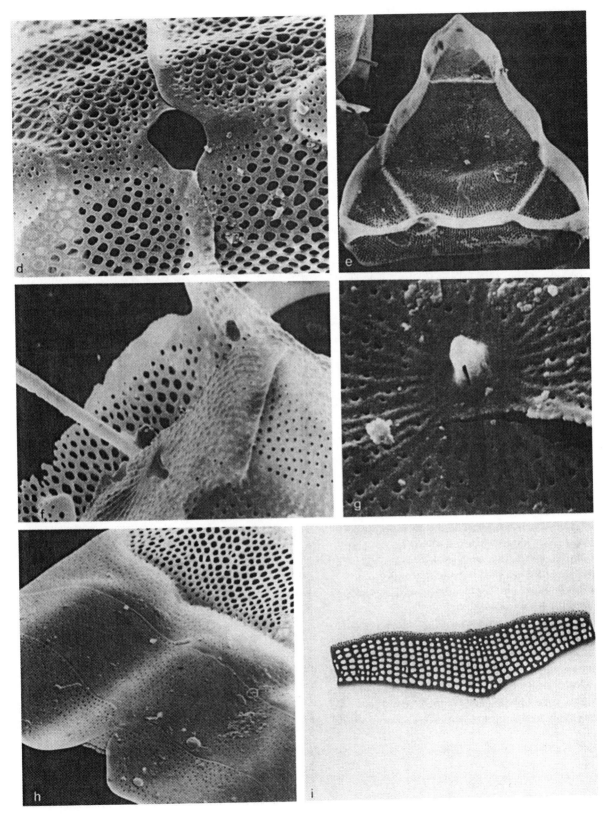

Ditylum L. W. Bailey 1861. Boston J. Nat. Hist. 7: 332

T: *D. trigonum* Bailey

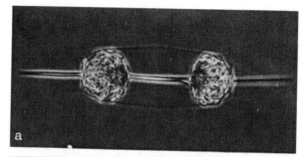

Cells free-living in the marine plankton, usually lying in girdle view, which is rectangular with a very conspicuous 'spine' arising from the centre of each valve.[a, b] In valve view the cell is usually triangular in outline,[c, d] but some populations of bipolar or quadripolar forms have been encountered. Numerous discoid chromatophores occur around the periphery. Rarely forming large populations but frequently found, especially in neritic plankton. Easily cultured and therefore used in several notable experimental studies.

Valves bi-, tri-, or quadrangular, with a conspicuous central tube arising from a small, clear central area. Rows of simple elliptical or round areolae occluded by rotae radiate from the central region to the valve mantle edge but are broken by a ridge bearing flattened spines, which often appear as two 'gutter' pieces fusing at the apex where further small spinules occur. In some forms the ridge is well-developed and deep but is perforated by slits,[f] as though the spines had fused. In our material of a bipolar form[k] the fimbriate ridge is circular rather than triangular. The region outside the ridge is valve mantle and here the rotae are attached at only two points, whereas on the valve face they are more complex.[g] Isolated simple pores occur scattered over both valve face and mantle;[g, h] these probably lack vela. Internally a single rimoportula opens at the base[h] of the central external tube and is unusual in having a closed central portion. Copulae numerous, scale-like, with ligulae on some.[i, j] Scanning electron micrographs often show longitudinal ridges and furrows[k] but we have not investigated these. Resting spores have been reported.

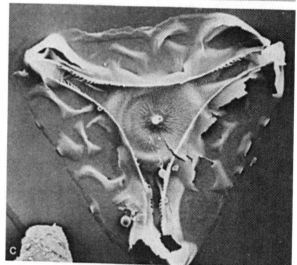

This genus was confused with the closely related *Lithodesmium* (p. 290) by earlier workers (see Hustedt, 1927–66). A small genus of which only *D. brightwelli* and *D. sol* are regularly recorded. There are forms or life cycle stages of this genus which require further investigation. A recent study has been made of valve formation (Li & Volcani, 1985a).

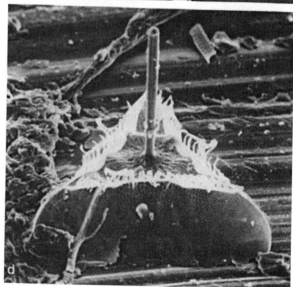

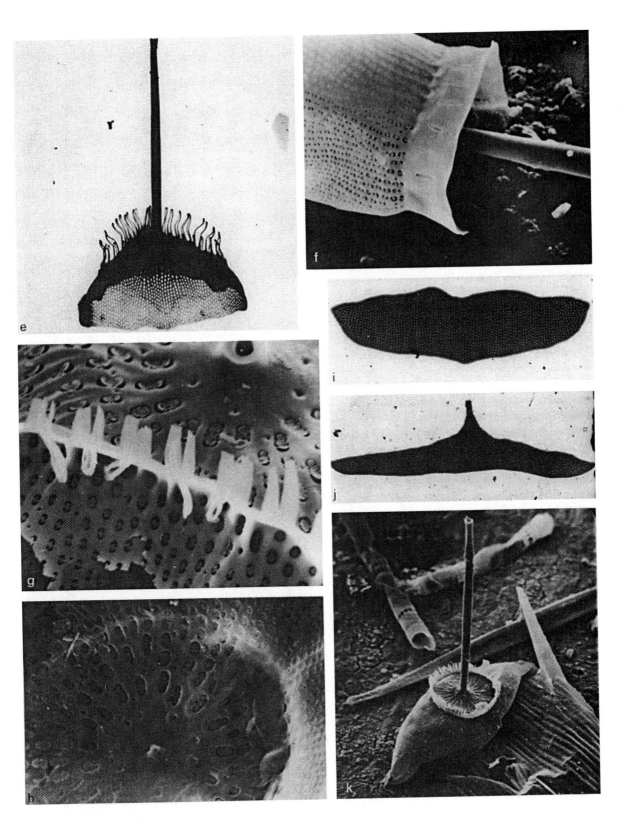

Corethron F. Castracane 1886. Rep. Sci. Res. Voy. 'Challenger', Bot. **2**: 85

T: *C. criophilum* Castracane

Cells solitary (reported rarely in short filaments), usually seen in girdle view, which is cylindrical with dome-shaped valves[a-f] and numerous discoid plastids. A cosmopolitan marine planktonic genus occurring in vast numbers, especially around Antarctica (see Fryxell & Hasle, 1971). The complex heterovalvy partly explains the perplexing variety of 'phases' illustrated by Hendey (1937) and first revealed with the EM by Fryxell & Hasle (1971).

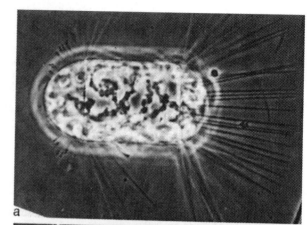

One valve is domed with a wide, upturned and scalloped brim.[b-d, g] Within this brim are deep sockets with peg-like projections at either side.[g, i] Into these sockets are inserted the bases of long, tapering, toothed spines which are T-shaped in section with the top of the T lying adjacent to the valve brim.[b-d, g, h] Hair-like filaments are often found on this valve.[b, c] In the intact cell[a] the spines on the valve just described point away from the girdle. The other valve[e, f] has a ring of spines of two types. Spines similar to those in the first described valve alternate with shorter spines that have distinctive claw-like tips[j] described in detail by Fryxell & Hasle (1971). Interestingly, there are right- and left-handed versions of these spines.[j] Near their base there is an expansion to form an oar-shaped blade before the spine enters the socket which itself is smaller than that of the long spine. Compare the two interiors.[d, f] The short spines point up and away from the valve rim but the long spine is inserted with the stem of the T towards the brim which is downturned in this valve. Consequently the large spines of both valves of an intact cell point in the same direction. When the spines are released from the constraint of the cingulum[a] as the cells divide, they swivel in their sockets to adopt an acute angle with the valve. *Corethron* is thus a unique diatom with moveable valve components. There are no tube-processes. Both the valve[i] and the scale-like girdle bands are finely areolate.

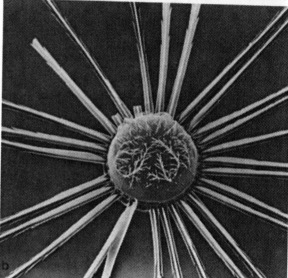

The scale-like nature of the bands led Fryxell & Hasle to place *Corethron* in the Rhizosoleniaceae. We cannot agree with this and equally see no justification for placing it in the Melosiraceae as Simonsen (1979) has done. We prefer to allocate this remarkable monospecific genus to its own family. Further detail is provided in Crawford & Round, 1989.

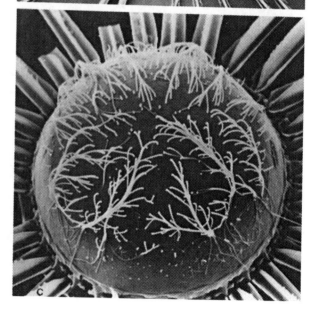

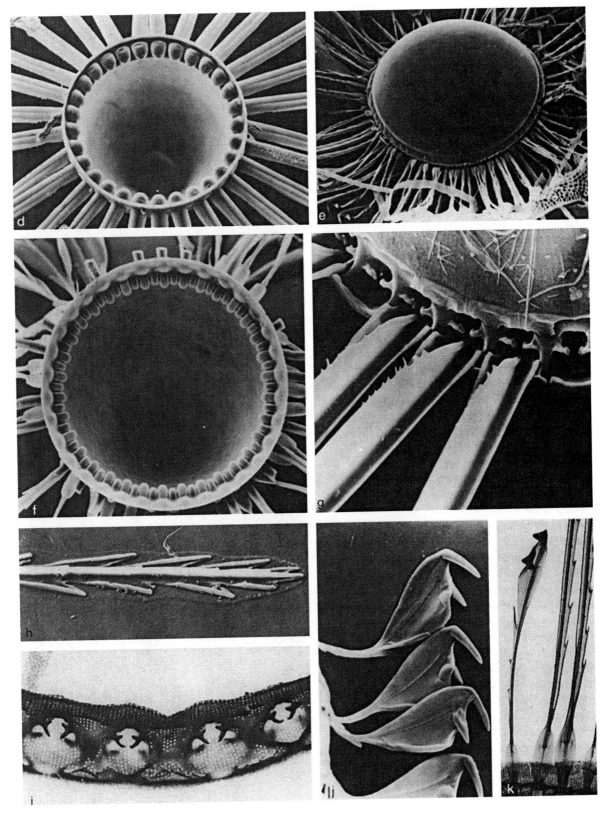

Cymatosira A. Grunow 1862. Verh. zool.-bot. Wien **12**: 377

T: *C. lorenziana* Grunow

A small genus of marine benthic diatoms forming filamentous colonies of rectangular cells,[a] the sides of which are curved so that the apices do not touch. Cells united by central interlocking spines which prevent separation of the cells.[b] Plastids 1 (2, 4) plate-like, lying at either end of the cell but details are not clear. Although often collected in inshore plankton it probably grows in the epipsammon; certainly it also occurs on sandy beaches and salt marshes.

Frustules heterovalvar. Valves broadly[h] to narrowly lanceolate[b, c, e, f] with pointed or blunt apices which terminate in ocelluli;[e-h] these are angled somewhat to the valve face and surrounded by a distinct rim. Hasle, von Stosch & Syvertsen (1983) report small ocelluli, except on the end (separation) valves of colonies where the ocellulus is larger on one valve. Areolae poroidal, arranged in lines across the valves. Sternum absent; in some there is a clear central region but when viewed from the inside (the best view to confirm a sternum) this is not thickened or distinct in any way and so does not appear to be a true sternum. Areolae occluded by cribra, which protrude from the outer valve surface[h, j] and have external spinules. Marginal spines conspicuous,[d, h] T-shaped, the upper parts of the Ts sometimes fused; with fine filigree (not illustrated, since it has been removed by our cleaning process) between the stems of the adjacent T-pieces. Complex spines forming a 'fence-like' structure are reported by Fryxell & Miller (1978). In *C. belgica* the T-shaped spines in the central interlocking region are larger than the remainder. Rimoportulae on one valve only, where there is one, placed off-centre, just internal to the row of spines along one edge or nearer the centre;[e, f] internally the rimoportula is sessile.[b, i] Copulae simple, open, with pores. Complex ligulae occur on the copulae and fimbriae on the most abvalvar band.

Fryxell & Miller (1978) and Hasle *et al.* (1983) give further details. The complicating feature in this genus lies in the two types of spines plus the fact that end (separating) cells also have a somewhat different morphology resulting in four valve types differing only slightly in morphology. The genus, however, clearly belongs in the Cymatosiraceae.

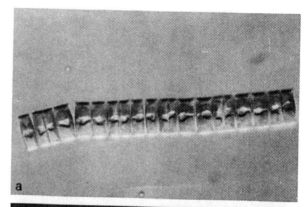

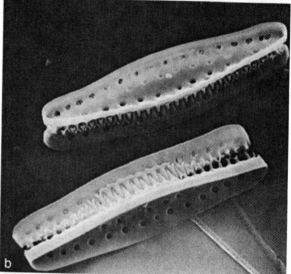

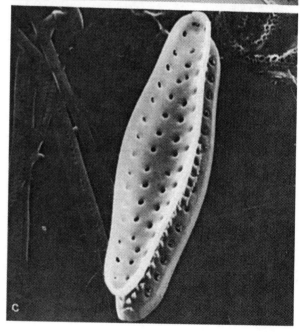

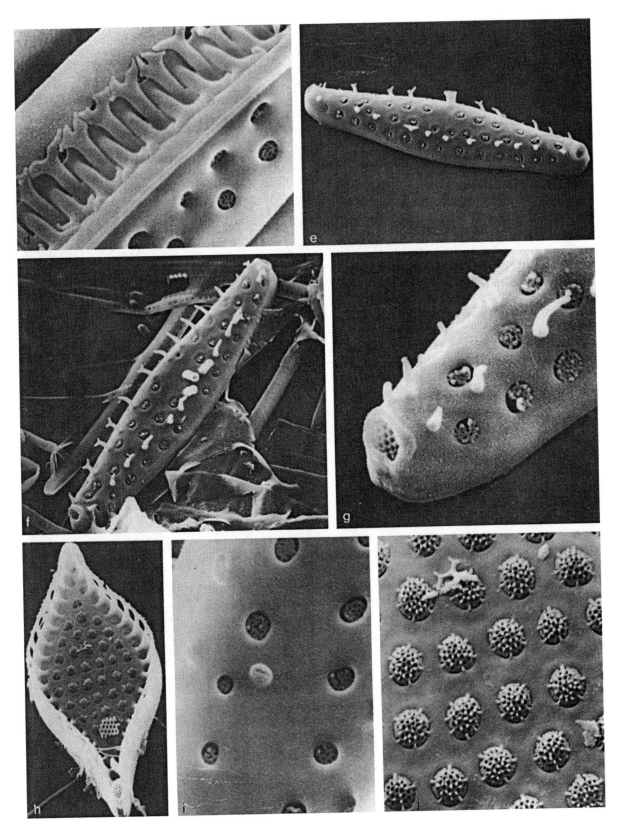

Campylosira A. Grunow ex H. Van Heurck 1885. Syn. Diat. Belg.: 157

T: *C. cymbelliformis* (A. Schmidt) Grunow ex Van Heurck (= *Synedra cymbelliformis*)

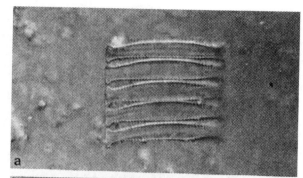

Cells narrowly rectangular in girdle view occurring in short filaments[a-c] joined by interlocking capitate spines. Plastids unknown. A common marine benthic genus usually associated with sand grains and detritus, and often covered with mucilage.

Frustules heterovalvar. Valves cymbiform,[d, h] with polar elevations.[a-e] One valve convex and the other either concave or straight (see Hasle *et al.* 1983). Areolae simple, round, irregularly spaced or in curved longitudinal striae[h] on the valve face; there may be a small central circular or transverse area without areolae.[i] Sternum absent. Apical ocelluli occur,[f] turned outwards towards the ventral side of the valve and, on the separation valves, with a spine angled towards the valve centre.[b, e] T- or Y-shaped linking spines [b, g] arise within a line of areolae at the junction of valve face and mantle, becoming simpler near the valve apices where they cannot interlink. There is a small rimoportula in the centre, near the curved dorsal margin.[i] The cingulum consists of four open copulae; the most abvalvar copula bears fine fimbriae according to Hasle *et al.* (1983). These workers also record separation valves in which the spines are simple, thus allowing fragmentation of the filaments.

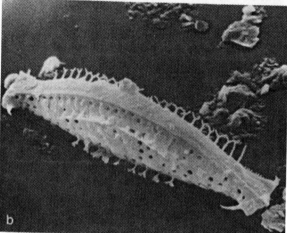

Additional illustrations and comments are to be found in the extensive study of the Cymatosiraceae by Hasle *et al.* (1983). *Campylosira* was one of the first genera to be described in this family and it is likely that many records (and of *Cymatosira*) are unreliable since they may refer to other, more recently described genera, e.g. *Plagiogrammopsis*. We thank the late Professor von Stosch for samples of the Cymatosiraceae.

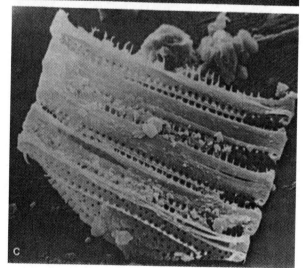

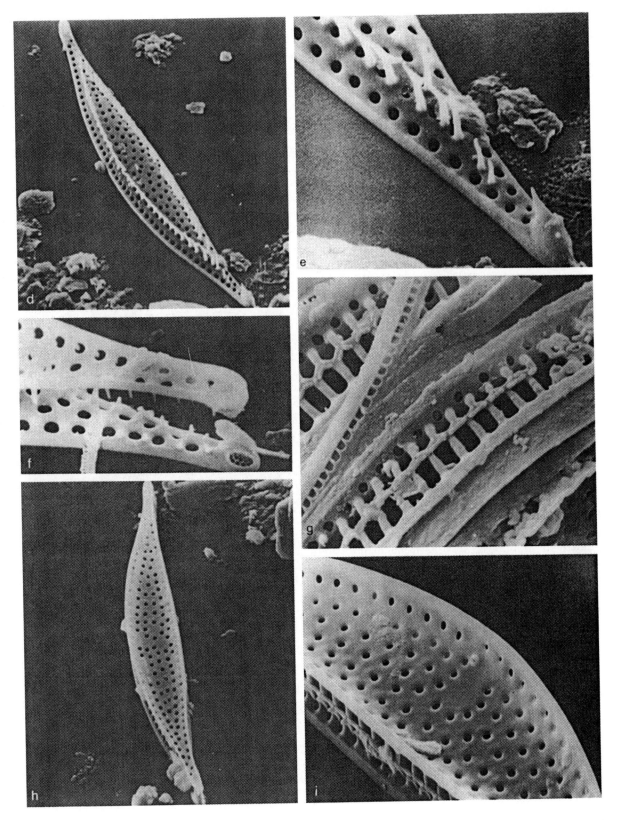

Plagiogrammopsis G. R. Hasle, H. A. von Stosch & E. E. Syvertsen 1983. Bacillaria 6: 30

T: *P. vanheurckii* (Grunow in Van Heurck) Hasle, von Stosch & Syvertsen
(= *Plagiogramma vanheurckii*)

Cells joined into short filaments in nature but long filaments in culture;[a] the cells are only loosely attached to each other and separate easily. Rectangular in girdle view, with apical elevations and a central expansion. A slight thickening of the valve centrally indicates the presence of a pseudoseptum. One plastid, lying along the girdle. Common in the marine benthos in shallow areas; possibly associated with sand grains.

Frustules heterovalvar, one valve bearing pili,[c–e, i] the other a rimoportula.[h] Valves lanceolate with rostrate to capitate apices,[b] which twist slightly to opposite sides.[i] Apices elevated,[e] bearing small ocelluli, which are somewhat lateral. Valve face slightly domed, curving into shallow mantles. Areolae large, circular;[f–i] closely spaced or sometimes in more-or-less transverse rows on the valve face; forming a single row on the mantle. Each areola is closed by a pegged cribrum,[f] which is raised slightly above the outer surface of the valve.[g] There is a plain area or transverse band (fascia) in the centre of the valve,[b, i, k] but no sternum. Long thin spines occur around the edge of the valve face;[c–f] some have flattened spathulate ends.[f] One valve has long barbed pili,[j] one arising near each ocellulus.[e] The other valve has a small rimoportula near the centre. Both have a prominent pseudoseptum[k] running across beneath the plain area or fascia in the middle of the valve. Girdle composed of 5–8 split copulae, which are ligulate and each have one row of fine pores.

This genus is related to *Cymatosira* and is distinguished from the superficially similar *Plagiogramma* by the presence of pili, long thin spines, a single pseudoseptum, rimoportulae, elevations bearing the ocelluli, and the absence of a sternum. Unlike some other members of the Cymatosiraceae, the degree of heterovalvy is slight.

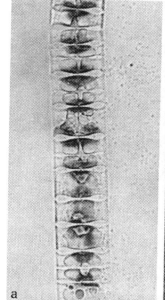
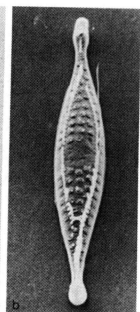
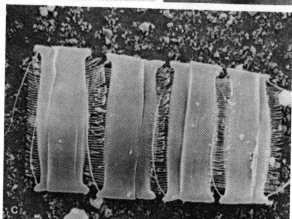
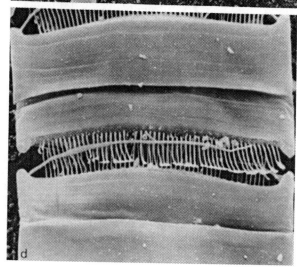

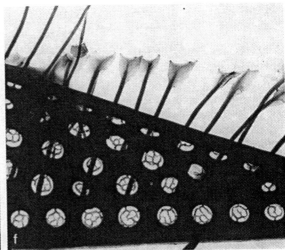

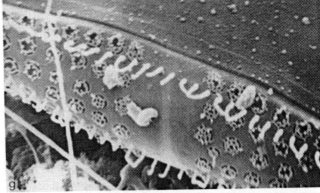

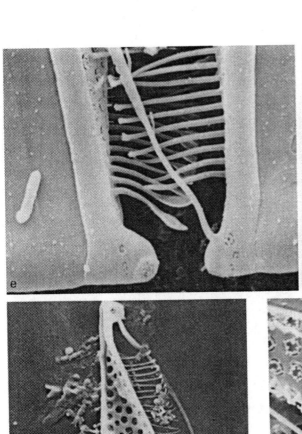

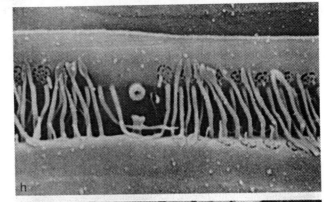

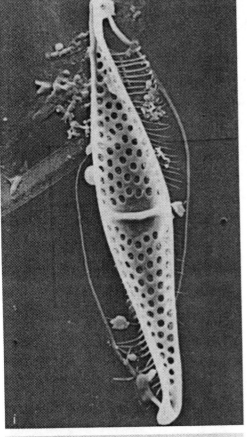

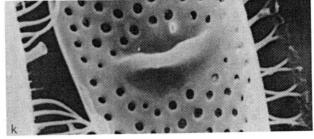

Brockmanniella G. R. Hasle, H.A. von Stosch & E. E. Syvertsen 1983. Bacillaria 6: 35

T: *B. brockmannii* (Hustedt) Hasle, von Stosch & Syvertsen (= *Plagiogramma brockmannii*)

Cells attached to form long chains[a] which may be twisted. Cells rectangular (oval in small cells) with slight apical and central expansions of the valve. Sibling valves touch at the centres and may also be in contact at the apices. One plastid, constricted in the centre and lying along one side of the cell. Common on the sediments of the North Sea and also appearing in the plankton. Probably widespread: our samples were cultures from around the British Isles but we have also found it off Florida.

Valves linear, lanceolate[b, f] to oval[c] with prominent offset ocelluli on apical elevations.[b-g] A central raised fascia[d-g] is devoid of areolae, except along the margin. There is no obvious sternum. The areolae are in transverse rows and are closed towards the outside by delicate vela.[f, h] A single row of spines occurs around the margin just above the outer ring of areolae and a few spines may occur scattered on the central area and elsewhere.[b, c, f, j] The valve surface may be finely granular.[c] Pili are absent. Ocelluli occur diagonally opposite one another and in a clockwise position; they have a slight rim. An offset process (rimoportula?)[a] occurs on one valve only.[arrow, f, g, j] Cingulum of numerous simple, slightly porous copulae, the last one having a row of granules close to the abvalvar edge (Hasle *et al.*, 1983).

This genus was erected for a species previously placed in *Plagiogramma* (*P. brockmannii*), based on its lack of pili and pseudosepta. As in other related genera the initial cells differ somewhat, e.g. they lack spines. We have not observed the concave and convex thecae recorded by Hasle *et al.* in post-auxospore material. The genus is quite close to *Cymatosira belgica* and to *Plagiogrammopsis vanheurckii* but we agree fully with Hasle *et al.* that it is preferable to maintain separate genera when there is a distinction involving such prominent features as pili and pseudosepta.

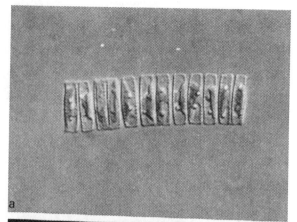

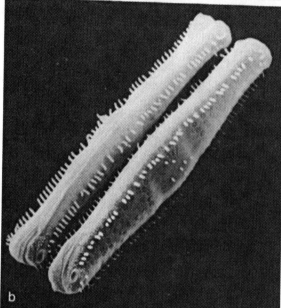

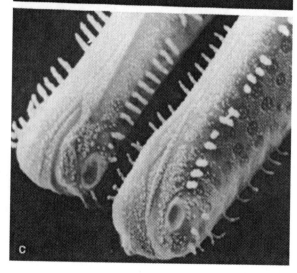

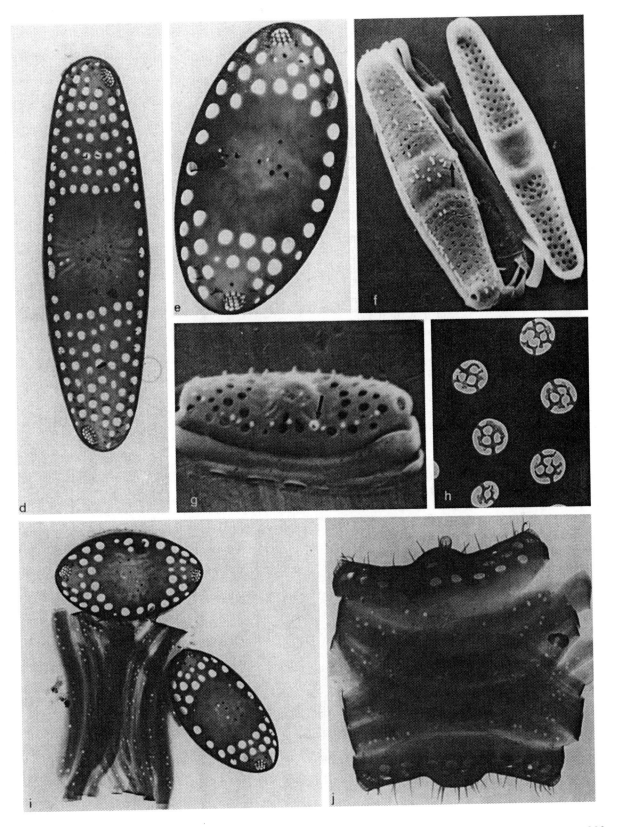

Minutocellus G. R. Hasle, H. A. von Stosch & E. E. Syvertsen, 1983. Bacillaria 6: 38

T: *M. polymorphus* (Hargraves & Guillard) Hasle, von Stosch & Syvertsen (= *Bellerochea polymorpha*)

Cells curved in girdle view (less so in small cells such as those we figure), growing as short chains, which fragment easily. One plastid. Marine, in the plankton of coastal waters; probably cosmopolitan but overlooked because of its small size.

Valves lanceolate, elliptical to subcircular;[a–f] mantle low, apices sometimes rostrate. One valve convex, with pili arising from opposite sides of the valve and directed towards the valve centre and a slightly raised apex bearing an ocellulus;[b, d] a row of marginal (mantle) areolae and others scattered[c, g] over the valve face (almost absent in some), the areolae being circular and occluded by an indistinct but porose velum. The pili have short branches near the base.[arrows, b] The ocelluli are placed diagonally and have a few porelli. A group of small papillae occur on the valve surface.[arrow, g] The other valve is concave, lacking pili[d–f] but with a small off-centre rimoportula; a small subapical spine occurs on this valve. Cingulum of 7–9 copulae (Hasle *et al.*, 1983), which are open and ligulate; a row of pores occurs on the advalvar bands, the others being plain.

The pictures we obtained in the SEM are poor. TEM yields better results – for an excellent detailed description of the three known species, see Hasle *et al.* (1983). Like other members of the Cymatosiraceae the valves of this genus become almost circular as the cells decrease in size in culture, the number of copulae increase and some of the valve detail disappears. The presence of pili distinguishes this genus from some others in the family and the absence of a marginal ridge on the valve distinguishes it from *Papilliocellulus*, *Arcocellulus* and *Leyanella*.

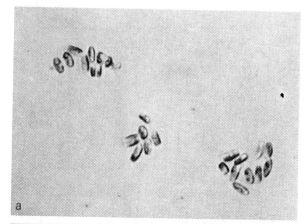

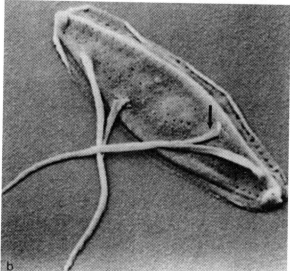

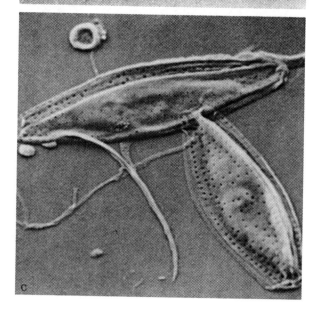

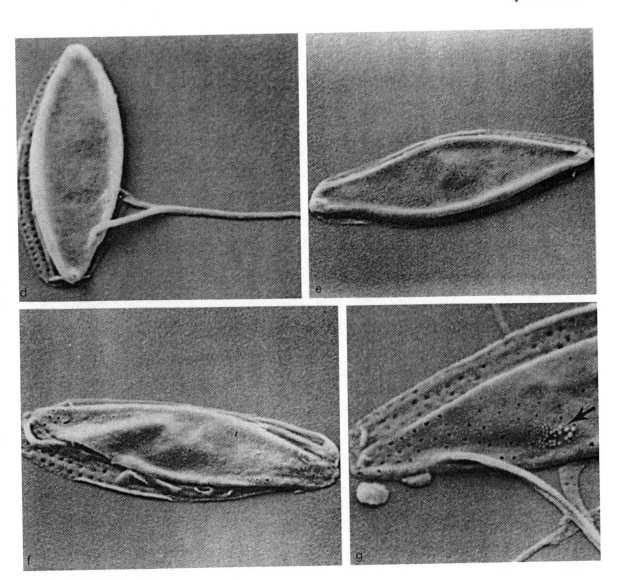

Leyanella G. R. Hasle, H. A. von Stosch & E. E. Syvertsen 1983. Bacillaria 6: 50

T: *L. arenaria* Hasle, von Stosch & Syvertsen

Cells curved in girdle view,[a] heterovalvar, attached to form long filaments; Hasle *et al.* (1983) also report solitary cells. One plate-like plastid. Marine, living amongst sand grains on coastal flats.

One valve of each cell concave, the other convex,[e, f] and bearing two long, hair-like projections (pili);[b, c, e] narrowly lanceolate to elliptical (after prolonged growth in culture). Mantles shallow. Areolae forming a single row along the margin of the valve face;[g] sometimes there are also a few scattered on the valve face itself. The areolae contain domed cribra at their external apertures. Ocelluli[c, d, g] occur at either end on slight elevations of the valve and are inclined, facing outwards.[h] The pili arise from near the ocelluli and the convex valve and each bears a reflexed branch near its base;[e] they are slightly displaced from the apical axis, lying diagonally with respect to it. For detail of the tip see Fig. e, inset. Beneath the tip of the reflexed branch the valve face has an area of small papillae. Along the junction between valve face and mantle there is a delicate upright membrane,[d-f, i-k] which is porous and lace-like and lies along the marginal row of areolae. The lower perforations through the membrane are triangular[j] and large; the upper ones are smaller and round, and contain fine porous plates. A small submarginal tubular process occurs at the centre of the concave valve.[g] The cingulum consists of eight or more open copulae; the most abvalvar has a row of small teeth along its abvalvar edge.

This genus of the Cymatosiraceae resembles *Arcocellulus* in its possession of a membranous ridge along the edge of the valve face but the valve structure is not the same and the pili occur on the convex valve in *Leyonella* but on the concave valve in *Arcocellulus*. For a more comprehensive description, see Hasle *et al.* (1983).

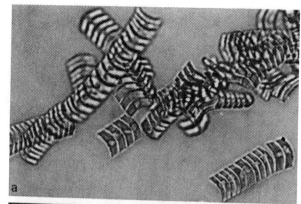

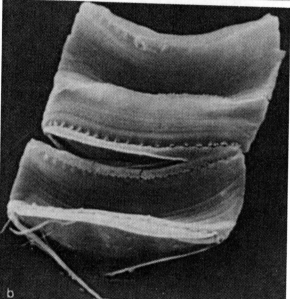

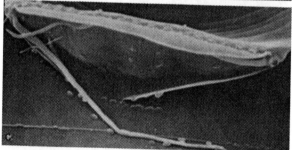

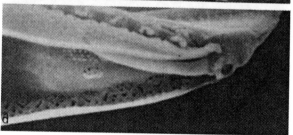

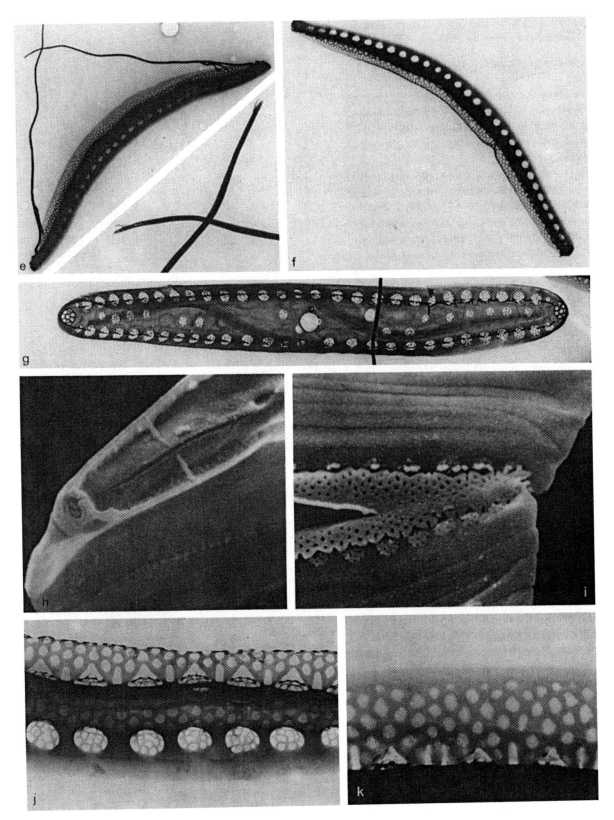

Arcocellulus G. R. Hasle, H. A. von Stosch &
E. E. Syvertsen 1983. Bacillaria 6: 54

T: *A. mammifer* Hasle, von Stosch & Syvertsen

Cells rectangular, genuflex[a] or falcate[e] in girdle view,
spindle-shaped in valve view[d] but tending towards
being round in small cells.[c] One plastid per cell.
Cells solitary or chain-forming, planktonic or
benthic, marine.

Valves elongate to oval, sometimes with slightly
capitate ends. Frustules heterovalvar, the concave
valve bearing two long pili[a, b, i] (spines) at least in
larger cells. Pili attached near the poles and with
reflexed branches[a, i] arising from their proximal
portion and short spines on the distal portion; small
areas of papillae[h] can also occur, usually towards
the bases of the pili. The valve surface may be
ornamented with low costae. Poroids more-or-less
circular.[d, e] A row of poroids lies along each margin
and occasional single poroids are scattered on the
valve face.[d, e, i, k] The vela are fine porous
membranes[h] or cribra. A shallow marginal ridge is
present which is better developed towards the
apices: it is formed by a series of loops of silica.[c, g, j]
A rimoportule occurs near the centre of the convex
valve.[f, k] Valve mantle narrow and plain. Copulae
10–15,[a, c, e, f, j] split and alternate, with their ends in
line with the valve poles. The copulae lack pores but
have more-or-less regularly arranged pits in the
surface.

The genus contains two species, which differ from
each other in the attitude of pili, the curvature of the
frustule in girdle view, relative valve length, velum
structure, and valve markings. It is closely related to
Leyanella, which has pili on the convex valves and a
more elaborate marginal ridge. *Arcocellulus* has been
thoroughly described by Hasle *et al.* (1983). The
tubular process is presumably a very much
simplified rimoportula, and this simplification is
also carried through to the ocelluli. The reduction of
the valve face to a network of costae is also
reminiscent of the reduction seen in some species of
the quite unrelated genus *Cylindrotheca* (but living
under similar ecological conditions).

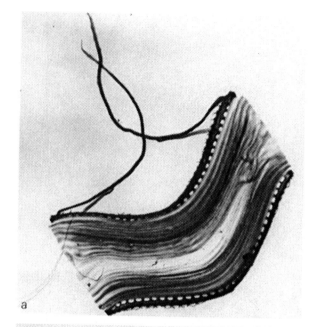

a

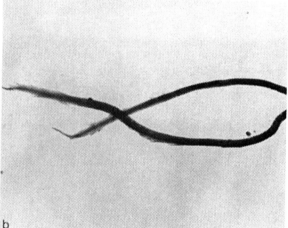

b

c

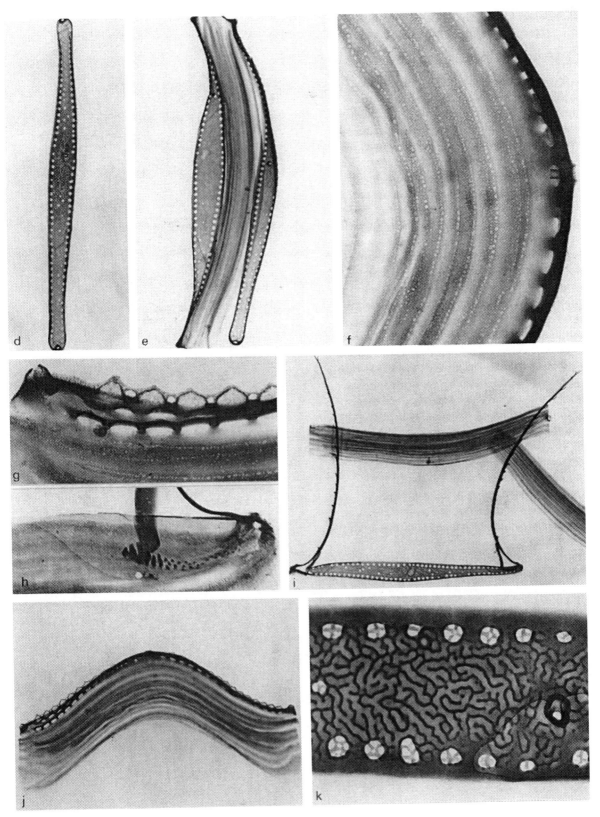

Papiliocellulus G. R. Hasle, H. A. von Stosch &
E. E. Syvertsen 1983. Bacillaria 6: 64

T: *P. elegans* Hasle, von Stosch & Syvertsen

Cells rectangular in girdle view,[a] and slightly
heterovalvar (with concave and convex valves)
joined to form loose filaments. A single plate-like
plastid occupies only half the cell. A marine genus
in coastal environments; possibly planktonic, but its
small size probably results in its being overlooked
and so the extent of its habitat is not known.

The valves are lanceolate to elliptical and most
unusual, built up from a framework of two rows of
thin transverse ribs, which join in a zig-zag manner
down the centre line.[b-f] At the centre there is a small
but distinct annulus. The transapical part of the
framework continues down the valve mantle, and at
the junction between face and mantle a wide
marginal frill (ridge) rises up, which exceeds the
height of the mantle.[d] The marginal frill may be
continuous or broken into scale-like structures and
consists of upright costae (finer than the valve
framework or cribra), which may bifurcate at their
tips and are interconnected by even finer costae.
Between the main costae of the valve face there
occur fine anastamosing siliceous filaments. The
basic framework of ribs is regarded by Hasle *et al.*
(1983) as delimiting areolae and if this is so then the
fine filaments form cribra. At the apices tiny
rimmed ocelluli occur, facing outwards,[arrow, b; c, f]
and on one valve a central tubular process is
present within the annulus (see Hasle *et al.*, 1983).
The cingulum consists of a number of non-porous
open bands.

This unusual diatom also has its cytoplasm
contracted into the region of the cell enclosed by the
cingulum, the space between this and the valve
being occupied by an organic, diatotepic layer
(Hasle *et al.*, 1983). Several features – the central
process, heterovalvar cells, marginal frill, etc. –
indicate that *Papilliocellulus* belongs within the
Cymatosiraceae, but as Hasle *et al.* remark, it
occupies a rather isolated position in it.

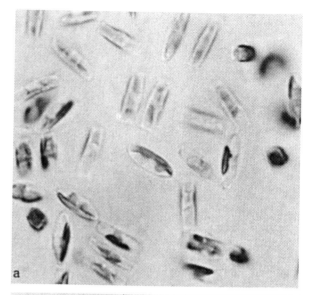

a

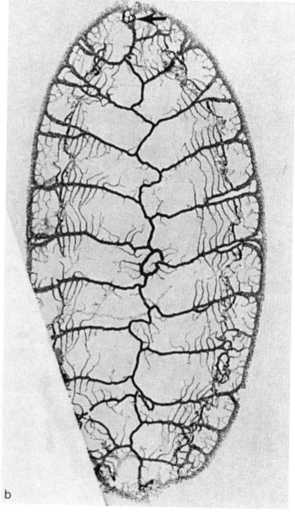

b

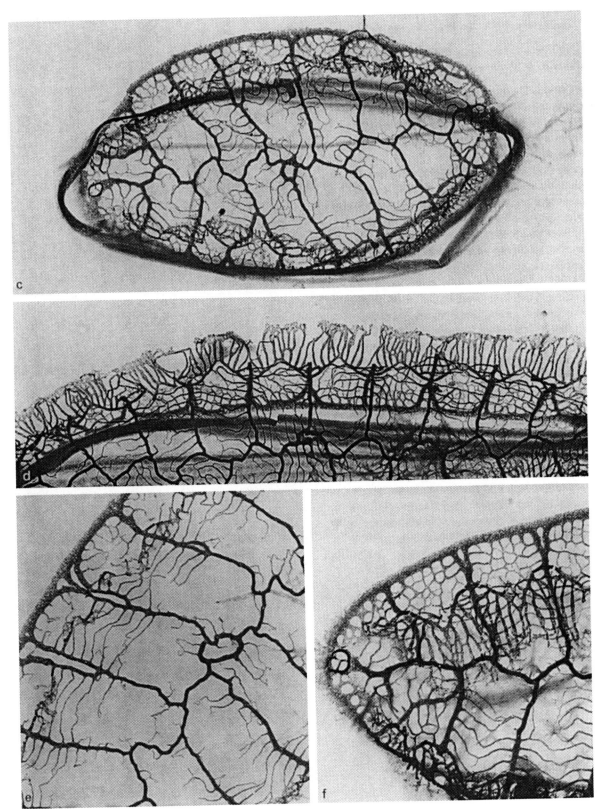

Extubocellulus G. R. Hasle, H. A. von Stosch & E. E. Syvertsen 1983. Bacillaria 6: 69

T: *E. spinifera* (Hargraves & Guillard) Hasle, von Stosch & Syvertsen (= *Bellerochea spinifera*)

Cells rectangular in girdle view, the pervalvar axis usually much exceeding the apical axis; joined in long chains.[a, b] One plastid, lying along the girdle. A marine coastal genus; exact habitat unknown but probably widespread.

Valves oval to almost circular;[c-g] valve face flat curving into an undifferentiated mantle; a few scattered spines sometimes occur on the valve face.[e, g] Areolae simple, circular, scattered over the valve. There were no obvious vela in our material but Hasle *et al.* report the areolae to be open or closed by fine cribra. Small raised ocelluli occur at the apices. No pili, marginal spines or marginal ridges are present. Cingulum consisting of many copulae,[c] the valvocopula having marginal fimbriae, the others being fragmented (quasifract of Hasle *et al.*) and bearing a single row of pores. No rimoportulae visible though occasional minute processes were found by Hasle *et al.*

This is an extremely small diatom (diam. <4 μm) and it is not obvious at first sight that the chains of cells are indeed diatoms. *Extubocellulus* has been placed in a subfamily of the Cymatosiraceae by Hasle *et al.* since the cells are not heterovalvate and do not have pili or marginal ridges/spines. The presence of ocelluli and the nature of the vela, the habit and habitat all clearly point to its inclusion in the Cymatosiraceae rather than the association with *Bellerochea* that had previously been proposed.

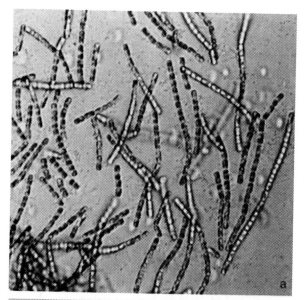

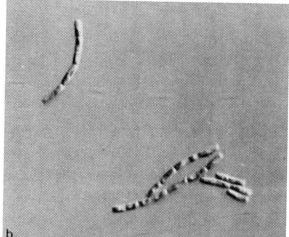

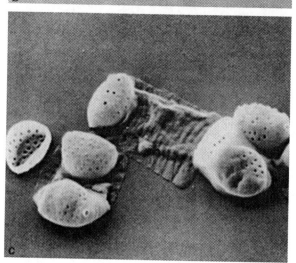

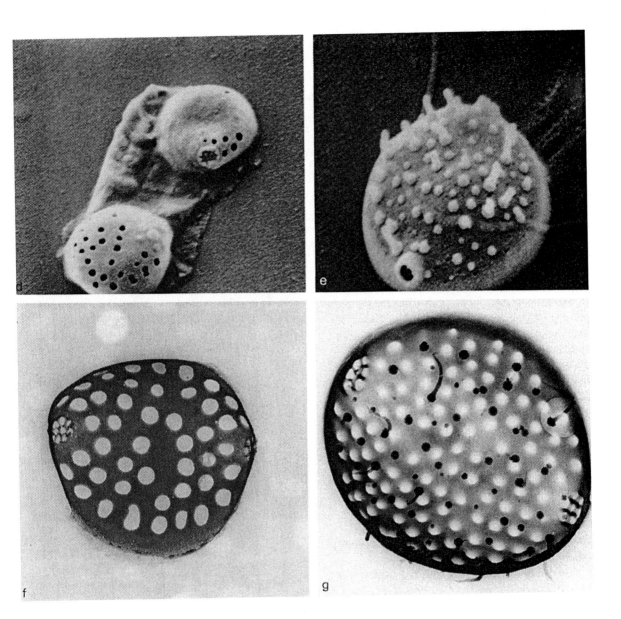

Rutilaria Greville 1863. Quart. J. Microsc. Sci. Ser. 3: 227

T: *R. epsilon* Kitton

Cells elongate and broadened at the centre. Sibling valves united by a pair of robust linking processes[a-c] (periplekta) at the valve centres to form chains of cells. This genus is known from Eocene to Miocene sediments.

Valves elliptical to spindle-shaped[a] as in the species illustrated. The margins of the mid-part of the valve parallel but tapering to slightly rostrate or capitate poles. The thick wall is perforated by radiating striae[a-e] of simple areolae: the closing plates are not preserved but we assume that the organisation was more complex than revealed so far. Striae terminate close to the well-marked margin which bears a single line of small spines[a-d] becoming indistinct towards the valve apices;[e] the spines are clearly linking in one species.[g, h] Mantle relatively shallow and lacking pores. The periplekton occurs at the centre of the valve and projects towards the sibling valve. At its apex, two branches wrap around the sibling process but not so tightly as to prevent sliding of the cells together and apart. Only one pair of branches is visible in the photographs; they belong to the (broken) upper valve and are wrapped around the lower periplekton. At the poles are ocelli which in some specimens[a, e, f] have not been preserved but in Fig. h they can still be seen. A view of the inside surface shows a small slit which in all probability is a rimoportula.[i] Consequently the periplekton is almost certainly a hollow tube.

The presence of an ocellus suggests a position in the Eupodiscaceae (see Ross, 1976). Complex linking processes of the type found in *Rutilaria* are absent from modern genera but present in other fossils (e.g. *Syndetocystis*). We thank P. A. Sims for material of this and the next genus.

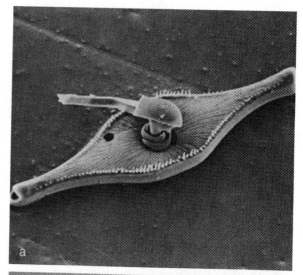

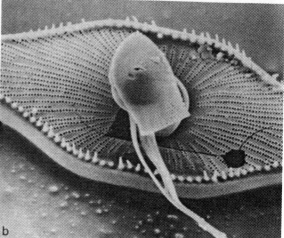

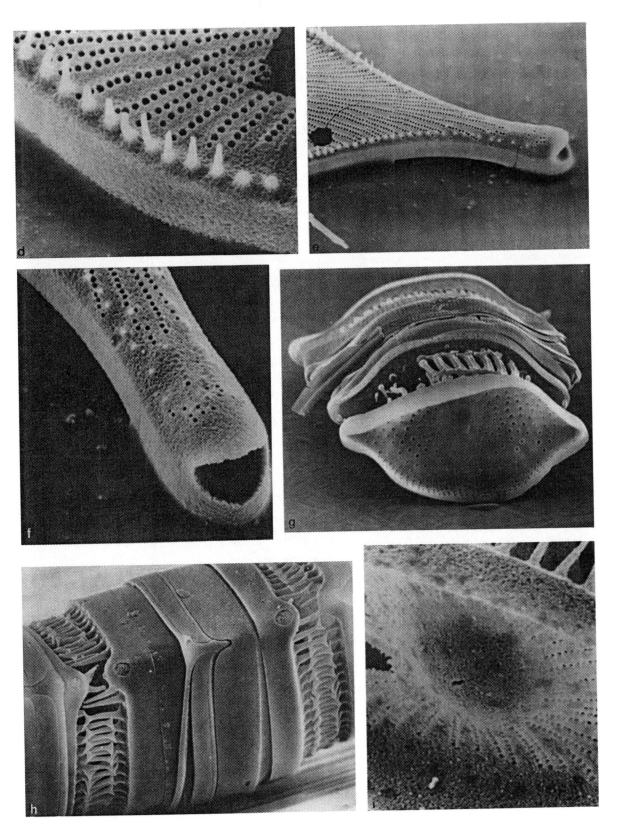

Syndetocystis J. Ralfs ex R. K. Greville 1866.
Trans. Microsc. Soc. Lond. Ser. 2, **14**: 125

T: *S. barbadensis* Walker & Chase

Cells in chains, linked by complex interlocking central processes;[a, b] in addition, having conspicuous apical elevations. A highly distinctive fossil genus, from Eocene to Oligocene deposits.

Valves elliptical.[c] Valve face depressed centrally, extended into rod-like elevations at the poles. At the junction of the valve face with the fairly shallow vertical mantles there is a marginal flange, which runs up onto the elevations at either end of the valve.[c, g] The elevations are spinose on the inner surface and plain on the outer. Lateral openings occur at the apices which must have been occluded in the live material but the structure has not been preserved: we presume that some form of ocellus was present. Valve face with scattered round areolae. From the slight central depression of the valve face there arises a long tubular process (periplekton); this expands at its tip into a ring, which clasps the stem of the process borne by the adjacent valve in the filament.[b, d-f] By analogy with other fossil genera (e.g. *Rutilaria*) this is probably the external extension of a central rimoportula, but we have not observed any portules, here or elsewhere on the valve. However, where the stem flattens out to form the ring there is an opening.[f] The valve mantle is plain. Copulae have not been observed.

In one photograph[a] the connecting processes have slid along each other so that the terminal rings lie against the sibling valves; whether this could happen *in vivo* is not known. As in so many fossil genera it is difficult to place *Syndetocystis* in a family. If an ocellus is proved then it will assist this: the complexity of the central process suggests a relationship to *Rutilaria*. The plain mantle, marginal ridge (spinose in *Rutilaria*), simple areolae, probable ocellus and the connecting mechanism of the periplekton all point to such a relationship.

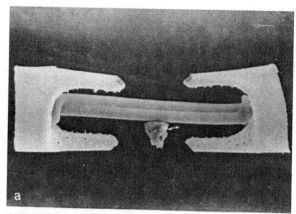

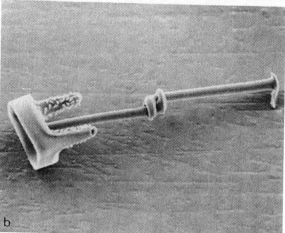

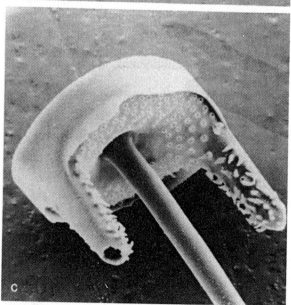

Rhizosolenia T. Brightwell 1858. Quart. J. Microsc. Sci. 6: 94 (nom. cons.)

T: *R. styliformis* Brightwell

Cells cylindrical short to very long,[a,d] straight or curved; free-living or forming long chains, and in some species spiral chains. Plastids numerous platelets, often in rows in cytoplasmic strands. A common genus in the marine plankton; also recorded (2 spp.) in freshwaters but these we have transferred to a new genus *Urosolenia* (see p. 324). The genus is easily recognisable by the single spined valves with a groove on each to accommodate the spine of the adjacent valve[b,e] and the numerous scale-like[k] girdle bands.

Valves (calyptrae) asymmetrical, cone-like,[b,c,g] drawn out into a 'spine', often with shoulders (otaria) at the base.[c,e-g] Areolae small, round to quadrate, arranged in vertical rows;[h] often indistinct when viewed from the outside because each areola is closed externally by a delicate plate penetrated by one to several slits or pores.[j,k] Hasle (1975) and Sundström (1984) refer to a loculate structure in some species but the loculi are at most very shallow. The valves are indented on one side where the spine of the adjacent cell fits;[e,f] the indentation sometimes extends onto the girdle[e] and the valve may be elongated in this region. A single rimoportula[b,c] is visible in the light microscope at the base of the spine, which indeed is actually the outer extension of the rimoportula. The cingulum is composed of intricately patterned scales,[i-l] which may link in pairs across the cingulum; or there may be many scales in each whorl. The scales are areolate and some are fimbriate and notched; the areolae have perforate plates similar to those on the valve.

Rhizosolenia is most distinctive in the structure of the valve and areolae. Several species have been illustrated by Hasle (1975). An earlier work of importance is Peragallo (1892). Several taxa (including *R. fragilissima, R. cylindrus, R. delicatula, R. stolterforthii*) have shorter cells and a relatively inconspicuous spine; these form a subgroup which will have to be removed from *Rhizosolenia*. Yet others under investigation by B. Sundström (personal communication) may also require new generic status.

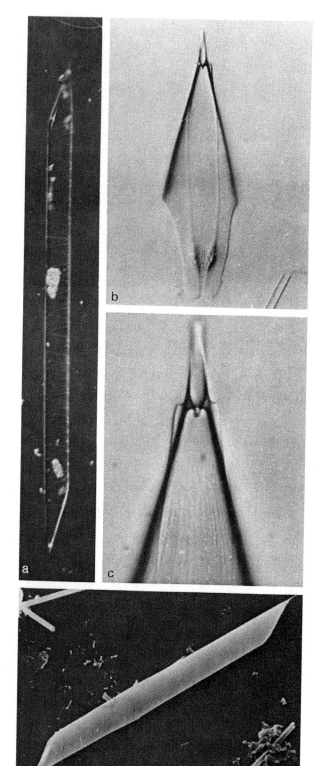

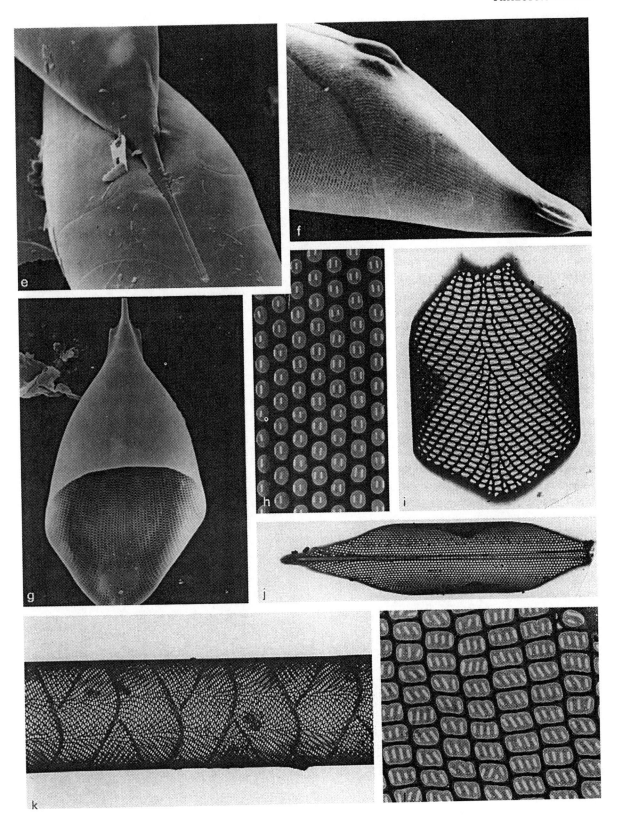

Proboscia B. G. Sundström. The marine diatom genus *Rhizosolenia*. A new approach to the taxonomy, p. 99. Lund, 1986. 117 pp.

T: *P. alata* (Brightwell) Sundström
(= *Rhizosolenia alata*)

Cells long, cylindrical,[a] usually solitary. Plastids almost certainly discoid. A widespread oceanic planktonic genus.

Valves conoid, tapering into a slightly curved proboscis.[b-f] A groove occurs at the base of the proboscis and is partially covered by two lateral flaps; the proboscis of the sister cell is held in this groove[h] prior to separation of the cells. No areolae obvious on the valve in SEM. TEM illustrations in Hasle (1975) and Sundström (1986) show rows of round areolae running to the apex of the proboscis and more irregular areolae on the basal part of the valve. The areolae have poroidal vela. Apex of proboscis obliquely truncate, surrounded by fine teeth. There is a rather inconspicuous groove at the apex[arrow, h] (see also figs 51, 56, 57 of Hasle, 1975; and in her fig. 52, the flat apex of the proboscis is closed except for a single off-central pore). The copulae are in two rows,[i, j] rhomboidal in outline but giving the appearance of triangular segments in SEM. They are finely porose.

This is a distinctive taxon which Hasle (1975) recognised as having several morphological features that set it apart from *Rhizosolenia*; of these the proboscis is the most useful in diagnosis. Several other species may need transfer into *Proboscia* but require detailed SEM studies. Previous attempts have been made to group the species of *Rhizosolenia* (see Hustedt, 1927–66) and several workers including Hustedt placed 'alata' in a subsection before Sundström described *Proboscia*.

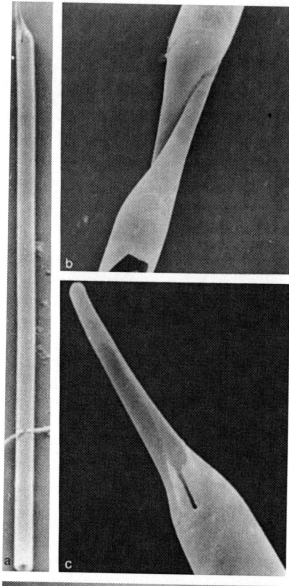

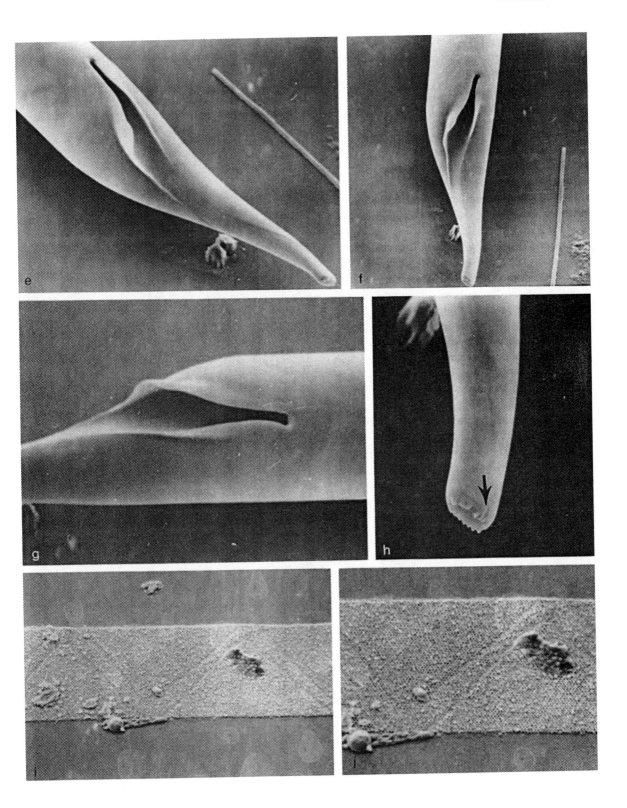

Pseudosolenia B. G. Sundström 1986. The marine diatom genus *Rhizosolenia*. A new approach to the taxonomy, p. 95. Lund, 1986

T: *P. calcar avis* (Schultz) Sundström (= *Rhizosolenia calcar-avis*)

Cells elongate,[a] cylindrical, usually solitary, each end with a sinuous spine. Plastids numerous, discoid. A widespread and common marine phytoplankton genus.

Valves (calyptrae) conical, drawn out into an asymmetric curving and narrowing process, which opens at the apex.[b-f] Indentation (= contiguous area of Sundström, 1986) forming a sigmoid groove[c, d] running down the process and across the valve. Areolae with large central aperture surrounded by four small pores, the latter opening to the outside.[j] A single rimoportula present at the base of the process;[b] the internal opening developed in a double ear-shaped structure which resembles the macro-rimoportula of some *Coscinodiscus* species. Figs 253, 254 of Sundström illustrate this structure. Cingulum composed of numerous scale-like (rhomboidal) copulae arranged in vertical rows;[g] copulae porate [h, i] with plain margins and fimbriate edges.

The distinguishing features of this genus are the coiled rimoportula, the asymmetrical valve and the form of the indentation of the valve (Sundström). We would add the lack of otaria, which are found in *Rhizosolenia* and *Proboscia*.

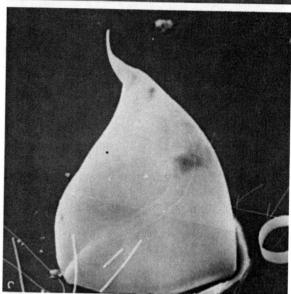

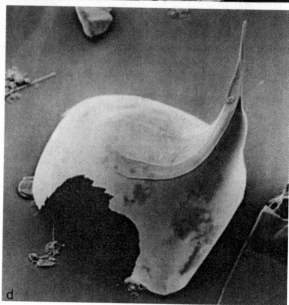

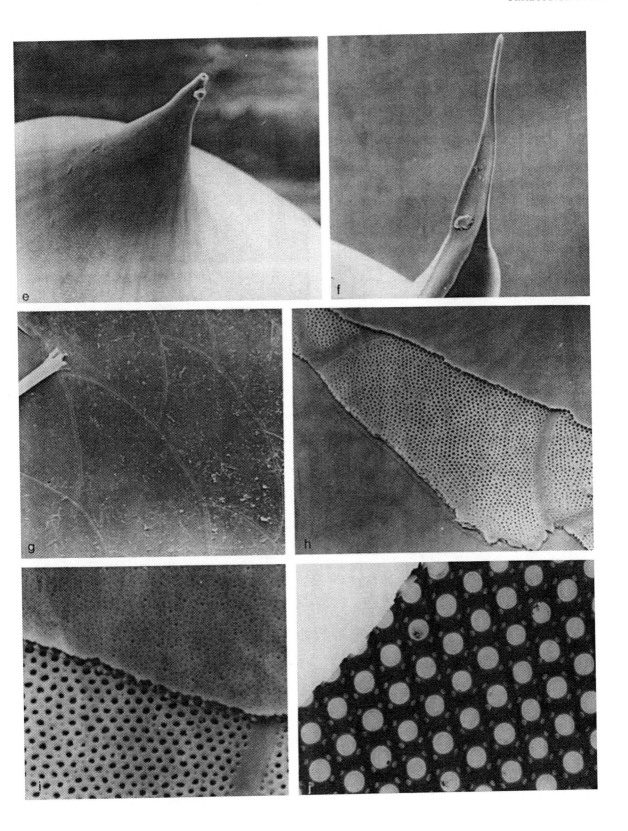

Urosolenia F. E. Round & R. M. Crawford, gen. nov.

T: *U. eriensis* (H. L. Smith) Round & Crawford (= *Rhizosolenia eriensis*)

Cells solitary, cylindrical, with a long fine hair-like extension. Plastids discoid, numerous. A widely distributed freshwater planktonic genus, although the cells are very delicate and easily overlooked.

Valves conical,[a-e] continuing into a long extension which terminates in a few small teeth.[g] The valve surface has irregular areolae, these becoming somewhat larger toward the base of the extension, which is plain. Rimoportulae absent. Copulae numerous, consisting of imbricating half-rings perforated by irregular openings.[e, f, h] One end of each ring lies slightly below the other end at the junction.[f]

The possession of a valve with a long extension and numerous copulae resulted in this genus being included within the wholly marine genus *Rhizosolenia*. Hustedt (1927–6) placed the freshwater (brackish) forms in a subsection Longisetae but the differences are sufficient to indicate a separate genus; the grooves on the valves of *Rhizosolenia sensu stricto*, the rimoportulae associated with the bases of the processes and the characteristic vela of *Rhizosolenia* are not present in the freshwater species.

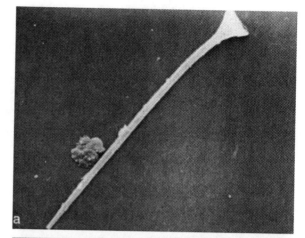

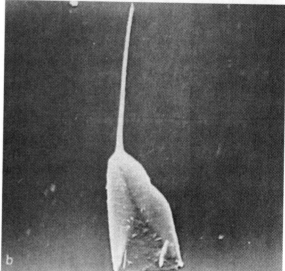

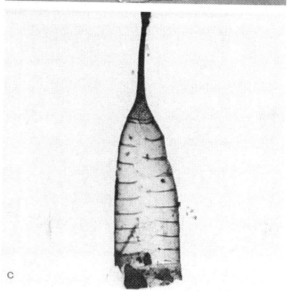

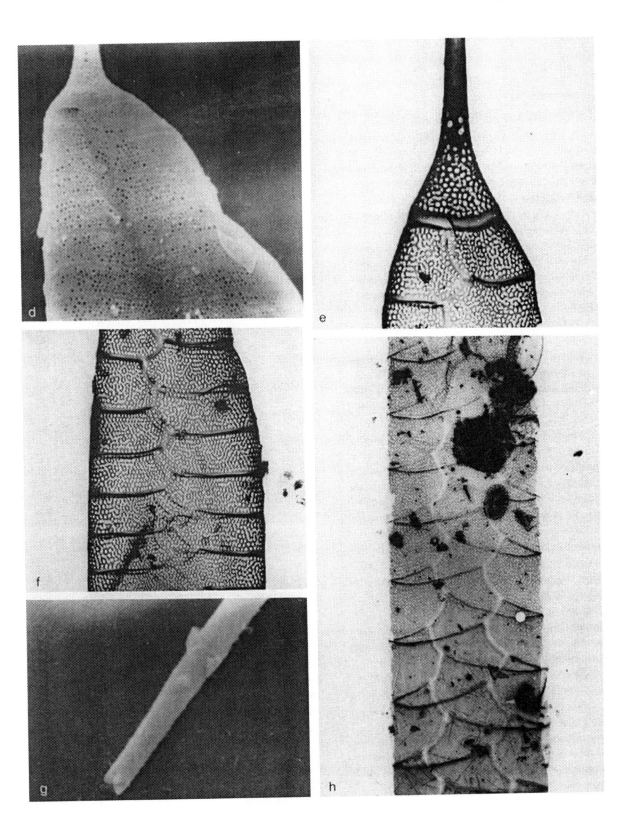

Guinardia H. Peragallo 1892. Le Diatomiste 1: 107

T: *G. flaccida* (Castracane) Peragallo (= *Rhizosolenia flaccida*)

Cells cylindrical,[a] united into filaments; these are fragile, often breaking up into single cells. Girdle view reveals a slight indentation on the valve face and many split copulae.[b] Numerous stellate plastids. A monotypic (but see below) very common, marine, planktonic genus.

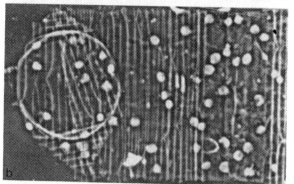

Valves circular,[c, d] almost hyaline, with a shallow valve mantle distinguished by more regular rows of areolae.[e] A small annulus (containing the single rimoportula) is located on the periphery[d] and from this rows of very fine areolae radiate, at first in a rather irregular fashion and then becoming more precisely arranged: this is shown nicely in fig. 81a of Hasle (1975). Directly inward from the rimoportula the areolae may be grouped 2-3 wide giving lenticular discontinuities[f-h] similar to those found near the swirls in *Bellerochea* (q.v.). No vela have been observed and none can be seen in the TEMs of Helmcke & Kreiger (1953-77) or of Hasle (1975). The external aperture of the single rimoportula is a raised curved tube, curling clockwise.[e] Fine ridges run from a point near the external aperture along the edge of the valve. Internally the rimoportula slit is positioned at right angles to the circumference and lies in a slightly raised unperforate region.[f, g] In some valves a lateral pore can be seen beneath the lips. Copulae numerous, split and with ligulae. The bands bear fine areolae in a rectangular arrangement; their apertures are smaller and the rows of areolae are more densely packed than on the valve face.

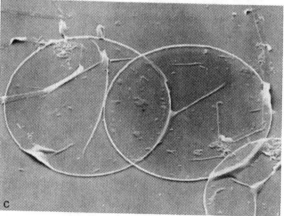

This genus is usually placed in the Rhizosoleniaceae and we feel this is correct. Its structure is only slightly reminiscent of *Rhizosolenia* as understood by references to species such as *R. styliformis*. Hasle (1975) considers that there is possibly only the one species since she has removed *G. blavyana* to *Dactyliosolen* which differs in the structure of the copulae and lack of external rimoportula process. However, as she pointed out, and we would agree, species such as *Rhizosolenia stolterfothii*, *R. fragilissima*, *R. delicatula* and possibly *R. cylindrus* and *R. tubiformis*, have much in common with *G. flaccida* and should be moved to *Guinardia*.

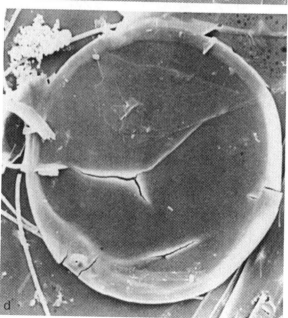

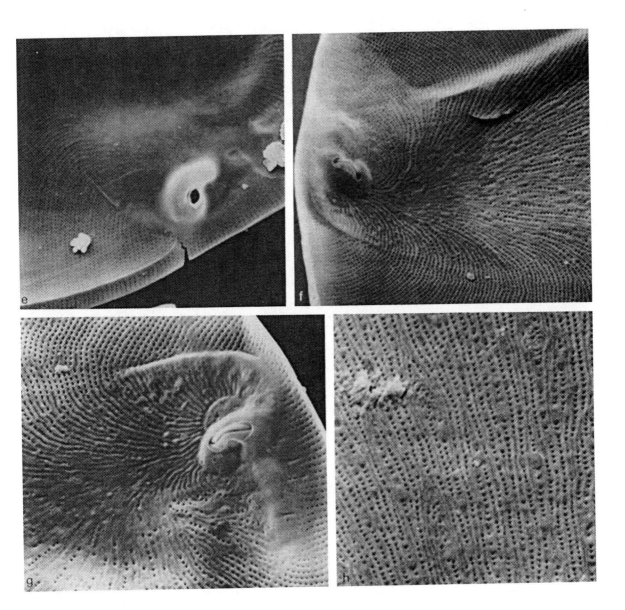

Dactyliosolen F. Castracane 1886. Rep. Sci. Res. Voy. 'Challenger', Bot. **2**: 75

T: *D. antarcticus* Castracane

Cells narrowly cylindrical, occurring in long filaments in the marine plankton. Girdle bands very numerous and the most obvious feature of the cells.[a, c-i] Plastids discoid. A small genus but quite widely distributed. Our figures are of *D. antarcticus*. Hasle (1975) recognised that *D. blavyanus* (= *Guinardia blavyana*) and *D. tenuijunctus* (= *Rhizosolenia tenuijuncta*) should be included in the genus.

Valves circular,[b] delicate, and from the outside showing only a rim of slightly more regular thickening and rather more distinctive pores. Almost no valve mantle distinguishable. Internally with delicate branching ribs radiating from an off-centre rimoportula. The rimoportula does not have surrounding elevations such as in *Guinardia flaccida*. The copulae are characterised by their thickness compared with the valve and by the single row of areolae on each; the areolae often appear unoccluded although in some material[d] it is possible to detect a fine net-like cribrum. The valvocopula is a complete, closed band, while the remaining girdle elements are crescent-like segments, with a flange on one end internally.[e-h] The ends of the segments are pointed and overlap one another so that the flanges of adjacent bands rest on one another.

Few workers have studied this genus but Hasle (1975) has made an excellent study of some of the type material. Whilst the valve is clearly very similar in form to that of *Guinardia*, the girdle bands are distinctive in possessing extremely large areolae and having internal flanges. We know of no other diatom with such an internal structure, which must affect the fitting together, or rather, sliding apart of the new and old bands. Also the presence of a cribrum has not been recorded in other taxa within the Rhizosoleniaceae.

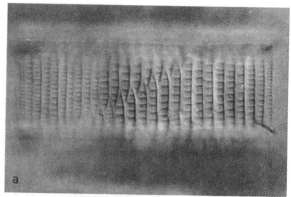

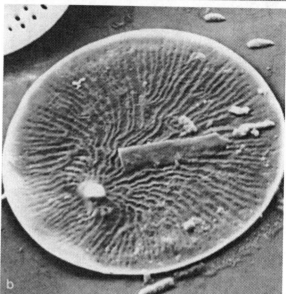

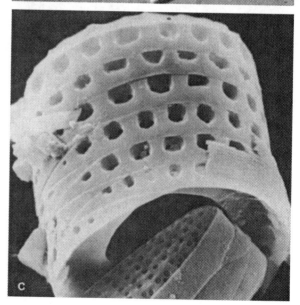

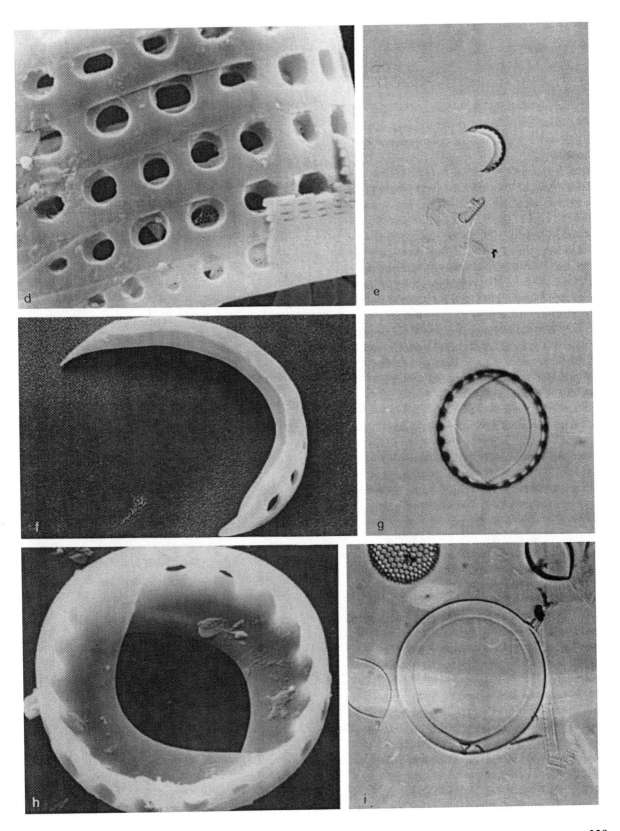

Pyxilla R. K. Greville 1865. Trans. Microsc. Soc. Lond. Ser. 2, 13: 1

T: *P. johnsoniana* Greville

Cells long in girdle view, each valve bearing a prominent central horn-like extension; probably forming filaments. A small genus of fossil diatoms, occurring in Eocene and Oligocene deposits. Habitat unknown.

Valves bell-shaped[a, b] with a long extension extending straight up from the centre. There is no valve mantle but a distinct flared rim is present around the edge of the valve. Lines of areolae, separated by narrow ribs, occur over the 'bell' and extend to the apex of the extension. Areolae large and round,[a, b, d, f] becoming elongate towards the apex of the extension;[e–g, i] vela apparently of the cribrate type.[d] At the apex of the extension there is a lateral flange subtended by a pore;[h] this structure fits closely into the sibling valve lateral projection.[c, g] Copulae with rows of small round pores are present (P. A. Sims, personal communication and see the Fig. d provided by Sims). The copulae are also figured by Nikolaev (1984).

The evidence from the linking of sibling valves suggests that filament formation might occur somewhat like that in *Rhizosolenia*, and indeed Gombos & Ciesielski (1983) illustrate such an attachment. The illustrations in Greville are of cells which are distinctly heterovalvate and with much shorter apical extensions. One illustration does, however, also show a flared valve margin, as in our specimens. Hendey (1969) erected the genus *Pyrgupyxis* for somewhat similar cells and he transferred several *Pyxilla* species to it, but not the type or the second species of Greville (*P. barbadensis*). The shape of the valves in Hendey's material is not close enough to the present material to warrant its inclusion in *Pyrgupyxis* though this cannot be entirely ruled out. Gombos & Ciesielski (1983) also maintain the two genera separate. There is some suggestion that the specimens are diatom resting spores and this is yet another point on which we cannot be certain. However, even if they are, they represent species which cannot be accommodated in any other known genus. The genus *Pyxilla* is generally included in the rhizosolenioid group, within the family Pyxillaceae containing only fossil genera. It would be interesting to determine whether or not there is a rimoportula internal to the lateral projection. The group of genera around *Pyxilla* all require further study.

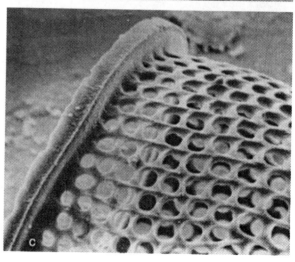

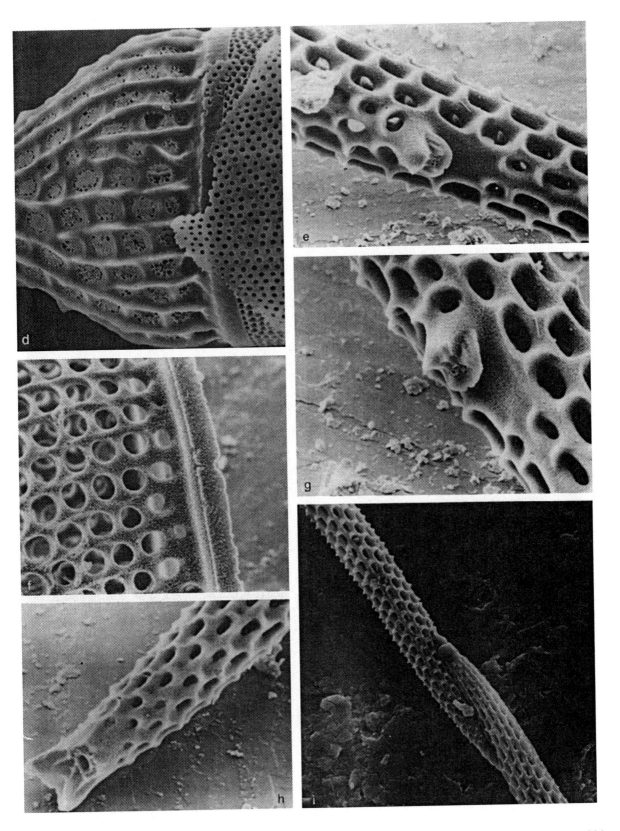

Chaetoceros C. G. Ehrenberg 1844. Ber. Bekanntm. Verh. Königl. Preuss. Akad. Wiss. Berlin, **1844**: 198

T: *C. tetrachaeta* Ehrenberg

Cells joined in straight,[a, c] curved or coiled filaments, rarely solitary; narrowly to broadly elliptical in valve view, rectangular in girdle view. Cells united by fusion or interlocking of setae produced from the valve. Some species have small 'prehensors' for linking of setae (Fryxell & Medlin, 1981). One or more small plate-like plastids. An important marine planktonic genus with many species. A few species have been recorded in freshwater.

Valves elliptical[e, g] to linear, with a flat or domed valve face, bearing one seta from near each pole. Valve a thin lamina with[e, g] or without a central annulus and rimoportula,[g] from which radiate rows of very fine areolae. Terminal valves[arrow, a] and setae differ morphologically. Solid spines may occur, either as a ring on the valve rim[d] or scattered over the valve surface. The setae[j-n] are built up in various ways and in some the cytoplasm with plastids penetrates along their length. Pore-plates have been found at the base of the setae.[e, h] The cingulum[i] consists of a number of copulae arranged in a scale-like pattern. Details of the cingulum are given by von Stosch and the distal or last-formed copula has a depression at both apices through which the setae of the new hypovalves project. Resting spores frequent.[arrow, b]

For details of life history including cell division, resting spore formation and sexual reproduction, see von Stosch *et al.* (1973). Two species, *Chaetoceros armatum* and *C. septentrionalis* are considered by Evensen & Hasle (1975) to have closer affinities with *Attheya decora* than with *Chaetoceros sensu stricto* and we have separated these into *Gonioceros*. An important recent contribution to the study of *Chaetoceros* is that of Rines & Hargraves (1988).

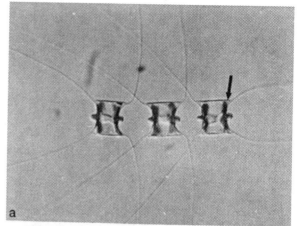

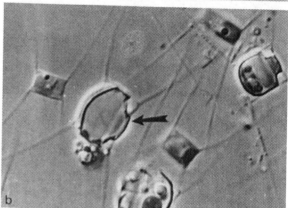

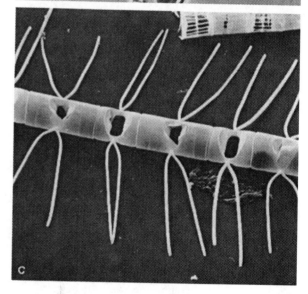

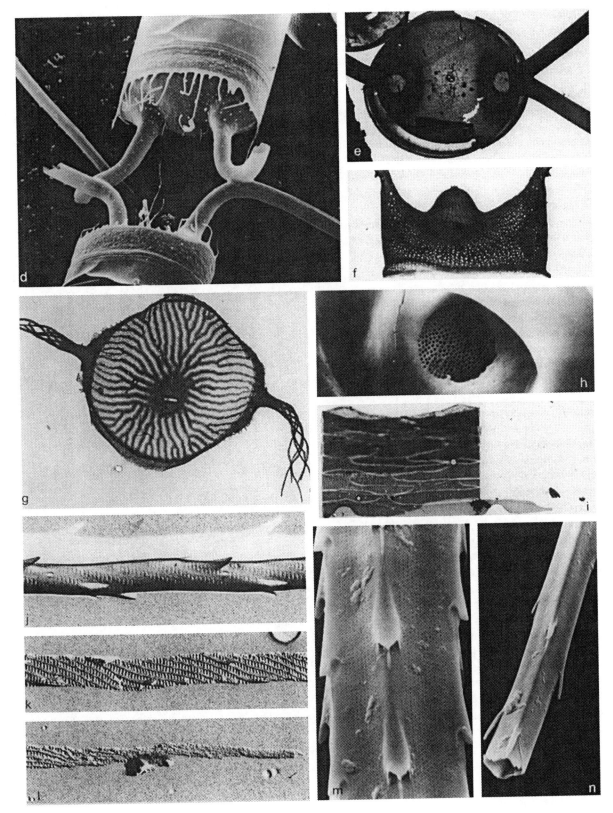

Gonioceros H. Peragallo & M. Peragallo, 1907
Diat. Mar. France: 468

T: *G. armatum* (T. West) H. & M. Peragallo
(= *Chaetoceros armatus*)

Cells rectangular, forming short chains or living singly (*G. septentrionalis*). In colonial forms the valve faces are appressed except around the margin, which results in the production of circular 'holes' at either side of the chain;[a, c] separation valves occur, with small spines. Two plate-like plastids. A marine genus living in the surf plankton (see Lewin, 1974) and formerly placed in *Chaetoceros* but more properly allied with *Attheya*. The filaments are often covered with silt particles presumably embedded in mucilage.

Valves elliptical to lanceolate[b, e-g] raised centrally,[c, d] apparently without any areolae or rimoportulae; mantle simple and shallow. Hollow setae arise from the apices of the valve.[a, c-f] They appear to be composed of spiral strips of silica,[e, f, h-j] but closer observation reveals a series of thin ascending elements, running the length of the seta, to which many thin transverse hoops are apparently attached. Towards the bases of the setae small outwardly facing spines occur;[d, f] in addition there is an internally directed group of stronger, longer spines.[c-f] Separation valves have a distinct central region of diffuse silica and a fine radiating outer layer; between the two regions is a ring of short 'thorn-like' processes.[f, g] Copulae numerous, finely porous (see Evensen & Hasle, 1975).

Hustedt (1927–66) placed *G. armatus* in *Chaetoceros* as an incompletely known form, and commented on the distinctive feature that the valve faces are appressed, whereas in almost all *Chaetoceros* spp. they are not. Evensen & Hasle (1975) also cast doubt on its position in *Chaetoceros* and noted several similarities to *Attheya*. They studied the copulae in detail and found that they differ from those of *Chaetoceros* species. Duke *et al.* (1973) described *C. septentrionalis* and this is clearly allied to *G. armatus*. The structure of the setae is perhaps the most striking feature of these diatoms, and indeed it was this that persuaded us to resurrect *Gonioceros*. The valves appear to be lacking pores but further investigation is required.

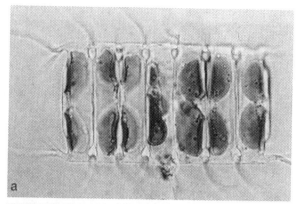

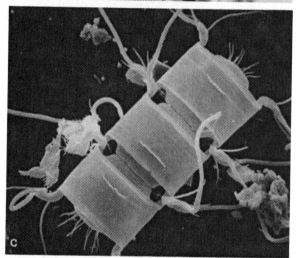

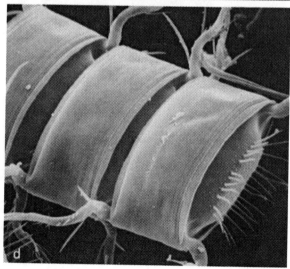

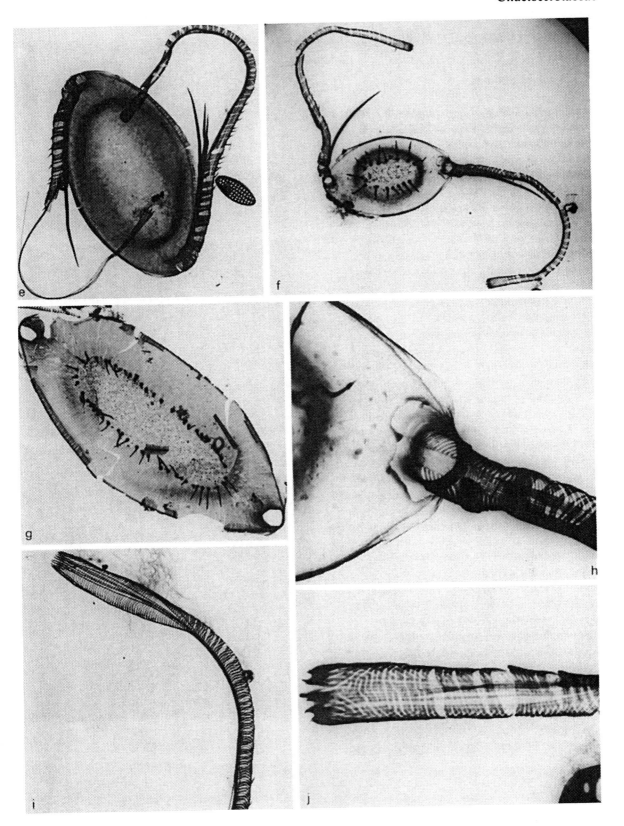

Bacteriastrum G. Shadbolt 1854. Trans. Microsc. Soc. Lond. Ser. 2, **2**: 14

T: *B. furcatum* Shadbolt

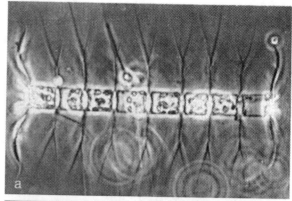

Cells cylindrical, linked to form filaments.[a] Each cell has several long, radiating setae which may be simple or bifurcate;[c] the setae from adjacent cells are fused. Separation valves occur at intervals in the filaments and have modified setae.[b] The colonies tend to lie in girdle view and the cells are slightly separated by the curvature of the basal part of the setae, leaving gaps between the cells. Plastids discoid. Marine, planktonic, and widely distributed.

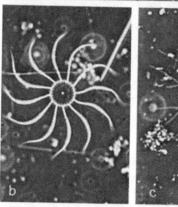
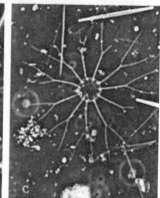

Valves centric with files of tiny, simple areolae radiating from a central annulus – we have been unable to photograph these adequately. Around the periphery hollow setae[f-h] arise and curve upwards before fusing with the setae of the sibling valve.[f] The fused setae then run out at right angles to the filament; the setae may be spinulose or variously ornamented.[h, j] The separation valve processes are unbranched, often thicker, curved and may have a few enlarged pores around the base.[i] Fusion of the setae is by their external walls but the two tubes maintain their integrity throughout. Towards their apices the setae bifurcate into two tubes again.[e] Internally the valves are simple, with the only obvious feature being the depressions leading into the bases of the setae. In the separation valves, but not in the others, the central annulus surrounds a single, simple rimoportula.[i] The copulae are porous and the more abvalvar are modified, with flaps which extend towards the adjacent cell, leaving gaps through which the paired setae project.[g]

There have been no comprehensive studies of this genus since that of Pavillard (1924). Boalch (1974) reports on the identity of the type species *B. furcatum* from Shadbolt's original slides and Fryxell (1978) provides ultrastructural detail. It is a small genus of less than 20 spp. yet it is usual to find cells in almost any plankton sample; only rarely does it seem to achieve dominance. *B. solitarium* is reported to exist as individual cells. This genus is always associated with *Chaetoceros* but differs in its radial symmetry and the lack of fenestration of the setae.

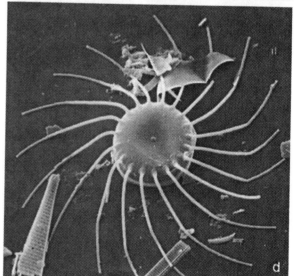

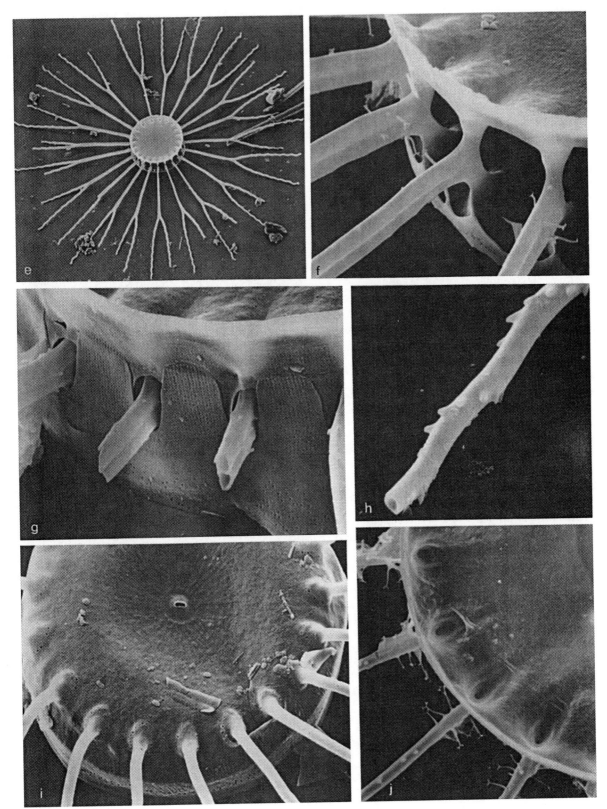

Acanthoceras H. Honigmann 1909. Arch. Hydrobiol. Planktonk. 5: 76.

T: *A. magdeburgense* Honigmann (= *Attheya zachariasi* J. Brun).

A freshwater form of very delicate structure, easily missed in planktonic samples. Cells solitary, usually seen in girdle view, rectangular with spines (setae) at each angle[a,b] and slight indentations at the centre of the valve faces; the median junction of the copulae visible with phase microscopy. Plastids reported as four small plates.

Valves moulded into two 'caps' connected by a thin 'bridge'.[c, d] The areolae are simple pores, arranged in longitudinal rows on the 'bridge' and radiating up the cap, stopping abruptly at the base of the spine. Spines tubular[e] and without areolae. No portules. The copulae are areolate, 'collar-like' structures.[g, f, h] Central copulae with few or no areolae.[g]

Acanthoceras was resurrected by Simonsen (1979) who suggested it be left in the Chaetocerotaceae. However, our electron microscope observations do not confirm this. The shape of the valve, the unperforate processes, the multiple girdle bands, the nature of the areolae, and the absence of portulae all suggest a distinct family and we propose the Acanthocerataceae for this monotypic genus. The illustrations in Honigmann (1909) do not really resemble a diatom (more likely a green alga, possibly a *Scenedesmus*). Planktonic samples subjected to acid-cleaning lose cells by solution. This diatom, with no portulae and exceedingly simple valves and girdle bands is almost as simple a structure as can be conceived for the diatom cell – the only elaboration being the production of spines. The copulae, observed on their own, could be mistaken for chrysophyte scales – whilst almost every other genus of diatoms has evolved such that we have no doubt of the divisional status of the group, this genus may indicate a link between the groups. The reproduction of *Acanthoceras* might provide an interesting insight into the problem. Hustedt (1927–66) merely records resting spores.

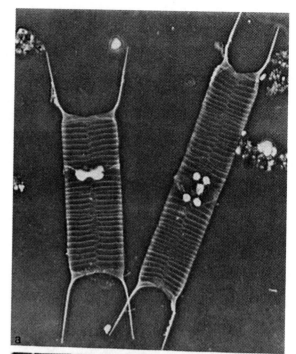

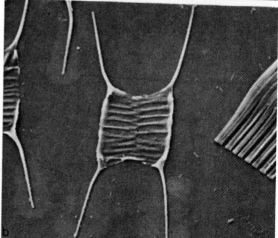

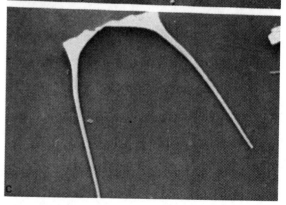

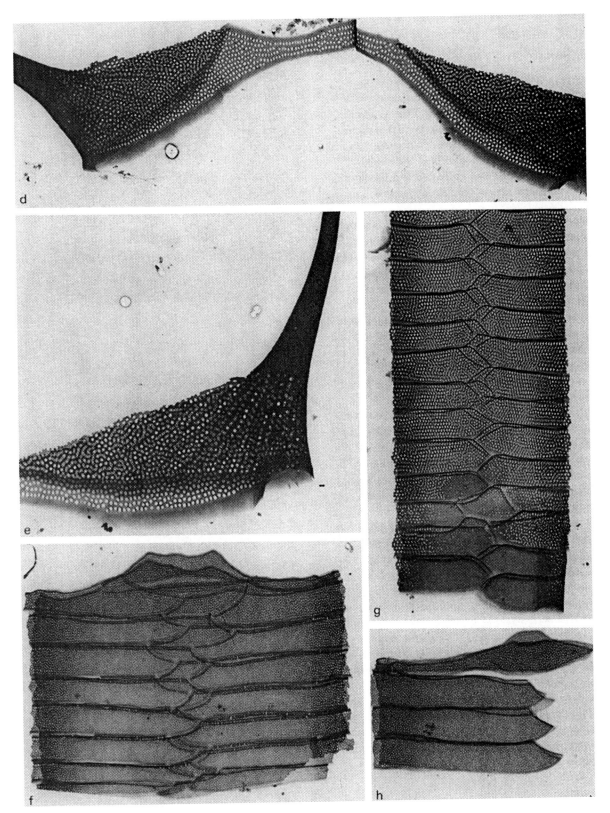

Attheya T. West 1860. Trans. Microsc. Soc. Lond. Ser. 2, 8: 152

T: *A. decora* West

Cells laterally compressed with short apical horns. Plastids spathulate, radiating from the centre with their broad ends directed outwards.[a] The cytoplasm extends into the horns, attaches to the valve centres at the rimoportulae and is also pressed against the girdle bands (see also Schnepf, Deichgräber & Drebes, 1980). *Attheya* lives in coastal marine waters and can be found on attached sand grains in the intertidal zone.

Valves elliptical, delicate, with an elongate central zone of closely spaced poroids[e, f, g] and, radiating from this, a delicate system of 'strands' of silica with occasional weakly developed connections (fewer towards the valve margin).[e, f] The siliceous strands extend out at the valve apices and curve in a loose spiral to form horns.[b, d] A single rimoportula occurs in an off-centre position.[e] Cingulum of many split copulae, all of approximately equal form;[b, c] openings occur at apices. Copulae porous on either side of the pars media, except in some of the more abvalvar bands, where there are plain areas.[h, i]

This is a monotypic genus now that the freshwater *Attheya zachariasi* has been removed to *Acanthoceras* by Simonsen (1979). The basic framework of the valves of *Attheya* has certain features in common with '*Chaetoceros*' *septentrionalis*, in particular the 'strands' forming the apical horns (*Attheya*) and setae (*Chaetoceros*); see also the discussion in Evensen & Hasle (1975). However, we have also decided that *C. septentrionalis* differs sufficiently from *Chaetoceros* to be allocated to a new genus (*Gonioceros*) so that there is now even less reason to suggest a transfer of *Attheya* to *Chaetoceros*. It is interesting that both *Attheya* and *Gonioceros* occur in the coastal surf zone – a very different habitat from the planktonic mode of both *Chaetoceros* and *Acanthoceras*. Evensen & Hasle also comment on the possible relationship of these taxa to families such as Hemiaulaceae, Biddulphiaceae and Eupodiscaceae rather than Chaetocerotaceae. Round & Crawford (1989) have proposed a separate family Attheyaceae but have left this within the Chaetocerotales for the present. Drebes (1977b) has described auxosporulation in *A. decora*.

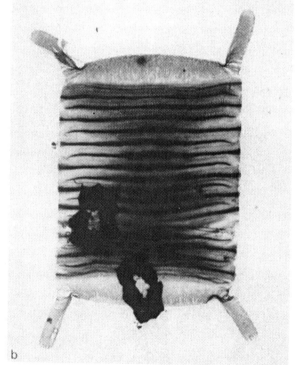

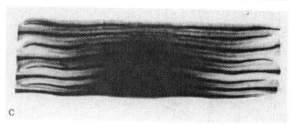

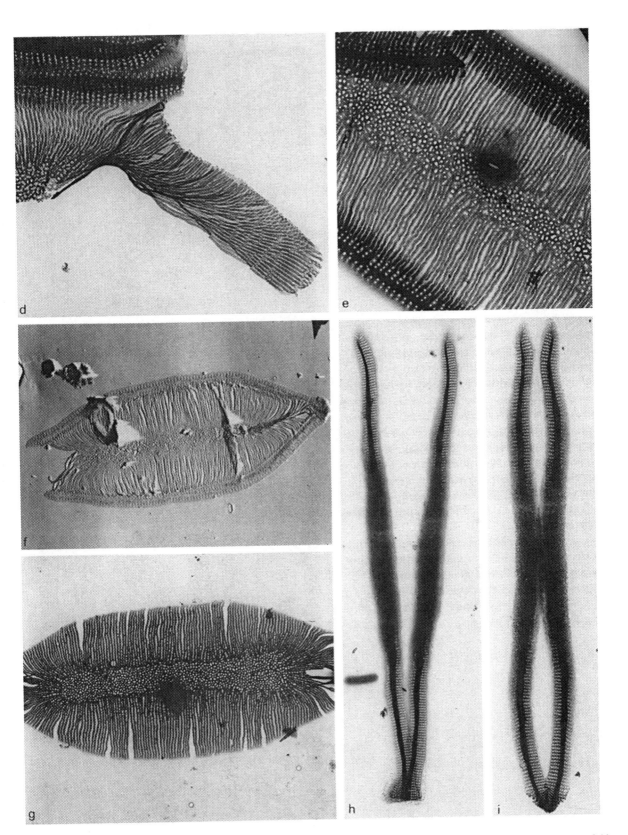

Leptocylindrus P. T. Cleve 1889. In C. G. J. Petersen, Vidensk. Udbytte Kanonbaaden 'Hauchs' togter i de Danske Have: 54

T: *L. danicus* Cleve

Cells narrow, cylindrical, united to form filaments.[a] Plastids 2 to many, plate-like, lying against the girdle. A common, delicate member of the marine plankton with less than five species, even if *Dactyliosolen mediterraneus* is transferred to it (Hasle, 1975); this is problematic and for the present our description of the structure of *Leptocylindrus* will be based on *L. danicus*.

Valves circular,[b-g] thin with a ring of projections around the margin of the flat valve face;[d, e] beneath this is a shallow valve mantle bearing vertical rows of areolae.[e-h] The marginal projections appear flap-like and between these are other blister-like markings.[arrows, d, e] Valve face with indistinct areolae in uniseriate striae;[f-h] these radiate from a central annulus within which is a central cluster of irregularly placed areolae.[f] There is also an off-centre pore (just outside the annulus) whose nature is uncertain and which has no parallel in any other genus we know. The flaps on the edge of the valve face have no structural counterpart on the inside of the valve, but in this region the radiating 'ribs' of the valve face become more distinct before running down the valve mantle; this feature is especially obvious in TEM. Copulae very numerous, consisting of tile-like segments,[i] each finely porous and with a wide truncated 'ligula'.

The organisation of the cell is very similar to that of many 'solenioid' diatoms and it certainly cannot be placed in the Melosiraceae as suggested, admittedly tentatively, by Simonsen (1979). We prefer Hasle's (1975) retention of Lebour's (1930) family Leptocylindraceae.

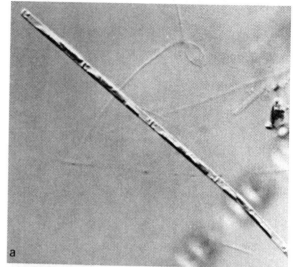

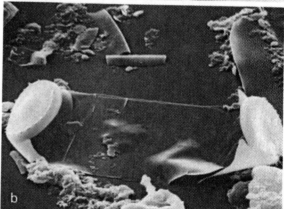

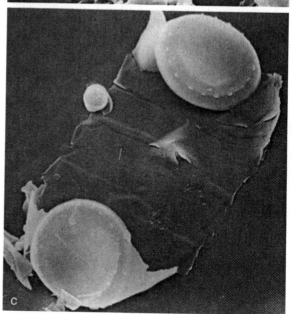

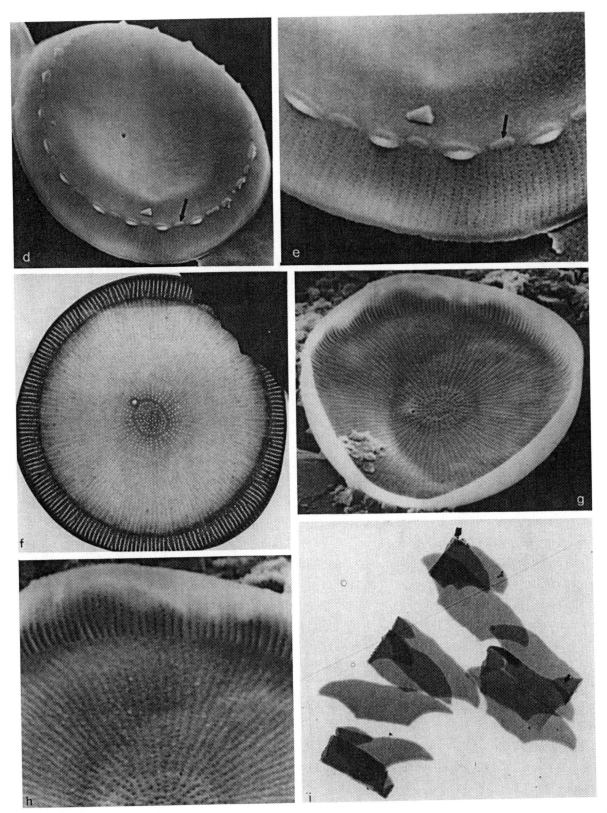

Araphid group

In many ways the classification of this group is the most difficult, since unlike the centric series their valve structure is rather simple, and unlike the raphid series the plastids and their arrangements have few distinguishing features. In addition, the valves tend to be linear and small, and it is not easy to discern their features, especially in the light microscope. Whilst the centric diatoms are predominantly planktonic and the raphid ones live mainly in the epipelon, the araphid diatoms are largely epiphytic/epizoic/epipsammic, and these habitats are less well studied than the others. All these aspects have led early workers to lump the genera; Hustedt (1927–66), for instance, had only seven subgroups. There are few details on plastids, nuclear movements or sexuality for the araphid genera and classification relies heavily on frustule structure. When all aspects have been studied we expect to see further adjustments to the system. All but one of the orders (the exception being the Fragilariales) are totally marine in distribution and the general impression is that evolution has led to much greater diversity of form in that habitat. In the Fragilariales there has been a tendency to insert the marine species into the freshwater genera simply on the basis of features such as overall form, e.g. *Asterionellopsis glacialis* was placed in *Asterionella* simply on colony morphology, while needle-like marine species were placed in *Synedra* even though, as recognised by Hustedt (1927–66), some of the taxa involved have the most complex loculate valves known in the araphid group. In cases like these the need for generic revision is clear. Other differences are more subtle, e.g. between the new genera clustered around *Rhaphoneis* (Andrews, 1975) and the splits of *Fragilaria* (Williams & Round, 1987) and *Synedra* (Williams & Round, 1986).

Fragilaria H. C. Lyngbye 1819. Tent.
Hydrophytol. Dan. 182 (1819) = *F. capucina*
Desmazières (1825) (see C. S. Boyer, Proc.
Acad. Nat. Sci. Philad. 78 Suppl. 182, 1927)

T: *F. pectinalis* (O. F. Müller) Lyngbye
(= *Conferva pectinalis*)

Cells joined to form ribbon-like colonies.[a-c] Cells in
nature seen in girdle view and either oblong or
swollen at the centre and then linked only there (*F.
crotonensis*). Plastids: 2 plates. Many species occur
on freshwater sediments and a few in the plankton.

Valves linear, linear-lanceolate,[i] elliptical, capitate,
sometimes with a slight central swelling. Sternum
linear or lanceolate, often expanded on one side at
the centre.[i] Areolae simple, arranged in transapical
uniseriate rows[f, g] passing over onto the valve
mantle; cribra delicate, often disc-like. Often a break
occurs in the rows at the valve face/mantle junction
where spines occur; spines may also extend around
the apices[f] and be simple, splayed[c] or branched.[d]
Apical pore plates present of a weakly-developed
ocellulimbus type[d, f] i.e. the apical pore plate lies
in a slight depression (see Williams, 1986).
Rimoportula openings circular or slit-like.
Characteristic small plaques[c-e] occur along the edge
of the valve mantle in some species. Internally the
areolar openings are simple,[h, j] sometimes lying
between rather well-developed costae. A single
rimoportula lies at one end of the valve[h] (often
laterally displaced). Copulae several;[d, e, k] 4 per valve
seems common; a single row of areolae present
along the advalvar edge but often obscured. The
valvocopula is usually split at one end though
populations with apparently complete valvocopula
have been seen.

Williams & Round (1987) restricted *Fragilaria* to
taxa that form colonies and have simple rows of
areolae and a single rimoportula. The genus is
freshwater and our investigations of marine
diatoms previously classed as *Fragilaria* suggest a
placing in other taxa. Lange-Bertalot (1980a) has
fused *Fragilaria* and *Synedra* (see also Poulin *et al.*
1986). We do not believe this is tenable since
characters can vary their state, e.g. *F. crotonensis* can
occur as single cells but is still recognisably *F.
crotonensis* – the colonial form of which is abundant
in thousands of lakes. The typification of *Fragilaria*
itself is a problem – it may have to refer to a marine
species. On practical grounds *Fragilaria* should be
conserved for freshwater species.

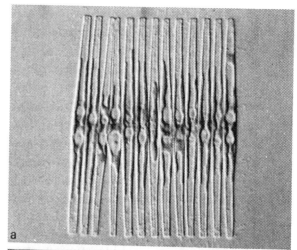

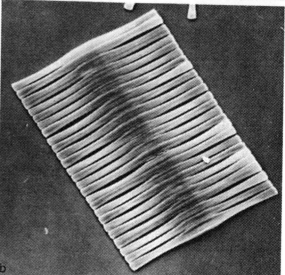

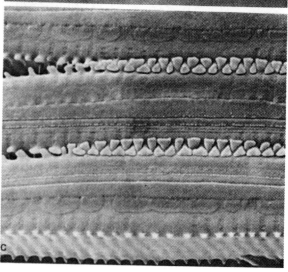

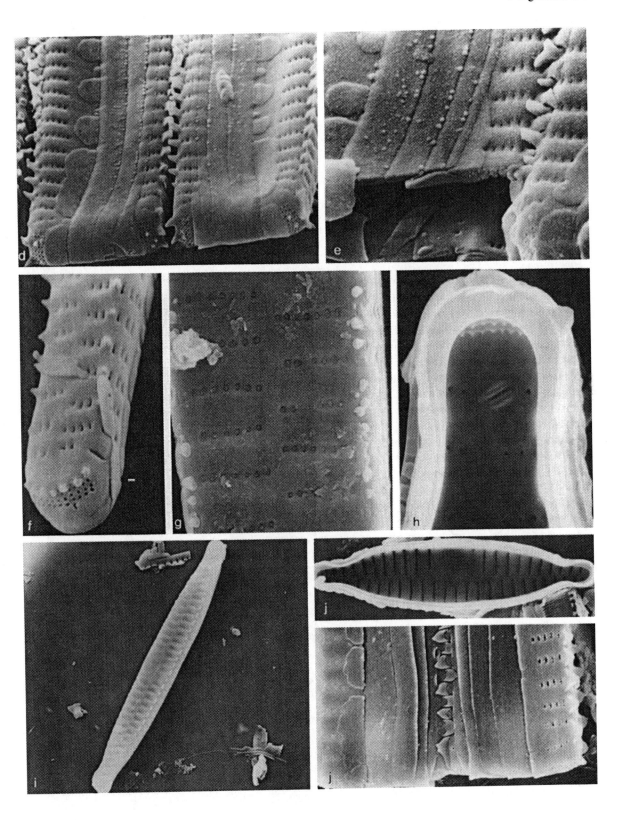

Centronella M. Voigt 1902. Forschungsber. Biol. Stat. Plön, 9: 41

T: *C. reicheltii* Voigt

Cells tri-radiate, with narrow arms.[a-c] Three small plastids are located at the bases of the arms and a tri-radiate nucleus lies in the central region (Schmid, 1978). The genus is found rather infrequently in the plankton of eutrophic lakes but can at times form large growths.

Valves tri-radiate with slight basal enlargements.[c-e] The arms may be of similar length or slightly unequal, and they may be twisted. Valve apices slightly capitate. A narrow sternum expands into a clear area in the centre.[d, e] Striae uniseriate, transverse to each arm, continuing down the valve mantle, which turns at a right angle to the valve face. Areolae slightly expanded laterally, often filled in towards the sternum. Small plaques (cf. *Fragilaria*, p. 346) occur along the edge of the valve mantle.[g] Spines occur at intervals along the junction between the valve face and mantle.[d, f, g] Small apical pore fields present and above these are two rather more prominent spines.[f] Single rimoportulae[h] occur near the apices of two of the arms only (Schmid, 1978). The cingulum consists of a valvocopula and a single copula; a row of areolae appears to be present on one band but the extent of this cannot be determined.

The status of the genus has been questioned by previous workers. The SEM pictures suggest *Centronella* is possibly a tri-radiate form of *Fragilaria*. The presence of the plaques and the finding of only two rimoportulae reinforces this view. We cannot detect any other feature, apart from the tri-radiate morphology, that might support separation of this form from *Fragilaria*; Schmid reached the same conclusion. Although occurring as single cells in the plankton it forms chains (cf. *Fragilaria*) in culture (Kreiger, 1927). How a tri-radiate morphology can arise from a bipolar form is unknown but there is always the possibility that it develops when an auxospore expands abnormally, though it is perhaps surprising that this yields a cell with arms almost always at approximately 120° to one another. Since *Centronella* has been widely recognised and since we cannot be sure that the tri-radiate shape is not genetically fixed within an ancient lineage of 'fragilarioid' diatoms, we illustrate it here and thank A.-M. M. Schmid for the illustrations.

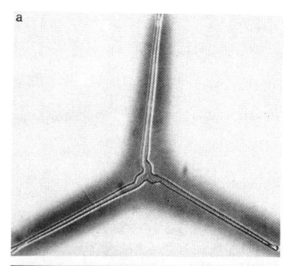

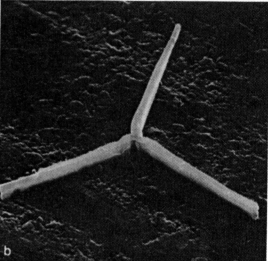

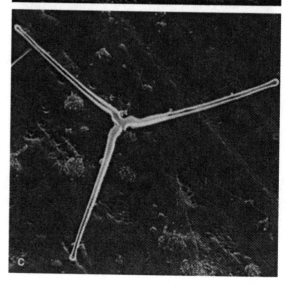

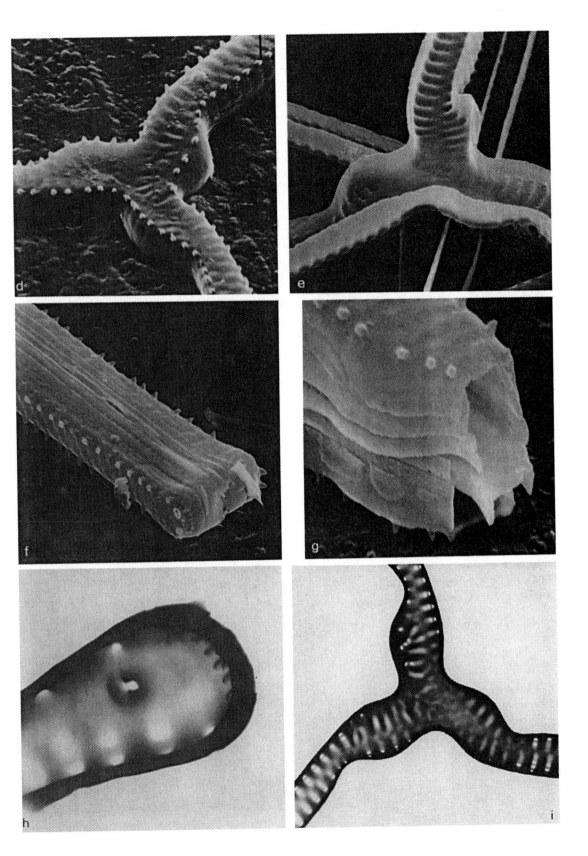

Asterionella A. H. Hassall 1850. Microscopic Examination of the Water. London.

T: *A. formosa* Hassall

Cells elongate joined to form stellate colonies.[a-c] Cells lie in girdle view and are slightly wider at the basal (attaching) pole than at the apical pole; in the colonies in fast-growing populations a second series of cells may be superimposed. Plastids: numerous small plates. A common freshwater, planktonic species often abundant in lakes. Frequently attacked by chytrid fungi seen as small colourless spheres along the cells.

Valves elongate, tapering slightly from the centre to the capitate apices;[d] the basal[f] larger than the apical pole.[e] Areolae uniseriate, sternum narrow. Vela indistinct. Apical pore fields present at either end of both valves.[e, f] A transversely orientated rimoportula at either end of both valves[g, h] opening externally via a pore somewhat larger than the areolae.[e, f] Spines occur along the edge of the valve mantle and above and below the apical pore field.[c, e, f] Copulae several, split; one or two rows of pores occur down the advalvar side.[i]

The cells are held together in the colonies by pads of mucilage. The ecology of *Asterionella* has been studied in detail by Lund (1949, 1950) who showed the close relationship between growth of the species and silicon content of the water. Several doubtful species have been described and a distorted form is often figured as var. *acaroides*. Jaworski *et al.* (1988) illustrate the extreme changes in cell morphology of *Asterionella* which can occur in cultures – the cells becoming almost oval and the stellate colony changing to a zig-zag form. The life cycle of *Asterionella* is discussed by Mann (1988b).

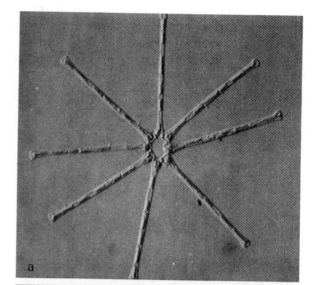

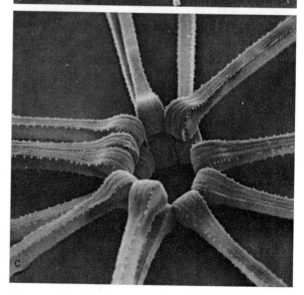

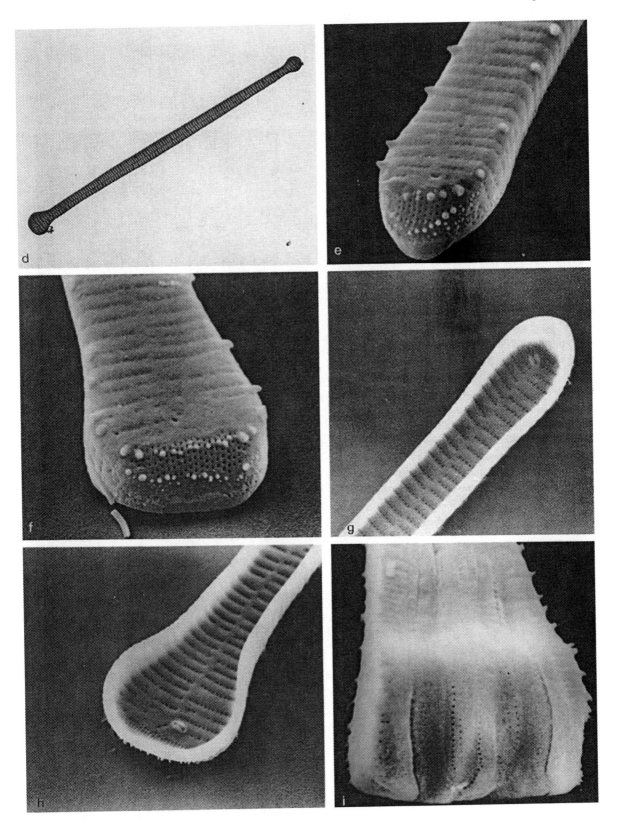

Staurosirella D. M. Williams & F. E. Round 1987. Diat. Res. **2**: 274

T: *S. lapponica* (Grunow in Van Heurck) Williams & Round (= *Fragilaria lapponica*)

Cells small, attached, and often forming short linear or zig-zag filaments. Plastids: two parietal plates adjacent to the girdle. A freshwater genus, often occurring attached to sand grains.

Valves linear, elliptical or cruciform,[a-c, e-g, m] with a wide sternum. Striae uniseriate, extending down the valve mantle[e-m] and containing slit-like areolae which are usually lost in preparation. The silica between the rows of areolae and between the areolae themselves may bear fine granules. Spines often complexly branched,[d] sometimes paired, situated at the valve face/mantle junction between the rows of areolae.[e, h-m] Apical pore fields present, consisting of few or many rows of poroids.[e, i-l] Mantle plaques usually present.[d] No rimoportulae. Cingulum of 8–10 open, ligulate, plain bands[k] with granulose abvalvar margins. Valvocopula wide and with a deeply crenulate advalvar edge. Copulae extremely curved at apices.

This genus encompasses the old *Fragilaria* species, *F. pinnata*, *F. africana* and *F. leptostauron* and differs from *Staurosira* in the nature of the areolae, apical pore plates and spine structure. The thick ribs separating the areolae are very characteristic of this taxon.

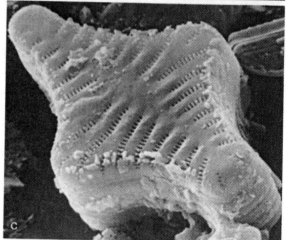

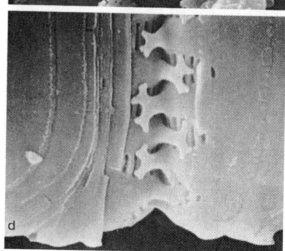

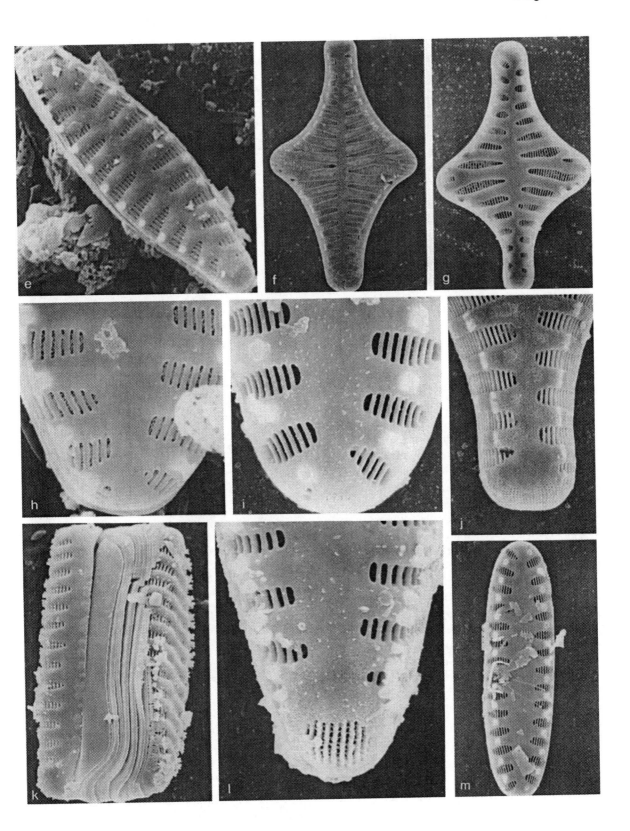

Staurosira (C. G. Ehrenberg) D. M. Williams &
F. E. Round 1987. Diat. Res. **2**: 272

T: *S. construens* (Ehrenberg) Williams & Round
(= *Fragilaria construens*)

Cells solitary or forming straight[b, c] or zig-zag
filaments of a few to many cells attached to one
another; attached to a substratum but often found
free-living. Plastids: a pair of plates lying along the
girdle. A common freshwater genus in lakes and
rivers, encompassing the old *Fragilaria construens*,
together with *F. elliptica* and probably several other
taxa still in *Fragilaria*.

Valves oval, elliptical, cruciform[d] or rarely
triangular, with widely spaced uniseriate rows of
areolae extending from the valve face onto the
curving mantle. Areolae circular or elliptical,
sometimes transapically extended, opening
externally into slight grooves in the valve face.[e–g]
Vela present but not yet studied in detail. Sternum
variable but never very narrow. No rimoportulae.
Marginal spines conspicuous, single or in pairs
between the areolae, spathulate or dichotomously
branched.[e–i] Small plaques sometimes occur along
the mantle margin[g] but these are relatively
inconspicuous compared with those in *Fragilaria*.
Apical pore fields variable, containing a few isolated
pores to several rows of them;[e, g, i] the pores often
appearing slightly rimmed. Internally with no
distinguishing features. Copulae 6–8 or possibly
more per valve;[c, e, j] usually extremely curved,
especially in small almost circular species where the
valve margin is also curved resulting in an enlarged
central mantle. Copulae almost certainly incomplete
(open) bands and without areolae.

This genus, formerly in *Fragilaria*, differs from
Fragilaria in the absence of rimoportulae, the non-
areolate copulae, the wide valvocopulae and
relatively narrow copulae. *Staurosira* differs from
Martyana (= *Opephora martyi*) since the latter has no
marginal linking spines, an apical pore field at one
end only, and a 'step' in the valve at the other end.

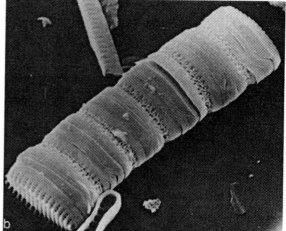

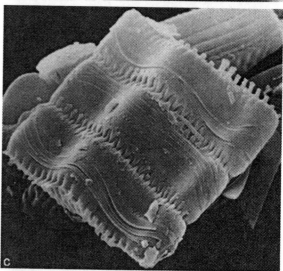

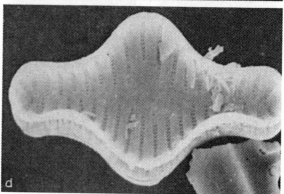

Pseudostaurosira (Grunow) D. M. Williams &
F. E. Round 1987. Diat. Res. **2**: 276

T: *P. brevistriata* (Grunow in Van Heurck)
Williams & Round (= *Fragilaria brevistriata*)

Cells joined tightly to form filaments. Plastids
probably parietal and plate-like. A common
freshwater genus, the species of which were
previously placed in *Fragilaria*.

Valves linear to elliptical, sometimes with
undulate margins,[a, c] or cruciform.[g, j] Sternum very
wide. Striae uniseriate, with only a few (1–4)
elongate areolae near the edge of the valve face[c–e]
and usually with a single areola on the mantle
below each spine.[d, f] Vela internal, mesh-like, though
in most preparations they are eroded. Spathulate[b] or
spathulate/branched spines[f–h] interrupt the striae at
the valve face/mantle junction. Plaques occur along
the mantle edge.[d–f] Small apical pore fields
sometimes present,[d] usually consisting of a few
poroids and sometimes with thickened rims.
Rimoportulae absent. Cingulum of several narrow
copulae,[b–d] which are not as narrowed at the apices
as in related genera. Copulae open, ligulate and
plain.

Apart from *P. brevistriata*, this genus also includes
some of the forms previously placed in *Fragilaria
construens* (e.g. the var. *bionodis*) and the recently
described *F. pseudoconstruens* (Marciniak, 1982) and
F. zeilleri Hérib. The most characteristic feature of
the genus is the sparse marginal areolae.

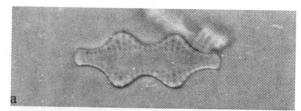

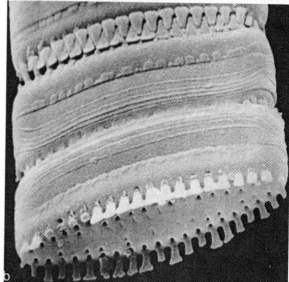

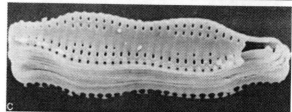

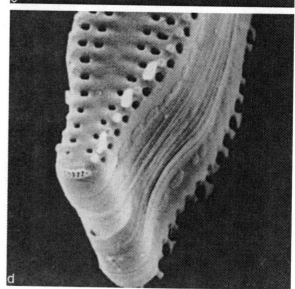

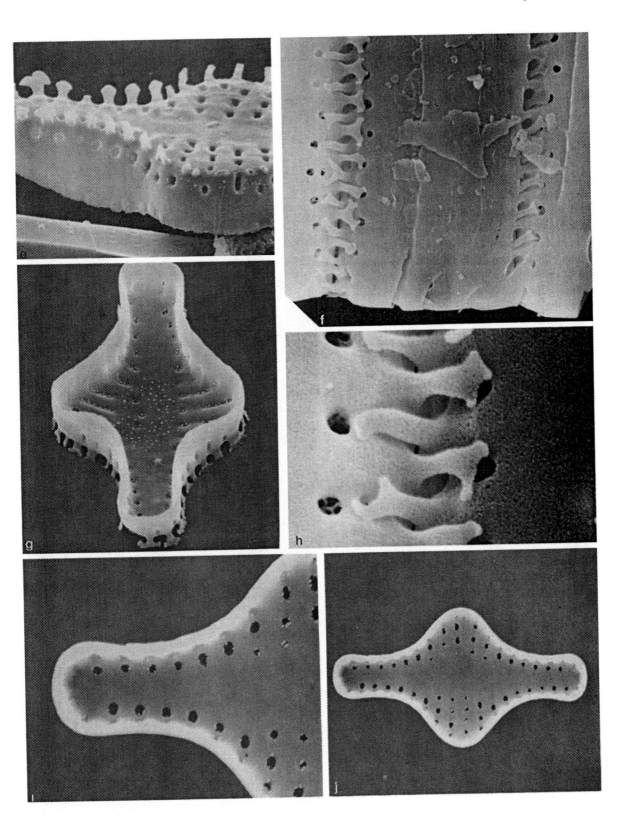

Punctastriata D. M. Williams & F. E. Round 1987. Diat. Res. **2**: 278

T: *P. linearis* Williams & Round

Cells forming short filaments or branching clusters. Plastids not known. A widespread freshwater genus. Difficult to identify by light microscopy but most records of 'Fragilaria pinnata' and 'Fragilaria var. lancettula' belong here. It has also been confused with *Martyana* (= *Opephora*).

Valves linear to elliptical, heteropolar,[a, b] with a slight depression at one end. Striae multiseriate,[b-h] ovoid, lying between raised ribs and curving onto valve mantle. A narrow central sternum separates the striae, which alternate. Each stria is interrupted across its centre by a plain region bearing short pyramidal spines which may bifurcate;[f, h] spines also occur on the ribs between the striae.[f] Apical pore field present at one end but difficult to detect. Valve mantle plaques present on *P. robusta*. Rimoportulae absent. Cingulum consisting of several bands, which are plain and narrow at the apices.[g, i]

This is a widespread genus, the species of which have previously been identified as *Fragilaria* – see illustrations in Rosen & Lowe (1981), Kobayasi & Yoshida (1984). Some perhaps have also been confused with those 'Opephora' spp. now referred to *Martyana*. It does share with *Martyana* the feature of a depression at one end of the valve, but the areolar structure is quite different. This also separates *Punctastriata* from *Fragilaria* and the other genera recently removed from *Fragilaria*. The detail of the valves makes *Punctastriata* a very distinctive taxon when viewed in the SEM. The spines are rather distinctive in shape and appear to be hollow; they contrast with those of other related taxa. Its exact microhabitat is not known. *P. ovalis* was found in the mucilage coating of a stone water trough.

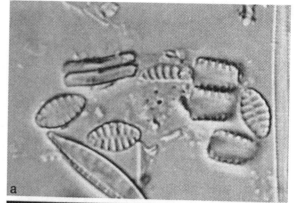

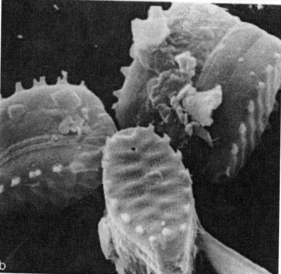

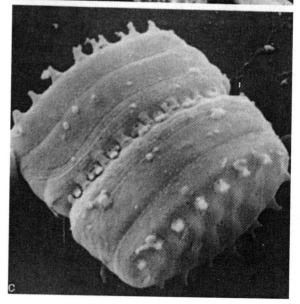

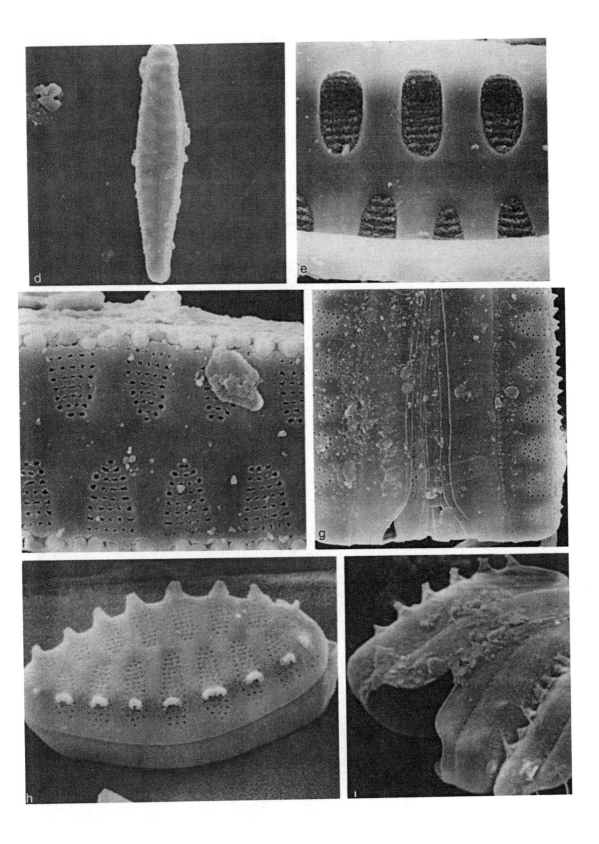

Fragilariforma (J. Ralfs) D. M. Williams & F. E. Round 1988. Diat. Res. 3: 265

T: *F. virescens* (Ralfs) Williams & Round (= *Fragilaria virescens*)

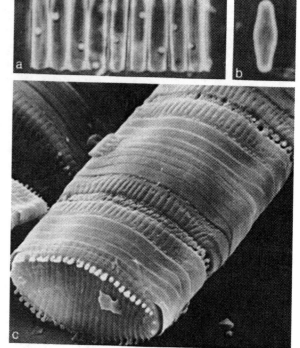

Cells forming linear[a] or zig-zag colonies. Plastids reported as several separate discs (Heinzerling, 1908). A new freshwater genus based on *Fragilaria virescens* (see Williams & Round, 1987).

Valves elliptical, lanceolate or linear, with tapering rostrate to capitate apices;[b-f] valves often centrally constricted. Spines occur marginally between the striae; when they break off, a small centrally depressed remnant is left.[f] Striae uniseriate, most extending from a very narrow central sternum (which may be indistinct when viewed from the outside), across the valve face and down the mantle without a break, but stopping short of the edge, where there is a row of plaques;[g] secondary uniseriate striae are usually intercalated between the longer ones.[i] Areolae small, circular, with an external, simple, plate-like velum with a single central opening. Apical pore fields well developed,[e, i] consisting of closely set unoccluded pores. A few isolated areolae occur between pore field and margin. One rimoportula present, situated near one apex;[h, i] its internal slit is aligned along a stria. Some material shows a minutely papillose region at the apical edge of the valve mantle.[h] Copulae 4–6, all incomplete, alternating and ligulate, with a single row of areolae on the advalvar side and in some a few additional scattered areolae.[c, g, k] The copulae are straight at their apices and the valvocopula has a slight apical septum.

These species are distinguished from *Fragilaria* by the structure and arrangement of the areolae, the simple apical pore plates and the distinctive arrangement of the copulae. Several of the species tend to be common in acidic waters but further work is needed on the whole range of species. We would include here *F. acidobiontica*, *F. bicapitata*, *F. cf. strangulata*, *F. constricta*, *F. lata*, *F. nitzschioides* and possibly *F. braunii*, *F. densestriata*, *F. sioli*, *F. obtusa* and *F. sublineata*.

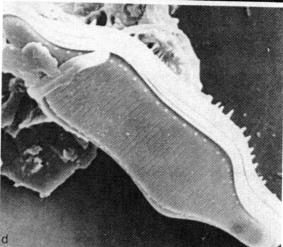

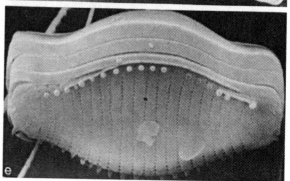

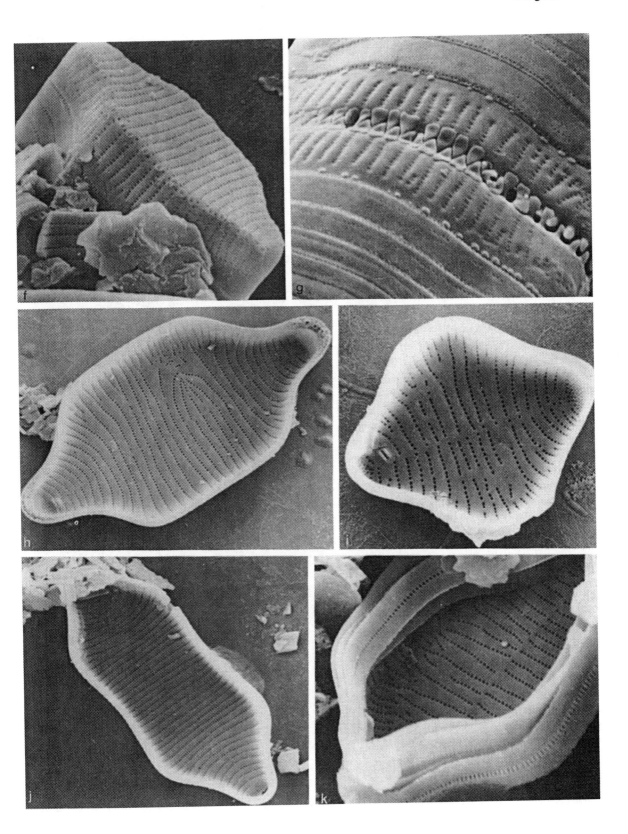

Martyana F. E. Round, gen. nov.

T: *M. martyi* (Héribaud) Round comb. nov.
(= *Opephora martyi* J. Héribaud, Les Diatomées
d'Auvergne, 1902, p. 43, 8/20)

Cells small, oblong to wedge-shaped in girdle view,
attached singly to sand grains by the narrow end.[a–d]
A freshwater genus based on '*Opephora martyi*'
which differs structurally from the original
Opephora, which is marine.

Valves ovate-elliptical,[c–e] becoming more-or-less
oval in small cells;[i] with a distinct depression[c–j]
('step') at the free head-pole of each cell. Striae
uniseriate, curving without a break down the valve
mantle which is not differentiated from the valve
face; sunken externally between distinct transapical
ridges; extending onto the 'step' at the head-pole
but stopping short of the base-pole, where there is
an apical pore field[h] through which mucilage is
secreted for attachment. Areolae slit-like, orientated
parallel to the apical axis. The margin is plain and
without the marginal plaques that are a feature of
Fragilaria and some other araphid genera. The valve
surface is plain with no linking spines. There are no
rimoportulae. Copulae up to 5 (or more) in number,
split, conspicuously curved and thin at the cell
apices. The valvocopula is very broad and the others
much narrower.

There is considerable confusion in the genus
Opephora which is based on marine material. The
problems have been caused because *Opephora* has
been used for almost any small ovate, araphid
diatom. The marine ovate forms have a rimoportula
at one end, unlike *Martyana*, which seems distinctive
enough in other respects too (e.g. in areola structure,
lack of marginal spines) to justify its separation. It is
possible that some species hitherto referred to
Opephora are simply form variants (i.e. heteropolar
populations) of other araphid taxa, e.g. Patrick &
Reimer record a non-ovate *Opephora americana*. It is
conceivable that some of the taxa usually included
in *Fragilaria* should be transferred to *Martyana*, but
even where 'fragilarioid' genera have areolae like
those of the '*Martyana*' type, they also have linking
spines, making it less desirable that they should all
be combined in one genus.

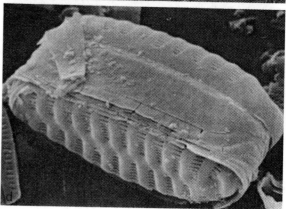

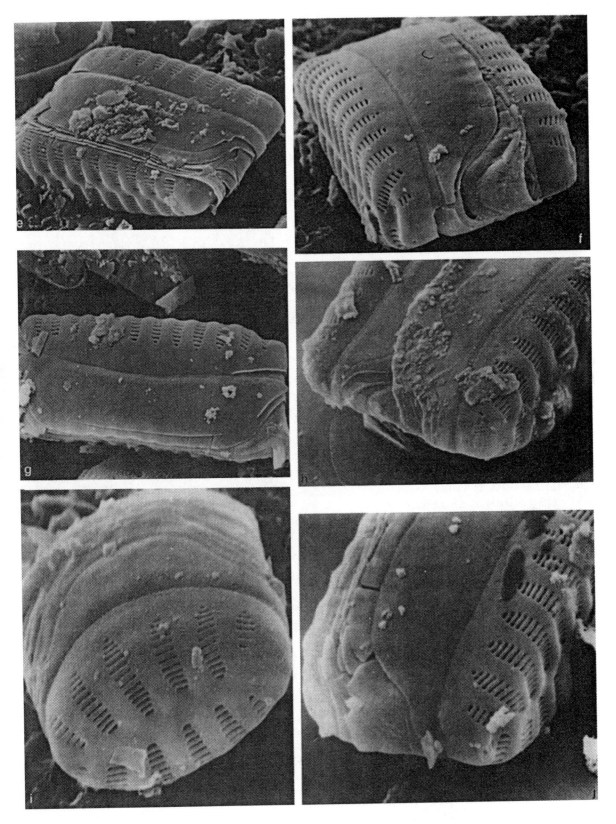

Diatoma J. B. M. Bory de St.-Vincent 1824.
Dict. Class. Hist. Nat. 5: 461 (nom. cons.)

T: *D. vulgare* Bory de St.-Vincent (typ. cons.)

Cells joined to form long ribbon-like colonies, or stellate/zig-zag colonies[a, c] (the planktonic species). Quadrate to oblong in girdle view. A strictly freshwater genus penetrating into some slightly saline sites. Plastids small; bacillar, plate-like or discoid.

Valves elliptical[b, e-g] to elongate, sometimes capitate,[d] with distinct bars of silica forming thickenings across the valve.[d, h, j] Striae uniseriate,[d-g] lying at right angles to a narrow, central sternum, becoming radiate at the apices. The striae tend to be in groups separated by transverse clear areas – the thickenings – and they continue down the vertical mantle.[e] At the apices, more closely spaced rows of pores occur, forming rather indistinct apical pore areas.[d-h] Spines occur along the edge of the valve face in some species[e, g] and in these the valves may also be spinulose. Internally the most conspicuous feature is the development of the transverse ribs[d, j] and the very distinct rimoportula,[h] which lies at one apex and has a transverse, slit-like opening externally.[d, g] Copulae split, with very characteristic Y-shaped ligulae,[i] two rows of areolae, and smooth margins. In some micrographs the proximal edge of the girdle band is waved over the internal ridges of the valves (cf. *Hannaea*). The outer surface of the bands of some species is finely granular.

Patrick & Reimer (1966) found that the *Diatoma* of de Candolle 1805 does not belong to the genus as now conceived. The monograph of *Diatoma* and *Meridion* (Williams, 1985) gives further details and discusses the subgeneric status of *Odontidium*.

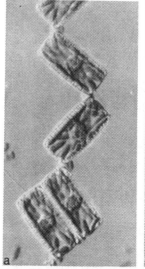
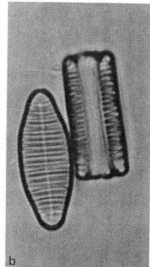
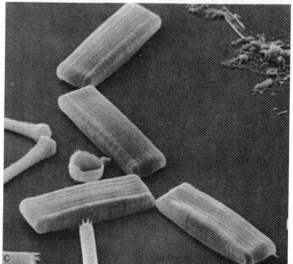
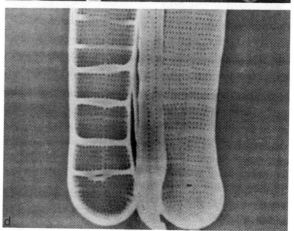

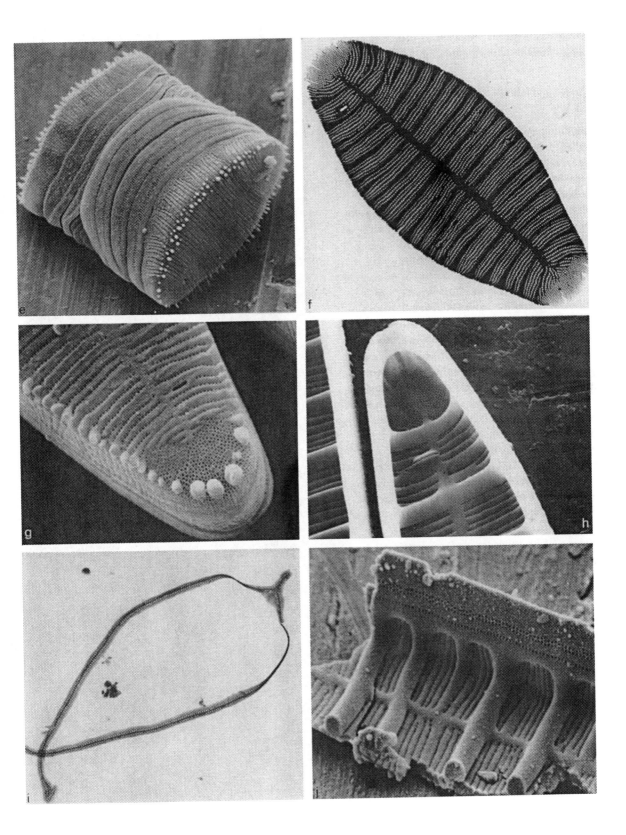

Hannaea R. Patrick 1966. In R. Patrick & C. W. Reimer, Monogr. Acad. Nat. Sci. Philad. **13** (1): 131

T: *H. arcus* (Ehrenberg) Patrick (= *Navicula arcus*)

Cells arcuate. Plastids probably 2 with lobes extending onto the dorsal side. Sometimes free-living, but often in clusters or short colonies; attached to rocks in freshwater streams and springs particularly in mountainous areas.

Valves linear and arcuate[a, c-e] with capitate poles and a slight swelling at the centre of the ventral side. Striae uniseriate, slightly sunken in valve framework, both internally and externally; orientated at right angles to a narrow central curved sternum. Areolae small, round, poroidal. Small spines occur along the valve face/mantle junction[f] and act as a loose interlocking mechanism; the files of areolae then continue down the valve mantle. Areolae absent at ventral swelling[e] but valve surface still transversely undulate forming 'ghost' striae; internally this central area is slightly depressed.[j] Apices with areas of closely spaced pores in vertical rows, forming apical pore plates.[g, h] Elongate plaques occur along the valve margins.[f] Rimoportulae very distinct internally,[i] one at each pole but occasionally lacking at one end, lying to one side of the sternum; internal lips and external slit-like aperture orientated parallel to the striae. Copulae split, with a 'scalloped' advalvar edge and with a row of areolae.[k]

This is a small genus, only *H. arcus* being recorded regularly. Patrick & Reimer (1966) discussed the early classification of this genus: *Ceratoneis* is the name that has generally been applied to it but the two spp. originally included in *Ceratoneis* are now allocated to *Nitzschia* and *Pleurosigma*, making another name necessary. The shape of the cell is almost the only feature distinguishing *Hannaea* from *Fragilaria* and the genus can be maintained only on its distinctive morphological character and possibly its very restricted ecological range.

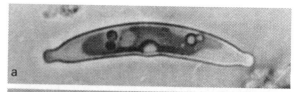

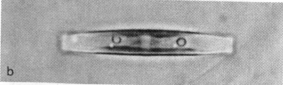

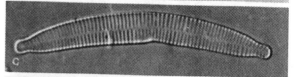

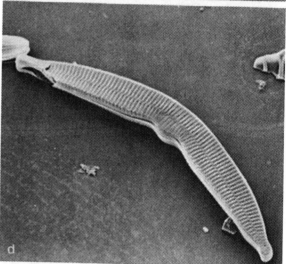

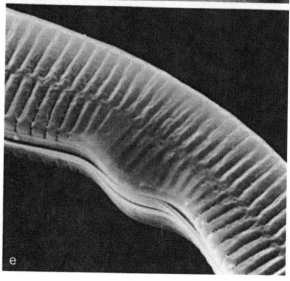

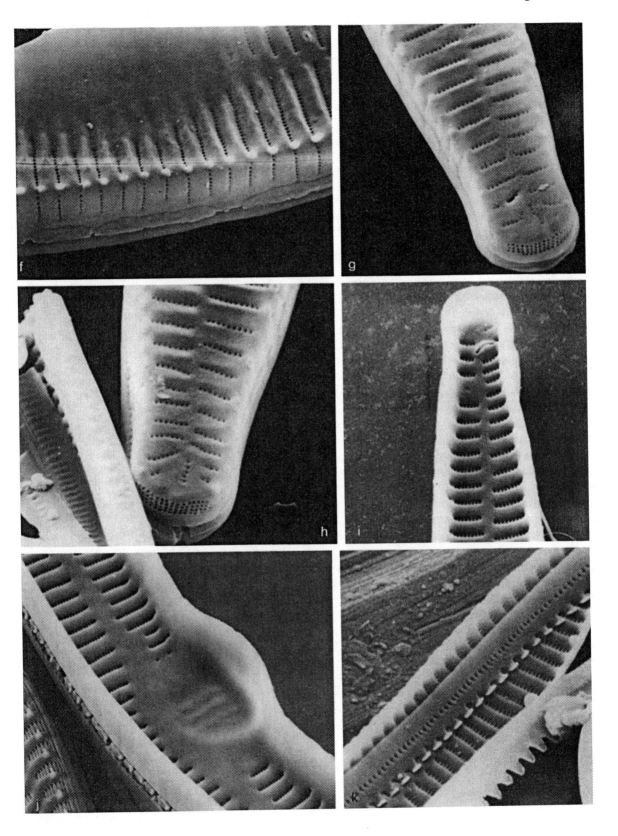

Meridion C. A. Agardh 1824. Syst. Alg. xiv: 2

T: *M. vernale* Agardh (= *M. circulare*)

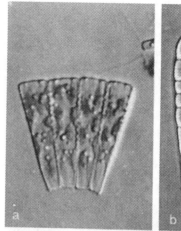
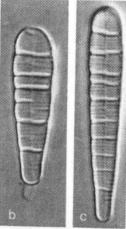

Cells attached closely by their valve faces to form fan-shaped colonies;[a, d, e] heteropolar, being cuneate both in valve and girdle view. Prominent ribbing present on valves, visible in both views. Plastids numerous, rather irregular, discoid, lying along the valve face. A completely freshwater genus found attached to stones and plants and in our experience always in flowing water; often abundant in springs.

Valves cuneate,[b, c] sometimes capitate. Valve face flat, curving fairly abruptly into mantles that diminish in height slightly from head-pole to base.[h] The mantle and copulae tend to be undulate and small regular or scattered spines occur along the valve face/mantle junction.[g] Striae uniseriate,[g, i] orientated at right angles to a narrow sternum, continuing down the mantle without a break but curving together thus leaving clear areas towards the mantle edge where series of plaques occur.[k] Areolae simple, circular to elliptical, and apparently without vela. At intervals the siliceous framework is enlarged to form thickened costae, which start at the mantle edge and run across half or the whole width of the valve.[h-j] At the base-pole there is an ill-defined area of closely spaced pores presumably involved in mucilage secretion. One rimoportula, lying off-centre near the head-pole, orientated transapically; sessile internally.[i] Copulae split, wider at the head-pole;[k] with one or more transverse rows of areolae.[g]

Apart from the cell shape and the development of thickened costae, this monotypic genus is very like *Fragilaria, Hannaea* and *Synedra*. It is apparently one of the few pennate genera in which resting spores are formed (see p. 52). Recently Williams (1985) has transferred *Diatoma anceps* to *Meridion*. If accepted this would mean that *Meridion* can be isopolar as well as heteropolar. Interestingly, this species also has a strong tendency to occur in running waters.

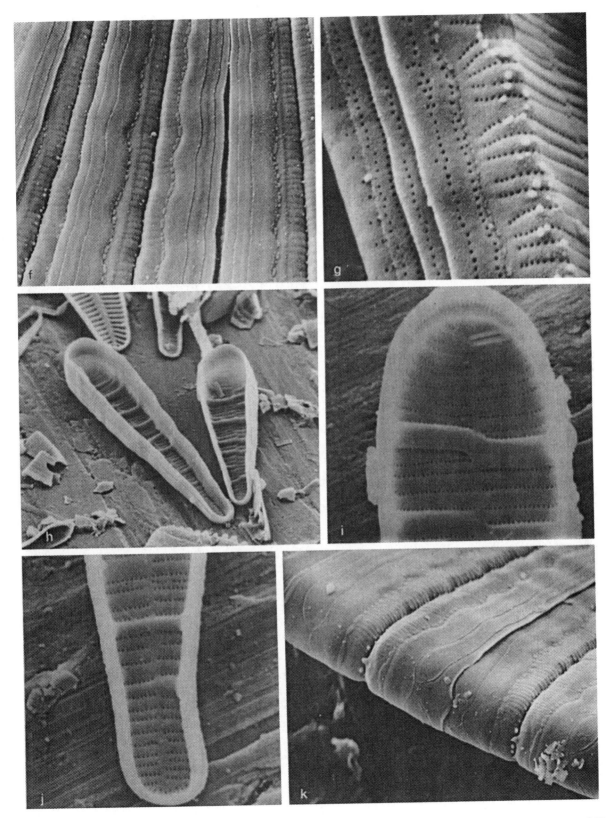

Synedra C. G. Ehrenberg 1830. Abh. Königl. Akad. Wiss. Berlin, 1830: 40

T: *S. ulna* Ehrenberg

Cells needle-like,[a-e] free-living or epiphytic; radiate colonies occur, in which the cells are attached to a common pad of mucilage, but only rarely are the cells joined valve face-to-face (*S. ungeriana*). Plastids: usually two long plates lying against the girdles and overlapping slightly on to the valve face. These may split up in unhealthy material, giving the impression of numerous discoid chromatophores. Freshwater in distribution.

Valves linear, some capitate, some centrally inflated. Striae perpendicular to the narrow sternum, sometimes absent or obscured from the central area.[j, k] Striae composed of rows of simple round or elongate areolae[f, g] (occasionally biseriate; this can occur in culture in some cells); continuing onto the valve mantle which is usually at right angles to the valve face. Each areola has its own individual opening internally. At the apices pore fields occur,[g, m] which are discrete, often depressed below the surface of the valve (ocellulimbus: Williams, 1986) and contain small round pores in a tight, orderly array. In some species two short horns project above the apical pore plates.[g] A rimoportula is located near each apex;[h, i] externally it has a slit-like opening which often lies in a slight depression; it is usually prominent internally. Copulae narrow, complete and with one row of areolae near the advalvar edge: the valvocopula often has a scalloped advalvar edge which fits on to the costal framework.[n]

Apart from the confusion of *Synedra* with *Fragilaria* (see p. 346) there were many species allocated to *Synedra* which have now been placed in new genera – see *Catacombas, Ctenophora, Hyalosynedra, Neosynedra* and *Tabularia* (Williams & Round, 1986). For a discussion and redefinition of *Synedra*, see Williams (1986).

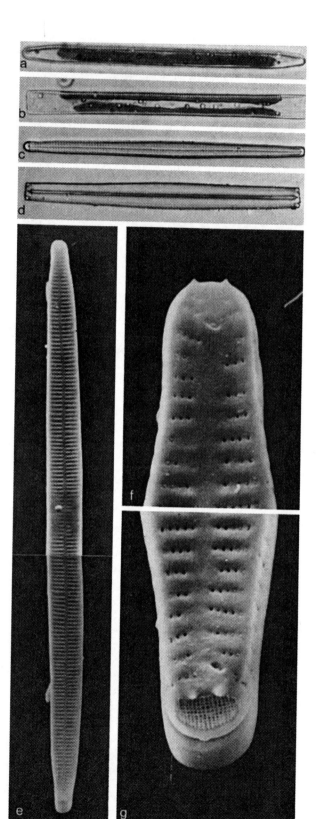

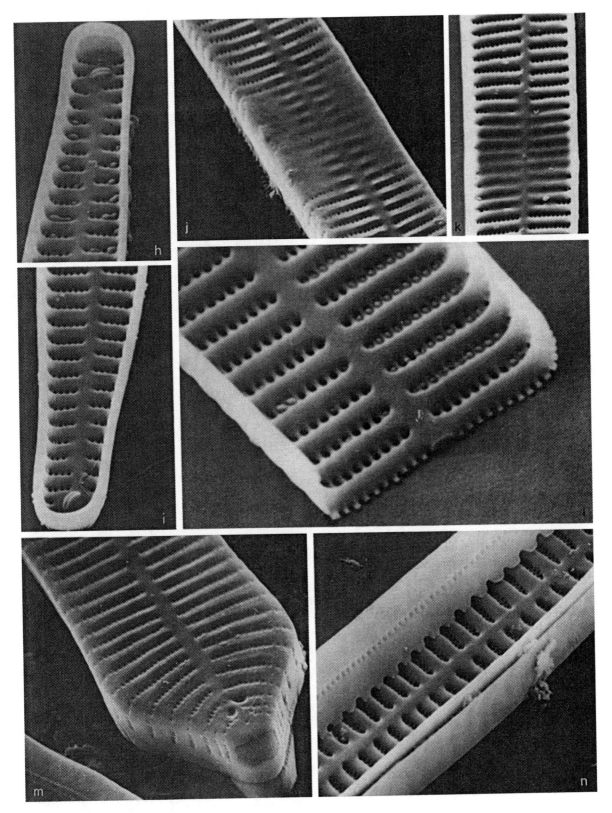

Ctenophora (A. Grunow) D. M. Williams &
F. E. Round 1986. Diat. Res. 1: 330 (= *Synedra*
subgen. *Ctenophora*)

T: *C. pulchella* (Ralfs ex Kützing) Williams &
Round (= *Synedra pulchella*)

Cells elongate, narrow, attached to substrata by
mucilage pads, usually in tufts. Two plate-like
plastids. A genus recorded frequently in brackish
water but also penetrating into a few fresh waters;
cosmopolitan.

Valves linear to linear-lanceolate,[a, b] with rounded
or slightly capitate ends. Valve face flat; mantles
distinct but shallow. Striae transverse,[c] uniseriate,
containing elliptical or more-or-less rectangular
poroids, which are closed externally by complex
cribra; absent centrally, where there is a broad plain
area (fascia).[b, c] Sternum narrow, virtually absent.
Internally the valve is prominently transversely
ribbed.[g, i] The central fascia is thickened internally
around its periphery and within it can be seen
'ghost' ribs and sternum.[i, j] This central area is very
distinctive in LM. At each pole there is an apical
pore field[d] set below the general surface of the valve
and containing small round pores (an 'ocellulimbus'
sensu Williams, 1986). In addition each pole has a
rimoportula, which opens by a simple pore
externally[d] and is diagonally orientated internally.[f, h]
Cingulum incompletely known but the valvocopula
is non-areolate and open, and its pars interior is
crenulate with a central modification to fit the
central thickenings of the valve. Other bands may
have a single row of pores.

At the moment this seems to be a monotypic
genus but further studies are needed. The
combination of the type of valve striation, the
cribrum structure, and the conspicuous plain area at
the centre of the valve, distinguish *Ctenophora* from
other related genera.

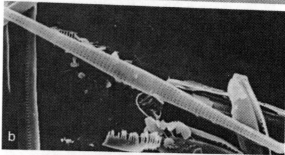

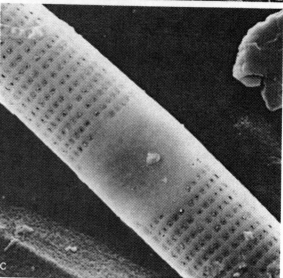

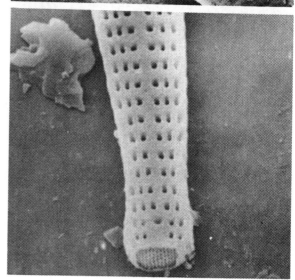

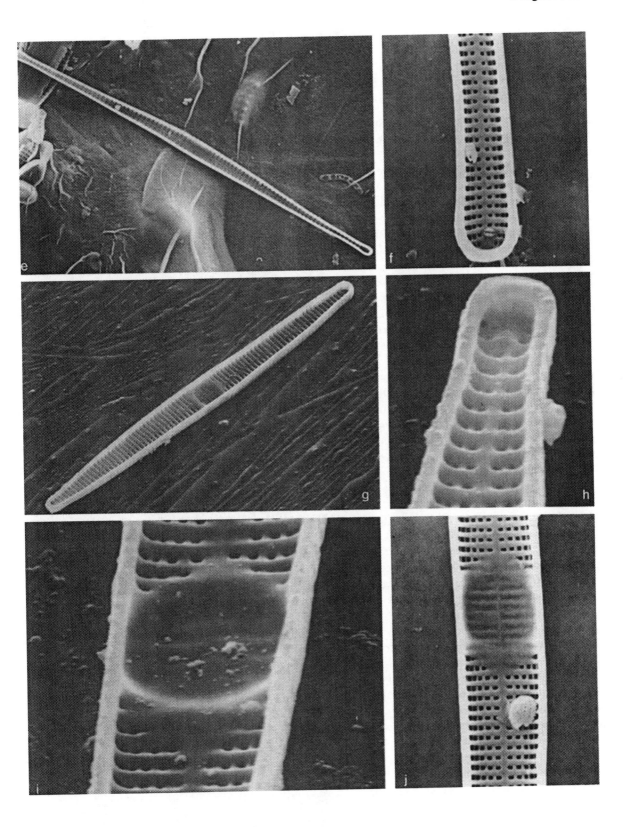

Neosynedra D. M. Williams & F. E. Round
1986. Diat. Res. **1**: 332

T: *N. provincialis* (Grunow) Williams & Round
(= *Synedra provincialis*)

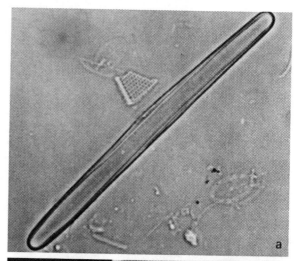

Cells elongate, attached, solitary. Plastids not
studied. A marine epiphytic genus of widespread
occurrence.

Valves elongate, linear;[a] slightly expanded at the
centre or uniformly undulate with somewhat
capitate, rounded apices.[c] Sternum inconspicuous or
moderately developed. Striae transverse,
uniseriate,[b, f, g] evenly spaced, continuing without a
break down the vertical valve mantle;[e] mantle edge
plain. Areolae elliptical or somewhat quadrate,[i, j]
each occluded by a cribrum lying near its external
aperture.[d, e] An apical pore field is present at each
end of the valve, composed of vertically extended
pores (slits) divided by intermittent cross bars.[d-g,i,j]
There is also a rimoportula at either end, usually
with 1 (2) incomplete rows of areolae between it and
the apical pore field.[i, j] The rimoportula is lateral to
the sternum and has a transapically elongate outer
aperture.[d-g] Internally the only notable point is the
form of the rimoportula which has a crescent-
shaped inner lip and a flap- or knob-like outer lip.[i, j]
Copulae split, ligulate, usually four per valve, each
with two rows of cribrate areolae.

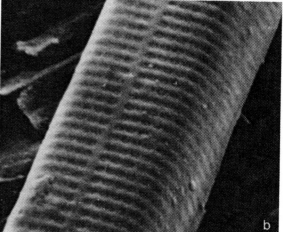

This genus is characterised by the slit-like pores
of the apical pore field and the form of the
rimoportula. *Cyclophora* has a similar pore field but
it is divided by a plain central region; *Cyclophora*
also differs in having complex central partitions and
a less distinctive rimoportula. The apical pore field
of *Protoraphis* also shows some similarity and it is
necessary to look into the interrelationships of these
three genera. The internal structure of the
rimoportula is reminiscent of the *Thalassionema/
Thalassiothrix* type but there the similarity ends.
Neosynedra is very difficult to study in the light
microscope; it is virtually impossible to make any
progress with species-level taxonomy in these
needle-like, extremely hyaline diatoms without EM
investigation. In natural samples, the species present
must first be determined with the EM, and then
subtle aspects of shape used to aid LM
differentiation. Cultures will no doubt help when
these are obtained. There is, however, no doubt
about the distinction of this genus.

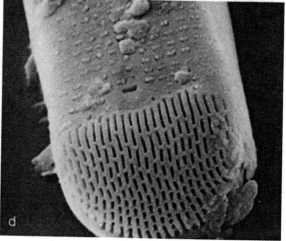

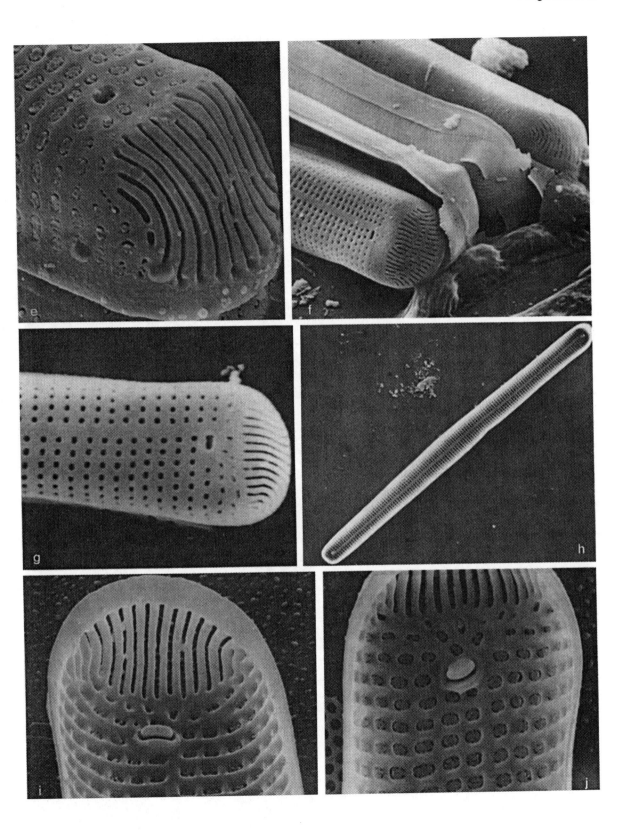

Tabularia (Kützing) D. M. Williams & F. E. Round 1986. Diat. Res. 1: 320 (= *Synedra* subgen. *Tabularia*)

T: *T. barbatula* (Kützing) Williams & Round (= *Synedra barbatula*)

Cells needle-like; solitary or in clusters but not attached to one another. Apparently with two plastids. A common marine or brackish, epiphytic and epilithic genus of world-wide distribution. The species were formerly placed in *Synedra* but this is an exclusively freshwater genus (see p. 370).

Valves elongate, linear or lanceolate;[a-c] sometimes capitate. Striae transverse; biseriate or consisting of one[c, f] or two[e] large areolae; reaching almost to the centre or leaving a very wide sternum; sometimes also broken by a hyaline area at the junction of valve face and mantle. Areolae round (in forms with biseriate striae) or transversely elongate, occluded externally by cribra with distinct cross bars; these are particularly obvious when viewed from the inside.[h, i] At each end there is a very distinct pore field set below the general valve surface[c, f] (the 'ocellulimbus' of Williams, 1986). At one end there is a single rimoportula, which is off-centre and aligned with the areolae or, in those species with a wide sternum, actually on the sternum;[j] the external opening is a simple pore.[e, f] At least 3 copulae per cingulum;[d] these are incomplete and ligulate. The valvocopulae are slightly waved along the advalvar margin and lack areolae; the other bands have a single advalvar row of pores.[c]

This is one of the commonest marine '*Synedra*' forms, and differs from the other genera in its family by virtue of the structure of the vela and presence of an 'ocellulimbus'. From the outside the cribra appear not unlike those in *Catacombas* but the internal structure is not chambered. Williams & Round (1986) include here *T. barbatula*, *T. investiens*, *T. tabulata* and *T. parva* but there are many other taxa that probably should also be transferred to *Tabularia*, which may ultimately need further splitting (see discussion in Williams & Round, 1986 and 1988 where 3 subgroups are discussed).

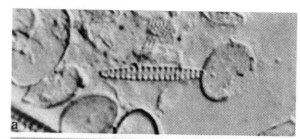

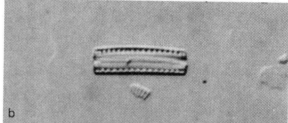

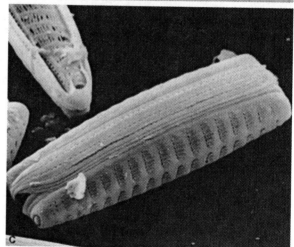

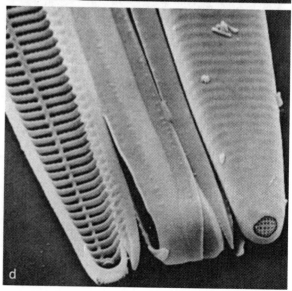

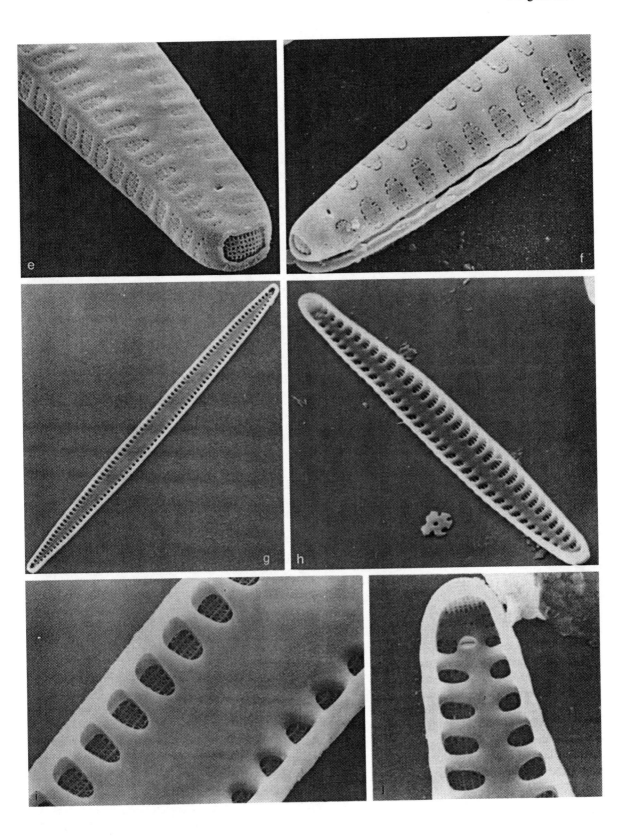

Catacombas D. M. Williams & F. E. Round
1986. Diat. Res. 1: 34

T: *C. gaillonii* (Bory) Williams & Round
(= *Navicula gaillonii*)

Cells robust, needle-like, more-or-less parallel-sided
in valve and girdle view. Attached by a basal
mucilage pad. Plastids numerous, small, discoid. A
cosmopolitan genus, living as an epiphyte on
seaweeds.

Valves linear, tapering slightly at the apices,[a, b]
with a narrow (*C. gaillonii*) or broad (*C. camtschatica*)
sternum. Striae transverse,[c] chambered (alveolate);
opening internally by an elliptical aperture on the
mantle;[e, f, h, i] opening externally via a complex
cribrum[c, d] or by a series of poroids each with their
own cribra: the striae are broken externally by a
hyaline area at the junction of valve face and
mantle. Mantles at right angles to the flat valve face.
A rimoportula opening is conspicuous at each end
and between this and the sunken apical pore field
or ocellus (ocellulimbus type);[d] there are a few
isolated simple pores on both the face and mantle.
Small spines may occur over the pore plate. The
large rimoportulae are located at the apices adjacent
to the last row of areolae; they have a wide base
internally but flare upwards.[f, g] Four massive
copulae per cell:[g, i] these are without areolae in *C.
gaillonii*, while in *C. camtschatica* the pair each bear
a single row of areolae.

C. gaillonii has always stood out from other
species classified in *Synedra sensu lato*. It is strictly
marine and has a set of features that distinguish it
from *Synedra sensu stricto*, namely, the structure of
the areolae and the chambering of the valve. It is
separated from *Ardissonea* by the structure of the
areolae, the presence of apical pore plates and
rimoportulae. In addition the junction of the valve
mantle and valve face is solid. It is close to
Hyalosynedra but differs in the structure of the
cribra, the longitudinal hyaline area at the valve
face/mantle junction and in the structure of the
rimoportula which has two large lips rather than a
crescent and knob form. Depending on the view
taken regarding the merit of the above distinctions,
the genus stands or must be fused with
Hyalosynedra: our view is that they should be kept
separate (see Williams & Round, 1986).

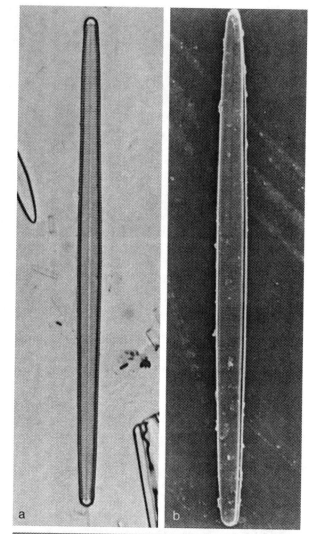

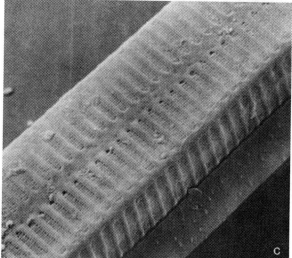

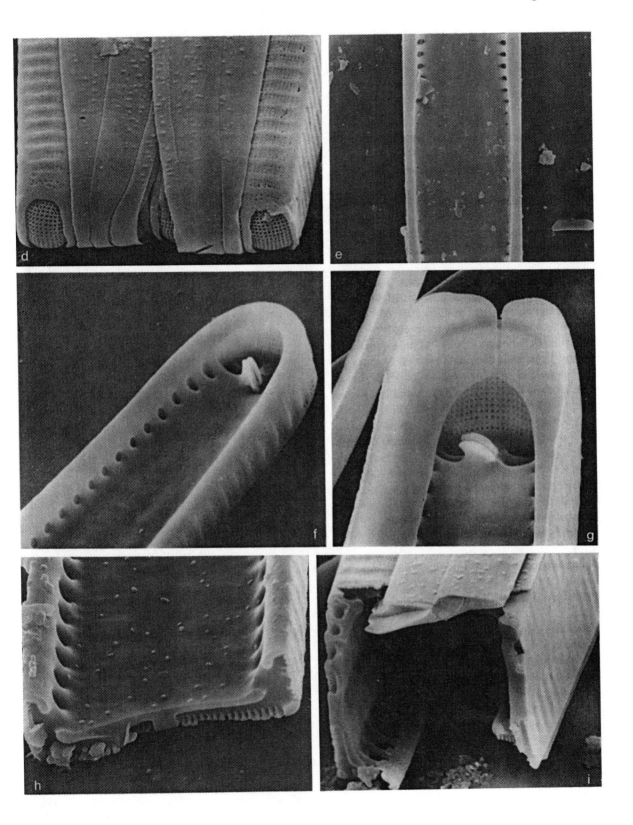

Hyalosynedra D. M. Williams & F. E. Round 1986. Diat. Res. 1: 316

T: *H. laevigata* (Grunow) Williams & Round (= *Synedra laevigata*)

Cells long, linear, living epiphytically. In girdle view also capitate/truncate. Plastids not yet studied. A very common marine epiphyte of world-wide distribution. This is a new genus for the species formerly classified as *S. laevigata*.

Valves linear, often with capitate poles,[a-c] with a very narrow sternum.[d-g] Valve face flat, curving into fairly deep mantles.[d] Striae mostly uniseriate, more-or-less alveolate, opening to the outside by a row of small pores and to the inside by a single round foramen[k] at the angle between valve face and mantle. The outer pores can appear paired but this is due to a bar across each; in some species the areolae reduce in size and form more distinct double rows on the valve mantle.[d] Mantle edge wide and plain. Valve ends each with clear plain areas and a rimoportula opening close to the point where the areolar rows cease.[d, e] An apical pore plate is present; it is distinctly sunken and with horizontal rows of pores.[d, e, j] The pores in this plate are larger than the areolae. A row of spines may occur above the apical pore plate.[d, e, j] Inner valve face featureless becoming slightly depressed at the apices by the pore plates.[i] A low, distinctive rimoportula occurs at the edge of the apical depression[i, k] (cf. *Neosynedra*): it has the same 'parrot beak-like' appearance internally as has been described by Hallegraeff (1986) in *Thalassiothrix*. External opening of rimoportula large, round or transversely elongate. The valves of this taxon tend to be thick due to the chambering of the wall. Copulae massive, split, non-porous.[j]

Hyalosynedra differs from *Synedra* in its chambered valve and from *Ardissonea* by the presence of both rimoportulae and apical pore plates; the chambering of *Ardissonea* is also much more complex. The structure of the areolae in *Hyalosynedra* is quite distinctive. This taxon seen in the light microscope is as Hustedt (1927–66) comments 'sehr hyalin und scheinbar strukturlos'; hence we proposed 'Hyalosynedra'. It is probable that *Synedra distinguenda* also belongs in this genus.

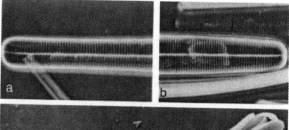

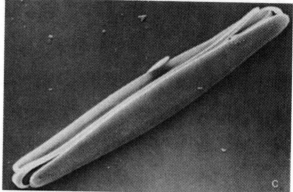

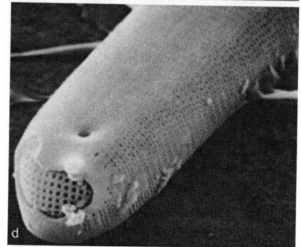

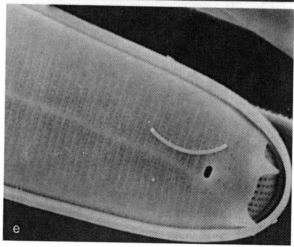

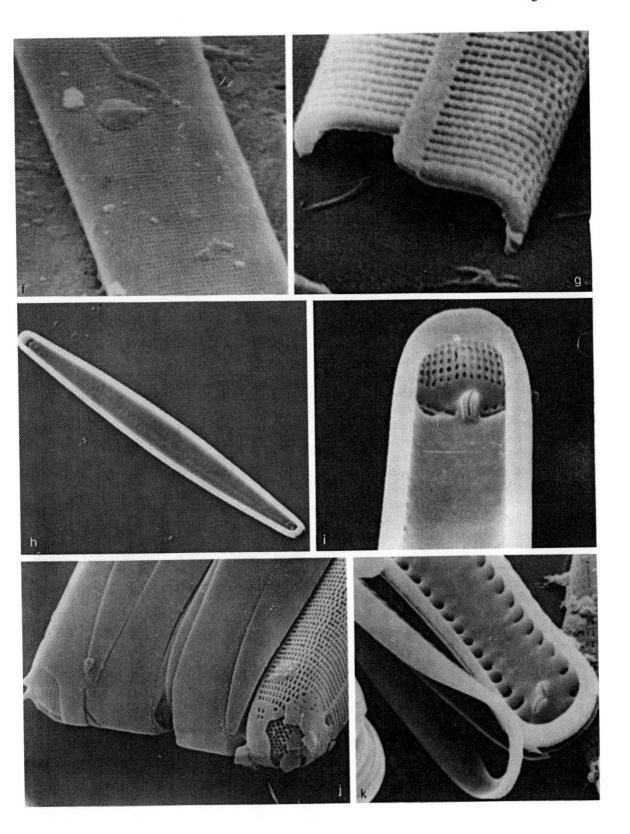

Opephora P. Petit 1888. Miss. Scient. du Cap Horn 1882–1885, Bot. 130

T: *O. pacifica* (Grunow) Petit (= *Fragilaria pacifica*)

Cells rectangular in girdle view, clavate to rhombic in valve view. Attached to sand grains and forming radiating colonies[a–c] but also growing epiphytically (Sullivan, 1979). Marine and not uncommon though there is considerable confusion as to what species should be placed in the genus.

Valves clavate[g–i] to rhombic, valve surface curved[d] to flat; sternum wide and may be laterally expanded at centre; mantle edge wide. Striae linear, occluded by complex vela set near the outer surface but bending inwards, at least in the cleaned material.[e, f] An apical pore field present at both poles.[g] Cingulum of 4 copulae;[j] the valvocopula is wide except at the apices; all the copulae are plain and split.

O. pacifica has been generally regarded as the type of the genus (Patrick & Reimer, 1966) and this is a distinctive species with linear striae occluded by complex vela. Sullivan (1979) thought that a structure at the head-pole might be a pseudonodulus but this does not appear to be the case. The valve margin does not bear spines and there are no rimoportulae. Other species that have been included in *Opephora*, e.g. *O. olsenii* have spines and vela with prominent cross-bars (Sundbäck, 1987, where the taxonomy is discussed in detail). For the moment these and several other similar taxa are excluded whilst further studies are in progress. As with genera such as *Synedra*, *Fragilaria*, *Navicula*, etc. there is a whole complex of species which has to be either included in a much enlarged *Opephora* or placed in several new genera. Frenguelli (1945) separated *O. schwartzii* into a new genus *Opephoropsis*, which now needs study in the SEM.

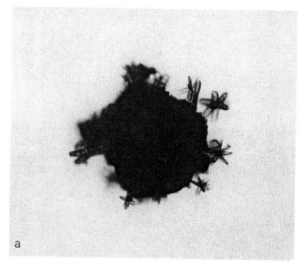

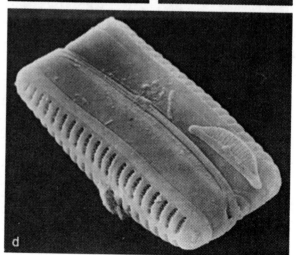

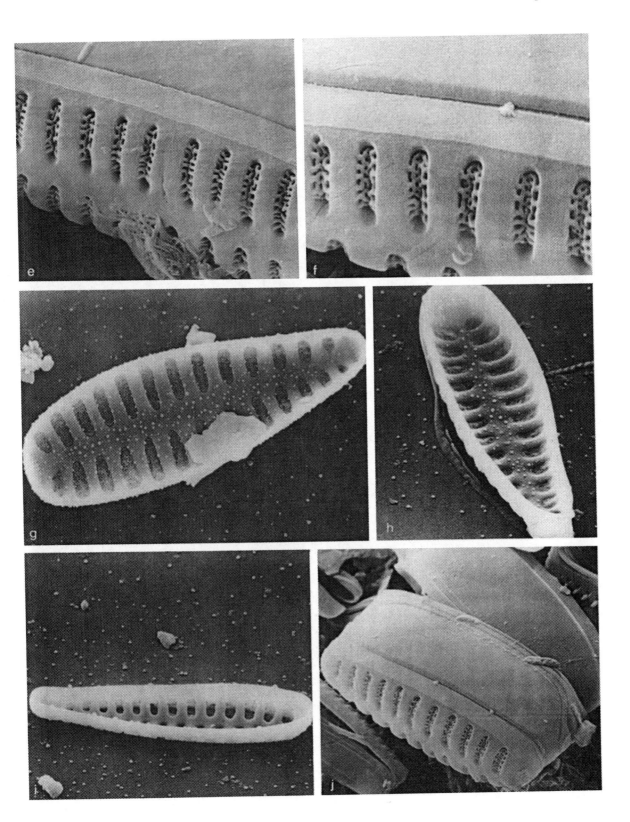

Trachysphenia P. Petit in Folin & Perier 1877.
Fonds de la Mer, 3: 190

T: *T. australis* Petit

Cells elongate, heteropolar in valve view, rectangular
in girdle view. Plastids not observed. A marine
genus epiphytic on algae (Hendey, 1964).

Valves lanceolate, heteropolar;[a-c] valve flat,
curving at its edges into shallow undifferentiated
mantles; sternum indistinct. Striae uniseriate.
Areolae circular to elliptical, containing branching
volate occlusions, which are located deep in the
areolae.[d-h] Pore fields occur at both apices,[d, e] each
consisting of a few small round pores which may
also be occluded by vela. No rimoportulae.
Cingulum not observed.

Only two species of *Trachysphenia* are recognised
in VanLandingham (1967-70) but the genus needs
further study. It is distinguished from other cuneate
marine genera by the combination of a particular
distinctive type of velum, the lack of rimoportulae
and the presence of apical pore fields. The pore
fields are unusual in that the pores appear to have
occluding membranes, whereas to our knowledge all
other genera have open pores. We thank K.
Sundbäck for allowing us to use her material, which
was collected from a sandy beach and may indicate
that this particular *Trachysphenia* occupied an
epipsammic habitat.

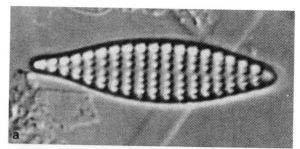

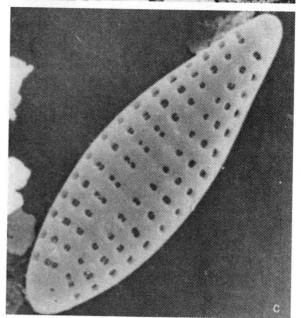

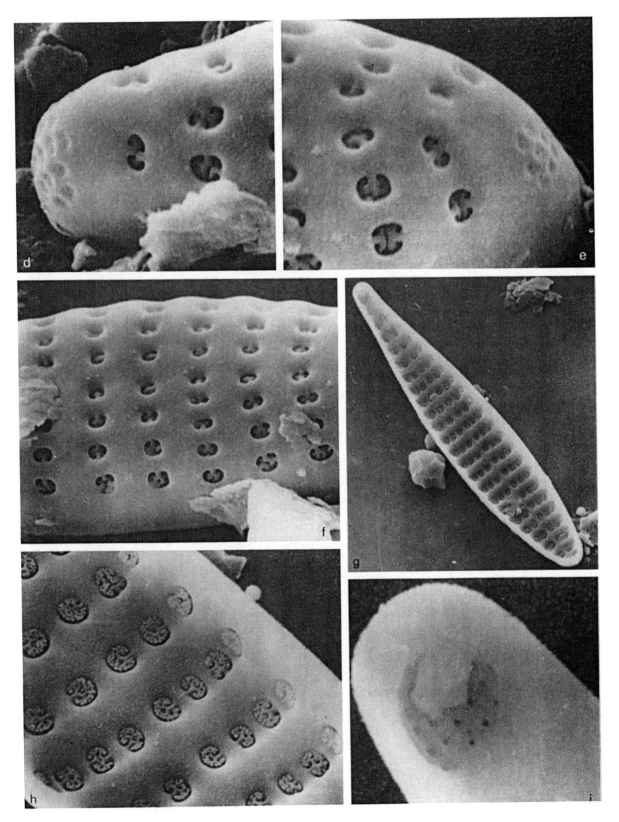

Thalassioneis F. E. Round gen. nov.

T: *T. signyensis* F. E. Round sp. nov.

Cells linear with central inflation. Plastids not known. Found in sea ice from Antarctica.

Valves linear, with rounded apices and central inflation.[a, b] Valve face flat, curving into shallow mantles. Striae uniseriate, widely separated and somewhat irregular;[b, c, e-h] areolae small, round, occluded externally by an indistinct cribrum; areolae opening internally via simple round pores. A slightly sunken apical pore field lies on the valve face;[d] the valve mantle below the pore field has a few scattered pores. A single rimoportula occurs on the central inflation,[e-h] slightly closer to one pole; the external opening is a small elongate, rimmed pore;[c] internally the opening is unstalked and aligned transapically. Copulae not observed.

Thalassioneis was frequent in samples from Signy Island and previously would have been placed in or close to *Fragilaria*. The new, restricted definition of *Fragilaria* (Williams & Round, 1987) relates to a freshwater taxon and *Thalassioneis* differs from this not only ecologically but in the nature of the striae, the position of the rimoportula, the form of the apical pore field and the lack of spines.

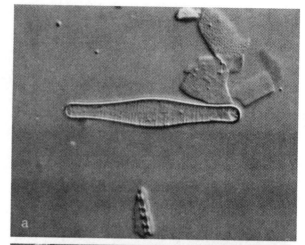

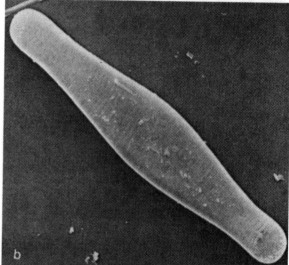

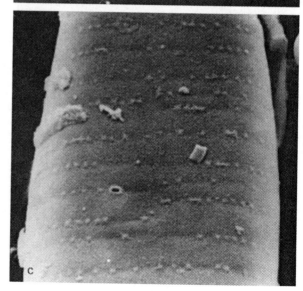

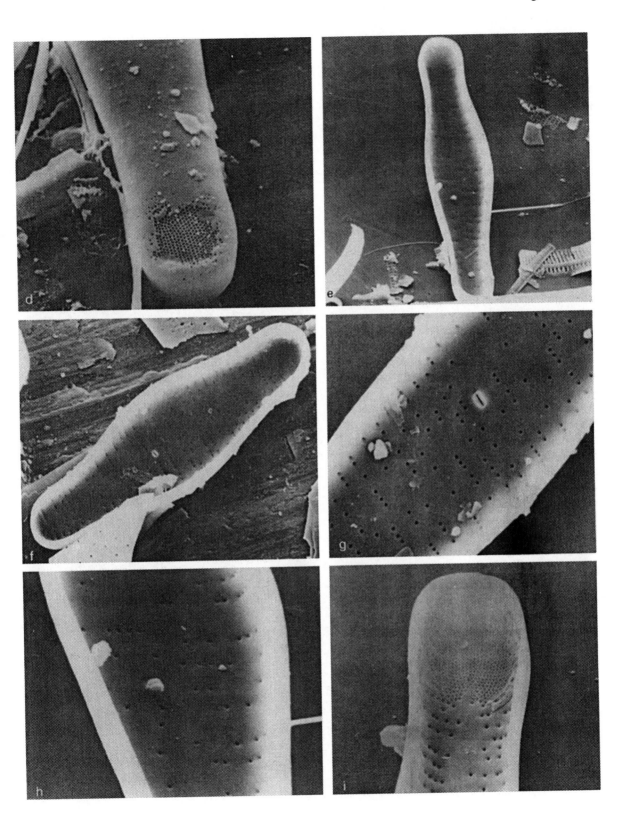

Falcula M. Voigt 1960. Rev. Algol., N.S. 5: 85

T: *F. rogallii* Voigt

Cells linear, lunate, probably attached singly or in clusters but live samples have not been studied. Plastids not known. A marine epiphytic/epizoic[a] genus; probably widespread but rarely reported.

Valves linear, lunate, tapering slightly to the apices; curved in transapical section.[b, c] Areolae in transverse uniseriate rows, broken by a slight sternum running on the convex side.[b–f] Vela not observed. Internally with slight or more pronounced cross-ribbing between the rows of areolae.[g, i, j] Ribbing absent around the apices, where the pores are smaller than those between the main ribs of the valve framework.[i, j] A transapically orientated rimoportula is present at one end on the convex margin.[d, h–j] At both poles there is also a series of slits (apical pore field) off-centre,[e, g, i, j] again on the convex side. When the slits of the apical pore field are viewed from the inside, the bars between are seen to be raised. Copulae several, split, and with two rows of areolae along their length.[e]

The genus was described by Voigt (1960a) but little detail can be seen in the LM. A new species *F. hyalina*, described by Takano (1983), has a much wider sternum and an apical pore field which is poroidal rather than with slits; this is a much smaller species than ours or than Voigt's and it may prove to be a separate genus – Figs a,d,h are of this species and are reproduced by kind permission of Dr Takano. The record of *F. hyalina* on the copepod *Acartia* adds another genus to the epizooic genera already recorded (*Pseudohimantidium* and *Protoraphis*). The slit-like opening at the apices of *Falcula* are in a similar position to the series of processes in *Pseudohimantidium*, but these latter processes have been interpreted as rimoportulae. More studies are required to determine the interrelationships of these genera.

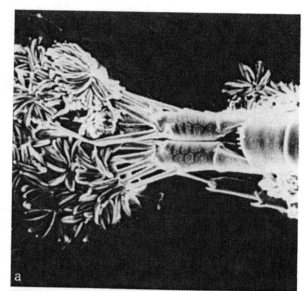

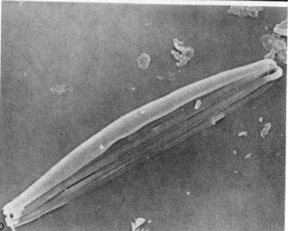

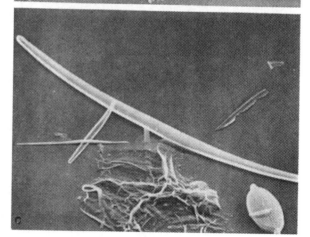

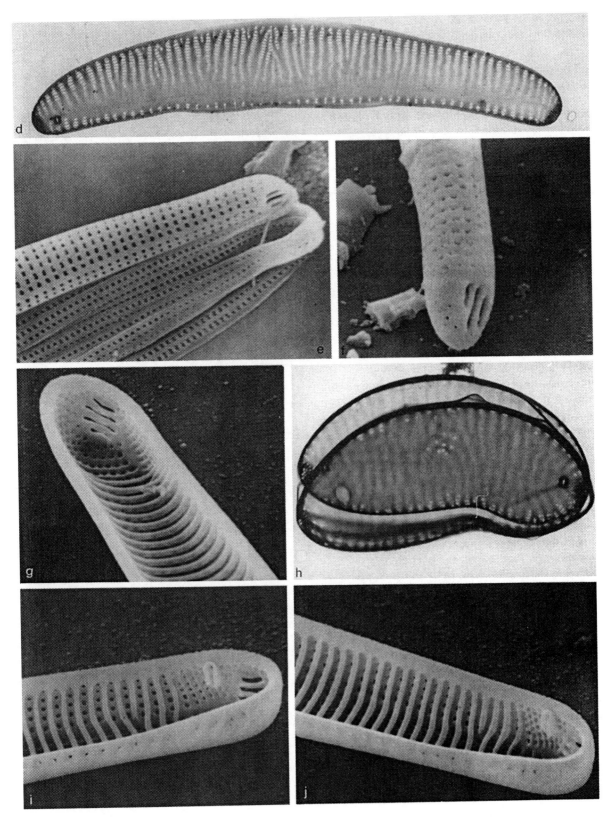

Pteroncola R. W. Holmes & D. A. Croll 1984.
Proc. 7th Int. Diat. Symp. (D. G. Mann, ed.):
267

T: *P. inane* (Giffen) Round comb. nov.
(= *Dimeregramma inane*)

Cells oblong in girdle view,[a-d] united to form zig-zag
colonies. Several plastids per cell (Hasle &
Syvertsen, 1981: as *Fragilaria hyalina*, though this
taxon needs reconsideration). A common epiphyte
on marine algae. The genus was in fact described
from marine diving bird feathers (Holmes & Croll,
1984), where it often formed 90–100% of the cells;
this interesting habitat for diatoms has only recently
become known and will undoubtedly repay further
study.

Valves linear to elliptical.[b-h] Valve face domed or
slightly flattened centrally, curving into deep
mantles. Striae chambered (alveolate), each chamber
opening to the outside by a row of small areolae
and to the inside by a small round foramen.[g, h]
Externally, the valve appears ribbed, small ridges
occurring between the rows of areolae, while
internally it is plain. A sternum is present, though
this is only obvious in TEM (Holmes & Croll, 1984).
An apical pore field occurs at both poles. Two
rimoportulae present, one near each apex, both
orientated transapically; external openings slit-like[e]
and difficult to distinguish from the stria slits;
internally small, sessile.[g, h] Copulae split, appearing
solid[i] but one row of areolae may occur, though
these tend to be obscured by the overlap between
bands.

This taxon can be the most abundant diatom on
seagrasses in the Gulf of Mexico (identified as
Fragilaria hyalina by Sullivan, 1979) and from our
collections we recognise that it is cosmopolitan. The
genus may have to receive other species after further
study since many records of marine *Fragilaria* are
suspect. '*P. marina* Holmes & Croll' is predated by
Giffen (1970) and hence we have used Giffen's
epithet after checking the type slide. We thank
Linda Medlin for pointing out this synonymy. The
black or white focus 'spots' of Holmes & Croll
(1984) are most likely small thickenings along the
valve mantle,[c] not parts of the valvocopula, and the
number of girdle bands is probably greater than
these authors quote.

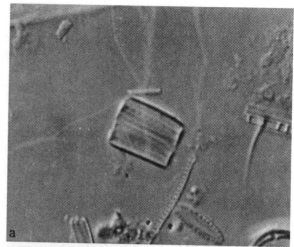

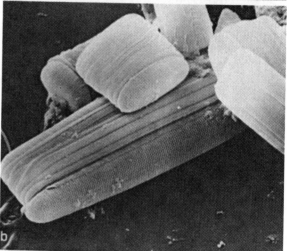

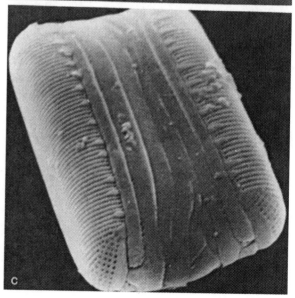

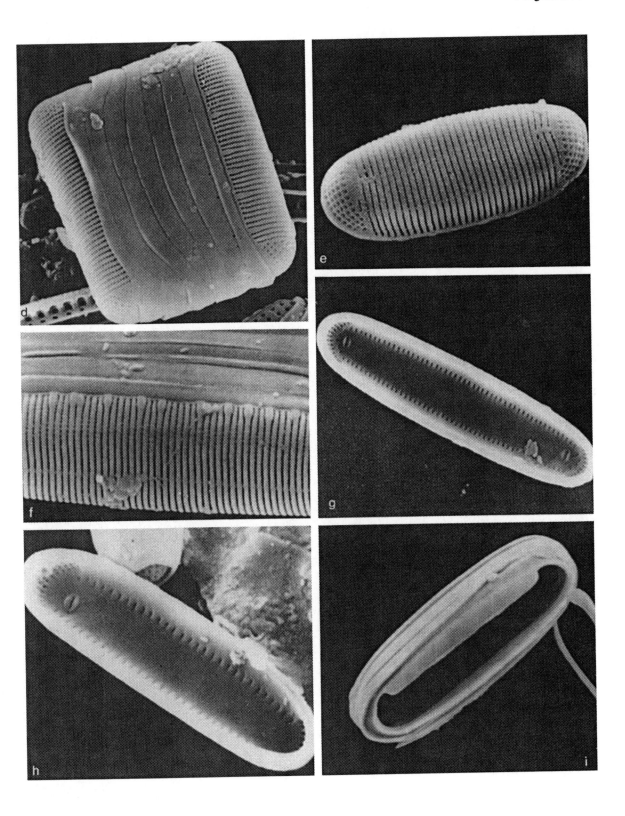

Asterionellopsis F. E. Round gen. nov.

T: *A. glacialis* (F. Castracane) F. E. Round, comb. nov. (= *Asterionella glacialis*)

Cells heteropolar;[a-c] in colonies joined by the valve faces of the expanded foot-pole (basal pole), so that viewed in the LM cells appear in girdle view. This genus is based on *Asterionella glacialis* Castracane (= *Asterionella japonica* Cleve: not validly published: Körner, 1970), which is a neritic marine species of world-wide distribution. Its affinity to *Asterionella* is simply its star-like colony. In fact it forms undulating, twisting filaments with two plastids located in the basal triangular swelling.[a]

Valves with an expanded rounded foot-pole, which narrows abruptly to a long thin extension often 3–4 times the length of the foot-pole: the extension enlarges only slightly at the apex.[c, h] Valve mantle very delicate and obvious only in the basal section[c] and along the extension.[f] In girdle view the symmetry is similar but the foot-pole is triangular in outline.[a] Sternum narrow; striae uniseriate, containing tiny round to rectangular poroids. In the extension the sternum is bordered by only a single row of areolae on either side: in this region the margins bear pointed spines externally.[f, g, i] Poroids with very fine closing cribra suspended by slightly thicker struts.[k] Oval 'ocelli' occur at either end of the valve, containing rows of elongate openings.[d-f, h] Silica framework very delicate, thickened only around the ocelli. A single rimoportula is present, lying adjacent to the 'ocellus' at the narrow pole. Several copulae,[g] with areolae that are smaller but similar to those of the valve; ligula and pars media present.[j]

Asterionellopsis differs from *Asterionella* in the shape of the valve, form of the areolae (which we regard as the best separating character), presence of distinctive 'ocelli' at both ends with elongate openings, the presence of a rimoportula at the narrow end of the valve, and the more keel-like spines along the valve rim. Körner (1970) recognised that *A. glacialis* differed from *A. formosa* in these details but did not separate them. Illustrations of *Asterionella kariana* Grun. in Körner and the TEM in Takano (1981) suggest that this too belongs in *Asterionellopsis* but *A. notata* does not and has been transferred to *Bleakeleya*. *A. socialis* (Lewin & Norris, 1970) should also be transferred to *Asterionellopsis*.

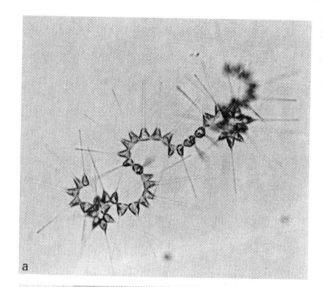

a

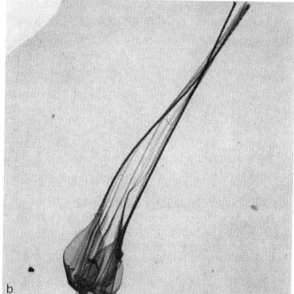

b

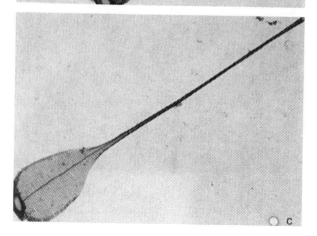

c

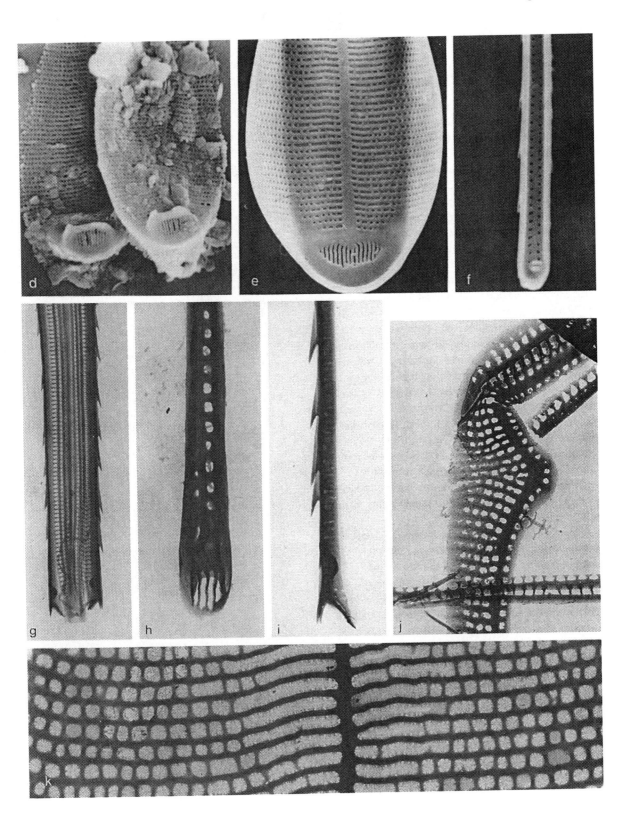

Bleakeleya F. E. Round gen. nov.

T: *B. notata* (Grunow in Van Heurck) Round comb. nov. (= *Asterionella bleakeleyi* var. *notata*)

Cells colonial, united by the flattened valve faces of the expanded end; thus cells lie in girdle view in the flat or twisted chains.[a, b] In valve view with a rounded, slightly inflated basal end, a narrow or slightly swollen shaft, and a narrow rounded apex.[b, c] Plastids: numerous small granules. A marine genus attached to seaweeds or free-floating – the habitat is not exactly clear; its rather rare occurrence may be due simply to its scattered occurrence in coastal plankton, which it may be contaminating from the benthos.

Valves linear, with one inflated end; with[b] or without a median inflation. A narrow or almost insignificant sternum[d, f] arises from a transverse bar across the basal inflation. Areolae in transapical rows, which continue down the valve mantle. Basal end with smaller pores than elsewhere, arranged in strongly radiating striae;[f, h] the valve face is here somewhat depressed.[e] The valve mantle striae continue around the basal pole. Spines occur along the valve face/mantle junction, all pointing towards the narrow apex and continuing around it.[b, d, e] No spines occur around the base pole though there may be granules. One rimoportula present,[g] slightly to one side of the sternum at the narrow pole;[a] external opening a simple hole. Girdle bands numerous, with rows of elongate pores.[i]

B. notata cannot be included in either *Asterionella* or *Asterionellopsis* without over-widening the circumscription of these and making imprecise genera. Hence we have separated it and chosen the name *Bleakeleya*, based on the specific epithet given to *Asterionella bleakeleyi* by Wm. Smith since this is the earliest synonym of *B. notata* (see Körner, 1970 and VanLandingham, 1967). Körner also described a var. *recticostata* and this has been re-found and is illustrated in Figs e–i. The structure of this variety at the SEM level suggests that this should be a species and not a variety of *B. notata*.

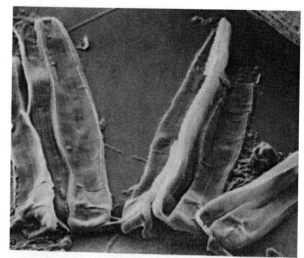

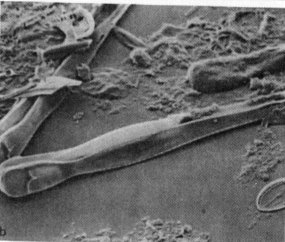

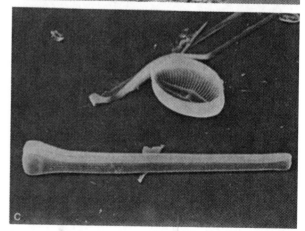

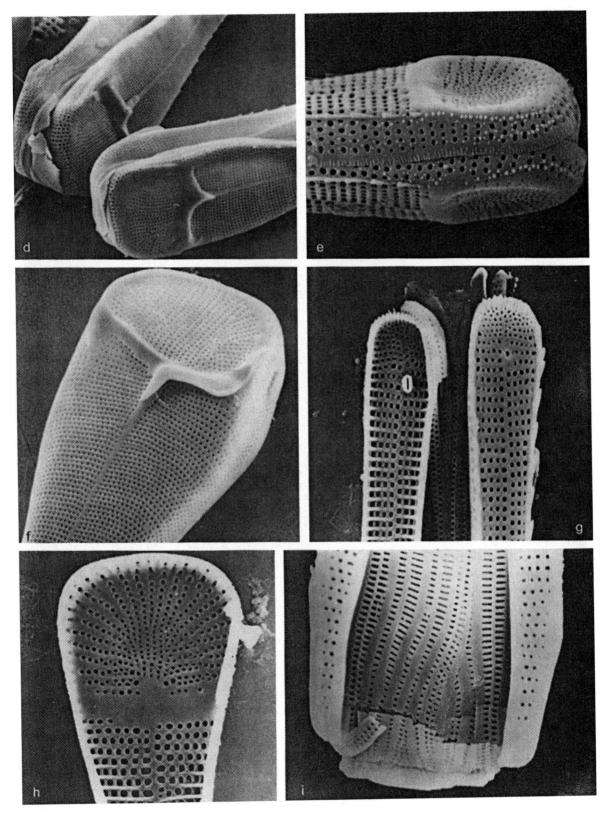

Podocystis J. W. Bailey 1854. Smithsonian Contr. Knowl. 7(3): 11 (nom. cons.)

T: *P. americana* Bailey

Cells epiphytic on marine algae, attached by a short mucilage stalk; heteropolar, cuneate in girdle view. Plastids numerous, small, plate-like. More common in tropical/subtropical waters than in temperate.

Valves obovate:[a, b] both ends rounded or the basal pole squared off.[c] Valve face flat; mantles fairly distinct but very shallow at the base-pole.[b, d-f] A distinct sternum is present.[a-h] Striae bi- to multiseriate,[b-h] transverse, continuing without a break down the valve mantle; separated in *P. adriatica* by well-developed costae:[e-g] these are especially thick on the mantles. Areolae circular, small, occluded by vela (we have no detail of their structure). A small isolated rimmed pore is present near the centre of some valves.[c] One rimoportula at each end,[f, g] just to one side of the sternum: external opening inconspicuous; internally sessile. No obvious apical pore field. Copulae split,[i] tapered to the basal end of the cell, ligulate. There is a conspicuous pars media running the length of the valvocopula; the areolae are spaced as on valves.[d] The copulae appear to lack a pars media and have smaller areolae.

Only two species are common: *P. americana* (= *P. adriatica*) and *P. spathulata* (= *Euphyllodium spathulatum* Shadb.). Two further species *P. javanica* and *P. ovalis* require checking. Additional studies are needed and a detailed comparison with other araphid genera required before its systematic position can be ascertained. It may be an isolated genus though on shape and girdle structure its nearest neighbour might be *Licmophora* which occupies a similar ecological niche.

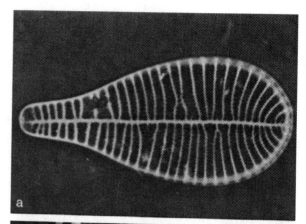

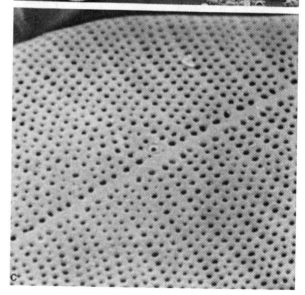

Tabellaria C. G. Ehrenberg 1840. Ber.
Bekanntm. Königl. Preuss. Akad. Wiss. Berlin,
1839:217

T: *T. flocculosa* (Roth) Kützing (= *Conferva
flocculosa*)

Cells joined together in long zig-zag,[a] partially
linear, or stellate (planktonic) colonies, with
prominent mucilage pads at the cell-to-cell
attachments and at the point of attachment to the
substratum. Cells usually lie in girdle view and
appear square to oblong in outline with distinct
septa on the copulae. Short strip-like plastids lie
between the septa. Nucleus central. A common
freshwater genus living attached to plants or stones
or, in the case of certain forms of *T. flocculosa*, in the
plankton (see Knudson, 1952, 1953a, b, for a full
discussion of the varieties). Distribution mainly in
acidic waters. A very distinctive genus.

Valves elongate, slightly capitate and equally or
more inflated in the central region.[b, c] Some
populations have a slight twist to the valve (cf.
Fragilaria crotonensis). Valve face flat; mantles
distinct. Sternum axial, narrow, expanded slightly in
the centre. Striae uniseriate,[c, d, g] transverse, more-or-
less irregularly spaced, containing small round
poroids; continuing over the edge of the valve face
onto the mantles. An area of small pores occurs at
each apex extending from the valve face onto the
mantle. Small spines occur on the valve margin
between the files of areolae;[d, e] the row of spines
continues through the apical pore area, where the
spines are often larger.[g] Koppen (1975) has reported
a population without spines. One rimoportula per
valve, placed in or near the central inflation, to one
side of the sternum; external opening slit-like,
transverse,[d] sessile internally.[i] Copulae of two types:
either complete with septa occluding almost half the
length,[f, h] or non-septate and split, with ligulae.[j] The
bands lack fimbriae. Occasional copulae appear to
have septa at both ends – rudimentary septa of
Knudson (1952). These have been considered an
important systematic character. The number of
copulae is very variable. Knudson (1952) used this to
distinguish *T. flocculosa* (many copulae) from *T.
fenestrata* and *T. quadriseptata*.

One of the few genera studied from the point of
view of micro-evolution, by Knudson (op. cit.). See
also Round & Brook (1959).

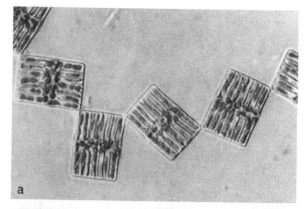

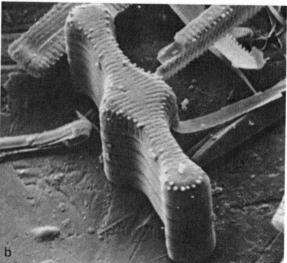

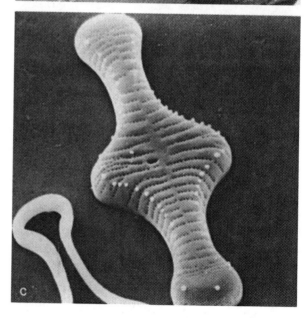

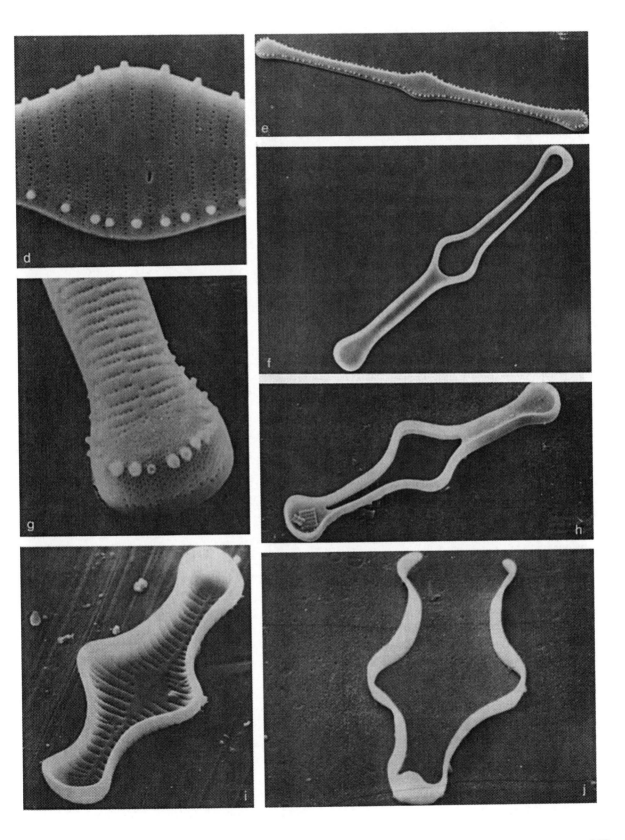

Tetracyclus J. Ralfs 1843. Ann. Mag. Nat. Hist.
12: 105

T: *T. lacustris* Ralfs

Cells in zig-zag chains held together by mucilage
pads at corners; square to oblong in girdle view,
with distinct septa. Occurs on freshwater sediments
in rather acid situations; also on bryophytes of rock
faces. Sometimes abundant but apparently absent
from large areas; tends to be northern/alpine in
distribution and its ecology requires further study.

Valves elongate to elliptical, often capitate and
often centrally expanded[a, b] (and also centrally
constricted in the form illustrated). Valve face flat;
mantles distinct and fairly deep. Striae transverse,
uniseriate,[b, c] *not* interrupted externally by the
underlying costae; continuing onto valve mantle,
where they often become rather disorganised.
Sternum narrow, rather irregularly defined by the
ends of the striae; this is rather unusual amongst
diatoms and may be a feature only of this sample.
Areolae small, simple, round; no vela observed. No
specialised apical region of areolae or pore field.
Internally the valves are strengthened by massive
transapical costae,[d-f] some running right across the
valve, others part way. The costae are fused with the
inwardly projecting valve margin, which is
somewhat expanded at the apices and in regions of
outward curvature of the valve. Valves may have 0, 1
or 2 rimoportulae; when present they lie to the side
of the sternum,[e, f] near the valve centre, and are
orientated transapically; their external openings are
not obvious and are simple slits. Copulae
numerous,[i] split and ligulate, provided with septa
across the ligulate ends;[g, h] they are not fimbriate but
have longitudinal rows of pores which seem to
combine internally to form large elongate openings.

For details on the systematics and inter-relations
of the forms, see Hustedt (1914) and Williams (1987).
Three species are recognised by Hustedt (1927–66)
but may all be forms of a single species, *T. lacustris*;
14 species are recorded by VanLandingham (1978).
The absence of a distinct apical pore field (plate)
contrasts with most araphid genera. *Tetracyclus* is
perhaps most closely related to *Tabellaria*.

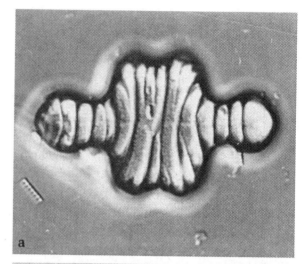

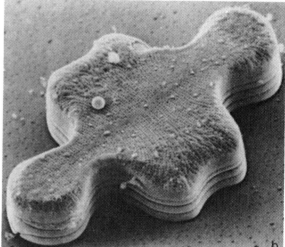

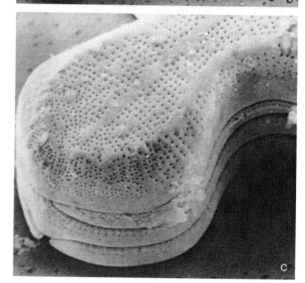

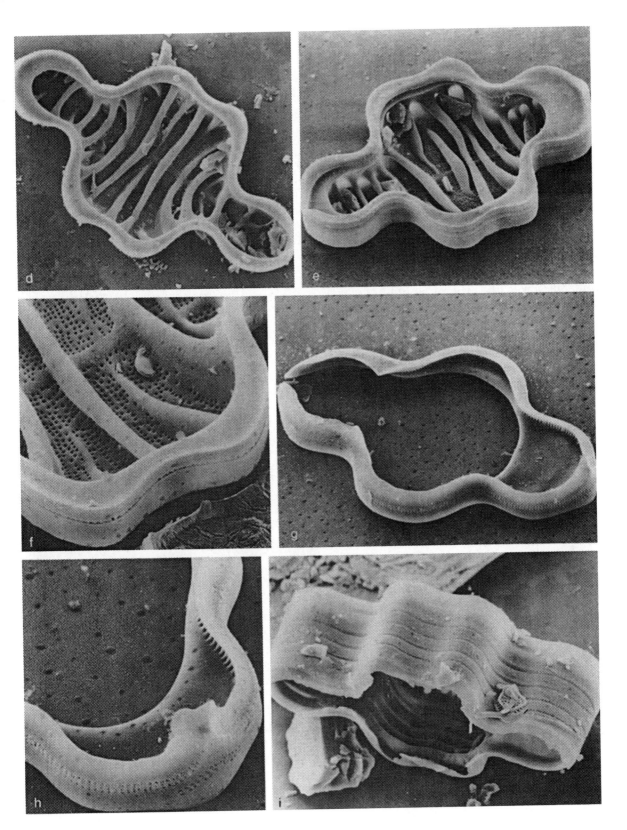

Oxyneis F. E. Round gen. nov.

T: *Oxyneis binalis* (C. G. Ehrenberg) F. E. Round, comb. nov. (= *Fragilaria binalis*)

Cells quadrate/oblong in girdle view, forming straight chains in the panduriform type (very rarely breaking) and zig-zag in the elliptical form. Plastids not observed. A genus confined to extremely acidic waters and a good indicator of them (see Battarbee, 1984; Flower, 1986).

Valves either elliptical[g, h] or panduriform.[a–e] Valve face flat, curving[a] into shallow mantles; junction of valve face and mantle bearing a row of spines.[c, g] Spines rather flat, bifurcating, and partially fused (Flower, 1986); interlocking between sibling valves. Striae uniseriate, containing round areolae; sternum narrow. Apical pore field present, consisting of rows of simple pores extending from the valve face down the mantle.[d, e, g, h] A single rimoportula occurs at one pole, at the end of an areola row adjacent to the apical pore field.[d, f, h] Cingulum of numerous open copulae,[i, j] each bearing a short terminal septum[j] and a row of areolae.

The taxon was formerly included in *Tabellaria* but it differs in several points – the greater complexity of the spine structure (small thorn-like spines occur in *Tabellaria*), the position of the rimoportula (central in *Tabellaria*), the difference in development of the septa (extending almost to the centre in *Tabellaria*) and the shape. Nevertheless, the genera are clearly very closely related.

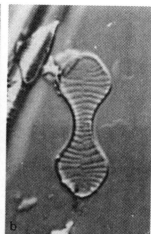
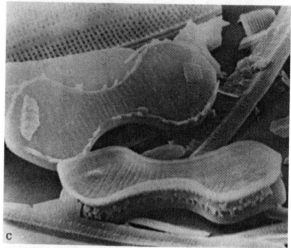
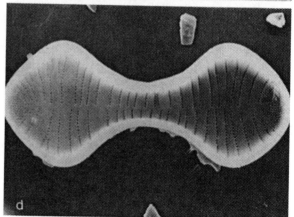

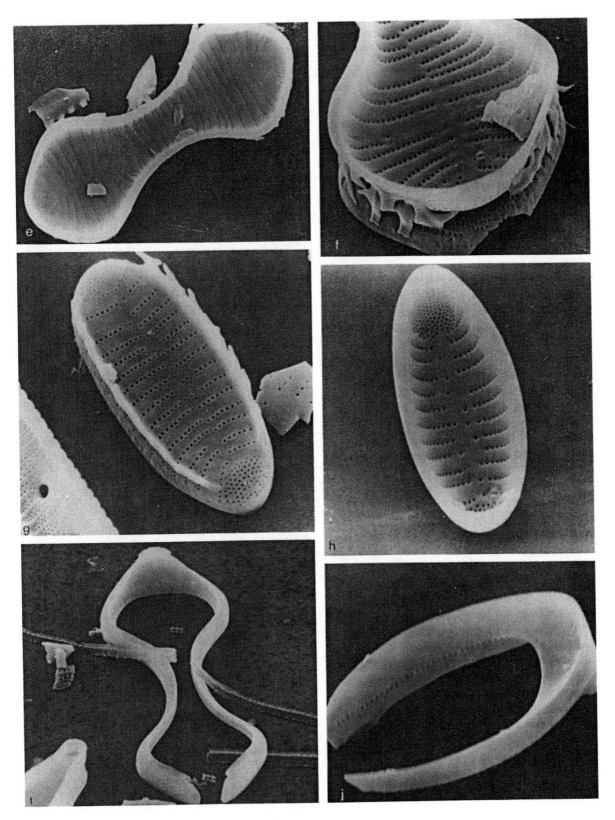

Licmophora C. A. Agardh 1827. Flora 10: 628 (nom. cons.)

T: *L. argentescens* Agardh (typ. cons.)

Cells heteropolar, cuneate in both views. Attached by the narrow basal pole, singly or in clusters on mucilage pads, and sometimes colonial on branched mucilage stalks.[a] Plastids of two types – either discoid or plate-like; this and other features suggest that the genus should probably be split. Cosmopolitan in coastal seas; always attached to seaweeds, marine angiosperms, shells, stones, etc. Also on the surface of whales and copepods.

Valves cuneate or clavate, often with the basal half extended and the basal pole somewhat capitate.[d, e] Valve face merging imperceptibly with fairly deep mantles.[b] Sternum present but sometimes indistinct. Striae uniseriate, parallel,[b, g] except at the apices, where they become radiate (especially at the broad pole).[g] Areolae elliptical or elongate[h] in the transapical direction, closed by cribra.[h, k] At the base-pole there is a row of slits,[d, f, i] through which the mucilage pads or stalks are secreted. Rimoportulae opening externally by slits. There is a single conspicuous, often fan-shaped rimoportula either near the basal pole[f] or near the apex.[c, g] The orientation and position (on valve face or mantle) appear to be variable. Copulae areolate[j] and tapering towards basal split. Septa present but may not be well-developed; borne internally around the apical ends and tapering off down each side of the copulae.

According to VanLandingham (1971) the type, *L. argentescens*, is synonymous with *L. flabellata*. Recently Wahrer *et al.* (1985) have investigated the genus *Campylostylus* and in a detailed comparison with 3 species of *Licmophora* have shown it to be merely a species of the latter genus. They also studied in detail the arrangement of the rimoportulae and came to the conclusion that the head-pole always has a rimoportula but we have examples of valves without one. There is a degree of heterovalvy in this genus. The genus *Licmosphenia* has not been studied but LM illustrations suggest a close relationship with *Licmophora* – it may even be congeneric.

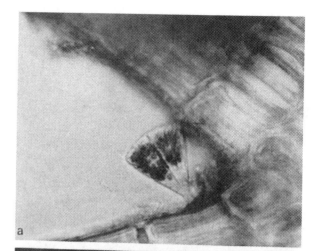

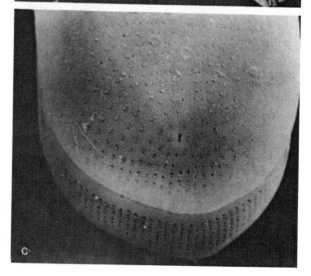

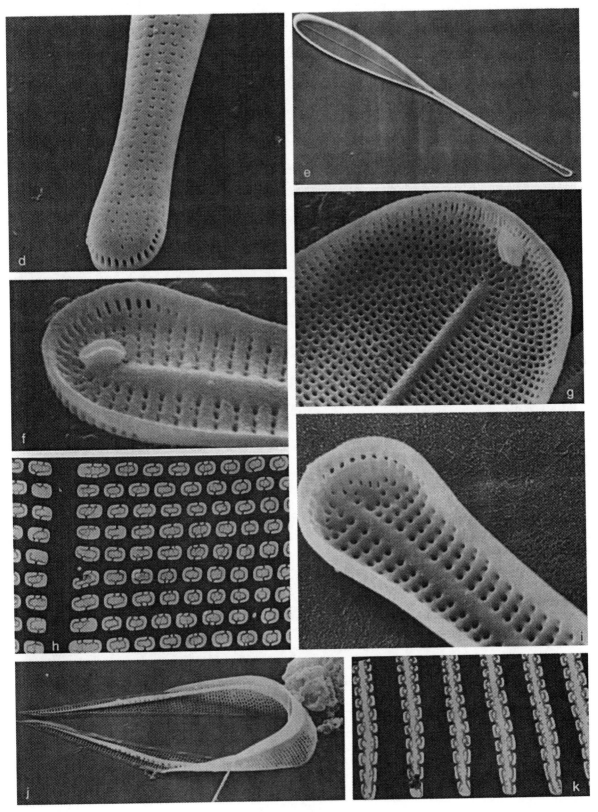

Rhaphoneis C. G. Ehrenberg 1844. Ber.
Bekanntm. Verh. Königl. Preuss. Akad. Wiss.
Berlin, 1844: 74

T: *R. amphiceros* (Ehrenberg) Ehrenberg
(= *Cocconeis amphiceros*: lectotype selected by
C. S. Boyer, 1927, Proc. Acad. Nat. Sci. Philad.
78 Suppl.: 190)

Cells solitary or forming short filaments (see Drebes,
1974); usually shallow in girdle view, broad in valve
view. Plastids numerous, small. Often occurring
attached to sand grains in shallow marine habitats;
common and widely distributed.

   Valves linear to subcircular, sometimes with
apiculate or capitate apices.[a-f] Valve face flat; valve
mantle shallow, often separated from the valve face
by a slight ridge;[c] face and mantle sometimes finely
granulose. Striae uniseriate, perpendicular to or
radiating from a narrow sternum. Areolae round or
elliptical, containing distinctive vela more-or-less
flush with the valve surface, with distinctive
concentric slits,[g] appearing like large perforate rotae.
A single row of areolae occurs on the mantle, below
the marginal ridge. Clusters of small pores without
vela occur at the apices,[d] and between these apical
pore fields and the remainder of the valve face is a
small rimoportula;[e, f, h] this is sessile internally and
opens by a simple pore externally. Copulae split but
no detail recorded; the valvocopula is sometimes
covered with small granules.

   This is a common genus with a good fossil record
(Andrews, 1975, 1978) but many species have been
removed to the genus *Delphineis* (Andrews, 1977).
Species allocated to *Delphineis* by Andrews have
small granules along the valve face/valve mantle
edge and lack apical pore fields. There are several
difficult taxa in the cluster of genera round the
classical *Rhaphoneis* (see *Neodelphineis*, *Diplomenora*,
Perissonoë and the new genus *Adoneis* of Andrews &
Rivera, 1987). For the moment *Rhaphoneis* is
retained for species with concentrically slit vela, a
rib but no spines or spinules on the margin of the
valve face, two rimoportulae and small apical pore
fields. The apical pore field is present only in
Rhaphoneis and *Perissonoë* amongst those genera we
place in the Rhaponeidaceae. Cells are often found
attached to the dead valves of centric diatoms, see
illustration in Drebes (1974).

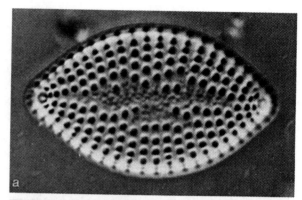

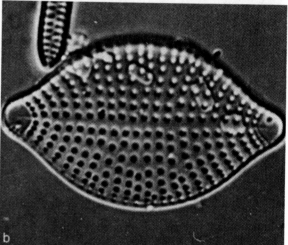

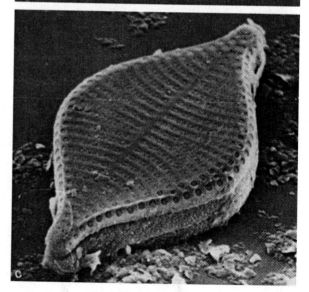

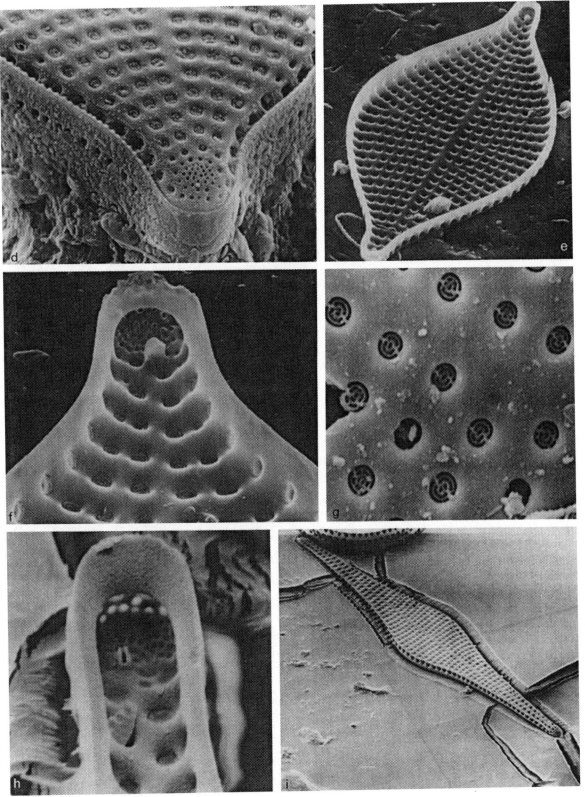

407

Diplomenora K. Blazé 1984. Morphology & taxonomy of *Diplomenora* gen. nov. Br. phycol. J. **19**, 217–225

T: *D. cocconeiformis* (Schmidt) Blazé
(= *Coscinodiscus cocconeiformis*)

Cells circular in valve view. Plastids unknown. A marine genus, living on sand grains. All records so far are from the southern hemisphere or from the coasts of Mexico and California.

Valves circular to elliptical,[a-f] with an extremely shallow mantle. Valve face flat, laminate-costate in structure. Striae radiating from an inconspicuous sternum and with more closely spaced groups of areolae at the poles of the sternum;[f-j] the areolae reduce in size near the mantle. Areolae with rotae suspended by 2–3 pegs near the outer face[e] (see Blazé, 1984). Internally the valve appears more costate than externally, with the areolae sunken. Rimoportulae[h-i] positioned near the margin, 1 or 2–10 in number, opening by simple external pores, sessile with outer and inner lips. Valvocopula split adjacent to the group of areolae at a pole. A row of very small pores occurs along the middle of the valvocopula; no other copulae observed.

This new genus is fully discussed by Blazé (1984) and its relationship to *Rhaphoneis*, *Psammodiscus* and *Detonia* pointed out. It belongs to a group of diatoms living primarily on sand grains and completely marine in distribution. They form a discrete cluster and we include them all in the Rhaphoneidaceae. Blazé (1984) described the genus with 2–10 rimoportulae but our example photographed in the SEM has a single or two rimoportulae; otherwise it is similar to the cells illustrated by Blazé. We believe that here and elsewhere it is possible to have a variable number of rimoportulae within a species or genus though in general there appears to be a 'normal' state to which the majority of valves conform.

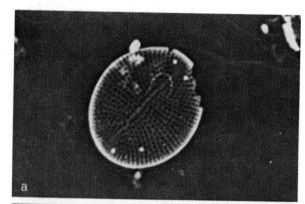
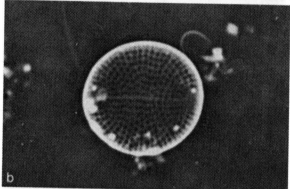
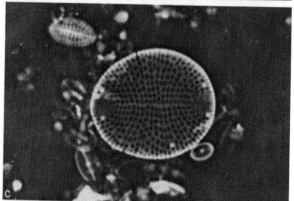
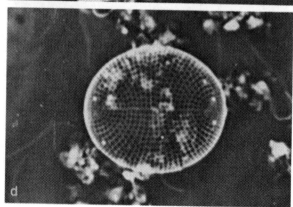

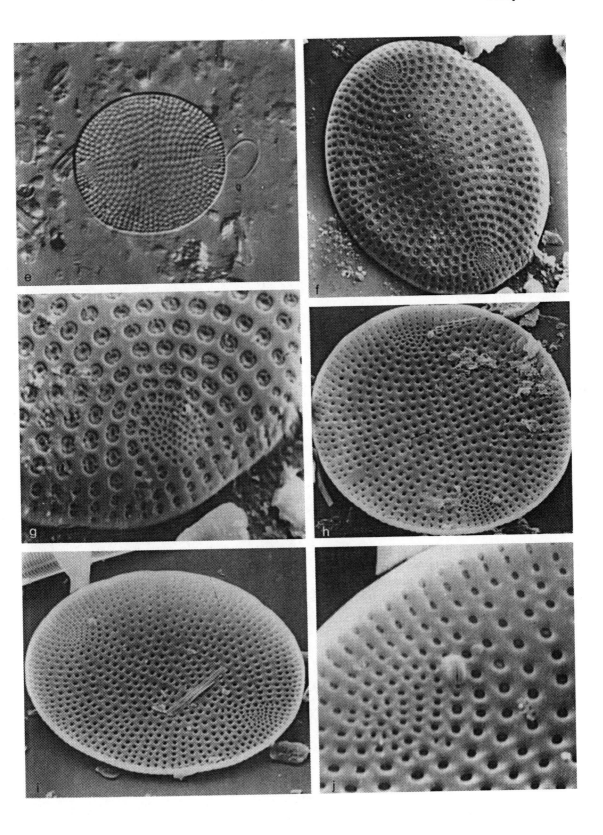

Delphineis G. W. Andrews, 1977. Nova Hedw.
Beih. 54: 249

T: *D. angustata* (Pantocsek) Andrews
(= *Rhaphoneis angustata*)

Cells shallow with flat valves, growing as long
chains in the plankton (*D. karstenii* – see Fryxell &
Miller, 1978) and probably as short chains in the
epipsammon;[b] marine. Several fossil species are also
recorded but their growth form is not known
(Andrews, 1981a, 1988). Plastids 4 to several
according to Fryxell & Miller.

Valves elongate, elliptical[a] or circular, with a flat
valve face, shallow mantles and a prominent clear
sternum,[a-g] which often enlarges near the poles.
Striae uniseriate, parallel to radial. Areolae closed
by rotae lying near the outer surface.[e, f] Along the
valve edge there are usually spines (*D. karstenii*)[f, g] in
the form of paired lipped structures, or short blunt
spines,[c, d] or closely placed granules;[e] these are
absent in some growth forms. There is usually one
row of areolae on the valve mantle below the spines,
and here the valve surfaces may be granular (see
Fryxell & Miller, 1978). In several species, plaques
occur along the edge of the valve mantle[c, d] (cf.
Fragilaria). At each apex there are two small pores[c- f]
located at the ends of the sternum and a single
rimoportula slightly off-centre.[h, i] The two
rimoportulae of one valve are diagonally placed
with respect to the sternum. The external
rimoportula opening may be a slightly raised ring in
a depression and in one example this elongates to
form a lipped structure;[e] in others it is difficult
to detect. Internally the rimoportulae are prominent
and the two small pores at the end of the sternum
seem to be unoccluded.[h, i] In some species each
rimoportula appears to have double lips on either
side of the slit. Copulae split and plain but require
further study.

This genus is characterised by the two small pores
at the ends of the sternum. The long or short
filaments are formed without interlocking of the
spines/granules on the valve edge. The small spines/
granules and lack of an apical pore field distinguish
this genus from *Rhaphoneis*. The grooved external
surface referred to by Andrews (1977) is not always
obvious in SEM but the slight enlargement of the
sternum at the apices and the ridge[e, f] around it are
often observed.

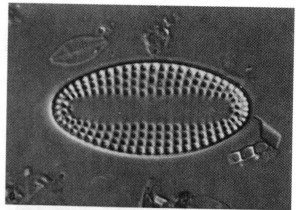

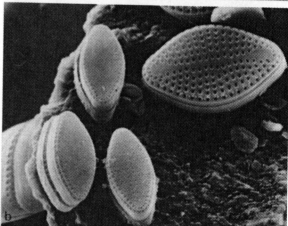

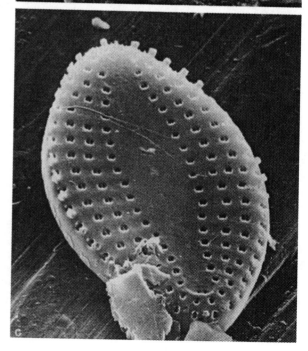

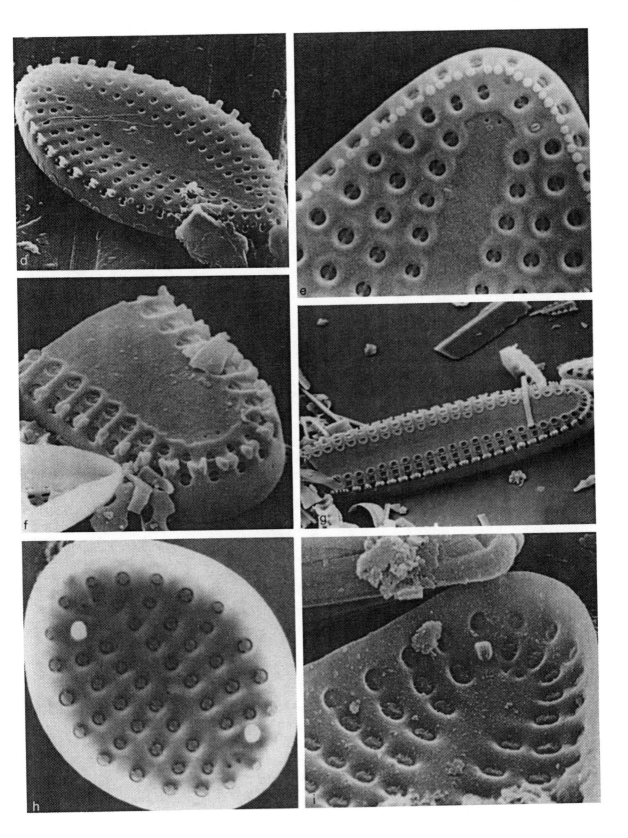

Neodelphineis H. Takano, 1982. Bull. Tokai
Fish. Res. Lab. **106**: 45

T: *N. pelagica* Takano

Cells elongate to elliptical, forming zig-zag colonies
in the neritic marine plankton. Plastids not known.
This is another genus which seems to be confined to
coastal waters. Originally described from Japan but
abundant along the Texas coast (L. Medlin,
personal communication) and along the coast of
Florida (Round unpub. obs.).

Valves linear,[a-c] becoming elliptical in smaller
cells. Valve face flat; mantles shallow. Sternum
narrow; striae uniseriate, transverse, sunk between
ribs externally, alternate.[d-f] Areolae circular or
elliptical, containing rotae; the two supports of each
rota are aligned parallel to the apical axis. Short
spines occur on the edge of the valve face between
the striae. There is a single row of areolae on the
valve mantle.[d] One rimoportula at each apex,[g-i] to
one side of the sternum; within a valve the
rimoportulae are placed diagonally.[g] External
opening of rimoportula simple; internally shallow
and almost circular. A small pore occurs adjacent to
each rimoportula, between it and the valve
apex.[arrow, h] There are no apical pore fields. Structure
of copulae unknown.

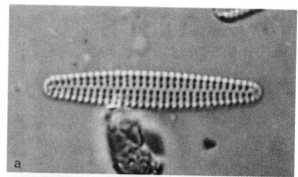

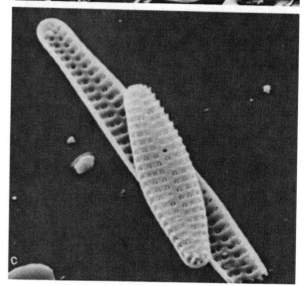

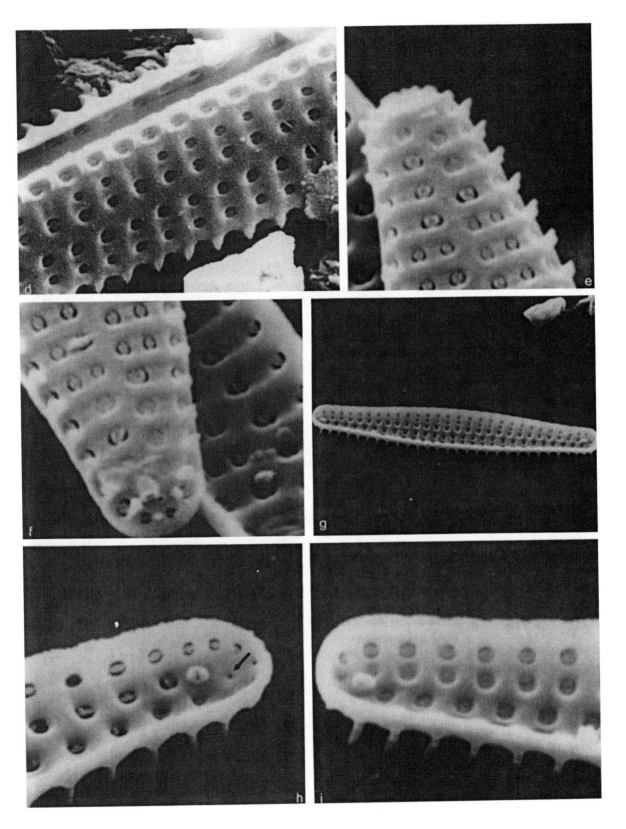

Perissonoë G. W. Andrews & V. A. Stoelzel
1984. In Proc. 7th Internat. Diat. Sympos.
(D. G. Mann, ed.): 226

T: *P. cruciata* (Janisch & Rabenhorst) Andrews
& Stoelzel (= *Amphitetras cruciata*)

Cells tablet-like, habit unknown. Plastids not
studied. A marine genus occurring in the littoral
zone and possibly more abundant in warm waters.
Our material was from a sample kindly supplied by
G. W. Andrews and collected from dead coral
rubble (to which it was attached) off Barbados. The
genus was formerly regarded as part of *Rhaphoneis*.

Valves quadrate or triangular, margins undulate
or almost straight.[a-f] Valve face planar, abruptly
differentiated from shallow mantles.[b] Ornamented
with small knobs on the valve face/mantle junction;
even smaller papillae are scattered on the valve face
but form a denser covering on the mantles.[b] Sterna
present, radiating from the centre but often not quite
forming a perfect cross.[c-e] The valve framework is
costate and the sterna are more distinct inside than
outside. Striae uniseriate, orientated at right angles
to the valve edge and curving slightly towards the
sterna; there are often a few incomplete striae at the
centre. A single row of slightly enlarged areolae
occurs on the mantle, even around the angles. The
areolae are round and sometimes open externally
into slight pits; they contain cribra, although these
might also be interpreted as perforate rotae.[g, i]
Clusters of small areolae (apparently without vela)
occur at each angle of the valve:[c-g, i] these are
termed pseudocelli by Andrews & Stoelzel (1984).
Valves heteromorphic, some without rimoportulae
and others with 1–4;[e, g, i] where two are present they
occur in an unexpected position, in adjacent angles.[e]
The rimoportulae are more-or-less central within
each angle and appear to be doubly lipped. Copulae
not seen by us: Andrews & Stoelzel report a rather
strange arrangement in *P. cruciata* of 4 segments
corresponding to the four sides.

Perissonoë appears from its valve structure and the
form of the areolae and rimoportulae, to be closely
related to *Rhaphoneis*. Andrews & Stoelzel pointed
out that it has features which are somewhat 'centric'
but pointed out its obvious affinities with
Rhaphoneis and *Delphineis*. The triangular species (*P.
trigona*) is exceptional among living species in also
being found back to the Miocene.

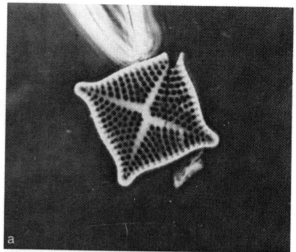

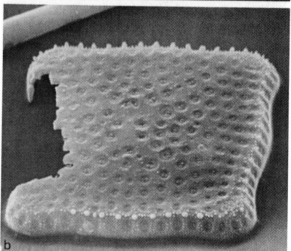

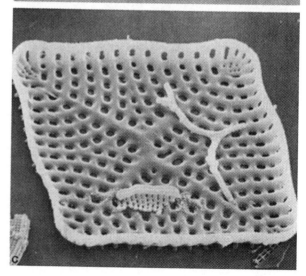

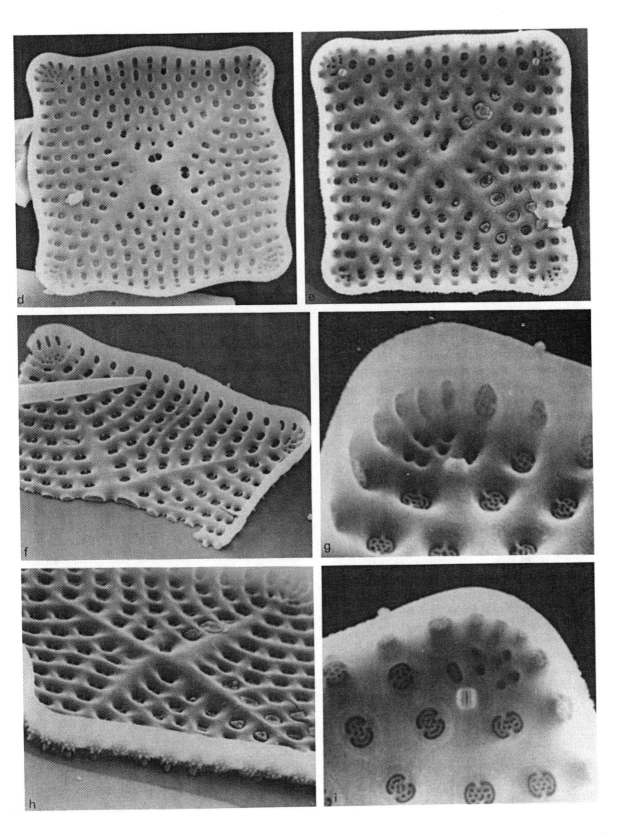

Sceptroneis C. G. Ehrenberg 1844. Ber.
Bekanntm. Verh. Königl. Preuss. Akad. Wiss.
Berlin, 1844: 264

T: *S. caduceus* Ehrenberg

Cells elongate, heteropolar; attached, probably to
seaweeds but rarely recorded live. Fossil forms better
known than the extant species. Plastids unknown. A
small genus (*c*. 10 spp.) requiring re-investigation.

Valves elongate and slightly curved,[a, b, f] with a
broad, slightly capitate head-pole and a narrow
base-pole; swollen in the central part. Valve mantle
sharply turned down, virtually at a right angle to the
flat valve face. Striae uniseriate, orientated at right
angles to a rather indistinct central sternum. Areolae
large, round or elliptical, containing perforate rotae[c, i]
across their external apertures, as in *Rhaphoneis*
(q.v.). Fossil material is often without occluding
plates. Apical pore fields present at both ends,[c-e]
extending onto the valve face. Within each field the
pores tend to radiate from the valve face; thus they
are not in vertical rows as in many other genera.
Our material is eroded but there appear to be traces
of a marginal ridge or spines[c, d] in a row around the
edge of the valve face, which continues through the
apical pore field. One rimoportula at each pole at
the end of the sternum; the external openings are
not obvious. Copulae not known.

This genus was placed by Peragallo & Peragallo
(1897–1908) in the family Raphoneidées and the vela
confirm this perceptive vision. As Hustedt (1927–66)
commented, *Sceptroneis* has very little in common
with *Synedrosphenia* and we are certain it is not in
any way related to *Synedra*. The form of the vela, the
two rimoportulae, the form of the apical pore field,
and the probable presence of a marginal ridge/
spines, are at the moment the main features
suggesting a position close to *Rhaphoneis*. The
position of the apical pore field is such that this is
one of the few genera in which it can be clearly seen
in light microscopy (cf. the excellent figure in
Hustedt, 1927–66).

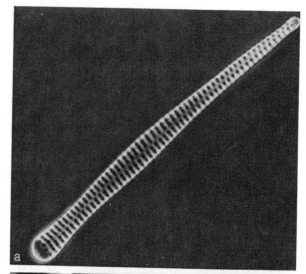

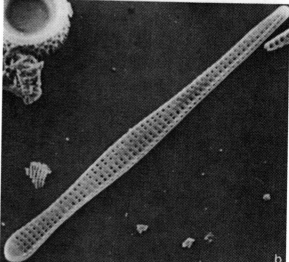

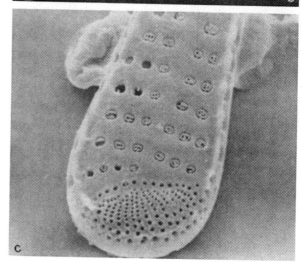

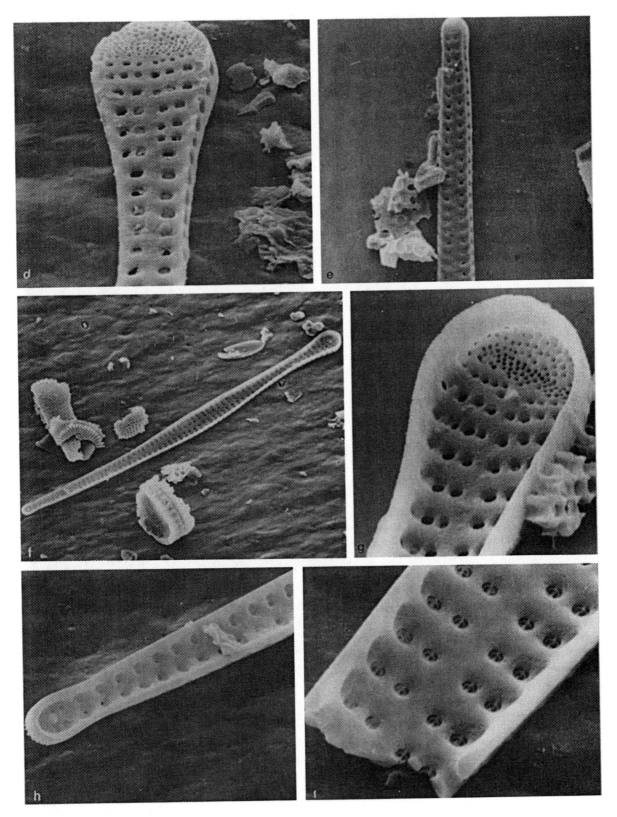

Psammodiscus F. E. Round & D. G. Mann
1980. Ann. Bot. **46**: 371

T: *P. nitidus* (Gregory) Round & Mann
(= *Coscinodiscus nitidus*)

Cells discoid,[a, b] solitary, attached by their valve faces to sand grains in the marine inter- and subtidal. Plastids unknown. World-wide in distribution.

Valves circular or elliptical. Valve face flat, bearing radiating rows of rather widely-spaced areolae,[a–c, f] which become smaller towards the mantles. Some valves have more closely spaced areolae and the radial pattern tends to be lost. Valves vary considerably in the density of the areolar rows. Valve mantle turned down sharply, with more closely spaced vertical rows of areolae which are smaller[b, c, e–g] than those on the valve face and tend to be oval in outline. The areolae are simple and occluded by rotae,[d, e] having two to several supporting pegs. In addition, a small pore is usually present near the centre of the valve face;[d] this is occluded internally by an irregular lobed structure,[i] whose significance is obscure. Some valves have a rimoportula. This is positioned near the centre,[f] at the end of one of the radial rows of areolae. It has a slightly rimmed external aperture and is very low and sessile internally.[h] Girdle bands rather narrow, open, the first having a row of pores, also occluded by rotae, though they are much smaller than those on the valve.

This is a very common epipsammic diatom which was originally placed in the genus *Coscinodiscus* as *C. nitidus* by Gregory (1857). Since *Coscinodiscus* has totally different valves and girdle bands, and is planktonic rather than benthic, this taxon was removed and placed in its own genus by Round & Mann (1980), where further details can be found. *Psammodiscus* is a further example of the care which must be taken when only overall form of a taxon is used to indicate affinity, since this virtually circular valved genus has probably to be allied with genera that are normally regarded as araphid pennate forms: several similarities exist between this and the araphid genera *Rhaphoneis*, *Delphineis* and *Diplomenora*.

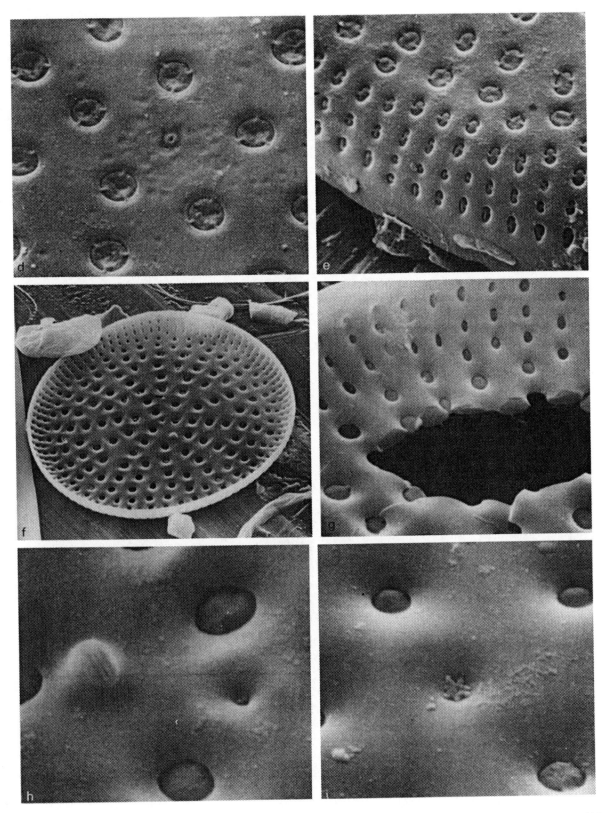

Ardissonea G. De Notaris 1870. Erbar.
Crittogam. Ital. ser. 2 no. **334**. G. De Notaris
1871, N. Giorn. Bot. Ital. **3**: 95

T: *A. robusta* (Ralfs in Pritchard) De Notaris
(= *Synedra robusta*)

Cells large (up to 0.75 mm long), linear,[a] attached at
one end onto massive branching, mucilage stalks.
Epiphytic on marine algae and angiosperms.
Plastids discoid but often touching and then
appearing as lobed plates.

Valves linear, often transapically undulate. Striae
uniseriate,[b, c] passing over onto the valve mantle
without change; sternum slight or absent. Apices
simple, with smaller areolae[b] in some species, but
apical pore plates absent. Valve structure complex,
chambered. Internally with a plain layer of silica in
which a central row, or 2 central and 2 marginal
rows, of apertures are located;[d-f] at the apices the
internal layer of silica fuses with the outer layer. The
internal rows of apertures and the external areolae
open into transapical chambers between the inner
and outer layers,[f] separated by solid ribs running
transapically[g] between the rows of areolae.
Rimoportulae absent. Copulae massive,[c, e, f] with
long fimbriae at the apices[e] and large flanges fitting
inside the valves.[f] The valvocopulae have notches at
the apices[b] in some species whilst the other copulae
are plain. None appear split. At the angle where the
valvocopula curves in under the valve mantle edge
there is a row of elongate openings and a similar
row on the next copula.[c, e, f]

This genus is obviously distinct from *Synedra* (see
also Round, 1979a) but may need further revision as
species are investigated in more detail. Hustedt
(1927-66) recognised that some species in his
subgenus had chambered valves and others did not.
The first mention of this genus is in the *exsiccata* set
with labels, on which there are diagnoses by De
Notaris (1870); the 1871 reference is the first
description in an available journal. Patrick &
Reimer (1966) followed Hustedt in recognising this
as a subgenus but we have no doubt of its merit as a
genus based on *A. robusta*, which has the internal
but perforated wall. It may prove best to place the
non-chambered species in yet another genus but we
require more data before further consideration is
given to this.

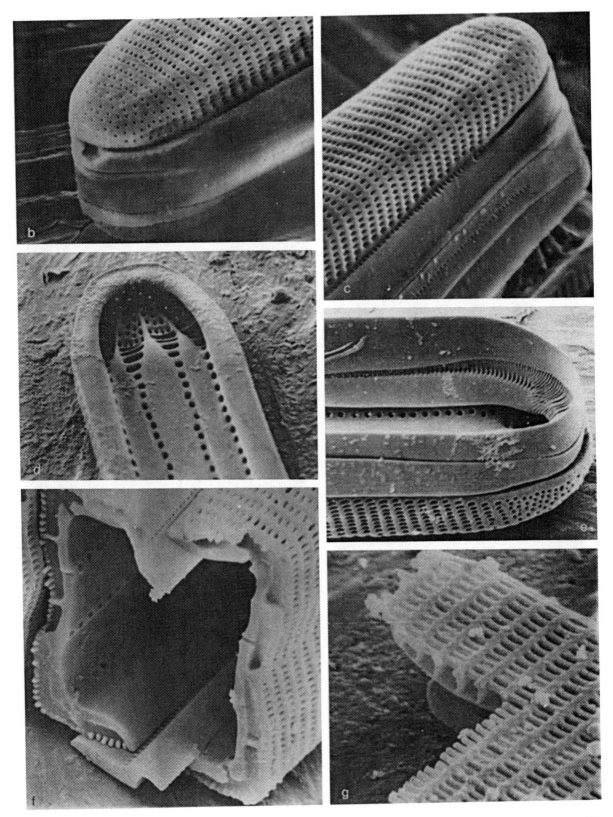

Toxarium J. W. Bailey 1854. Smithsonian Contrib. Knowl. 7 (3): 15

T: *T. undulatum* Bailey

Cells needle-like[a-c] in valve and girdle view, sometimes curved or undulate. Plastids unrecorded. Growing attached or perhaps only intermingled with other epiphytes on marine algae and angiosperms. Common, especially in tropical/subtropical waters.

Valves up to 1 mm long, slightly expanded at both apices and at the centre;[b] slightly curved or with undulate margins.[j] Valve face flat; mantles distinct but shallow. No distinct sternum.[a, d, f] Areolae scattered over much of the valve face but in fairly well-defined rows along the edge of the valve face and on the mantle; simple and round internally and externally. Vela are present deep within the areolae, but we have no detail of their structure. There is no rimoportula nor any apical pore plate but a slight development of a flange (septum?) at the apices of *T. hennedyanum*. Copulae of *T. hennedyanum* of two distinct types: valvocopula narrow with a single row of pores next to the valve edge; second band broader and thinner with similar row of pores next to the valvocopula and then a broad area bearing rows of pores which run towards the opposite edge of the band. Occasionally pores are joined to form 'tears' in the band.

This genus was placed in *Synedra* soon after its initial description and Hustedt (1927–66) was not prepared to recognise it as a separate genus. In fact it probably has little relationship to *Synedra*: Hustedt was not aware, of course, of the absence of rimoportulae (Round, 1979a) although he noted the absence of a sternum. This latter is a rather rare condition in pennate genera and coupled with it is a distinctive semi-circular arrangement of the areolae at the apices. The only species other than the type is the non-undulate *T. hennedyanum* but taxa in VanLandingham (1978) need checking. '*Synedra reinboldii*' was also placed in *Toxarium* by Mills (1933–5) but this has a very distinct rimoportula and cribra (see p. 428) – it has been placed in a new genus *Trichotoxon* (Reid & Round, 1987).

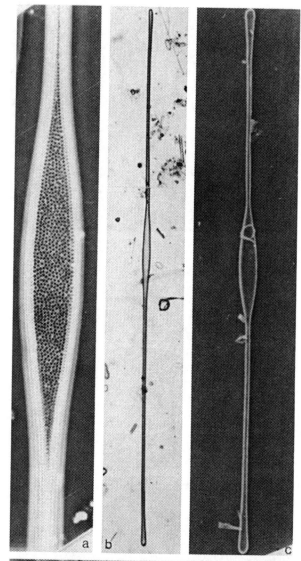

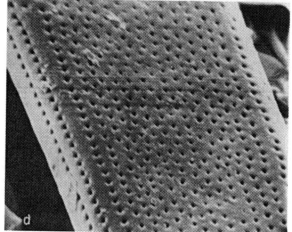

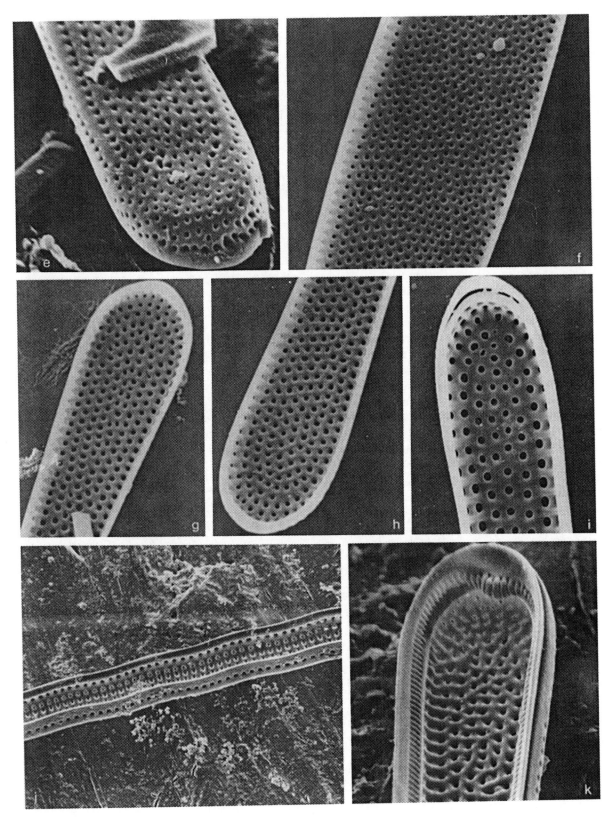

Thalassionema A. Grunow ex F. Hustedt 1932
in Rabenhorst Krypt.-Flora 7 (2): 244

T: *Thalassiothrix nitzschioides* (Grunow)
Grunow ex Hustedt (= *Synedra nitzschioides*)

Cells linear, isopolar,[a] forming stellate, zig-zag[b, c] or
fan-like colonies joined by mucilage pads. Plastids
discoid. A common marine planktonic genus of 3
species (see below).

Valves elongate, linear, their ends tapering slightly
and rounded. Valve face flat centrally, curving down
into indistinct shallow mantles. The areolae seem to
be reduced to a marginal series; across each there is
a raised simple or Y-shaped bar externally,[d, g] which
almost obscures the opening. Small spines were
reported by Hasle & Mendiola (1967) but these are
only the bars seen in side view. Internally the
areolae are simpler, with circular openings[i] which
are much narrower than the external openings. At
each pole there is a single, slightly raised large pore
externally, which is the opening of a rimoportula. It
is somewhat unusual in the thickening around it,
which may extend into a short spine around the
opening.[f-h] At one end the siliceous bars across the
areolae become confluent.[e] The other end retains the
individuality of the bars,[f] and whilst the structure is
somewhat modified it does not have such prominent
apical pores and the opening of the rimoportula is
not always so obvious. Girdle bands non-porous,
split.

The genus has often been regarded as monotypic
(*T. nitzschioides*): two other species have been
reported (VanLandingham, 1978) although these
were synonymised by Hasle & Mendiola (1967) into
T. bacillaris. Hallegraeff (1986) has made a detailed
study of the genus and concluded that *T.
nitzschioides*, *T. bacillaris* and *T. frauenfeldii* all belong
in *Thalassionema*.

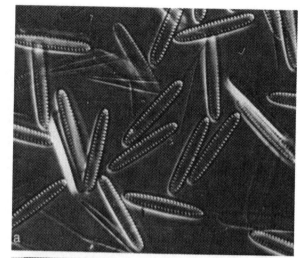

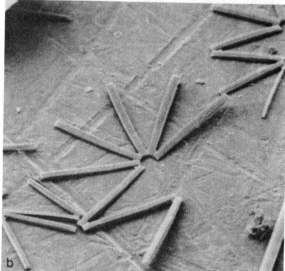

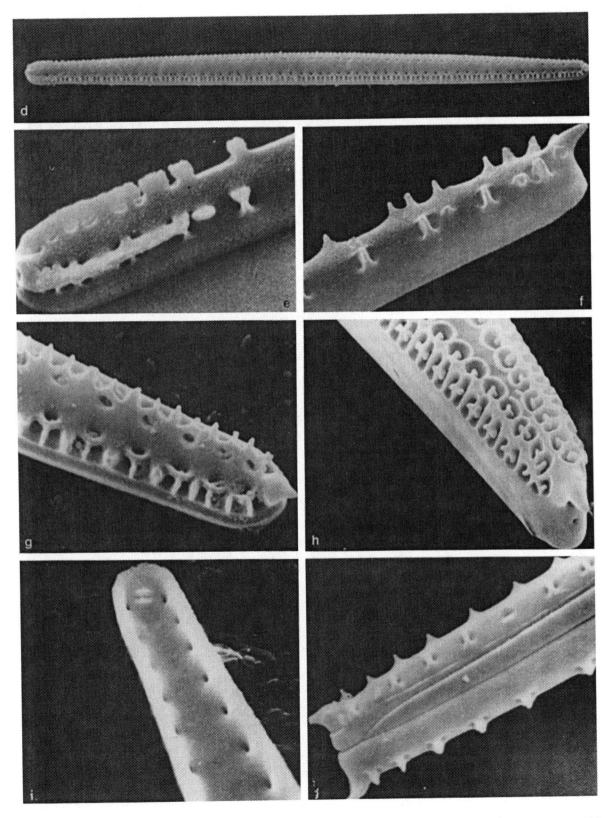

Thalassiothrix P. T. Cleve & A. Grunow 1880. Kongl. Svenska. Vetensk.-Akad. Handl. Ser. 2, **17** (2): 108

T: *T. longissima* Cleve & Grunow

Cells needle-like,[a] often somewhat curved and its ends twisted; free-living or forming radiating, stellate or zig-zag colonies. The cells of this genus may be the longest known amongst diatoms – records of cells 4 mm long are reported. Numerous small plate-like plastids. An oceanic, planktonic genus with 13 species recognised in VanLandingham (1978); some of these belong in *Thalassionema*. We have examined Antarctic and tropical material.

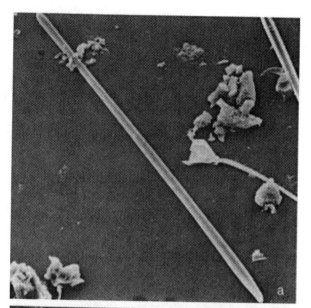

Valves extremely long, slightly heteropolar;[a] foot-pole rounded,[b, c] head-pole narrower and more pointed with two prominent spines.[d, f] Sternum wide, except near apices; areolae often forming two rows on either side of the sternum closed by a porate cribrum[b, c, e] (often eroded).[d, f] Spines arise between the two rows of areolae on each side of the valve and point to the head pole.[b, d-f] Internal openings of areolae circular or slightly elongate.[g, h] Rimoportulae at both ends opening externally by simple pores; internally often with unequal lips – one rather upright and the other curving around it. Copulae without areolation or with a single row of poroids.[i]

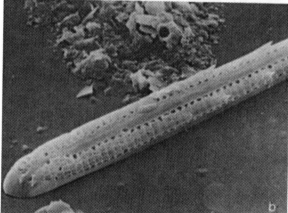

This genus is distinguishable from *Thalassionema* by the spines along the margin and the cribrate areolae. The cells tend to be heteropolar with two projections developed at one end in contrast to the one in *Thalassionema*. In the distinction between *Thalassionema* and *Thalassiothrix* heteropolarity has been mentioned – it is, however, only a matter of degree, being very obvious in *Thalassiothrix* and less so in *Thalassionema*; *Thalassiothrix* is more likely related to the old *Synedra reinboldii* now *Trichotoxon* (see p. 428). The spines are a characteristic feature of *Thalassiothrix* and reports in the literature of their loss are, we believe, not tenable since the spine is a composite structure extending between 2–3 areolae. In one sample from Lagos Lagoon the spines cannot be seen in the LM but are easily resolved in SEM. Semina (1981) reports *T. longissima* lacking spines and with terminal serrated ledges but it is not clear whether this taxon should be regarded as *T. longissima*. Her statement that it is not present in the Antarctic is incorrect – we have found it and so has Hallegraeff (1986), whose account together with that of Hasle & Semina (1987) provide a comprehensive survey of the genus.

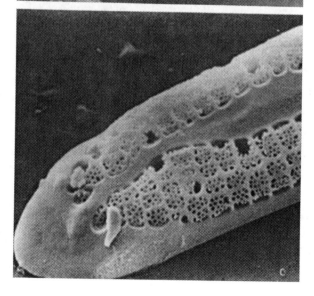

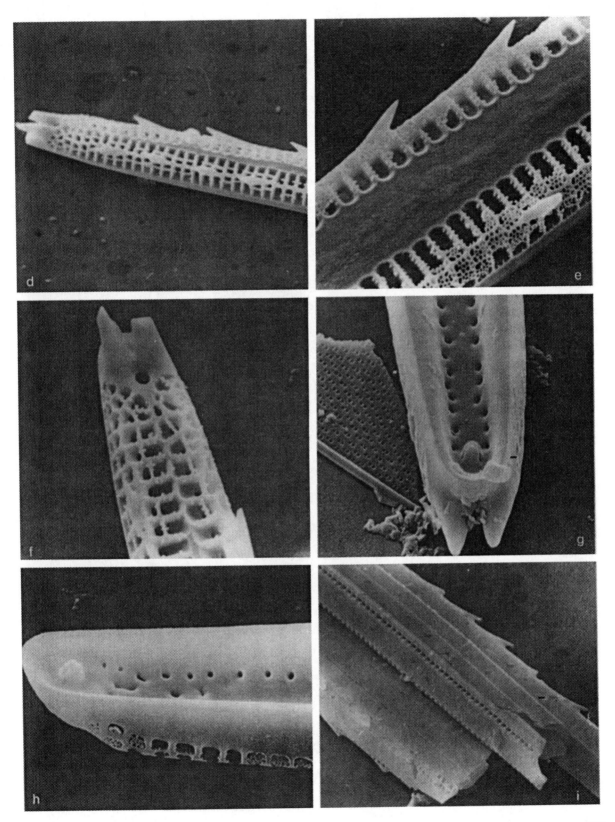

Trichotoxon F. M. Reid & F. E. Round 1987. Diat. Res. **2**: 224

T: *T. reinboldii* (Van Heurck) Reid & Round (= *Synedra reinboldii*)

Cells thin, extremely long, and bow-shaped;[arrow, a] solitary or in bundles. Plastids discoid, numerous. A marine planktonic genus probably confined to the southern hemisphere: abundant around Antarctica and displaced in water currents flowing northwards.

Valves linear, arcuate,[a] slightly inflated at the centre and apices, shallower at the poles than elsewhere. Valve face flat; mantles distinct. Sternum wide. Striae occupying the mantles and extending slightly over the junction between valve face and mantle; a few areolae can be found scattered on valve face.[d, e] Areolae elliptical to quadrate, with external cribra and opening internally by simple pores along the valve face and mantle junction;[f, g] the areolae form fairly large shallow cavities in the valve, the base of each cavity opening to the interior by one of the small simple pores. Internal views often reveal transverse indentations which may reflect the chambering. Rimoportulae 2, one at each pole, opening externally through simple pores;[c, d] internally large and sessile.[h, i] Cingulum consisting of several plain bands.

This genus has been erected for a taxon which was clearly out of place in *Synedra* and cannot reasonably be placed in any other needle-like marine taxon (see *Thalassionema* and *Thalassiothrix*). Other workers, e.g. Semina (1981) and Hasle & Semina (1987), also comment on the individuality of this taxon. Neverthless it undoubtedly belongs in the same family as these genera.

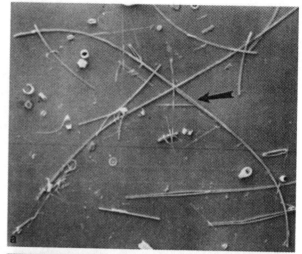

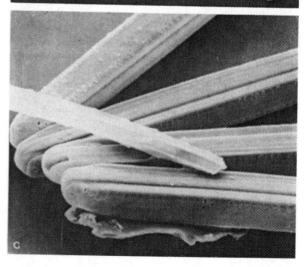

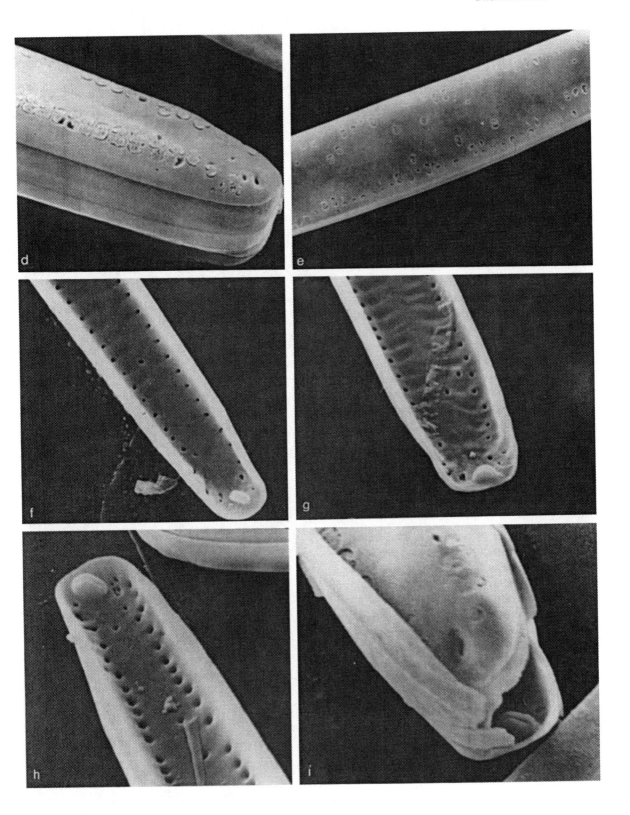

Rhabdonema F. T. Kützing 1844. Kiesl. Bacill. Diat.: **126** (nom. cons.)

T: *R. minutum* Kützing

Cells attached by their valve apices to form tabular or zig-zag colonies;[a] rectangular in girdle view.[a–d] Plastids one to many; stellate with a central pyrenoid. Epiphytic on marine plants around all coasts.

Valves linear to lanceolate. Valve face flat; mantles deep. Sternum narrow or inconspicuous externally, but massive internally and joined to equally massive transapical costae.[f, h, j] Striae uniseriate, containing round or elliptical areolae,[e] which are occluded at their outer ends by rotae or cribra, and which open internally into deep grooves between the costae. At the apices there are large apical pore fields (ocelli), which are quite distinct from the main framework of the valve;[e, g] there are also a few scattered simple pores around the valve mantle. The large rimoportulae are exceptional in their positioning, occurring along the sides of the sternum;[f, j] they have slit-like external openings, which are easily distinguished from the areolae. Cingulum deep and complex. There are several complete copulae, bearing several rows of pores with vela similar to those of the valve; in *R. adriaticum* the arrangement of the pores gives a false impression of interlocking segments.[d] These copulae bear septa[i] of various kinds and may also be loculate and have marginal flaps overlapping adjacent, more advalvar copulae.[k] The last-formed copulae are simplified and *R. arcuatum* has a series of 'scales' at the apices.

This is a very complex genus which von Stosch (1958b) considered from its reproduction and auxospore structure might be intermediate between the centric and araphid pennate genera, a view we cannot dispute from ultrastructural features. *R. minutum* is the third commonly recorded species and appears to have a simpler structure than the others; the genus as presently constituted is heterogeneous and has no obvious affinities to other diatoms. See also Pocock & Cox (1982) for further details of valve structure.

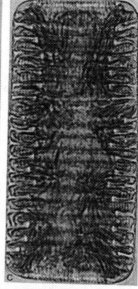

Striatella C. A. Agardh 1832. Consp. Crit. Diat.: 60

T: *S. unipunctata* (Lyngbye) Agardh
(= *Fragilaria unipunctata*)

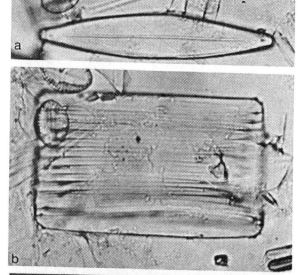

Cells secreting a stalk to which they are attached at one corner (appearing like a flag on a pole); tabular in girdle view with slightly truncated corners. Plastids rod-like, radiating from central pyrenoids. This is a fairly abundant epiphytic marine genus, but not well documented as to distribution, precise habitat, etc.

Valves lanceolate.[a, c, f] Valve face flat in the centre but merging imperceptibly with the mantles. Striae uniseriate,[a, e] orientated at right angles to a narrow sternum;[d, e, g] the areolae are staggered, giving a diamond-shaped pattern. In some valves the apical striae change direction and run towards the poles.[d, e] Areolae simple, tranversely elongate, opening externally by a narrow slit. A distinct sunken apical pore plate (ocellulimbus), with vertical rows of small circular pores (porelli), occurs at each apex surrounded by a rim of plain silica.[d, e] A few areolae occur below this ocellus and above is the slit opening of a rimoportula which is in line with the sternum.[e, f] In girdle view the corners of the cell appear cut off [b] – this is due to the positioning of the apical pore plate with its slight rim. Internally the sternum is raised and the rimoportula is a very broad-based, unstalked structure lying in the apical plane. Copulae numerous,[b, c, i, j] with short or long septa and slightly elongate areolae.

S. unipunctata, illustrated here, differs from the other species placed in this genus and we propose that for the moment *Striatella* be regarded as a monospecific genus. An illustration in Hasle (1974b) shows most distinctly how the mucilage stalk arises from the apical pore plate.

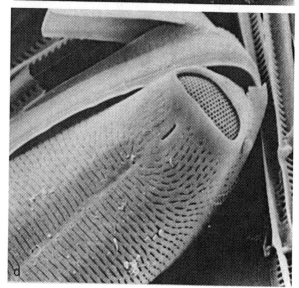

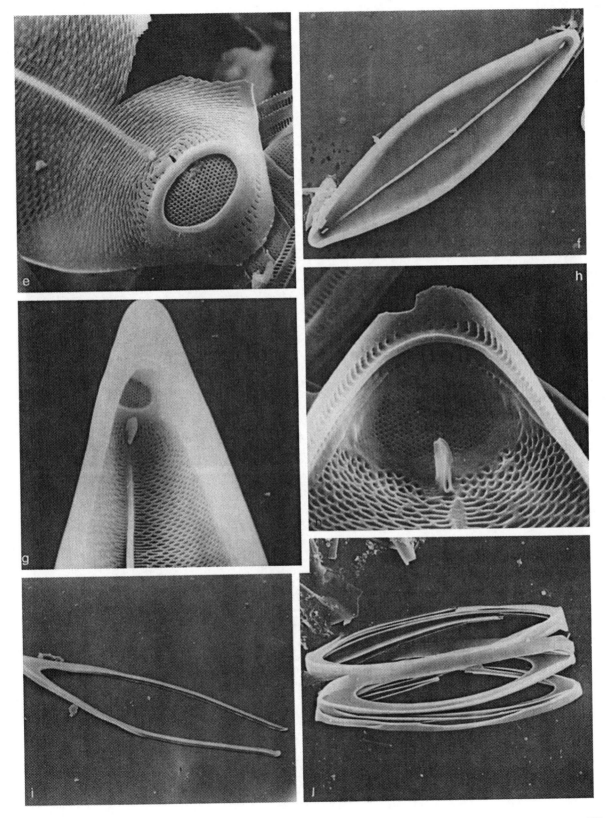

Microtabella F. E. Round, gen. nov.

T: *M. interrupta* (Ehrenberg) Round, comb. nov.
(= *Tessella interrupta*)

Cells attached at their poles, forming zig-zag colonies or attached along the length of the valves. In girdle view oblong, valve view elongate, slightly expanded centrally rounded or with capitate ends. Plastids 4, located in the centre of the cell. A widespread marine genus but little known ecologically.

Valves elongate with no or slight development of a sternum and areolae in parallel rows continuing down the mantle without change or with slight modification, e.g. doubling.[a, b, e] Valve mantle often increasing in depth in centre section. In some species, small spines occur, often flowing together to form plates or a ridge along the valve edge.[d, i] Large apical pore areas are present[f, j] and a slightly elongate exit pore of the rimoportula is sometimes distinguishable amongst the areolae near the apex. The connecting mucilage is extruded through the apical pore area. Internally the costae of the framework run smoothly across the valve;[g, h] in some species the costae are slightly expanded. Vela have not been observed. The slightest remnant of a sternum may occur near the apices where a rimoportula is located. Copulae numerous, with rows of elongate areolae. Last-formed copulae (central) tend to have fewer areolae. Septa either short or extending almost to the centre[a] of the copulae and may continue as a slight internal longitudinal ridge down the band.

Microtabella differs from *Striatella* in the form and arrangement of the areolae, virtual lack of a sternum, form of the apical pore plate and the presence of 4 independent plastids. The apical pore areas of *Microtabella* species are never sunken with a conspicuous plain rim as in *Striatella*. There are many taxa of small quadrate, epiphytic diatoms that remain to be studied and further splitting and additions of species/genera must be expected in this area.

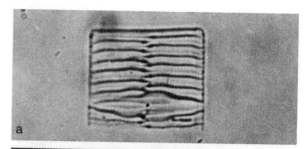

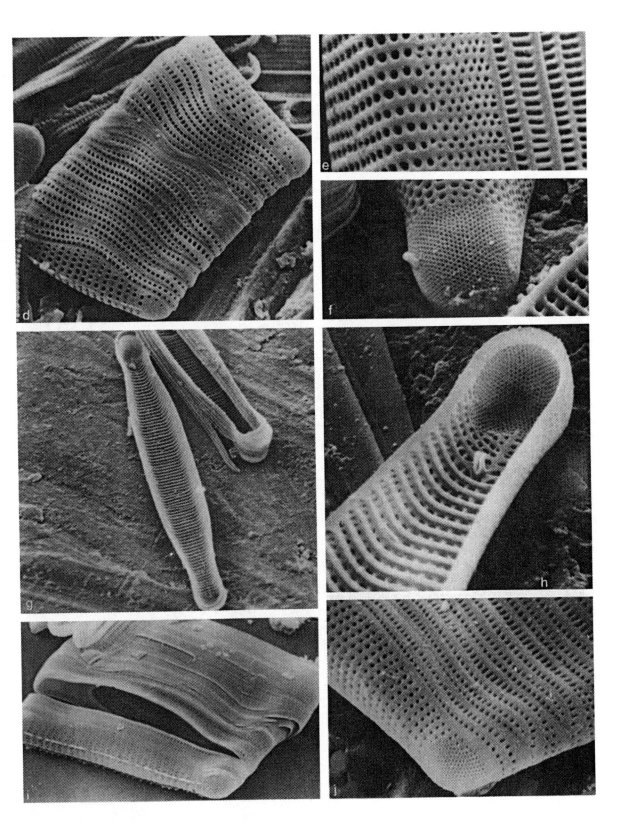

Grammatophora C. G. Ehrenberg 1840. Ber. Bekanntm. Verh. Königl. Preuss. Akad. Wiss. Berlin, 1840: 161

T: *G. angulosa* Ehrenberg (lectotype selected by C. S. Boyer, Proc. Acad. Nat. Sci. Philad. **78** Suppl.: 154)

Cells joined in zig-zag colonies,[a, b] living attached to marine macrophytes and other substrata. Square to rectangular in girdle view, with conspicuous undulate septa on the girdle bands.[a] Chromatophores 4 per cell, central with lobes filling the spaces between the septa. A small genus (less than 50 spp.) of unmistakable form, confined to the marine littoral and world-wide in distribution.

Valves elongate, sometimes undulate or even arcuate with a narrow, usually indistinct sternum,[f, g] and no distinct central area. Striae transverse, uniseriate,[e, f] leaving an area at each apex which is occupied by an apical pore field. Small spines[c] (and sometimes 2 large spines) occur on the apical mantle and in some species along the valve face edge. The striae areolae extend down the 'mantle' which curves without a break to the valve edge, which is often broad and plain. Rows of areolae also continue around the apices below the apical pore fields.[e, f] A slight development of the margin into pseudosepta occurs at the apices.[g] Areolae simple, poroidal. Two rimoportulae[h] per valve, each lying at an apex, internal[g] to the apical pore field. Copulae numerous,[c] either plain or with rows of areolae; all split except the valvocopula, which is a complete ring with prominent areas of elongate areolae around apices.[j] Each valvocopula bears two complex undulate septa,[j] which end in central thickenings or curve back towards the poles; at their apices the advalvar edges have complex flanges which fit around the valve mantle pseudoseptum. In some species the abvalvar edge of the valvocopula has plaques lying on it[c] (cf. those in *Fragilaria* except that there they are on the valve edge).

This genus is the marine counterpart of the freshwater *Tabellaria*, each being araphid, and having septa and zig-zag colonies. They occupy similar niches but do not appear to be closely related.

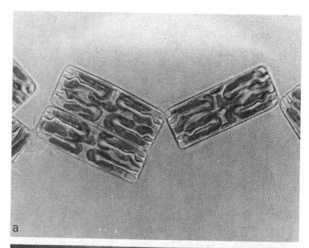

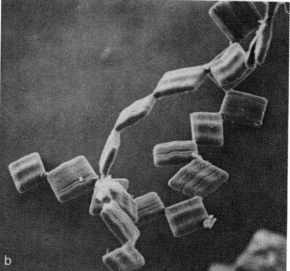

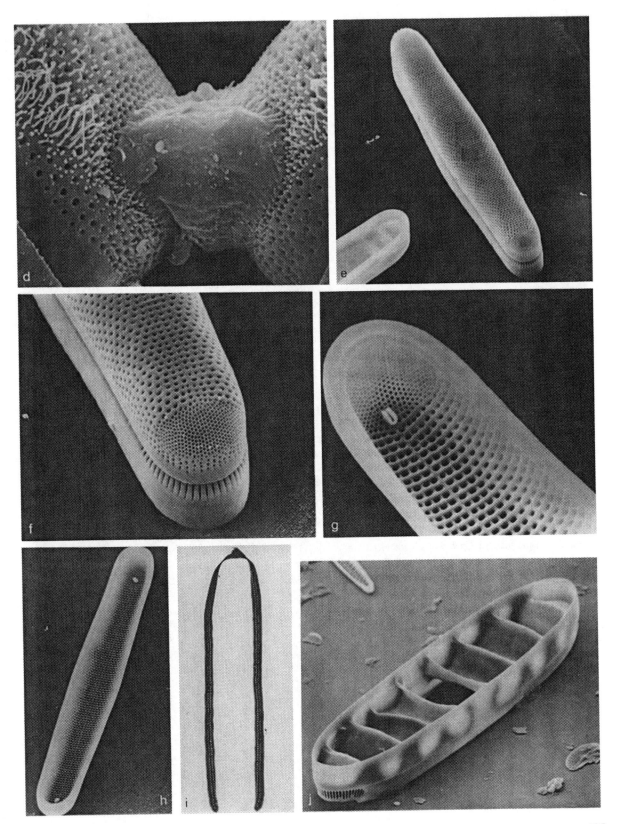

Cyclophora A. F. Castracane 1878. Atti dell'
Acad. Pont. de Nuovi Lincei, 31: 186

T: *C. tenuis* Castracane

Cells heterovalvar, forming zig-zag filaments
attached to the substratum by a terminal mucilage
pad. Plastids elongate, radiating from around the
centre (Hustedt, 1927–66). A rarely recorded marine
genus.

Valves linear with slightly capitate poles.[b-e] Valve
face flat centrally, curving into deep mantles.[f] One
valve has a central deepening of the mantle, while
the other has a complementary depression.[e-f] The
valve with the mantle expansion bears an almost
circular, cup-like internal thickening;[a, i, k] the other is
plain. Striae uniseriate containing simple poroidal
areolae. We have not been able to study the vela
properly but they appear to be located at the outer
apertures of the areolae.[g] Internally the rows of
areolae are separated by slight ribbing and the
narrow sternum is slightly raised.[i-k] At the apices
there are elongate curved slits, arranged in a
distinctive manner;[g] these appear to be unoccluded
and probably represent the pathway by which
mucilage is secreted. Rimoportulae with elongate
and slightly raised external openings and positioned
to one side of the sternum near the poles;[j] their
internal slits are transapically aligned. Copulae
numerous, bearing rows of elongate areolae[l] which
are most obvious in internal view.

The heterovalvy is an unusual feature in araphid
diatoms. The form of the apical pore field resembles
those in *Neosynedra* and *Protoraphis*. The
arrangement of plastids in *Cyclophora* (see Hustedt,
1927–66) has not been seen in other araphid genera,
though in *Striatella* the rod-like plastids radiate from
a central point. Further studies are required to
establish the affinity of *Cyclophora*. The SEM details
confirm Castracane's original view that this
distinctive genus is a member of the Fragilarioideae;
this was earlier maintained also by Hustedt (1927–
66) against other views.

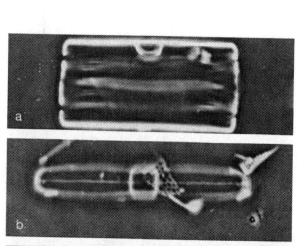

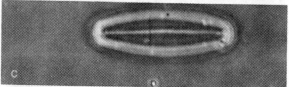

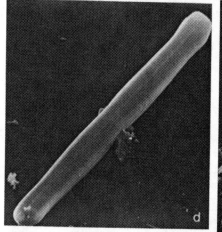

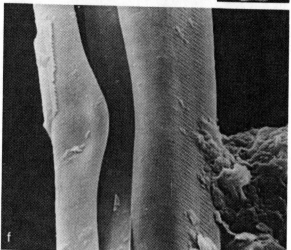

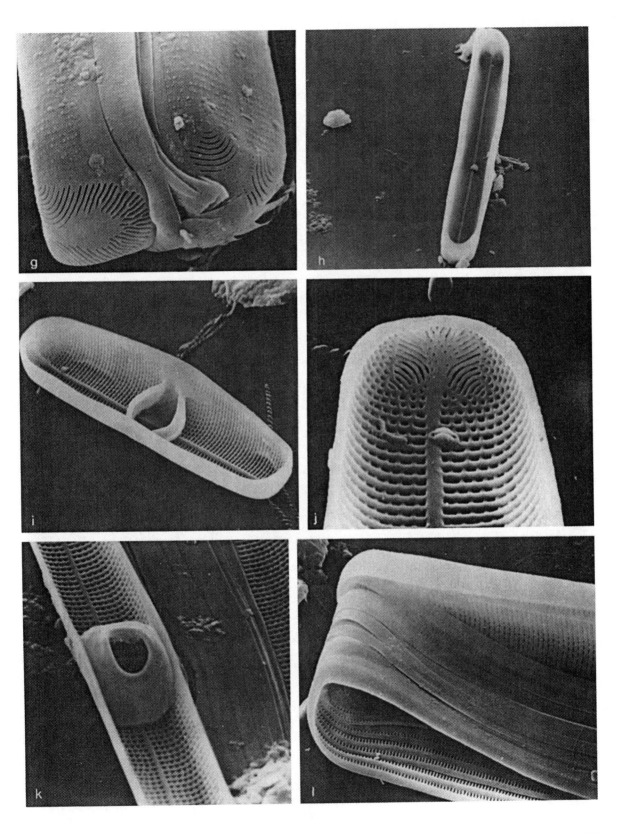

Gephyria G. A. W. Arnott 1858. Quart. J. Microsc. Sci. 6: 163.

T: *G. telfairiae* Arnott

Cells arcuate in girdle view[a, b] and thus with one concave and one convex valve; straight and isopolar in valve view. Plastids not observed. Occurs as an epiphyte on tropical (?) marine algae. Collected once only by the authors from *Sargassum* in Hawaii. Rarity and distribution unknown.

Valves elongate,[c, d] linear, apices rounded, with a fine sternum from which transverse striae run without a break across the valve face and down the curved mantle. Valves chambered. Externally there are biseriate striae[e, f] of small round pores; towards the valve margin the striae become triseriate.[f] These pores open internally into chambers separated by thick walls.[k] The chambers themselves open internally by large, somewhat irregularly-shaped apertures,[g–i] which form a more-or-less zig-zag row along each side of the valve. Extensive apical pore fields occur at each end of the concave valve but not the convex. The line of the sternum continues through two-thirds of the apical pore plate. One rimoportula is present on the edge of the pore field at each end of the concave valve, both lying on the same side of the axial area. Rimoportulae also occur on the convex valve in similar positions. Internally, each rimoportula is a large spade-shaped structure.[h] The copulae are split rings, chambered like the valves, with a single row of chamber apertures on both external and internal[j] surfaces; the internal apertures are very much smaller than those of the valve chambers. Cingulum composed of at least 2 copulae.

This is a most distinctive genus and, except for *Entopyla* (which we have not been able to investigate) the only bilaterally symmetrical 'araphid' genus with this arcuate type of morphology. Six valid genera listed by VanLandingham (1971) but only *G. media* has ever been regularly recorded. Other LM/EM can be found in John (1986). The chambering of the valves and the massive, chambered copulae are reminiscent of *Ardissonea* but *Gephyria* differs in the presence of rimoportulae and apical pore fields.

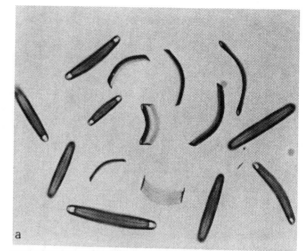

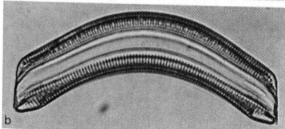

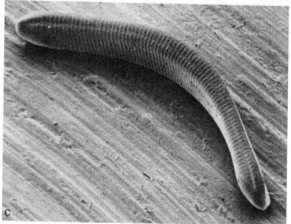

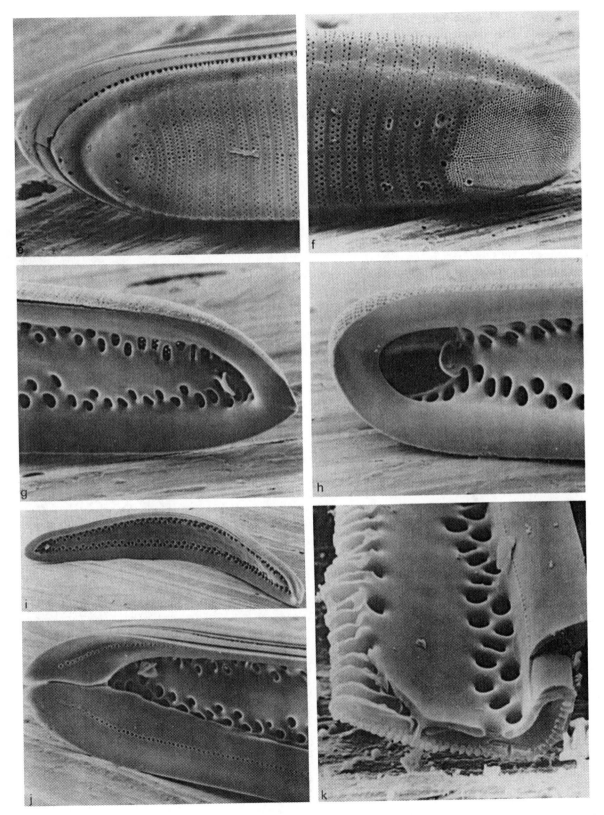

Climacosphenia C. G. Ehrenberg 1843. Abh. Königl. Akad. Wiss. Berlin 1841: 401

T: *C. moniligera* Ehrenberg

Cells clavate in valve[a-e] and girdle view, attached by branched mucilage stalks arising from the basal pole. Plastids numerous, small, plate-like. Epiphytic or epilithic in marine coastal habitats. Predominantly tropical or subtropical, extending into cooler water along some coasts.

Valves clavate,[a-e] apices rounded in valve view; without a median sternum, but with two faint lateral lines which are continuous around the apex;[f, g] internally these are seen as distinct costae.[h, j] The rows of areolae meet along the centre line of the valve forming a 'fault line'[f] (see Mann, 1984b). Valve mantle deep and at right angles to valve face. Areolae in straight uniseriate rows, which continue down the vertical valve mantle and radiate at the apices. Small spines occur on the valve face/mantle junction at the broader head pole.[g] Internally the areolae may show a slight discontinuity down the centre line[j] ('fault-line') though the thickenings between the rows of areolae tend to be continuous over much of the internal valve face.[j] On the internal, apical part of the valve mantle there are a series of irregular siliceous thickenings.[j] No portules or apical pore plates occur. Cingulum composed of a fimbriate valvocopula[m] bearing transverse septa[k, l] and two copulae. All bear rows of areolae as on the valve, but the areolae are more distinct, being larger and more widely spaced. On the valvocopula the pars media bears the septa whilst on the other copulae there is simply a break in the rows of areolae. The septa are solid, complete bars towards the head pole but grow together as ball-and-socket[i] or finger-like[l] joints towards the base-pole. Sometimes these lower joints are covered by a porous layer of silica.[k]

A common tropical diatom of which only two very closely related species, *C. elongata* and *C. moniligera* are reliably recorded: see Round (1982a) for greater detail and discussion. The genus is one of a small number of araphid genera that lack any obvious wall organelle involved in secreting the thick mucilage stalks. The occurrence of septa in this genus and in the naviculoid *Climaconeis* is a good example of convergence.

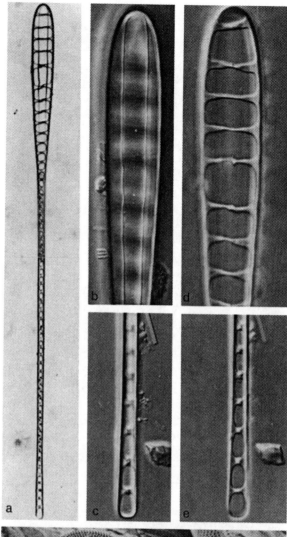

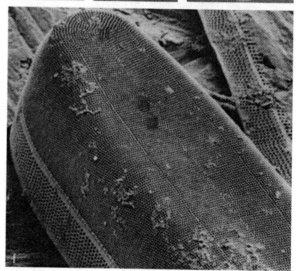

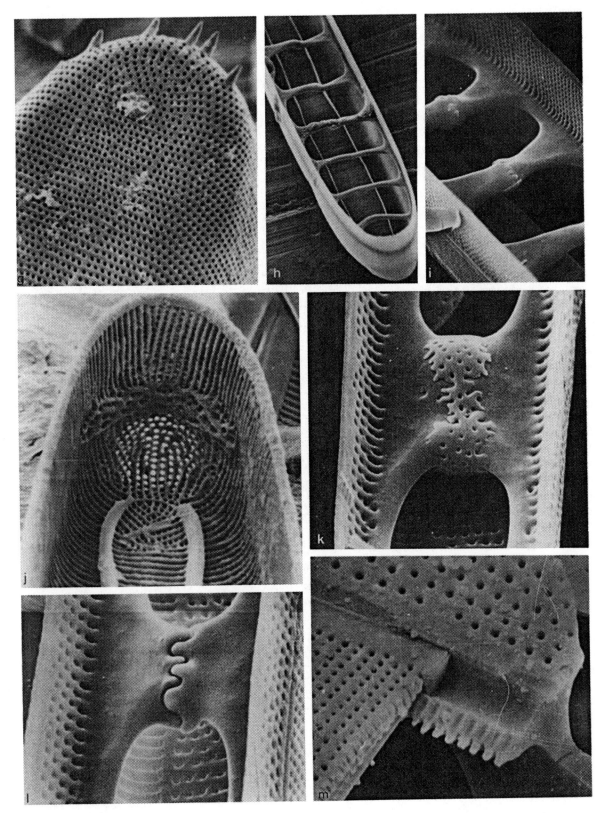

Synedrosphenia (Peragallo) F. Azpeitia Moros
1911. Asoc. Esp. Prog. Ci. Congr. 1908, **4** (2):
220 (= *Synedra* subgen. *Synedrosphenia*)

T: *S. giennensis* Azpeitia

Cells elongate, heteropolar in valve view;[a] not
observed in girdle view. We have only been able to
study SEM illustrations provided by D. Williams
and thus have no personal knowledge of plastid
structure or ecology. The genus is reported as
marine and from its form we would suspect it to be
epiphytic. VanLandingham recognises only 4 good
species and 5 unconfirmed – the genus is in urgent
need of re-investigation.

Valves cuneate, tapering to the foot-pole.[a–e] Valve
face transapically undulate,[c, d] fairly sharply
differentiated from the mantles. Striae uniseriate,
apparently subtended by 2 lateral sterna, one lying
at each edge of the valve face; as a result 2 opposing
sets of striae meet in an irregular manner down the
centre line. Internally the striae open between robust
transverse ribs, which may fork and fuse,[h, i] or
simply meet along the midline.[f, g] Near the head-
pole, at the widest part of the valve, the ribs are
somewhat irregular and further towards the head-
pole they are radiate. Internally the lateral sterna are
especially prominent, appearing as solid bars of
silica separating the ribs of the valve face from those
of the mantle[f] (cf. the situation in *Climacosphenia*).
Further investigations are required to determine the
nature and consistency of the bifurcation, fusion,
etc. of the rib system. The internal ribbing dies out
at the very tip of the head-pole[f] and foot-pole,[i, j]
leaving areas of simple pores (apical pore fields?).
There appears to be a slight pseudoseptum at the
valve apices. Areolae small, simple, circular.
Rimoportulae apparently absent. The cingulum has
not been examined in detail but the bands appear to
have vertical rows of pores which are smaller[d] than
those on the valve.

We have identified this material as *Synedrosphenia*
purely by comparison of the SEM illustrations with
the illustration in Hustedt (1927–66). We have
not detected any portules but the head-pole is
obscured somewhat by detritus. If it proves that
there are no rimoportulae *Synedrosphenia* would be
rather similar to *Climacosphenia*, differing in shape,
the absence of septae on the copulae, and the fact
that the copulae have smaller pores than the valve
(vice versa in *Climacosphenia*).

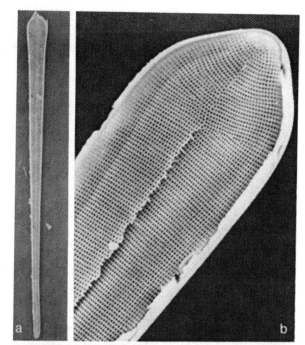

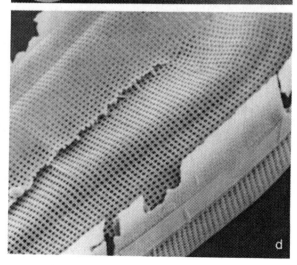

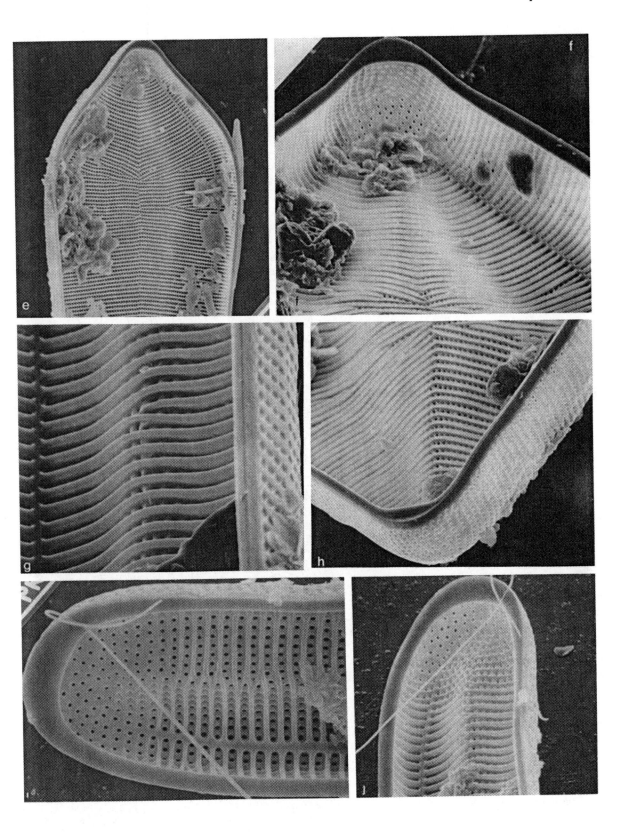

Pseudohimantidium F. Hustedt & G. Krasske
1941. Arch. Hydrobiol. 38: 272

T: *P. pacificum* Hustedt & Krasske

Cells lunate,[a] markedly dorsiventral, secreting stalks[b]
which attach the diatom to marine copepods of the
genera *Corycaeus*, *Farranula* and *Euterpina*. Division
of the cells results in the formation of branching
colonies. Plastids numerous (Gibson, 1979) but their
form is not clear. The genus is only known from the
epizooic habitat but it is clearly widespread in
temperate and tropical waters.

Valves lunate, with a flat valve face that curves
gently into the mantles; mantles higher on the
convex dorsal side[d-f] than on the ventral side.
Sternum prominent internally, off-centre,[d] ending in
a differentiated region at both apices. Striae
uniseriate; composed of transversely elongate
areolae; the vela have not been studied. At the
apices there are curved slits externally,[c] which
represent the common opening of 4–9
rimoportulae;[d-g] these are rather open-lipped
internally, their openings lying at right angles to the
row. In one valve[g] seen from the inside, between the
row of rimoportulae and the valve mantle there
appears to be a rimoportula placed parallel to the
row. The valves are heteropolar in that one end has
more rimoportulae than the other (8–9 or so, as
opposed to 3–4). Small scattered pores occur on
either side of the rows of rimoportulae and these are
the most likely points from which the mucilage
stalks arise – a view confirmed by sections in
Gibson's (1979) paper. Copulae split, porous, wider
on the dorsal side.

The taxonomy of this genus has been most
confused. Early workers, using light microscopy,
placed it in *Licmophora*, *Cymbella* or *Amphora*, and
later workers in *Hormophora* or *Sameioneis*, but it is
clear that Hustedt & Krasske were correct in
assigning it to a new genus (see Gibson, 1978, for
detailed discussion and early references). The
ultrastructure of Florida Current material of *P.
pacificum* has been described by Gibson (1979).
Simonsen (1970) proposed the family
Protoraphidaceae for this genus and *Protoraphis*.
This was perhaps an unfortunate name since it
implies a link between this group and the raphid
diatoms, and also an origin for the raphe that at
present cannot be proved.

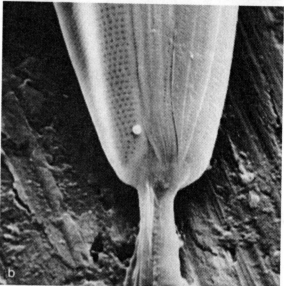

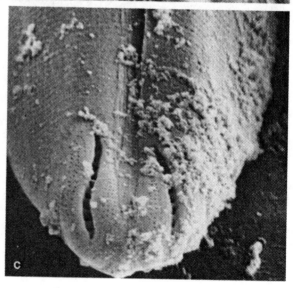

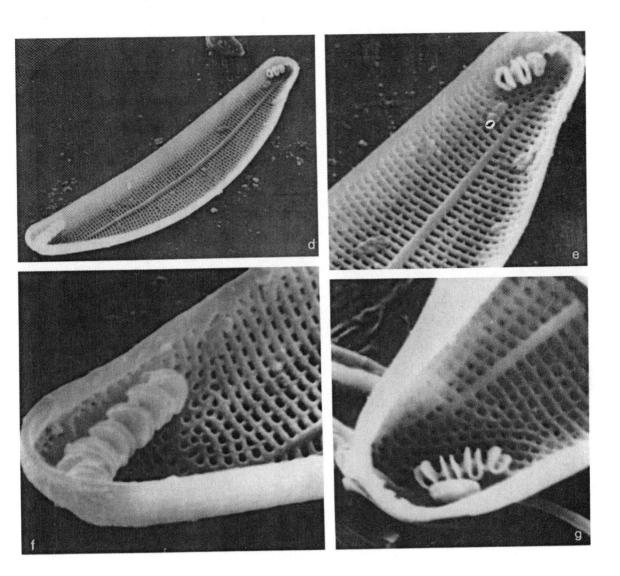

Raphid group

Our account of the raphid diatoms differs a great deal from the classifications given in the major floras (see p. 124). The changes involve the description of new genera (*Aneumastus, Luticola, Petrodictyon, Petroneis, Psammodictyon, Scolioneis, Seminavis*); the adoption of recently described new genera (e.g. *Gomphonemopsis, Pseudogomphonema*) or of revised concepts of old genera (e.g. *Berkeleya, Placoneis*); and the abandonment of some genera that appear to be superfluous. One major change from older classifications is that we do not recognise the monoraphid genera as a special group but merely an order alongside the many others in the raphid section — as did Hendey (1964) and Patrick & Reimer (1966). Details of our new or revised genera are of course written into the individual accounts. However, some general remarks are necessary since, if the *reasons* for change are not known and appreciated, the changes themselves will not gain acceptance.

Our researches have dealt not only with the morphology and anatomy of the silica frustule but also with the protoplast. Wherever possible we have determined the arrangement and structure of the plastids, how the plastids divide and when, and what kinds of movements the nucleus makes during the cell cycle. In many cases we have made observations of sexual reproduction or made reference to the extensive work done by Geitler (summarised 1973, 1984) in order to check the validity of conclusions drawn from frustule and protoplast data. Only a fraction of this information can be incorporated into the generic descriptions and unfortunately much of what we have discovered remains unpublished. An example may give an idea of the data we have attempted to gather and how it has guided our attempts to classify the diatoms. *Craticula* was originally established because of the production of peculiar skeletal internal valves but in the light microscope the structure of its normal valves appears very similar to *Navicula* and hence for over 100 years it has been included in *Navicula*. Not only are the valves apparently similar, but so is the interphase plastid arrangement; in both cases there are two plate-like plastids, one against each side of the girdle. More detailed examination reveals some further similarities. Furthermore, both genera produce a 1:2 ratio of *cis:trans* frustules, as a result of strict oscillation during consecutive cell cycles (p. 80 and Mann & Stickle, 1988). But in other ways *Craticula* and *Navicula* differ. We have noted in the generic descriptions that there are differences between them in the structure of valve areolae and raphe, in the porosity of the girdle bands, and in the form of the pyrenoid. We did not have space to add that in *Craticula* plastid division occurs later in the cell cycle than in most *Navicula* species and that rotation of the daughter plastids begins before they have separated fully; in *Navicula* rotation generally occurs well after separation. We also could not point out that there are differences in sexual reproduction. In the species of *Navicula* that have been investigated, the supernumerary nuclei produced in each gamete (as a result of the fact that only 1 or 2 gametes are formed following each meiosis) survive for a long time before degenerating; young auxospores are therefore often multinucleate (Karsten, 1896; Geitler, 1952a, 1952c, 1958b; Mann & Stickle, 1989). In *Craticula*, however, the supernumerary nuclei degenerate immediately after meiosis II. Then again, *Navicula* gametangia remain closely associated throughout meiosis, whereas *Craticula* gametangia lie somewhat separated within a wide mucilage envelope; and in *Navicula* the auxospores expand parallel to the apical axes of the gametangia, whereas in *Craticula* there is no fixed relationship. So for all these reasons, together with the original one, that *Craticula* produces special internal valves, we separate *Craticula* from *Navicula*. And it is because we also have information on sexual reproduction and protoplast dynamics in *Stauroneis* that we suggest that *Craticula* may in fact be more closely related to *Stauroneis* than to *Navicula*.

All the information upon which our judgements are based will have to be published before fellow phycologists will be able to see fully why our classification is as it is. We hope nevertheless that we have given enough information both to convince other workers that our changes are necessary and to enable them to use the classification. We hope too that our classification will be be tested by other taxonomists, using new characters and different methods of analysis. We must emphasise that in the raphid diatoms it is essential (and in fact very easy and enjoyable) to study the protoplast: the morphology of the frustule, even when thoroughly investigated, often does not yield enough characters to allow confident assessment of relationships (see also Mann 1989b).

We do not pretend that revision of the raphid genera is complete. We have separated many genera from *Navicula* and intend that *Navicula* should in future contain only the lineolate species. Unfortunately we have been unable to allocate all the *non*-lineolate species to our new genera. Some species remain unassigned because we do not yet know enough about them even to guess where they should go; we urge other workers to study these, using frustule *and* protoplast characters, so that all the necessary transfers can be made as soon as possible. Other naviculoid taxa are probably unrelated to any of the genera we recognise and further new genera will be needed. We believe, however, that the genera we describe cover by far the majority of the '*Navicula*' species that have been described. In other genera or families the situation is far less satisfactory. *Amphora* in particular requires attention and we intend to give work on this genus a high priority. Within *Amphora* there is a great diversity of valve, girdle and protoplast structure; some *Amphora* groups may prove to be asymmetrical versions of unrelated naviculoid taxa, as we have suggested here for *Seminavis* and *Biremis*. *Caloneis*, as presently circumscribed, with *C. amphisbaena* as type, cannot be distinguished from *Pinnularia* and we have not attempted to do so. Apart from a few misplaced species, this complex is clearly a natural group, with its characteristic alveolate valve structure, but further work will probably show that it could be split into a number of segregate genera, if this is considered desirable.

Eunotia C. G. Ehrenberg 1837. Ber. Bekanntm. Verh. Königl. Preuss. Akad. Wiss. Berlin, **2**: 44

T: *E. arcus* Ehrenberg (lectotype selected by C. S. Boyer 1927, Proc. Acad. Nat. Sci. Philad. **78** suppl.: 215)

Cells solitary or united into band-like colonies or occasionally attached by mucilage pads. Cells lunate or at least laterally asymmetrical in valve view and hence dorsiventral; usually with a relatively wide girdle and often lying in girdle view. Usually two,[a] but occasionally more (Geitler, 1958), elongate plastids, one lying against each valve and the part of the ventral girdle adjacent to the valve. The extant species are restricted to freshwater and are particularly abundant in the epiphyton and metaphyton of oligotrophic waters. The occasional cells found in marine fossil deposits dating back to the Eocene are surprising (possibly washed in).

Valves always asymmetrical about the apical plane,[b] often strongly curved. Valve face and mantle usually clearly differentiated,[d, f] rarely with short spines at their junction.[d] Striae often rather irregularly spaced and arranged about a narrow sternum, which lies near or at the edge of the valve face.[i] Striae uniseriate, containing small round poroids that apparently lack hymenes or other occlusions. Raphe slits very short, one lying at each pole.[h] Raphe not coincident with the sternum but to one side, on the ventral mantle;[e, h–j] little of it is visible in valve view in LM except an internal thickening corresponding to the helictoglossa. Raphe often continuing past the helictoglossa externally as a short terminal fissure, which curves towards the dorsal margin or back towards the centre of the valve.[h, j] Rimoportulae present, usually one per valve, lying near one apex;[k] rimoportulae of epivalve and hypovalve at opposite ends of the cell. Rimoportulae short internally and opening externally by a simple pore that is difficult to distinguish from the areolae. Cingulum containing several (6–10 in the species examined) similar, open, porous bands;[d–g] the band width and the number of transverse rows of poroids diminish from valvocopula to cingulum edge.

Von Stosch & Fecher (1979) have described the formation and characteristics of resting spores in *E. soleirolii*. Supernumerary thecae are formed by various species. A large genus in need of revision at the species level.

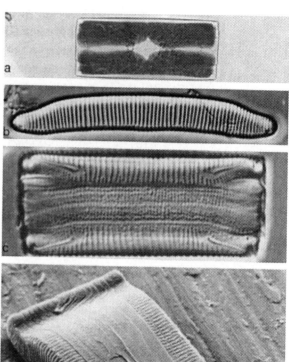

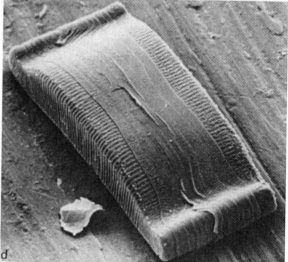

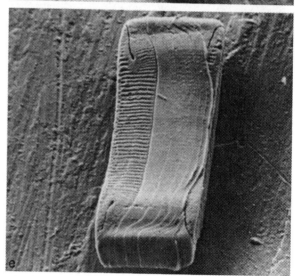

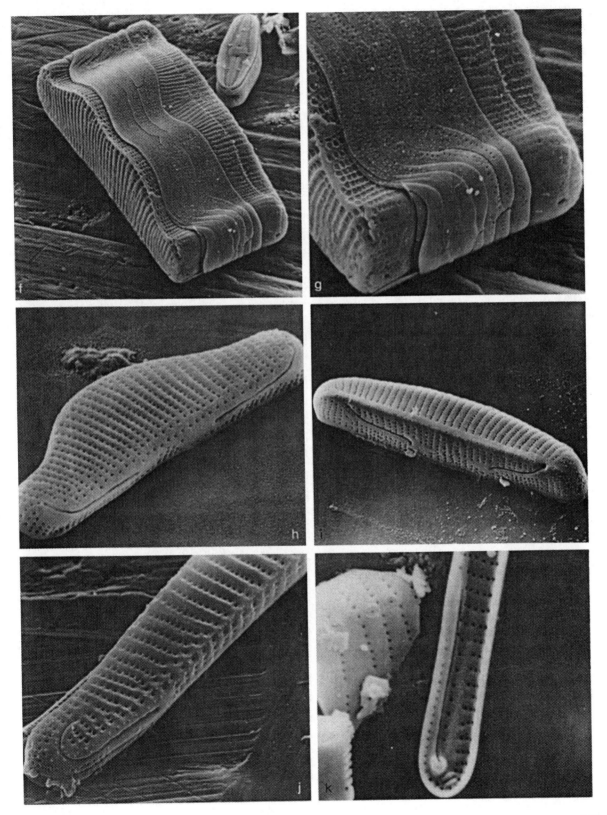

Actinella F. W. Lewis 1864. Proc. Acad. Nat. Sci. Philad. **15**: 343 (nom. cons.)

T: *A. punctata* Lewis

Cells solitary or clustered, slightly curved, expanded at one end (head pole end) in valve view,[a, c] attached to solid substrata by the narrow end. Free cells tend to lie in valve view because of the curvature of the valve. Plastid number and arrangement unknown. A small and rather rare genus found in extremely acid, humic waters; mainly tropical in distribution but also, for instance, extending up the east coast of the United States (Patrick & Reimer, 1966). Our material was collected in New Jersey.

Valves elongate, asymmetrical about the apical plane, swollen at one end and here with an apical indentation.[a, c, d] Junction of valve face and mantle furnished with a row of prominent spines. Striae uniseriate, containing small round poroids. Raphe slits short, lying near the poles on the ventral side; commencing on the valve face and turning to run along the concave side of the valve, ending close to the edge of the mantle.[d, e] Internally the raphe slits open only onto the mantle[g] so that the part of the raphe visible on the valve face externally is blind and comparable with the terminal fissures of other raphid diatoms. Rimoportulae present, one at each pole,[f, g] lying in the apical part of the valve mantle; external apertures of the rimoportulae not easily distinguishable from those of the areolae. Narrow and broad poles also furnished with narrow 'sterna' or breaks in the striae, which run along the valve parallel to the raphe. Girdle composed of open bands,[j] which bear two or more transverse rows of small round poroids;[i] where the bands open at the apical end of the cell, they do so at the most pointed part.[h]

The position and structure of the raphe slits, rimoportulae, stria pattern and overall form clearly ally this genus to *Eunotia*, as does the ecological restriction to extremely acid waters. It differs from *Eunotia* in its heteropolarity. Species with crenulate (undulate) margins have been reported (Schmidt 1874–1959) but need investigation.

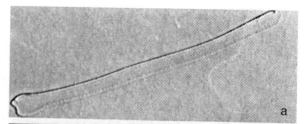

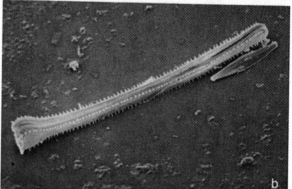

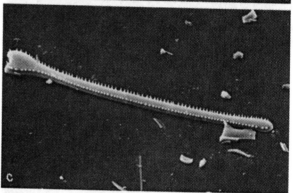

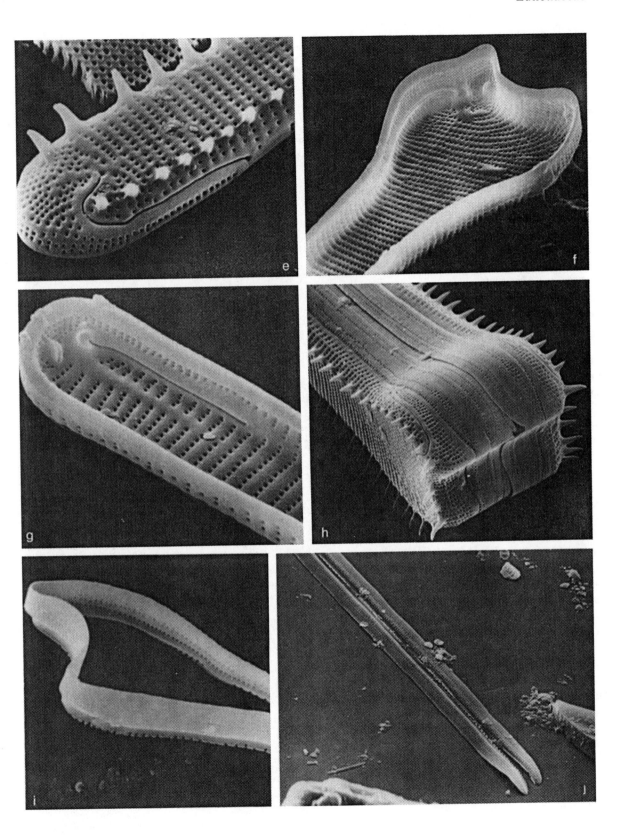

Semiorbis R. Patrick 1966. In R. Patrick & C. W. Reimer *The Diatoms of the United States* 1: 162

T: *S. hemicyclus* (Ehrenberg) Patrick
(= *Synedra hemicyclus*)

Cells lunate, attached to freshwater filamentous algae. The curvature of the cells means that in most instances only the curved valve view will be obvious. Plastids not observed. This is a monotypic genus first described as a *Synedra* and then as *Eunotia*, *Ceratoneis*, *Pseudoeunotia* and *Amphicampa*. It is relatively rarely recorded and seems to be confined to mountain habitats, often in places where *Eunotia* is common.

Valves lunate,[a-c] shallow, with conspicuous transverse ridges which extend upwards into spines along both sides of the valve. The ridges interdigitate at cell division, but do not appear to function as linking structures. Striae uniseriate[d-f] or partially biseriate (especially near the dorsal margin), lying between the ridges. The areolae are small round poroids; along the centre of the valve they are rather sparse. Short raphe slits are present at the poles[d-f] as in *Eunotia*; the terminal fissures turn across the valve and end on the convex side. Internally the raphe slit is visible only along the concave (ventral) edge and ends in a helictoglossa.[h] No rimoportula is present. Girdle bands have not been studied.

This small genus is clearly a member of the Eunotiales. It has been described in detail and compared with *Eunotia* by Moss *et al.* (1978) who proposed that the genus *Semiorbis* be retained; its ecology still requires study. More recently, Morrow, Deason & Clayton (1981) have suggested that *Semiorbis* should once more be included in *Eunotia* but we believe this is premature and should not be done until *Eunotia* itself has been revised. The genus was proposed by Patrick & Reimer (1966) but placed in the Fragilariaceae (though they admitted that it would have to be transferred to the Eunotiales if a raphe was confirmed) and separated from *Amphicampa*: we have not had the opportunity to study this genus.

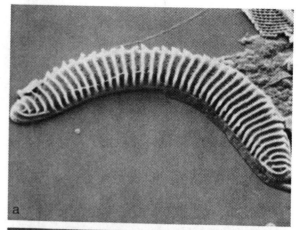

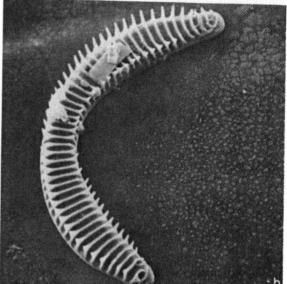

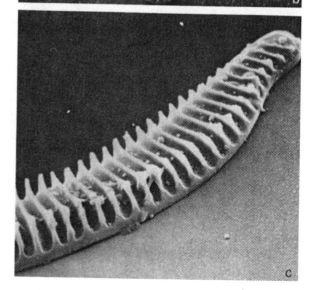

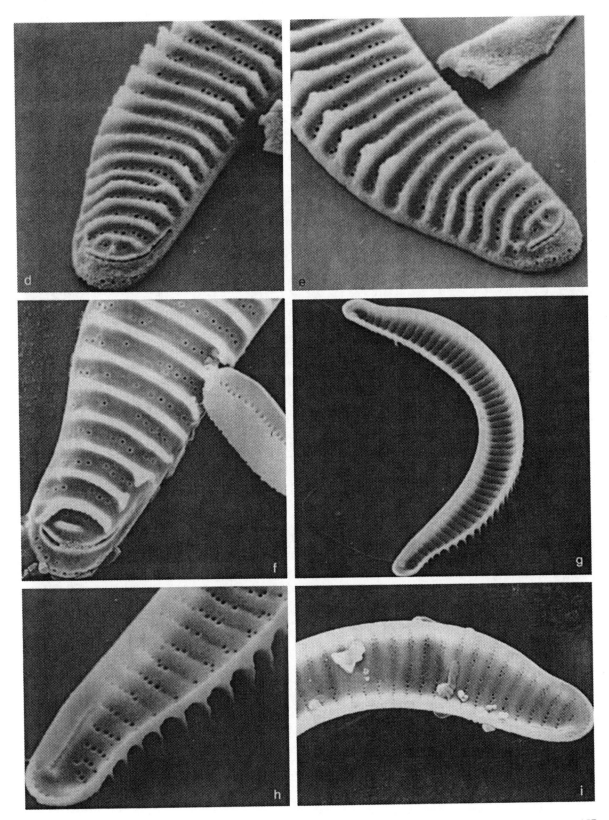

Peronia A. de Brébisson & G. A. W. Arnott ex
F. Kitton 1868. Quart. J. Miscrosc. Sci. Ser. 2
(8): 16 (nom. cons.)

T: *P. fibula* (Brébisson ex Kützing) Ross
(= *Gomphonema fibula*; = *P. erinacea* Brébisson
& Arnott ex Kitton nom. illeg.)

Cells solitary, cuneate in valve and girdle views,[a-d]
attached to aquatic plants by mucilage stalks. Intact
cells usually lie in girdle view. Plastids apparently
two laterally placed plates. An infrequently recorded
genus but abundant in acidic oligotrophic fresh
waters.

Valves linear, heteropolar[a-d] and often slightly
capitate[a] at the broader end. The uniseriate striae
are somewhat irregularly spaced and contain simple
poroids. They are transverse, becoming radial or
irregular at the apices.[f, g] Along the secondary side of
the valve the rows tend to be more disorganised.[f]
Small plaques occur along the edge of the mantle[e]
(cf. *Fragilaria*). As in *Eunotia* there is a very narrow
sternum between the raphe slits; in *Peronia*, however,
this is central.[f, i, k-n] Around the rim of the valve face
is a row of conical spines.[c-e] Where these have fallen
off, small round projections with a central
indentation are left.[g] *Peronia* exhibits a type of
heterovalvy similar to that in *Rhoicosphenia* although
these genera are not closely related. One valve of
each frustule (the R-valve) has a well-developed
raphe system occupying two-thirds of the valve
length.[m] In the other (the P-valve) the raphe slits are
diminutive and often missing at the head-pole.[j] The
R-valve has a rimoportula at each pole between the
helictoglossa and the valve apex,[k, n] while in the
P-valve there is only one rimoportula, lying at the
base-pole. The raphe is simple and lacks terminal
fissures. The 'central' endings are expanded and
curved slightly to one side.[f, i] Apical pore fields are
apparently absent. The cingulum consists of at least
five porous open bands.[e] These are narrower and
have fewer transverse rows of poroids than in the
related *Eunotia*.

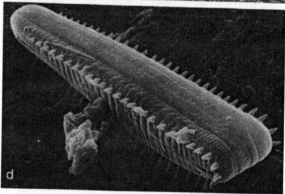

Peronia is closely related to *Eunotia* although the
raphe system of the R-valve is more complex than
anything found in that genus. It is of considerable
interest that these two genera are often abundant in
the same highly acidic waters. Patrick & Reimer
(1966) stated that the cells of *P. fibula* occur in
curved filaments but we have never observed this.

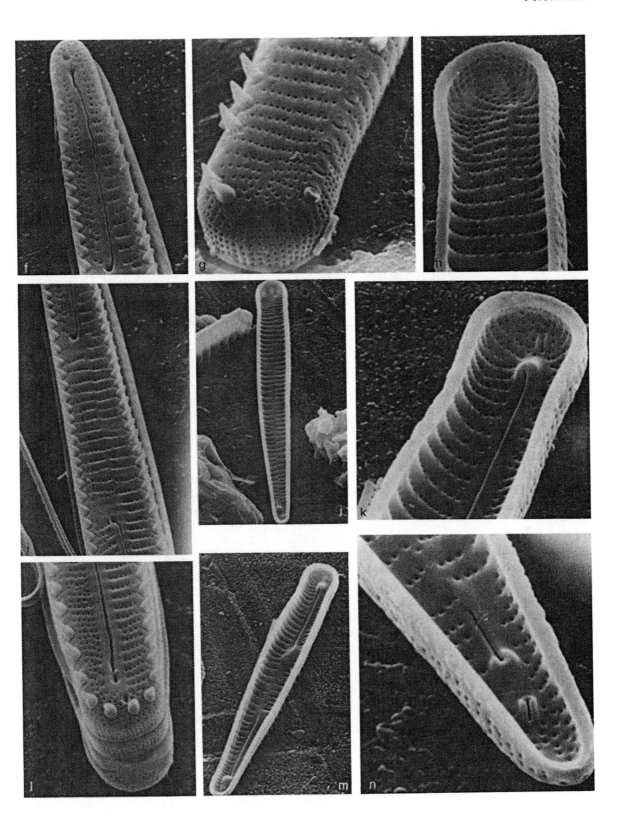

Lyrella N. I. Karajeva 1978. Bot. Zh. **63**: 1595

T: *L. lyra* (Ehrenberg) Karajeva (= *Navicula lyra*)

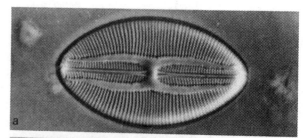

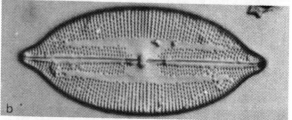

Cells solitary, naviculoid, usually lying in valve view.[a-d] Where known, 2 or 4 plastids per cell. If two plastids, each usually composed of two lobed plates, one against each valve, connected by a narrow isthmus lying near the central nucleus; or two lobed plates as in *Petroneis*. If four plastids, two lying against each valve, disposed symmetrically about the median transapical plane. A marine group, commonest in the epipelon of sandy sediments.

Valves heavily silicified, broadly linear or lanceolate, with bluntly rounded or broadly rostrate poles. Valve face flat or slightly undulate,[c, d] curving into relatively shallow mantles. Striae more-or-less coarse, uniseriate, containing round poroids occluded internally by siliceous flaps;[j] hymenes absent. Striae interrupted by a lyre-shaped, non-porous, thickened area of silica,[c-f] which is often slightly depressed below the general level of the valve face and may be ornamented externally with ribs or warts. Raphe system central, straight.[a-d] Terminal fissures present, turned towards the secondary side.[d, g, h] At the external centre, the raphe fissures open into oblanceolate grooves;[e] the endings themselves are simple. Internal central endings hooked or T-shaped.[j] Girdle containing a few open bands (where known, 3 per cingulum), each bearing one transverse row of poroids near the junction of pars exterior and pars interior.[g, h]

Lyre-shaped 'hyaline' areas appear to have arisen independently in several raphid groups, including *Lyrella*, *Cocconeis* and *Mastogloia*. *Lyrella* is undoubtedly closely related to *Petroneis* but the plain lyre-shaped area (sternum) and details of the areola structure serve to separate them. The smaller lyrate naviculoids allied to '*Navicula*' *pygmaea* do not belong here, differing in valve, areola, plastid and raphe structure; these are separated into *Fallacia* (q.v.). *Lyrella* species are known from sediments dating back to the Eocene and are probably preserved well because of their heavy silicification. The genus was established for the type, *L. lyra*, by Karayeva (1978a), but few other combinations appear to have been made until now.

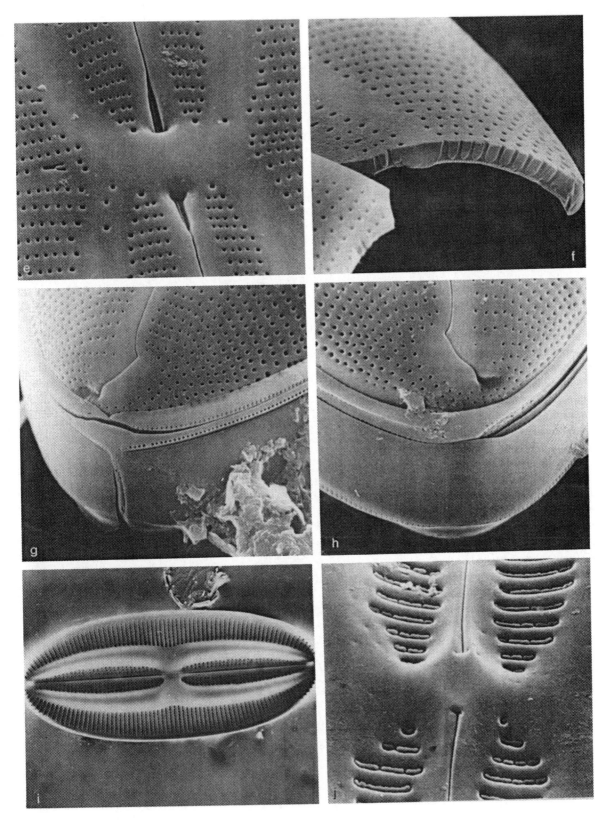

Petroneis A. J. Stickle & D. G. Mann, gen. nov.

T: *P. humerosa* (Brébisson) Stickle & Mann, comb. nov. (= *Navicula humerosa*)

Cells solitary, naviculoid, usually lying in valve view. Where known, two large butterfly- or X-shaped plastids per cell, one against each valve; plastid margins often lobed. Plastid containing curved, elongate pyrenoids. Marine, epipelic; particularly common on sandy sediments.

Valves linear to elliptical, broad, often with rostrate poles;[a-c] heavily silicified and coarsely patterned. Valve face flat, curving gently into shallow mantles.[b, i] Striae uniseriate,[b, c] usually more-or-less radial, containing large round or transapically elongate poroids. Poroid occlusions complex, volate;[f-h] frequently the small flaps (volae) are borne on elaborate cribra.[g] No hymenes present. Raphe-sternum central,[a-c] straight, expanded centrally into a round or rectangular area.[e, f] External raphe fissures opening centrally into an oblanceolate, spathulate or T-shaped groove;[c, d] external central raphe endings themselves simple or slightly expanded terminally. Terminal fissures present, turned towards the secondary side.[i] Internal central raphe endings shaped like a shepherd's crook,[f] both curved to the same side. Girdle containing a few open bands, each bearing one or two transverse rows of large poroids.

A widespread and ancient group, common in sandy sediments in temperate or tropical seas. Most closely related to *Lyrella* but differing from it in the absence of a lyre-shaped plain area, and in details of raphe and areola structure. Species of *Petroneis* have usually been classified in *Navicula* in the section *Punctatae*, but they differ from *Navicula sensu stricto* in almost every respect except the central position of the raphe.

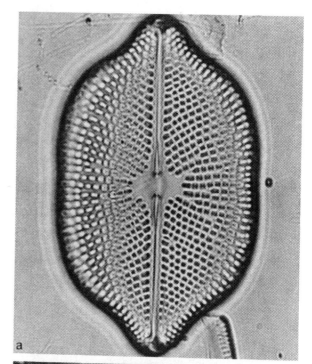

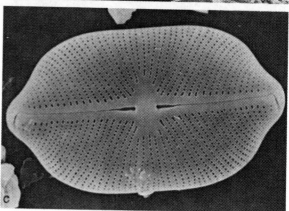

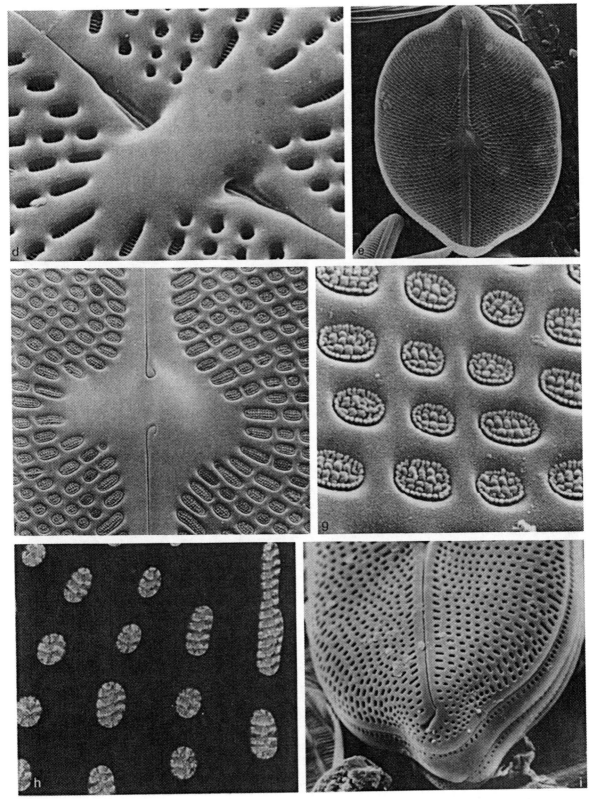

Aneumastus D. G. Mann & A. J. Stickle, gen. nov.

T: *Aneumastus tusculus* (C. G. Ehrenberg) Mann & Stickle, comb. nov. (= *Navicula tuscula*)

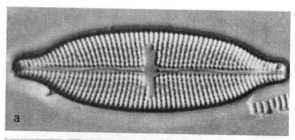

Cells solitary, usually lying in valve view, with 2 plastids arranged one on either side of the median transapical plane. Plastids H-shaped in girdle view, consisting of 2 plates appressed to the valves, connected by an isthmus containing a large pyrenoid; plates invaginated beneath the raphe. A small genus occurring in the epipelon of basic freshwaters.

Valves naviculoid,[a, b] lanceolate, often with rostrate or somewhat capitate poles. Valve face flat,[b-d] fairly abruptly demarcated from the shallow mantles. Striae uniseriate throughout, or becoming biseriate near the edge of the valve face. Areolae complex, opening into deep pits internally[f-i] via sieves of small pores; occluded externally by flaps of silica[b-d] borne by the areola walls. Valve margin thickened, rib-like.[h] Raphe-sternum fairly narrow, especially externally, but widening at the centre to form an elliptical or rectangular central area. External raphe fissures sinuous.[b] External central endings expanded and then turned slightly towards the opposite side of the valve from the terminal fissures, which curve around, ending close to the valve margin.[d] Girdle narrow, consisting of open bands. First band (valvocopula) modified to fit around the thickened valve margin. Its pars interior is finely porous.[i] A series of tiny chambers is formed in it around the edge of the cell, which open to the outside by a single row of round pores in the pars exterior.

Aneumastus has usually been included in *Navicula* (lately as the subgenus *Tuscula*), but the valve and raphe structure, together with the kind of plastid arrangement, show clearly that its affinities lie with *Mastogloia* as was recognised by Hajós (1973). *Aneumastus* lacks the complexity of the system of bulbous chambers found in the girdle of *Mastogloia* but otherwise these genera are very similar.

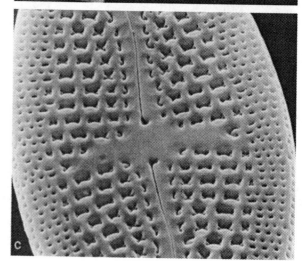

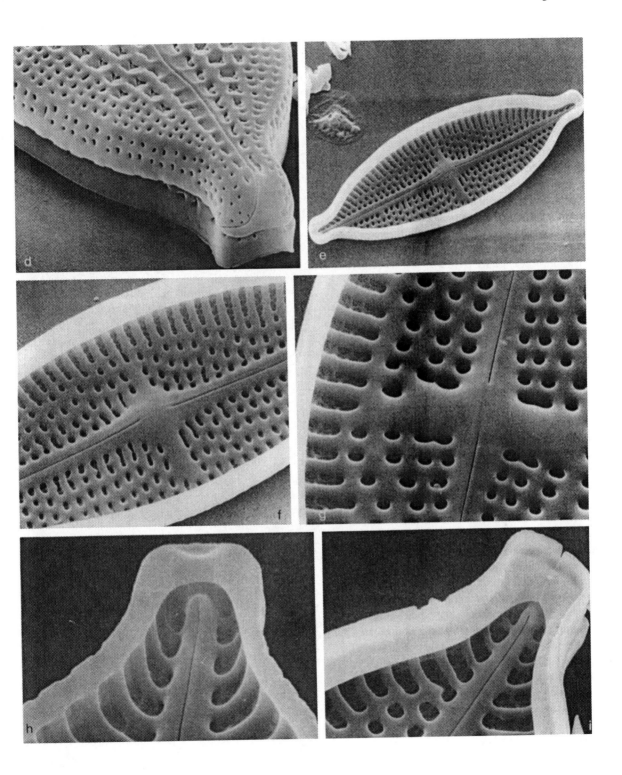

Mastogloia G. H. K. Thwaites ex W. Smith
1856. Syn. Brit. Diat. 2: 63

T: *M. dansei* (Thwaites) Thwaites ex W. Smith
(= *Dickieia dansei*; lectotype selected by C. S.
Boyer 1928, Proc. Acad. Nat. Sci. Philad. **79**
Suppl.: 327)

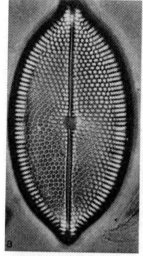
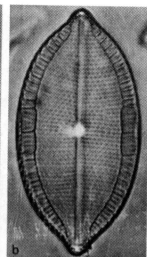

Cells naviculoid, solitary or forming gelatinous,
bubble-like colonies. Single cells lying in valve or
girdle view. Two plastids, placed one towards each
pole, consisting of two valve-appressed plates
connected by a bridge containing a large pyrenoid.
An epipelic or epiphytic genus containing many
species; principally marine but extending into
brackish and fresh waters.

Valves isopolar or slightly heteropolar, linear to
elliptical, sometimes with apiculate apices.[a–c] Valve
face flat, curving into fairly shallow mantles;
conopea sometimes present. Valve margins often
thickened, especially at the poles. Striae uni–[c] or
biseriate, often changing in structure across the
valve[e, f] and sometimes interrupted by lyre-shaped
lateral sterna.[h] Areola structure extremely variable
and complex, often loculate. Areolae closed by
cribra[k] or volae, without hymenes. Raphe-sternum
central,[a] narrow, sometimes extended into flaps
externally which overlap the raphe such that the
external raphe fissures (or even the whole raphe) are
often sinuous.[d] External central endings often
expanded; polar endings with or without hooked
terminal fissures. Raphe-sternum often thickened
internally;[h–j] central endings usually straight and
simple, sometimes ending in helictoglossae.[g, h]
Girdle consisting of open bands and having a
unique structure. The first band (valvocopula) bears
one to many large chambers (partecta), which open
externally via apertures close to the valve margin.[c, e, f]
The aperture can be distant from its chamber, in
which case the two are connected by a duct (see also
Novarino, 1987). The chambers have porous
internal walls and are concerned with the
production of mucilage strands (see Figs 22b, c). The
bands may also bear one or more rows of simple
pores.

Several species have been described by Stephens
& Gibson (e.g. 1979a, b, 1980) and the architecture
of the partecta by Stoermer *et al.* (1964). *Mastogloia* is
very distinctive and natural, but extraordinarily
variable in areola structure. The chambered
valvocopula is paralleled only by the much simpler
structure in *Aneumastus*.

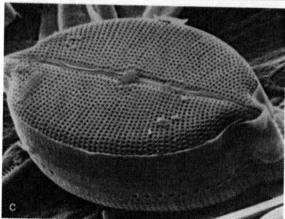

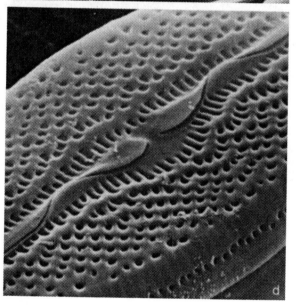

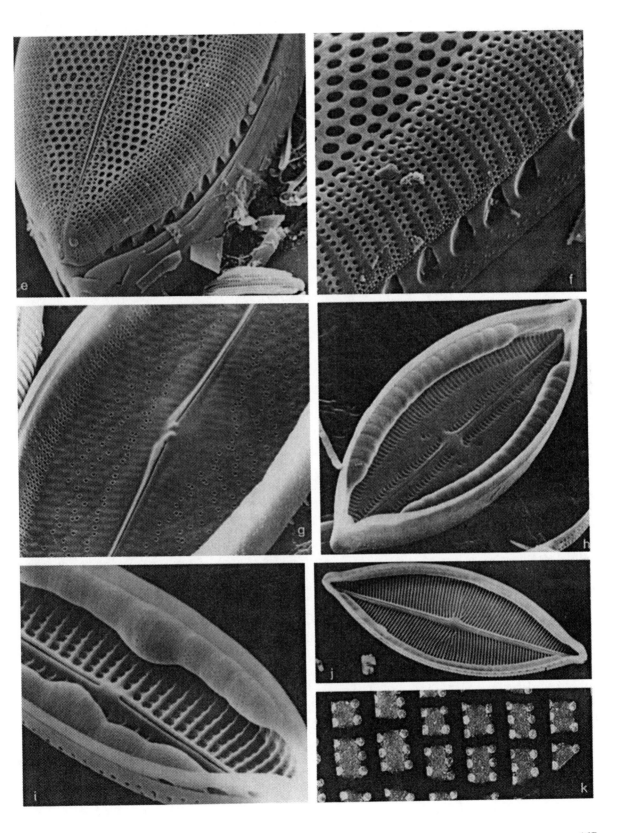

Dictyoneis P. T. Cleve 1890. Diatomiste 1: 14

T: *D. marginata* (F.W. Lewis) Cleve (= *Navicula marginata*; lectotype selected by C. S. Boyer 1928, Proc. Acad. Nat. Sci. Philad. **79** Suppl.: 343)

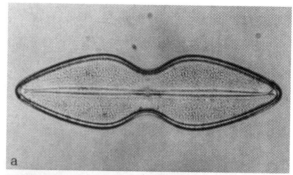

Cells solitary, naviculoid, usually lying in valve view. Plastids unknown. A marine epipelic genus, apparently more widespread in the tropics than in temperate seas. Only the type is recorded at all frequently; other species rare or fossil.

Valves panduriform,[a, b] with acutely rounded poles. Valve face curving gently into fairly deep mantles. Valve framework loculate.[h, j] Inner layer of valve penetrated by uniseriate striae containing small round poroids. This layer is exposed externally near the central endings[c, g] and poles[d] but elsewhere is overlain by a more coarsely structured layer. This outer layer is also penetrated by uniseriate striae containing round or oval pores, but the striae are more distantly spaced and the pores are larger than in the inner layer;[c–e] near the margin they are often expanded into longitudinal slits or grooves. Fractures, such as the one we illustrate[j] may give a false impression of a longitudinal canal near the margin of the valve. Internally, the small poroids of the inner layer are difficult to distinguish because of the strong development of the ribs separating one row of poroids from the next.[h–k] Raphe-sternum narrow, central, thickened internally[h, i] and slightly expanded in the middle. Raphe slits bordered externally by thin ridges of silica.[c, f] Terminal fissures present, deflected to opposite sides at the two poles so that the raphe is effectively slightly sigmoid. External and internal central endings simple or slightly expanded. Girdle composed of open porous bands, bearing small indentations externally.[c–e]

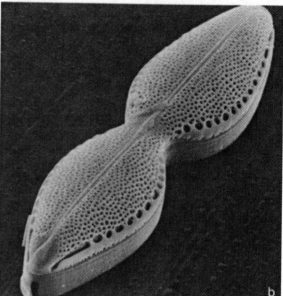

The relationships of this very distinctive but infrequently recorded genus are unknown. The valve structure is not like that in any other raphid genus, except perhaps *Mastogloia*. We have reflected its apparently isolated position by putting it in a monotypic order, the Dictyoneidales.

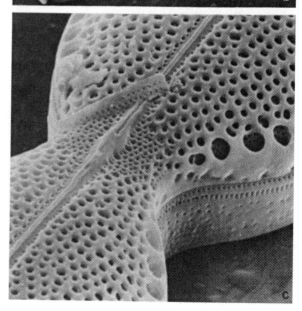

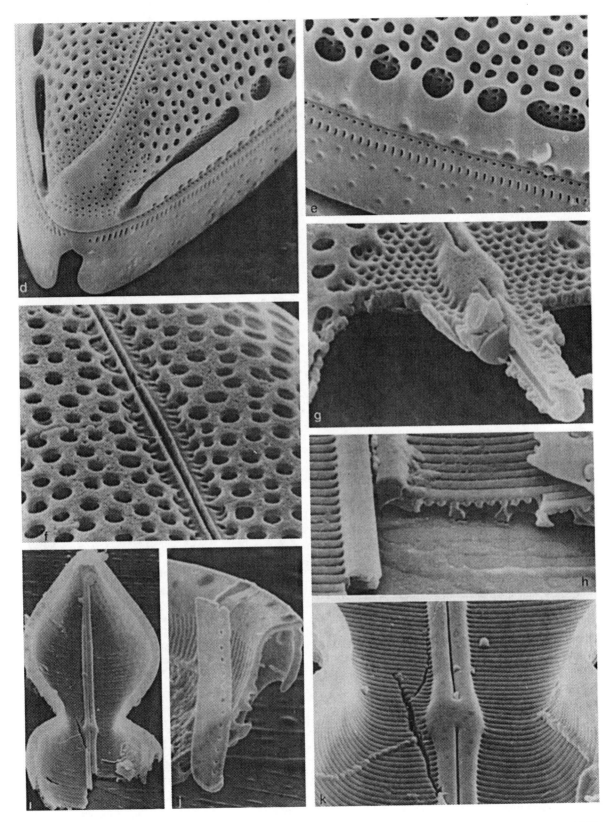

Rhoicosphenia A. Grunow 1860. Verh. zool.
-bot. Ges. Wien **10**: 511

T: *Rh. curvata* (Kützing) Grunow
(= *Gomphonema curvata*)

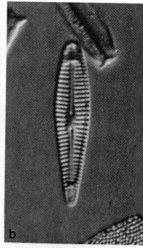

Cells usually colonial, stalked; when detached,
usually lying in girdle view. Heteropolar[h] or
isopolar, flexed in girdle view.[a] One lobed plastid,
lying with its centre against one side of the girdle
and extending beneath both valves and part of the
opposite girdle. A small epiphytic and epilithic
genus of marine and fresh waters.

Valves dissimilar in shape and structure,[d, e]
heteropolar or isopolar, linear to linear-lanceolate.
Striae uniseriate or biseriate, containing apically
elongate poroids closed by hymenes. At one or both
poles of one or both valves the striae are much more
closely spaced and contain unoccluded pores,[c]
through which the mucilage stalks are secreted.
Valve margins thickened and forming inwardly
projecting pseudosepta at the poles.[i] Raphe system
differently developed on the two valves. The
concave, ventral valve has a full raphe system;[d] the
convex, dorsal valve has a reduced system, with
short raphe slits lying close to the poles.[e, g] Internal
raphe endings similar in both valves: at the poles
are helictoglossae, while the central endings are
hooked. External central endings slightly expanded.[d]
Slightly curved terminal fissures present,[f] except in
the heteropolar species, where the broader pole of
the convex valve often has no terminal fissure.
Girdle consisting of open bands (3–5 per
epicingulum), which each bear one transverse row of
poroids or are plain.[j–l] The valvocopula is often
modified[k] to fit around the thickened valve margin
and interlocks with it strongly *in vivo*.

This genus has been studied in detail by Mann
(1982a, b, 1984d) and Medlin & Fryxell (1984a, b).
According to Lange-Bertalot (1980b) *Rh. curvata* is a
later synonym of *Rh. abbreviata* but there are races
within this group which may require recognition at
species level; hence we retain *Rh. curvata* for the
present.

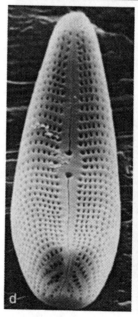
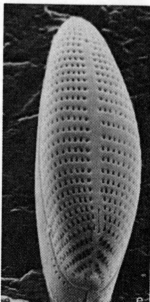

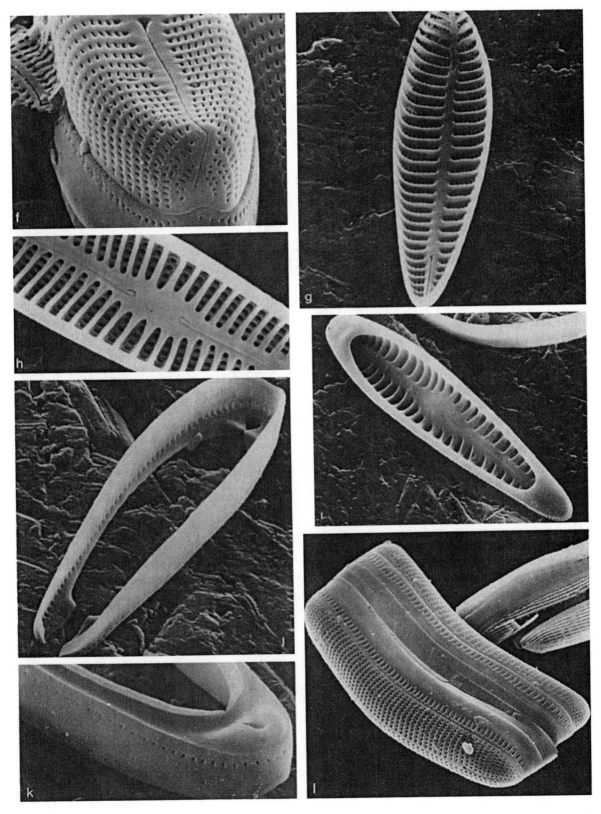

Campylopyxis L. Medlin 1985. Br. phycol. J. **20**: 318

T: *C. garkeana* (Grunow) Medlin (= *Rhoikoneis garkeana*)

Cells solitary, usually lying in girdle view[a, b] and then appearing flexed; attached. Plastids unknown. A small marine genus, found growing on seaweeds.

Valves linear to linear-lanceolate with bluntly rounded apices.[c] One valve with a strongly convex valve face;[d] the other valve with a concave valve face. In neither is the valve face sharply differentiated from the mantles. Valve margin everywhere very strongly thickened[i] but not forming true pseudosepta at the poles. Striae uniseriate,[d-h] continuing around the valve apices,[g, h] containing round poroids occluded internally by hymenes; no pore fields present at the poles. Raphe-sternum central.[c] Raphe-slits running the full length of both valves; central raphe endings not widely separated.[f] External central raphe endings expanded, pore-like; internal endings hooked towards the primary side.[j] Terminal fissures present only on the ventral valve,[h] deflected. Cingula containing four open bands,[k] each with one transverse row of elongate poroids.

The only known species, *C. garkeana*, was originally allocated to *Rhoikoneis* but later transferred to *Navicula* sect. *Microstigmaticae* (which has now been removed in part into *Parlibellus*). The valve, raphe and girdle structure show, however, that *Campylopyxis* is most closely related to *Rhoicosphenia*; the only characters separating these genera are the greater heterovalvy of *Rhoicosphenia* and the absence of a true pseudoseptum. *Campylopyxis* is described and discussed in detail by Medlin (1985). The superficial similarity to *Achnanthes* and *Achnanthidium*, and the true affinity with isopolar *Rhoicosphenia* species is likely to cause confusion unless great care is taken with the identification of all marine 'achnanthoid' species. Many old records of these forms are probably wrong.

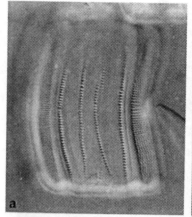

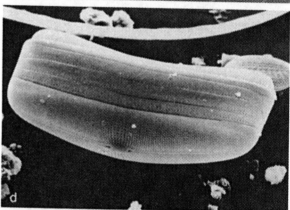
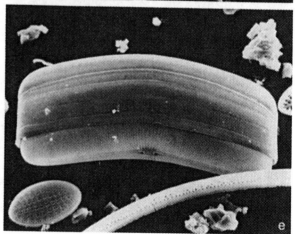

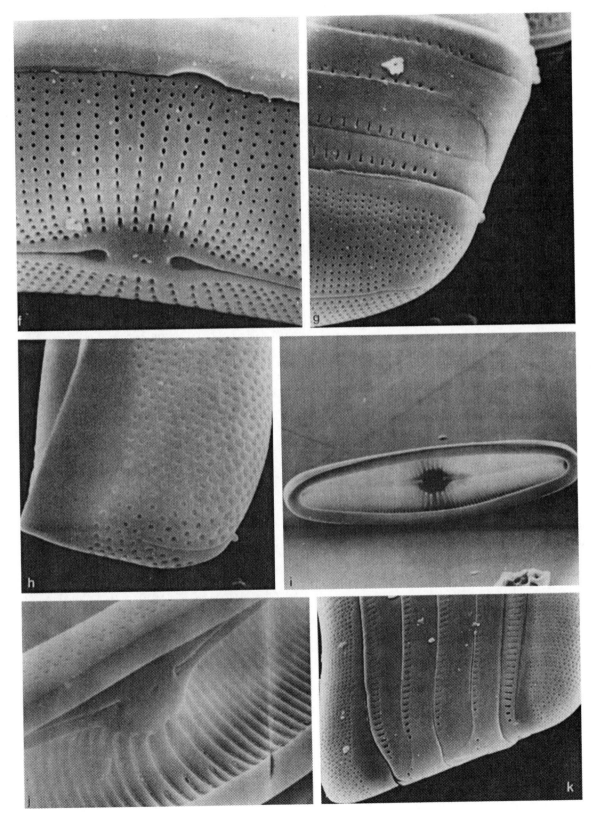

Cuneolus M.H. Giffen 1970. Bot. Mar. 13: 90

T: *C. skvortzowii* (Nikolaev) Medlin
(= *Gomphonema skvortzowii*)

Cells solitary, naviculoid in valve view;[a] weakly
cuneate and straight, or perhaps slightly bent in
girdle view.[b] Plastids unknown. An epiphytic (?)
diatom genus, whose only species was first described
from Russia and later from S. Africa; marine.

Valve slightly heteropolar,[c, d] lanceolate, with a
central stauros[e, h] and radial striae. Valve mantle
produced into slight pseudosepta at either end,[h, k]
the pseudoseptum at the head-pole being somewhat
better developed. Raphe slits unequal in length, the
slit lying towards the deeper, head-pole being the
shorter.[a] Raphe-sternum rib-like, slightly raised
internally.[h, j] Central internal raphe endings hooked
towards the primary side;[h-j] external endings
expanded, pore-like.[d-g] Terminal fissures hooked,[e-g]
meeting a transverse furrow near the valve margin
and producing a 'T' piece[e, f] at the foot-pole, but
only curved to one side at the head-pole.[g] Cingulum
consisting of 3 or more open bands. The wide
valvocopula bears one transverse row of round
poroids[c, d] and a distinct septum at the head-pole.
The remaining narrow bands are not porous.

As Giffen (1970) pointed out, this very small-
celled species has certain apparent affinities to
Gomphonema and *Rhoicosophenia*. At the moment
only the type species is known to belong to the
genus. Nikolaev (1969) described it as *Gomphonema
skvortzowii* and we believe this is synonymous with
Cuneolus minutus Giffen (1970). This small species
has almost certainly been overlooked; it has a wide
distribution, with records from Oregon, S. Africa
and Japan. *Cuneolus* is most closely related to the
Rhoicospheniaceae, particularly *Rhoicosphenia
genuflexa*. Owing to its extremely small size (less
than 10 μm in length) it has probably often been
considered simply as a small *Navicula*, like *N. diserta*.

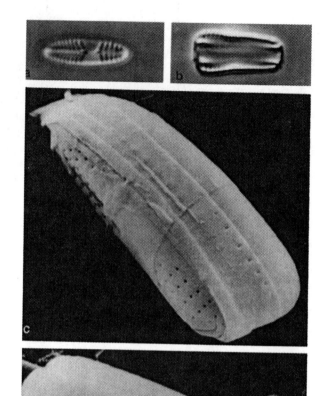

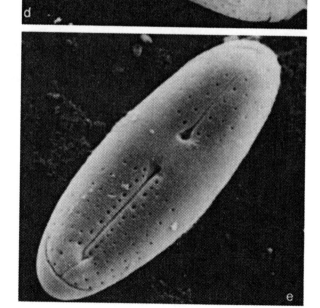

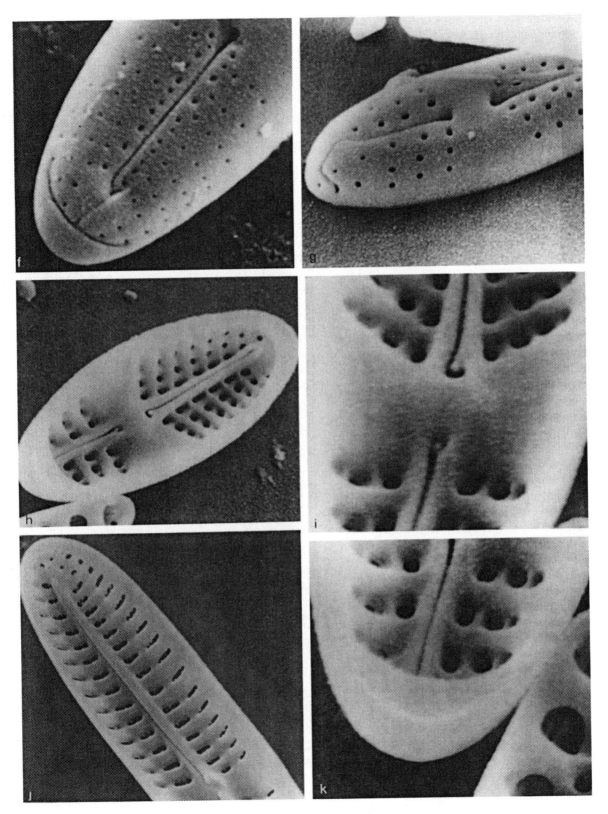

Gomphoseptatum L. Medlin in L. Medlin & F. E. Round 1986. Diat. Res. **1**: 212

T: *G. aestuarii* (Cleve) Medlin (= *Gomphonema aestuarii*)

Cells solitary, heteropolar,[a, d] attached by a single mucilaginous stalk; usually lying in girdle view. Plastids unknown. A monotypic genus, growing on seaweeds in the marine littoral. Widespread but apparently avoiding the coldest seas.

Valves linear to linear-lanceolate, heteropolar, with rounded poles.[a, c, d] Valve face flat, moderately well differentiated from a shallow mantle. A well developed pseudoseptum occurs at the foot-pole.[g] Striae uniseriate, separated into a face and a mantle series. Each half of the stria is either composed of a single areola constricted into several sections by short projections; or it is wholly subdivided into pores. Towards the foot-pole the marginal parts of the striae are reduced to single round poroids.[c, d] The foot-pole has a cluster of densely packed round unoccluded (?) pores[c, d, f] through which the stalk is secreted; head-pole without cluster of pores.[e] Raphe central, straight; the primary half of the raphe-sternum is thickened internally,[g, h] so that the raphe opens slightly laterally. External central raphe endings expanded, pore-like; internal endings slightly hooked. Terminal fissures present, bent; the valve is unusually heavily silicified beside the terminal fissure on the primary side at the foot-pole. Girdle composed of several open copulae,[i, j] each with one transverse row of poroids. The first band (valvocopula) bears a prominent septum, mimicking the valve pseudoseptum, but at the head-pole.

The presence of a pseudoseptum on the valve and a septum on the valvocopula, together with the fine detail of pore and raphe structure serve to separate this genus from other marine gomphonemoid genera, such as *Gomphonemopsis* and *Cuneolus*. Details of the structure and relationships of *Gomphoseptatum* are discussed by Medlin & Round (1986).

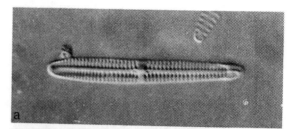

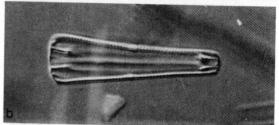

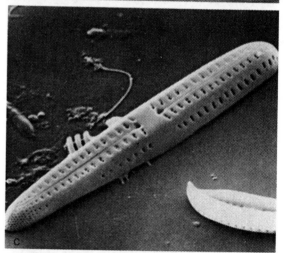

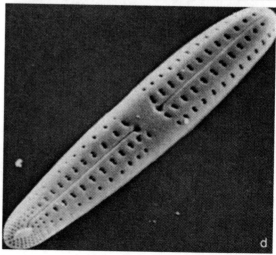

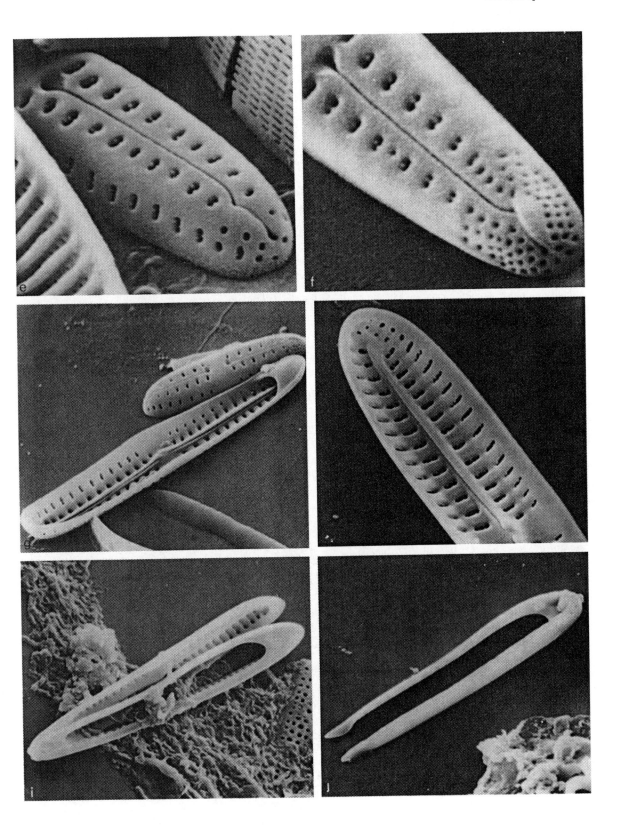

Gomphonemopsis L. Medlin in L. Medlin &
F. E. Round 1986. Diat. Res. **1**: 207

T: *G. exigua* (F.T. Kützing) Medlin
(= *Gomphonema exiguum*)

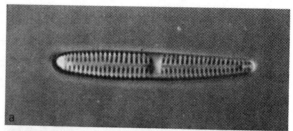

Cells solitary, attached to the substratum by a single
unbranched stipe, heteropolar, usually lying in
girdle view. Plastids unknown. A small marine,
epiphytic genus, widespread but not abundant in
temperate to tropical seas.

Valves linear to linear-lanceolate, heteropolar,[a–d]
with a bluntly rounded head-pole and an acutely
rounded foot-pole. Valve face flat, sharply
differentiated from the mantle, with a strip of
imperforate silica externally, separating the striae
into two portions. Striae uniseriate, containing
transapically elongate areolae[d, e] occluded near their
external apertures by a delicate siliceous membrane,
which is perforated by one or a few small pores[i]
(which may be equivalent to the poroids in many
other raphid diatoms). Foot-pole with a few
scattered, apparently unoccluded pores[d, g] or a row
of such pores (sometimes slit-like), through which
mucilage is presumably secreted. Pseudosepta
absent. Raphe central, straight. Central endings
straight, simple or slightly expanded externally,
simple internally. Terminal fissures absent, except at
the base-pole of some species, where a small groove
is present connecting the two basal areolae,[g] giving
the appearance of a T-shaped terminal fissure.
Cingulum, where known, containing 5 open bands,[j]
each bearing one transverse row of round or
elongate poroids over at least part of their length.

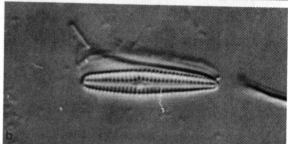

A number of species previously included in
Gomphonema have been transferred to this genus;
these include *G. exiguum*, *G. pseudoexiguum*, *G.
littorale* and an undescribed form wrongly included
in the taxon, *Gomphonema abbreviatum*. The habitat
of this genus is usually dominated by the similarly
shaped *Licmophora*, and *Gomphonemopsis* has
probably often been overlooked for this reason.

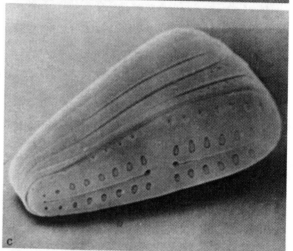

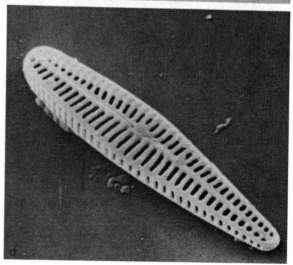

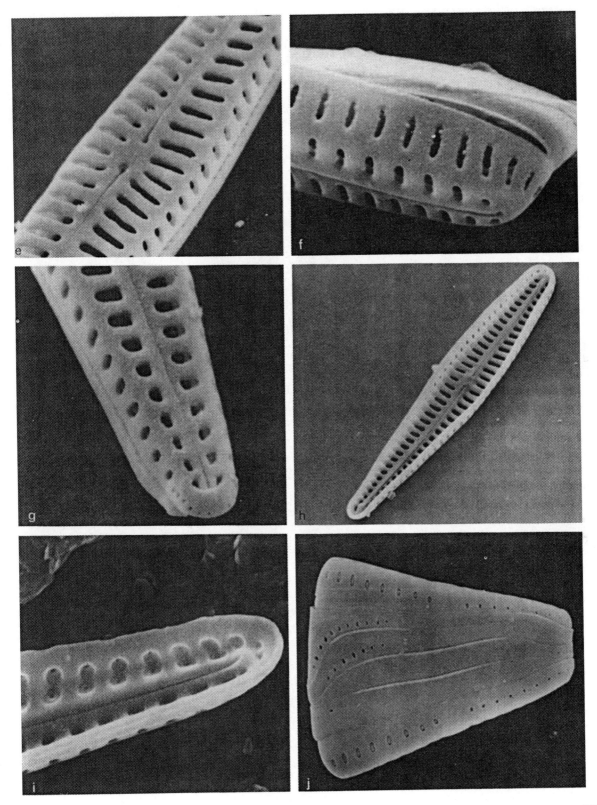

Anomoeoneis E. Pfitzer 1871. Bot. Abh.
Morphol. Physiol. **2**: 77

T: *A. sphaerophora* (Kützing) Pfitzer (= *Navicula sphaerophora*)

Cells solitary, usually lying in valve view. One highly lobed plastid, with its centre close to one side of the girdle; deeply invaginated beneath each raphe slit and also centrally along the mid-line of the girdle. One large, more-or-less spherical pyrenoid lies at the centre of the plastid. Epipelic, living on sediments in freshwater and in waters of high conductivity; extremely resistant to fluctuations in the osmotic potential of its environment.

Valves lanceolate,[a] often with rostrate or capitate poles. Valve face flat,[b] curving into fairly deep mantles, which abruptly become shallower near the poles. Striae uniseriate, containing small, oval poroids, which are occluded by hymenes. Striae interrupted by a solid area near the junction of the mantle and valve face,[c, d] but then continuing down the mantle as rows of smaller poroids. The poroids form an orderly line on either side of the raphe-sternum but elsewhere are distantly spaced and arranged in rather irregular longitudinal files. In many places the areolae appear to be filled in, producing shallow pits externally ('ghost areolae').[c-f] Raphe central, with hooked terminal fissures.[d] External central endings turned towards the secondary side;[e, f] internal endings hooked towards the primary side.[g] Girdle bands open, porous.

The removal of several taxa to *Brachysira* by Round & Mann (1981) leaves this as a very small genus with *A. sphaerophora* and *A. costata* the only commonly recorded species. The removal of *Brachysira* solved one of the mysteries of *Anomoeoneis sensu* Hustedt (1927–66), which contained some species restricted to low-conductivity, usually acidic water (some *Brachysira* spp.) and others characteristic of high-conductivity or even hypersaline waters (*Anomoeoneis sensu stricto*). *Anomoeoneis* appears to be most closely related to *Staurophora, Gomphonema, Cymbella, Placoneis* and their allies; the plastid and raphe structure are very similar in all, despite the different valve shapes and habit.

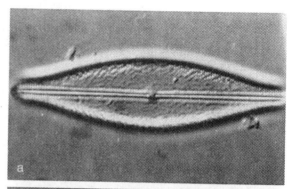

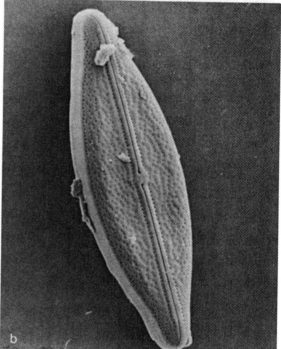

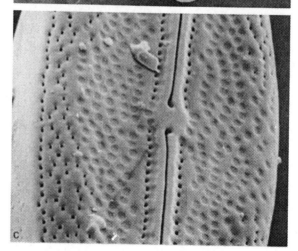

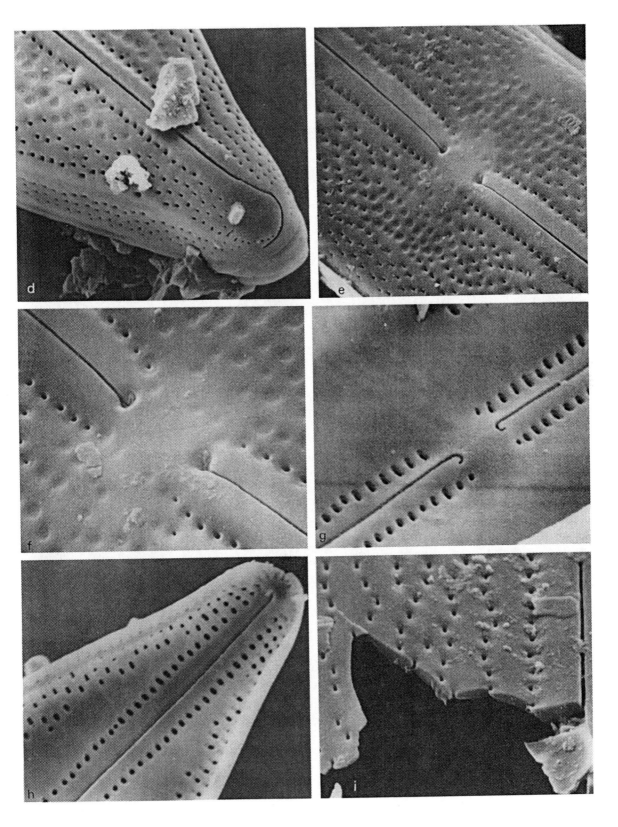

Staurophora C. Mereschkowsky 1903. Beih. Bot. Centralbl. **15**: 20

T: *S. amphioxys* (Gregory) D. G. Mann, comb. nov. (lectotype selected by D. G. Mann = *Stauroneis amphioxys*)

Cells solitary, lying in valve[a] or girdle view with approximately equal frequency. One plastid, lying with its centre against one side of the girdle;[b] from here lobes extend beneath the valves, each lobe being longitudinally indented along the raphe. The centre of the plastid contains one or two prominent rounded pyrenoids, which therefore lie to one side of the cell, the nucleus being displaced towards the other side. Brackish or marine, epipelic.

Valves linear or lanceolate, sometimes with slightly rostrate poles. Valve face gently curved in transapical section, merging gradually into the mantle;[c] at least in some species, the mantle becomes shallower towards the poles. Striae uniseriate,[c-e] containing small round poroids occluded internally by hymenes.[h] An internally prominent but flat-topped stauros is present centrally,[f, g, i] although it does not reach to the margin of the valve. Raphe-sternum narrow.[f] External central raphe endings bordered by lips and lying in a spathulate groove,[c, d] which is deflected slightly to one side; terminal fissures hooked towards the secondary side of the valve.[c] Internal central raphe endings turned or hooked towards the primary side of the valve.[g] Girdle consisting of several open bands, of which some at least bear two rows of tiny round poroids.

Staurophora is easily distinguished from *Stauroneis sensu stricto* by its plastid structure and arrangement, while details of the raphe structure (e.g. the central raphe endings) support the conclusion that these two genera are not so closely related as the light microscopical image of the valve might suggest. Thickened central strips have evolved independently in several groups of raphid diatoms (e.g. *Staurophora, Stauroneis, Proschkinia, Haslea*) and the grouping together of all forms possessing a stauros is quite artificial. The nearest relative of *Staurophora* appears to be *Anomoeoneis*, which has a similar plastid, valve and raphe structure.

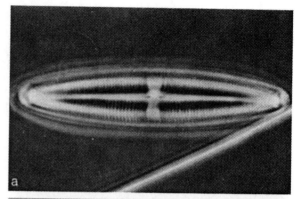

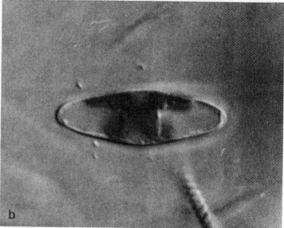

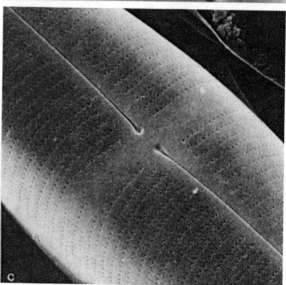

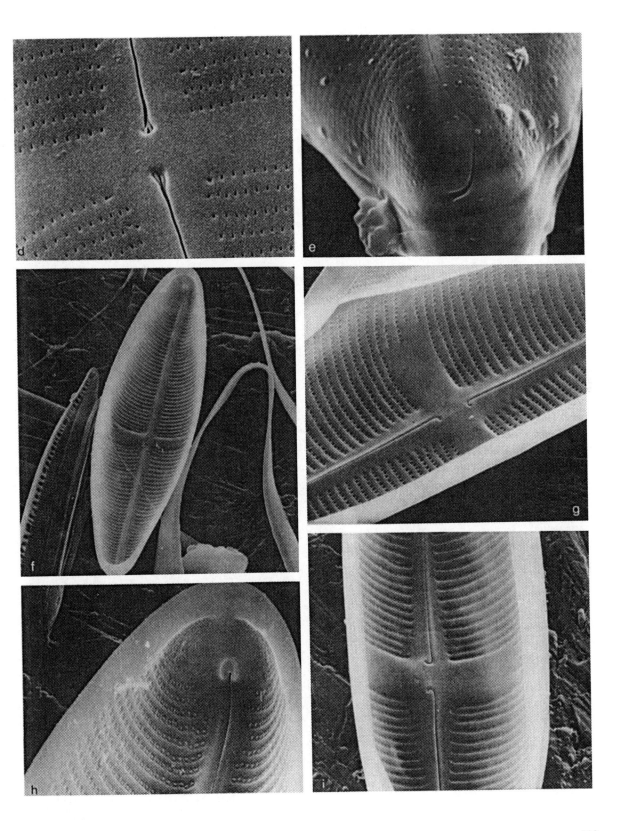

Placoneis C. Mereschkowsky 1903. Beih. Bot. Centralbl. **15**: 3

T: *P. gastrum* (Ehrenberg) Mereschkowsky (= *Pinnularia gastrum*)

Cells solitary, naviculoid, usually lying in valve view. The single, large elaborate plastid is divided into two X-shaped plates lying one against each valve, connected by a broad column. This column, which contains a large flat pyrenoid, either lies centrally in the cell or is displaced towards one side of the girdle; the nucleus lies near the opposite side of the girdle. Freshwater and perhaps marine, epipelic.

Valves lanceolate to linear, often with rostrate or capitate poles.[a-d] Striae uniseriate, containing small round poroids[c-f] closed by volate occlusions.[g, i] Valve face flat, fairly sharply differentiated from the mantles; these become much shallower near the poles.[h] Internal central raphe endings hooked towards the primary side of the valve. External central endings slightly expanded. External polar endings hooked towards the same or opposite sides. Girdle consisting of open bands, at least the most advalvar of which bear one transverse row of poroids apiece.[e-f]

A small genus, which was for many years included within *Navicula* sect. *Lineolatae*. *Placoneis* differs from *Navicula*, however, in virtually every character of frustule and protoplast, and is much more closely related to *Cymbella* and *Gomphonema*. The method of sexual reproduction also supports this conclusion. *Placoneis* has been reviewed by Cox (1987), who lists the species known to belong to it.

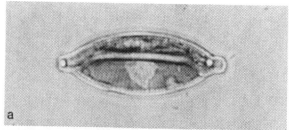

a

b

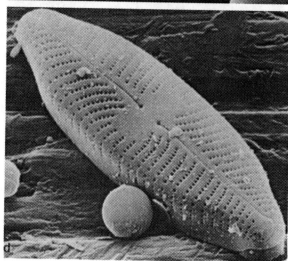

c

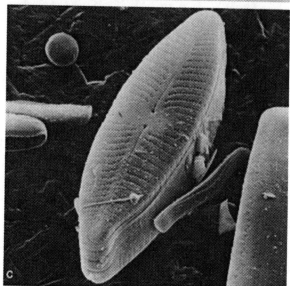

d

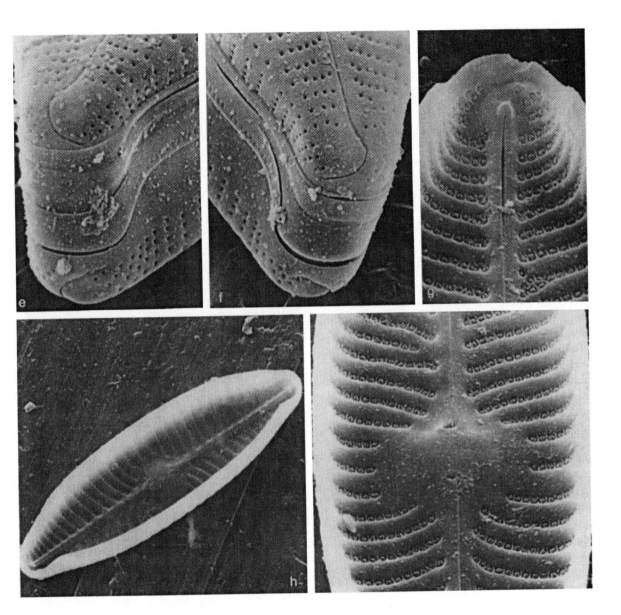

Cymbella C. Agardh 1830. Consp. Crit. Diat. p. 1 (nom. cons.)

T: *C. cymbiformis* Agardh (typ. cons.) (Håkansson & Ross, 1984. Taxon 33: 525)

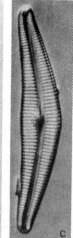

Cells slightly to strongly dorsiventral,[a-d] colonial and forming branched mucilage stalks, or solitary. One plastid, consisting of two H-shaped valve-appressed plates connected towards the dorsal side of the girdle by a massive bridge containing the pyrenoid; pyrenoid large, usually almost isodiametric. Plastid margins usually entire but lobed in *C. lanceolata* which is also aberrant in other respects. Epiphytic, epilithic or epipelic; freshwater.

Valves almost naviculoid to strongly arcuate; poles rounded, rostrate or capitate. Mantles more-or-less equal in naviculoid forms but progressively more unequal with increasing asymmetry of valve outline; valve face planar,[d-f] rarely ridged near dorsal mantle. Striae uniseriate, containing poroids. External apertures of poroids round, slit-like[d-g] or dendritic. Internally, poroids unoccluded or with volae. In stalked species, both apices with more-or-less discrete areas of small, round, unoccluded pores,[g, j] through which stalk material is secreted. Raphe system lying along or near the midline of the valve, curved in strongly dorsiventral forms. External raphe fissures often sinuous,[d-f] ending centrally in expanded pores or in a hook pointing towards the ventral margin; terminal fissures turned towards the dorsal margin.[d, g] Internal central endings almost always covered and obscured by a nodular or flap-like development from the primary side of the raphe-sternum.[h, i] One or more stigmata,[d-f] with convoluted internal occlusions[h, i] often present in this central part of the raphe-sternum. Cingulum composed of 4 open bands, the second being reduced to little more than a ligula; bands usually bearing one transverse row of poroids.

Closely related to *Encyonema* but distinguished by the dorsal plastid and ventral nucleus, and the structure and orientation of the raphe system. *Cymbella* as understood here is a large and diverse genus, and corresponds to Krammer's (1982) subgenera *Cymbella* and *Cymbopleura*.

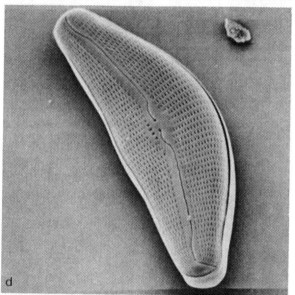

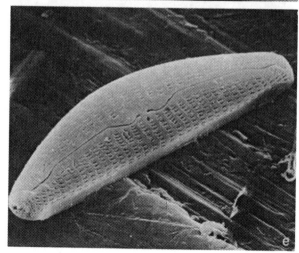

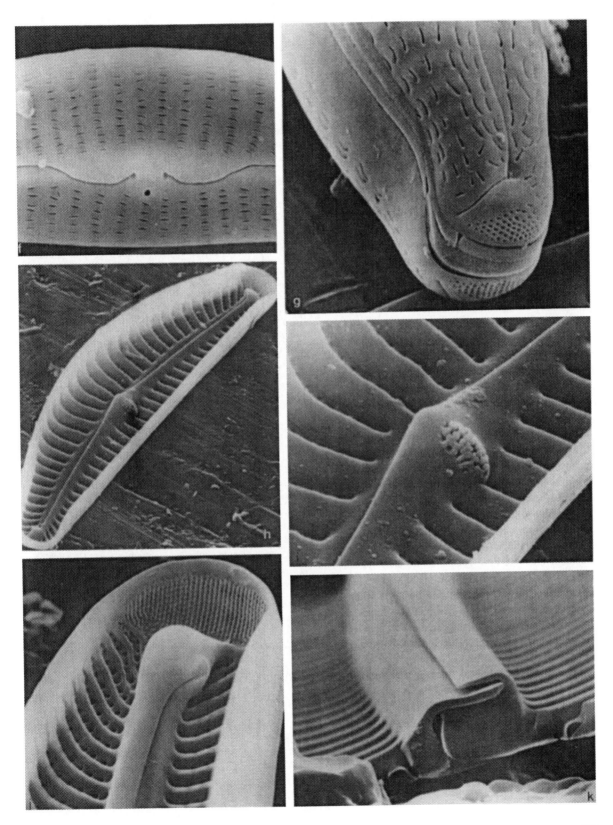

Brebissonia A. Grunow 1860. Verh. zool.-bot. Ges. Wien **10**: 512 (nom. cons.)

T: *B. boeckii* (Ehrenberg) O'Meara
(= *Cocconema boeckii*)

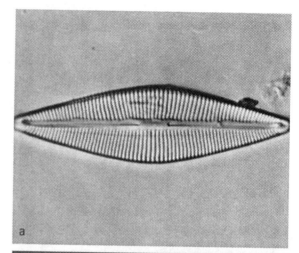

Cells naviculoid,[a, b] colonial, producing branched mucilage stalks by which they attach to solid substrata. One plastid lying with its centre against one side of the girdle and extending beneath both valves and part of the opposite girdle; deeply indented under the raphe system of both valves and along the midline of the girdle; containing a prominent central pyrenoid. A very small haptobenthic genus of brackish or marine waters.

Valves rhombic-lanceolate or lanceolate, isopolar.[a, b] Valve face flat, curving gently into shallow mantles.[c] Striae biseriate,[c, d] containing small, apically elongate poroids, except on either side of the raphe system centrally, where they are transapically elongate.[c] Transapical ribs robust.[a, f-i] Near either pole the striae are more closely spaced and contain small round unoccluded pores,[d, e] through which the mucilage stalks are secreted. Valve margins developed into small pseudosepta at the poles.[h] Raphe system central. Raphe-sternum thickened internally,[f-i] narrower in the centre of the cell than elsewhere (because of attenuation on one side). Raphe-slits widely separated at the centre. Internal central endings hooked towards the primary side; external endings expanded.[c] Helictoglossa prominent, slightly oblique;[h, i] raphe-sternum continuing beyond the helictoglossa, containing a long terminal fissure which is kinked on the external valve face above the helictoglossa.[d, e] Girdle containing open bands; number and detailed structure unknown but the first and perhaps other bands appear to have one transverse row of areolae.

Brebissonia is probably closely related to *Cymbella*: this is suggested by the isopolarity, plastid structure and raphe structure, and by the formation of mucilage stalks. The bilateral symmetry and stria structure separate them, as does the separation of the central raphe endings and the habitat. *Brebissonia* is not a common genus but seems to occur in quantity when discovered. Mahoney & Reimer (1986) have described the frustule structure of the type species, which they consider should be called *B. lanceolatum*.

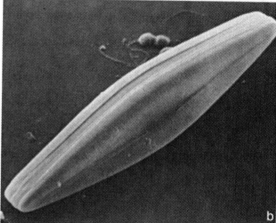

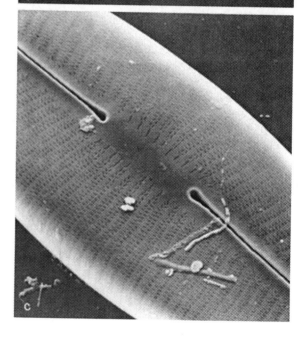

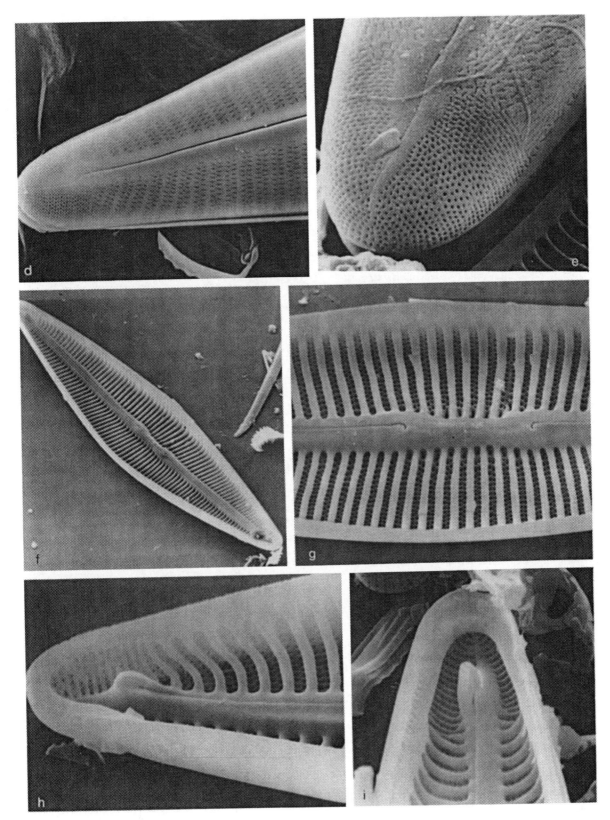

Encyonema F. T. Kützing 1833. Linnaea 8: 589

T: *E. paradoxum* Kützing

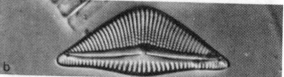

Cells strongly dorsiventral;[a-c] colonial, forming mucilage tubes or rarely solitary. One plastid, lying with its centre against the ventral side of the girdle and extending beneath both valves to the dorsal side. Plastid margins strongly indented beneath both raphe systems and along the midline of the ventral side; pyrenoid more-or-less inconspicuous, central. Freshwater; epiphytic, epilithic, often metaphytic.

Valves asymmetrical about the apical plane, with a more-or-less straight ventral margin and a strongly convex dorsal margin; poles acutely or bluntly rounded, sometimes rostrate. Dorsal mantle deeper than ventral mantle;[c] valve face planar. Striae uniseriate, containing poroids which open externally by narrow, apically elongate slits[c-g] (rarely small round apertures). Internally the slit-like areolae lie between prominent transapical ribs. No apical pore fields. Raphe system parallel to ventral margin.[d, e] External raphe fissures slightly sinuous,[d] ending centrally in expanded pores that are deflected towards the dorsal margin;[c-f] terminal fissures curved towards ventral margin.[c-e, g] Central internal endings always visible, hooked towards the dorsal side.[i, j] Helictoglossae deflected toward the ventral side. No true stigma, but one of the poroids near the central raphe endings may be slightly modified in shape. Girdle composed of open bands, some of which are reduced and do not extend around the whole cell circumference; at least some bands have one tranverse row of small round poroids.

A small group recognised by Krammer (1982) as a subgenus within *Cymbella*. We prefer to separate at the generic level. Live cells are easily identified because of the ventral plastid and dorsal nucleus. The whole cell interior and the orientation of the raphe system are opposite in *Encyonema* and *Cymbella* relative to the dorsiventrality of the cell (Mann, 1981a; Geitler, 1981; Krammer, 1982).

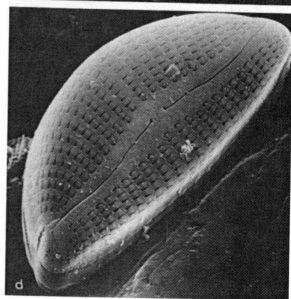

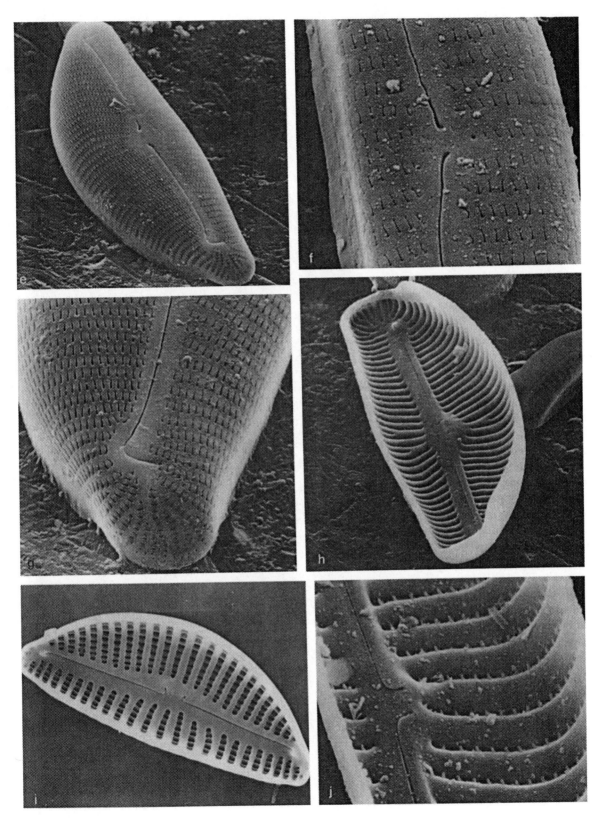

Gomphocymbella O. Müller 1905. Bot. Jahrb. Syst. 36: 145

T: *G. vulgaris* (Kützing) Müller

Cells free-living or attached, asymmetrical with respect to both the median transapical plane and the apical plane (both 'cymbelloid' and 'gomphonemoid'). Plastids unknown but probably as in *Cymbella* and *Gomphonema*. A small freshwater genus.

Valves heteropolar[a] and also asymmetrical with respect to the apical plane, lanceolate, with slightly rostrate or capitate poles.[b, c] Valve face flat, curving fairly quickly into the mantles; dorsal mantle higher than ventral. Mantles becoming shallower near poles.[i] Striae uniseriate or partly biseriate, interrupted at the top of the mantle,[c] containing oval or almost slit-like poroids; poroids smaller and nearly circular on the mantles.[c, i] Striae separated by very robust transapical ribs.[f–i] An area of small, round unoccluded pores is present at the narrower pole (this has broken off in the specimens illustrated); by analogy with *Cymbella* and *Gomphonema* this area is presumably responsible for the secretion of a mucilage stalk. Raphe lying more-or-less along the midline of the cell. External raphe fissures slightly sinuous, deflected towards the dorsal (primary) margin centrally.[b] Terminal fissures turned towards the ventral margin. External central raphe endings expanded into pores; internal endings abruptly reflexed towards the primary (dorsal) side.[f, g] A simple stigma with an elongate internal slit is present on the dorsal side, near the central raphe endings.[f, g] Girdle unknown.

This genus is very closely related to *Cymbella* and *Gomphonema*. The areola structure is closer to *Cymbella*, while the stigma closely resembles that of some *Gomphonema* species. The presence of a pore field at only one end is also reminiscent of *Gomphonema*. We thank T. Watanabe for the material on which the SEM account is based.

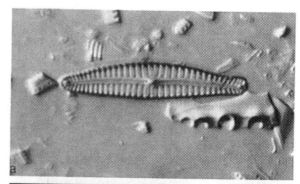

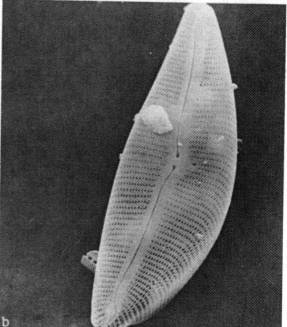

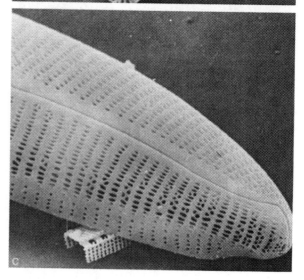

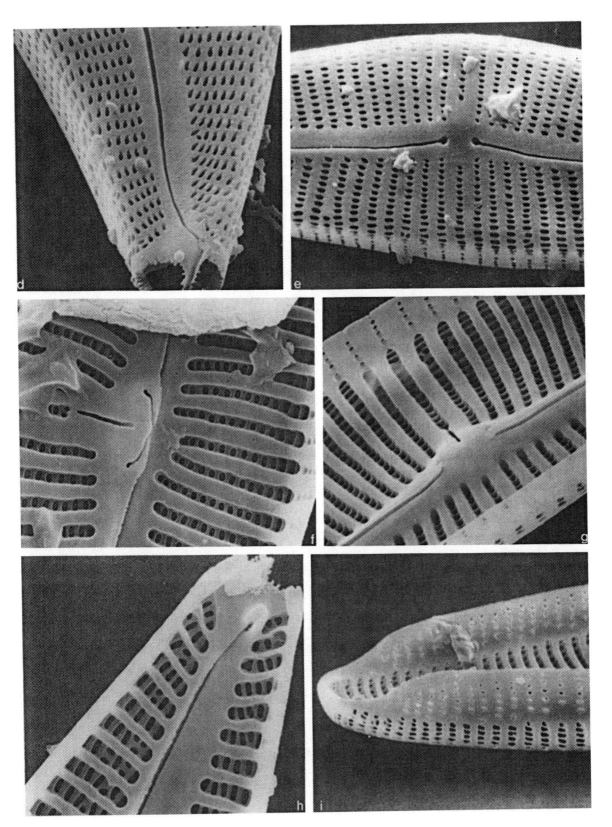

Gomphonema C. G. Ehrenberg 1832. Abh. Königl. Akad. Wiss. Berlin, **1831**: 87 (nom. cons.)

T: *G. acuminatum* Ehrenberg (typ. cons.)

Cells colonial, forming branched mucilage stalks, attached to solid substrata. Cells heteropolar,[a, b] wedge-shaped in girdle view.[c] One plastid, lying against one side of the girdle and extending beneath both valves and part of the opposite side of the girdle. Plastid deeply indented longitudinally beneath the raphe systems of both valves and along the midline of the girdle; containing one central pyrenoid. Very common in freshwater haptobenthic communities.

Valves linear or lanceolate, heteropolar,[d] with a narrow base pole and a wider, sometimes rostrate or capitate head-pole. Striae uniseriate (occasionally biseriate), containing poroids.[e] Each poroid is usually occluded by a single reniform vola, which leaves a crescent-shaped slit linking cell exterior and interior. Discrete areas of small, round unoccluded pores are present at base-pole,[f, i] through which the mucilage stalks are secreted. Raphe-sternum straight, central. External raphe fissures straight or slightly sinuous. Internal central endings hooked towards primary side.[j] External central endings usually expanded and straight; terminal fissures slightly curved. Raphe-slits unequal in length; upper being the shorter. Stigma sometimes present, much more simply constructed than in *Cymbella*. Girdle consisting of a few (often 4) open bands, usually bearing one or two transverse rows of poroids; second band often reduced to a ligula at the base-pole.[f]

A few marine species have been reported but some (see Medlin & Round, 1986), and perhaps all, can be separated from *Gomphonema* as defined here (e.g. see *Gomphonemopsis, Gomphoseptatum*). *Gomphonema* is fairly closely related to *Cymbella, Gomphocymbella, Encyonema* and *Brebissonia*, less closely to *Anomoeoneis, Staurophora* and *Rhoicosphenia*; all have the same type of plastid and resemble each other in aspects of raphe and girdle structure.

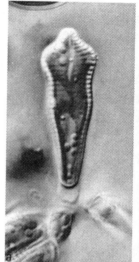

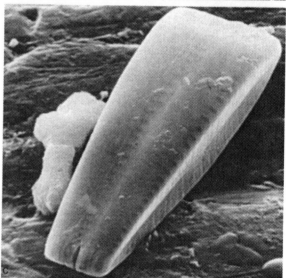
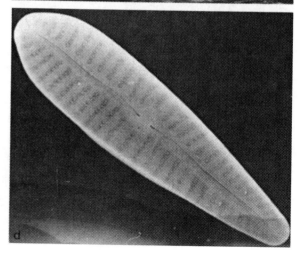

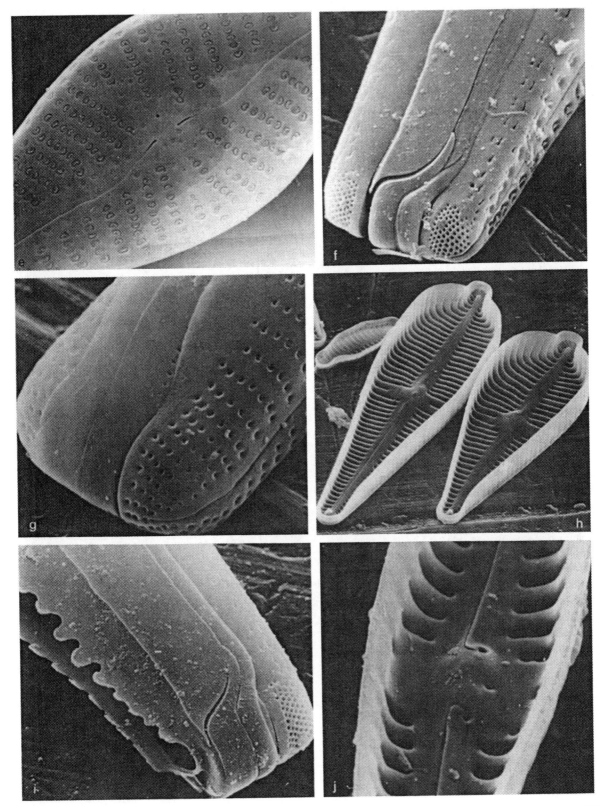

Didymosphenia M. Schmidt 1899. In A. Schmidt, Atlas Diat.: 214 (nom. cons.)

T: *D. geminata* (Lyngbye) M. Schmidt (= *Echinella geminata*)

Cells large, colonial, forming branched mucilage stalks[a] by which they are attached to a substratum; large colonies forming macroscopic grey growths. Frustule heteropolar in valve and girdle view. A small freshwater epiphytic or epilithic genus.

Valves robust, heteropolar, usually capitate at the poles.[b, c] Valve face flat, sometimes with a marginal ridge at the junction with the mantles,[c, e] which are deep; the ridges often end in spines at the head-pole.[e] Striae uniseriate, containing large poroids.[d-g] Poroids occluded by several volae leaving a dendritic slit for communication between inside and outside;[g] opening internally between transverse ribs.[h- k, m] Raphe system central, straight. Raphe-sternum, transapical costae and margin sometimes ornamented externally with shallow pits, warts and papillae.[d, f, l] The narrower base-pole has an area of very small, unoccluded, round pores through which the mucilage stalks are secreted.[f] Raphe-sternum relatively narrow, pierced centrally by several stigmata,[d, i] which have simple round apertures externally but open internally through convoluted spongy bosses of silica.[j] Terminal fissures present, hooked.[f, h] External central raphe endings expanded into pores.[d] Internal central endings obscured by a nodular outgrowth of silica on the primary side of the raphe-sternum;[i, j] nodule sometimes indented at its centre[i, j] but sometimes entire. Helictoglossa simple or deflected to one side.[k, m] Girdle composed of open bands (4 per epitheca) with plain partes exteriores but apparently having a transverse row of poroids in the pars interior.

Dawson (1973a, b) has described 2 species in detail. *Didymosphenia* is similar to *Gomphonema* but distinguished by the cell size, pore structure, stigma structure, and the much-branched transapical costae of the valve centre. Kociolek & Stoermer (1988a) consider that *Didymosphenia* is more closely related to *Cymbella* and *Encyonema* than to *Gomphonema* and *Gomphoneis*. Before accepting this conclusion we would wish to see further analysis of the evolution of heteropolarity and investigation of reproductive characteristics.

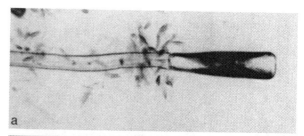

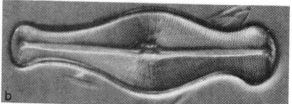

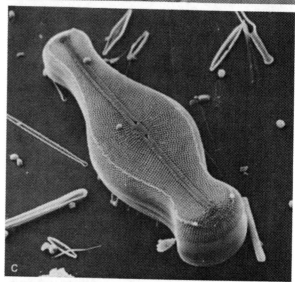

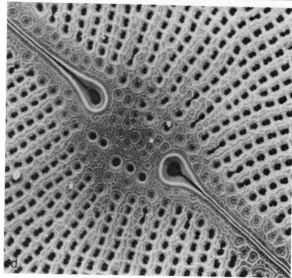

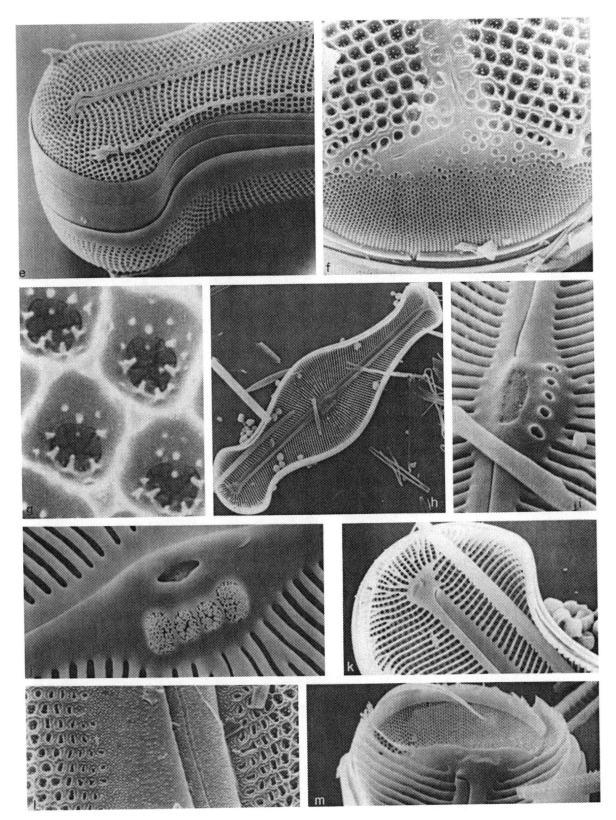

Gomphoneis P. T. Cleve 1894. Kongl. Svenska Vetensk.-Akad. Handl. Ser. 2, **26** (2): 73

T: *G. elegans* (Grunow) Cleve (= *Gomphonema elegans*; lectotype selected by C. S. Boyer 1928, Proc. Acad. Nat. Sci. Philad. 79 Suppl.: 299)

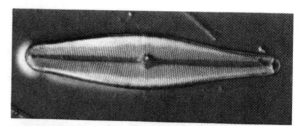

Cells large, colonial, forming branched mucilage stalks. Frustule heteropolar,[a] wedge-shaped in girdle view.[b] Plastids unknown. A small group of attached freshwater diatoms, apparently reaching its maximum development in N. America, but apparently also present in L. Baikal (U.S.S.R.) (Kociolek & Stoermer, 1988b).

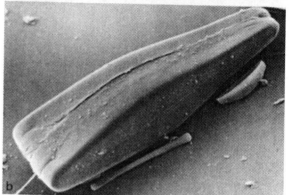

Valves heteropolar, linear to lanceolate. Head pole bluntly rounded[e] or occasionally apiculate. Valve face flat, curving gently into relatively deep mantles, which become markedly shallower near both poles.[j, k] Two spines sometimes present near the head-pole, one on either side of the external raphe ending. Valve margin thickened, especially at the head-pole, where a pseudoseptum is formed.[f, i] Striae alveolate. Each alveolus, or chamber, opens to the outside by two rows[c] of poroids occluded by flap-like or rim-like volae, and to the inside by a transapically elongate aperture. The inner aperture is only about half the length of the stria[g] so that in the LM longitudinal lines can be seen crossing the striae, as in many *Pinnularia* spp. At the narrower pole there is a discrete area of very fine, unoccluded pores,[d] through which the mucilage stalks are secreted. External raphe fissures straight or gently curved. Central external raphe endings expanded;[c] terminal fissures deflected at the head-pole, straight at the base-pole.[d] Internal central endings lying in a central nodule;[g] hooked towards the primary side. Raphe-sternum pierced centrally by one, or sometimes several stigmata, which open externally by a simple pore[c] and internally by a narrow slit.[arrow, g] Girdle consisting of open, porous bands.

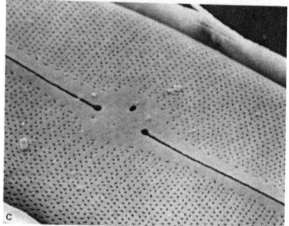

Dawson (1974) transferred some small *Gomphonema* species into *Gomphoneis* because they had biseriate striae. These species do not appear to be closely related to *Gomphoneis sensu stricto* (Kociolek & Rosen, 1984) and should perhaps be left in *Gomphonema* unless other characters can be found to ally them with large celled *Gomphoneis*. As understood here, *Gomphoneis* spp. can be recognised and distinguished from *Gomphonema* by their alveolate striae.

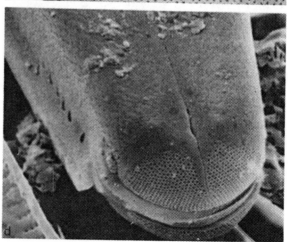

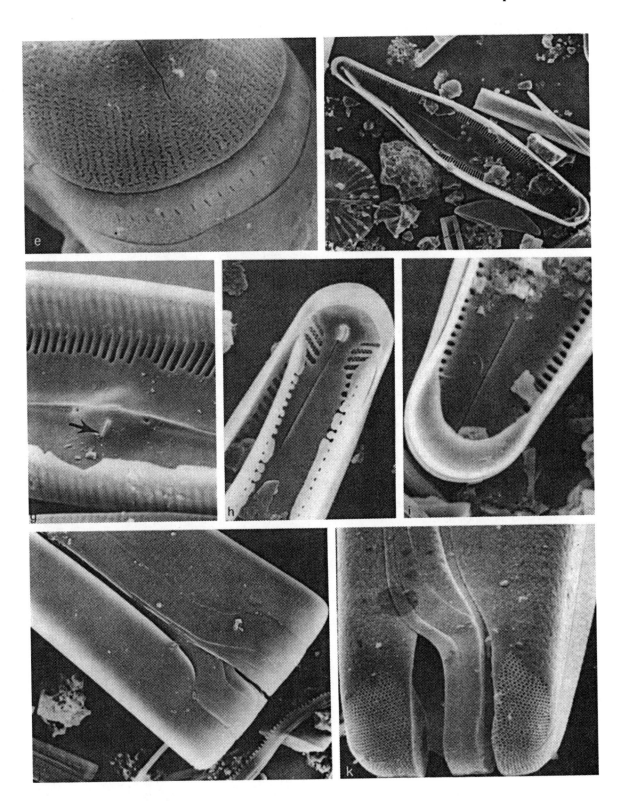

Reimeria J. P. Kociolek & E. F. Stoermer 1987. Syst. Bot. **12**: 457

T: *R. sinuata* (Gregory) Kociolek & Stoermer (= *Cymbella sinuata*)

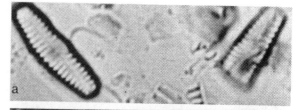

Cells slightly dorsiventral, with a slight unilateral expansion;[a,b] free-living, motile. One much-lobed plastid. Freshwater, associated with stone surfaces particularly in rivers.

Valves linear or linear-lanceolate, subcapitate,[d,e] asymmetrical about the apical axis: dorsal side convex, ventral less convex or straight, but with an expanded central convexity. Valve face flat, curving fairly abruptly into the mantles.[b,c] Mantles deep, except at the poles. Striae distant, biseriate, or uniseriate, containing tiny round poroids;[b-d] absent from the ventral swelling, opening internally between prominent ribs.[e-g] Mantles with a wide plain margin. An apical pore field lies at each end on the ventral mantle, and shorter striae on the dorsal apical mantle. Raphe system more-or-less central, straight.[d-g] External fissures slightly sinuate; central endings expanded; terminal fissures present, curved towards the ventral (secondary) side. Internally with central raphe endings turned to the dorsal (primary) side; helictoglossae prominent.[e-g] Between the central raphe endings, or slightly to the dorsal side of them, is a single isolated pore (stigma), which is unoccluded both internally and externally. Cingulum apparently containing 4 open bands, at least some of which bear a single row of pores.[c,h-j]

This recently erected genus differs from *Cymbella* in the unusual shape of the valve and the ventrally displaced apical pore field. The simple stigma, central internal raphe endings and pore fields are also very reminiscent of many *Gomphonema* species (see Kociolek & Stoermer, 1987, for a detailed discussion).

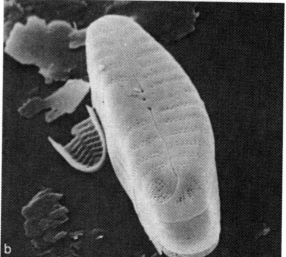

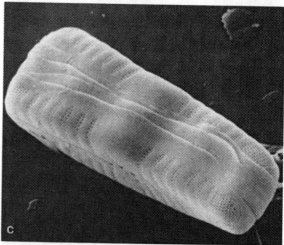

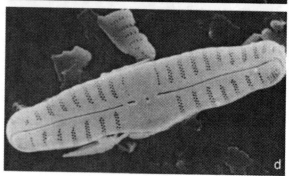

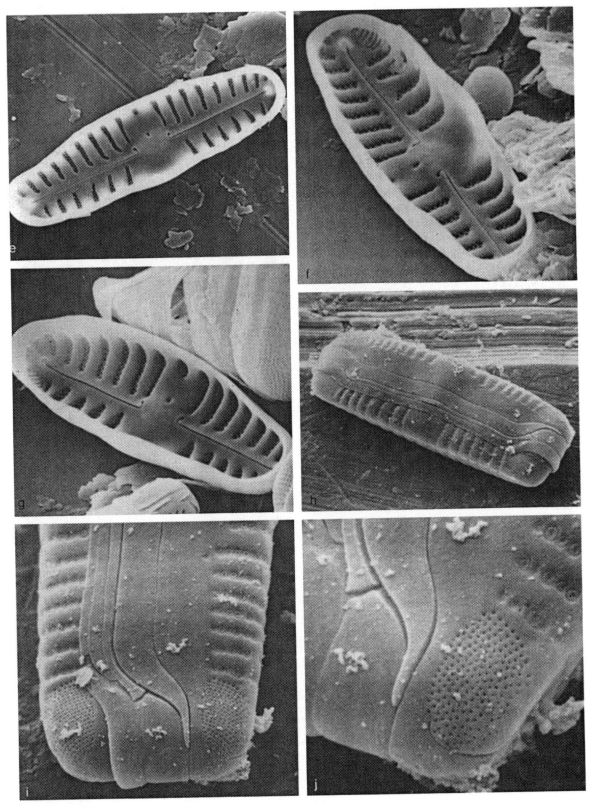

Achnanthes J. B. M. Bory de St.-Vincent 1822.
Dict. Class. Hist. Nat. 1: 79

T: *A. adnata* Bory

Cells heterovalvar[a,b] with one raphid (R-)[c] and one
rapheless (P-)[d] valve. Solitary or forming short
bands attached to substrata by a mucilage stalk[b]
secreted from the raphe of one R-valve; more rarely
forming chains, by the attachment of cells to each
other by very short stalks, again produced from part
of the raphe near one pole. Cells bent about the
median transapical plane (with the R-valve on the
inside of the shallow V),[e] lying in girdle or valve
view. In *A. longipes*, many small plastids per cell;[b] in
other species, 2 large plastids, one on either side of
the median transapical plane, each H-shaped in
valve view and consisting of 2 plates connected by a
central pyrenoid bridge. Usually haptobenthic,
predominantly marine but a few in freshwater where
they are mainly subaerial.

Valves linear to lanceolate. Mantles more-or-less
sharply differentiated from the concave valve face in
the P-valve;[f, j] merging gradually with the convex
valve face in the R-valve.[e] Striae uniseriate or, in
A. longipes, bi- or triseriate, containing poroids
occluded by complex cribra bearing volae:[i] there are
no hymenes. R-valve usually with a fascia or stauros
and a central raphe-sternum.[g] P-valves without
plain, transverse areas, with a rapheless sternum [h]
(pseudoraphe) often displaced towards one margin;
this sternum is always narrower than the raphe-
sternum. External central raphe endings straight,
expanded; internal endings coaxial and simple or
slightly hooked to the same side. Terminal fissures
curved. Cingulum containing 3–7 open bands
bearing one or 2 transverse rows of large poroids.

Although long associated with it, *Achnanthes*
differs from *Achnanthidium* in areola, raphe, girdle
and plastid structure. These genera are possibly
distantly related, but there are similarities in the
flexure of the frustule and the method of stalk
formation. These may result from convergence. The
traditional use of the name '*Achnanthidium*' e.g. by
Hustedt (1927–66), for some members of this group
(e.g. *A. brevipes*, *A. coarctata*, *A. inflata*) is wrong.
Hustedt used '*Achnanthes*' as a subgenus to refer
only to *A. longipes*. The remaining species he placed
in the subgenus *Microneis* (our *Achnanthidium*).
Some aspects of morphogenesis are described by
Boyle *et al.* (1984).

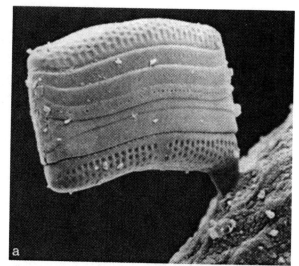

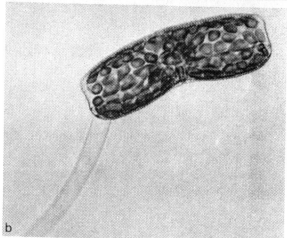

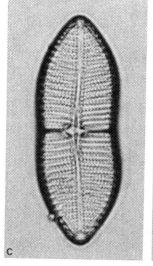

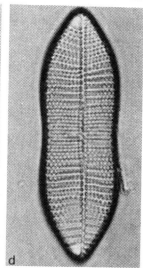

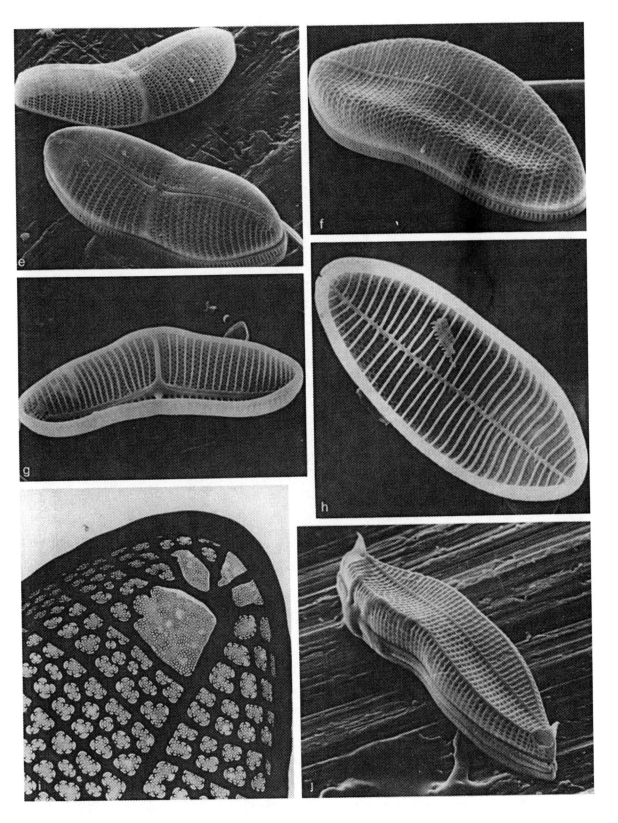

Cocconeis C. G. Ehrenberg 1837. Abh. Königl. Akad. Wiss. Berlin, **1835**: 173

T: *C. scutellum* Ehrenberg (lectotype selected by C. S. Boyer, 1928, Proc. Acad. Nat. Sci. Philad. **79** Suppl.: 242)

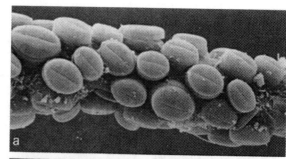

Cells solitary, heterovalvar. One valve (R-valve)[b] with a raphe-sternum, the other (P-valve)[c] lacking a raphe but with a corresponding rapheless sternum ('pseudoraphe'). Valves and girdle shallow, so that cells or valves almost always lie in valve view. One plastid, which is flat and C-shaped, and may be simple in outline or elaborately lobed; containing one to several elongate pyrenoids. Freshwater to marine, living on plants,[a] rocks, etc. to which the cells are attached, via the R-valve, by mucilage.

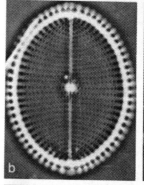

Valves elliptical or almost circular. R-valve usually less convex than P-valve. In *C. pediculus* and some others the R-valve is markedly concave and here there is a tendency for the P-valve to be correspondingly convex, and for the raphe to be raised on a light keel.[g] Mantles shallow, usually sharply distinguished in the R-valve but less so in the P-valve. Striae usually uniseriate,[g-i] containing small round poroids, but multiseriate striae[e] and loculate areolae are present in some. P-valve often more complex in structure than R-valve (e.g. *C. pediculus*: Gerloff & Rivera, 1979). Poroids closed by hymenes with linear perforations. In the R-valve the striae are sometimes interrupted by a submarginal rim of silica,[k] which also projects internally. Raphe-sternum and 'pseudoraphe'[e] are usually similar in shape though the latter is often wider;[f] both are central, and straight or slightly sigmoid. Terminal fissures absent.[g] External central raphe endings simple or slightly expanded; internal endings non-coaxial, deflected towards opposite sides.[h, i] Cingulum consisting of a few narrow, non-porous bands, of which the valvocopula at least is often closed and often bears simple or complex projections,[i, j] which fit exactly beneath the valve ribs and pores.

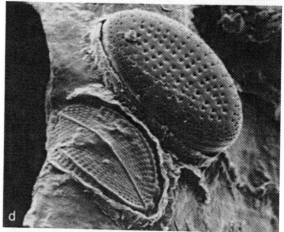

In *C. pediculus* the pseudoraphe is formed by filling in of the raphe primordium during valve silicification (Mann, 1982a). Further SEM details of some species are given by Holmes *et al.* (1982). This is a complex genus which is being divided into smaller genera, e.g. by Holmes (1985), who has separated off two groups of species living on whale skins into the new genera *Bennettella* and *Epipellis*.

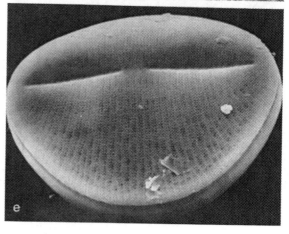

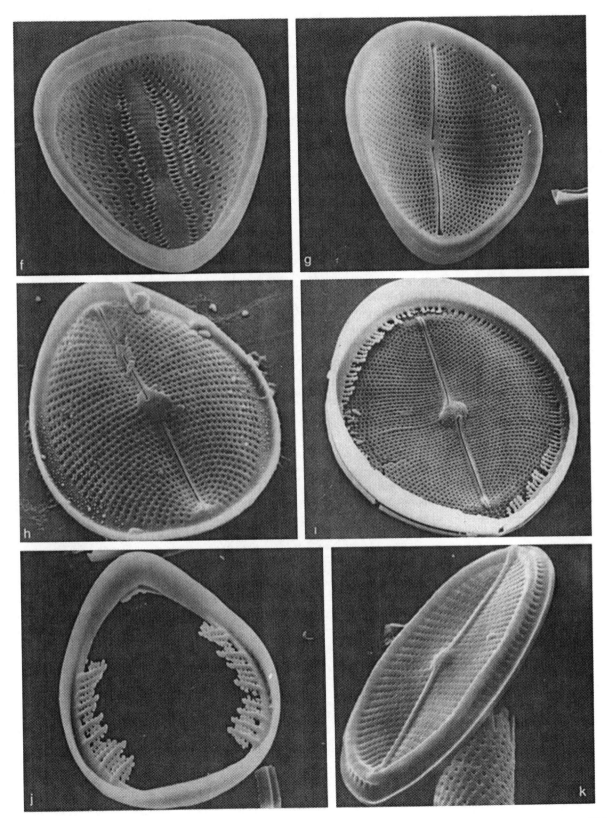

Campyloneis A. Grunow 1862. Verh. zool.-bot. Ges. Wien **12**: 429

T: *C. argus* Grunow

Cells solitary, heterovalvar, with R- and P- valves[a-d] as in *Cocconeis*; attached. Usually lying in valve view. Two plate-like plastids (Hendey, 1964). A fairly common genus found attached to marine algae; probably monotypic.

Valves oval or almost circular. Valve face of R-valve concave,[e] P-valve[f] complementary, with a concave central area though the overall form of the valve is convex.[f] R-valve with uniseriate striae containing small round poroids.[h-k] Raphe (R-valve only!) central, restricted to valve face.[e] External raphe endings straight and simple at centre and pole; internal central endings slightly non-coaxial, deflected towards opposite sides.[h, i] P-valve more complex with cribrate areolae opening into shallow chambers[f, l, m] which themselves open to the inside by small simple pores.[r, s] The areolae are larger near the pseudoraphe and regularly square around the perimeter of the valve.[f, l]

Internally, the frustule is sufficiently complex to warrant extra illustration. The valvocopulae (no other bands have been observed) are extraordinary structures with a very narrow pars exterior but a pars interior that extends a web of flat ribs over much of the internal surface of the valves. They are very closely pressed to both R- valve[e, n-q] and P-valve[g, m] and reflect the heterovalvy in their different morphologies. The advalvar surface of the R-valvocopula[n, o] is smooth without pores corresponding to those in the valve. The abvalvar face[u-w] of the P-valvocopula does show pores, a rather fuller mesh and a circular foramen at the centre.[u] When in place on the P-valve the outline of the band can only be distinguished from the underlying valve at the centre[r, t] and at a few other places.[arrows, s] The 4 components of the frustule are held together by 2 rings of clustered facets (e.g. on the R-valve)[h, j, k] and rugose bosses (e.g. on the abvalvar face of the P-valvocopula).[r, s] Fig. x shows the two valvocopulae attached to one another.

Some of the girdle characteristics occur in *Cocconeis* (Holmes *et al.*, 1982), though in less pronounced form, suggesting a relationship. Hustedt (1927–66) comments on the very variable development of the 'septum' which is the striking and diagnostic feature of *Campyloneis*.

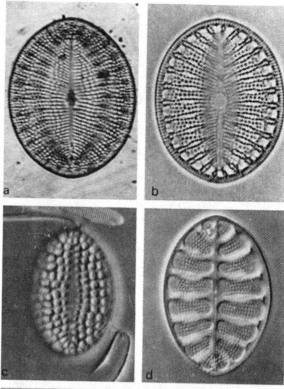

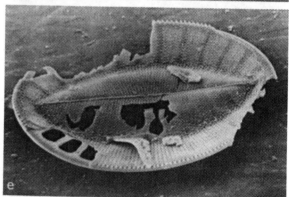

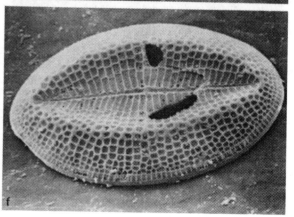

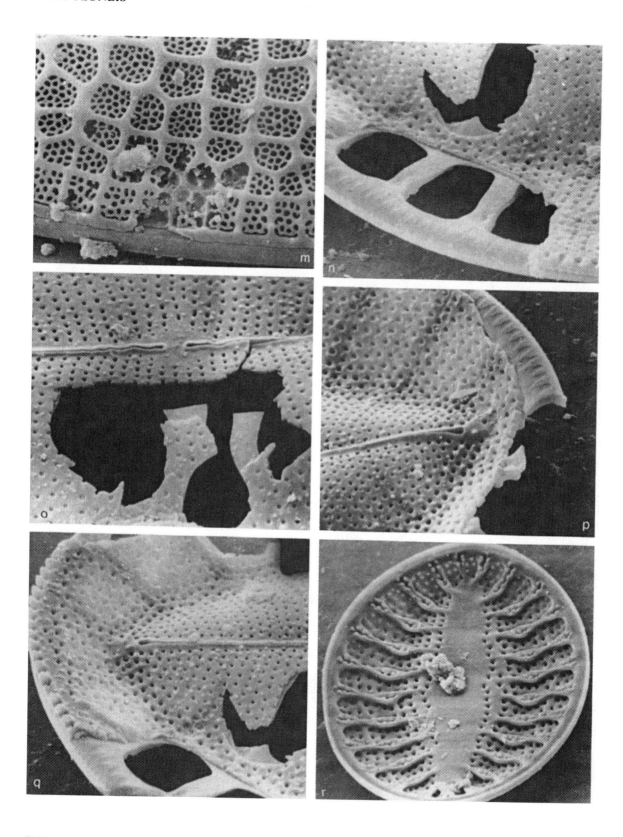

Anorthoneis A. Grunow 1868. Reise Fregatte 'Novara' Bot. 1 (1): 9

T: *A. excentrica* (Donkin) Grunow (= *Cocconeis excentrica*)

Cells solitary, disc-like, almost always lying in valve view; heterovalvar with a raphe-sternum on one valve (R-valve)[c] and a plain, rapheless sternum or pseudoraphe on the other (P-valve).[a, b] One flat, C-shaped plastid. Often reported free-living but probably usually attached to sand grains; marine.

Valves elliptical, often almost circular. Valves very shallow with no true mantles. Valve surface undulate,[c, d] the form of the R-valve complementing that of the P-valve. Striae uniseriate, containing round or oval poroids; these are occluded by hymenes perforated by elongate slits. The hymen is often supported by 1–4 pegs[i] and these give the poroids the appearance of containing rotae.[i] Raphe-sternum and araphid sternum short,[e–h] not reaching the poles, eccentric. Raphe without terminal fissures. Internal central endings non-coaxial, being curved in opposite directions as in *Cocconeis*. External central endings straight, slightly expanded. Sternum of P-valve as wide or wider than the raphe-sternum, sometimes very much so (*A. hyalina*). For confident identification both valves must be observed. Girdle incompletely known but apparently containing non-porous bands as in *Cocconeis*.

A small genus with only 8 spp. listed by VanLandingham (1967). *Anorthoneis* is clearly very closely related to *Cocconeis*, but the eccentricity of the raphe system is diagnostic. Comparable variations in structure have been used by Holmes (1985) to split genera from *Cocconeis*, and in the light of his changes the maintenance of *Anorthoneis* seems reasonable.

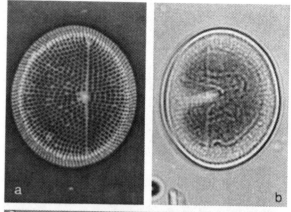

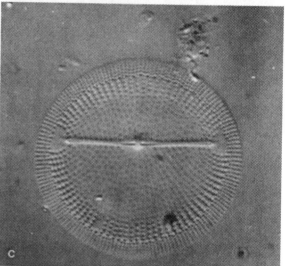

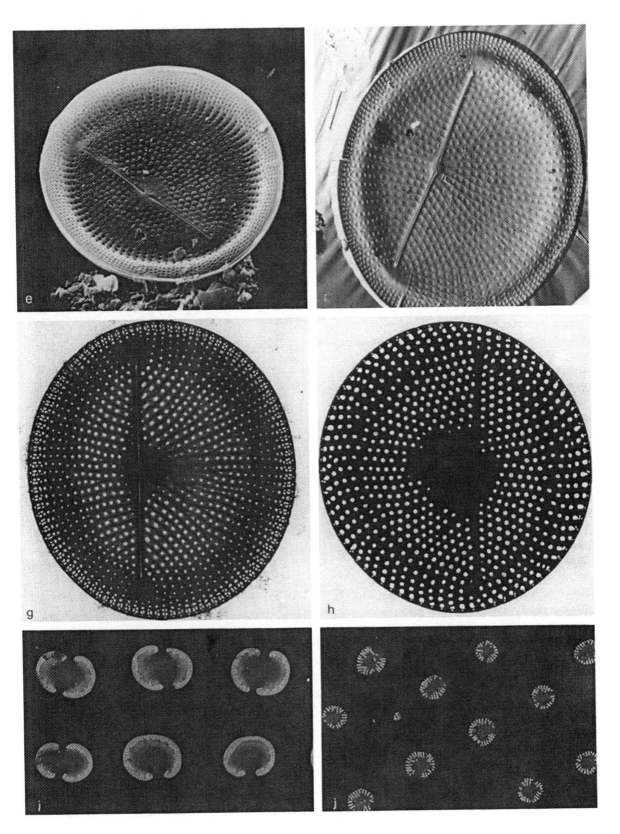

Achnanthidium F.T. Kützing 1844. Kies. Bacill. Diat. 75

T: *A. microcephalum* Kützing (lectotype selected by L. Rabenhorst 1853. Süssw.-Diat. Freunde Mikr. 25 & 26

Cells solitary or in short chains,[a] heterovalvar,[c, h] with R- and P-valves as in *Cocconeis*. Attached to substrata by a mucilage stalk secreted from one end of the R-valve. Cells bent[d] about the median transapical plane so that each is shaped like a shallow 'V' in girdle view, with the R-valve usually on the inside of the V. One plastid, lying against one side of the girdle but also extending beneath one or both valves. Freshwater, haptobenthic; a few species are recorded as marine but these appear to have a more complex valve structure and need investigation.

Valves linear, lanceolate or elliptical,[b, e] with rounded or capitate poles. Valves fairly shallow with well differentiated mantles. Valve face usually slightly convex in the R-valve and concave in the P-valve. Striae uniseriate to multiseriate, containing simple poroids closed by hymenes; these have round or slightly elongate perforations. In *A. lanceolata* and its allies, the P-valve has a clear area on one side of the centre, which is sometimes partly roofed over by a thin hood,[g] the *hufeisenformige Fleck* of Hustedt (1927–66). The R- and P-valves usually have a similar structure but the rapheless sternum (pseudoraphe) of the P-valve is sometimes wider than the raphe-sternum or has a different shape centrally. Occasionally the striation density differs. Both valves must be observed for confident identification. Raphe usually without terminal fissures. External central raphe endings simple or slightly expanded; internal endings non-coaxial, deflected in opposite directions.[f] Girdle shallow, containing open, non-porous bands.[c, d]

When, as here, *A. microcephala* and its allies are considered as a separate genus, the correct name for them is *Achnanthidium*, not *Achnanthes* Bory or *Microneis* Cleve. '*Achnanthidium*' is often applied, wrongly, to the *Achnanthes brevipes* group (e.g. Hustedt, 1927–66). Further investigation is required to determine whether the *A. lanceolata* cluster should be recognised as a separate genus.

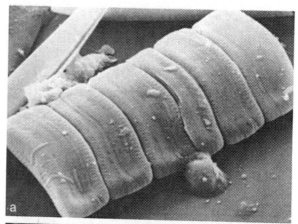

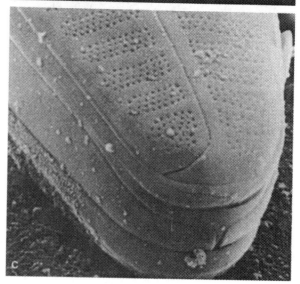

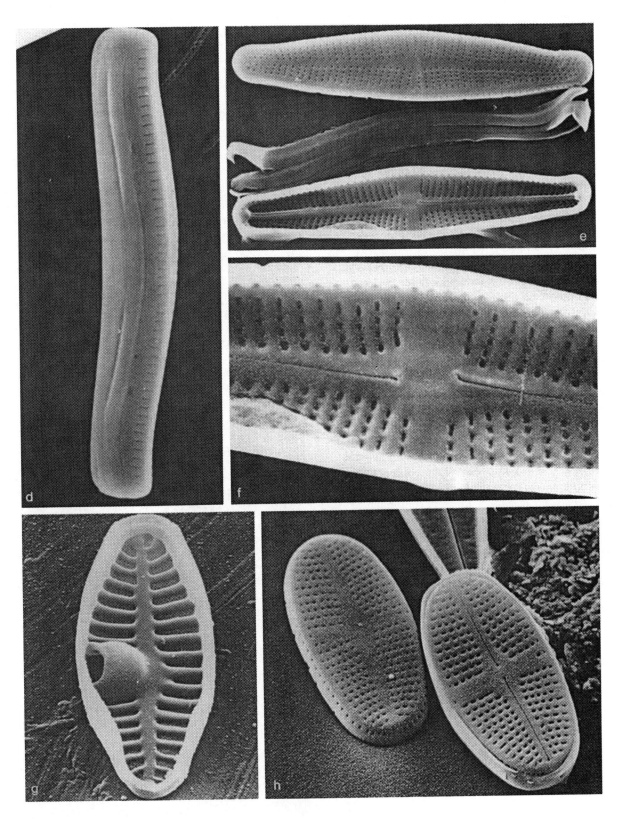

Eucocconeis P. T. Cleve 1895. Köngl. Svenska Vetensk.-Akad. Handl. **27** (3): 173

T: *E. flexella* (Kützing) P. T. Cleve (= *Cymbella flexella*; lectotype selected by D. G. Mann)

Cells solitary, heterovalvar, with R- and P-valves as in *Cocconeis*. Frustule bent about the median transapical plane as in *Achnanthes* but also twisted about the apical axis.[a-c] Cell usually lying in valve view. One plate-like plastid. A small genus found mostly in oligotrophic freshwaters.

Valves linear, lanceolate or more-or-less elliptical, with round, rostrate or somewhat truncate poles. Valve face well differentiated from relatively deep mantles. Valve face markedly convex in R-valve, correspondingly concave in P-valve.[c, e] Striae fine, radiate, uniseriate,[d, f] containing simple but deep, round poroids closed internally by hymenes.[j] Valve margin wide. Junction of valve face and mantle often with a plain region and rows of pores slightly less regular on the mantle.[c, d] Raphe-sternum and rapheless sternum[e] (pseudoraphe) central, sigmoid, expanded at their centres. Terminal fissures absent. External central raphe endings expanded into 'pores', rimmed in some cases;[d] internal endings non-coaxial,[j] deflected towards opposite sides. Girdle containing narrow, non-porous bands.

Our concept of *Eucocconeis* is narrower than that of Cleve (1894–5) but is near that of Hustedt (1930). Several authors (e.g. Hustedt, 1927–66) have rejected *Eucocconeis* because of an apparent continuum of variation in raphe curvature between *Achnanthidium sensu stricto* and *Eucocconeis*, from straight to strongly sigmoid. This requires further investigation but the shape of the frustule and the valve structure suggest that the genus is probably worth retention, although it is clearly closely related to both *Achnanthidium* (q.v.) and *Cocconeis*.

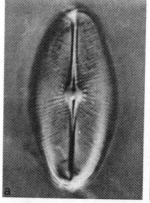
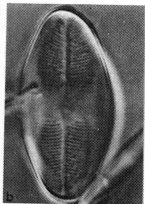
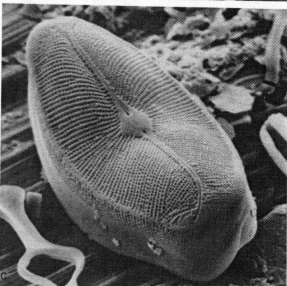

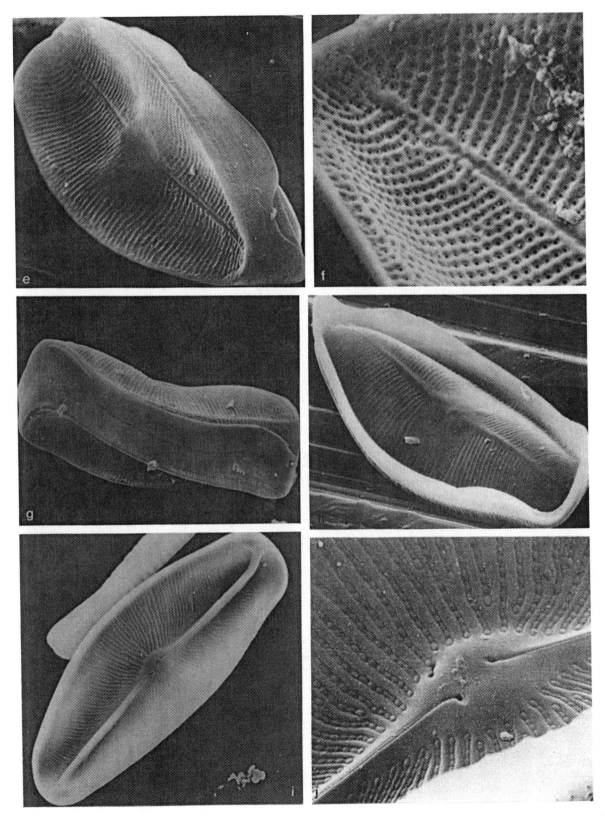

Parlibellus E. J. Cox 1988. Diat. Res. 3: 9–38

T: *P. delognei* (Van Heurck) Cox (= *Navicula delognei*)

Cells naviculoid, often forming mucilage tubes, in which the cells live and move. Single cells also occur and usually lie in girdle view. Two butterfly-shaped plastids, lying one against each side of the girdle. Marine; epilithic or epiphytic.

Valves lanceolate or linear, with bluntly or acutely rounded apices.[a, c, d] Valve face curved,[b, c, e] merging gently into the mantles, or slightly flattened. Striae somewhat more distantly spaced at the centre, uniseriate, containing small round poroids occluded by hymenes.[g, h] One or a few isolated pores are sometimes present centrally,[e] but these do not appear to penetrate through to the valve interior.[h] External central raphe endings simple or slightly expanded. Very short terminal fissures are present, turned towards the same side of the valve.[c] Central internal endings simple, or raphe fissure continuing into a narrow rib-like or helictoglossa-like structure.[g] Girdle composed of many open bands, each bearing two transverse rows of round poroids.[j]

A widespread group, formerly included within *Navicula*. Their rather simple raphe system may reflect evolutionary reduction in relation to the endotubular habit. The combination of plastid and raphe is sufficient to separate this genus from other naviculoid taxa with a similar, simple valve structure (e.g. *Sellaphora*), while the raphe and valve structure is quite different from that in *Navicula*, which has a similar plastid arrangement. The nomenclature and frustule structure of two species have been discussed by Cox (1978). There are numerous species in this genus which is extremely abundant in collections from coastal sites; the allocation of species to *Parlibellus* is dealt with in detail in Cox (1988). The affinities of *Parlibellus* are unclear. We are led to place it in the Berkeleyaceae by the simple valve and raphe structure, the porous girdle bands, and the capacity to produce mucilage tubes.

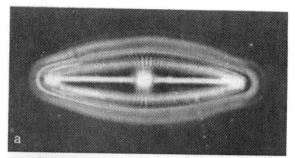

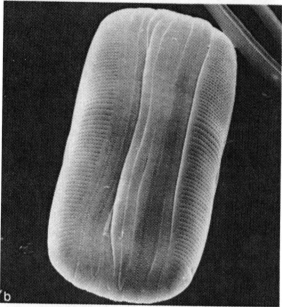

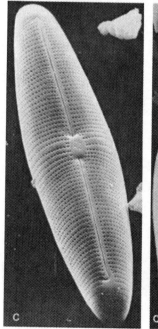

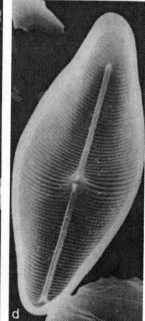

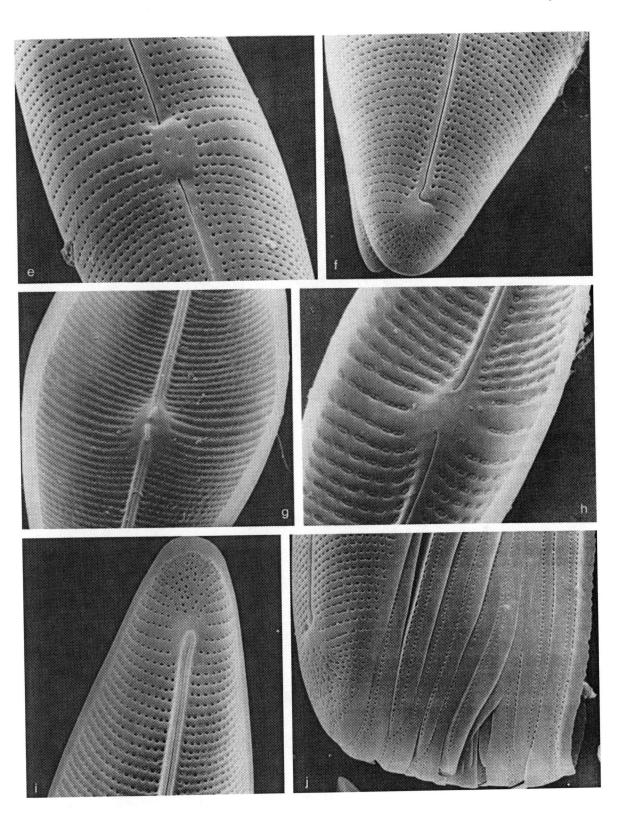

Berkeleya R. K. Greville 1827. Scott. Crypt. Fl. 5: 294

T: *B. fragilis* Greville

Cells solitary, or forming mucilage tubes[a] or films; as wide or wider along the pervalvar axis than along the transapical so that girdle views are frequent. One plastid, consisting of two girdle-appressed plates connected by a narrow isthmus, which is often not central but slightly nearer to one pole. Brackish to marine, abundant world-wide; epilithic and epiphytic, often coating the rock surface in intertidal pools or on coralline red algae.

Valves linear to linear-lanceolate with bluntly rounded or slightly capitate poles.[b, c, f, g] The valve face grades into the mantle, the latter being fairly shallow. Striae uniseriate, consisting of simple poroids closed by hymenes. The raphe-slits are short in *B. rutilans*,[f, g] longer in *B. fragilis* and *B. micans*, and separated by a narrow axial sternum.[b-e] Only the primary half of each raphe-sternum is continuous with this central sternum so that the raphe slits appear to be lateral to the pattern centre in a manner reminiscent of *Eunotia* and *Peronia*. Immediately adjacent to the sternum or raphe-sternum the poroids are often much wider transapically than elsewhere.[d, e] Sometimes these poroids open externally into a shallow, longitudinal groove. External raphe endings straight or slightly deflected toward the secondary side. There are no ribs bordering the raphe internally and the central internal endings are straight, sometimes with small helictoglossa-like lips (Cox, 1975b). The mature cingulum consists of at least 5 open bands,[i] each of which bears 2 rows of round or oval poroids.

Giffen (1970) proposed that *Berkeleya* should be restored for species placed in *Amphipleura*. Later *Berkeleya* was treated systematically by Cox (1975a, b, 1979a) who pointed out that *Amphipleura pellucida* was distinct and should be kept separate. *Berkeleya* is probably most closely related to *Climaconeis* (for which see p. 520 and Cox, 1982) and the papers by Cox should be consulted for a fuller discussion.

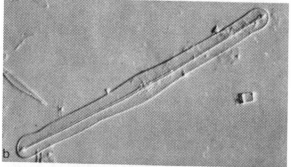

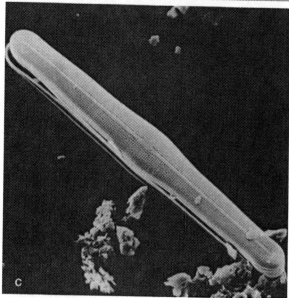

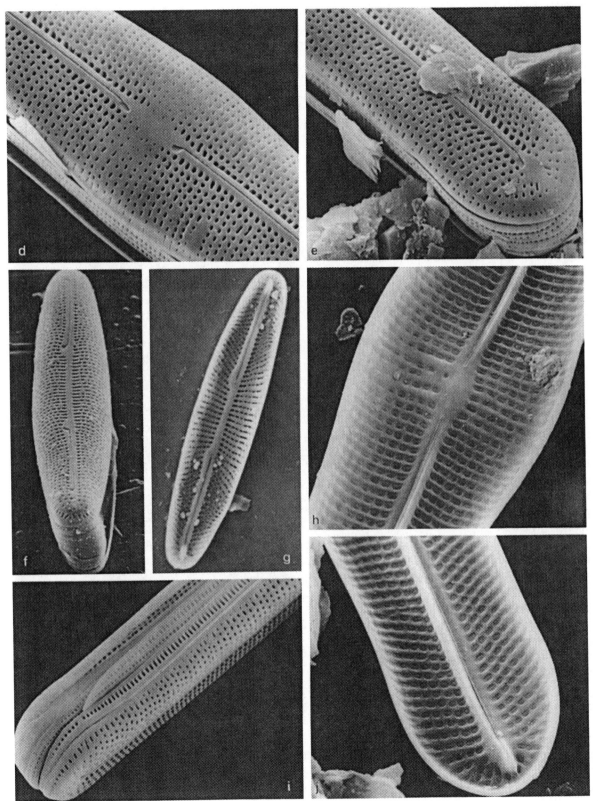

Climaconeis A. Grunow 1862. Verh. zool.-bot. Ges. Wien, **12**: 421

T: *C. lorenzii* Grunow

Cells solitary, elongate, usually lying in valve view. Four to 20 H-shaped plastids in a line along the cell,[a] each composed of two girdle-opposed plates connected by a central bridge containing a pyrenoid. Marine, epipelic.

Valves narrowly linear, linear-lanceolate or lunate, with rounded or slightly capitate poles; centre sometimes inflated.[b] Valve face flat, curving into fairly shallow mantles.[c] Striae uniseriate, containing round or oval poroids, which are closed internally by hymenes: poroids adjacent to the raphe-sternum often separated from the adjacent ones[c, d] or larger than elsewhere. Striae continuing around the valve apices.[d, g, i] A stauros is present in some species.[l] Raphe central and straight, or biarcuate and sometimes eccentric. Raphe-sternum thickened internally.[e-g, j-l] Terminal fissures absent.[d, i] External central raphe endings simple or slightly expanded,[c] sometimes deflected towards the ventral margin in lunate forms; internal central endings simple,[f] polar endings with helictoglossa.[g] Girdle bands numerous,[d] each bearing two transverse rows of poroids. The valvocopulae sometimes bear 'craticular bars'[h, j, k] which grow out from both sides of the band and interdigitate in the centre (cf. *Climacosphenia*); pars interior sometimes developed into an inwardly directed flange.[j, k]

This is one of several cases where early data on plastid structure (Mereschkowsky, 1901a), which supported the separation of a genus from *Navicula*, were ignored by later workers. *Climaconeis* is closely related to *Berkeleya*. There are many nomenclatural problems with *Climaconeis* species, since many have at times been referred to other genera, including *Stictodesmis*, *Climacosphenia*, *Denticula*, *Amphipleura*, *Frustulia*, *Okedenia* and *Navicula*. The 'craticular bars' bear no relation to the internal valves formed by *Craticula*. The genus is excellently reviewed by Cox (1982).

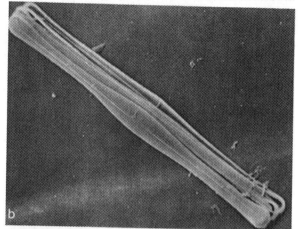

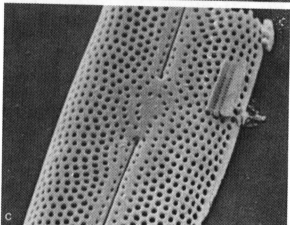

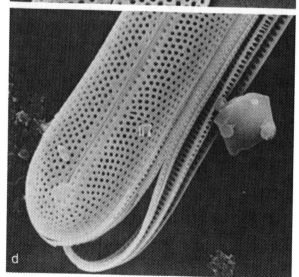

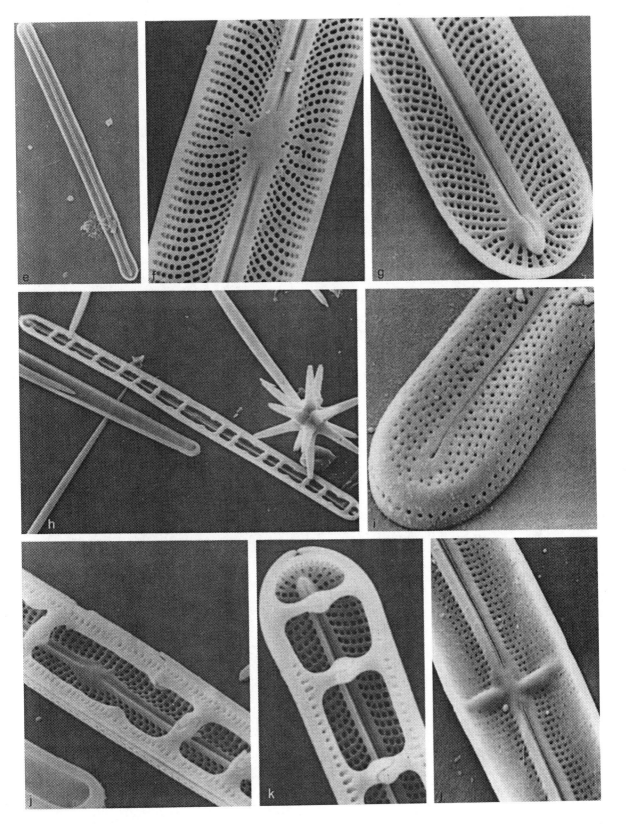

Stenoneis P. T. Cleve 1894. Kongl. Svenska
Vetensk.-Akad. Handl. Ser. 2, **26** (2): 123

T: *S. inconspicua* (Gregory) P. T. Cleve
(= *Navicula inconspicua*)

Cells solitary, bacillar.[a] Plastids unknown. A small
genus, of which only one species is commonly
recorded, occurring in benthic marine habitats.

Valves linear,[a, b] with bluntly rounded poles. Valve
face flat,[d] curving down at its edges into extremely
shallow mantles.[d] Striae uniseriate,[c-e] containing
small round poroids. The arrangement of the
poroids is often rather disorderly; towards the centre
of the valve the striae are interrupted by irregular
clear areas,[c] which are wider centrally than towards
the poles, and whose long axes are orientated
parallel to the raphe. Raphe-sternum thickened
internally to form two prominent longitudinal ribs,[g-i]
one on either side of the raphe, from which the
transapical ribs arise. At the centre the raphe is also
bordered externally by ribs, though these are of
limited extent.[c] External central raphe endings
markedly expanded;[f] terminal fissures more-or-less
absent, the external fissures ending in expanded
pores, which are deflected slightly to one side.[d, e]
Internal central endings turned to one side;[g]
helictoglossa simple, not forming a compound
structure with the ribs of the raphe-sternum[h, i]
(contrast *Frustulia, Berkeleya, Climaconeis*). Girdle
not studied in detail.

The validity and relationships of this genus
require further study. *Climaconeis* has a similar valve
structure and some features of the raphe system are
also alike; it may prove that *Climaconeis*, as
modified by Cox (1982), cannot be separated from
Stenoneis, in which case *Stenoneis*, as the later
synonym, would have to be abandoned.

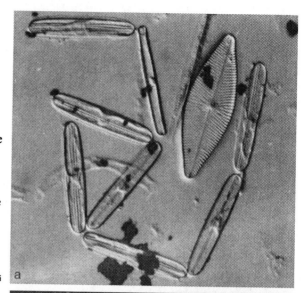

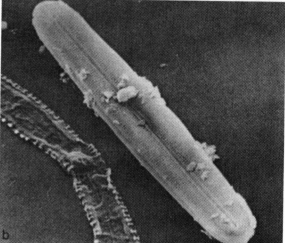

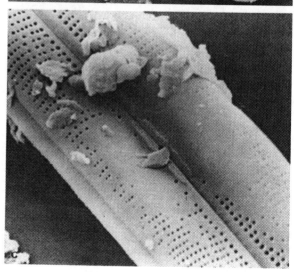

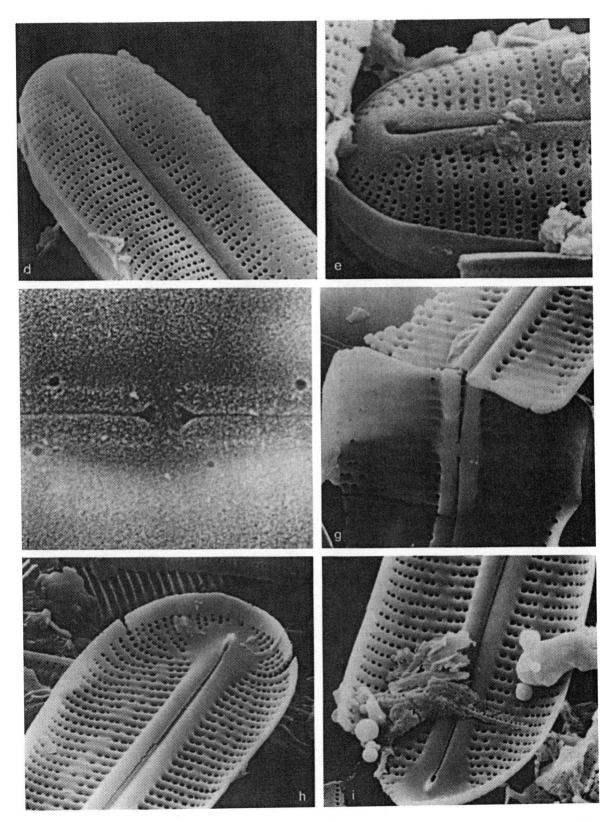

Cavinula D. G. Mann & A. J. Stickle gen. nov.

T: *C. cocconeiformis* (Gregory ex Greville)
Mann & Stickle (= *Navicula cocconeiformis*)

Cells solitary, naviculoid, usually lying in valve view.
One or two plastids; if two, one towards each pole.
Each plastid H-shaped in girdle view, composed of
two plates lying against the valves, connected by a
thick bridge containing the pyrenoid; the plates are
usually highly and irregularly lobed but can be
simpler, with indentations beneath the raphe. A
small genus apparently restricted to freshwater,
occurring in the epipelon of oligotrophic lakes or in
shady, moist subaerial habitats.

Valves linear-lanceolate or rhombic-lanceolate[a] to
more-or-less elliptical,[b] sometimes with slightly
rostrate poles. Valve face flat, curving fairly abruptly
into shallow mantles. Striae uniseriate, fine, radiate,
containing small round poroids[b-e] (somewhat
transversely elongate near the central raphe
endings[b, d]). Poroids are occluded internally by
hymenes (probably eroded in Figs f, h) which within
a stria are more-or-less confluent,[i] forming a strip
along the inside of the stria. Raphe-sternum central,
slightly thickened internally.[f, h] External central
raphe endings expanded, pore-like. Terminal
fissures present, abruptly bent to one side[c, e] but
sometimes very short;[b, d] fissures turned in opposite
directions at the two poles so that externally the
raphe is slightly sigmoid. Internal central endings
unexpanded, straight,[h] sometimes accompanied by
small siliceous papillae.[i] Girdle composed of narrow
open bands, each bearing two transverse rows of
tiny poroids.

This genus, like *Sellaphora*, *Craticula*, *Lyrella*,
Placoneis and others, has traditionally been placed
in *Navicula*. However, it exhibits no resemblance to
Navicula sensu stricto beyond the overall naviculoid
plan of the valve. No close relatives of *Cavinula* are
yet known, although these may be detected once
some of the smaller and rarely encountered
'*Navicula*' species are re-investigated with the EM
and in the living state.

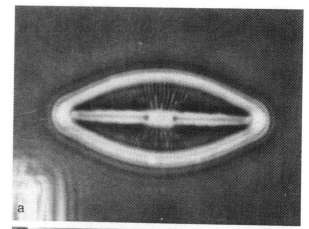

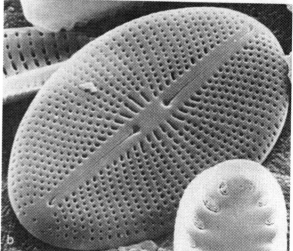

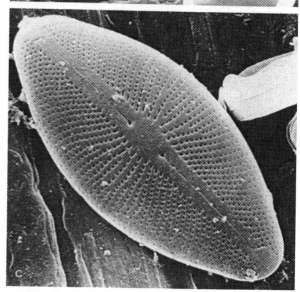

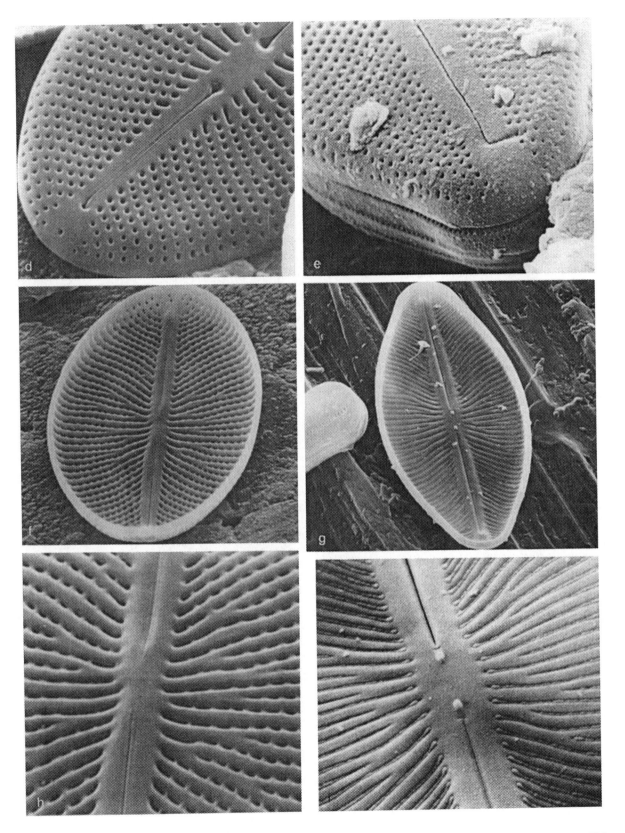

Cosmioneis D. G. Mann & A. J. Stickle gen. nov.

T: *C. pusilla* (W. Smith) Mann & Stickle (= *Navicula pusilla*)

Cells solitary, naviculoid, lying in valve[a] or girdle view.[b, c] Two H-shaped plastids, one against each valve. A small genus, including the type species and probably 2 or 3 other species previously included in *Navicula*. The type species is commonest in subaerial, calcareous, freshwater habitats.

Valves lanceolate or elliptical, with rostrate or strongly capitate poles.[a, d, e] Valve face flat, curving fairly abruptly into the deep mantles, which, however, become much shallower near the poles.[d, f, j] Striae radial, uniseriate, containing small round poroids occluded by hymenes;[h] the hymenes are internal and more-or-less confluent within a stria. Raphe-sternum slightly thickened internally, expanded centrally to form an elliptical 'central area'[d, e, g, h] Central internal raphe endings anchor-shaped[h]; external endings greatly expanded,[d, e, g] forming ± conical depressions. External raphe fissures slightly expanded above the helictoglossa;[d, f, k, l] terminal fissures curved towards the secondary side of the valve. Cingulum[i–k] containing at least 7 open bands, each bearing one transverse row of small round poroids. Band 1 (valvocopula) is perforated by a large, apparently unoccluded hole, which lies next to the valve margin at one pole of the cell;[j] at the opposite pole there is a similar but smaller hole in band 2.[i]

Cosmioneis is superficially similar to *Petroneis* and both have traditionally been classified together in the same section (sect. *Punctatae*) of *Navicula sensu lato*. There are several differences between *Cosmioneis* and *Petroneis*, each apparently insignificant on its own, but combining with the others to make generic separation necessary. They differ with respect to pore occlusions, central raphe endings, coarseness of the valve pattern, cingulum structure and mantle depth, and in whether the mantle becomes shallower near the poles. Furthermore, the plastids of *Cosmioneis* are not constricted in the median transapical plane and the nuclei do not always divide on the same side of the cell, so that both *cis* and *trans* frustules are formed (*Petroneis* produces *cis* frustules only). The two genera are not closely related.

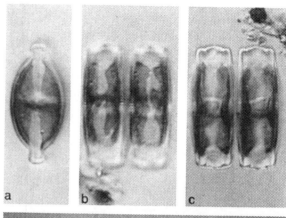

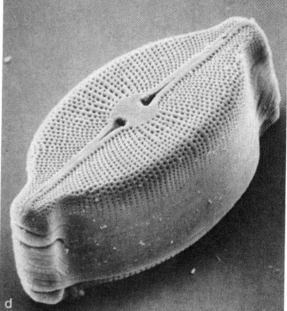

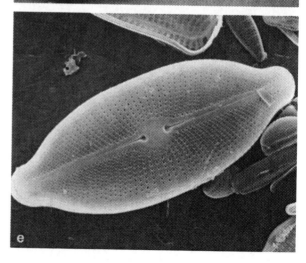

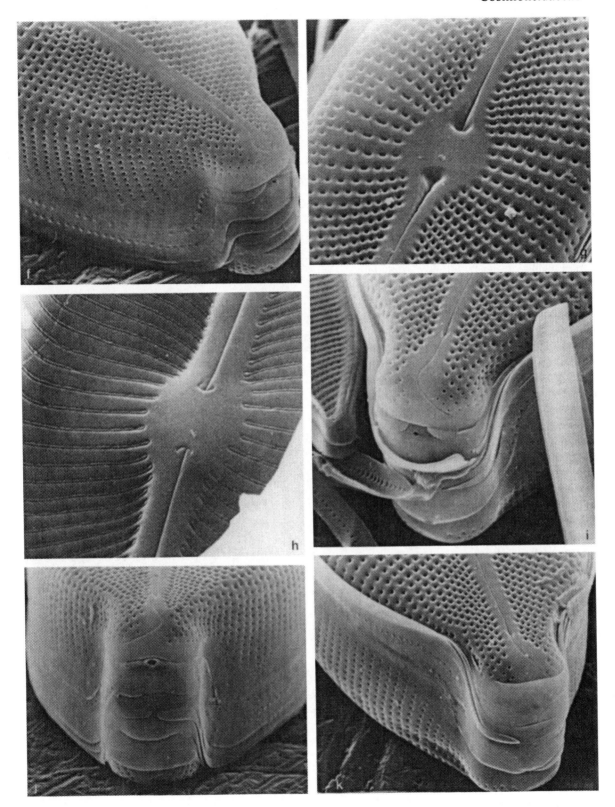

Scolioneis D. G. Mann, gen. nov.

T: *S. tumida* (Brébisson) Mann
(= *Navicula tumida*)

Cells solitary, twisted about the apical axis,[b, c] usually lying in girdle view. Two H-shaped plastids, one against each valve; the plastids sometimes appear to be connected on one side of the cell, across the girdle, by a narrow isthmus. Each plastid contains several rounded pyrenoids. A small genus restricted to brackish water, silts and sands.

Valves linear with acutely or bluntly rounded poles; twisted and hence the raphe system apparently sigmoid.[a, c, d] Valve face curving smoothly into deep mantles. Striae uniseriate,[e, f] containing small round poroids, which open externally by linear or forked slits and are occluded by hymenes. Valves everywhere one-layered (contrast *Scoliopleura, Scoliotropis*). Raphe-sternum usually narrow, expanded slightly in the centre.[f] External central raphe endings simple; terminal fissures straight, continuing to the valve margin. Internal central endings T-shaped;[h] helictoglossae elongate.[i] Girdle composed of several open bands, of which the most advalvar bears one transverse row of poroids and the others two.[g] Valvocopula toothed along the advalvar edge.[d, e]

Formerly included in *Scoliopleura* but lacking the offset central raphe endings and longitudinal canals of that genus. Distinguished from *Scoliotropis* by having fewer plastids, which lie against the valves rather than the girdle, by the simple uniseriate striae and by the raphe structure.

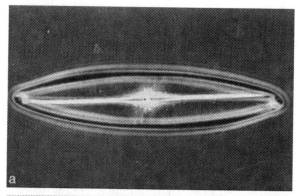

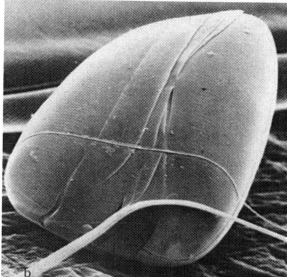

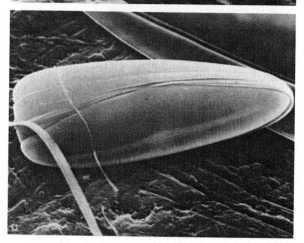

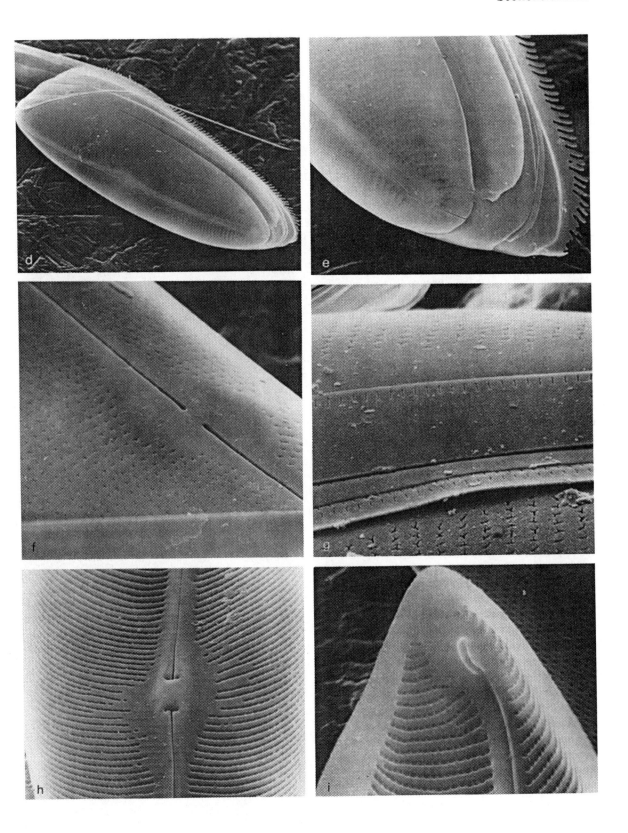

Diadesmis F. T. Kützing 1844. Kies. Bacill.
Diat.: **109**

T: *D. confervacea* Kützing

Cells small (usually <20 μm), often more-or-less
bacillar;[a, b] sometimes solitary but often forming
band-like colonies[c, e] in which the cells are
connected by their valve faces. Individual cells lie in
valve or girdle view with approximately equal
frequency. One simple or slightly lobed plastid,
lying against one side of the girdle and one valve;
sometimes also extending under the other valve. A
small freshwater genus, virtually restricted to
subaerial habitats, e.g. damp moss or rock.

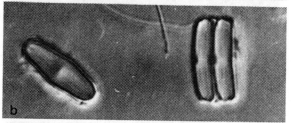

Valves linear or linear-lanceolate, with bluntly
rounded poles. Valve face flat, sharply differentiated
from fairly shallow mantles.[c-g] Junction of valve face
and mantle often bearing a ridge of silica,[d] or a row
of short spines or projections.[e, f] In some species the
projections of sibling valves interdigitate (cf. some
Fragilaria); elsewhere the integrity of colonies is
maintained by the close fit of the valve faces,
polysaccharide secretion, and girdle overlap. Striae
uniseriate, containing round or (especially on the
mantles) transapically elongate poroids,[c-g] which are
occluded by hymenes. The hymenes lie more-or-less
flush with the inner surface of the transapical ribs,
making the valve interior appear somewhat
featureless;[h, j] only when the hymenes have been
eroded away do the poroids become obvious.
Raphe-sternum central, relatively broad. Raphe-slits
sometimes filled in (cf. the P-valves of
Achnanthidium or *Cocconeis*), producing valves that
superficially resemble those of fragilarioid genera.[f]
Central raphe endings and external polar endings
all simple or T-shaped. Cingulum consisting of a
few to many open bands, each with one or 2
transverse rows of round or slit-like poroids.[d, k]

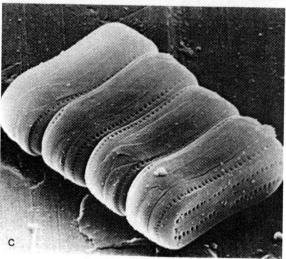

A genus fairly easily separated from other
naviculoid diatoms by the combination of stria and
raphe structure, valve shape and (often) colonial
habit. *Luticola* is often found in the same habitats
but has a different raphe structure, a stigma and a
more elaborate plastid. Nevertheless *Diadesmis* and
Luticola appear to be closely related. *Diadesmis* also
shows some similarities to *Brachysira*. The girdle
structure of *D.* (= *Navicula*) *confervacea* has been
studied in detail by Rosowski (1980), and
polymorphism in *D. gallica* by Granetti (1977, 1978).

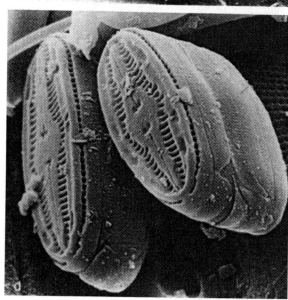

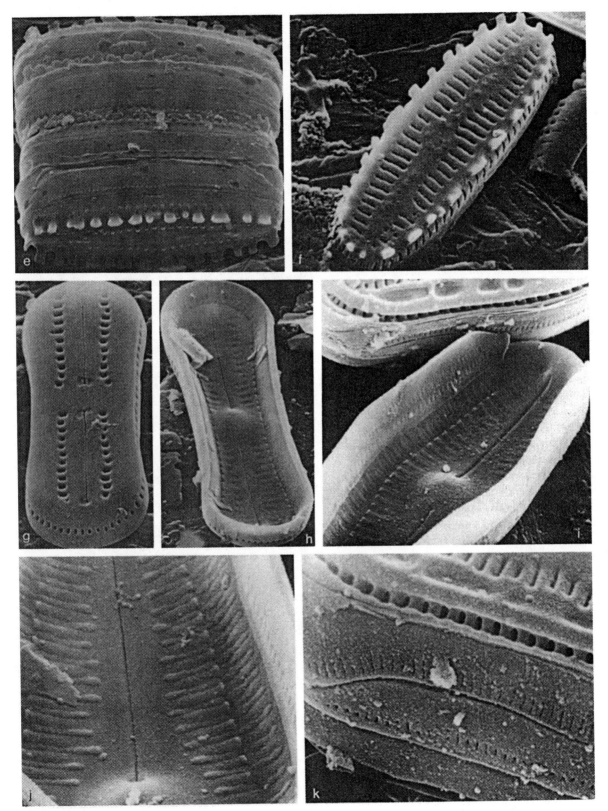

Luticola D. G. Mann, gen. nov.

T: *L. mutica* (Kützing) Mann (= *Navicula mutica*)

Cells solitary or (rarely) forming chains, naviculoid, usually lying in valve view.[a, b] One plastid, lying with its centre against one side of the girdle; from here 2 lobes extend beneath each valve, one on either side of the median transapical plane. The lobes are longitudinally indented beneath the raphe. There is a single, central pyrenoid. A fresh or slightly brackish water genus, commonest in soils or subaerial habitats and in estuaries.

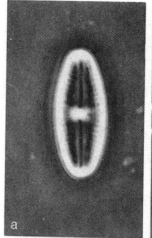

Valves linear, lanceolate or elliptical, with bluntly rounded to capitate poles;[c, d] valve face flat, fairly sharply differentiated from the mantles; junction of valve face and mantle sometimes bearing spines. Valve margin notched approximately half-way between centre and pole in each quadrant of the valve.[c-f] Striae uniseriate, each containing several more-or-less round poroids on the valve face and one round pore on the valve mantle. The poroids of the valve face are occluded by hymenes,[g, i] which are confluent within a stria, forming a transapical strip across the valve; the mantle pores have their own hymenes. Where valve face and mantle join, there is a longitudinal canal within the thickness of the valve, slightly reminiscent of the canal in *Neidium*. On one side of the valve, opposite the stigma, the inner wall of the canal is sometimes elaborated into a flap-like structure extending in towards the centre of the valve.[f, g, i] Raphe-sternum fairly narrow, expanded and thickened centrally to form a short stauros, penetrated on one side by a single stigma; this opens internally by a curved, lipped slit[f-i] and externally by a simple pore.[c, d] Central internal raphe endings straight; simple or slightly lipped. Central external endings deflected, curved or abruptly bent away from the stigma. External polar endings usually curved in the opposite direction from the central endings.[c, d] Girdle containing open bands, each normally with one or two rows of tiny round poroids;[e] band 1 bears a series of coarse projections.[j]

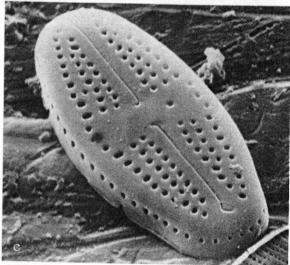

The combination of raphe and pore structure, and the unique type of stigma, serves to separate *Luticola* from *Diadesmis*. There is little risk of confusion with any other genus, except perhaps *Placoneis*, providing live cells are examined. Phipps & Rosowski (1983) and Mayama & Kobayasi (1986) have described some aspects of valve and girdle structure.

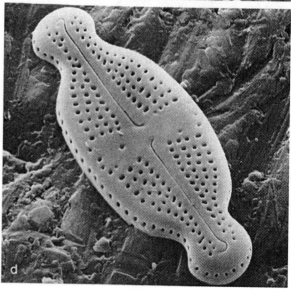

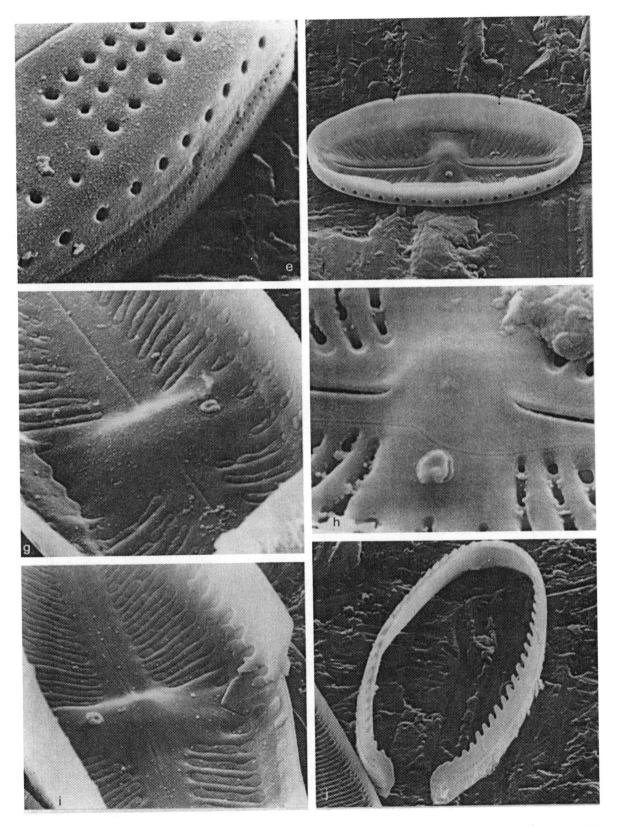

Frickea H. Heiden 1906. A. Schmidt Atlas Pl. 264, Fig. 1.

T: *F. lewisiana* (R. K. Greville) Heiden
(= *Navicula lewisiana*)

Cells naviculoid. Habit, girdle and plastid structure are unknown. Brackish. A rarely recorded monotypic genus.

Valves broadly linear.[a] Valve face and mantles poorly differentiated.[b] Straie fine relative to the size of valve, uniseriate, containing small round areolae.[b-d] There is a tendency towards the formation of loculi through the narrowing of each poroid towards the inner and outer faces of the valve.[i] The striae are exactly parallel over most of the valve and run unbroken externally from the extremely narrow raphe-sternum to the margin. Some of the areolae, however, are blocked internally by the prominent ribs[f-h] running parallel to the raphe. Raphe system straight,[a] central; stopping well short of the pole.[d] External polar and central raphe endings similar and T-shaped.[b-c] Central internal endings simple.[g] At each pole internally there is a very long helictoglossa[f, h] which is very obvious in the LM. This is not fused to the longitudinal ribs, which at first run parallel to the helictoglossa and then run unbroken from one pole to the other with only a slight change of direction centrally, to accommodate the prominent nodule between the central raphe endings. Each rib is separated from the raphe by about 2 rows of poroids which represent the first 2 poroids of each transapical stria. This contrasts markedly with *Frustulia* where the ribs are borne by the raphe-sternum itself.

The structure of *Frickea*, in particular the raphe system, the fineness of the valve structure and the elongate helictoglossa, suggests a relationship with *Frustulia*. There are differences, however, in the position of the longitudinal ribs and the lack of fusion between these and the helictoglossa. We are indebted to P. A. Sims who provided us with material – the fragmentary nature of the material is due to our mis-treatment!

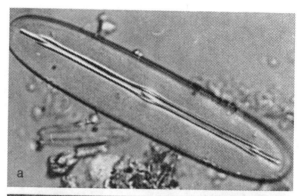

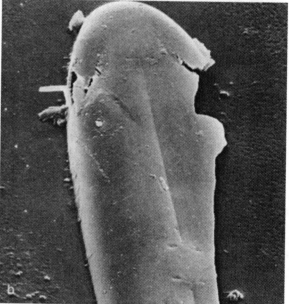

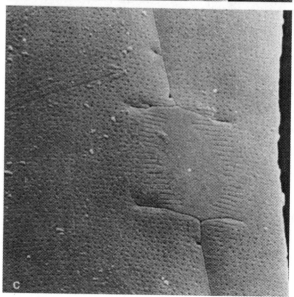

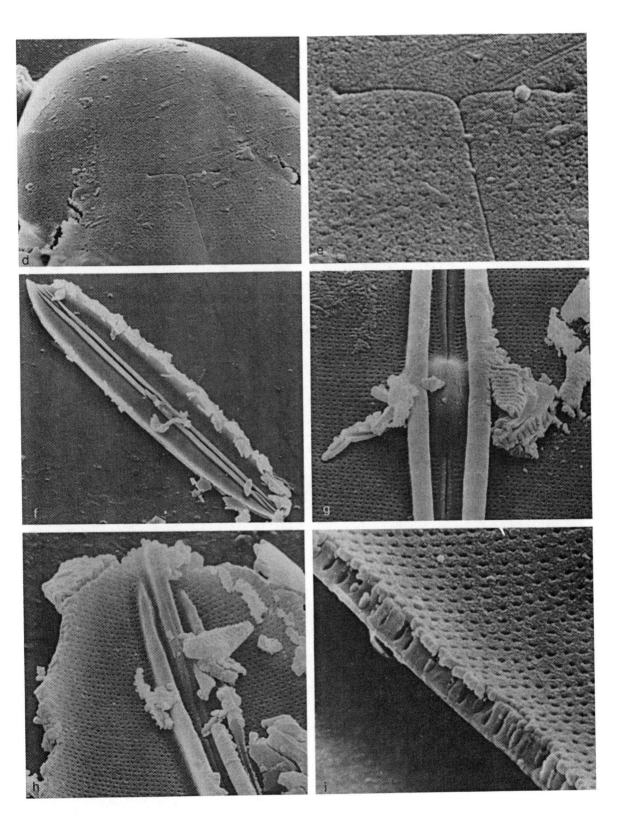

Amphipleura F. T. Kützing 1844. Kies. Bacill. Diat.: **103**

T: *A. pellucida* (Kützing) Kützing (= *Frustulia pellucida*; lectotype selected by P. T. Cleve 1894. Kongl. Svenska Ventensk.-Akad. Handl. 26 (2): 106)

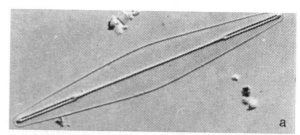

Cells solitary, naviculoid, usually lying in valve view. One central H-shaped plastid lying against one valve and extending beneath the girdle, containing a central pyrenoid. Freshwater; epipelic and loosely associated with filamentous algae around plants.

Valves linear-lanceolate or linear with rounded or rather acute poles.[a, e] Valve flat, curving into very shallow mantles; valve margin thickened and rib-like.[f-j] Valve structure fine and delicate (hence the use of *A. pellucida* as a test object for microscope lenses since the striae are 37–40 in 10 μm). Striae uniseriate, at right angles to sternum and raphe, containing closely and very regularly spaced poroids.[c, f] Each poroid opening to the exterior by a narrow, apically orientated slit and to the interior by a circular aperture, across which is a hymen.[j] Raphe-slits short and restricted to near the poles;[a, e, g, i] sometimes longer but slits always separated by a long, narrow axial sternum, which projects internally as a relatively massive rib.[f] This, together with the two ribs that run alongside each raphe-slit, is the most obvious feature of the valve. The two ribs enclosing the raphe are joined to the sternum and are also fused to the helictoglossa[g-j] to form the characteristic 'porte-crayon' shape described by Cleve (1894–5): this referred to the predecessor of the modern propelling pencil. The internal and external central raphe endings, and the external polar endings, are alike, being simple, straight and very slightly expanded. Girdle incompletely known, but apparently consisting of open bands, each with two transverse rows of poroids.

A small genus most closely related to *Frustulia*. Cox (1975a, b) has shown that several brackish or marine species formerly included in *Amphipleura* (*A. micans, A. rutilans*) should be placed in *Berkeleya* The type is the only commonly recorded species; its ecological requirements are poorly known but it is usually found in small numbers and when abundant it is tolerating slightly alkaline or somewhat acid conditions in ponds, streams or bogs.

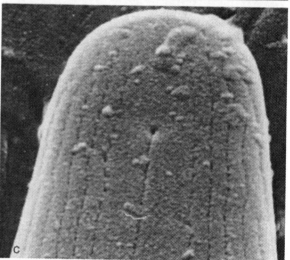

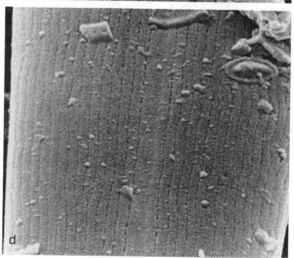

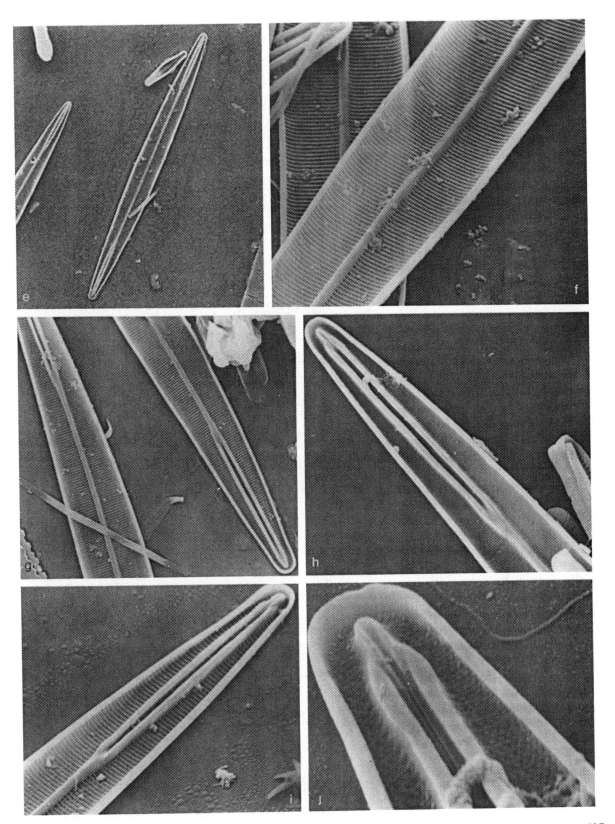

Frustulia L. Rabenhorst 1853. Süssw.-Diat.
Freunde Mikr.: **50** (nom. cons.)

T: *F. saxonica* Rabenhorst (typ. cons.)

Cells solitary or within mucilage tubes; single cells
usually lying in valve view. One H-shaped plastid
lying against one valve and extending beneath both
sides of the girdle. Freshwater to slightly brackish,
epipelic or loosely associated with macrophytes; the
F. rhomboides complex is particularly characteristic
of acid, peaty waters.

Valves linear-lanceolate to lanceolate,[a, h]
sometimes with capitate poles. Valves with flat valve
faces and ill-defined, shallow mantles; margins
sometimes slightly thickened. Striae closely spaced,
parallel, uniseriate,[c-e] containing poroids closed by
hymenes at their inner apertures;[f-i] outer apertures
of poroids slit-like or circular. Raphe system straight
or slightly biarcuate. Raphe-sternum bearing
prominent ribs of silica internally,[f-i] one on either
side of the raphe, running parallel and close to it.
At the poles these longitudinal ribs are fused to the
helictoglossa to give a 'porte-crayon' ending;[f, g] the
ribs are sometimes also fused centrally to each
other[f, h] and to the nodule that occurs between the
central raphe endings in these forms. Central
internal raphe endings simple.[h] External endings,
both polar[e] and central,[c, d] usually T- or Y-shaped;
occasionally the external fissures are extended into
blind grooves centrally, which are deflected in the
same direction (*F. weinholdii*). Girdle containing
open bands bearing one or two transverse rows of
poroids. In *F. rhomboides* band 1 (the valvocopula),
and perhaps others, is folded into a tube;[j, k] this
opens to the valve interior by poroids and to the
outside by a slit, which is expanded centrally and *in
vivo* lies by the valve margin.

Ross & Sims (1978) erected *Berkella*, considering
this to be intermediate between *Frustulia* and
Berkeleya. There is a little evidence that *Berkella* is
closely related to *Berkeleya*; its resemblance to *F.
vulgaris* suggests either that *Frustulia* must be
remodelled to exclude *F. vulgaris* (which would then
have to be transferred to *Berkella*) or, as we prefer,
that *Berkella* should be combined with *Frustulia*.
Frustulia is most closely related to *Amphipleura*.

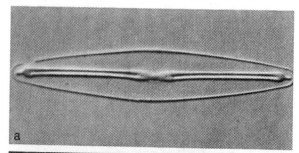

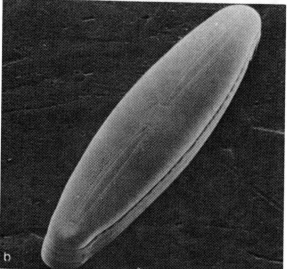

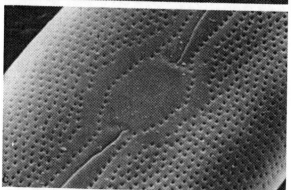

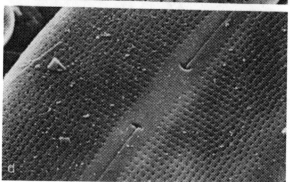

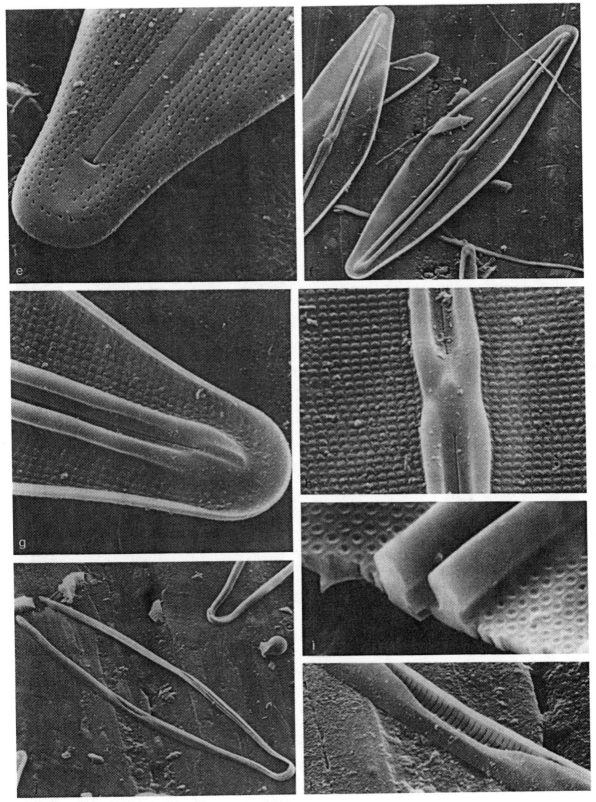

Brachysira F. T. Kützing 1836. Algarum Aquae Dulcis German. **16**: 153

T: *B. aponina* Kützing

Cells solitary, naviculoid, lying in valve or girdle view. One plastid. Freshwater or marine (1 sp.) to hypersaline; epipelic. Especially common in oligotrophic lakes and bogs.

Valves linear, lanceolate or rhombic, with bluntly rounded or capitate poles.[a-d] Valve face flat, usually ornamented with warts,[b, d, e] spines or longitudinal ribbing.[c, f] A prominent marginal ridge or clear zone runs around the whole valve face.[b-h] Striae uniseriate, containing transapically elongate poroids, which are closed by hymenes across their inner apertures.[i] Raphe-sternum narrow but sometimes expanded centrally, bearing longitudinal ribs externally;[b-h] these are especially prominent in Fig. e. Central internal and external endings simple, straight. No terminal fissures but raphe sometimes ending in a T-shaped depression over which a flap of silica extends.[d, g, h] Cingula composed of open copulae.[g, j, k] Valvocopula folded inwards at its junction with the valve margin to form a long chamber, which opens to the outside by a narrow crack,[k] and to the inside by a row of round or elongate poroids. The more abvalvar bands sometimes also each bear one transverse row of poroids.

This genus was re-established by Round & Mann (1981) when material of *B. aponina* was discovered. A group of freshwater species usually referred to *Anomoeoneis* (*vitrea, serians, styriaca, zellensis*) resemble *B. aponina* very closely but are quite different from the type of *Anomoeoneis, A. sphaerophora*. They must therefore be put in *Brachysira*. It is a remarkable genus in that we have found *B. aponina* in marine samples from all over the world; the other species are equally widely dispersed but only in freshwater. Some are virtually confined to and abundant in acidic (low conductivity) freshwaters and others are reported from more alkaline sites. It is hard to see how such disjunct distributions can have arisen.

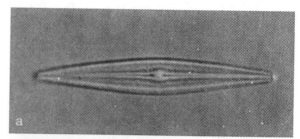

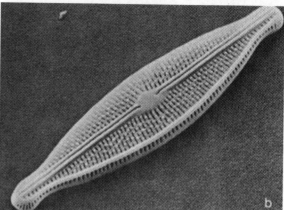

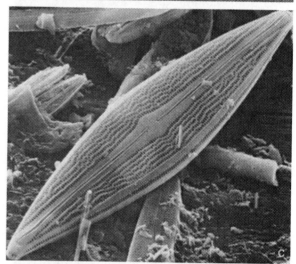

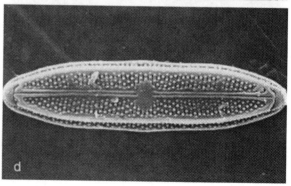

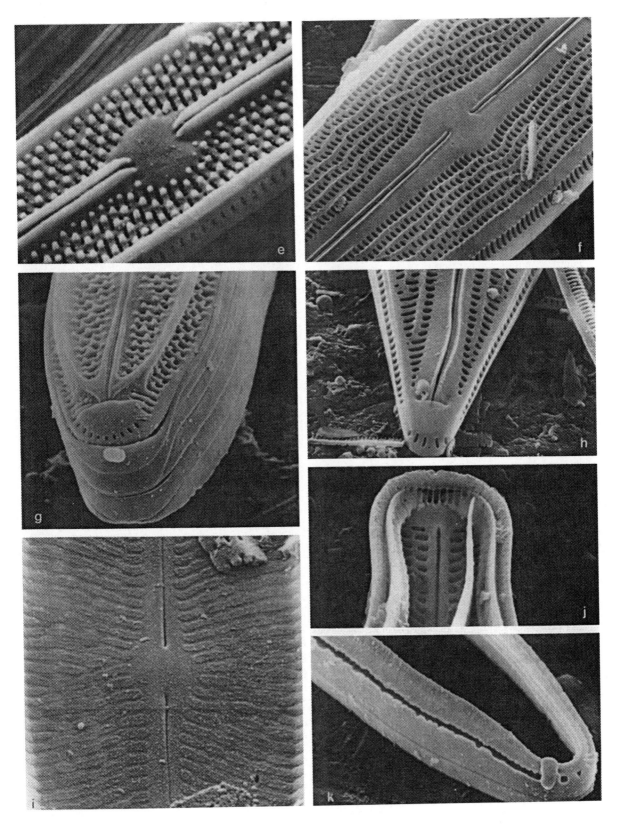

Neidium E. Pfitzer 1871. Bot. Abh. Morphol. Physiol. **1**: 39

T: *N. affine* (Ehrenberg) Pfitzer (= *Navicula affine*; lectotype selected by C. S. Boyer 1928, Proc. Acad. Nat. Sci. Philad. **79** Suppl.: 320)

Cells solitary, naviculoid, lying in valve or girdle view. Usually 4 girdle-appressed plastids, disposed symmetrically, one in each quadrant of the cell; occasionally 2 plastids lying against the epivalve. An exclusively freshwater genus of the epipelon, very widely distributed but rarely abundant.

Valves linear to lanceolate,[a, b] with blunt or rostrate poles, and sometimes constricted centrally. Valve face flat, fairly well differentiated from the deep mantles;[b] mantles becoming shallower towards the poles. Striae uniseriate,[c, d, f] appearing in the LM as lines of round or transapically elongate pores, which sometimes cross the valve at a slight angle to the transapical axis. Valve structure complex, consisting of two layers of porous silica.[h, i] These are usually furthest apart at the junction of valve face and mantle, where the space between them forms a large longitudinal canal;[h, i] nearer the raphe or valve margin the space diminishes and the layers are connected by vertical struts. Each pore visible in the LM represents a loculate areola, partially connected laterally to nearby areolae. The areola opens to the outside by an unoccluded round aperture and to the inside by a round or transapically elongate pore, which is closed by a hymen lying flush with the inner surface of the valve.[e, g] The internal raphe fissures end in helictoglossae at both centre and pole:[e, g] the two central helictoglossae form a compound structure.[f, g] The polar raphe endings are characteristically forked externally into two long, straight terminal fissures,[c] while the central external endings are almost always curved or deflected in opposite directions.[d] Girdle apparently consisting of non-porous, open bands.

A very distinctive genus, set apart from all others by the combination of valve, raphe and plastid structure. The mode of sexual reproduction suggests a link with *Amphipleura* and *Frustulia* (Mann, 1984c).

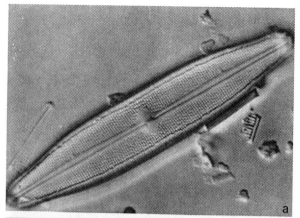

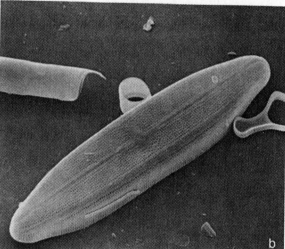

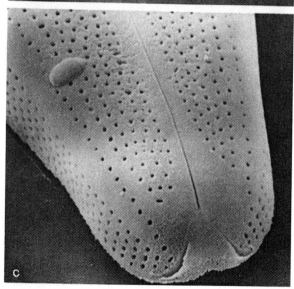

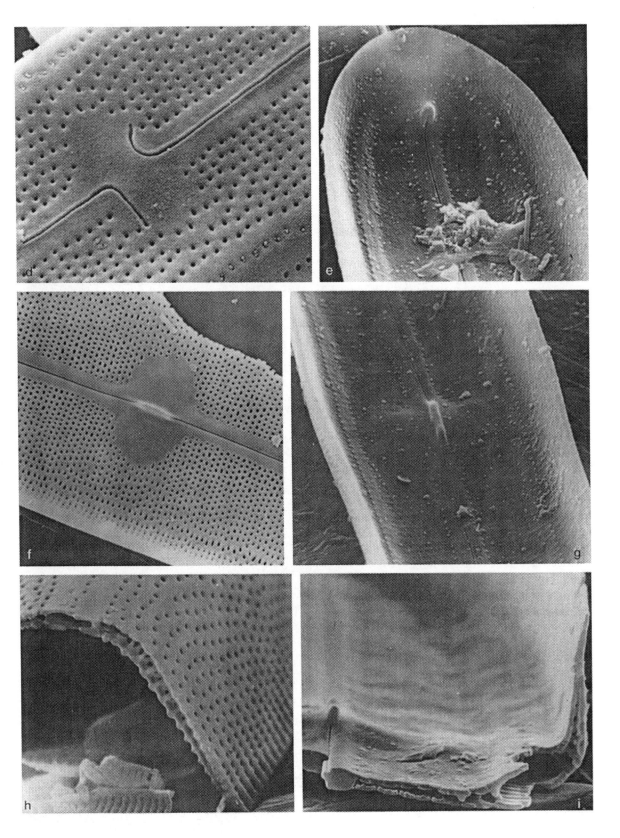

Scoliopleura A. Grunow 1860. Verh. zool.-bot. Ges. Wien **10**: 554

T: *S. peisonis* Grunow (lectotype selected by C. S. Boyer 1928. Proc. Acad. Nat. Sci. Philad. **79** Suppl.: 361)

Cells solitary, lying in girdle view or valve view.[a, b] Frustule slightly twisted about the apical axis. Plastids unknown (reports of plastids in *Scoliopleura* refer to *Scoliotropis*). A small genus occurring in saline lakes (Neusiedler See, Caspian Sea) and brackish waters; epipelic.

Valves linear, with bluntly round[b, c] or somewhat apiculate apices. Valve face curving more-or-less uniformly into fairly deep mantles. Striae loculate, each consisting of a continuous transapical chamber[g, i] that opens to the inside via a finely porous strip and to the outside by a single row of larger round pores. Near the raphe the inner wall of the chamber bulges into the cell,[g, i] here the walls between adjacent striae probably disappear, as in *Scoliotropis* (q.v.), so that a longitudinal canal is formed within the valve framework. Above the canal there is a row of larger pores communicating with the exterior.[b, d, e] Raphe system sigmoid.[b] Central external raphe endings turned to opposite sides;[d] terminal fissures forked.[e] Internal raphe fissures ending centrally in a double helictoglossa.[g, i] Girdle containing open bands.

Very similar to *Scoliotropis* and clearly related to it. There are slight differences in the stria structure and in the raphe (see under *Scoliotropis*), but further investigation may show that these genera are not distinct. Both are quite different, however, from *Scolioneis*. *Scoliopleura sensu stricto* appears to have an interesting biogeography, occupying several of the old landlocked basins of eastern Europe and S.W. Asia.

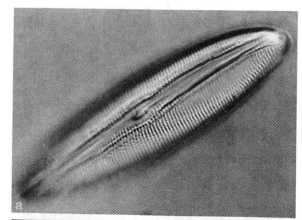

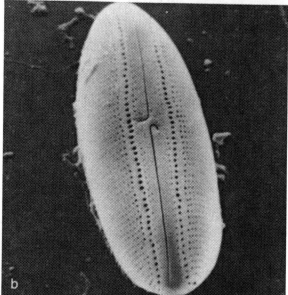

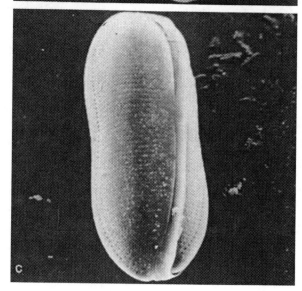

Scoliotropis P. T. Cleve 1894. Kongl. Svenska. Vetensk.-Akad. Handl., Ser. 2, **26** (2): 72

T: *S. latestriata* (Brébisson) Cleve
(= *Amphiprora latestriata*)

Cells solitary, lying more often in girdle view. Frustule slightly twisted about the apical axis.[a, b] Four girdle-appressed plastids, arranged symmetrically about the median transapical and apical planes. A small genus of brackish waters; epipelic on silty sediments.

Valves linear with slightly acute, but rounded poles.[a, b] Valve face curving uniformly into deep mantles.[b, d] Striae loculate, each consisting of a continuous transapical chamber that opens to the outside by 2 rows of small round pores,[d-f] and to the inside by a fine sieve of tiny holes.[g-j] In addition each chamber has 2 large external openings, one adjacent to the raphe-sternum,[b-e] the other adjacent to the margin.[f] Next to the raphe the inner wall of the chamber bulges into the cell and here the walls between adjacent striae disappear, so that a longitudinal canal[h-j] is formed on either side of the raphe-sternum. Raphe very slightly sigmoid.[a, b] Terminal fissures slightly deflected, continuing to the valve margin.[d] Central external endings expanded, lying in shallow depressions;[c, e] internal endings straight, simple, lying in a prominent double helictoglossa.[h] Girdle deep, consisting of open, porous bands,

Similar in shape and habitat to *Scolioneis* but differing in almost all aspects of valve, raphe and plastid structure. *Scoliotropis* is very closely related to *Scoliopleura*, both genera having loculate striae and longitudinal canals near the raphe-sternum. There are minor differences in the positions of the longitudinal canals, in the raphe endings, and in the structure of the striae (*Scoliotropis* has biseriate outer striae; *Scoliopleura*, uniseriate outer striae). Whether these are enough to justify separation is questionable, but we keep the two genera apart since we have insufficient information on some aspects of the cell that may provide important points of contrast, e.g. plastid structure and division.

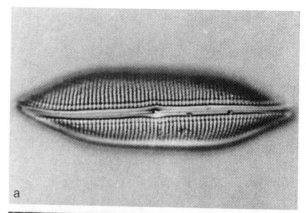

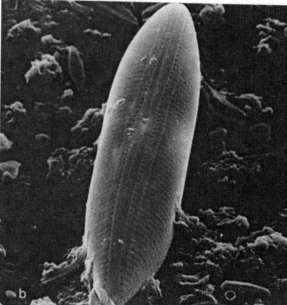

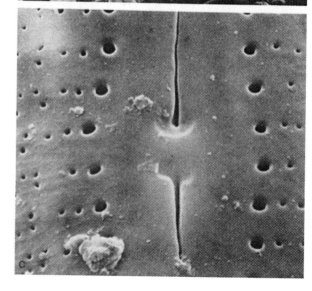

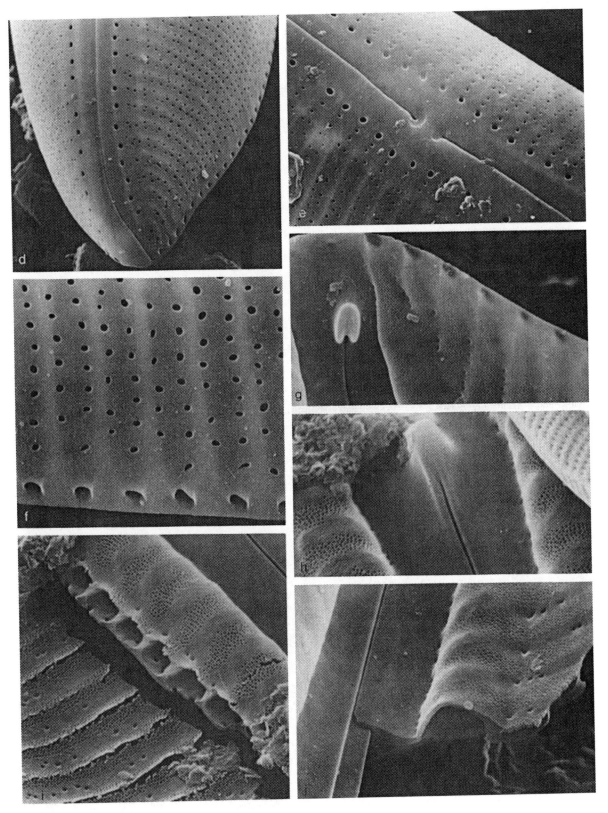

Biremis D. G. Mann & E. J. Cox gen. nov.

T: *B. ambigua* (P. T. Cleve) Mann
(= *Pinnularia ambigua*)

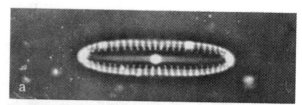

Cells solitary, slightly to markedly dorsiventral, usually lying in girdle view. Plastids 2, arranged one on either side of the median transapical plane; each consisting of 2 plates appressed to the girdle, connected by a bridge containing a large pyrenoid. Common in the epipelon of sandy marine sediments.

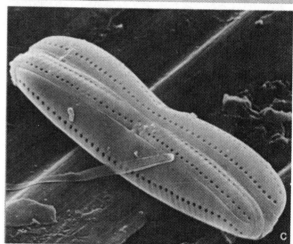

Valves linear, almost bilaterally symmetrical[a] to strongly dorsiventral,[c] when they resemble the valves of *Amphora* (q.v.); deep. Valve face curved, not differentiated from the mantles. Stria structure complex: each stria consists of 2 incompletely separated, transapically elongate chambers, which are bounded internally by a delicate porous strip (cribrum);[i] near the margin and near the raphe-sternum there is a narrow crescent-shaped slit,[g, i, j] providing further avenues of communication between valve interior and chamber. Externally, the chambers open by circular or slit-like foramina, which form 2 longitudinal rows,[c-f] one adjacent to the raphe-sternum, one near the valve margin (it is this pattern that suggested the generic name). Raphe central to eccentric, straight or biarcuate, lying in a wide raphe-sternum. External central endings simple, expanded;[d, e] internal endings forming a double helictoglossa.[g, i] Polar helictoglassa narrow,[j] terminal fissures curving off smoothly and turning back along the valve margin, becoming aligned with the marginal row of foramina.[f] Girdle consisting of several to many open, porous bands; these are at least sometimes wider dorsally, as in *Amphora*.

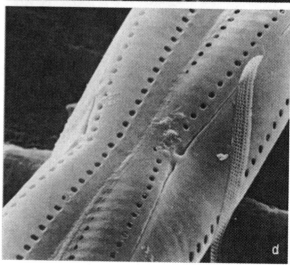

Members of this genus have been classified in *Amphora* (*digitus*, *mülleri*, *ridicula*) or *Pinnularia* (*ambigua*). *Amphora* is a heterogeneous group needing much further revision; *Biremis* is not closely related to the type group of *Amphora* (*A. ovalis* and its allies) and the valve structure sets it apart from those '*Amphora*' species with a similar plastid arrangement. The closest relatives of *Biremis* seem to be *Scoliotropis*, *Scoliopleura* and *Progonoia*. The pattern of adaptive radiation in these genera resembles that in the *Navicula* group, where again a particular combination of protoplast, raphe and valve structure has been retained despite great changes in symmetry and shape.

Progonoia H. Schrader 1969. Nova Hedw., Beih. **28**: 58

T: *P. didomatia* Schrader

Cells solitary, naviculoid, usually lying in valve view. Where known, cells with 4 plastids, arranged in pairs, one pair lying against each valve; within a pair the 2 plastids are placed symmetrically about the median transapical plane, one lying towards each pole. A small and uncommon genus, apparently restricted to the marine littoral.

Valves broad and panduriform,[a, b] with acute apices. Striae restricted to marginal areas, the central part of the valve being occupied by a broad strip of non-porous ('hyaline') silica. Striae complex in structure, each consisting of a chamber that opens to the exterior by a few large pores[b-d] and which is bounded internally by a delicate (?porous) siliceous membrane[e-g] (eroded in 2 of our figures).[h, i] Around the edge of this membrane, where it abuts onto the valve margin[i] or the central 'hyaline' area, there appear to be tiny holes giving access from the cell interior into each chamber. Each chamber is subdivided by a partition, which is continuous between the chambers, forming a longitudinal rib.[f, g] This structure is reminiscent of that in *Diploneis* and especially in *Scoliotropis*, but unlike these genera, there do not appear to be true longitudinal canals. Externally the raphe endings are simple at the centre[c] and turned to the same side at the poles.[d] Internally the raphe slits end in a small 'beak' resembling 2 helictoglossae back-to-back.[f] Girdle apparently consisting of open bands bearing many rows of small poroids.

An ancient genus, recorded from Eocene deposits, whose extant species are poorly known. Valve structure is described by Schrader (1971). *Progonoia* occupies a rather isolated position within the raphid group, though there seem to be links with *Biremis*, *Scoliopleura* and *Scoliotropsis*.

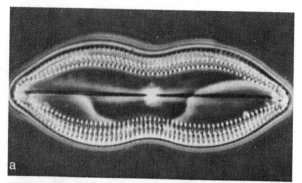

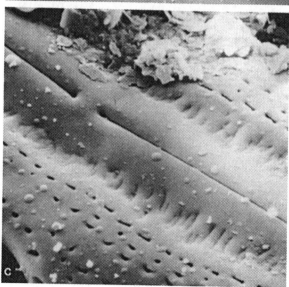

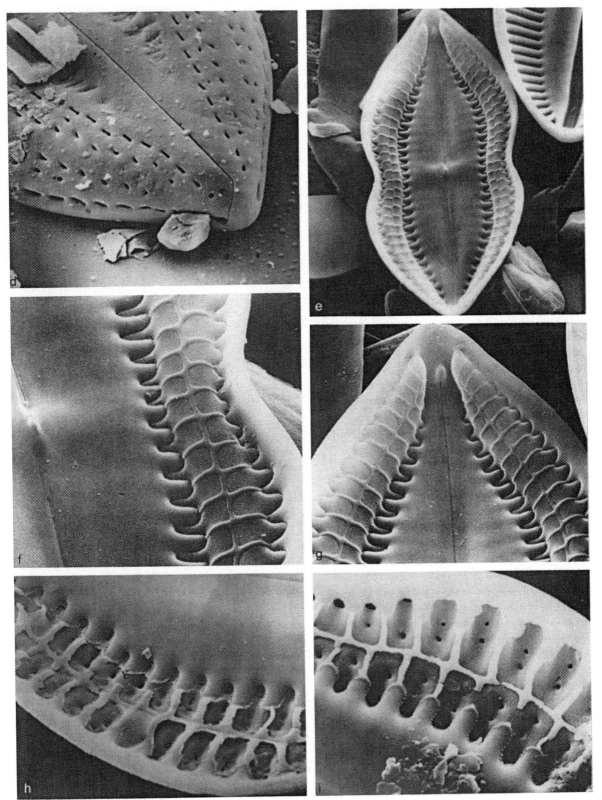

Sellaphora C. Mereschkowsky 1902. Ann. Mag. Nat. Hist. Ser. 7, **9**: 186

T: *S. pupula* (Kützing) Mereschkowsky (lectotype selected by D. G. Mann, 1989b; = *Navicula pupula*)

Cells solitary, usually lying in valve view. A single H-shaped plastid lies with its centre against the epivalve and extends beneath both sides of the girdle and often onto the hypovalve. Plastid containing one, often polyhedral pyrenoid,[a] positioned near the epivalve, usually to one side of the narrow bridge joining the 2 halves of the plastid, or in the bridge itself. A major group in freshwater and also extending into brackish and probably into marine sites; epipelic.

Valves linear to lanceolate or elliptical,[a-d, f] usually with bluntly rounded or capitate poles. Valve face flat, curving fairly gently into shallow or moderately deep mantles; often grooved near the raphe externally. Striae uniseriate,[e-h] containing small round poroids occluded near their internal apertures by hymenes.[j] A non-porous conopeum is sometimes present externally.[f] Transapical, bar-like thickenings occur at the poles in some species. Raphe system central, straight. Terminal fissures usually present, deflected or hooked.[g] Central external endings expanded, slightly deflected towards the primary side.[e] Central internal endings also turned or deflected towards the primary side.[j] Cingulum consisting of a few (in *S. bacillum* apparently 4) open, usually non-porous bands.[g]

A large genus, containing many of the small naviculoid diatoms placed by Hustedt (1927-66) in *Navicula* sects. *Bacillares* and *Minusculae*. Examples of species transferred to *Sellaphora* are *S. pupula, S. bacillum, S. laevissima, S. seminulum, S. disjuncta* (Mann, 1989b). Plastid structure must be examined before a species can be reliably allocated using LM to this genus, or indeed to the other genera we have split from *Navicula. Sellaphora* is easily separated from *Navicula sensu stricto* by the plastid, areola and raphe structure. Plastid and cell division have been described in detail by Mann (1984a, 1985). Crossing experiments have shown that some of the apparently minor form variants of *S. pupula* in Hustedt (1930) are not interfertile and might therefore be considered separate species (Mann, 1984e, 1988a, 1989c).

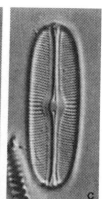
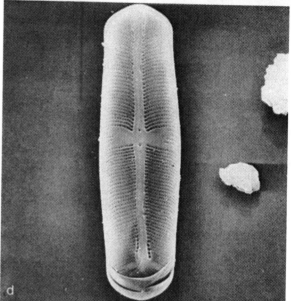
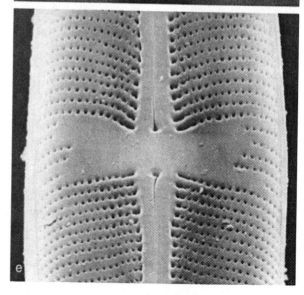

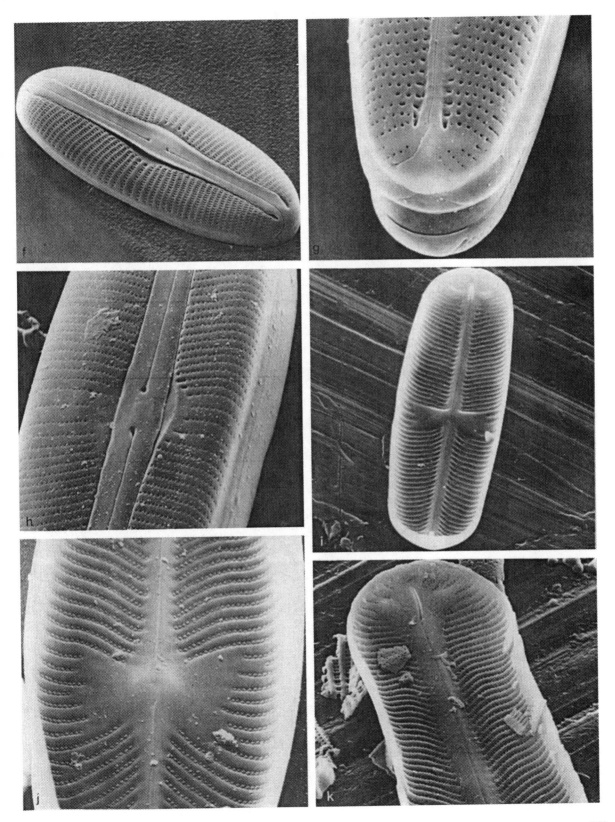

Fallacia A. J. Stickle & D. G. Mann, gen. nov.

T: *F. pygmaea* (Kützing) Stickle & Mann
(= *Navicula pygmaea*)

Cells solitary, usually lying in valve view, containing a single plastid. Plastid basically H-shaped, consisting of 2 girdle-appressed plates connected by a narrow isthmus lying against the epivalve; there may also be narrow lobes parallel to the raphe, extending out from the isthmus, and from one of the lateral plates. One or two invaginated pyrenoids present. A large group of marine and freshwater epipelic diatoms.

Valves naviculoid;[a–e] linear, lanceolate to elliptical, usually with bluntly rounded poles. Valve face flat, curving down at its edges into shallow mantles. Striae uniseriate[b] (rarely partly biseriate), interrupted by lateral sterna,[a, d, h] which are depressed below the general level of the valve and often appear in LM as a hyaline 'lyre' (cf. *Lyrella*). The striae are partially or completely covered externally by finely porous conopea,[b, e, i] which can have entire margins,[e] or bear finger-like extensions[j] along the striae; the conopea can have free edges or be so tightly associated with the rest of the valve as to constitute a skin-like covering over it.[c, f] A lyre-shaped canal[i] is formed between the depressed lateral sterna and the conopea; this may open to the outside around the whole edge of the conopeum, or just at the poles,[b, k] or only via the tiny pores of the conopeum itself. Areolae round, closed by hymenes.[i] Raphe-sternum narrow. Internal central raphe endings deflected towards the primary side.[h, i] External central endings straight, expanded[b, c, e, f]; terminal fissures deflected, bent or hooked.[j, k] Girdle incompletely known but apparently consisting of a few plain, open bands, of which the first is by far the deepest.[k]

Plastid movements and structure are very similar in *Fallacia* and *Sellaphora*, and these genera are closely related. *Fallacia* is probably also related closely to *Rossia* Voigt (1960b). *Fallacia* (Latin for 'trickery') bears a very strong resemblance to *Lyrella*, because of the lyre-shaped markings. However, *Lyrella* lacks conopea, has volate rather than hymenate pore occulsions, and has a completely different plastid structure and arrangement; furthermore, its nuclei always divide on the same side of the cell, whereas in *Fallacia* there is an oscillation (Mann & Stickle, 1988). *Fallacia* includes many of the smaller species formerly placed in *Navicula* sect. *Lyratae*.

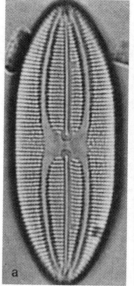
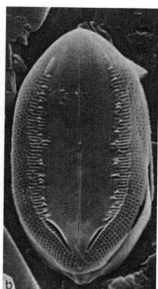

a b

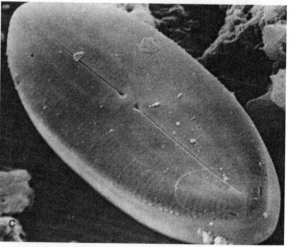

c

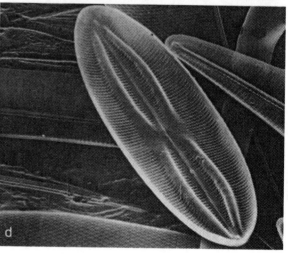

d

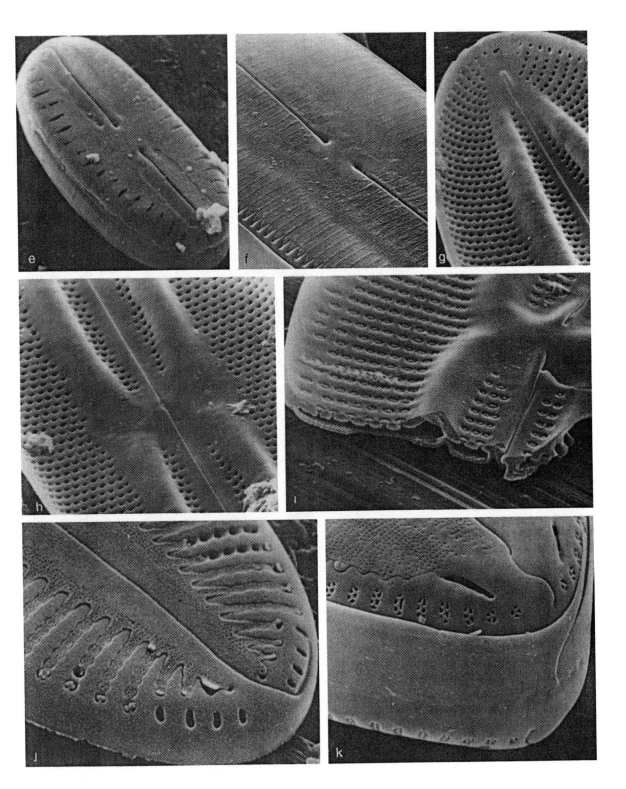

Pinnularia C. G. Ehrenberg 1843. Ber.
Bekanntm. Verh. Königl. Preuss. Akad. Wiss.
Berlin, 1843: 45 (nom. cons.)

T: *P. viridis* (Nitzsch) Ehrenberg (= *Bacillaria
viridis*) (typ. cons.)

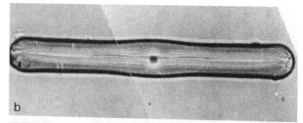

Cells solitary, naviculoid, lying in valve or girdle
view; very rarely, forming band-shaped colonies (see
Round, 1988). Two girdle-appressed plate-like
plastids, sometimes with highly dissected margins;
or the two plates united beneath the hypovalve by a
very narrow central isthmus to form a single H-
shaped plastid. Pyrenoids sometimes invaginated. A
very large epipelic genus; freshwater to (rarely)
marine, the species in the latter habitat sometimes
being strongly dorsiventral.[j, k]

Valves linear,[a] lanceolate or elliptical, sometimes
with rostrate or capitate poles and sometimes with
undulate margins.[a, b] Valve face flat or curving
smoothly into relatively deep mantles.[c] Valve surface
often ornamented.[e, f] Striae basically multiseriate but
usually also chambered.[d, h] Each chamber (alveolus)
has as its outer wall a porous plate containing many
rows of small round poroids occluded by hymenes.
The inner wall consists of a plain siliceous plate
perforated by one large transapically elongate
aperture,[h] or by 1 or 2 small round holes; rarely this
inner wall is missing and the striae are simply
multiseriate. The poroids are usually invisible in the
LM and the striae appear structureless, apart from
the internal apertures of the alveoli. Occasionally
the striae are interrupted by lateral sterna. Raphe
system usually central.[c] External central endings
expanded; long hooked terminal fissures present at
the poles.[c, e] Internal central endings turned towards
the primary side, where there is a prominent nodule;[g]
or internal fissures continuous.[i] Outer and inner
raphe fissures not superimposed and sometimes
flexed in different directions, resulting in sinuous
curves, which sometimes cross over each other along
the raphe (as seen in LM). Girdle consisting of a
few open bands. The first band is usually the widest
and bears one row of elongate poroids.

We are unable to find a satisfactory basis for the
traditional separation of *Pinnularia* from *Caloneis*.
We have investigated many species, including the
type of *Caloneis*, *C. amphisbaena*, and conclude that
if *Pinnularia* is ever split, it will not be along the
traditional boundary between these genera.

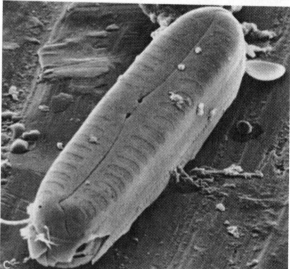

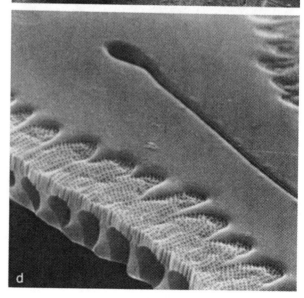

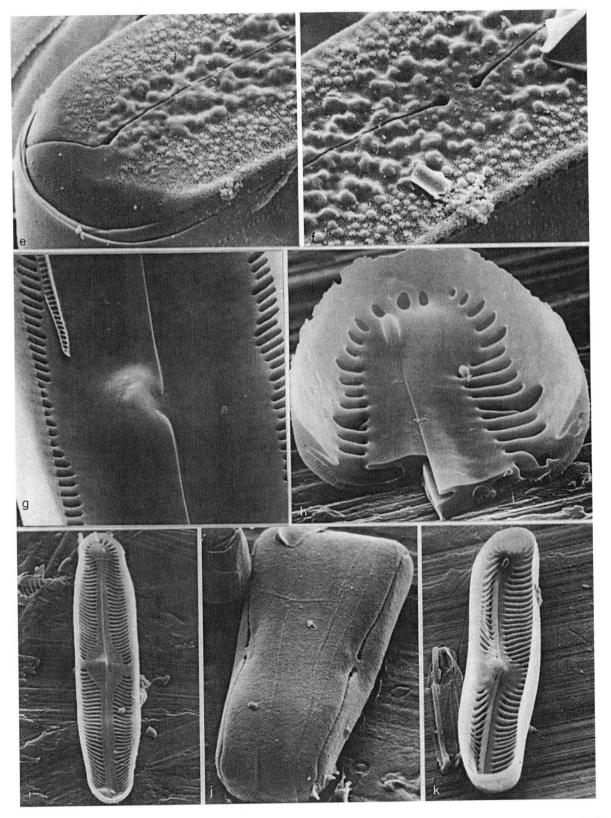

Diatomella R. K. Greville 1855. Ann. Mag. Nat. Hist. Ser. 2, **15**: 259 (nom. cons.)

T: *D. balfouriana* (W. Smith) Greville (= *Grammatophora balfouriana*)

Cells forming chains or zig-zag colonies; single cells usually lying in girdle view[h] because of the depth of the mantles and girdles. Plastids unknown. An uncommon freshwater genus, found mainly in subaerial habitats, e.g. in moss clumps, and also in mountain streams.

Valves linear to elliptical,[a] with obtusely to acutely rounded poles. Valve face flat,[c] abruptly or gently curving into the deep mantles.[c-f] Striae essentially biseriate,[e-g] alveolate. Outer wall of each alveolus perforated by two rows of small round poroids; alveolus opening internally by a single oval or elongate aperture located near the junction of valve face and mantle. Poroids sometimes more distant from each other at the edge of the valve face.[d-f] Valve margins sometimes bearing small blebs or warts[g] of silica. Raphe-sternum central, often broad; central raphe endings relatively widely separated. External central raphe endings expanded,[d, e] straight or deflected to one side (apparently the primary side); terminal fissures curving first in the opposite direction to the central endings, then back to the midline of the valve and then continuing along this to near the valve margin.[f] Internal central raphe endings deflected like the external endings, unexpanded. Mature cingula consisting of 4 open, non-porous bands. Bands 1 and 3 are wide, 2 and 4 narrow. Internally band 1 bears two pairs of massive projections (septa) which meet and interdigitate in the middle of the cell.[h, j] The pars interior must also interlock very strongly with the valve margin, since fully formed valves are scarcely ever found separated from the first band with its septa.

Two *Diatomella* spp. have been studied in the SEM by Le Cohu (1983); they resemble the species illustrated here (*D. balfouriana*). *Diatomella* shows few similarities to other genera, but its alveolate structure and deflected, expanded central raphe endings suggests a relationship to *Pinnularia*.

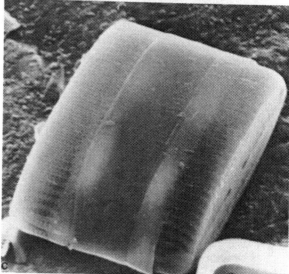

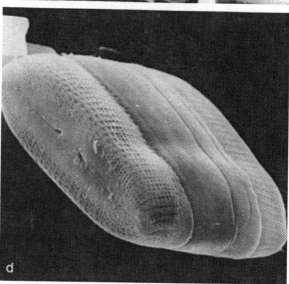

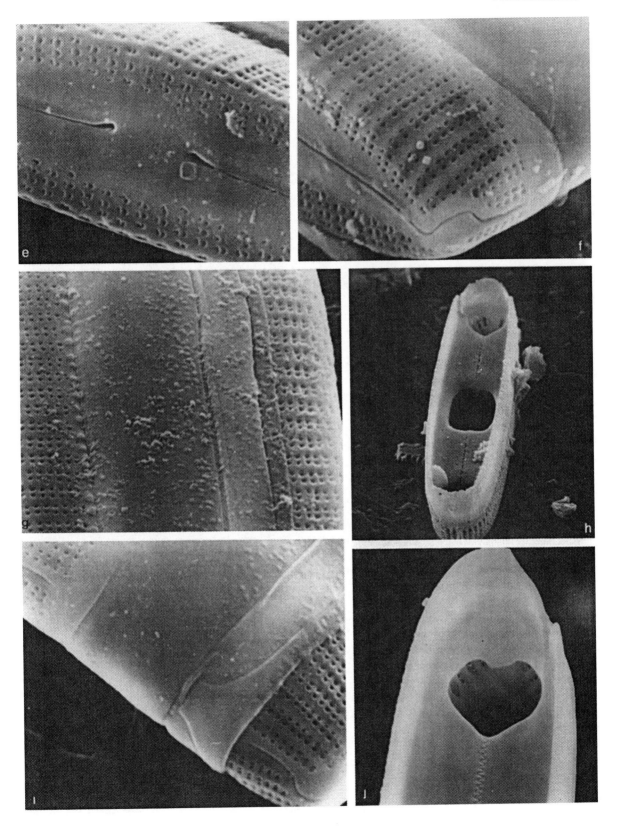

Phaeodactylum Bohlin 1897. Zur Morphologie und Biologie einzelliger Algen. Ofvers. Vetensk.-Akad. Forh., Stokh. **54**: 507

T: *P. tricornutum* Bohlin

Cells unicellular with one plastid and existing in one of three forms; triradiate,[a] fusiform[b] and oval.[b] Marine littoral, planktonic and benthic according to form. One species.

The siliceous valve is naviculoid and occurs solely in the oval cell form and then on one side of the cell only.[f-i] Girdle bands have not been reported. The other half of this cell is covered by an organic 'valve'. Fusiform[d] and triradiate[e] cells are also covered by organic valves, whose structure has been examined in detail for the fusiform cell by Reimann & Volcani (1968). There are three layers: a thin (3 nm) electron dense layer immediately outside the plasmalemma, a less densely stained middle layer (4–6 nm) and an outer dark layer (7–10 nm). The outer layer is composed of sets of 2–5 or more ridges 7 nm high. In a set each ridge lies parallel to the others. Lewin, Lewin & Philpott (1958) have shown the organic 'valves' to be composed of polymers of xylose, mannose, fucose and galactose.

The fusiform and triradiate forms are common when grown in liquid media but oval forms have been observed by us only on the sides of the culture vessel. Lewin *et al.* (1958) noted oval forms when *Phaeodactylum* was grown on an agar surface. Triradiate and fusiform[c] cells can then be found giving rise to oval forms.[arrow, c] This suggests that the oval form is benthic whilst fusiform and triradiate forms may be planktonic. Lewin *et al* (1958) report that the fusiform cells are more buoyant than the oval cells.

The taxonomic position of *Phaeodactylum* is uncertain. Lewin (1958) compared the siliceous valve with that of *Cymbella* on account of the slight asymmetry[f] yet 'in view of the unique characters' established a new suborder Phaeodactylineae. Excellent studies of the different morphotypes were made by Borowitzka, Chiappino & Volcani (1977) and Borowitzka & Volcani (1978).

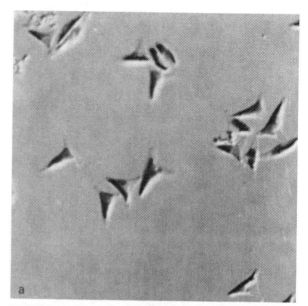

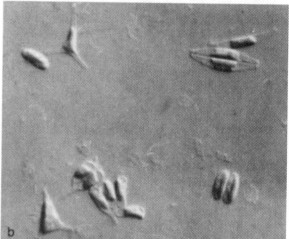

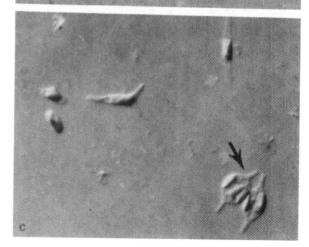

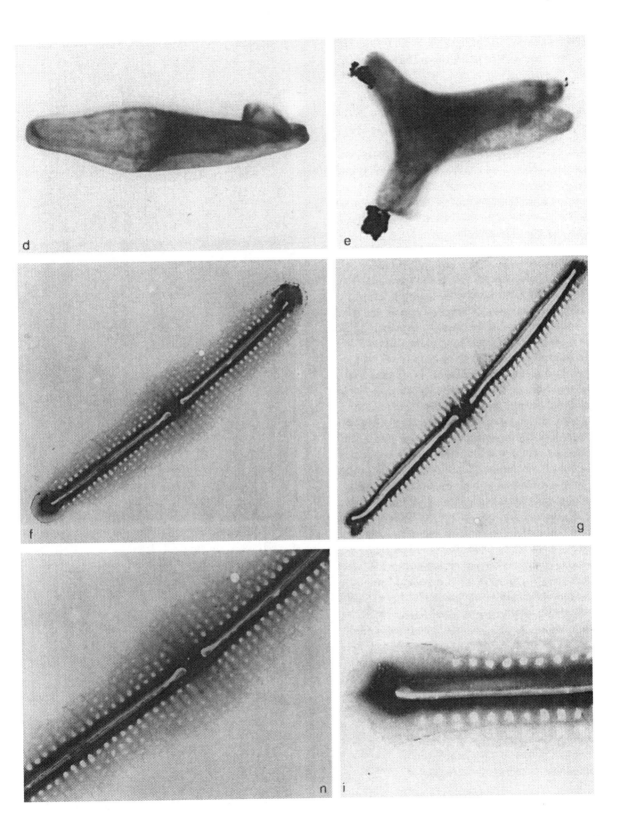

Diploneis C. G. Ehrenberg ex P. T. Cleve 1894. Kongl. Svensk Vetensk.-Akad. Handl. Ser. 2, 26 (2): 76

T: *D. didyma* (Ehrenberg) Cleve (= *Navicula didyma*; lectotype selected by C. S. Boyer 1927, Proc. Acad. Nat. Sci. Philad. 79 Suppl.: 345)

Cells solitary, naviculoid, usually lying in valve view. Two plastids, one on either side of the apical plane, each containing one pyrenoid. Each plastid lies against the girdle but bears lobes extending beneath the valve. Predominantly marine, but with a few freshwater species; epipelic.

Valves linear to elliptical,[h] or panduriform,[a–c] with bluntly rounded poles. Valve face flat, curved or undulate (undulations parallel to the apical axis),[c] poorly or not differentiated from mantles. Striae very complex, changing in structure across the valve and containing loculate areolae. On each side of the raphe there is a continuous longitudinal canal within the valve structure,[arrow, i] whose inner wall projects into the cell. The canal is not open to the cell interior[g] but opens externally by one or a few longitudinal rows of pores.[b, f] The areolae open to the outside by large round or transversely elongate apertures,[b, c] or by groups of pores, or by pores occluded by complex cribra;[d, e] they open to the inside by a finely porous siliceous membrane which may form a continuous strip along the whole stria or occlude individual pores through the inner wall layer. Raphe-sternum central.[b, c, f, h] Terminal fissures deflected or hooked. Internal central raphe endings simple, straight. External central endings simple or expanded; straight, deflected or hooked to one side. Girdle consisting of a few open copulae of which the most advalvar (valvocopula) is by far the widest.

A large and diverse genus, easily recognisable by the longitudinal canals and coarse appearance in the LM. It represents a natural group but may nevertheless be divisible into smaller genera. The various types and arrangements of the areolae and ornamentation, etc. of the valves are amongst the most complex in the raphid diatoms, except perhaps for *Mastogloia*, and we have only been able to illustrate a small fraction of the diversity. See Idei & Kobayasi (1989) for further details of the ultrastructure of the valves and copulae.

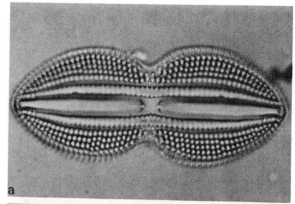

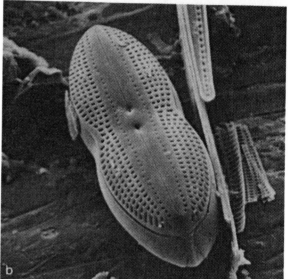

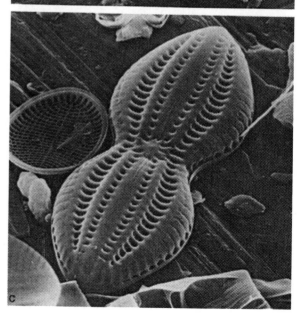

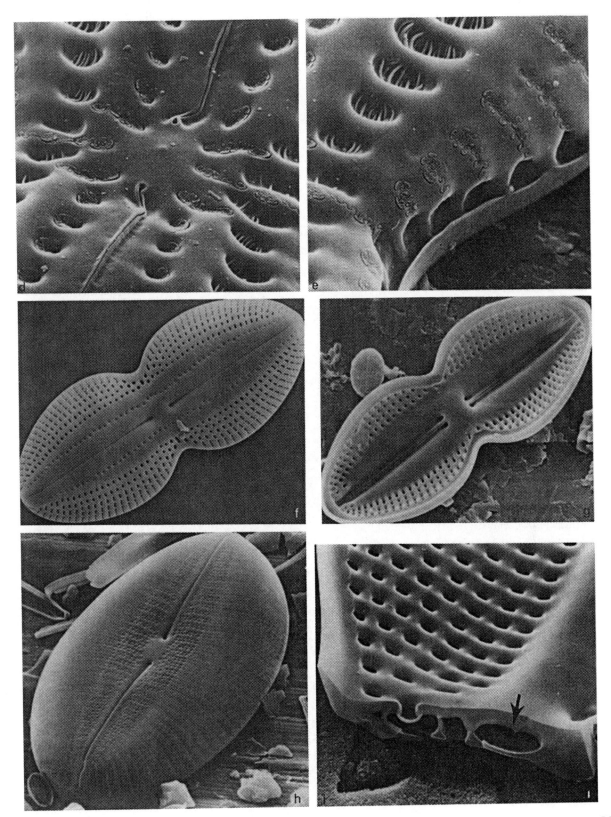

Raphidodiscus H. L. Smith in T. Christian 1887, Microscope (Ann Arbor) 7: 65

T: *R. marylandicus* Christian

Cells solitary, discoid. A small extinct genus found world-wide in marine Miocene deposits.

Valves elliptical to more-or-less circular.[a-d] Valve face undulate, curving gently into the extremely shallow mantle, which bears a circumferential ridge close to the valve margin. The structure of the striae changes across the valve. In the lanceolate, concave central region the valve bears prominent radial ribs externally, which occasionally branch, but internally it is plain[f] (cf. *Diploneis*). A honeycomb-like structure can sometimes be observed in craters in the central elliptical area[h] but these are probably merely erosions, revealing the complexity of the valve structure. Around the central region is a zone of more-or-less uniseriate striae, which are separated internally by prominent transverse or radial ribs[f-h] and which contain large round poroids. Outside this the striae become narrower and more closely spaced, and their structure becomes simpler. In our eroded material the areolae are more-or-less simple holes, but fine pegs and strands remain,[e] suggesting that cribra or other coarse occlusions were present originally. Raphe-sternum central, short, and slightly elevated; internally the raphe slits open into deep grooves. All raphe endings simple, straight. Girdle poorly known but apparently containing narrow plain bands.

Raphidodiscus has been considered in some detail by Andrews (1974), who suggests that it should be placed in a group of its own, the Raphidodiscoideae. Among extant raphid diatoms it seems most closely related to *Diploneis*, although this suggestion is based on little more than the general appearance of the valve. *Raphidodiscus* has a very restricted range stratigraphically and is for that reason useful as a marker fossil (Andrews, 1974).

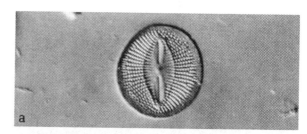

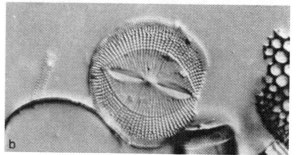

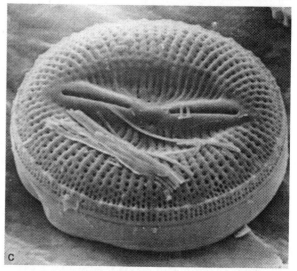

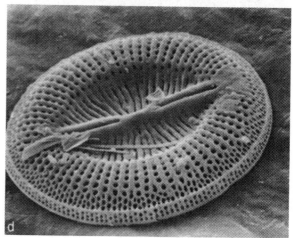

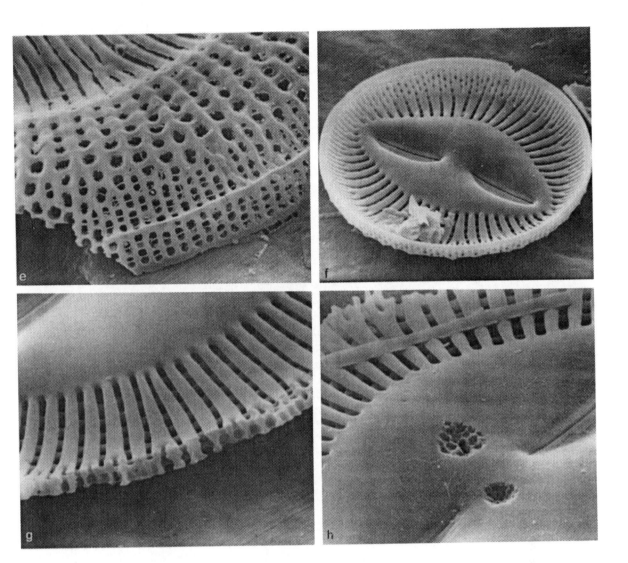

Navicula J. B. M. Bory de St.-Vincent 1822.
Dict. Class. Hist. Nat. 2: 128

T: *N. tripunctata* (O. F. Müller) Bory (= *Vibrio tripunctatus*)

a

b

Cells solitary, naviculoid![a-d] Most species lie in valve view but a few are strongly compressed laterally and lie in girdle view. Two girdle-appressed plastids, one on either side of the apical plane, each containing an elongate rod-like pyrenoid. Freshwater or marine, epipelic. Extremely common; hardly a sample can be taken of epipelon without encountering this genus.

c

Valves lanceolate to linear, with blunt, rostrate or capitate apices. Valve face flat or curved, usually curving gently into the mantles. A short conopeum is sometimes present.[j, k] Striae uniseriate or rarely biseriate, containing apically elongate, linear poroids, which are closed by hymenes[m] at their inner apertures. The poroids of adjacent striae are aligned with each other, so that straight or gently curving longitudinal striations are usually visible in the LM. Striae interrupted by lateral sterna in a few species. Raphe-sternum thickened,[f, h, i, l] especially on the primary side, so that the internal fissures open laterally, except at centre and poles; the primary side usually bears in addition an accessory rib running the whole length of the valve internally, parallel to the raphe. Central internal raphe endings straight and unexpanded, lying in a small oval nodule;[i] sometimes the internal fissures are continuous across the central nodule; rarely a double helictoglossa is formed. External central endings simple, expanded into pores or hooked, usually towards the secondary side.[c] External polar endings simple to strongly hooked.[g] Girdle composed of several open, usually plain bands.

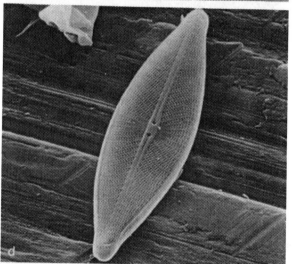

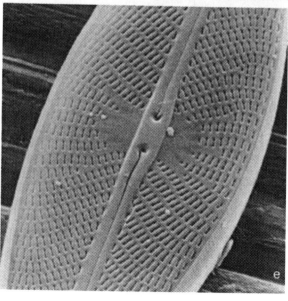

Navicula has traditionally been a dump for all bilaterally symmetrical raphid diatoms lacking particularly distinctive features. If such features were discovered the taxa concerned were usually removed into separate genera, such as *Neidium*, *Anomoeoneis*, etc. Even with such removals, however, the genus remained heterogeneous. *Navicula* should be used only for the natural group described here, which corresponds to the sect. *Lineolatae* recognised by most authors. There will be nomenclatural difficulties until all the taxa failing to satisfy the above description have been found their proper homes. The type species has been described by Cox (1979b).

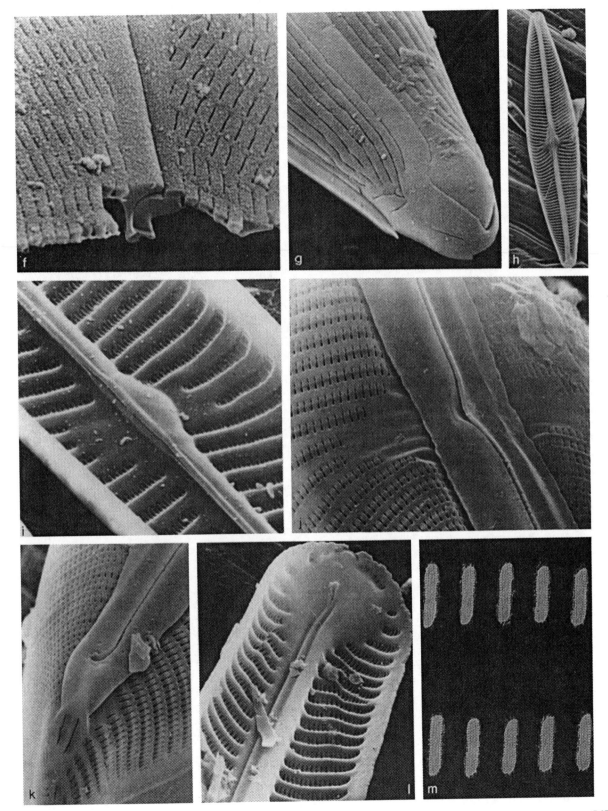

Trachyneis P. T. Cleve 1894.

T: *T. aspera* (C. G. Ehrenberg) Cleve
(= *Navicula aspera*; lectotype selected by C. S.
Boyer, 1928. Proc. Acad. Nat. Sci. Philad. **79**
Suppl.: 428)

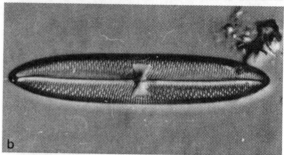

Cells solitary, naviculoid, lying in valve or girdle
view. Two girdle-appressed plastids, which often
have lobed margins. A distinctive small genus of the
marine littoral, where it lives epipelically on sand.

   Valves linear to linear-elliptical, with acutely or
bluntly rounded poles.[a-c] Valve face curved, merging
imperceptibly into deep mantles.[e, i] Striae basically
uniseriate, but partitioned into several alveoli. Each
alveolus opens to the inside by a single round hole[f-h]
and to the exterior by a series of slit-like poroids,[d, e]
which are closed by hymenes. The slits are often
aligned longitudinally to produce 'lineolate striae' as
in *Navicula* (q.v.). Valve structure often interrupted
centrally by a broad oval or bow-tie shaped plain
area.[h] Raphe-system central.[a, b, f] The two sides of the
raphe-sternum are very unequally developed: the
primary side of the raphe-sternum is thicker than
the secondary (so that the raphe opens laterally,
except near the centre of the valve) and with a
prominent flange, which curves over beneath the
raphe, towards the secondary side.[f-i] Central internal
raphe endings straight, very slightly expanded, lying
in a transversely elongate nodule.[g-i] External raphe
fissures often sinuous near the centre, the central
endings themselves being expanded and hooked
towards the secondary side.[d] Terminal fissures
hooked.[e] Sides of valve fusing subapically to delimit
a pore[h] as in *Rhoikoneis* (q.v.). Girdle consisting of
open, non-porous bands, of which the most advalvar
is by far the widest.[j]

   The valve structure, with its partitioned alveolate
striae is very distinctive. Nevertheless *Trachyneis* is
clearly allied to *Navicula* by its plastid structure,
apically elongate poroids and simple internal raphe
endings; furthermore, some *Navicula* species have
hooked external central endings. *Trachyneis* is also
closely related to *Rhoikoneis* and *Pseudogomphonema*.

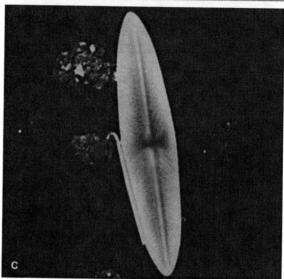

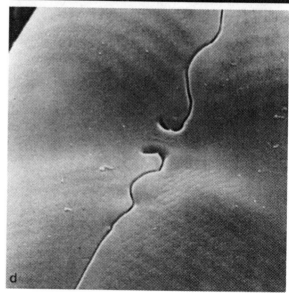

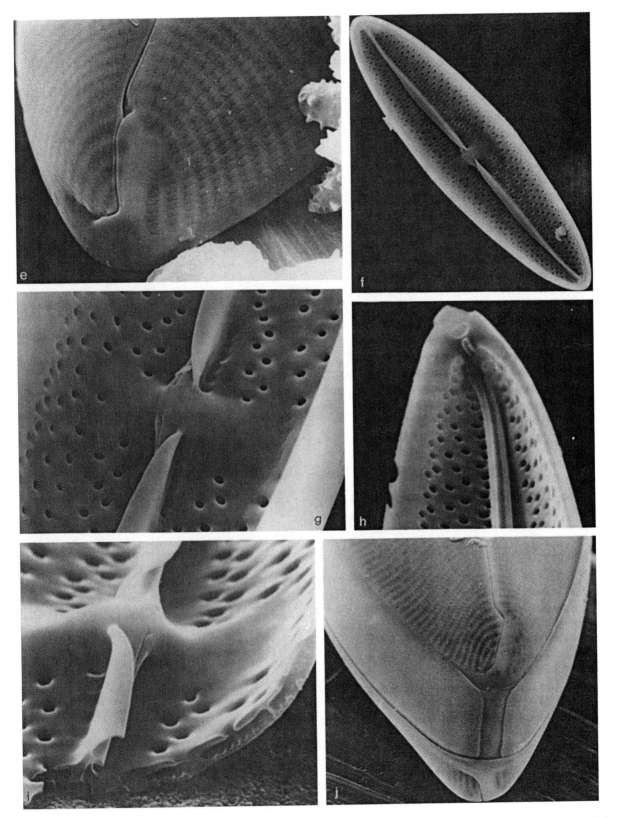

Pseudogomphonema L. Medlin 1986. Diat. Res.
1: 214

T: *P. kamtschaticum* (Grunow) Medlin
(= *Gomphonema kamtschaticum)*

Cells solitary, attached by a single short stalk or
pad, heteropolar; usually lying in girdle view.[b]
Plastids unknown. A marine, epiphytic genus,
primarily of polar seas.

Valves linear to linear-lanceolate, with broadly
rounded or slightly rostrate head poles and more
acutely rounded foot-poles.[a] Valve face flat in the
centre, or curving uniformly into relatively deep
mantles.[b-e] Striae uniseriate, containing apically
elongate, slit-like poroids occluded by hymenes;
opening internally into grooves between strongly
developed transapical ribs.[g, h] Valve margin
thickened at the foot-pole and fused to the sides of
the helictoglossa, leaving an apical chamber[g] which
opens to the outside by a row of unoccluded (?)
slits.[f] Raphe-sternum central, more strongly
developed on the primary side internally, so that the
raphe opens laterally, except at the centre and poles;
this also has the effect (in LM) that the raphe
appears asymmetrically placed in relation to the
striae. External central raphe endings expanded,
pore-like;[d, e] internal endings straight,[h] simple, closer
together than the external endings. Terminal fissures
more-or-less absent at the head-pole,[c] hooked at the
base-pole.[f] Cingulum [i, j] containing 4 open, non-
porous bands of which bands 2 and 4 are reduced
to small segments (ligulae) lying at the foot-pole;
furthermore, in bands 2 and 4 the pars interior is
modified to clasp the thickenings of the valve
margin.

This genus belongs to the cluster of genera that
includes *Rhoikoneis, Trachyneis* and *Navicula sensu
stricto*. The valve, raphe and areola structure of
Pseudogomphonema are quite different from those of
Gomphonema and there is no close relationship
between these genera, despite the similar shape and
habit of the cells. Medlin & Round (1986) give a
full discussion of *Pseudogomphonema*, including
details of the four known species (*P. arcticum,
P. groenlandicum, P. kamtschaticum* and
P. septentrionale).

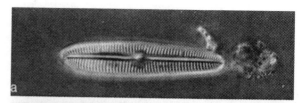

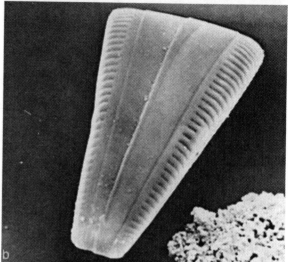

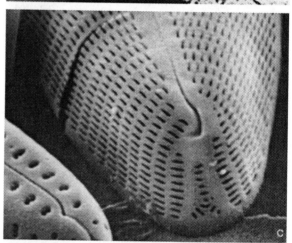

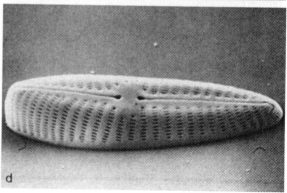

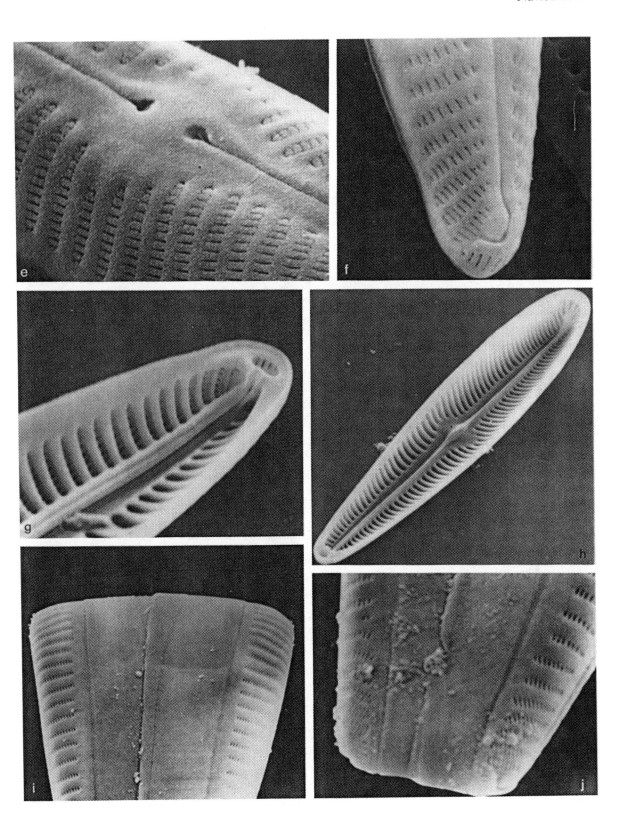

Seminavis D. G. Mann, gen. nov.

T: *S. gracilenta* (Grunow ex A. Schmidt) Mann (= *Amphora angusta* var. *gracilenta*)

Cells solitary, strongly dorsiventral,[a-c] usually lying in girdle view, containing two girdle-appressed plastids; the dorsal plastid is larger than the ventral. A small genus of the marine epipelon and epiphyton.

Valves semi-lanceolate, with a straight or slightly curved ventral margin and an evenly curved dorsal margin.[a, b] Valve face curving into a deep dorsal mantle; ventral mantle absent.[b-d] Striae uniseriate,[d, e] opening internally between prominent transapical ribs,[f, h] containing apically elongate areolae, which are closed internally by hymenes and open externally by narrow slits. Raphe-sternum wider dorsally than ventrally.[b, d] Internally the ventral (primary) side is extended over towards the dorsal side;[f] the raphe therefore opens laterally, on the crest of a narrow ridge, except at the centre and poles. Internal central raphe endings simple,[c, f] unexpanded. External central endings slightly expanded and deflected towards the ventral side,[d] or hidden by flap-like structures[e] extending across from the ventral side; terminal fissures hooked towards the dorsal side.[b, g] Girdle wider dorsally than ventrally, though less so than in *Amphora*, apparently consisting of 3 plain, open bands: a wide first band, a wide segmental band occupying the gap left at one pole by the ends of band 1, and a very narrow abvalvar element.[g]

The plastid arrangement, stria and raphe structure, and some aspects of girdle structure indicate that *Seminavis* has been derived from diatoms allied to *Navicula sensu stricto* and is not closely related to *Amphora*. It is part of a group that has diversified much in cell shape and symmetry, producing gomphonemoid, sigmoid, naviculoid, amphoroid and achnanthoid forms (*Pseudogomphonema*, *Gyrosigma*, *Navicula* and *Trachyneis*, *Seminavis*, and *Rhoikoneis*, respectively: for discussion of some of these, see Medlin, 1985, and Medlin & Round, 1986).

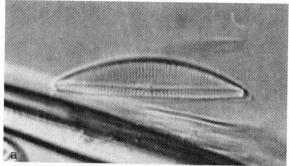

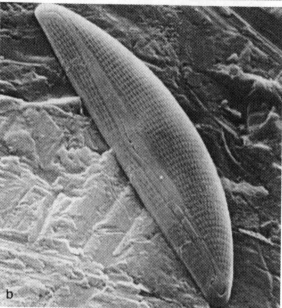

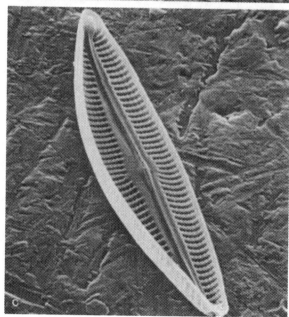

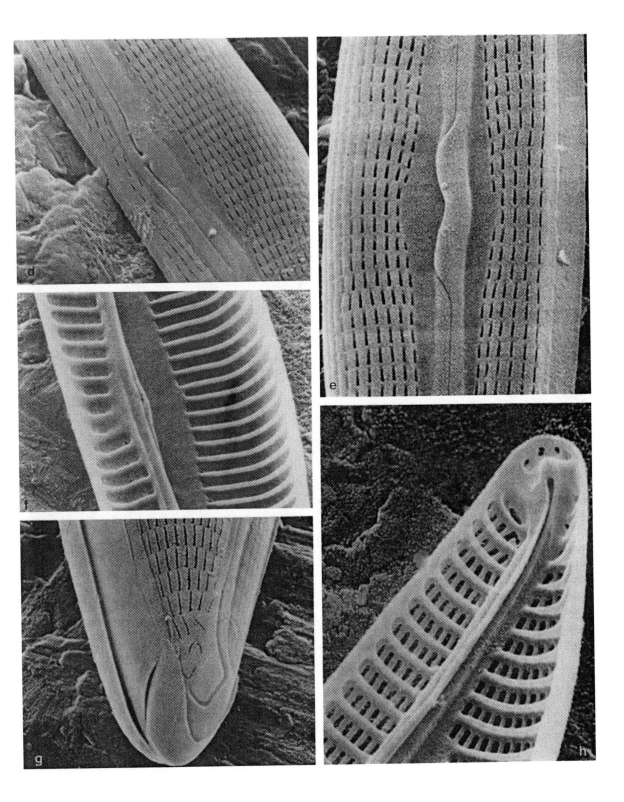

Rhoikoneis A. Grunow 1863. Verh. zool.-bot. Ges. Wien **13**: 147

T: *Rh. bolleana* Grunow

Cells usually solitary; naviculoid and isopolar,[b, c] but curved in girdle view.[a, d] Plastids unknown. Epiphytic on seaweeds and apparently widespread, though confused in the literature since Grunow's genus has been misunderstood (Medlin, 1985).

Valves structurally similar but one is convex and the other concave. Valve face curving smoothly into the mantles. Striae uniseriate,[c-f] slightly chambered[h, i] (alveolate), containing apically elongate poroids[e, f] probably closed by hymenes. Transapical ribs strongly developed, running from the raphe-sternum to a point near the valve edge where they fuse with a solid sheet of silica extending in from the valve margin;[h, i] here the valve is chambered (a faint line[b] marks the internal edge of this chambering when valves are viewed by LM). The silica sheet continues around the apices of the valves and at the apex there is a hole in it[i] – the isolated vertical slits seen below the raphe endings on the outside of the valve[c, j] open into a small chamber at this point. Raphe central. Internal fissures obscured by a flap of silica extending across (apparently) from the primary side of the raphe-sternum;[h, i] fissures more-or-less continuous centrally.[g] External central endings expanded and slightly deflected towards the primary side; terminal fissures present, bent. Girdle consisting of plain, open bands, some of which are reduced to elements occupying much less than the full circumference of the cell.[j]

Rhoikoneis is allied to *Navicula sensu stricto*, but differs from it in the alveolate striae, more strongly asymmetrical raphe-sternum and heterovalvy. It is also related to *Trachyneis*, but in *Trachyneis* the cell is straight in girdle view and the alveoli are partitioned. We thank L. Medlin for the illustrations.

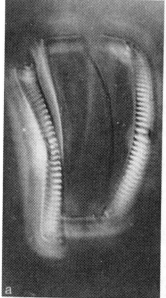
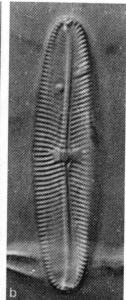
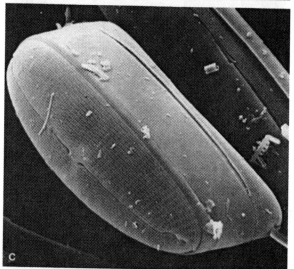
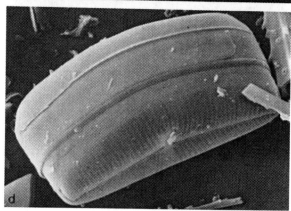

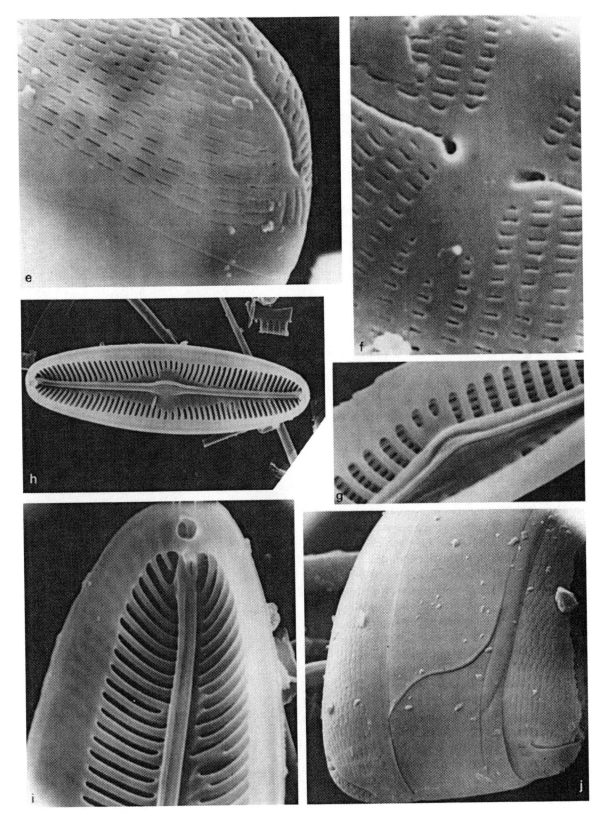

Haslea R. Simonsen 1974. 'Meteor'
Forschungsergebnisse, Reihe D **19**: 46

T: *H. ostrearia* (Gaillon) Simonsen (= *Navicula ostrearia*)

Cells solitary or forming mucilage tubes, naviculoid, lying in valve or girdle view. Where known, there are either 2 plastids, one against each side of the girdle, or many small plastids. *H. ostrearia* is the 'blue diatom', so named because of the blue colour of its vacuoles; it is often abundant in oyster ponds. A fairly common member of the marine epipelon and also planktonic in warm and tropical waters.

Valves lanceolate[a, b] with somewhat acute poles; the ends are sometimes slightly twisted in cleaned material mounted for the LM, perhaps as a result of drying of the delicate valves. Valve face curved in section, merging imperceptibly with the shallow mantles.[c, i, j] Striae uniseriate, containing square[i] or rectangular[j] poroids, and overlain externally by longitudinal strips,[c-g] many of which are continuous from pole to pole. The poroids are closed internally by hymenes.[i, j] In some species the central 2 or 3 transapical costae on each side of the valve are thickened, forming stauros-like structures.[h, i] Raphe system central. External raphe fissures turned to the same side centrally[c, d] and extended into bent or hook-like terminal fissures at the poles.[c-g] Internal central raphe endings simple, straight, unexpanded.[i] Raphe-sternum with a prominent ridge internally on one side of the raphe,[j] as in *Navicula* and *Trachyneis*; this is fused to the thickened central ribs (where present)[i] and hides the internal raphe fissure over much of the length of the valve; the internal fissures themselves open laterally,[j] as in *Navicula*. The ridge ends close to,[j] but not at the apex and it may be the absence of the ridge apically that allows the twisting noted above. At the centre there may be a short second ridge parallel to the first,[h, i] but on the other side of the raphe, fused to the other thickened ribs. The cingulum contains a few open, plain bands.[g]

Simonsen (1974) transferred several *Navicula* species to this genus. He included the epipelic or tube-dwelling species *H. crucigera* and most of our illustrations are of this. The planktonic species lack the enlarged central costae. Related to *Navicula sensu stricto* and *Gyrosigma*, but separated by a combination of valve shape, raphe structure and areola structure. Three taxa are described in detail by von Stosch (1985).

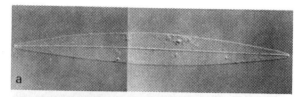

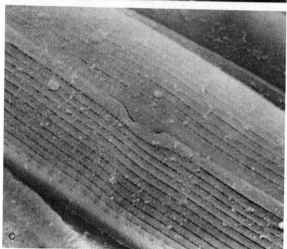

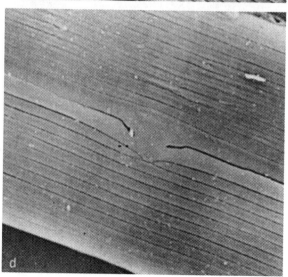

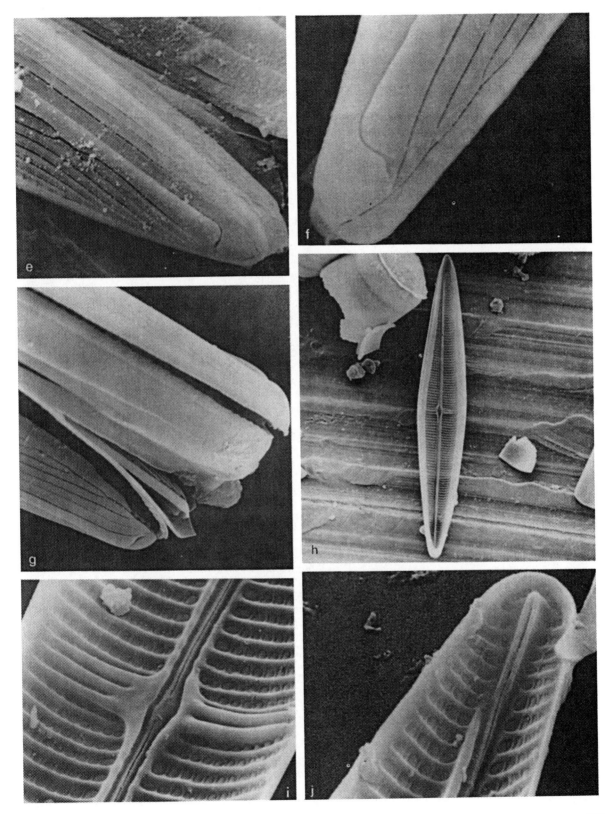

Cymatoneis P. T. Cleve 1894. Kongl. Svenska Vetensk.-Akad. Handl., Ser. 2, **26** (2): 75

T: *C. sulcata* (Greville) Cleve (= *Navicula sulcata*, lectotype selected by C. S. Boyer, Proc. Acad. Nat. Sci. Philad. 79 Suppl.: 306)

Cells solitary, lying in valve or girdle view. Plastids unknown. A small infrequently recorded genus of the marine epipelon, apparently more frequent in subtropical or tropical seas, in sandy sediments.

Valves linear-lanceolate to elliptical, sometimes with undulate margins; poles bluntly rounded or apiculate.[a, b] Valve face thrown into a series of folds, so that the raphe is raised on a keel;[b-f] bearing one or more prominent ridges parallel to the raphe. Mantles shallow, indistinct. Raphe system often slightly sigmoid and eccentric, possibly reflecting a need for sibling valves to have complementary shapes; the longitudinal ridging of the valve face is also not symmetrical with respect to the apical plane (e.g. 2 ridges on one side and 3 on the other in the specimen illustrated). Striae uniseriate,[b-i] containing apically elongate poroids; in well-preserved specimens the poroids seem to be occluded by hymenes and to open externally by slits (cf. *Navicula*); poroids also aligned longitudinally. Raphe-sternum very narrow.[f] External central raphe endings expanded;[b, f] internal endings forming a small beak-like double helictoglossa.[h, i] External raphe fissures sometimes bordered by ridges, especially towards the poles; no terminal fissures. Girdle not studied; apparently consisting of non-porous bands (Cleve, 1894–5).

An interesting genus requiring further study, especially in relation to its plastids. The areola structure suggests a relationship to *Navicula sensu stricto*, and a double helictoglossa can be found in some marine *Navicula* species. No other genus suggests itself as close kin. We thank Klaus Kemp for finding and mounting material of this genus so that we could examine it in the SEM.

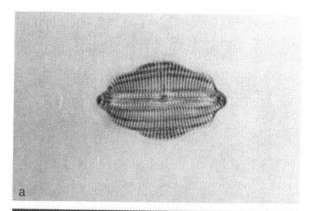

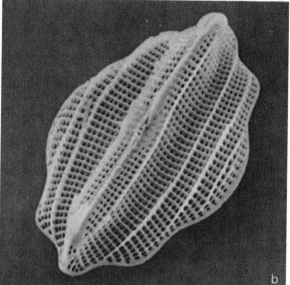

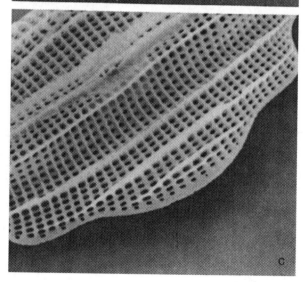

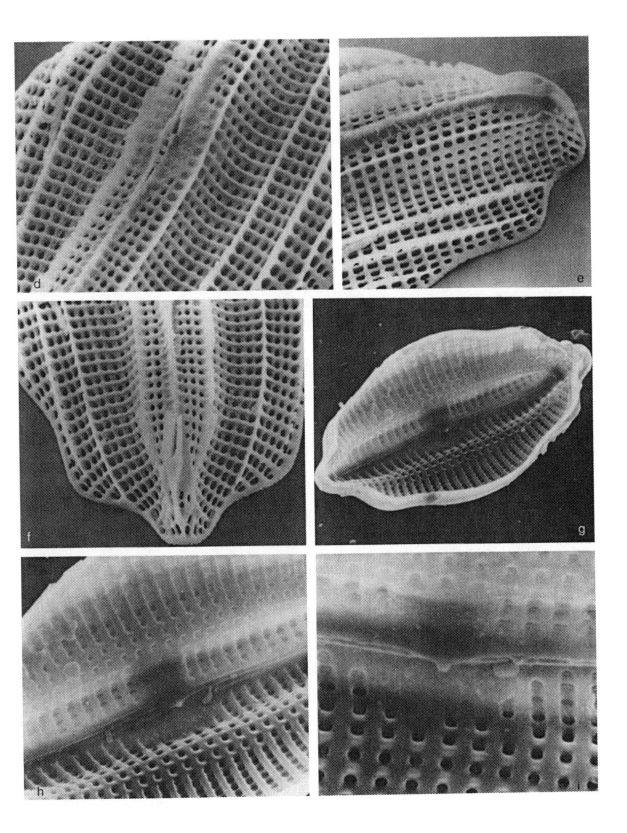

Pleurosigma W. Smith 1852. Ann. Mag. Nat. Hist., Ser. 2, 9: 2 (nom. cons.)

T: *P. angulatum* (Quekett) W. Smith (= *Navicula angulata*)

Cells solitary, usually lying in valve view and then sigmoid.[a, b] Two or four ribbon-like plastids, which follow a convoluted but orderly pattern from pole to pole, or pole to centre; rarely with many small discoid plastids. A large genus of marine and brackish waters; usually epipelic on sand or silt but occasionally planktonic. Occasionally also in high-conductivity freshwaters.

Valves sigmoid, though sometimes only very slightly; linear-lanceolate to lanceolate or rhombic. Valve face flat, curved or slightly angled at the raphe;[c–e] mantles scarcely distinguishable and valve usually very shallow. Areolae loculate, arranged in decussate rows.[e–h] Each areola opens to the outside by an apically elongate slit[e] and to the inside by one or two poroids, which are sometimes rimmed[f, h] and which are closed by hymenes. The areolae communicate with each other laterally by pores, so that a continuous space is formed within the structure of the valve. Raphe system sigmoid, often more so than the valve itself. Raphe-sternum narrow, slightly thickened internally;[f, g] thickening more-or-less equal on either side of the raphe (contrast *Gyrosigma*). Central internal raphe endings expanded, lying in a small oval nodule flanked on either side by short curved ridges.[g] Helictoglossa sometimes turned to one side.[f] External central raphe endings sometimes slightly expanded, both turned towards the same side of the valve;[c, d] terminal fissures hooked[e] towards opposite sides. Girdle narrow, containing open, non-porous bands.

Pleurosigma was last monographed in 1891, by H. Peragallo, and urgently needs revision. It is very closely allied to *Toxonidea* and *Rhoicosigma*, which have very similar valve and plastid structure. It is more distantly related to *Gyrosigma*, which differs not only in the arrangement of the loculate areolae, but also in the structure of the plastids and raphe-sternum. The separation of *Pleurosigma* and *Gyrosigma* is discussed by Stidolph (1988), but with no mention of the clear difference in plastid morphology.

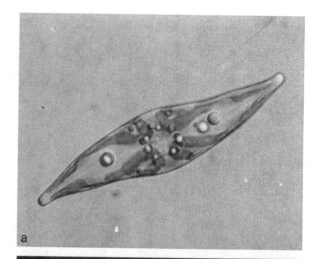

a

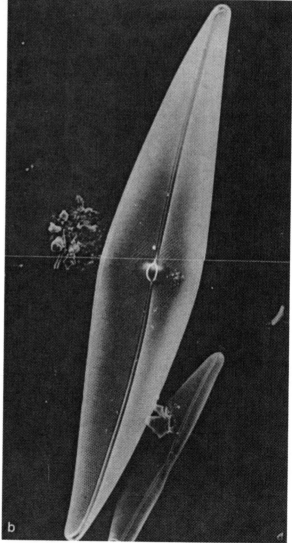

b

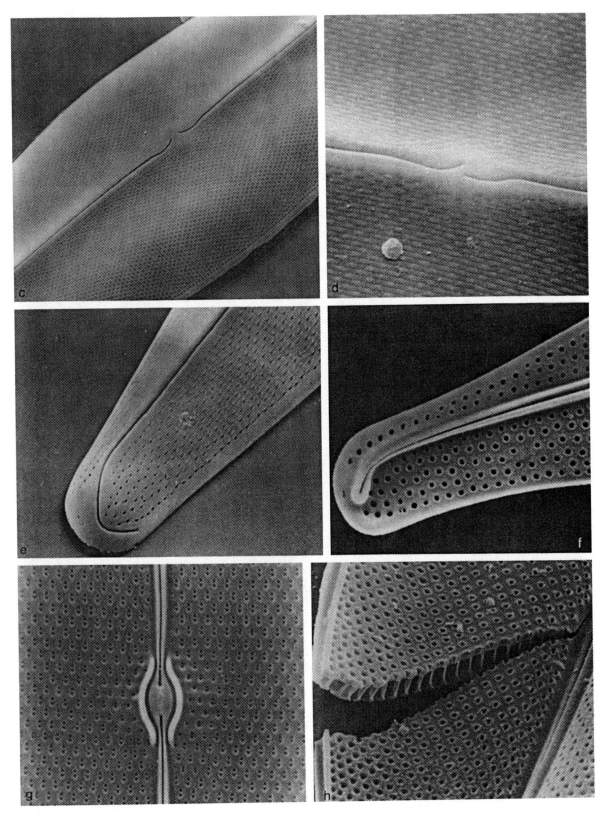

Toxonidea A. Donkin 1858. Trans. Microsc. Soc. Lond. Ser. 2, 6: 19

T: *T. gregoriana* Donkin (lectotype selected by C. S. Boyer 1928. Proc. Acad. Nat. Sci. Philad. 79 Suppl.: 476)

Cells solitary, arcuate, lying in valve view.[a, b] Two or four ribbon-like plastids; these contain numerous rectangular pyrenoids and follow a tortuous but orderly path around the periphery of the cell as in *Pleurosigma*. A small genus of the marine epipelon, especially on sandy substrata.

Valves lanceolate, arcuate. Valve face flat, curving gently into the very shallow, scarcely distinguishable mantles. Areolae loculate[e] and arranged in decussate rows,[c, d] as in *Pleurosigma*. Each areola opens to the inside by large rimmed pores[e-i] occluded by hymenes, and to the outside by small round or apically elongate apertures.[c, d] Raphe system biarcuate, lying nearer the concave margin of the valve.[a, b, f] Raphe-sternum slightly thickened on one side, but not bearing ridges (contrast *Gyrosigma*). Central internal endings straight, expanded, lying in a small nodule and bordered on either side by a crescent-shaped ridge.[f, h] External central endings bent or hooked towards the convex dorsal margin of the valve;[d] terminal fissures present, also bent towards the dorsal margin.[c] Girdle shallow, consisting of open, plain bands.

Toxonidea is very closely related to *Pleurosigma*, but the shape of valve and raphe set them apart. Shape and symmetry appear to have been remarkably labile in this group of genera, producing sigmoid, biarcuate, more-or-less naviculoid and even bent 'achnanthoid' forms (classified in *Rhoicosigma*, which we do not illustrate). All these have a very similar valve and plastid structure, and they occupy similar habitats. Nevertheless we feel that there are here several groups of species that are sufficiently distinct to be recognised as separate genera within a natural family, Pleurosigmataceae.

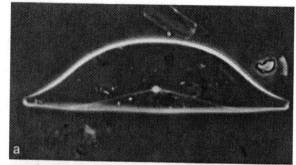

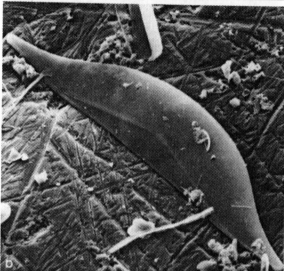

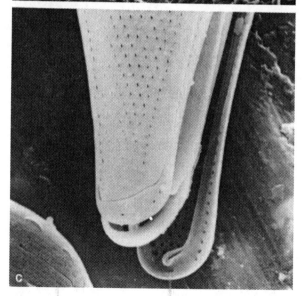

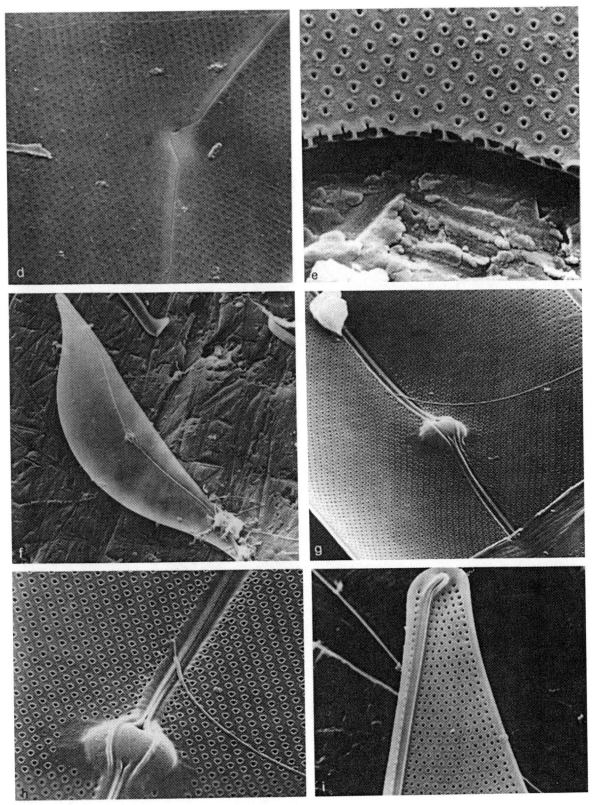

Donkinia J. Ralfs in A. Pritchard 1861. Hist. Infus., Ed. **4**: 920

T: *D. carinata* (Donkin) Ralfs (= *Pleurosigma carinata*; lectotype selected by Cox 1983, Bot. Mar. **26**: 573)

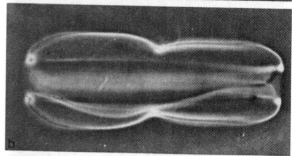

Cells solitary, lying in valve[a] or more often girdle view.[b] Two or 4 girdle-opposed plastids with elaborately lobed margins and containing several short, bar-shaped pyrenoids. A small genus of the marine epipelon.

Valves linear-lanceolate or linear, slightly sigmoid, with bluntly or acutely rounded poles.[d] Valves fairly deep but with no real differentiation into valve face and mantle. Valve structure loculate.[j] Each chamber opens to the outside by a simple longitudinally orientated slit[e, f] and to the inside by an oval,[g] hymen-occluded pore; this is sometimes subdivided by a transverse bar. Chambers usually in longitudinal and transverse rows as in *Gyrosigma*, occasionally in decussate rows. Raphe system more-or-less keeled towards the poles, strongly sigmoid, veering from central in the middle of the valve to submarginal near the poles.[d] Terminal fissures short, deflected, sometimes expanded into 'pores'.[f] External central endings expanded, straight or deflected towards one margin;[e] internal endings simple, or slightly expanded, lying in a flat, fusiform nodule which is bordered on each side by a short lunate ridge.[g] Primary side of raphe-sternum without a ridge. Girdle deep, consisting of non-porous open copulae.

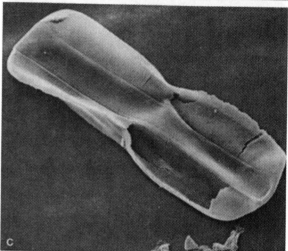

Some authors do not separate *Donkinia* from *Gyrosigma* and *Pleurosigma*. These genera are certainly closely related and it may prove difficult to distinguish *Donkinia* from *Gyrosigma*; the plastid and valve structure are very similar. Detailed studies of *Donkinia* have been made by Cox (1981c, 1983a, b).

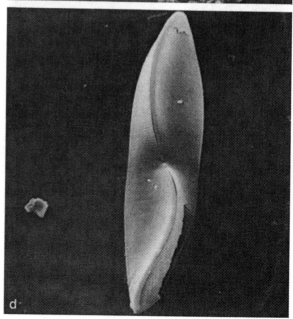

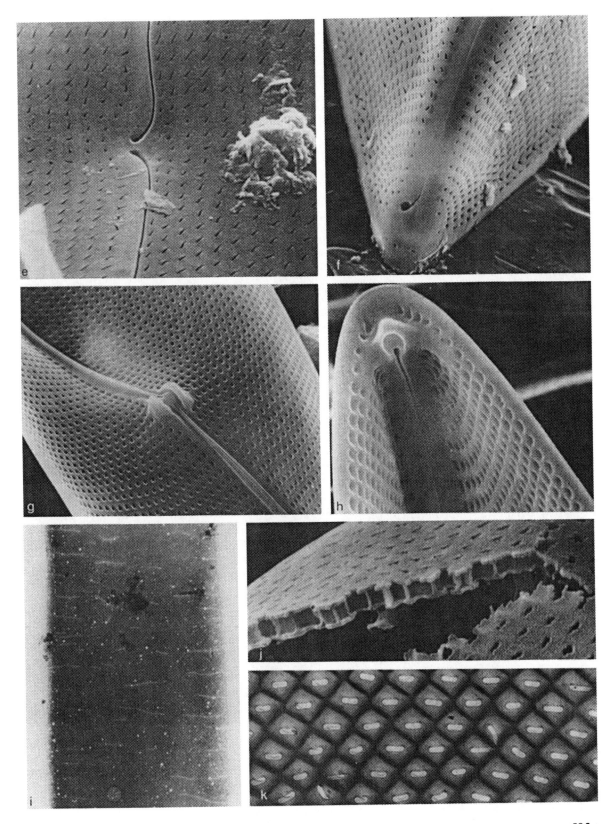

585

Gyrosigma A. Hassall 1845. Hist. Brit. Freshw. Algae 1: 435 (nom. cons.)

T: *G. hippocampus* (Ehrenberg) Hassall (= *Navicula hippocampa*)

Cells solitary or occasionally within mucilage tubes, usually lying in valve view.[a] Two large plate-like plastids per cell, one against each side of the girdle; sometimes extending beneath the valves and then their margins often highly lobed and dissected. Principally epipelic, living in brackish habitats but extending into marine habitats; a few species are common in freshwater.

Valves sigmoid,[b] linear to lanceolate, occasionally with rostrate poles. Valve face flat, curving smoothly into very shallow mantles. Valves composed of 2 layers held a short distance apart by struts placed in the interstices between the pores.[i] Inner layer penetrated by round to oval pores,[e-i] which are arranged in longitudinal and transverse striae; pores closed by hymenes; outer layer penetrated by slits. In most species, every internal pore lies below one of the apically orientated slits[c] in the outer layer of the valve; occasionally each outer slit lies above several pores. Raphe system usually along or close to the midline of the valve and hence sigmoid. Terminal fissures present,[d] turned in the same direction as the valve pole. The external raphe fissures are extended in the centre of the valve into blind grooves which are usually turned in opposite directions;[c] these 'central fissures', however, are occasionally T-shaped or turned in the same direction. Raphe opening internally onto the crest of a ridge,[e-h] which, except at the centre, is turned over towards the secondary side; central internal endings simple and expanded, or T-shaped, lying in a small, flat, fusiform nodule, which is flanked by lunate ridges.[e, f] On the primary side, the lunate ridge is continuous with a low ridge,[e-h] or row of small teeth borne on the raphe-sternum and running parallel to the raphe. Girdle containing open, non-porous bands.

Closely related to *Donkinia*, more distantly to *Pleurosigma* and *Toxonidea*, which have a different plastid structure and areola arrangement, but a similar raphe system and (in *Pleurosigma*) valve shape.

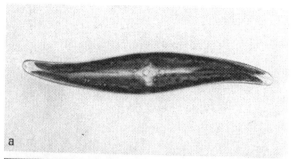

a

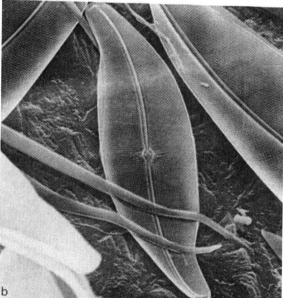

b

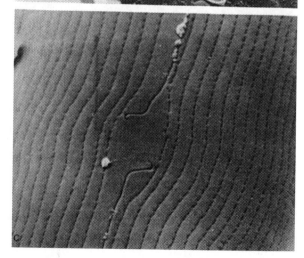

c

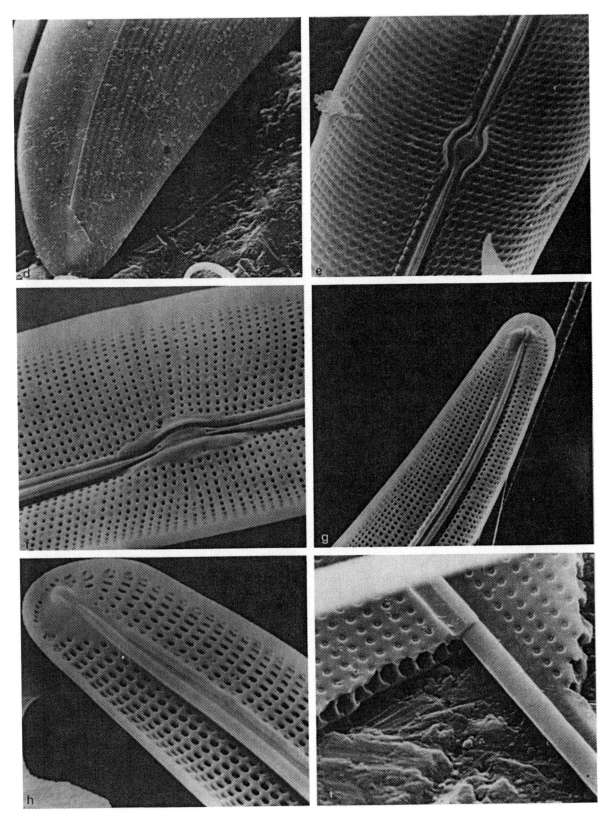

Plagiotropis E. Pfitzer 1871. Bot. Abh. Morphol. Physiol. **2**: 93

T: *P. baltica* Pfitzer

Cells solitary, almost always lying in girdle view unless moving actively. Four (sometimes 3 or 2) plastids disposed symmetrically about the apical and median planes, and lying against the girdle. Each plastid complexly lobed, usually with a prominent unilateral constriction; constrictions of the plastids appear superimposed in girdle view since the linking sections both lie against the same (valve) side of the cell. In the space left by these constrictions a large, complex volutin granule can be seen. Brackish or marine, epipelic.

Valves variable in form, sometimes compressed laterally. The raphe lies at the top of a high eccentric keel, which sometimes decreases in height or is even absent centrally.[a–c] Striae uniseriate.[e] Areolae often loculate, aligned longitudinally so that there appear to be longitudinal as well as transverse striae.[h, i] Shapes of inner and outer apertures of areolae variable: one or other aperture is often apically elongate. Hymenes present, apparently near the outer surface of valve. Raphe endings variable externally:[d, e] simple, pore-like, forked (polar endings only) or turned to one side. Internal central endings simple but lying in a small flat, oval nodule,[c] which is often accompanied on either side by short, curved ridges of silica (cf. *Pleurosigma*). Helictoglossa sometimes elaborated into a large funnel-like structure.[f, g] Girdle consisting of open bands.

A large distinctive group found most often in sandy or silty marine sediments. Unjustly neglected taxonomically. Probably related most closely to the *Pleurosigma–Gyrosigma* group, rather than to *Entomoneis* as previously thought. Reimer in Patrick & Reimer (1975) explains the reason for rejecting the more familiar name *Tropidoneis* in favour of *Plagiotropis*. The most commonly recorded species is *P. lepidoptera* and the var. *proboscidea* is reported from alkaline streams in the United States. Paddock (1988) has been studying this group in detail and considerable changes, involving the creation of several new genera, have been made. Unfortunately we have not been able to consider these here; reference should be made to Paddock (op. cit.). Plastid studies are urgently necessary.

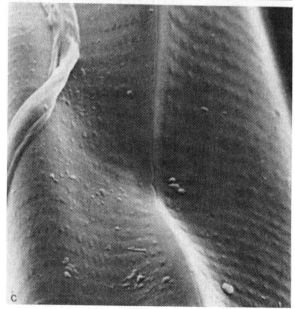

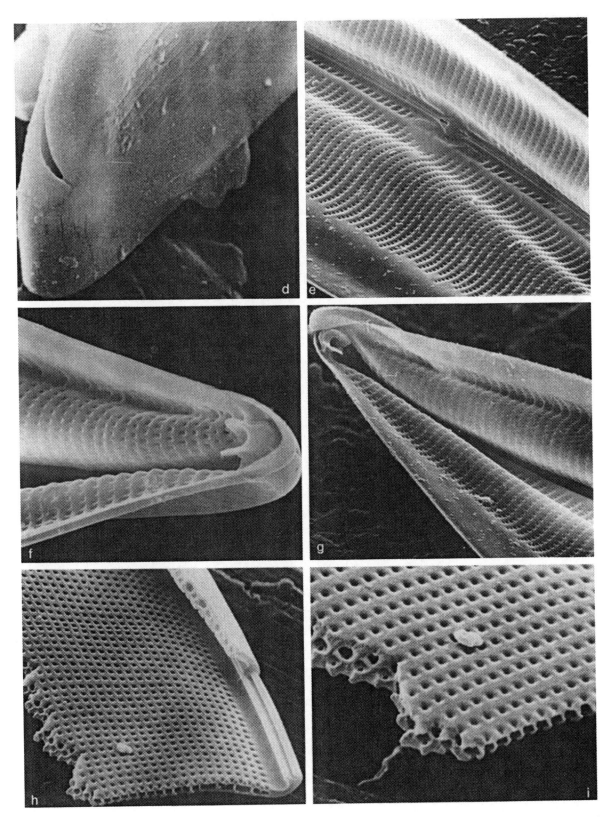

Stauropsis A. Meunier 1910. Duc d'Orléans Campagne Arctique de 1907, Bot. Microplankton: 318

T: *S. membranacea* (Cleve) Meunier (= *Navicula membranacea*)

Cells connected by their valve faces into band-like colonies; deeper (along the pervalvar axis) than wide, so that isolated cells tend to lie in girdle view.[a] Four folded, ribbon-like plastids, disposed symmetrically about the apical and median transapical planes; in valve view, each occupying a quadrant of the cell. Each plastid follows a tortuous path across the girdle, from one valve to the other and back. Marine, planktonic.

Frustule delicate, collapsing and becoming distorted if dried.[b] Valves lanceolate with acute apices; vaulted and more-or-less keeled near the poles, with no clear distinction between valve face and mantles. Striae fine, uniseriate,[c, d] containing small round poroids, each occluded at its external aperture by a thin membrane perforated by a single slit aligned parallel to the apical axis;[h] hymenes not yet demonstrated. Raphe-sternum narrow, central,[b] connected at the middle of the valve with a robust rib-like stauros, which is T-shaped in cross-section. Near the poles the raphe-sternum bears short flanges of silica, which appear in the LM as small refractile 'knobs' at the corners of the cell. Central raphe endings straight and simple, both externally and internally. Internal polar raphe endings unknown; externally the raphe fissure ends in or close to a large rimmed pore.[b, g] Girdle incompletely known, but containing open, very finely porous bands.[e, f]

The only species known with certainty to belong to *Stauropsis* is the type (Paddock, 1986), although several other species were assigned to the genus by Meunier (1910). There are resemblances between *Stauropsis* and *Pachyneis* (Simonsen, 1974), in the delicate valve structure and the presence of small flanges on the raphe-sternum. Large terminal pores are present in some members of the *Plagiotropis* complex, which also have complex plastids, though they are not ribbon-like as in *Stauropsis*. *Stauropsis* is not closely related to *Stauroneis*, in spite of the presence in both of a stauros. The habitat, cell shape, plastids and girdle structure all differ.

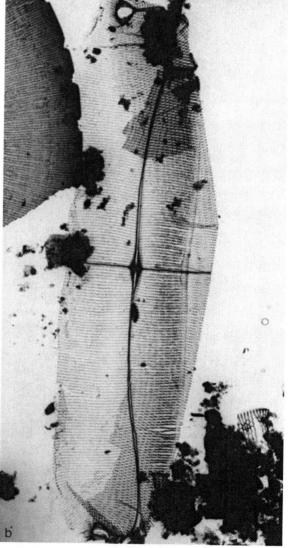

Stauroneis C. G. Ehrenberg 1843. Ber. Bekanntm. Verh. Königl. Preuss. Akad. Wiss. Berlin, 1843: 45

T: *S. phoenicenteron* (Nitzsch) Ehrenberg (= *Bacillaria phoenicenteron*; lectotype selected by C. S. Boyer 1928, Proc. Acad. Nat. Sci. Philad. 79 Suppl.: 420)

Cells naviculoid, solitary, or more rarely in small colonies (*S. acuta*).[b] Two plastids, one against each side of the girdle;[a] occasionally the plates are connected by a narrow central isthmus. A freshwater epipelic genus, with some subaerial forms in soil and moss.

Valves lanceolate to elliptical,[a, c, d] often capitate; sometimes with a slight ridge at the junction of the flat valve face and the mantle. Valve mantle often reducing in depth near apices; margin sometimes thickened at poles to form pseudosepta.[b, d] Striae uniseriate,[c–g] containing small round poroids, the external apertures of which can be transversely elongate; poroids closed internally by hymenes.[i] Pattern interrupted centrally by a thick stauros,[h, i] which extends out from the almost equally thickened raphe-sternum. Raphe-sternum sometimes bearing longitudinal ribs internally,[j,k] in a way reminiscent of *Frustulia* (q.v.). External central raphe endings strongly expanded, often deflected towards one side; internal endings simple or slightly deflected. Terminal fissures curved towards the secondary side. Girdle composed of several (at least 4 in one species) open porous bands.

Stauroneis should be understood to contain only the freshwater species with the plastid structure described above. Other '*Stauroneis*' spp. must be transferred elsewhere, e.g. to *Stauropsis* and *Staurophora* and a number of new genera will have to be erected. A preliminary survey of plastid arrangement and how these change during the cell cycle has been made by Stickle & Mann (1988). As Andrews (1981b) comments 'the genus has no phylogenetic unity but is made up of diatoms from diverse lineages, the only common feature being a stauros'. *Stauroneis sensu stricto* is probably most closely related to *Craticula*.

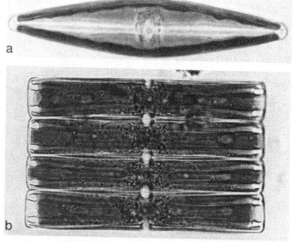

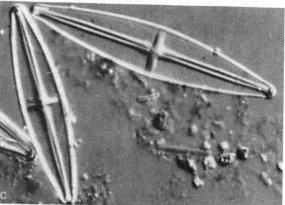

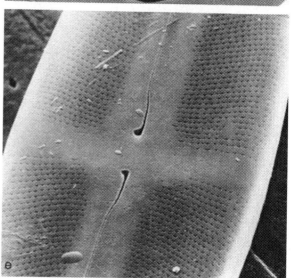

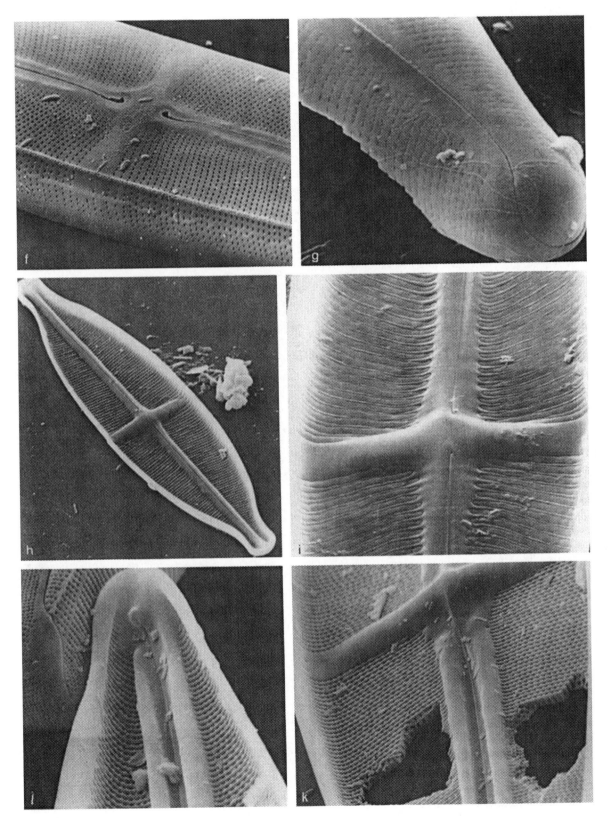

Craticula A. Grunow 1868. Reise Fregatta
'Novara' Bot. 1(i): 20

T: *C. perrotettii* Grunow

Cells solitary, naviculoid, usually lying in valve
view.[a, b] Two elongate, simple plate-like plastids per
cell, lying one against each side of the girdle. One to
several pyrenoids per plastid; pyrenoids not long
rods as in *Navicula*. Polymorphic, producing
characteristic reduced valves (craticulae), or internal
valves with changed stria pattern, in response to
osmotic stress. Freshwater to brackish, epipelic.

The valves are lanceolate,[c, d] with narrow rostrate
or capitate poles. Striae of normal valves more-or-
less strictly parallel,[e, f] uniseriate and consisting of
small round or elliptical poroids occluded by
hymenes at their internal apertures.[i, j] The poroids
and the frets separating them are aligned
longitudinally so that there appear to be
longitudinal[e–g] as well as transverse striae.
Furthermore the longitudinal elements formed by
the aligned frets are often thickened, so that
externally the valve appears as if covered by
longitudinal ribs of silica.[e–g] Craticular valves
consisting of a raphe-sternum and a system of
distantly spaced stout transverse bars (Schmid,
1979b). In the internal valves the striae are radial,
the poroids are not obviously aligned and no
external ribbing is present (see Schmid, 1979b).
Raphe-sternum thick in relation to valve face[e, j, k] but
raphe not accompanied by accessory ribs (contrast
Navicula). Internal central raphe endings simple or
slightly curved;[j, k] external central endings expanded
into 'pores' and hooked or turned towards the
primary side of the valve.[e, f] Terminal fissures
hooked, ending very close to the valve margin.[g]
Girdle composed of open, porous bands bearing one
transverse row of poroids apiece.

A small genus, distinguished from *Navicula* on the
basis of its different areola, raphe, girdle and
pyrenoid structure, and its valve polymorphism. The
internal valves probably function as protective
casings analogous to the resting spores of centric
species – Schmid (1979b) terms the whole complex
of valves/craticula/dormant cell/outer mucilage
envelope, a resting spore. From plastid behaviour,
sexual reproduction and some aspects of valve
structure this genus appears to be more closely
related to *Stauroneis* than to *Navicula*.

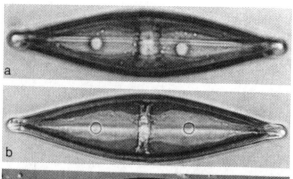

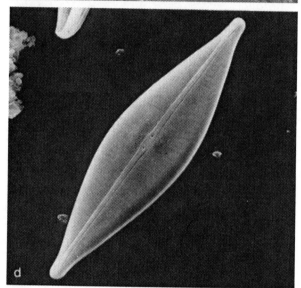

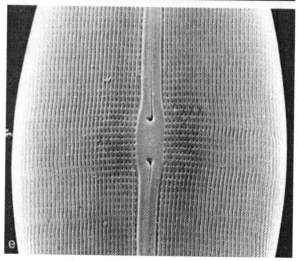

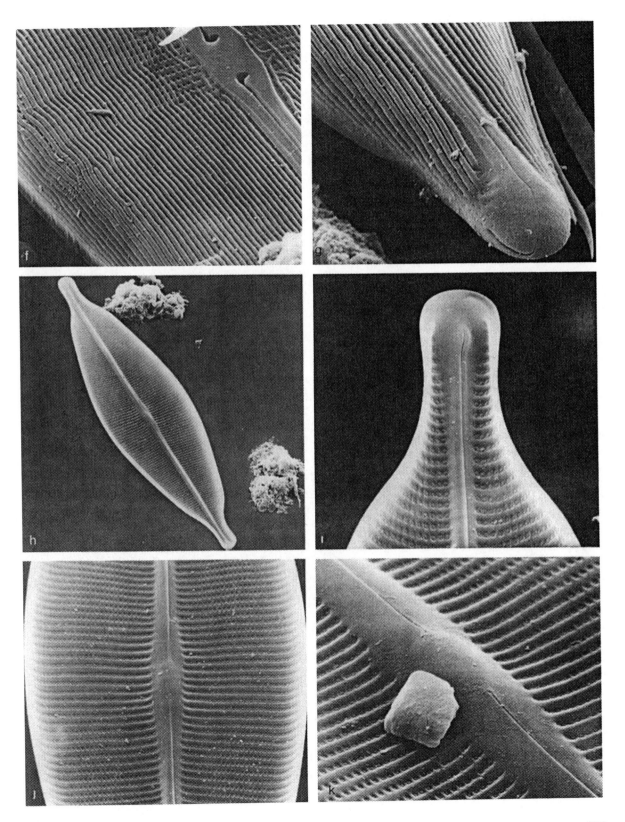

Proschkinia N. I. Karayeva 1978. Bot. Zh. **63**: 1748

T: *P. bulnheimii* (Grunow) Karayeva

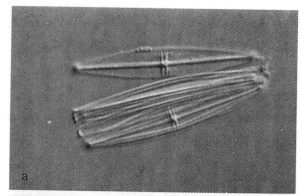

Cells solitary, usually lying in girdle view as a result of the great depth of the girdle. One plastid, divided into 2 plates lying on opposite sides of the girdle, each approximately half the length of the cell and displaced so that the plastid is diagonally symmetrical about the median transapical plane; plates connected by a narrow isthmus which crosses the cell at the central cytoplasmic bridge containing the nucleus. Marine; planktonic and benthic.

Valves narrowly lanceolate or linear,[a-c] with acute to slightly capitate apices. Valve face flat, grading into often very shallow mantles. Striae uniseriate,[h] containing more-or-less rectangular poroids closed by hymenes. The poroids are aligned between continuous longitudinal ribs of silica[c-e] (cf. some *Craticula*). At the centre of the valve the striae are more distant, especially on the primary side of the valve, and the costae separating them are thickened, producing stauros-like structures.[d, e] Near the central raphe endings, there are often small stigmata on the primary side of the valve.[g, h] The raphe system is sometimes central but often slightly eccentric, displaced towards the secondary side of the valve. Externally it is accompanied by one to several longitudinal ribs,[d, e] one of which often partially encloses the central raphe endings.[d] The central endings themselves are expanded externally into large 'pores'. The external polar endings are hooked towards the same side of the valve. Internally the raphe opens to one side of the raphe-sternum, which is considerably thicker than the remainder of the valve.[g,h] Centrally the fissures end internally in a structure resembling a double helictoglossa.[g] Girdle consisting of many open, porous, grooved, and more-or-less chambered bands,[i] together with some plain abvalvar bands.

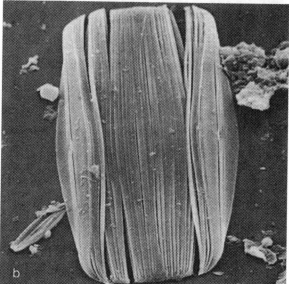

A genus apparently misunderstood in its original description, in that the raphe system was described as being like that of the genus *Nitzschia*. The creation of a new suborder by Karayeva (1978b) cannot be substantiated. The genus requires much further study (but see Brogan & Rosowski, 1988). There are superficial similarities to *Craticula* and *Navicula* but the raphe and girdle structure in particular make a close relationship unlikely.

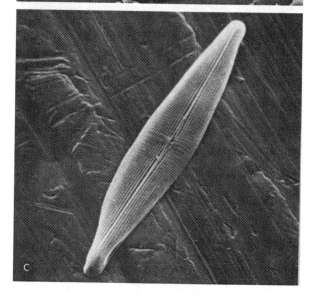

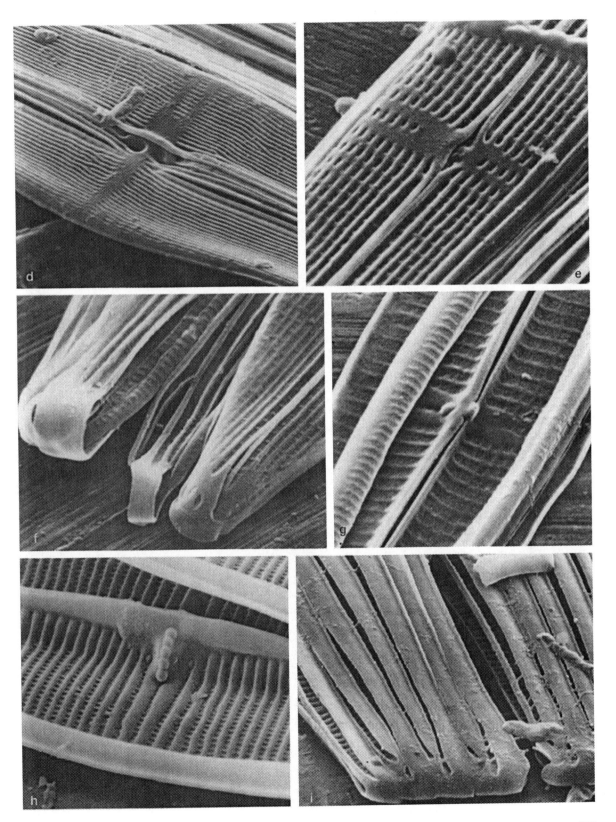

Catenula C. Mereschkowsky 1903. Scripta Botanica Horti Universitatis Petropolitanae, **19**: 97

T: *C. pelagica* Mereschkowsky (lectotype selected by C. S. Boyer 1928, Proc. Acad. Nat. Sci. Philad. **79** Suppl.: 492)

Cells dorsiventral, attached by their valve faces to form ribbon-like filaments.[a] Valve faces parallel (contrast *Amphora*). One plastid, lying against the ventral side of the cell, constricted along the midline of the girdle. A marine genus, living attached to or associated with the sediments (*C. adhaerens*), or in the plankton (*C. pelagica*).

Valves strongly asymmetrical about the apical plane,[b, c, i] with a more-or-less straight ventral margin and a convex dorsal margin; apices sometimes slightly rostrate. Valve face flat,[e] sharply differentiated from the shallow mantles. Striae uniseriate but apparently filled in on the valve face, so that true pores are found only around the dorsal and ventral mantles, the valve face being plain or with slight transapical ribbing. Mantle pores transapically elongate.[d-h] Raphe strongly eccentric,[e, i] lying at the junction of valve face and ventral mantle. Central raphe endings distant, simple internally and externally. Polar endings more-or-less distant from the poles.[f] Helictoglossa prominent in both LM[b] and SEM;[i, j] indeed the central and polar raphe endings are the only features of the valves obvious in the LM. External polar raphe endings simple, unexpanded. Cingulum consisting of several (at least five in *C. adhaerens*) open, non-porous bands.[c, d, g, h]

The two species originally described by Mereschkowsky are still the only species known. Salah's (1955) *Amphora salyii* seems to be conspecific with *C. adhaerens* (Simonsen, 1962, and observations of the type slide by K. Sundbäck, personal communication); this species is considered in detail by Sundbäck & Medlin (1986). *Catenula* is similar to some *Amphora* species in its plastids, valve shape and simple raphe structures. However, the heterogeneity of *Amphora* suggests that it would at present be unwise to combine these two genera.

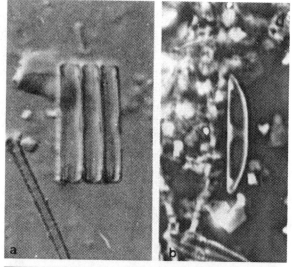

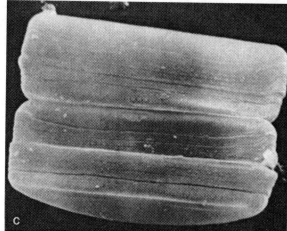

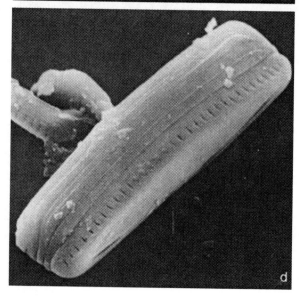

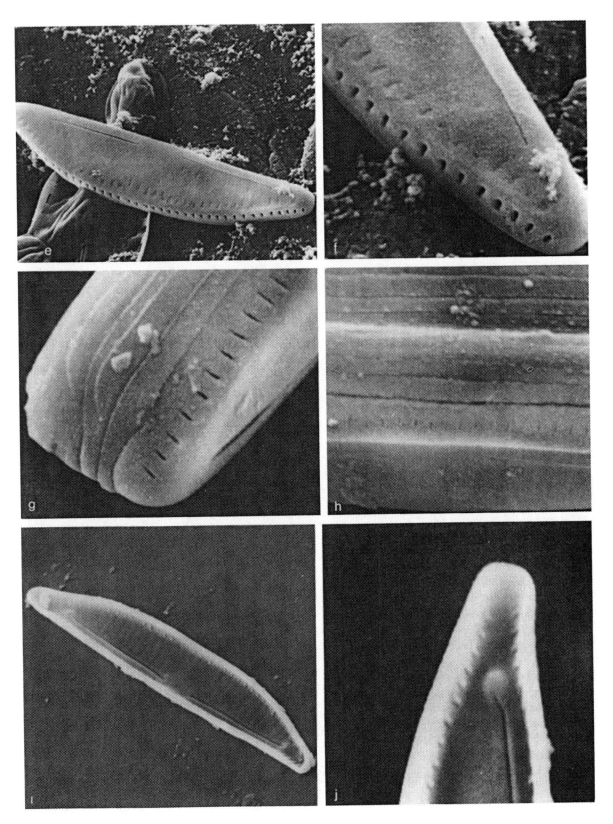

Amphora C. G. Ehrenberg ex F. T. Kützing 1844. Kiesel. Bacill: 107

T. *A. ovalis* (Kützing) Kützing (= *Frustulia ovalis*: lectotype selected by C. S. Boyer, 1928, Proc. Acad. Nat. Sci. Philad. 79 Suppl.: 253)

Cells solitary, sometimes sessile but usually motile, almost always lying in girdle view,[a–f] and then appearing elliptical or lanceolate, with truncate ends. A whole cell or frustule is like 'a third of an orange' (Hendey, 1964). Valves so arranged that both raphe systems lie on the same (ventral) side of the cell.[g–i, k, l, o] Plastids usually 1 or 2, sometimes many; extremely diverse in position, shape and structure. Epiphytic, epilithic or epipelic. A very large mainly marine genus with relatively few representatives in freshwater.

Valves asymmetrical about the apical plane ('cymbelloid'),[q, w] and sometimes constricted centrally or near the poles. Valve face sometimes well differentiated from mantles, sometimes merging imperceptibly with them. Ventral mantle always shallower than dorsal mantle and sometimes absent. Striae uni- or biseriate, containing areolae that may be anything from simple, round poroids[r–t] to complex loculate structures.[n] Areolae occluded by hymenes.[k, x] Raphe system eccentric, lying nearer the ventral margin,[f–i, k, l, w] but biarculate in some and then approaching the dorsal margin near the poles.[q, s] Raphe–sternum often extended over the valve externally as a short conopeum, especially on the dorsal side of the raphe.[g–i, l] External central raphe endings straight or deflected towards either margin,[g, i, k, l] usually expanded; internal endings simple[t] or hidden by a flap of silica extending from the ventral side.[r, u] Terminal fissures usually present, short, diverse in form. Copulae numerous,[j, l–n] open, each one wide on the dorsal side and narrow on the ventral; plain or perforated by one or more rows of areolae. Girdle structure is important for correct identification.

There is a study of some freshwater species by Krammer (1980), and an important series of papers by Schoeman & Archibald (see references in Schoeman studied the type species and revealed much infraspecific variation; *A. copulata* has also been studied (Lee & Round, 1988). The genus needs revision, based on frustule *and* protoplast characters.

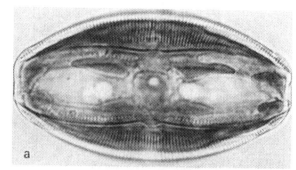

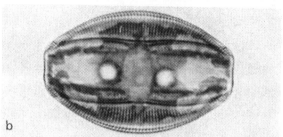

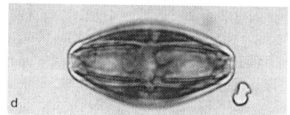

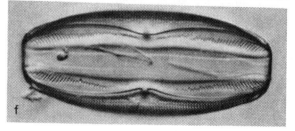

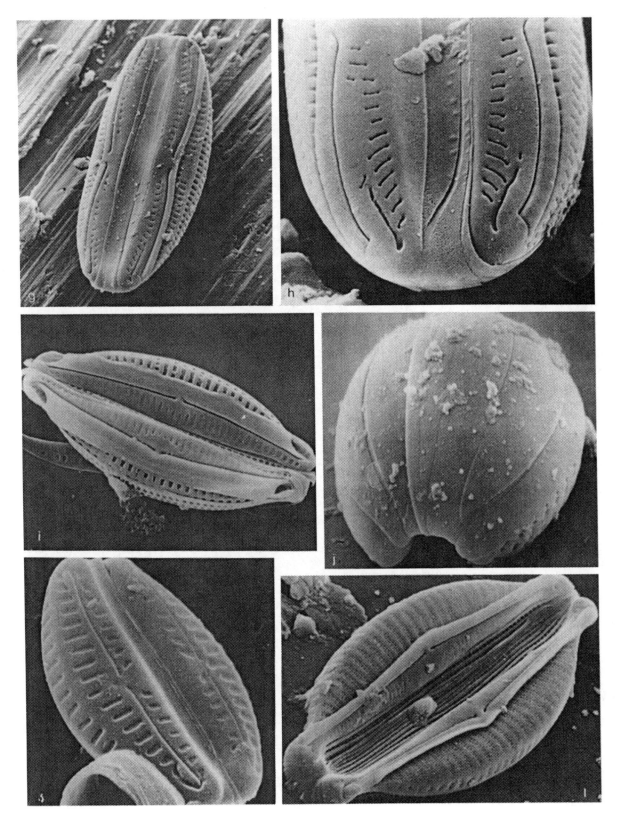

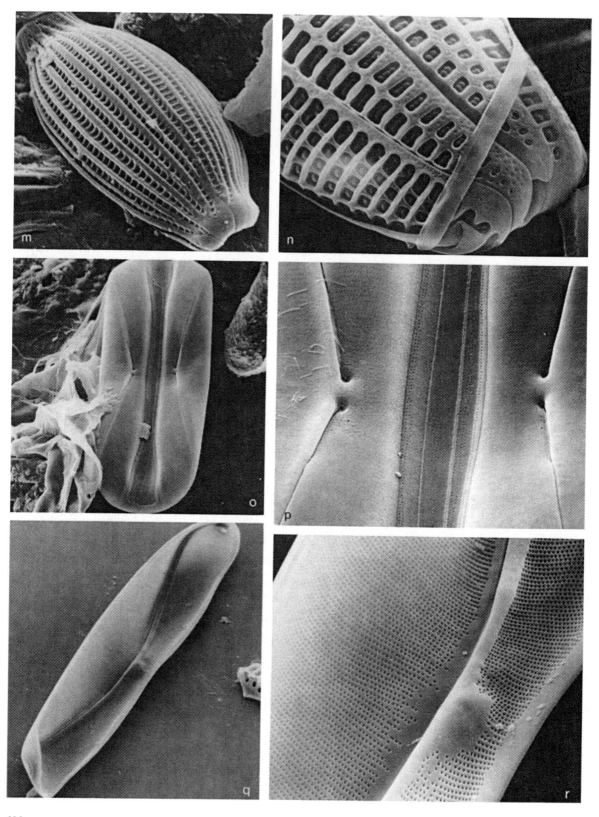

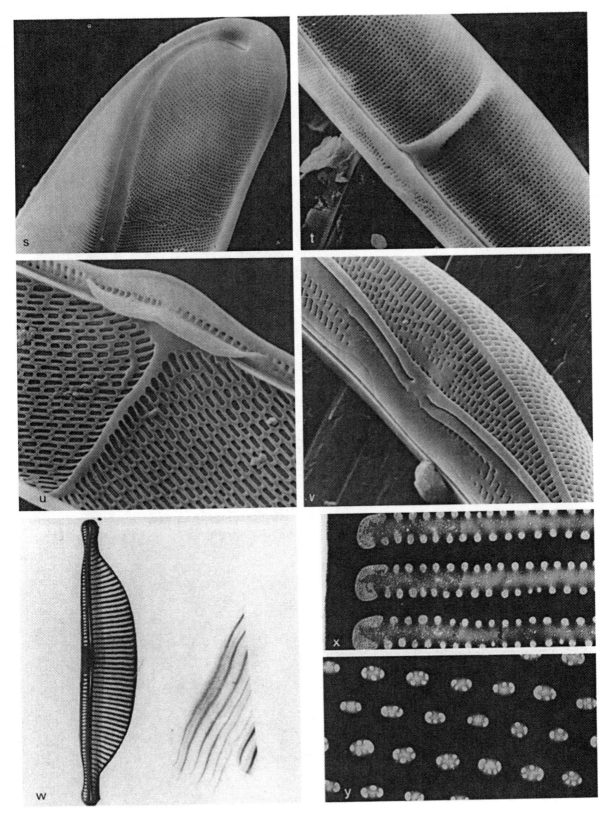

Undellata T. B. B. Paddock & P. A. Sims 1980.
Bacillaria 3: 169

T: *U. lineata* (Greville) Paddock & Sims
(= *Amphiprora lineata*)

Cells solitary, strongly dorsiventral, with very narrow valves and numerous girdle bands.[a] Usually lying in girdle view and then rectangular but constricted centrally.[a] Plastids: many and discoid in *U. quadrata*, unknown elsewhere. Marine, epipelic.

Valves narrow, with a biarcuate ventral margin and a sinuous or straight dorsal margin.[b, c] Striae uniseriate,[h] containing small round poroids. Pattern sometimes interrupted centrally by a massive stauros.[h] Raphe system raised on a narrow, eccentrically placed keel, lying near the biarcuate ventral margin.[b, d] Raphe usually subtended by short, bar-like fibulae.[a, h] Internal[h] and external central raphe endings simple. External raphe fissure deflected towards the dorsal margin at the poles.[f] Girdle elements numerous[a] and unusual, in that they are tubular with flanges on either side to allow for overlap.[d, i, k] The tubes bear 1 or 2 transverse rows of areolae together with some other scattered areolae.[i] Towards the apices, and sometimes elsewhere, where the ventral and dorsal sides of the girdle approach each other closely, there are often peg-like projections from either side of the bands, which interdigitate[g] or fuse[d, j] in the midline of the cell.

The genus was erected by Paddock & Sims (1980) for several species formerly placed in *Amphora*, *Amphiprora* or *Auricula*. They discuss the possible affinities of this genus and give a detailed description of valve and girdle structure. Their 1981 paper should also be consulted. We thank T. B. B. Paddock and P. A. Sims for the material.

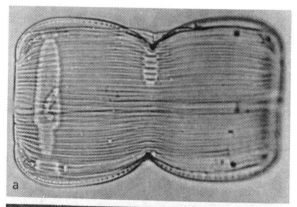

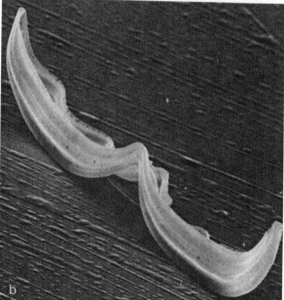

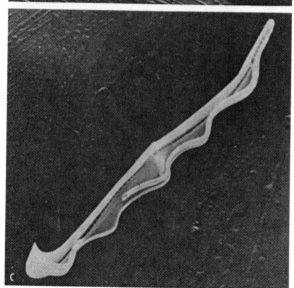

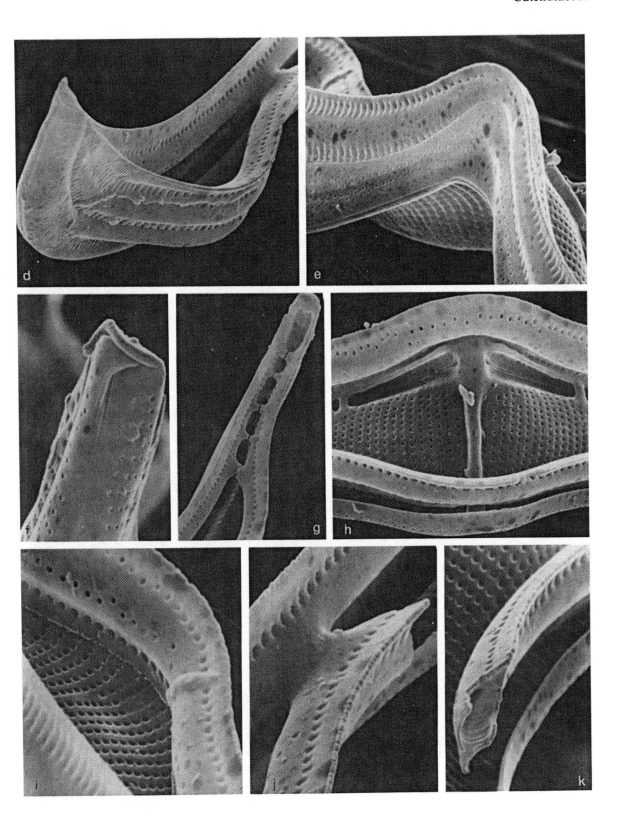

Thalassiophysa P. Conger 1954. Smithsonian Misc. Coll. **122** (14): 1

T: *T. hyalina* (Greville) Paddock & Sims
(= *Amphiprora hyalina*; = *T. rhipidis* Conger, nom. illeg.)

Cells solitary, strongly dorsiventral, almost always lying in girdle view.[a] One axial plastid per cell, lying near the ventral side of the girdle. A monotypic genus of the marine littoral.

Valves compressed,[b] with a strongly keeled, markedly biarcuate raphe system. Ventral side of valve more-or-less planar; dorsal side arched. Striae uniseriate, containing small round poroids[f, h] occluded at their external apertures by hymenes. Raphe system fibulate, lying on top of the keel near the poles but curving onto the ventral side of the valve nearer the centre. Raphe forming a narrow 'V' centrally;[c, d, i] central raphe endings lying by the ventral valve margin. Polar raphe endings close to valve margin;[e] no terminal fissures. Near the centre, the raphe is accompanied internally by a siliceous ridge on each side:[f] the ridges combine with the raphe-sternum centrally to form a characteristic beak-like structure[c, f, g, i] (termed a 'mucro' by Paddock & Sims, 1980). Fibulae of two types: arched clasps[f] and more-or-less straight struts[h] ('adventitious fibulae'). The clasps can be seen in the central region of the raphe system linking the 2 ridges noted above, but cease some distance from the central raphe endings.[f] The clasps also occur in the keeled region, but are hidden and we have not been able to investigate them in detail. The struts link the sides of the narrow, high keel and occur at various levels beneath the raphe.[h] Girdle consisting of open bands bearing striae similar to those of the valve.

A very distinctive diatom, widespread in temperate and tropical seas. The valve structure is covered in detail by Paddock & Sims (1980) where the genus is called *Proboscidea*. This name was illegitimate, as recognised by Paddock & Sims (1981) in a review of fibulate diatoms. The plastid structure and fibulae suggest a possible relationship to *Entomoneis*. However, we prefer to link it to *Undatella* and *Amphora sensu lato*.

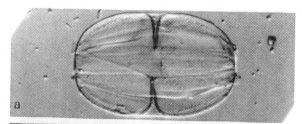

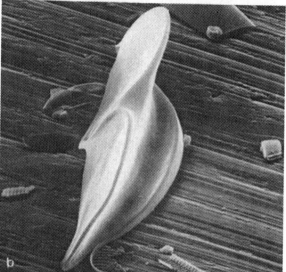

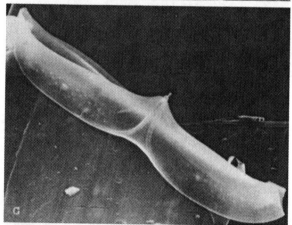

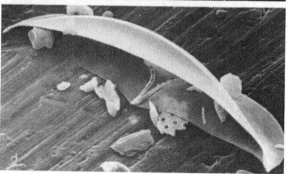

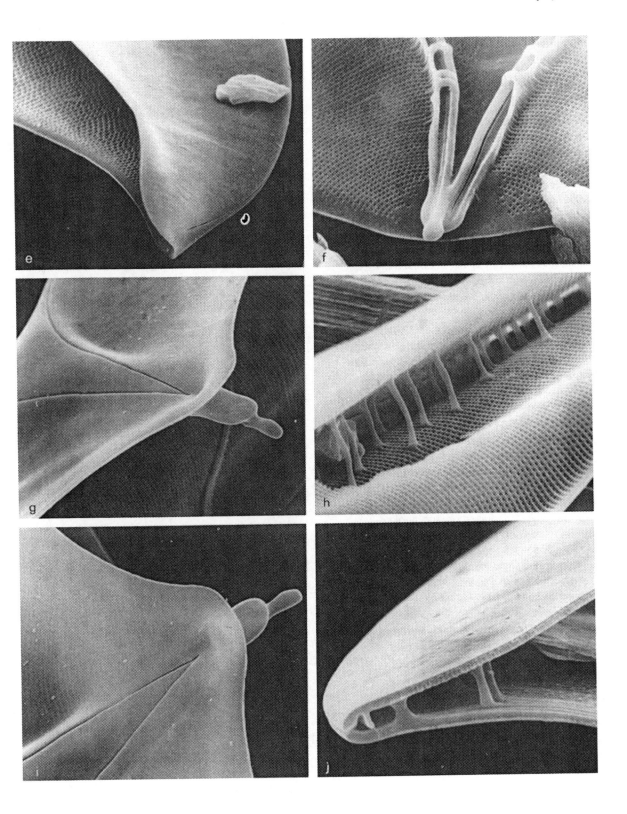

Bacillaria J. F. Gmelin 1791. Syst. Nat., Ed. 13, 1 (6): 3903

T: *B. paxillifer* (O. F. Müller) Hendey (= *Vibrio paxillifer*)

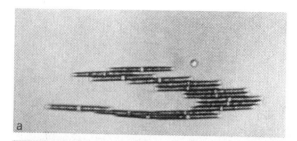

Cells elongate, forming a unique type of motile colony[a, b] in which individual cells slide to and fro with respect to each other, held together by interlocking ridges and grooves developed on the raphe-sterna. Colonies extending to form linear arrays (by pole-to-pole contact only) and then retracting into tabular arrays, with all the cells side by side. Usually 2 plastids, one towards each pole, as in *Nitzschia*. A small epipelic genus of marine, brackish, and occasionally high-conductivity fresh waters. Often found in the plankton of shallow seas where it has presumably been swept up from the sediments.

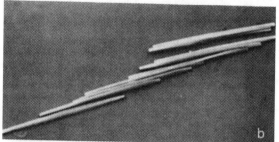

Valves linear or linear-lanceolate;[c, d] poles rostrate or capitate. Valve face more-or-less flat, curving into extremely shallow mantles. Irregular to reticulate ridges of silica are sometimes present near the junction of valve face and mantle.[f] Striae uniseriate[i, j] or (rarely) biseriate,[h] containing small round poroids occluded by hymenes. Raphe system central or nearly so, slightly keeled,[c, f] fibulate.[d, h-j] Raphe-sternum bearing a prominent ridge externally,[c, e-g] on one side of the raphe; this is involved in the cell-cell interlock (Drum & Pankratz, 1965b). Raphe continuous from pole to pole. External polar endings pore-like,[g] T-shaped or with hooked terminal fissures. Fibulae rib-like, arching into the cell.[h-j] Girdle consisting of many open bands, which often have tiny warts externally and may be porous.

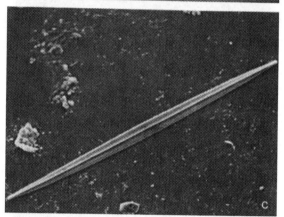

A distinctive genus, which is also the earliest to have been described. Attachment of cell to cell to form colonies does occur in other members of the Nitzschiaceae (e.g. *Fragilariopsis*, some *Nitzschia* spp.) but these never exhibit the motility evident in *Bacillaria* colonies. A genus in urgent need of investigation at the species level.

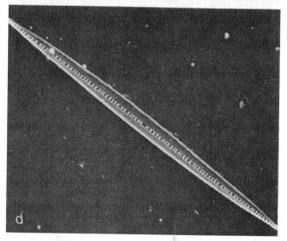

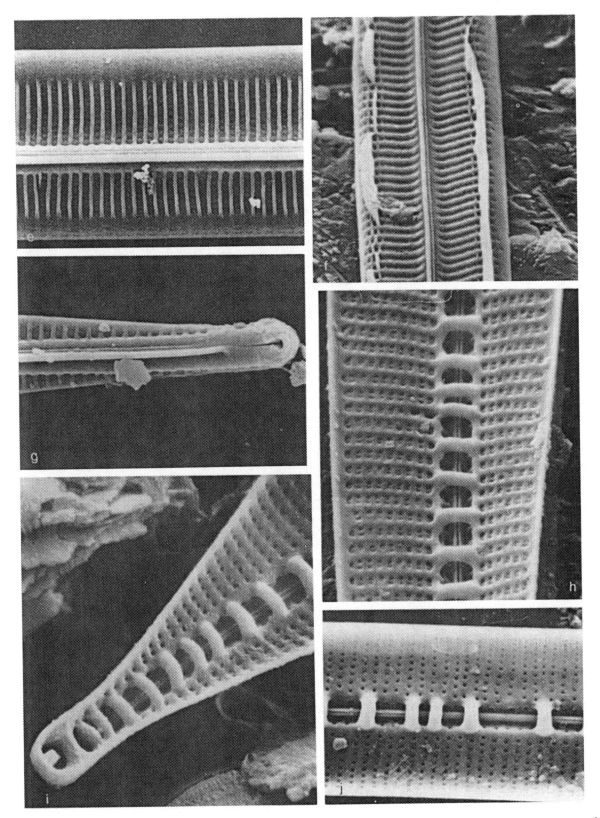

Hantzschia A. Grunow 1877. Mon. Microsc. J.
18: 174 (nom. cons.)

T: *H. amphioxys* (Ehrenberg) Grunow
(= *Eunotia amphioxys*) (typ. cons.)

Cells solitary, straight or sigmoid, usually lying in
girdle view,[a] markedly dorsiventral. Two (rarely 4)
plastids, lying one on either side of the median
transapical plane. Plastids simple or complexly
lobed and lying against the ventral side of the cell;
or consisting of two girdle-apposed plates connected
by a prominent central pyrenoid. Widely distributed
in the marine and freshwater epipelon, especially in
intertidal sand, but extending into subaerial habitats
(e.g. *H. amphioxys*, which is almost universally
present on soils).

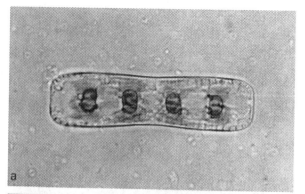

The valves are asymmetrical[b] with respect to the
apical plane, or sigmoid. The raphe is eccentrically
placed on the valve[c, f, h] and both raphes are on the
same, usually less convex, ventral side of the cell, i.e.
the frustule is mirror-symmetrical about the valvar
plane ('hantzschioid symmetry'). Valve structure
simple, with uniseriate or biseriate rows of round or
reniform poroids containing hymenes; cribra may
also be present. Marginal ridges occasionally
present. Valve face quite distinct in some[d] but in
others grading imperceptibly into the mantles.
Raphe continuous from pole to pole or interrupted
centrally, often biarcuate (closer to the ventral side
centrally than nearer the poles). Where present,
central raphe endings simple[e] or partially enclosed
dorsally by a flap of silica; internally the raphe
fissures are continuous or deflected to opposite
sides[g] (cf. *Cocconeis*). Raphe subtended internally by
fibulae,[f-i] which are massive, or slender and rib-like.
Polar raphe endings simple or hooked towards the
dorsal side.[c] Girdle complex, containing open or
closed bands, of which at least some bear two or
more rows of poroids.[j]

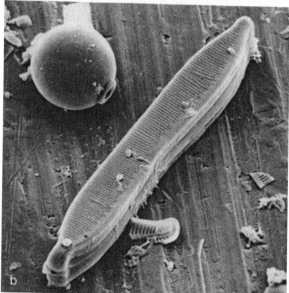

Although undoubtedly a natural group, *Hantzschia*
is difficult to define, since few characters are
common to all *Hantzschia* spp. but absent from all
Nitzschia spp. All *Hantzschia* spp., however, seem to
have a type of division in which cells of
hantzschioid symmetry always give rise to two
daughter cells *both* of which have hantzschioid
symmetry: this contrasts with the situation in
Nitzschia. The genus has been discussed by Round
(1970) and Mann (1977, 1980a, b, 1981b).

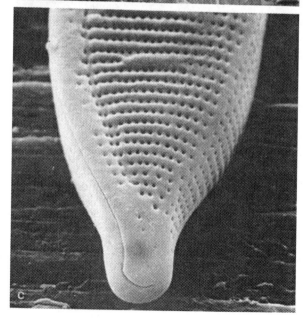

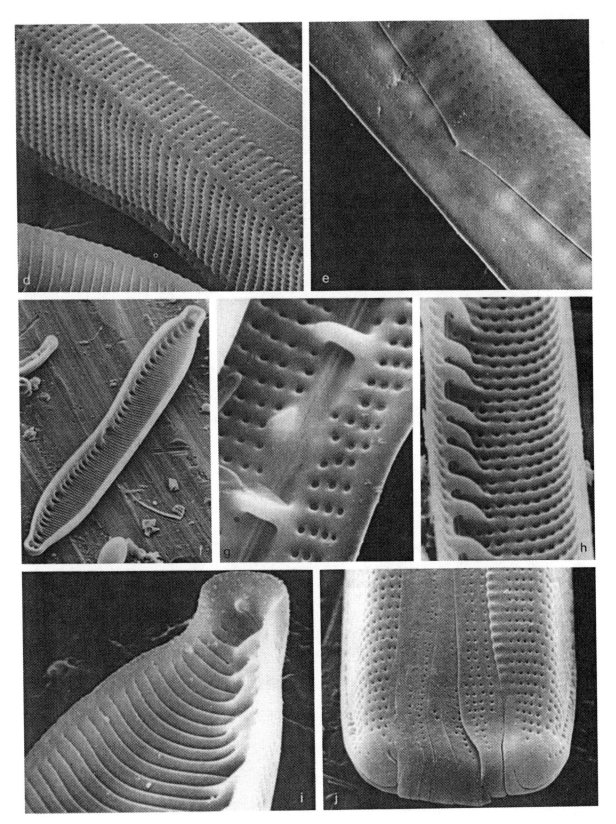

Psammodictyon D. G. Mann gen. nov.

T: *P. panduriforme* (Gregory) Mann (= *Nitzschia panduriformis*)

Cells solitary, usually lying in valve view.[a] Frustules diagonally symmetrical about the median valvar plane (nitzschioid symmetry).[c] Two plastids, one towards each pole; the gap between them is often oblique to the transapical plane (contrast *Nitzschia*). A marine, epipelic genus, particularly widespread on sandy substrata.

Valves panduriform or broadly linear, with apiculate or bluntly rounded poles.[a, b] Valve face bounded by the keeled raphe system on one side and merging imperceptibly with a shallow mantle on the other;[b] in transverse section usually sinusoidal.[f] Valve structure almost always loculate over at least part of the valve face.[f] Each chamber opens to the valve interior by one[g, h] to several small pores, which are occluded by hymenes, and to the exterior by a single round aperture;[d] chambers in hexagonal array.[a] In non-chambered parts of the valve face or on the proximal mantle, the transapical striae are uni- or multiseriate, containing simple poroids. Striae often interrupted by an axial sternum.[a, d] Raphe system submarginal, keeled, fibulate.[f, h, i] External raphe fissures usually accompanied by a ridge on either side; ridges very closely appressed except near the central raphe endings. Central external endings slightly expanded or deflected and separated by an oval nodule;[c] internal fissures ending in a double helictoglossa.[h] Terminal fissures hooked[c] or deflected. Raphe subtended internally by simple rib-like or bar-like fibulae.[h, i] Girdle consisting of several open bands; the most advalvar band usually bears two to several transverse rows of small round poroids; the other bands are narrower and non-porous.[e]

A natural group closely related to *Nitzschia* and *Tryblionella*, but separable by the combination of valve structure, raphe and fibula structure, and to a lesser extent by habitat, valve shape and plastid position. Formerly included in *Nitzschia* as the sect. *Panduriformes*.

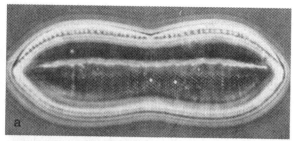

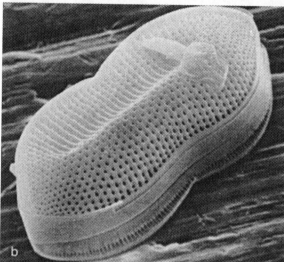

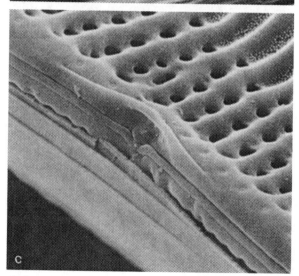

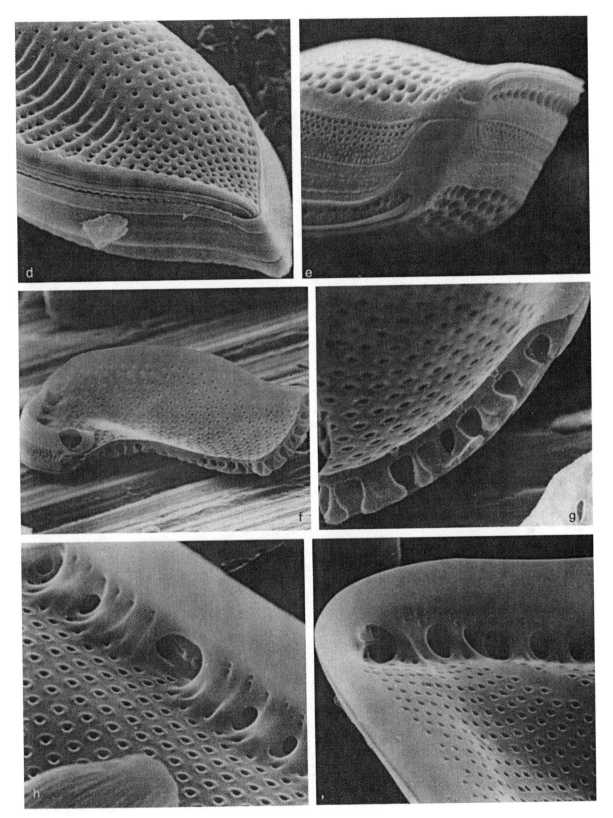

Tryblionella W. Smith 1853. Syn. Brit. Diat. 1: 35

T: *T. acuminata* W. Smith (lectotype selected by D. G. Mann)

Cells solitary, usually lying in valve view.[a, b] Frustules diagonally symmetrical about the median valvar plane ('nitzschioid' symmetry). Two plastids, one on either side of the median transapical plane. A fairly large epipelic genus, widespread but not often common in brackish and marine sediments; also present in high-conductivity fresh waters.

Valves robust, broad; elliptical, linear or panduriform, with bluntly rounded or apiculate poles. Valve face often bearing warts or ridges externally,[c, e] undulate,[c] bounded on one side by the keeled raphe system; on the other side often bearing a marginal ridge at its junction with the extremely shallow distal mantle.[f, h, k] Striae uniseriate to multiseriate,[e, g] usually interrupted by one or more sterna,[d, g] and containing small round poroids occluded by hymenes; rarely, alveolate (cf. *Pinnularia*). Raphe system strongly eccentric, keeled, fibulate.[g, l] Central external raphe endings close together,[i] slightly expanded or deflected; occasionally absent. Internal fissures ending centrally in a double helictoglossa. Terminal fissures short, deflected.[h] Fibulae squat,[g, l] often as broad or broader apically than transapically, sometimes striate like the valve[l] and then sometimes hollow. Girdle narrow, containing plain or sparsely porous, open bands.[k]

Closely related to *Psammodictyon* and *Nitzschia*. Difficult to define monothetically but nevertheless a natural cluster, including most species traditionally placed in *Nitzschia* sects. *Tryblionella, Circumsutae, Apiculatae* and *Pseudotryblionella*. Common diagnostic combinations are: massive apically elongate fibulae, high marginal ridge, uniseriate striae interrupted by one or two sterna, tiny poroids; *or* small, stubby fibulae, low marginal ridge, multiseriate striae, thickened axial sternum; *or* valves panduriform, strongly undulate valve face, uniseriate striae, sterna present or absent, bar- or plate-like fibulae (this for the group closest to *Nitzschia* and *Psammodictyon*); *or* central raphe endings absent, thickened axial sternum, multiseriate alveolate striae. The broad, shallow valves, undulate valve face and strongly eccentric raphe system are useful supporting characters.

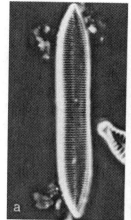

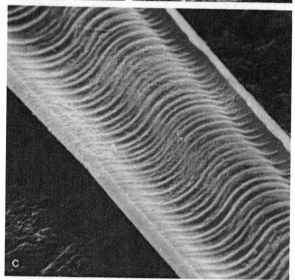
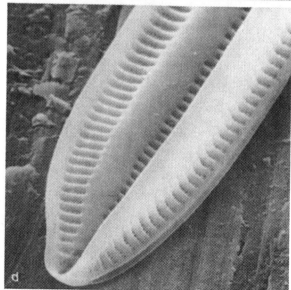

Cymbellonitzschia F. Hustedt 1924. In A.
Schmidt, Atlas Diat., T. 352, figs 12, 13

T: *C. minima* Hustedt

Cells solitary or forming short chains; individual
cells lie in valve or girdle view with approximately
equal frequency. In valve view the cells are
asymmetrical about the apical plane; in girdle view
the cells are rectangular, since the valve faces are
parallel (contrast most *Cymbella* spp.). Two simple,
plate-like plastids, one on each side of the mean
transapical plane (cf. *Nitzschia*). A small genus of
fresh and brackish waters.

   Valves with a straight ventral margin and a
convex dorsal margin.[a-c] Valve face flat, fairly
sharply differentiated from the mantles. Striae for
the most part uniseriate, containing small round
poroids occluded by hymenes; the hymen is external
and often more-or-less domed.[b-e] Near the raphe, in
the walls of the subraphe canal, the striae become
biseriate (*C. diluviana*) and here the poroids are
much smaller than elsewhere. Raphe system
fibulate, eccentric, lying at the junction of valve face
and one mantle.[c-e] The raphe systems of the two
valves are on the same side of the cell
('hantzschioid' symmetry): this may be the ventral
side (*C. diluviana*, *C. hossamedinii*) or the dorsal side
(*C. minima*: but see Simonsen, 1987). The remainder
of this description applies to *C. diluviana*, the only
species known in detail. External raphe fissures
accompanied on one side (nearer the valve face) by
a low, longitudinal ridge of silica.[c-e] Central raphe
endings fairly distant; straight and simple, internally
and externally. Terminal fissures always curved
towards the ventral margin.[c, e] Raphe subtended
internally by short, bar-like fibulae, whose bases are
connected by longitudinal ridges.[f-n] Cingulum
consisting of a few open bands,[i] some of which bear
a single transverse row of tiny poroids.

   We illustrate the commonest species, *C. diluviana*.
C. minima, the type, may prove to be quite different,
since here the raphe system is on the convex dorsal
side. *C. diluviana* is very similar to some species of
Nitzschia sect. *Lanceolatae*, but the difference in valve
shape and frustule symmetry must reflect significant
changes in the control of auxospore development
and valve ontogeny (the raphe systems of sibling
valves must move in the same direction during
silicification as in *Hantzschia*).

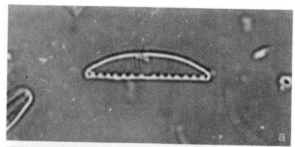

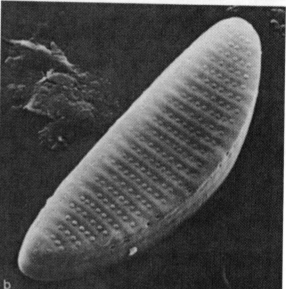

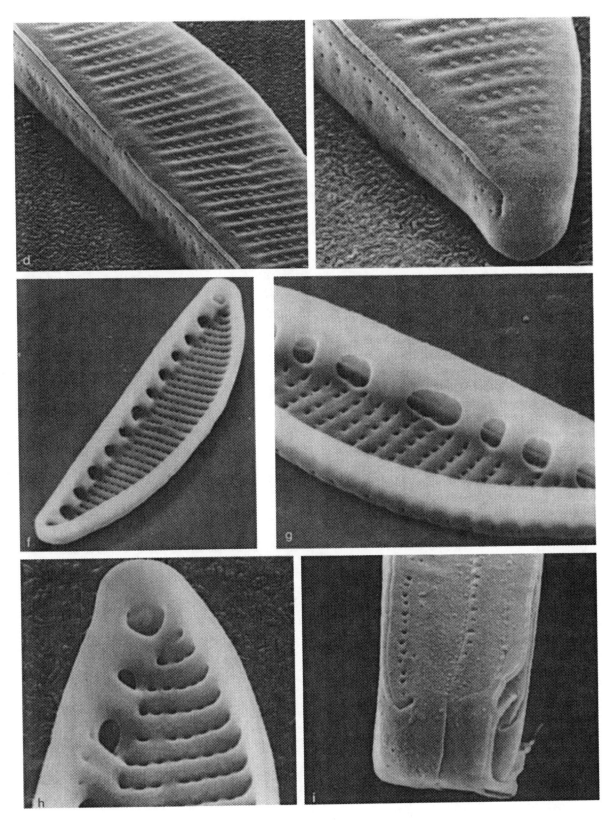

Gomphotheca N. I. Hendey & P. A. Sims 1982. Bacillaria 5: 199

T: *G. sinensis* (Skvortzow) Hendey & Sims
(= *Nitzschia chinensis*)

Cells very long, ?colonial, heteropolar in valve and girdle view. Plastids unknown. A small genus, known to include only 2 brackish to marine, rarely recorded species; apparently tropical and subtropical.

Valves narrowly linear,[a] with a bluntly or acutely rounded head-pole. Valve face curved, merging into the mantles, so that the valve is almost semi-circular in transverse section.[f, i, j] Striae fine, uniseriate, containing small but deep poroids.[c, d] Near the junction between valve face and raphe-sternum, some poroids of each striae open into a transapically elongate groove.[b-d] Striae shorter near the centre, leaving a plain central area.[b] At intervals, very conspicuous/robust ribs extend internally across the valve from near the raphe to the valve margin, which is also thickened and rib-like. Raphe system fibulate.[e-h] Raphe-sternum more-or-less central or eccentric, massive, raised on a slight keel;[c, d, f, i, j] bearing two grooves, one on either side of the raphe slit and parallel to it. Central raphe endings simple or slightly expanded, internally and externally. Terminal fissures present, more-or-less straight, continuing almost to the edge of the valve. Raphe subtended by massive fibulae, which connect with one or two of the thickened transapical ribs.[e-h] Fibulae linked at their bases by longitudinal ridges; the ridges and fibulae together delimit round or oval holes (portulae), which connect the subraphe canal[f, i, j] to the valve interior. Girdle incompletely known. Valvocopula apparently a very robust, chambered structure,[i, j] with many small pores externally and a few (3) rows of larger pores internally; pars interior with scalloped and hooked edges which fit over the expanded valve margin.

The two known species were formerly classified in *Gomphonitzschia* but separated from it following light and electron microscopical investigation by Hendey & Sims (1982). Valve and fibula structure are critical for the separation of *Gomphonitzschia* and *Gomphotheca*, which resemble different sections of *Nitzschia*. Special thanks are due to P. A. Sims for providing us with material of this genus.

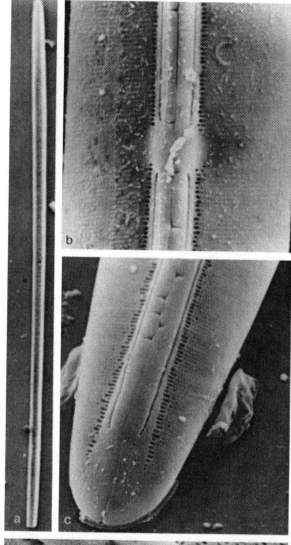

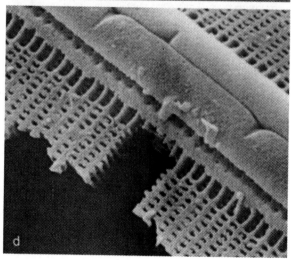

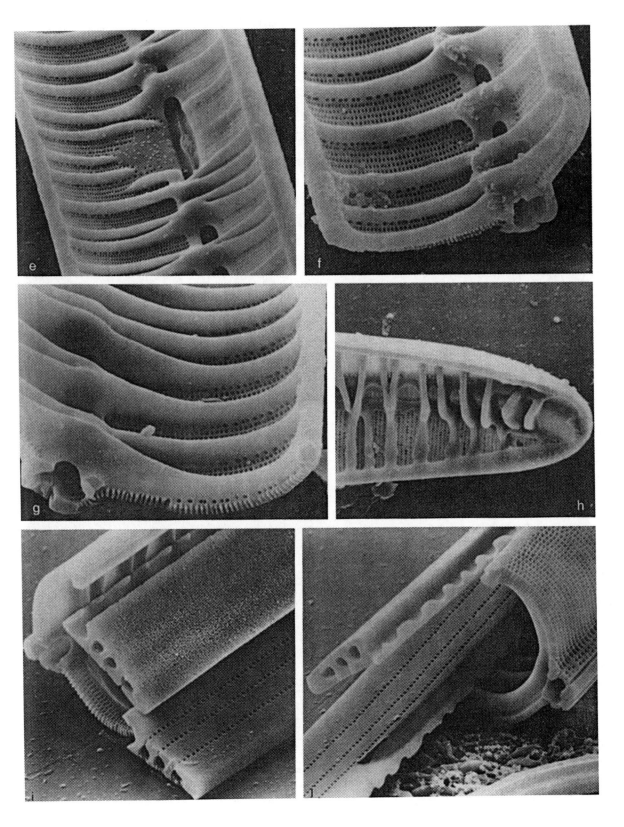

Nitzschia A. H. Hassall 1845. Hist. Brit. Freshw. Algae 1: 435 (nom. cons.)

T: *N. sigmoidea* (Nitzsch) W. Smith (= *Bacillaria sigmoidea*; as *N. elongata* nom. illeg. in Hassall 1845)

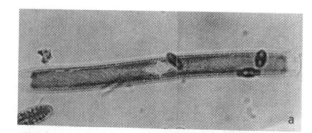

Cells solitary, or forming chain-like or stellate colonies, or living in mucilage tubes. Usually straight and needle-like, sometimes sigmoid, lying in valve or girdle view. Two plastids, one at each end of the cell;[a] rarely, many small discoid plastids. Plastids simple or complexly lobed, containing one to many rod-like pyrenoids. Freshwater to marine; usually epipelic or planktonic.

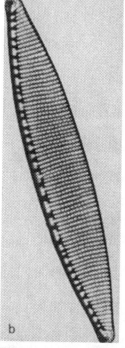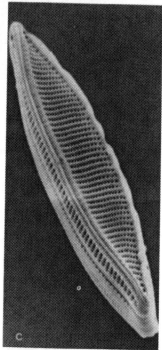

Valves straight or sigmoid, narrow; linear, lanceolate or elliptical, sometimes expanded centrally;[h] more-or-less symmetrical in outline with respect to the apical plane but often strongly asymmetrical in structure. Poles various, often rostrate or capitate. Striae usually uniseriate[b-f] not interrupted by lateral sterna, containing small round poroids occluded by hymenes and sometimes by cribra as well; areolae occasionally loculate in the larger sigmoid species. Conopea[g, k] or marginal ridges[c-f] sometimes present. Raphe system slightly to strongly eccentric, closer to the proximal margin (by definition), fibulate.[h-k] Raphe systems of the two valves on the same ('hantzschioid' symmetry) or opposite sides ('nitzschioid' symmetry) of the frustule. During cell division the raphe systems of the sibling valves are formed on opposite sides of the parent cell; hence hantzschioid cells do not breed true (contrast *Hantzschia*). Central raphe endings present in some species,[d] raphe continuous from pole to pole in others. Where present, central internal raphe endings simple, lying in a small or large double helictoglossa; external endings simple or deflected towards the distal margin. Terminal fissures usually present, turned or hooked towards the proximal or distal side.[e, f] Fibulae very diverse, sometimes extended across the valve. Girdle bands open, very variable in number, usually with one to several transverse rows of poroids apiece.[l]

A difficult and large genus, split into several sections by Cleve & Grunow (1880). The type species and its close allies are discussed by Mann (1986b). We separate *Fragilariopsis*, *Psammodictyon* and *Tryblionella* from *Nitzschia* and expect that others will follow.

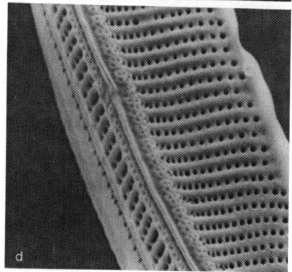

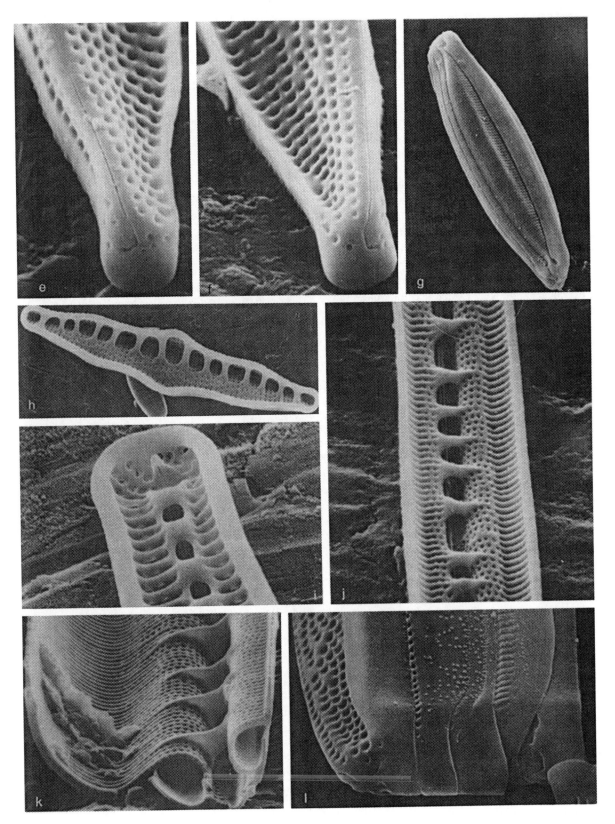

Denticula F. T. Kützing 1844, Kies. Bacill. Diat.: 43

T: *D. elegans* Kützing (lectotype selected by C. S. Boyer, 1928. Proc. Acad. Nat. Sci. Philad. **79** Suppl.: 530)

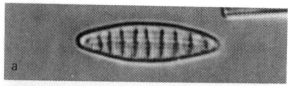

Cells small; solitary or forming short chains; usually lying in girdle view. Two simple plastids per cell, arranged symmetrically about the median transapical plane. Freshwater and marine, benthic.

The valves are linear or lanceolate with blunt or slightly rostrate poles.[a-d] Striae uniseriate or biseriate, containing round poroids occluded by centrally placed hymenes; a delicate cribrum is also present in some. The fibulate raphe system is nearly central to moderately eccentric.[c] The raphe may be borne on a wide, low keel, in which case the valve face also bears a groove[c-e] (which accommodates the keel of the sibling cell during division), or the valve face may be simply curved. Within a cell the raphe systems of the valves are diagonally opposite (nitzschioid symmetry). The raphe system is subtended internally by massive fibulae.[f-h] These also extend across the valve face, forming partitions running from margin to margin. The fibulae widen at their bases (near the raphe) so that only small apertures[f, h] ('portulae') are left connecting cell interior with subraphe canal. Raphe with or without[c] central raphe endings. Where present, central raphe endings simple internally and externally. Polar raphe endings hooked towards the secondary side of the valve, which may be the narrower[c] or broader[d, e] side. Girdle consisting of a few open bands or half-bands (Figs i, j show the two ends of a single frustule), the more advalvar of which sometimes bear one transverse row of poroids apiece. The valvocopula often extends beneath the fibulae to form septum-like structures.

Closely related to *Nitzschia* sect. *Grunowia*, from which it is difficult to separate, except on the extent of the fibulae. *Denticula* formerly included some marine species now placed in *Denticulopsis* (a fossil genus with only one extant species) and in addition some freshwater forms that belong in or near *Epithemia*. The only marine, true *Denticula* of which we are aware is the newly described *D. neritica* growing on the feathers of marine birds (Holmes & Croll, 1984).

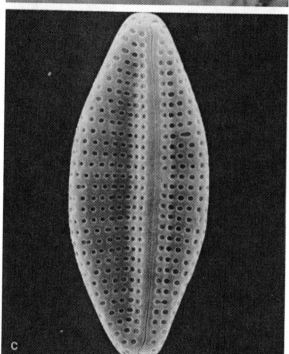

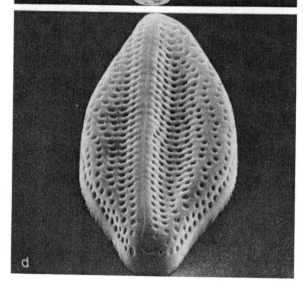

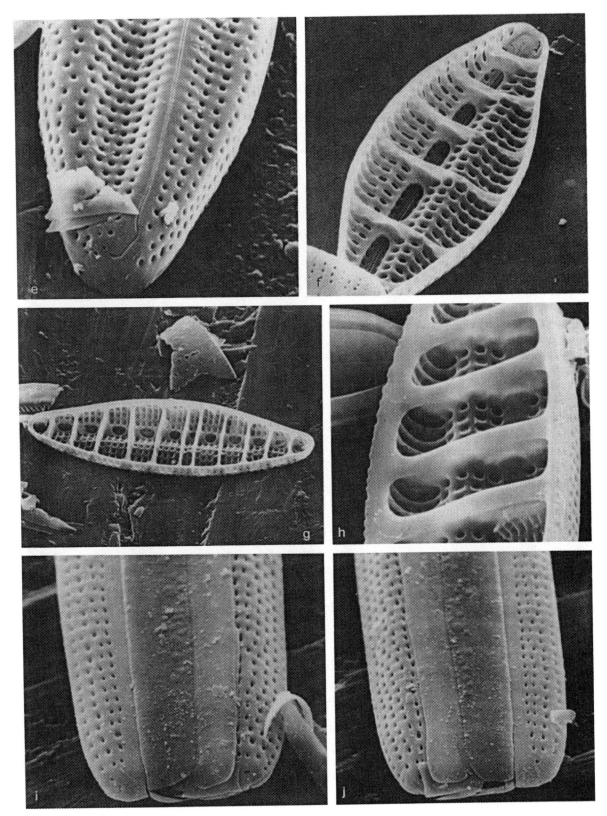

Fragilariopsis F. Hustedt 1913. A. Schmidt Atlas Diat., T. 299, figs 9–14

T: *F. antarctica* (Castracane) Hustedt
(= *Fragilaria castracane*)

Cells bacillar, usually united into flat, ribbon-like colonies.[a, b] Single cells lying in valve or girdle view. Two plate-like plastids, placed symmetrically, one on either side of the median transapical plane. Marine, planktonic or benthic; particularly abundant in Antarctic waters.

Valves relatively wide; linear, linear-lanceolate or elliptical;[b, c] isopolar or heteropolar. Valve poles usually bluntly rounded. Valve face flat, sharply differentiated from the shallow mantles.[d–f] Striae parallel throughout except at the poles, where they may become radiate; lying between robust transapical ribs;[g–i] usually bi- to multiseriate, containing small round poroids closed by hymenes. Parts of striae frequently irregularly porous or even plain; striae interrupted by a plain longitudinal strip at the junction of valve face and distal mantle.[d] Raphe system strongly eccentric, lying at the junction of valve face and proximal mantle; fibulate.[h, i, k] Raphe with or without central endings. All raphe endings simple and straight; no terminal fissures.[d, j] Fibulae small, inconspicuous in the LM, shaped like very short bars; often not positioned exactly in relation to the transapical costae separating the striae. Girdle containing several open bands, some of which bear one or more transverse rows of poroids.

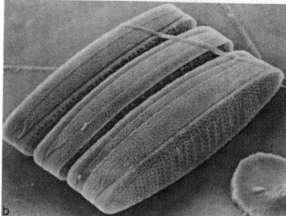

Recent authors, following Hasle (1972b), have treated *Fragilariopsis* as a section of the genus *Nitzschia* because some of the species are not colonial. On the basis of valve shape, raphe structure and stria structure, *Fragilariopsis* appears to be a natural group. This being so, it is a matter of taste whether this group is called a subgenus, a genus or a subfamily. We prefer generic status in view of the enormous diversity already present in *Nitzschia*. The genus has an interesting geographical distribution, apparently with its centre in Antarctica where the majority of species occur, often in great abundance.

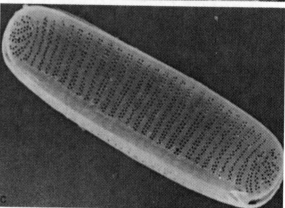

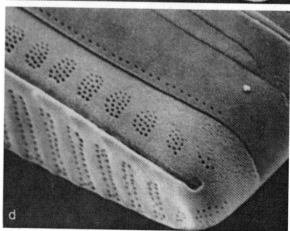

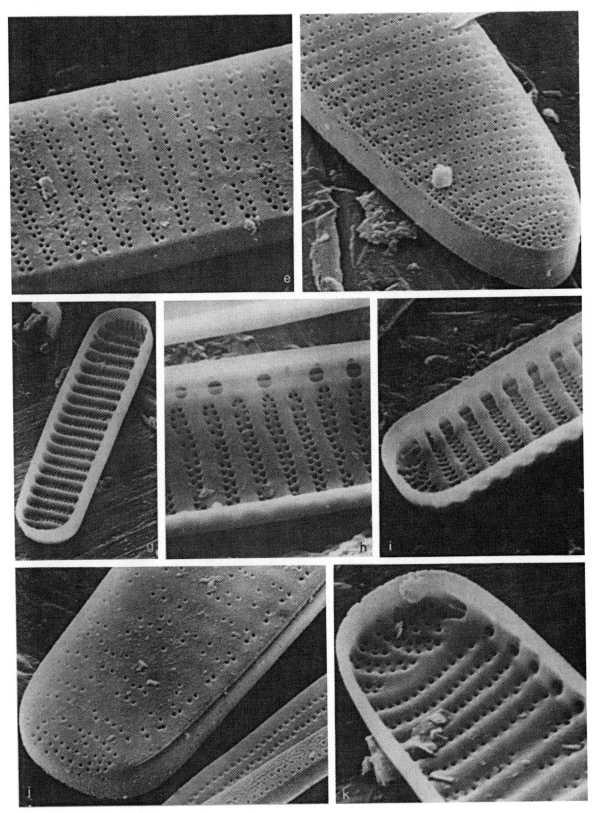

Cylindrotheca L. Rabenhorst 1859. Algen
Sachsens no. 801

T: *C. gerstenbergeri* L. Rabenhorst

Cells solitary, needle-like, straight or arcuate.[a-g]
Frustules usually strongly twisted about the apical
axis,[f] so that valve and girdle describe a spiral course
around the cell; consequently the cells rotate as they
move through the sediment. Plastids two to many,
plate-like or discoid. A small genus but widely
distributed in the epipelon of marine habitats; rarely
in freshwater. Most species are permanently epipelic
and only *C. closterium* is regularly stirred up and
spends time in the plankton.

Valves long and narrow and only lightly or
partially silicified; often reduced to strips of material
bordering the raphe (see Reimann & Lewin, 1964).
Striae irregular if present.[i] In species where the valve
is most highly developed (*C. closterium*) delicate
sinuous costae extend out from the raphe-sternum
and these are separated by areas of small pores.
Raphe subtended by numerous thin, rib-like
fibulae,[h, i] which are relatively heavily silicified.
Central raphe endings present[h] or absent, in which
case the raphe is continuous. When present the
endings are straight and very slightly expanded.
Polar endings simple. Girdle bands narrow and
numerous,[f] consisting of lightly silicified plain strips.

This genus is best observed in the TEM. *C.
closterium* is still placed by many authors in
Nitzschia, in spite of the detailed studies of
Cylindrotheca species made by Reimann & Lewin
(1964). *Nitzschia* and *Cylindrotheca* are closely related
but the reduced and lightly silicified valves, the
spiral twist of the frustule (present in *C. closterium*
but only near the poles), the fibula structure and the
girdle together serve to separate *Cylindrotheca* from
all other 'nitzschioid' genera. This is a small genus
(4 spp. in VanLandingham, 1969) and very
abundant in coastal waters world-wide. The valves
are so delicate that they are usually destroyed by
acid-cleaning.

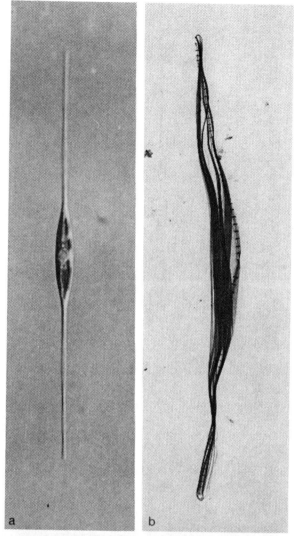

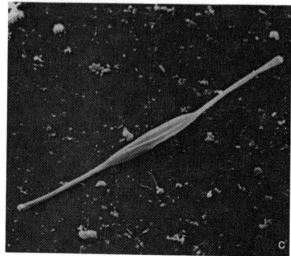

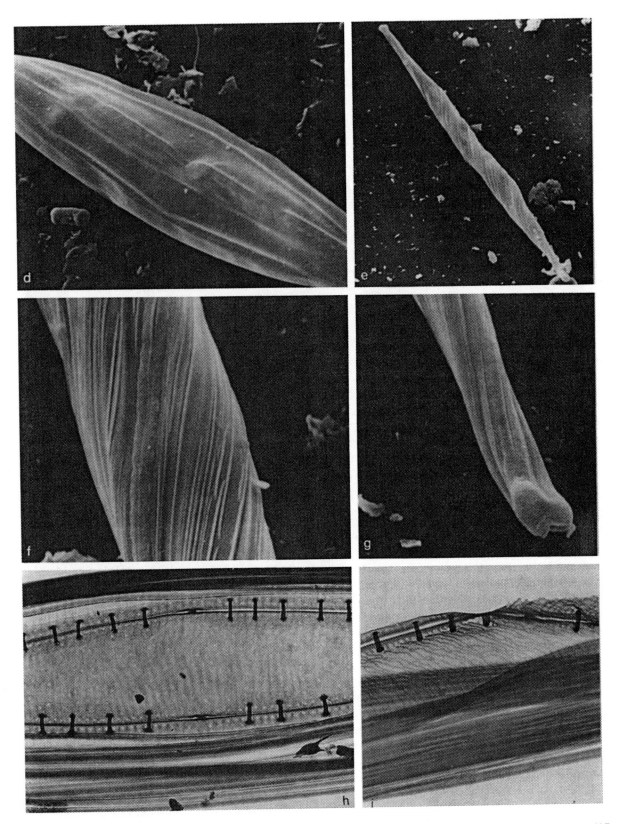

Epithemia F. T. Kützing 1844. Kies. Bacill. Diat.: 33

T: *E. turgida* (Ehrenberg) Kützing (= *Eunotia turgida*; lectotype selected by C. S. Boyer 1928. Proc. Acad. Nat. Sci. Philad. 79 Suppl.: 488)

Cells solitary, strongly dorsiventral, usually lying in girdle view and then linear to almost elliptical. One large plate-like plastid with highly lobed margins lies against the girdle on the ventral. All cells contain a few small endosymbiotic cyanophytes. An exclusively freshwater epiphytic and epipelic genus, widely distributed in base-rich habitats.

Valves strongly asymmetrical about the apical plane,[a] often arcuate, with blunt to broadly capitate poles. Valve face usually flat, sometimes bearing warts externally, sometimes with a marginal ridge at the junction with the dorsal mantle. Striae apparently uniseriate,[d] but areolae often very complicated in structure,[e-g] making their identity and boundaries difficult to determine; sometimes loculate.[i] Areolae occluded externally by flap-like occlusions[e] and opening only by narrow crescent-shaped slits.[g] Transapical costae robust,[h, i] some being additionally thickened internally and running from margin to margin, acting as fibulae beneath the raphe. Raphe system eccentric, biarcuate,[b] opening internally into a canal, which communicates with the cell interior by small round or oval holes (portulae) lying between the major transapical ribs (fibulae). External central[b-d] and polar raphe endings[g] simple or slightly expanded; no terminal fissures. Internally the raphe slits are more-or-less continuous with each other at the centre[i] and end in small helictoglossae[j] at the apices. Girdle sometimes wider on the dorsal side, complex, consisting of open and closed plain bands. The more advalvar bands combine to form a system of inwardly pointing projections[k] (often erroneously called a 'septum'), which in the intact frustule clasp tightly around the major transapical ribs.

The type species has been studied in detail by Sims (1983). *Epithemia* is closely related to *Rhopalodia* but not to *Denticula senus stricto*, even though they have often been classified together (e.g. by Krammer & Lange-Bertalot, 1988). Some species traditionally classified in *Denticula*, however, (e.g. *D. vanheurckii*) are undoubtedly close to *Epithemia*, on the basis of valve and protoplast structure; they are only superficially like true *Denticula*.

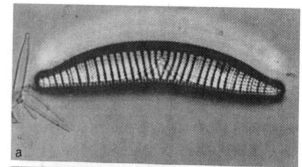

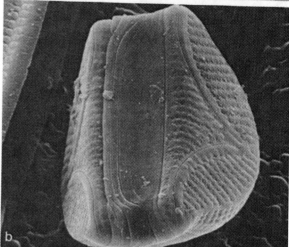

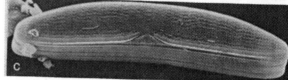

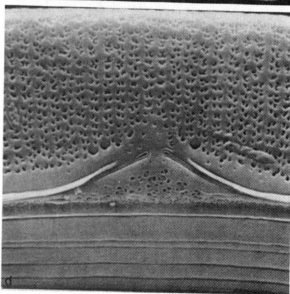

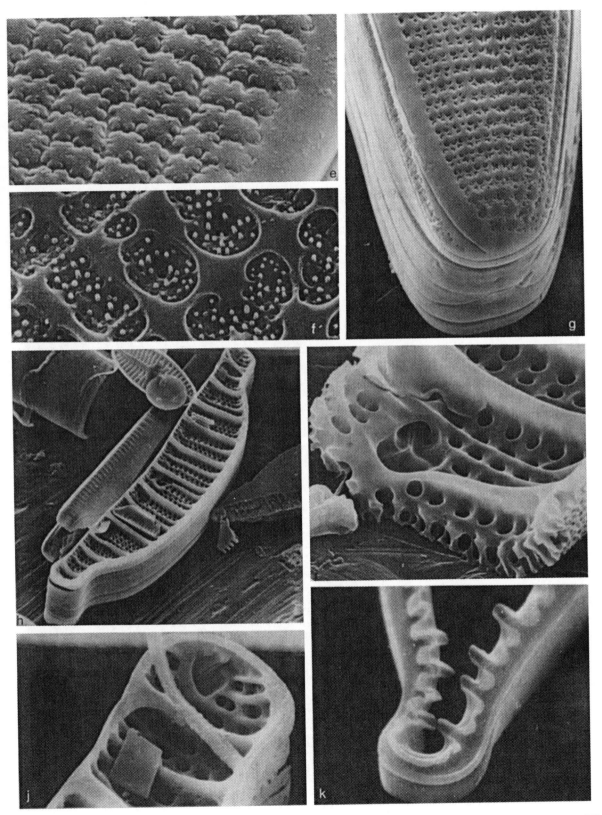

Rhopalodia O. Müller 1895. Bot. Jahrb. **22**: 57 (nom. cons.)

T: *Rh. gibba* (Ehrenberg) O. Müller (= *Navicula gibba*; lectotype selected by C. S. Boyer 1928, Proc. Acad. Nat. Sci. Philad. 79 Suppl.: 491)

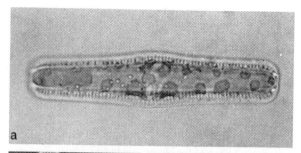

Cells solitary, dorsiventral, attached or free, and usually lying in girdle view. Frustules linear, lanceolate or more-or-less elliptical in girdle view, heteropolar or isopolar. One large plate-like plastid with lobed margins lying on the ventral side of the cell.[a] At least in some species the cells contain a few small endosymbiotic cyanophytes. Epipelic and epiphytic, freshwater to marine.

Valves linear or arcuate,[b-d] often shaped like an orange-segment,[c] strongly asymmetrical about the apical plane. It is very difficult to distinguish a discrete mantle on either side, unless the whole of the valve dorsal to the raphe is taken as a mantle. Valve sometimes bearing warts externally. Striae uni- to multiseriate, containing poroids occluded by one or a few volae.[c, g] Transapical costae often robust, some being additionally thickened and running from margin to margin, acting as fibulae beneath the raphe.[h] Raphe system eccentric, closer to the dorsal margin,[c, e] often raised on a keel. Raphe fissures often bordered externally by siliceous flanges or ridges. External central raphe endings expanded, sometimes deflected slightly towards the ventral side;[e] internal endings simple.[h] External polar endings simple.[f] Raphe opening internally into a canal, which communicates with the cell interior by small round or oval holes[i] (portulae) lying between the major transapical ribs (fibulae). Girdle wider dorsally than ventrally, complex, consisting of open and closed bands which are usually porous but plain in some species.[j] No elaborate interlock formed with the valve, in contrast to *Epithemia*.

An interesting genus, much more diverse in structure than its close relative *Epithemia*, and found in a wider range of habitats. Frustule shape, girdle structure and the course of the raphe system serve to separate *Rhopalodia* from *Epithemia*. Krammer (1988a, b) has made a very detailed study of the Gibberula group of *Rhopalodia*.

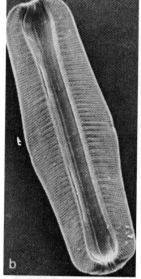

Entomoneis C. G. Ehrenberg 1845. Ber. Bekanntm. Verh. Königl. Preuss. Akad. Wiss. Berlin, **1845**: 71

T: *E. alata* Ehrenberg

Cells solitary, twisted about the apical axis, usually lying in girdle view[a, b] and then appearing bilobate. The torsion of the cell means that valves or whole frustules can present a great variety of aspects depending on exactly how they lie relative to the observer. One axial, plate-like plastid;[a] or two plastids, one on each side of the median transapical plane. A fairly large epipelic genus, of brackish marine sediments; occasionally in freshwater.

Valves lanceolate or linear with acute poles, often strongly compressed laterally[e–h] and bearing a high, narrow keel,[b, h] which becomes lower (and may disappear) in the centre of the valve, and also decreases in height towards the poles. Discrete mantles usually absent. Outside of valve sometimes bearing warts; valve margins and raphe-sternum sometimes with longitudinal ribbing. Striae usually biseriate[i] or multiseriate,[g] containing small round poroids occluded by hymenes; or the whole stria consisting of little more than a delicate, porous siliceous membrane (in which case the perforations are usually still in two lines, next to the transapical costae). Raphe system fibulate,[i, j] or the sides of the keel fused beneath the raphe so that the subraphe canal is connected to the rest of the cell lumen only near the central and polar raphe endings. In fibulate forms, fibulae may occur at many levels beneath the raphe, especially where the keel is high; the fibulae are short, bar-like struts and are borne on the transapical costae. Central raphe endings (internal and external)[g] and external polar endings all similar: straight, and not or only slightly expanded. The external raphe fissure ends very close to each pole. Girdle bands numerous,[c] open and porous.[d]

A complex genus, with a variety of pore and raphe structures. It is possible that *Entomoneis* could be split into several smaller genera. Nevertheless, the shape of the valve and frustule, the tendency towards fusion of the two sides of the keel, the typically biseriate structure of the striae and the relatively simple raphe endings, suggest that *Entomoneis* represents a natural group. Some aspects of the keel structure are discussed by Paddock & Sims (1981): the more familiar name *Amphiprora* was abandoned in favour of *Entomoneis* by Reimer (in Patrick & Reimer, 1975).

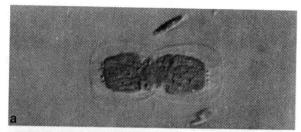

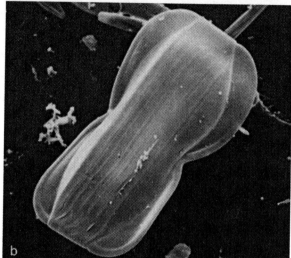

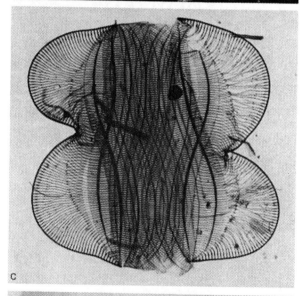

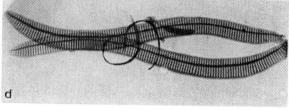

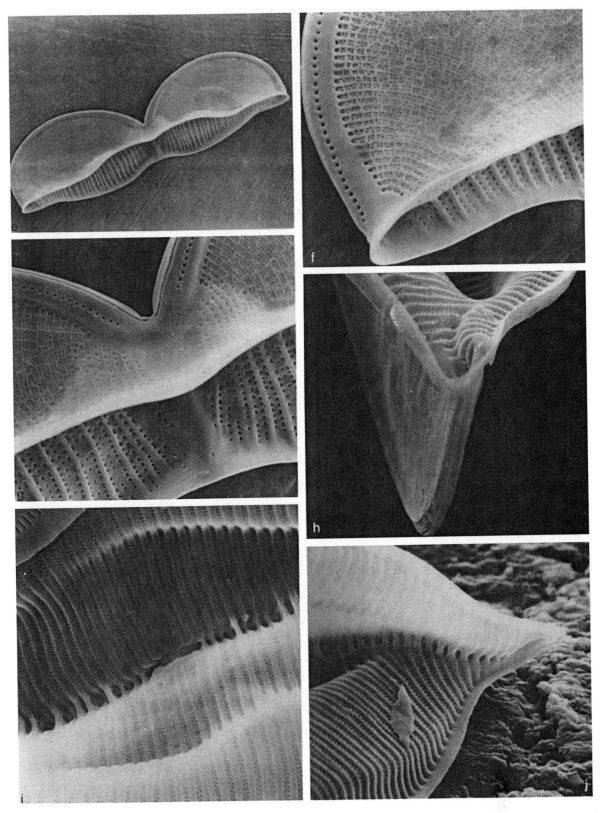

Auricula F. Castracane 1873. Atti Acad. Pontif. Sci. Nuovi Lincei **26**: 407

T: *A. amphitritis* Castracane

Cells solitary, almost always lying in girdle view.[a] One large central plastid, with a central elongate pyrenoid. A small marine genus, living in the epipelon of coastal sediments.

Valves crescent-shaped or bilobate, with bluntly rounded poles[b-c] and a strongly eccentric raphe system. Valve face flat or slightly undulate. Raphe in an excentric position leaving a wide valve face on one side and a narrower one on the other.[c] Striae biseriate, each stria consisting of a delicate strip of silica which contains small round poroids close to the thin costae separating the striae.[i,j] Raphe system biarcuate[d], more-or-less keeled,[e-g,j,k] fibulate.[i,j] Raphe accompanied internally by two ridges, one on either side of the raphe,[i] from which the transapical costae extend out, converging and fusing on the wide valve face. The half of the raphe-sternum on the valve face side is wider than that on the (proximal) mantle. Raphe system often extending around more than half of the valve circumference, sometimes so much so that the 'polar' raphe endings lie fairly close to each other on the distal margin. Internal and external central raphe endings straight, simple, indented into the cell. Terminal fissures absent, the polar endings being very close to the valve margin. Fibulae short, rib-like, arising from the two ridges flanking the raphe internally. Copulae numerous, simple.

Auricula has been considered in detail by Paddock & Sims (1980), who provide the only modern account of the genus and supplied us with material. *Auricula* shows similarities to *Entomoneis* and *Rhopalodia*, and has been confused with these and with *Amphora* in the past.

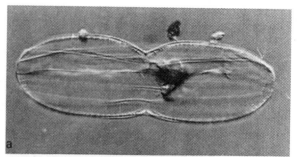

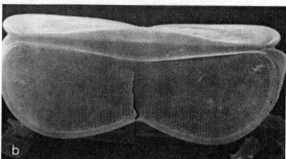

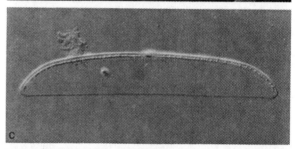

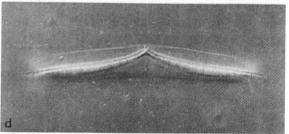

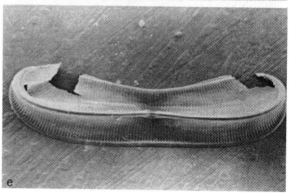

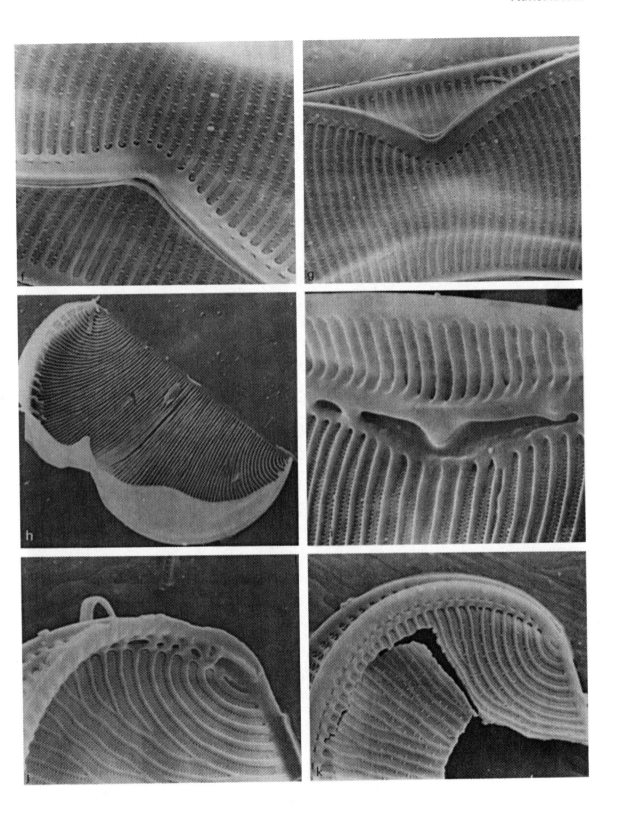

Hydrosilicon J. Brun 1891. Mém. Soc. Phys. Genève **32** (2): 31

T: *H. mitra* Brun

Cells solitary; rounded and centrally constricted (panduriform) in valve view, similar but truncated in girdle view. Plastid structure unknown. Epipelic in marine coastal sediments. A rarely encountered genus containing two species according to VanLandingham (1971).

Valve slightly constricted centrally on both sides, otherwise linear-elliptical.[a-c] Valve face slightly undulate, separated from the mantles by the keeled raphe system which runs around the whole circumference of the valve face. Striae biseriate,[e-j] containing small round poroids,. Striae radiating from the raphe system, both towards the valve margin, and also inwardly, towards the middle of the valve; here they meet and fuse to form a series of ribs, which mimic the pattern centres (sterna) of araphid diatoms.[e,f] Raphe system fibulate;[g] raphe consisting of two slits, each of which runs from one side of the central constriction of the valve,[d] around the valve pole, to the other side of the constriction. The major structural axis of the cell (between the two sets of raphe endings) is thus at right angles to the major morphological axis of the cell (between the two poles): cf. *Surirella*. Raphe flanked on each side internally by a ridge of silica, which bears the fibulae and delimits the subraphe canal. External raphe endings simple or slightly expanded;[d] internal fissures apparently continuous across the nodule formed by the central constriction of the valve and keel.[h-j] The fibulae and internal longitudinal ridges combine to produce circular or oval holes (portulae) which give access into the subraphe canal from the cell interior; fibulae variable in size but large in relation to the portulae. Girdle unknown.

Hydrosilicon has a raphe system similar to that in *Surirella*, *Cymatopleura* and their allies. The interrelationships of *Hydrosilicon* and similar genera are discussed by Paddock & Sims (1977). We thank T. B. B. Paddock for material.

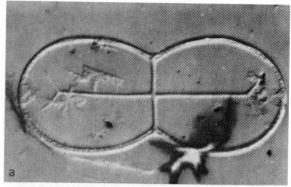

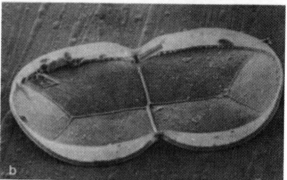

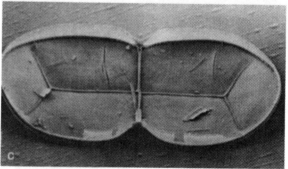

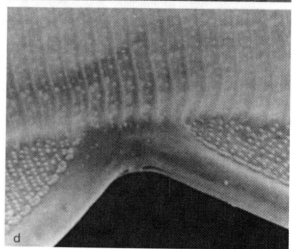

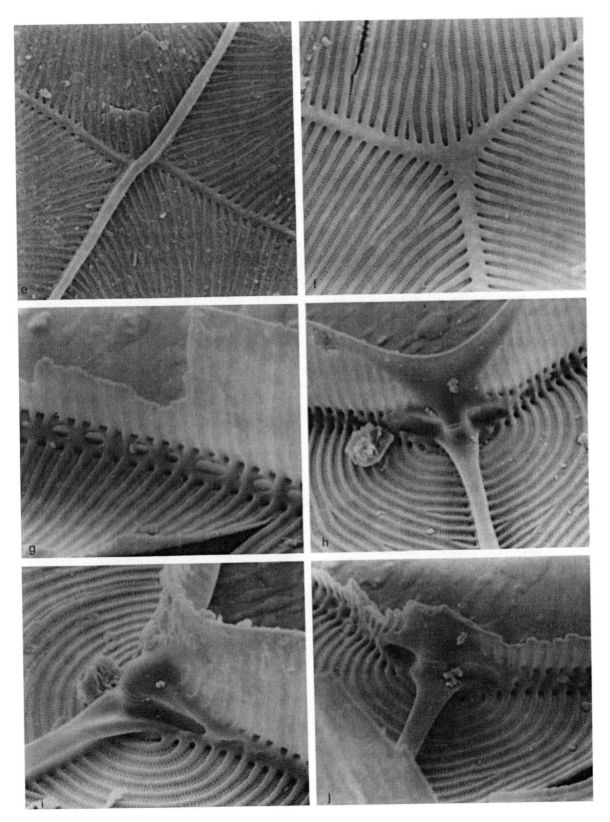

Petrodictyon D. G. Mann, *gen. nov.*

T: *P. gemma* (Ehrenberg) D. G. Mann
(= *Surirella gemma*)

Cells solitary, lying in valve or girdle view; valve
and girdle heteropolar. One plastid, folded at the
base pole, expanding above this into 2 valve-
appressed, lobed plates. Brackish or marine,
epipelic.

Valve oval-cuneate,[a-c, g] with a raphe system
around its whole circumference. Valve face smooth,
slightly undulate, more-or-less grooved along its
midline and slightly indented above each thickened
internal transapical costa. Raphe system raised on a
shallow keel.[b-f] Striae uniseriate, [d, e] containing
relatively large, somewhat quadrate poroids, which
are closed by complex cribrum-borne volae.
Transapical costae fusing in the midline of the valve
to form a narrow longitudinal sternum.[g] Some
transapical costae are strongly thickened and extend
from the margin to the midline of the valve[g, h]
(which is developmentally a valve margin!); these
thus act as fibulae and where they cross beneath the
raphe they are especially deep, like angle irons.[h-j] In
addition, there are other fibulae, also borne on
transapical costae, which are not so extensive[h] and
between which circular to oval openings (portulae)
give access to the subraphe canal. 'Central' raphe
endings (at the broader pole) elaborate, forming
nostril-like structures externally,[d] and deep-lipped
structures internally.[j] External 'polar' raphe endings
(at the narrow pole) simple or slightly expanded;[e]
internal endings slightly lipped[i] but not as massively
developed as at the 'central' endings. Girdle
containing several open bands.

The simple valve structure (with its regular
alternation of thin transapical ribs and uniseriate
striae), costate thickenings and the characteristic
'central' raphe endings are sufficient to separate this
genus from other surirelloid taxa. *Petrodictyon
gemma* and some other taxa referrable to this genus,
e.g. *P. foliatum, P. gemmoides* and *P. contiguum* have
been described by Paddock (1978); this account
gives further illustrations, detail and discussion of
the relationship with *Plagiodiscus*.

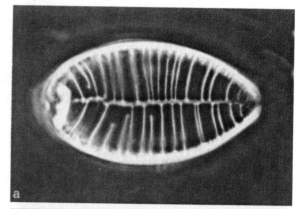

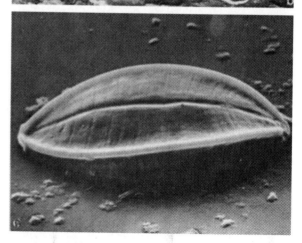

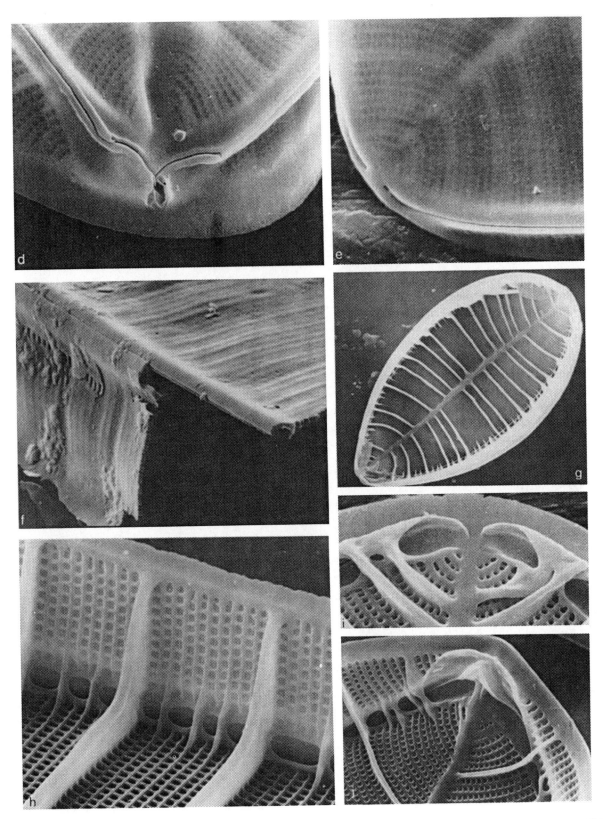

Plagiodiscus A. Grunow & T. Eulenstein 1867.
Hedwigia 6: 8

T: *P. nervatus* Grunow

Cells solitary, disc-like, reniform in valve view.
Plastids unknown. A marine, coastal genus
containing a few species, which have sometimes
been classified in *Surirella*. Recorded so far only
from tropical or subtropical waters.

Valves broad, with a central indentation on one
side.[a, b, e] Valve face flat or slightly undulate, slightly
ribbed externally near the raphe system, often with
granules externally. Valve face separated from
mantles by the slightly keeled raphe system,[b-d, f]
which runs around the whole perimeter of the valve
as in *Surirella*. Mantle shallow on the ventral
(indented) side,[c] but rising along the dorsal side
from either end towards the centre.[b, d] Striae bi- to
multiseriate, containing small round poroids.[e, g, j]
Valve interior with conspicuous ribs, which cross
beneath the raphe system and there thicken to form
angle irons; they also function as fibulae. The ribs
thin towards the centre or fuse with a curved rib
along the midline of the valve. Hollow invaginations
of the valve mantle sometimes present, forming flat
porous[i] sacs (palmulae)[e, g] within the cell; these are
connected to the mantle by short, hollow stalks.[h]
Raphe endings located in the central indentation[c]
and on the convex margin diametrically opposite
this,[d] not at the poles. External raphe endings
slightly expanded; on the ventral side, the internal
raphe endings are produced into a beak-like
projection[k] similar to the mucro in *Thalassiophysa*.
Raphe opening internally into a cylindrical
canal, which connects with the cell lumen via
circular to oval apertures (portulae). Girdle
unknown.

Plagiodiscus is a distinctive genus, which has been
treated in detail by Paddock (1978). It has clear
affinities with other surirelloid genera, especially
Petrodictyon, but is distinguished by its symmetry,
possession of palmulae, and with the raphe endings
at opposite ends of the short axis of the cell, not the
long axis as in *Surirella*, *Cymatopleura* or
Stenopterobia.

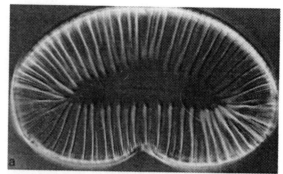

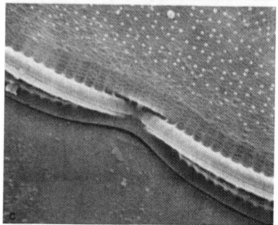

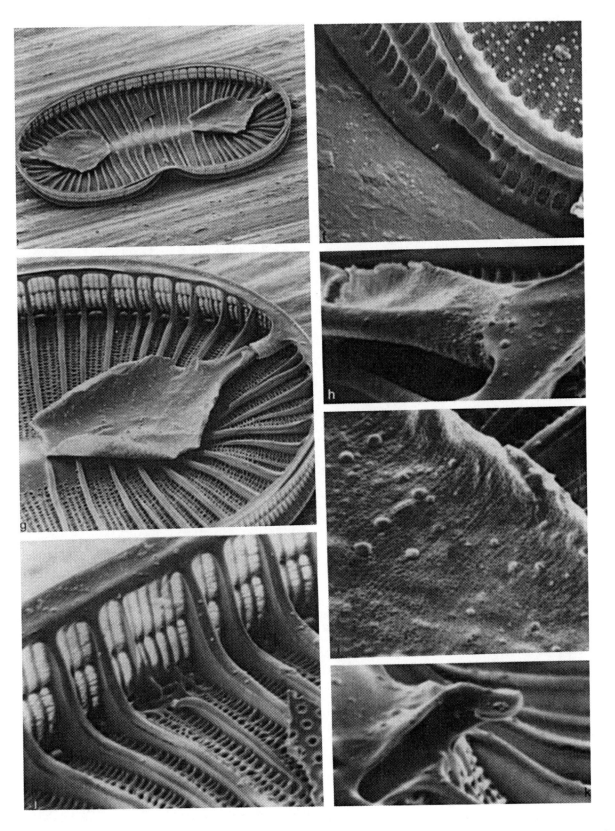

Stenopterobia A. de Brébisson ex H. Van Heurck 1896. Treat. Diat.: 374

T: *S. sigmatella* (Gregory) Ross (lectotype selected by D. G. Mann)

Cells solitary, sigmoid[a, b] or straight, lying in valve or girdle view. One plastid per cell, divided into two plates lying one against each valve, connected by a very narrow isthmus near one pole. Freshwater, epipelic; apparently restricted to acid oligotrophic lakes and ombrogenous mires.

Valves sigmoid or straight, narrow, linear. Valve face slightly undulate,[c] with a fibulate raphe system around the whole circumference. The fibulae[d] are formed in part by the infolding and fusion of parts of the valve face and mantle;[i] this has the effect of raising the raphe system and the canal beneath it above the general level of the valve.[c-i] The fusion of valve face and mantle is sometimes so extreme that holes are produced beneath the raphe system, so that the raphe system is connected to the rest of the cell only by a series of narrow tubes.[h] Striae multiseriate,[d] containing tiny round poroids. No hymenes. Transapical costae often bearing mushroom-like projections[c] or warts externally; meeting in the midline of the valve and fusing to form a narrow longitudinal rib or sternum[d] which is sometimes perforated by a few scattered poroids. Raphe endings simple or slightly curved externally; simple internally. Girdle unknown.

This small genus is often understood to include only the remarkable sigmoid species *S. sigmatella* (= *S. 'intermedia'*). Brébisson's original concept, which we adopt here, was broader and included narrow, straight surirelloid forms, e.g. *S. delicatissima* (we illustrate this species as well as *S. sigmatella*). The relatively simple structure of valve and raphe system, the elongate shape and the ecological restriction to acid waters make this a natural group, separable from *Surirella*. *Stenopterobia* and *Surirella*, however, are obviously very closely related.

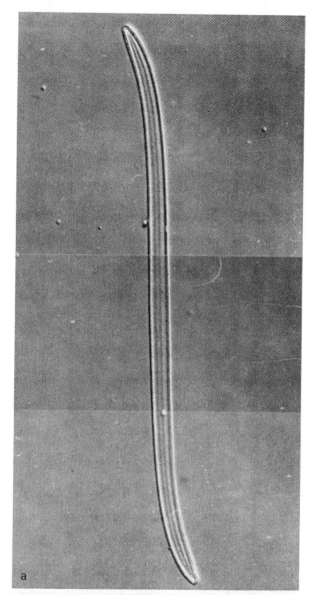

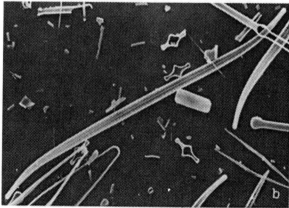

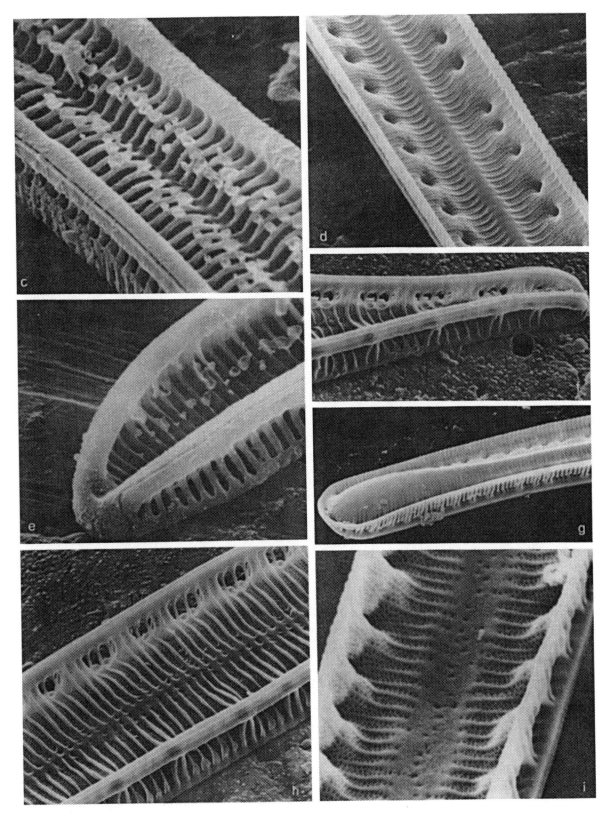

Surirella P. J. F. Turpin 1828. Mém. Mus. Hist. Nat. **16**: 363

T: *S. striatula* Turpin

Cells solitary, lying in valve or girdle view. Frustule isopolar or heteropopolar and wedge-shaped, occasionally twisted about the apical axis. One plastid, consisting of 2 large valve-appressed plates connected by a very narrow isthmus near one pole of the cell (the narrower pole in heteropolar species); or 2 plastids. Plastid margins often highly lobed. A large freshwater to marine, epipelic genus.

Valves usually strongly silicified; linear to elliptical or obovate,[a-c] sometimes panduriform,[d] with a raphe system running around the whole perimeter of the valve face. Valve face planar, or often concave, and then frequently with undulations parallel to the apical axis (contrast *Cymatopleura*); sometimes ornamented with siliceous warts or ridges, and occasionally with spines along the midline of the valve (two very large subapical spines in *S. capronii*). Valve structure not obviously costate (contrast *Petrodictyon*, *Hydrosilicon*). Striae frequently multiseriate,[g, i] containing very small round poroids occluded by volae; often interrupted by sterna, especially near or along the midline of the valve. Raphe system raised on a shallow or deep keel, whose walls are often crimped together and sometimes fused.[f, g] In extreme cases holes may be formed beneath the raphe system so that the subraphe canal communicates with the rest of the cell interior only by a series of narrow tubes. In addition to any such structures the raphe may be subtended internally by rib- or plate-like fibulae.[g, h] Raphe simple with straight unexpanded endings internally and externally;[e, f] internal fissures sometimes continuous.[h] Raphe endings located at the poles.[e, f] Cingulum consisting of several copulae; at least in one species, the splits are staggered and not aligned beneath the raphe endings.[j]

Cell division studies show that one end of the cell (the broader end in heteropolar species) is homologous with the centre of naviculoid valves, while the other end is homologous with the two poles of naviculoid valves. The nucleus migrates to the 'central' end for mitosis, and valve silicification begins from here in the daughter cells (see also Schmid, 1979). Krammer & Lange-Bertalot (1987) discuss the complexities of the *S. ovalis* group.

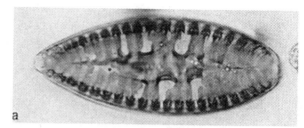

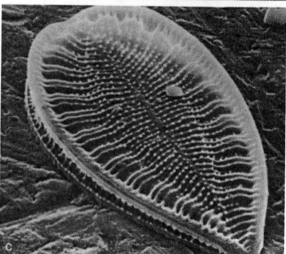

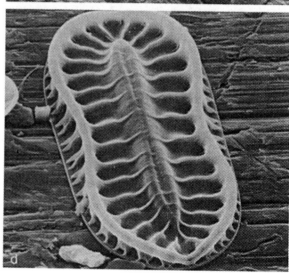

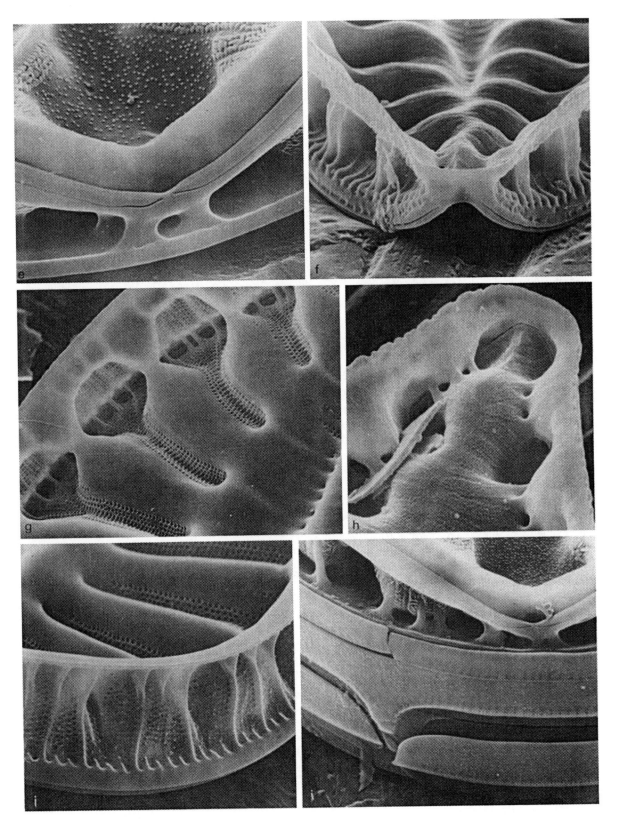

Campylodiscus C. G. Ehrenberg ex F. T.
Kützing 1844. Kies. Bacill. Diat.: 59

T: *C. clypeus* (Ehrenberg) Ehrenberg ex Kützing
(= *Cocconeis? clypeus*: lectotype selected by C. S.
Boyer, 1928, Proc. Acad. Nat. Sci. Philad. **79**
Suppl.: 548).

Cells solitary, usually lying in valve view and then
appearing (sub) circular.[a] Cells saddle-shaped;[b, c]
valves similar but set at right angles to each other.
One plastid, divided into two valve-appressed plates
with lobed margins, linked by a narrow isthmus.
Epipelic in freshwater, brackish and marine
habitats. A large genus with only one species
recorded frequently from freshwater.

Valves (sub)circular, convex along the 'apical
plane' (the plane intersecting both sets of raphe
endings), concave along the 'transapical plane'.[c]
Apical planes of the two valves set at right angles.
Valve face often with warts and ridges externally,
and much folded.[c, e-h] Striae bi- to multiseriate,[h- j]
interrupted by sterna and usually containing small
round poroids; poroids sometimes larger, cribrate.[d]
Raphe system submarginal, raised on a keel,[e-h] and
occupying the whole circumference of the valve face;
subtended internally by small rib-like fibulae[i, j] and/
or larger structures formed by the fusion of the two
sides of the keel. Fusion in some cases leads to the
formation of holes in the keel beneath the raphe-
canal so that the raphe-canal is connected to the
rest of the cell only by a series of narrow, distantly
spaced tubes and bars.[g, h] All raphe endings usually
simple or slightly expanded, internally and
externally.[f] Girdle composed of open bands.

A very distinctive genus, yet perhaps not natural.
Parallels between groups within *Surirella* and
Campylodiscus (e.g. see Paddock, 1985) suggest that
the characteristic crossed axes and circular shape of
Campylodiscus may have evolved more than once,
just as spiralling of the frustule must have evolved
independently in *Surirella* and *Cymatopleura*.

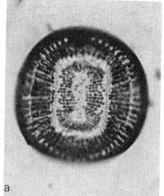

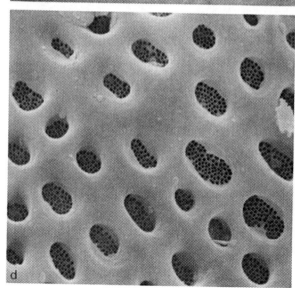

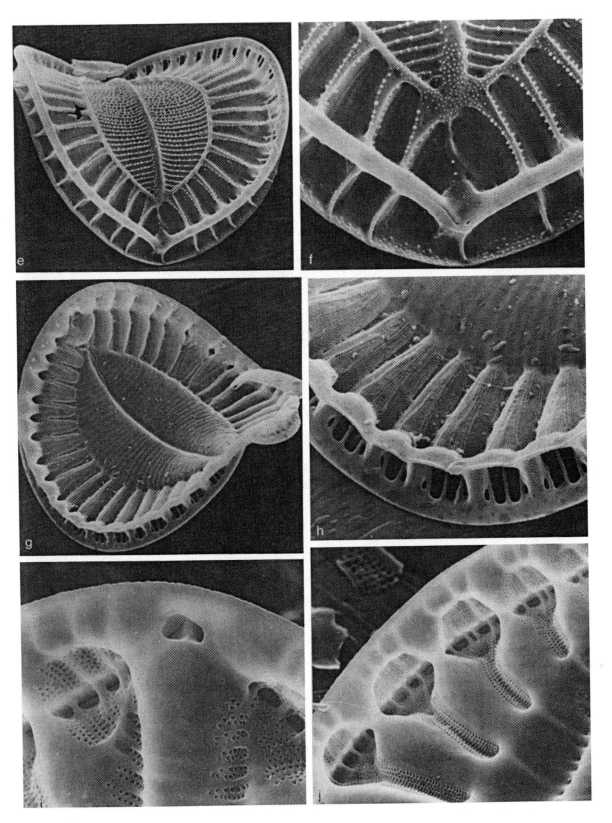

Cymatopleura W. Smith 1851. Ann. Mag. Nat. Hist. Ser. 2, 7: 12 (nom. cons.)

T: *C. solea* (Brébisson) W. Smith (= *Surirella solea*)

Cells solitary, lying in valve or girdle view.[a, b] Frustule isopolar, occasionally twisted about the apical axis. One plastid,[a, b] consisting of two large, valve-appressed plates connected by a very narrow isthmus, which lies near one end of the cell. The plastid margins are often highly lobed and extend onto the girdle.[b] Freshwater, epipelic, with a tendency to be more abundant in high-conductivity (alkaline) waters.

Valves panduriform[c,d] or linear to elliptical,[h] with a raphe system running around the whole of the valve circumference. During early stages of deposition the unsilicified or partially silicified valve face is planar but later it becomes thrown into a series of waves,[d, h] often shallower at the centre. The waves of one daughter valve are exactly complemented by waves of its sibling. Valves strongly silicified, often bearing costate and reticulate thickenings externally,[f-i] but smooth internally. Raphe system raised on a shallow keel, whose walls are crimped below the raphe-sternum.[i] Striae uniseriate, meeting to form a fault line along the middle of the valve,[k] containing small round poroids, which are often rimmed internally[j] (the remains of a delicate velum?). Raphe simple, with straight or slightly curved endings at both poles externally[f, g] and a continuous groove beneath these internally.[j] Girdle with open, slightly verrucose bands. The open ends of the bands are laterally positioned,[e] not at the poles.

Although the *Cymatopleura* frustule is apparently isopolar, the cell within is heteropolar. Before mitosis the nucleus always moves to the same end of the cell (always the end away from the plastid bridge), where it divides. It moves to this end too at meiosis, while the cell copulates via the opposite end. Thus one pair of raphe endings is apparently homologous to the central endings of naviculoid diatoms, and the other pair to the polar endings. A particularly distinct and well-defined, yet small genus having affinities with *Surirella* and *Campylodiscus*. There is a problem concerning the correct name of the type: Mann (1987, 1989c) considers that *C. librile* is not necessarily synonymous with *C. solea* as has been assumed by some workers. We have not therefore followed the Index Nominum Genericorum in this instance.

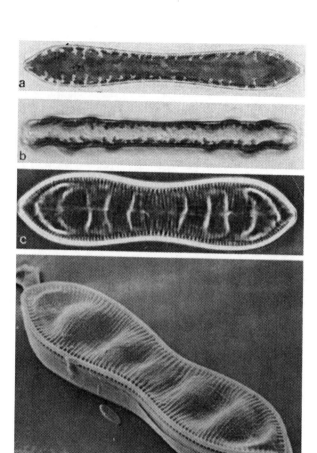

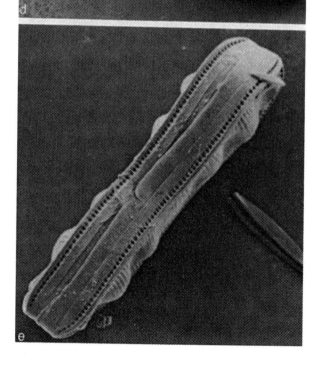

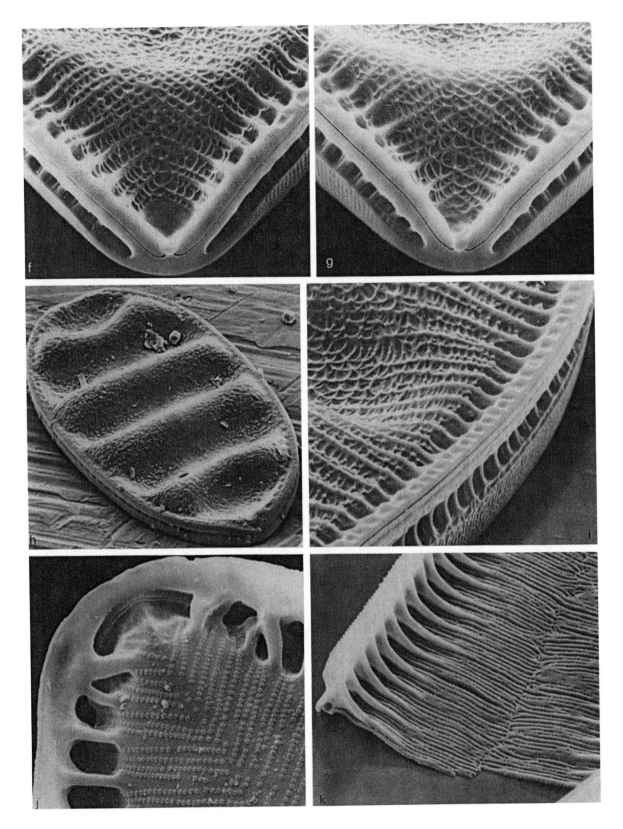

Appendix I: New taxa

CLASS DESCRIPTIONS
Coscinodiscophyceae F.E. Round & R.M. Crawford, *class. nov.*

Striae primariae radiantes circum annulum ordinatae.

Fragilariophyceae F.E. Round *class. nov.* **et Fragilariophycidae F.E. Round** *subclass. nov.*

Ordinatio striarum primarium pennata; sterna 1–2 praesentia.

Raphe carens.

Bacillariophyceae Haeckel 1878 emend. D.G. Mann

Ordinatio striarum primarium pennata. Raphe et sternum praesens.

SUBCLASS DESCRIPTIONS
Bacillariophycidae D.G. Mann, *subclass. nov.*

Raphe praesens. Ab Eunotiophycidis absentia absoluta rimoportularum distinguendae.

Biddulphiophycidae F.E. Round & R.M. Crawford, *subclass. nov.*

Anguli valvarum prominentiis in ocellos pseudocellosve terminantibus instructi.

Chaetocerotophycidae F.E. Round & R.M. Crawford, *subclass. nov.*

Valvae setis cavis ex polis extensis instructae.

Corethrophycidae F.E. Round & R.M. Crawford, *subclass. nov.*

Peripheria unius valvae frustuli processibus unguiformibus munita.

Coscinodiscophycidae F.E. Round & R.M. Crawford, *subclass. nov.*

Cellulae disciformes vel cylindricae, solitariae vel catenas facientes. Chromatophora disciformia. Valvae circulares, subcirculares vel semicirculares. Areolae loculatae, unaquaque extra cribris occlusa, intra per foramen unicum aperta. Rimoportulae plerumque praesentes. Copulae multae apertae. Plantae praecipue marinae, aliquot generibus aquam dulcem habitantibus.

Cymatosirophycidae F.E. Round & R.M. Crawford, *subclass. nov.*

Cellulae plerumque heterovalvares, rimoportulis solum in valva una, ocellulis et pilis introrsum versus diagonaliter dispositis instructae.

Eunotiophycidae D.G. Mann, *subclass. nov.*

Et raphe simplex et rimoportulae plerumque praesentes.

Fragilariophycidae F.E. Round, *subclass. nov.*

Vide descriptionem Fragilariophycearum.

Lithodesmiophycidae F.E. Round & R.M. Crawford, *subclass. nov.*

Valvae processu unico centrali tubuloso longo, ad juncturam frontis cum limbo ala vel spinis instructae.

Rhizosoleniophycidae F.E. Round & R.M. Crawford, *subclass. nov.*

Valvae calyptriformes, extensionibus lateralibus ad valvas germanas conjunctis.

Thalassiosirophycidae F.E. Round & R.M. Crawford, *subclass. nov.*

Valvae et rimoportulis et fultoportulis instructae.

ORDINAL DESCRIPTIONS
Achnanthales Silva 1962

Cellulae solitariae vel coloniales, interdum catenas breves facientes. Chromatophora varia: 1 laminiforme; 2 H-formia ex laminis duabus ad cincturam appressis isthmo lato pyrenoidem continenti connexis constans; multa disciformia. Nucleus vegetativus centralis vel excentricus. Frustula (semper?) cis-conformata heterovalvaria. Areolae simplices vel loculatae, a volis vel hymenibus occlusae. Raphovalva saepe concava systemate raphis centrali (rarissime excentrico) praedita. Areovalva saepe convexa sterno centrali vel excentrico praedita. Extrema centralia interna raphis unciformia vel a se ex adverso flexae. Fissurae terminales nullae vel latus idem versus flexae. Cingulum ex taeniis apertis vel clausis, porosis vel non porosis, constans.

Typus: *Achnanthes* Bory 1822

Anaulales F.E. Round & R.M. Crawford, *ord. nov.*

Cellulae elongatae arcuatae solitariae. Valvae bipolares rectae vel arcuatae, costis transversis internis. Rimoportula unica centralis. Plantae solum maris habitantes.

Typus: *Anaulus* Ehrenberg 1844

**Arachnoidiscales F.E. Round, *ord. nov.* &
Arachnoidiscaceae F.E. Round, *fam. nov.***

Cellulae disciformes. Frons plana costis
validissimis radiantibus intra praedita; limbus
verticalis a fronte distinctus. Striae in sectoribus
radiantibus ordinatae, ad marginem frontis
interruptae sed tum in limbum continuantes.
Areolae grandes volis occlusae. Annulus
centralis rimoportulis elongatis circumcinctus.
Copulae inornatae. Plantae marinae
epiphyticae.

Typus: *Arachnoidiscus* Deane ex Pritchard 1852

**Ardissoneales F.E. Round, *ord. nov.* &
Ardissoneaceae F.E. Round, *fam. nov.***

Cellulae lineares interdum ad centrum polosque
aliquantum dilatatae, polis rotundatis. Frons
plana vel undulata. Striae alveolatae, extra per
series areolarum simplicium intra per foramina
pauca apertae. Apices valvae seriebus pororum
minorium instructi, sed sine areis distinctis
porellorum. Rimoportulae nullae. Copulae
clausae unaquaque serie unica areolarum
praedita. Plantae marinae epiphyticae.

Typus: *Ardissonea* De Notaris 1870

**Asterolamprales F.E. Round & R.M. Crawford,
*ord. nov.***

Cellulae aspectu valvae subcirculares vel ovales,
solitariae. Chromatophora multa disciformia.
Valvae radiis tubularibus extra membranis
siliceis subtilibus obtectis. Areolae loculatae,
unaquaque extra cribro occlusa, intra per
foramen unicam aperta. Rimoportulae ad
extremitates exteriores radiorum et interdum
etiam prope centrum positae. Copulae paucae
inornatae. Plantae marinae planctonicae.

Typus: *Asterolampra* Ehrenberg 1844

**Aulacoseirales R.M. Crawford, *ord. nov.* &
Aulacoseiraceae R.M. Crawford, *fam. nov.***

Cellulae cylindricae a spinis circum peripheriam
frontis positis conjunctae, catenas ita facientes.
Areolae in fronte interdum nullae, in limbo in
seriebus spiralibus vel rectis verticalibusque
ordinatae, cribris occlusae. Collum "Ringleiste"
infra marginem valvae positum rimoportulis
perforatum. Valvae speciales rupturam catenae
permittentes interdum factae. Plantae
planctonicae aquam dulcem habitantes.

Typus: *Aulacoseira* Thwaites 1848

Bacillariales Hendey 1937 emend D.G. Mann

Cellulae solitariae vel coloniales, interdum
catenas facientes. Chromatophora 2
laminiformia, uno utrumque polum versus
posito; raro chromatophora multa disciformia.
Nucleus vegetativus plusminusve centralis.
Frustula isovalvaria cis- vel trans-conformata,
vel solum cis-conformata. Areolae simplices vel
loculatae hymenibus occlusae. Systema raphis
semper fibulatum plerumque excentricum
carinatum. Extrema centralia interna raphis
recta saepe in helictoglossam duplicem
terminantia aut extrema centralia nulla.
Cingulum ex taeniis apertis vel clausis constans.

Typus: *Bacillaria* Gmelin 1791

**Chaetocerotales F.E. Round & R.M. Crawford,
*ord. nov.***

Cellulae aspectu valvae circulares vel ellipticae,
aspectu cincturae rectangulares, coloniales, raro
solitariae. In coloniis, cellulae inter se per setas
duas vel multas affixae; valvae speciales
rupturam catenae permittentes interdum factae.
Chromatophora 2–multa. Rimoportulae
centrales, interdum solum in valvis specialibus
praesentes. Plantae marinae planctonicae.

Typus: *Chaetoceros* Ehrenberg 1844

Chrysanthemodiscales F.E. Round, *ord. nov.*

Cellulae orculiformes, extremis convexis, inter se
per frontes affixae, catenas ita facientes. Frons
tholiformis, annulo centrali. Striae uniseriatae
ex annulo radiantes, inter costas ramificantes
subtiles positae. Portulae nullae. Copulae
multae clausae, unaquaque seriebus pluribus
areolarum praedita. Plantae marinae
epiphyticae.

Typus: *Chrysanthemodiscus* A. Mann 1925

**Climacospheniales F.E. Round, *ord. nov.* &
Climacospheniaceae F.E. Round, *fam. nov.***

Cellulae elongatae clavatae. Frons plana; limbus
profundus. Striae uniseriatae a sternis
lateralibus interruptae, in media valvae inter se
incidentes inordinationem longitudinalem hic
facientes. Areolae simplices. Rimoportulae
nullae. Areae apicales porellorum nullae.
Cingulum ex copulis duabus constans.
Valvocopula seriebus areolarum et projecturis
internis validis complanatis vel costiformibus
instructa; copula similis est, sed sine projecturis.
Plantae marinae epiphyticae.

Typus: *Climacosphenia* Ehrenberg 1843

Corethrales F.E. Round & R.M. Crawford, *ord. nov.*

Cellulae solitariae heterovalvares. Valvae tholiformes annulum peripheralem spinorum serratorum munitae, una valva processibus unguiformibus etiam praedita, poris simplicibus dispersis. Portulae nullae. Plantae marinae planctonicae.

Typus: *Corethron* Castracane 1886

Coscinodiscales F.E. Round & R.M. Crawford, *ord. nov.*, et Coscinodiscaceae Kützing emend. F.E. Round & R.M. Crawford.

Cellulae disciformes vel breviter cylindricae vel hemidisciformes, solitariae. Chromatophora disciformia. Valvae loculatae; loculi extra cribris complexis occlusi, intra per foramina aperti. Rimoportulae in annulo in limbo positae aut in centrum frontis dispersae vel aggregatae. Macrorimoportulae interdum praesentes. Copulae multae apertae. Plantae marinae planctonicae.

Typus: *Coscinodiscus* Ehrenberg 1838

Cyclophorales F.E. Round & R.M. Crawford, *ord. nov.* et Cyclophoraceae F.E. Round & R.M. Crawford, *fam. nov.*

Cellulae tabulares heterovalvares, una valva costa annulari centrali praedita. Areae apicales porellorum elongatorum praesentes. Copulae multae simplices apertae. Plantae solum maris habitantes.

Typus: *Cyclophora* Castracane 1878

Cymatosirales F.E. Round & R.M. Crawford, *ord. nov.*

Cellulae coloniales heterovalvares, una valva rimoportula praedita, pilis et ocellulis ad propeve apices.

Typus: *Cymatosira* Grunow 1862

Cymbellales D.G. Mann, *ord. nov.*

Cellulae coloniales vel solitariae saepe dorsiventrales. Chromatophorum unicum, ex laminis duabus ad valvas appressis secus raphem indentato isthmo lato excentrico pyrenoidem continenti connexis constans. Nucleus vegetativus excentricus. Frustula cis-conformata, isovalvaria vel heterovalvaria. Valvae unistratae aut striae alveolatae. Areolae plusminusve simplices volis vel hymenibus intra striam raro confluentibus occlusae. Systema raphis centrale vel excentricum. Extrema

centralia interna raphis unciformia aut fissurae internae trans centrum continuae. Fissurae terminales latus idem versus vel a se ex adverso flexae. Cingulum ex taeniis apertis saepe porosis constans.

Typus: *Cymbella* Agardh 1830

Dictyoneidales D.G. Mann, *ord. nov.*

Cellulae solitariae. Frustula isovalvaria. Structura valvae complexa ex strato exteriori irregulariter grosseque areolato et strato interiori subtiliter striato constans. Systema raphis centrale. Extrema centralia interna externaque raphis recta. Fissurae terminales a se ex adverso flexae. Cingulum ex taeniis apertis porosis constans.

Typus: *Dictyoneis* Cleve 1890

Ethmodiscales F.E. Round, *ord. nov.* & Ethmodiscaceae F.E. Round, *fam. nov.*

Cellulae cylindricae. Frons plana vel tholiformis. Striae ex centro non areolato radiantes. Areolae loculatae. Rimoportulae typicae circum centrum valvae dispersae; rimoportulae minores in limbo positae. Copulae apertae latae confertim areolatae. Plantae marinae planctonicae.

Typus: *Ethmodiscus* Castracane 1886

Eunotiales Silva 1962

Cellulae solitariae vel catenas facientes, plerumque dorsiventrales. Chromatophora 1-multa laminiformia. Nucleus vegetativus saepe excentricus prope latus dorsale positus. Frustula cis-conformata isovalvaria. Valvae unistratae. Areolae simplices volis (?) occlusae. Systema raphis in sternum haud vel non perfecte conflatum, fissuris raphis brevibus. Extrema 'centralia' interna raphis recta. Rimoportulae 1-2 plerumque praesentes. Cingulum ex taeniis apertis simplicibus porosis constans.

Typus: *Eunotia* Ehrenberg 1837

Fragilariales Silva 1962, emend. F.E. Round

Cellulae longae vel breves, tum aspectu valvae ellipticae, coloniales vel solitariae. Chromatophora 2-multa. Valvae sterno simplici instructae, marginibus saepe spinulosis. Areolae simplices. Rimoportulae plerumque praesentes prope polos positae, sed interdum nullae. Areae apicales porellorum plerumque praesentes. Copulae multae apertae. Plantae aquam dulcem habitantes vel marinae, planctonicae vel solum habitantes.

Typus: *Fragilaria* Lyngbye 1819

Hemiaulales F.E. Round & R.M. Crawford, *ord.*
nov.

Cellulae circulares vel bipolares, coloniales, ocellis
parum elevatis vel prominentiis altis ocellis
redactis spinisque/ve instructis conjunctae.
Chromatophora disciformia. Rimoportula unica
centralis vel marginalis, aut rimoportulae
multae. Plantae marinae praecipue planctonicae
saepe fossiles.

Typus: *Hemiaulus* Heiberg 1863, *nom. cons.*

Leptocylindrales F.E. Round & R.M. Crawford,
ord. nov.

Cellulae cylindricae coloniales, in dimensione
pervalvari longae. Valvae cylindricae, ad
juncturam frontis cum limbo annulo
prominentiarum cristiformium et
pustuliformium alternantium. Valvae sine
processibus complexis, sed poro excentrico
praeditae. Copulae squamiformes. Plantae
marinae planctonicae.

Typus: *Leptocylindrus* Cleve 1889

Licmophorales F.E. Round, *ord. nov.*

Cellulae aspectu valvae cincturaeque clavatae,
polo basali ad substratum per mucum affixo.
Copulae septa apicalia brevia saepe ferentes.
Plantae marinae epiphyticae.

Typus: *Licmophora* Agardh 1827

Lithodesmiales F.E. Round & R.M. Crawford,
ord. nov.

Cellulae solitariae vel catenas breves facientes.
Chromatophora disciformia. Valvae plerumque
tri- vel quadripolares, areolis rotis simplicibus
vel complexis occlusis. Rimoportula unica
centralis extra tubo longo prominenti, intra
rimis duabus. Margo frontis serie spinorum
erectorum vel membranis aliformibus munitus.
Copulae segmentatae plurimae. Plantae
marinae planctonicae.

Typus: *Lithodesmium* Ehrenberg 1839

Lyrellales D.G. Mann, *ord. nov.*

Cellulae solitariae. Chromatophora 2 vel 4, ex
laminis ad valvas appressis secus raphem
indentatis constantia. Nucleus vegetativus
centralis. Frustula cis-conformata isovalvaria.
Valvae unistratae. Areolae simplices volis vel
volis cribrisque occlusae. Systema raphis
centrale. Extrema centralia interna raphis
unciformia vel T-formia. Fissurae terminales

latus idem versus (interdum a se ex adverso?)
flexae. Cingulum ex taeniis apertis porosis
constans.

Typus: *Lyrella* Karayeva 1978

Mastogloiales D.G. Mann, *ord. nov.*

Cellulae solitariae. Chromatophora 2 aspectu
cincturae H-formia, utrumque ex laminis
duabus secus raphem indentatis isthmo lato
centrali pyrenoidem continenti connexis
constans. Nucleus vegetativus centralis. Frustula
cis- vel trans- conformata, isovalvaria. Structura
valvae complexa, areolis cribris vel volis
occlusis sine hymenibus. Systema raphis
centrale. Extrema centralia interna externaque
raphis recta. Fissurae terminales latus idem
versus flexae. Cingulum ex taeniis apertis
porosis constans; taenia prima complexa
loculata.

Typus: *Mastogloia* Thwaites 1856

Melosirales R.M. Crawford, *ord. nov.*

Cellulae sphaericae vel cylindricae, inter se
binatim affixae vel catenas longas facientes.
Chromatophora disciformia. Valvae loculatae,
intra areolis poroidibus rotis occlusis, extra
foraminibus vel laminis porosis praeditae. Modi
conjunctionis varii: per projecturas tubulares
aut spinas aut mucum. Rimoportulae minimae,
prope limbum saepe positae aut aggregatae
interdum dispersae. Copulae multae apertae.
Plantae marinae vel salsuginosae vel aquam
dulcem habitantes.

Typus: *Melosira* Agardh 1824, *nom. cons.*

Naviculales Bessey 1907 emend. D.G. Mann

Cellulae solitariae vel catenas facientes, interdum
dorsiventrales. Chromatophora 1, 2 vel 4
laminiformia, raro multa disciformia vel H-
formia. Nucleus vegetativus centralis raro
excentricus. Frustula isovalvaria cis- vel trans-
conformata raro solum cis-conformata. Areolae
simplices vel loculatae hymenibus occlusae.
Systema raphis nunquam fibulata, centrale vel
aliquantum excentricum, rectum vel
sigmoideum, rarissime biarcuatum. Extrema
centralia interna raphis recta vel unciformia vel
T-formia vel in helictoglossam duplicem
terminantes. Fissurae terminales plerumque
latus idem versus raro a se ex adverso flexae.
Cingulum ex taeniis apertis constans.

Typus: *Navicula* Bory 1822

Orthoseirales R.M. Crawford, *ord. nov.* &
Orthoseiraceae R.M. Crawford, *fam. nov.*
Cellulae cylindricae catenas breves facientes.
Striae radiantibus trans frontem limbumque
extensae. Areolae grandes. Carinoportulae 2–5
ad centrum frontis positae. Valvae germanae
dissimiles, una gradu in limbo praedita, altera
limbo recto habenti. Copulae incrassationibus
characteristicis praeditae. Plantae saepe
subaeriae praecipue inter bryophyta aquam
dulcem habitantes.

Typus: *Orthoseira* Thwaites 1848

Paraliales R.M. Crawford, *ord. nov.*
Cellulae valdissimae heterovalvares, inter se
affixae catenas ita facientes. Stratum internum
valvae crassissimum, seriebus tuborum
longorum perforatum, tubis in loculos non
profundos extra aperientibus; loculi lamina
porosa obtecti. Valvae germanae plerumque
dissimiles, una gradu in limbo praedita, altera
limbo recto habenti. Ordo et familia
tuboprocessibus complexis propriis. Solum
maris et aquam dulcem habitant.

Typus: *Paralia* Heiberg 1863

Rhabdonematales F.E. Round & R.M. Crawford,
***ord. nov.* & Rhabdonemataceae F.E. Round &**
R.M. Crawford, *fam. nov.*
Cellulae tabulares. Valvae loculatae. Copulae
validae septis complexis. Plantae solum maris
habitantes.

Typus: *Rhabdonema* Kützing 1844.

Rhaphoneidales F.E. Round, *ord. nov.*
Cellulae solitariae vel catenas breves facientes.
Valvae circulares vel bipolares vel multipolares.
Areolae simplices rotis occlusae. Rimoportulae
plerumque ad apices, raro circum peripheriam
positae. Areae apicales porellorum praesentes
sed indistinctae. Copulae plures apertae.
Plantae solum maris habitantes.

Typus: *Rhaphoneis* Ehrenberg 1844

Rhizosoleniales Silva 1962, emend . F.E. Round
& R.M. Crawford.
Cellulae cylindricae, coloniales vel solitariae.
Chromatophora multa parva. Areolae simplices
vel loculatae, extra per rimas apertae, loculis
intra per foramina apertis. Rimoportula unica
ad basim extensionis valvae posita. Copulae
multae apertae, dimidiae vel segmentatae,
areolatae quam in valva. Plantae fere semper

marinae planctonicae.

Typus: *Rhizosolenia* Brightwell 1858, *nom. cons.*

Rhopalodiales D.G. Mann, *ord. nov.*
Cellulae solitariae dorsiventrales.
Chromatophorum unicum laminiforme axiale.
Nucleus vegetativus excentricus prope latus
dorsale cincturae positus. Frustula cis-
conformata isovalvaria. Areolae simplices vel
loculatae, volis occlusae. Aliae costae
transapicales valde incrassatae, aliae non
incrassatae. Systema raphis fibulatum
excentricum biarcuatum vel carinatum. Extrema
centralia interna raphis recta aut fissurae
internae trans centrum continuae. Cingulum ex
taeniis apertis vel clausis constans.

Typus: *Rhopalodia* O. Müller 1895

Stictocyclales F.E. Round, *ord. nov.* &
Stictocyclaceae F.E. Round, *fam. nov.*
Cellulae cylindricae. Frons plana a limbo verticali
distincta. Striae uniseriatae ex annulo centrali
radiantes, ordinatione earum in fronte limboque
costis vel cristis internis radialibus incrassatis
secta. Rimoportulae in centro frontis dispersae,
extra per poros rimiformes apertae.
Pseudonodulus unicus prope marginem frontis
positus. Copulae clausae latae porosae. Plantae
marinae epiphyticae.

Typus: *Stictocyclus* A. Mann 1925

Stictodiscales F.E. Round & R.M. Crawford, *ord.*
nov.
Cellulae circulares vel multiangulatae solitariae.
Superficies externa valvae ad centrum
incrassationibus irregularibus praedita,
marginem versus radiatim incrassata vel costata.
Limbus non profundus. Areolae simplices.
Portulae nullae.

Typus: *Stictodiscus* Greville 1861

Striatellales F.E. Round, *ord. nov.*
Cellulae aspectu cincturae oblongae. Valvae
lineares vel lanceolatae, marginibus interdum
undulatis, polis rotundatis interdum capitatis.
Frons plana vel curvata. Sternum angustum
interdum apparenter nullum. Striae uniseriatae.
Areolae simplices. Areae apicales porellorum
distinctae bene effectae. Rimoportulae ad apices
vel secus sternum positae. Copulae multae
plerumque septa ferentes. Plantae marinae
epiphyticae.

Typus: *Striatella* Agardh 1832

Surirellales D.G. Mann, *ord. nov.*

Cellulae solitariae interdum dorsiventrales. Chromatophorum unicum laminiforme axiale vel ex laminis duabus ad valvas appressis isthmo angusto prope polum unum cellulae posito connexis constans; rarissime chromatophora 2 laminiformia, uno utrumque polum versus posito. Nucleus vegetativus centralis vel prope latus dorsale cincturae positus. Frustula isovalvaria. Areolae simplices volis vel hymenibus occlusae. Systema raphis semper fibulatum carinatum interdum sigmoideum plerumque valde excentricum plus quam dimidium peripheriae frontis circumiens. Extrema centralia interna raphis recta saepe in helictoglossam duplicem terminantia aut fissurae internae trans centrum continuae. Cingulum ex taeniis apertis constans.

Typus: *Surirella* Turpin 1828

Tabellariales F.E. Round, *ord. nov.*

Cellulae aspectu cincturae rectangulares. Valvae ellipticae usque elongatae lineares, plerumque ad centrum polosque dilatatae. Frons plana spinis ad marginem; limbus non profundus. Striae uniseriatae, areolis simplicibus. Uterque polus valvae area porellorum praeditus. Rimoportula unica centralis vel apicalis. Copulae multae apertae ligulatae, unaquaque septo instructa. Plantae epiphyticae aquam dulcem habitantes.

Typus: *Tabellaria* Ehrenberg ex Kützing 1844

Thalassionematales F.E. Round, *ord. nov.* & **Thalassionemataceae F.E. Round,** *fam. nov.*

Cellulae aciculares saepe longissimae, interdum et arcuatae et centraliter apicaliterque dilatatae. Sternum plerumque latum, striis brevibus in limbum extensis. Areolae loculatae circulares vel transapicaliter elongatae, unaquaque extra cribris complexis occlusae intra per foramen parvum circulare aperta. Junctura frontis cum limbo ad utramque apicem spinis prominentibus 1–2, alibi spinis minoribus munita. Rimoportulae 2, una ad utrumque polum posita. Copulae plures, unaquaque serie unica areolarum praedita. Plantae marinae planctonicae, solitariae vel coloniales.

Typus: *Thalassionema* Grunow ex Hustedt 1932

Thalassiophysales D.G. Mann, *ord. nov.*

Cellulae solitariae valde dorsiventrales, raro catenas facientes. Chromatophora varia: 1 laminiforme axiale; 2 varie formata (laminiformia, H-formia, etc.); multa disciformia; etc. Nucleus vegetativus plerumque excentricus. Frustula cis-conformata isovalvaria. Areolae plerumque simplices hymenibus occlusae. Systema raphis excentricum, rectum vel biarcuatum, interdum fibulatum. Extrema centralia interna raphis plerumque recta; fissurae internae trans centrum interdum continuae. Fissurae terminales nullae vel latus idem versus flexae. Cingulum ex taeniis apertis constans.

Typus: *Thalassiophysa* Conger 1954

Toxariales F.E. Round, *ord. nov.* **& Toxariaceae F.E. Round,** *fam. nov.*

Cellulae aciculares ad centrum polosque aliquantum dilatatae, marginibus valvarum rectis vel undulatis. Areolae circulares simplices, pro parte maxime dispersae sed prope marginem frontis et in limbo non profundo in seriebus transversalibus dispositae. Portulae nullae. Areae porellorum nullae. Copulae ?clausae porosae. Plantae marinae epiphyticae.

Typus: *Toxarium* Bailey 1854

Triceratiales F.E. Round & R.M. Crawford, *ord. nov.*

Cellulae bi- vel multipolares. Chromatophora disciformia. Structura valvae varia. Ocelli rimoportulaeque praesentes. Copulae multae. Plantae marinae planctonicae vel solum habitantes.

Typus: *Triceratium* Ehrenberg 1839

SUBORDINAL DESCRIPTIONS

Diploneidineae D.G. Mann, *subord. nov.* et **Diploneidaceae D.G. Mann,** *fam. nov.*

Cellulae solitariae. Chromatophora 2 laminiformia, uno ad utrumque latus cincturae appresso, pyrenoidibus bacillaribus vel rotundatis interdum indentatis. Frustula circum planitiem valvarem medianam, planitiem transapicalem medianam et planitiem apicalem symmetrica. Valvae plerumque ellipticae vel lineari-ellipticae vel panduriformes, polis obtuse rotundatis. Structura valvae complexa, canalibus longitudinalibus duabus ad raphem parallelis, uno in utroque latere raphis, striis vel areolis loculatis. Areolae vel stria tota intra hymenibus vel laminis subtiliter porosis occlusae, extra per foramina circularia vel elliptica vel rimiformia aut per poros minores aggregatos apertae. Raphosternum angustum. Extrema centralia interna raphis recta non dilatata; extrema centralia externa recta vel flexa vel unciformia. Fissurae terminales flexae vel unciformes utraque latus idem versus curvata. Cingulum e taeniis apertis non porosis constans, taenia prima (valvocopula) lata taeniis ceteris angustis.

Typus: *Diploneis* Ehrenberg ex Cleve 1894

Naviculineae Hendey

Cellulae plerumque solitariae. Chromatophora 2 vel 4 (raro 1) laminiformia ad latera cincturae appressa, interdum lobata vel taeniiformia, pyrenoidibus saepe bacillaribus, non indentatis. Nucleus vegetativus centralis, in statu dividenti idem latus cellulae quam in divisione proximo fere nunquam occupans, frustulis cis- et trans-conformatis in ratione 1:2 igitur praesentibus. Valvae lineares, lanceolatae, ellipticae vel sigmoideae, raro circum planitiem apicalem asymmetricae, polis variis. Striae plerumque uniseriatae, interdum alveolatae. Areolae simplices vel loculatae saepe apicaliter elongatae, hymenibus intra striam non confluentibus occlusae. Raphosternum plerumque centrale plerumque rectum, interdum sigmoideum vel biarcuatum. Extrema centralia interna raphis recta saepe in nodulo parvo interno cristis brevibus limitato posita. Extrema centralia externa varia: recta dilatata, a se ex adverso vel latus idem flexa, unciformia, etc. Fissurae terminales plerumque unciformes, a se ex adverso vel latus idem versus flexae, interdum nullae. Cingulum ex taeniis apertis parce vel non porosis constans; taenia prima quam taeniae aliae saepe latior.

Typus: *Navicula* Bory 1822

Neidiineae D.G. Mann, *subord. nov.*

Cellulae solitariae vel tubos mucosos catenasve facientes. Chromatophora varia pyrenoidibus non indentatis. Nucleus vegetativus plerumque centralis, in statu dividenti idem latus cellulae quam in divisione proximo interdum vel semper, raro nunquam occupans. Valvae lineares, lanceolatae vel ellipticae, raro circum planitiem apicalem asymmetricae, polis variis.

Striae plerumque uniseriatae simplices, interdum loculatae. Areolae simplices vel loculatae plerumque circulares vel transapicaliter elongatae, hymenibus intra striam saepe confluentibus occlusae. Raphosternum plerumque centrale rectum, raro excentricum. Extrema centralia interna raphis simplicia rectaque, vel T-formia, vel in helictoglossam duplicem terminantia. Extrema centralia externa varia: recta dilatata, T-formia, a se ex adverso vel latus idem versus flexa, etc. Fissurae terminales similiter variae. Cingulum ex taeniis apertis porosis inter se plusminusve similaribus constans.

Typus: *Neidium* Pfitzer 1871

Sellaphorineae D.G. Mann, *subord. nov.*

Cellulae solitariae rarissime catenas facientes. Chromatophorum unicum aspectu valvae H-formia ex laminis duabus ad cincturam appressis isthmo angusto ad hypo- vel epi-valvam posito connexis constans, aut chromatophora 2 laminiformia ad latera cincturae appressa. Pyrenoides saepe indentatae. Nucleus vegetativus centralis vel excentricus, in statu dividenti idem latus cellulae quam in divisione proximo fere nunquam occupans, frustulis cis- et trans-conformatis in ratione 1:2 igitur praesentibus. Valvae lineares, lanceolatae vel ellipticae, polis obtuse rotundatis interdum subcapitatis vel capitatis. Striae uniseriatae aut alveolatae. Areolae vel pori exteriores alveolorum circulares, hymenibus intra striam non confluentibus occlusae. Raphosternum centrale rectum. Extrema centralia interna raphis latus primum versus flexa aut fissurae internae trans centrum continuae; extrema centralia externa dilatata latus primum versus aliquantum flexa. Fissurae terminales plerumque unciformes latus secundum versus curvantes. Cingulum ex taeniis apertis parce vel non porosis constans; taenia prima quam taeniae aliae latior.

Typus: *Sellaphora* Mereschkowsky 1902

FAMILY DESCRIPTIONS

Acanthocerataceae R.M. Crawford & F.E. Round, *fam. nov.*

Cellulae aspectu cincturae oblongae, valvis projecturis polaribus praeditis. Valvae ex galeris duobus polaribus areolatis fascia tenui areolata connexis constantes; utrumque galerum in projecturam longam inornatam tubularem productum. Copulae multae colliformes areolatae (copulis abvalvaribus exceptis quae inornatae). Plantae planctonicae aquam dulcem habitantes.

Typus: *Acanthoceras* Honigmann 1909

Achnanthaceae Kützing 1844 emend. D.G. Mann

Cellulae solitariae vel coloniales, plerumque pedunculatae. Chromatophora 2 aspectu valvae H-formia, utroque ex laminis duabus ad cincturam appressis, isthmo lato per lumen cellulae perducto pyrenoidem continenti connexis constans; rarius chromatophora multa disciformia. Frustula circum planitiem valvarem medianam asymmetrica, circum planitiem transapicalem medianam et planitiem apicalem plusminusve symmetrica; heterovalvaria. Raphovalva introflexa, fronte in sectione transapicali convexa, raphosterno centrali. Areovalva retroflexa, fronte in sectione transapicali concava, sterno angusto excentrico. Limbi profundi. Striae uni-, bi- vel multi-seriatae. Areolae grandes subquadratae cribris volas ferentibus occlusae. Extrema centralia interna raphis recta vel unciformia; extrema centralia externa recta dilatata. Fissurae terminales latus idem versus curvantes. Cingulum ex taeniis apertis porosis constans.

Typus: *Achnanthes* Bory 1822

Achnanthidiaceae D.G. Mann, *fam. nov.*

Cellulae solitariae vel catenas breves facientes. Chromatophorum unicum laminiforme. Frustula heterovalvaria cis-conformata aspectu cincturae flexa, circum planitiem transapicalem medianam et planitiem apicalem symmetrica, raro circum axem apicalem torta. Valvae lineares vel lanceolatae vel ellipticae, polis rotundatis vel apiculatis vel subcapitatis, una (raphovalva) raphe centrali praedita altera (areovalva) sine raphe sed sterno centrali praedita. Limbus non profundus. Striae uni-, bi- vel multiseriatae. Areolae parvae simplices circulares hymenibus occlusae. Extrema centralia interna raphis (raphovalva!) a se ex adverso leviter flexa; extrema centralia externa recta. Fissurae terminales breves vel nullae. Cinctura angusta, cingulo e taeniis apertis

paucis constanti. Ab Achnanthaceis differunt velis, positione sterni (in areovalva), structura et dispositione chromatophororum, extremibus centralibus internis.

Typus: *Achnanthidium* Kützing 1844

Anomoeoneidaceae D.G. Mann, *fam. nov.*

Cellulae solitariae. Chromatophorum unicum ex laminis duabus ad valvas appressis secus raphem indentatis isthmo laterali valido pyrenoidem continenti connexis. Nucleus excentricus, semper in latere eodem cellulae dividens, frustulis omnibus igitur cis-conformatis. Frustula circum planitiem valvarem medianam, planitiem transapicalem medianam et planitiem apicalem symmetrica. Valvae lanceolatae vel ellipticae polis rotundatis vel capitatis. Frons plana; limbus plusminusve profundus prope polos deminutus. Striae uniseriatae. Areolae interdum semotae et in seriebus longitudinalibus sinuatis ordinatae, semper hymenibus occlusae. Raphosternum centrale, ad centrum valvae saepe valde expansum. Stauros interdum praesens. Extrema centralia externa raphis dilatata aliquantum flexa. Extrema centralia interna unciformia, latus primum versus curvata. Fissurae terminales unciformes. Cingulum e taeniis apertis constans.

Typus: *Anomoeoneis* Pfitzer 1871

Arachnoidiscaceae F.E. Round, *fam. nov.*

Vide descriptionem Arachnoidiscalium.

Ardissoneaceae F.E. Round, *fam. nov.*

Vide descriptionem Ardissonealium.

Attheyaceae R.M. Crawford & F.E. Round, *fam. nov.*

Cellulae aspectu cincturae oblongae, valvis projecturis polaribus brevibus praeditis. Valvae cymbiformes, area centrali elongata areolata, parte marginali et projecturis apicalibus ex filis tenuibus constantes, filis in projecturis in spiras contortis. Rimoportula unica excentrica interdum praesens. Copulae multae apertae, in utroque latere partis mediae areolatae. Plantae marinae ad grana arenae affixae.

Typus: *Attheya* West 1860

Aulacoseiraceae R.M. Crawford, *fam. nov.*

Vide descriptionem Aulacoseiralium.

Bellerocheaceae R.M. Crawford, *fam. nov.*

Cellulae aspectu cincturae oblongae catenas

facientes. Valvae bi- vel multi-polares, ex reticulo costarum subtilium respectu annuli centralis lateralisve ordinatarum constantes. Junctura frontis cum limbo crista fimbriata praedita. Rimoportula unica intra annulum posita. Unusquisque polus ocello debiliter effecto instructus. Copulae subtiles. Plantae marinae planctonicae.

Typus: *Bellerochea* Van Heurck 1885

Berkeleyaceae D.G. Mann, *fam. nov.*

Cellulae solitariae vel tubos mucosos facientes. Chromatophora 1-multa; unumquidque ex laminis duabus ad cincturam appressis isthmo connexis constans. Frustula circum planitiem valvarem medianam, planitiem transapicalem medianam et planitiem apicalem symmetrica; raro circum planitiem apicalem asymmetrica quod valvae arcuatae. Valvae lineares vel lanceolatae, polis rotundatis vel acutis. Striae uniseriatae. Areolae simplices circulares vel subquadratae, hymenibus occlusae. Raphosternum centrale intra saepe incrassatum, raro biarcuatum. Extrema centralia externa raphis dilatata latus unum versus interdum parum flexa; extrema centralia interna recta non dilatata. Fissurae terminales nullae vel breves flexaeque. Cinctura plerumque profunda. Cingulum e taeniis apertis pluribus vel multis porosis constans, unaquaque taenia seriebus duabus pororum saepe praedita; raro taenia prima (valvocopula) serie projecturarum geminarum costiformium in medio valvae conjunctarum intra praedita.

Typus: *Berkeleya* Greville 1827

Brachysiraceae D.G. Mann, *fam. nov.*

Cellulae solitariae. Chromatophorum unicum. Mores nuclei vacillantes, frustulis et cis- et trans-conformatis igitur factis. Frustula circum planitiem valvarem medianam, planitiem transapicalem medianam et planitiem apicalem symmetrica. Valvae lineares vel lanceolatae. Striae uniseriatae ad juncturam frontis cum limbo crista externa prominenti interruptae. Frons plana plerumque spinosa vel verrucosa vel longitudinaliter costata. Areolae transapicaliter elongatae, in seriebus (striis) longitudinalibus sinuatis ordinatae, intra hymenibus occlusae. Raphosternum angustum ad centrum expansum. Extrema externa raphis

ad polos centrumque simplicia recta non dilatata; fissurae externae raphis utrinque cristis limitatae. Extrema interna centralia raphis recta non dilatata. Cingulum e taeniis apertis constans, partibus exterioribus earum non porosis; pars interior taeniae primae (valvocopulae) sulcata porosa.

Typus: *Brachysira* Kützing 1836

Cavinulaceae D.G. Mann, *fam. nov.*

Cellulae solitariae. Chromatophora 1-2, utrumque ex laminis duabus ad valvas appressis isthmo plusminusve centrali connexis constans, isthmo lato pyrenoidem conspicuum continenti. Frustula circum planitiem valvarem medianam, planitiem transapicalem medianam et planitiem apicalem symmetrica. Valvae ellipticae, lanceolatae vel fere lineares, polis obtusis rotundatis. Frons plana; limbus plerumque non profundus ad polos non abrupte diminutus. Striae uniseriatae radiatae. Areolae circulares vel ellipticae hymenibus intra occlusae. Extrema centralia externa raphis recta expansa; extrema centralia interna recta non expansa vel T-formia, interdum in papillas vel cristas minutissimas transientia. Fissurae terminales nullae vel a se ex adverso flexae. Cingulum e taeniis apertis angustis porosis constans, unaquaque seriebus duabus pororum parvorum instructa.

Typus: *Cavinula* D.G. Mann, *gen. nov.*

Climacospheniaceae F.E. Round, *fam. nov.*

Vide descriptionem Climacospheniarum.

Coscinodiscaceae Kützing emend. F.E. Round & R.M. Crawford

Vide descriptionem Coscinodiscalium.

Cosmioneidaceae D.G. Mann, *fam. nov.*

Cellulae solitariae. Chromatophora 2 laminiformia secus raphem indentata, uno eorum ad utramque valvam appresso. Frustula cis- vel trans-conformata, circum planitiem valvarem medianam, planitiem transapicalem medianam et planitiem apicalem symmetrica. Valvae lineares vel lanceolatae, polis saepe rostellatis vel capitatis. Frons plana; limbus profundus sed prope polos abrupte diminutus. Striae uniseriatae. Areolae circulares hymenibus intra occlusae, hymenibus secus strias saepe confluentibus. Extrema centralia externa raphis valde expansa poriformia; extrema interna

T-formia. Fissurae terminales unciformes latus secundum valvae versus flexae, supra helictoglossas dilatatae. Cingulum e taeniis apertis multis constans, unaquaque serie transversali pororum circularium parvorum instructa; centrum taeniae primae et taeniae secundae foramine circulari perforatum.

Typus: *Cosmioneis* D.G. Mann, *gen. nov.*

Cyclophoraceae F.E. Round & R.M. Crawford, *fam. nov.*

Vide descriptionem Cyclophoralium.

Diadesmidaceae D.G. Mann, *fam. nov.*

Cellulae parvae solitariae vel per frontes inter se affixae ita catenas facientes. Chromatophorum unicum. Nucleus semper in latere eodem cellulae dividens, frustulis omnibus igitur cis-conformatis. Frustula circum planitiem valvarem medianam, planitiem transapicalem medianam et planitiem apicalem symmetrica. Valvae lineares vel lanceolatae vel ellipticae, marginibus interdum sinuatis. Striae uniseriatae ad frontis cum limbo juncturam interruptae, junctura hac cristis vel spinis interdum praedita. Areolae circulares vel transapicaliter elongatae hymenibus intra occlusae, hymenibus secus strias plerumque confluentibus. Systema raphis simplex extremis externis polaribus centralibusque inter se saepe similaribus, flexis vel rectis vel T-formibus. Extrema centralia interna raphis recta vel T-formia. Stigma interdum praesens. Cingulum e taeniis apertis porosis constans.

Typus: *Diademis* Kützing 1844

Dictyoneidaceae D.G. Mann, *fam. nov.*

Cellulae solitariae. Frustula circum planitiem valvarem medianam, planitiem transapicalem medianam et planitiem apicalem symmetrica. Valvae panduriformes polis rotundatis. Structura valvae complexa extra irregulariter ex grosse areolata intra costis transapicalibus angustis et striis uniseriatis praedita. Raphosternum angustum incrassatum. Extrema centralia raphis et intra et extra recta. Fissurae terminales a se ex adverso flexae.

Typus: *Dictyoneis* Cleve 1890

Diploneidaceae D.G. Mann, *fam. nov.*

Vide descriptionem Diploneidinearum.

Endictyaceae R.M. Crawford, *fam. nov.*

Cellulae cylindricae catenas breves facientes.

Valvae validae loculatae. Loculi intra per poros parvos a volis occlusos extra per foramina aperti. Rimoportulae in annulo positae, per cristam exiguam ad juncturam frontis cum limbo aperientes. Copulae non visae. Solum maris verisimiliter habitant.

Typus: *Endictya* Ehrenberg 1845

Ethmodiscaceae F.E. Round, *fam. nov.*

Vide descriptionem Ethmodiscalium.

Gosleriellaceae F.E. Round, *fam. nov.*

Cellulae disciformes. Frons plana; limbus non profundus. Striae ex centro non areolato rimoportulam unicam continenti radiantes. Areolae simplices parvae. Valvocopula areolata, circulo ad epivalvocopulam affixo setis siliceis radiantibus munito; aliae setae simplices singulares, aliae cohaerentes setas compositas crassas ita facientes. Plantae marinae planctonicae.

Typus: *Gossleriella* Schütt 1893

Hyalodiscaceae R.M. Crawford, *fam. nov.*

Cellulae lenticulares sphaericae vel compressae, solitariae vel binatim affixae. Valvae loculatae. Loculi intra volis occlusi, extra per aliquot poros aperti. Rimoportulae dispersae vel aggregatae. Solum orarum maris habitant, ad algas grandes vel grana arenae affixae.

Typus: *Hyalodiscus* Ehrenberg 1845

Lauderiaceae (Schütt) Lemmermann, emend. F.E. Round & R.M. Crawford

Cellulae breviter cylindricae, in catenas conjunctae. Chromatophora multa disciformia. Valvae subtiliter loculatae, fultoportulis multis peripheriam valvae versus positis, aliquot fultoportulis in centro. Fultoportulae extra per tubos longos aperientes. Processus occlusi tubulares etiam praesentes. Rimoportula unica prope marginem valvae posita. Copulae multae loculatae. Plantae marinae planctonicae.

Typus: *Lauderia* Cleve 1873

Lithmodesmiaceae F.E. Round, *fam. nov.*

Cellulae multipolares. Frons undulata ad juncturam eius cum limbo profundo crista spinosa vel integra areolataque instructa. Striae radiantes, areolis volis occlusis. Rimoportula unica centralis, extra in tubum robustum longum prolongata. Copulae multae squamiformes. Plantae marinae planctonicae.

Typus: *Lithodesmium* Ehrenberg 1839

Lyrellaceae D.G. Mann, *fam. nov.*

Cellulae solitariae fere semper marinae.
Chromatophora duo vel quattuor e laminis ad
valvas appressis secus raphem indentatis
constantia, laminis interdum trans cellulam per
lumen connexis. Nucleus semper in latere
eodem cellulae dividens, frustulis omnibus
igitur cis-conformatis. Frustula valida circum
planitiem valvarem medianam, planitiem
transapicalem medianam et planitiem apicalem
symmetrica (raro circum planitiem apicalem
asymmetrica). Striae uniseriatae interdum (in
Lyrella) a sternis lateralibus interruptae. Areolae
grandes volis vel cribris volisque occlusae.
Extrema centralia externa raphis recta in sulcos
vel foveas spatulatas vel cuneiformes aperientia.
Extrema centralia interna unciformia vel
T-formia. Fissurae terminales plerumque
unciformes. Cinctura non profunda. Cingulum
e taeniis apertis paucis constans, unaquaque
seriebus 1–2 pororum praedita.

Typus: *Lyrella* Karayeva 1978

**Melosiraceae Kützing 1844 emend. R.M.
Crawford**

Cellulae sphaericae, inter se binatim affixae vel
catenas longas facientes. Chromatophora
disciforma. Valvae loculatae, intra areolis
poroidibus rotis occlusis, extra laminis porosis
praeditae. Modi conjunctionis per spinas aut
mucum. Rimoportulae minimae, prope limbum
saepe positae aut aggregatae interdum dispersae.
Plantae marinae vel salsuginosae vel aquam
dulcem habitantes, planktonicae vel affixae.

Typus: *Melosira* Agardh 1824, *nom. cons.*

Naviculaceae Kützing 1844 emend. D.G. Mann

Cellulae solitariae. Chromatophora 2
laminiformia, uno ad utrumque latus cincturae
appresso, pyrenoidibus bacillaribus. Frustula
plerumque circum planitiem valvarem
medianam et planitiem transapicalem
medianam et planitiem apicalem
symmetrica, sed interdum circum planitiem
valvarem medianam aut planitiem apicalem
asymmetrica. Valvae lineares vel lanceolatae,
interdum heteropolares vel semilanceolatae,
polis variis. Striae uniseriatae rarissime
biseriatae, interdum alveolatae. Areolae vel pori
exteriores alveolarum simplices apicaliter
elongati hymenibus occlusi. Raphosternum

plusminusve rectum, centrale vel excentricum;
pars prima eius costam ad raphem parallelam
saepe ferens, plerumque latus dorsale versus ita
producta, ut fissurae internae raphis laterales
fiant, orificiis earum in apice cristae angustae
positis. Extrema centralia interna raphis recta
non dilatata, plerumque in nodulo parvo
circulari rarius in helictoglossam duplicem
terminantia. Extrema centralia externa varia:
dilatata, unciformia, alis e parte secunda
raphosterni productis obtecta, etc. Fissurae
terminales unciformes vel flexae latus idem
versus curvantes, raro nullae. Cingulum ex
taeniis apertis plerumque non porosis constans.

Typus: *Navicula* Bory 1822

Orthoseiraceae R.M. Crawford, *fam. nov.*

Vide descriptionem Orthoseiralium.

Pinnulariaceae D.G. Mann, *fam. nov.*

Cellulae solitariae (raro catenas breves facientes).
Chromatophora 1–2: si 1 nunc H-forme ad
cincturam valvasque appressum isthmo ad
hypovalvam; si 2 nunc ad cincturam appressa
uno in utroque latere cellulae. Pyrenoides saepe
indentatae. Nucleus in statu dividenti latus
idem ac in divisione proxima nunquam (vel
rarissime?) occupans, moribus eius ita
oscillantibus. Frustula circum planitiem
valvarem medianam, planitiem transapicalem
medianam et planitiem apicalem fere semper
symmetrica. Valvae lineares vel lanceolatae
subinde ellipticae, polis obtuse rotundatis
interdum capitatis vel subcapitatis vel breve
productis. Striae plerumque ex alveolis 1–2
constantes, raro simpliciter multiseriatae. Pori
externi alveolorum circulares parvuli
hymenibus occlusi; foramen internum circulare
vel ellipticum vel transapicaliter elongatum.
Extrema centralia externa raphis dilatata saepe
poriformia; extrema centralia interna latus
primum versus flexa vel fissura interior trans
centrum continua. Fissurae terminales
unciformes. Cingulum e taeniis apertis paucis
constans, taenia prima (valvocopula) lata serie
unica areolarum elongatarum praedita.

Typus: *Pinnularia* Ehrenberg 1843

Plagiotropidaceae D.G. Mann, *fam. nov.*

Cellulae solitariae. Chromatophora 2 vel 4 lobata
unoquoque sinu grandi laterali. Grana
magnopere manifesta composita e materia dicto

anglice 'volutin' constata saepe (semper?) praesentia, uno in utroque dimidio (polari) cellulae posito. Frustula circum planitiem valvarem medianam et planitiem transapicalem symmetrica. Valvae compressae lanceolatae circum planitiem apicalem asymmetricae. Systema raphis carinata eccentrica non fibulata. Striae uniseriatae. Areolae simplices vel loculatae saepe apicaliter elongatae, hymenibus occlusae. Extrema centralia externa raphis recta vel latus idem versus flexa; extrema centralia interna recta cristis brevibus interdum concomitata. Fissurae terminales nullae vel aliquantum flexae interdum dilatatae poriformes. Helictoglossae interdum grandes infundibuliformes.

Typus: *Plagiotropis* Pfitzer 1871

Proschkiniaceae D.G. Mann, *fam. nov.*

Cellulae solitariae profundae aspectu cincturae plerumque visae. Chromatophorum unicum ex laminis duabus ad cincturam appressis isthmo centrali connexis constans, una lamina utrumque polum versus posita, protoplasto circum planitiem transapicalem medianam igitur diagonaliter symmetrico. Valvae lineares vel lanceolatae polis acutis. Striae uniseriatae parallelae centrum versus valvae distantiores. Areolae hymenibus intra occlusae extra per rimas longitudinales apertae, rimis inter costas rectas longitudinales positis. Raphosternum angustissimum centrale vel aliquantum excentricum intra incrassatum. Extrema centralia externa raphis valde dilatata latus unum versus parum flexa; extrema centralia interna recta non dilatata. Stigmata interdum (semper?) praesentia. Fissurae terminales unciformes. Cingulum e taeniis apertis multis similibus sulcatis porosisque constans. A *Naviculaceis* structura chromatophororum raphisque cincturaeque distinguendae.

Typus: *Proschkinia* Karayeva 1978

Psammodiscaceae F.E. Round & D.G. Mann, *fam. nov.*

Cellulae disciformes. Valvae circulares vel ellipticae. Frons plana; limbus non profundus. Striae uniseriatae et longae et breves radiantes, in cellulis parvis plusminusve irregulariter ordinatae. Areolae circulares simplices rotis occlusae in limbo minores. Rimoportula unica fere centralis interdum praesens. Plantae marinae ad grana arenae affixae.

Typus: *Psammodiscus* Round & Mann 1980

Rhabdonemataceae F.E. Round & R.M. Crawford, *fam. nov.*

Vide descriptionem Rhabdonematalium.

Rocellaceae F.E. Round & R.M. Crawford, *fam. nov.*

Valvae circulares, areolis maximis circum regionem centralem rimoportulam atque porum unicum simplicem continentem ordinatis.

Typus: *Rocella* Hanna 1930

Scolioneidaceae D.G. Mann, *fam. nov.*

Cellulae solitariae circum axem apicalem tortae. Chromatophora 2 laminiformia, uno ad utramque valvam appresso secus raphem indentato. Valvae lineares vel lanceolatae, profundae, ubique unistratae. Striae uniseriatae. Areolae simplices hymenibus occlusae extra per rimas interdum furcatas apertae. Extrema centralia externa raphis recta; extrema centralia interna T-formia. Cingulum e taeniis apertis constans, unaquaque seriebus 1–2 pororum praedita.

Typus: *Scolioneis* D.G. Mann, *gen. nov.*

Skeletonemataceae Lebour, emend. F.E. Round

Cellulae processibus e fultoportulis exorientibus conjunctae. Valvae loculatae, loculis intra cribris occlusis extra apertis. Rimoportula unica praesens. Copulae multae apertae. Plantae marinae planctonicae, raro aquam dulcem habitantes.

Typus: *Skeletonema* Greville 1865

Stauroneidaceae D.G. Mann, *fam. nov.*

Cellulae solitariae vel catenas breves facientes. Chromatophora 2, uno ad utrumque latus cincturae appresso pyrenoidibus raro bacillaribus. Nucleus non semper in latere eodem cellulae dividens; mores nuclei aut oscillantes aut vacillantes. Frustula circum planitiem valvarem medianam, planitiem transapicalem medianam et planitiem apicalem symmetrica. Valvae lanceolatae plerumque non profundae stauro interdum praeditae; valvae internae speciales interdum factae. Striae uniseriatae. Areolae circulares vel ellipticae hymenibus intra occlusae. Raphosternum sine costis vel cristis internis. Extrema centralia externa raphis dilatata; extrema interna recta simplicia. Fissurae terminales unciformes. Cingulum e taeniis apertis constans, unaquaque serie unica pororum plerumque praedita. Per meiosem gametangia separata in massa mucilagina inclusa. Isogametae 2 ante conjunctionem in gametangio circumductae, una in utroque latere planitiei transapicalis medianis denique disposita. Angulus inter axem apicalem alterutrius gametangii et axem alterutrius auxosporae non constans.

Typus: *Stauroneis* Ehrenberg 1843

Stictocyclaceae F.E. Round, *fam. nov.*

Vide descriptionem Stictocyclalium.

Streptothecaceae R.M. Crawford, *fam. nov.*

Cellulae aspectu cincturae quadratae, catenas planas vel tortas facientes. Valvae bipolares angustae laminares. Frons disperse perforata, junctura eius cum limbo crista exigua praedita. Limbus minime profundus. Rimoportula unica centralis vel marginalis. Ocelli debiliter effecti, uno ad utrumque polum posito. Copulae disperse perforatae, ut in valvis. Plantae marinae planctonicae.

Typus: *Streptotheca* Shrubsole 1890

Thalassionemataceae F.E. Round, *fam. nov.*

Vide descriptionem Thalassionematalium.

Thalassiophysaceae D.G. Mann, *fam. nov.*

Cellulae solitariae valde dorsiventrales. Chromatophorum unicum axiale laminiforme. Frustula circum planitiem valvaram medianam et planitiem transapicalem medianam symmetrica. Valvae angustae circum planitiem apicalem valde asymmetricae. Striae uniseriatae. Areolae parvae circulares vel ellipticae hymenibus extra occlusae. Systema raphis valde excentricum carinatum biarcuatum fibulatum. Fibulae costiformes in carina tota dispersae. Extrema centralia interna raphis in rostrum producta; fissurae externae ad centrum V-formes sine extremis. Fissurae terminales nullae. Fibulae costiformes in carina tota dispersae. Cingulum e taeniis apertis seriebus multis pororum parvorum praeditis constans.

Typus: *Thalassiophysa* Conger 1954

Toxariaceae F.E. Round, *fam. nov.*

Vide descriptionem Toxarialium.

GENERIC DESCRIPTIONS & NEW COMBINATIONS

ANEUMASTUS D.G. Mann & A.J. Stickle, *gen. nov.*

Cellulae solitariae, aspectu valvae plerumque visae. Chromatophora 2, uno utrumque polum versus; utrumque chromatophorum ex laminis duabus ad valvas appressis isthmo connexis secus raphem indentatis constans, isthmo lato pyrenoidem conspicuum continenti. Valvae lanceolatae, polis saepe rostratis vel aliquantum capitatis. Frons valvae plana, a limbo non profundo bene distincta. Striae raphosternum versus valvae uniseriatae sed prope margines biseriatae, vel omnino uniseriatae. Areolae circulares vel quadratae, structura tortuosa, intra in foveas profundas per cribra subtilia apertae, volis extra occlusae; sine hymenibus. Margo valvae incrassatus costiformis. Raphosternum angustum, praecipue extrinsecus, sed ad centrum dilatatum, aream centralem ellipticam vel oblongam faciente. Fissurae externae raphis sinuatae, extremis centralibus expansis latus unum versus curvatis; fissurae terminales latus alterum versus recurvatae, prope marginem valvae terminantes. Extrema centralia interna raphis recta non expansa. Cinctura angusta ex taeniis apertis constans. Series loculorum minutorum in parte interiori taeniae primae (valvocopulae) praesens, loculis intra per partem interiorem subtiliter porosam extra per seriem pororum majorum circularium in parte exteriore aperientibus.

Genus novum *Mastogloiae* Thwaites affine, a qua absentia partectorum genuinorum praecipue differt.

Holotypus: *Aneumastus tusculus* (Ehrenberg) D.G. Mann & A.J. Stickle, *comb. nov.*

Navicula tuscula C.G. Ehrenberg 1840. Ber. Bekanntm. Verh. Preuss. Akad. Wiss. Berlin, 1840: 215

Aneumastus stroesei (Østrup) D.G. Mann, *comb. nov.*

Navicula tuscula var. *stroesei* E. Østrup 1910. Danske Diat.; 84

Aneumastus tusculoides (Cleve-Euler) D.G. Mann, *comb. nov.*

Navicula tusculoides A. Cleve-Euler 1953. K. svenska VetenskAkad. Handl., Fjärde Ser. 4(5): 119

ASTERIONELLOPSIS F.E. Round, *gen. nov.*

Cellulae inter se per partes basales frontium affixae colonias longas tortas formantes, heteropolares, parte basali dilatata triangulata chromatophora duo continenti. Valvae heteropolares, ex polo basali rotundato, scapo elongato tenui et polo apicali parum dilatato constantes. Sternum angustum. Striae uniseriatae, areolis circularibus membranis subtiliter porosis occlusis. In scapo frontis cum limbo junctura spinis praedita. Uterque polus ocello costas erectas continenti instructus. Rimoportula unica ad polum apicalem posita. Copulae plures porosae ligulatae. Coloniae cellulaeque planctonicae marinae.

Holotypus: *Asterionellopsis glacialis* (Castracane) F.E. Round, *comb. nov.*

Asterionella glacialis F. Castracane 1886. Rep. Sci. Res. H.M.S. Challenger Bot. 2: 50

Asterionellopsis kariana (Grunow in Cleve & Grunow) F.E. Round, *comb. nov.*

Asterionella kariana Gronow 1880. In P.T. Cleve & A. Grunow, K. Svenska Vetensk.-Akad. Handl. **17**(2): 110.

BIREMIS D.G. Mann & E.J. Cox, *gen. nov.*

Cellulae solitariae vix vel valde dorsiventrales, plerumque aspectu cincturae visae. Chromatophora 2, uno utrumque polum versus; utrumque chromatophorum ex laminis duabus ad cincturam appressis isthmo connexis constans, isthmo lato pyrenoidem conspicuum continenti. Valvae lineares profundae, bilateraliter plusminusve symmetricae vel valde asymmetricae, tum valvas *Amphorarum* simulantes; poli rotundati. Frons valvae in sectione transapicali curvata a limbo non vere distincta. Striae plusminusve uniseriatae, unaquaque ex loculis duobus imperfecte sejunctis transapicaliter elongatis constanti; uterque loculus lamina porosa (cribro) tenui intra limitatus, etiam per rimas duas angustas lunares, una prope marginem valvae altera prope raphosternum posita, cum lumine cellulae conjunctus, extra per foramen circulare vel rimiforme apertus. Foramina in seriebus duabus in utroque latere valvae igitur disposita; est proprietas haec, quae nomen generis mihi subjecit. Raphe recta vel biarcuata in raphosterno lato centraliter vel lateraliter posita,

extremis centralibus externis rectis expansis, extremis centralibus internis rectis helictoglossam duplicem facientibus; helictoglossa apicalis angusta; fissurae terminales recurvatae, extremis cum foraminibus marginalibus coordinatis. Cinctura ex taeniis apertis porosis constans; pars dorsalis cincturae interdum, fortasse semper, latior quam pars ventralis, ut in *Amphora*.

Holotypus: *Biremis ambigua* (Cleve) D.G. Mann, *comb. nov.*

Pinnularia ambigua P.T. Cleve 1895. K. Svenska Vetensk.-Akad. Handl. 27(3): 94

Biremis? baculus (Hustedt) D.G. Mann, *comb. nov.*

Mastogloia baculus F. Hustedt 1933. Rabenhorsts Krypt. Fl. Deutschl. Österr. u. Schweiz, 7, 2(4): 564

Biremis digitus (A. Schmidt) D.G. Mann, *comb. nov.*

Amphora digitus A. Schmidt 1875. Atlas Diat.: taf. 26, fig. 30

Biremis fritschii (Salah) D.G. Mann, *comb. nov.*

Pinnularia fritschii M.M. Salah 1953. J. Roy. microsc. Soc. 72: 166

Biremis ridicula (Giffen) D.G Mann, *comb. nov.*

Amphora ridicula M.H. Giffen 1976. Bot. Mar. 19: 382

BLEAKELEYA F.E. Round, *gen. nov.*

Cellulae coloniales ad polos expansos frontibus complanatis conjunctae, catenas planas vel tortas facientes, igitur aspectu cincturae plerumque visae. Chromatophora numerosa parva graniformia. Valvae anguste lineares, polo uno leviter dilatato, polo altero angusto curvato, scapo ad centrum interdum expanso. Frons plana ad polum basalem aliquantum depressa. Sternum angustum vel fere indistinctum prope basin valvae transtro exoriens. Striae uniseriatae transapicales, ad polos exceptae ubi radiales, in limbum prolongatae etiam ad polum basalem. Areolae circulares, ellipticae vel subquadratae, simplices. Junctura frontis cum limbo praeter ad polum basalem spinis apicem angustam versus flexis munita; limbus granulis instructus. Rimoportula unica ad polum angustum apicalem in latere uno valvae prope sternum posita, apertura externa simplici circulari. Taeniae numerosae seriebus pororum elongatorum instructae.

Holotypus: *Bleakeleya notata* (Grunow) F.E. Round,

comb. nov.

Asterionella bleakeleyi var. *notata* A. Grunow 1867,
 Hedwigia 6:2

BRACHYSIRA: new combinations

Brachysira atacamae (Hustedt) D.G. Mann, *comb.*
 nov.

   Navicula atacamae F. Hustedt 1927. Arch.
    Hydrobiol. 18: 243

Brachysira longirostris (Hustedt) D.G. Mann, *comb.*
 nov.

   Anomoeoneis longirostris F. Hustedt 1942. Int. Rev.
    ges. Hydrobiol. 42: 48

Brachysira rhomboides (Hustedt) D.G. Mann, *comb.*
 nov.

   Anomoeoneis rhomboides F. Hustedt 1942. Int. Rev.
    ges. Hydrobiol. 42: 48

CAVINULA D.G. Mann & A.J. Stickle, *gen. nov.*
   Cellulae solitariae, aspectu valvae plerumque
    visae. Chromatophora 1 vel 2, si 2 uno
    utrumque polum versus; utrumque ex laminis
    duabus ad valvas appressis isthmo plusminusve
    centrali connexis constans, isthmo lato
    pyrenoidem conspicuum continenti; margines
    chromatophororum multum lobati vel solum
    secus raphem indentati. Valvae ellipticae vel
    lanceolatae raro fere lineares, polis obtusis
    rotundatis. Frons plana; limbus plerumque non
    profundus, ad polos non abrupte diminutus,
    margine incrassato. Striae uniseriatae radiatae
    saepe arcte dispositae. Areolae simplices
    circulares vel ellipticae, intra hymenibus
    occlusae, hymenibus secus strias confluentibus.
    Raphosternum centrale, ad centrum interdum
    expansum. Extrema externa centralia raphis
    recta expansa; extrema interna recta non
    expansa vel T-formia, interdum in papillas vel
    cristas minutissimas transientia. Fissurae
    terminales nullae vel a se ex adverso flexae.
    Cinctura non profunda. Cingulum ex paucis
    taeniis apertis angustis porosis constans,
    unaquaque seriebus duabus pororum parvorum
    circularium instructa.
   Structura et forma areolarum *Cosmioneidis* similis,
    a qua fissuris terminalibus a se ex adverso
    flexis, limbo prope polos non diminuto,
    conformatione et dispositione
    chromatophororum, etc. distinguenda. Ab
    Aneumasto, cuius chromatophora eis *Cavinulae*
    similia sunt, differt structura areolarum

raphisque cingulique.

Holotypus: *Cavinula cocconeiformis* (Gregory ex
 Greville) D.G. Mann & A.J. Stickle, *comb. nov.*

   Navicula cocconeiformis W. Gregory 1855. Ex R.K.
    Greville, Ann. Mag. Nat. Hist. 2nd Ser., 15: 256

Cavinula jaernefeltii (Hustedt) D.G. Mann & A.J.
 Stickle, *comb. nov.*

   Navicula jaernefeltii F. Hustedt 1942. Arch.
    Hydrobiol. 39: 111

Cavinula lacustris (Gregory) D.G. Mann & A.J.
 Stickle, *comb. nov.*

   Navicula lacustris W. Gregory 1856. Quart. J.
    microsc. Sci. 4: 6

Cavinula pseudoscutiformis (Hustedt) D.G. Mann &
 A.J. Stickle, *comb. nov.*

   Navicula pseudoscutiformis F. Hustedt 1930. Bacill.
    (Pascher SüsswasserFl.): 291

Cavinula scutiformis (Grunow ex A. Schmidt) D.G.
 Mann & A.J. Stickle, *comb. nov.*

   Navicula scutiformis A. Grunow 1881. Ex A.
    Schmidt, Atlas. Diat.: taf. 70, fig. 62

Cavinula variostriata (Krasske) D.G. Mann, *comb.*
 nov.

   Navicula variostriata G. Krasske 1923. Bot. Archiv.
    3: 197

COSMIONEIS D.G. Mann & A.J. Stickle, *gen. nov.*
   Cellulae solitariae profundae. Chromatophora 2
    laminiformia secus raphem indentata, uno
    eorum ad utramque valvam appresso. Nucleus
    in statu vegetativo centralis, in statu dividenti
    latus cellulae idem ac in divisione proxima non
    semper occupans, frustulis et cis- et trans-
    conformatis igitur factis. Valvae lineares vel
    lanceolatae, polis saepe rostellatis vel capitatis.
    Frons plana; limbus profundus sed prope polos
    abrupte deminutus. Striae uniseriatae. Areolae
    simplices circulares, intra hymenibus occlusae,
    hymenibus secus strias saepe confluentibus.
    Raphosternum centrale, centrum versus
    aliquantum expansum. Extrema centralia
    externa raphis valde expansa, poriformia;
    extrema interna T-formia. Fissurae terminales
    unciformes latus secundum versus, supra
    helictoglossas dilatatae. Cingulum ex taeniis
    apertis multis inter se plusminusve similaribus
    constans, unaquaque serie transversali pororum
    circularium parvorum instructa; centrum
    taeniae primae (valvocopulae) et taeniae
    secundae foramine circulari perforatum.

Cosmioneis Petroneidis primo adspectu similis est, sed ab hac differt moribus protoplasti, velis areolarum, deminutionibus abruptis limbi profunditate cincturae, copia taeniarum, etc.

Holotypus: *Cosmioneis pusilla* (Wm. Smith) D.G. Mann & A.J. Stickle, *comb. nov.*

Navicula pusilla Wm. Smith 1853. Syn. Brit. Diat. I: 52

Cosmioneis delawarensis (Grunow ex Cleve) D.G. Mann, *comb. nov.*

Navicula delawarensis A. Grunow 1893. Ex P.T. Cleve, Diatomiste 2: 13

Cosmioneis? grossepunctata (Hustedt) D.G. Mann, *comb. nov.*

Navicula grossepunctata F. Hustedt 1944. Ber. dt. Bot. Ges. **61**: 271

Cosmioneis lundstroemii (Cleve in Cleve & Grunow) D.G. Mann, *comb. nov.*

Navicula lundstroemii P.T. Cleve 1880. In P.T. Cleve & A. Grunow, K. Svenska Vetensk.-Akad. Handl. **17**(2): 13

CRATICULA: new combinations

Craticula accomoda (Hustedt) D.G. Mann, *comb. nov.*

Navicula accomoda F. Hustedt 1950. Arch. Hydrobiol. **43**: 446

Craticula ambigua (Ehrenberg) D.G. Mann, *comb. nov.*

Navicula ambigua C.G. Ehrenberg 1843. Abh. Akad. Wiss. Berlin, **1841**: 417

Craticula cuspidata (Kützing) D.G. Mann, *comb. nov.*

Frustulia cuspidata F.T. Kützing 1833. Syn. Diat., Linnaea 8: 549

Craticula halophila (Grunow ex Van Heurck) D.G. Mann, *comb. nov.*

Navicula cuspidata var. *halophila* A. Grunow 1885. Ex H. van Heurck, Syn. Diat. Belg.: 100

Other species belonging to the genus:

Craticula perrotettii Grunow 1868

DIADESMIS: new combinations

Diadesmis aerophila (Krasske) D.G. Mann, *comb. nov.*

Navicula aerophila G. Krasske 1932. Hedwigia **72**: 111

Diadesmis brekkaensis (Petersen) D.G. Mann, *comb. nov.*

Navicula brekkaensis J.B. Petersen 1928. Bot. Iceland 2: 389

Diadesmis contenta (Grunow ex Van Heurck) D.G. Mann, *comb. nov.*

Navicula contenta A. Grunow 1885. Ex H. van Heurck, Syn. Diat. Belg.: 109

Diadesmis laevissima (Cleve) D.G. Mann, *comb. nov.*

Fragilaria laevissima P.T. Cleve 1898. Bih. Kongl. Svenska Vetensk.-Akad. Handl. **24**, 3(2): 9

Diadesmis perpusilla (Grunow) D.G. Mann, *comb. nov.*

Navicula perpusilla A. Grunow 1860. Verh. zool.-bot. Ges. Wien 10: 552

Other species belonging to the genus:

Diademis confervacea Kützing 1844

D. gallica Wm. Smith 1857

ENCYONEMA: new combinations

Encyonema alpinum (Grunow) D.G. Mann, *comb. nov.*

Cymbella alpina A. Grunow 1863. Verh. zool.-bot. Ges. Wien 13: 148

Encyonema braunii (Hustedt.) D.G. Mann, *comb. nov.*

Cymbella braunii F. Hustedt 1955. Abh. naturw. Ver. Bremen 34: 57

Encyonema brehmii (Hustedt) D.G. Mann, *comb. nov.*

Cymbella brehmii F. Hustedt 1912. Arch. Hydrobiol. 7: 695

Encyonema elginense (Krammer) D.G. Mann, *comb. nov.*

Cymbella elginensis K. Krammer 1981. Bacillaria 4: 136

Encyonema formosum (Hustedt) D.G. Mann, *comb. nov.*

Cymbella formosa F. Hustedt 1955. Abh. naturw. Ver. Bremen 34: 57

Encyonema grossestriatum (O. Müller) D.G. Mann, *comb. nov.*

Cymbella grossestriata O. Müller 1905. Englers bot. Jahrb. 36: 154

Encyonema javanicum (Hustedt) D.G. Mann, *comb. nov.*

Cymbella javanica F. Hustedt 1938. Arch. Hydrobiol., Suppl. 15: 424

Encyonema juriljii (Hustedt) D.G. Mann, *comb. nov.*

Cymbella juriljii F. Hustedt 1955. Abh. naturw. Ver. Bremen 34: 55

Encyonema lacustre (Agardh) D.G. Mann, *comb. nov.*

Schizonema lacustre C.A. Agardh 1824. Syst.Alg.: 10

Encyonema latens (Krasske) D.G. Mann, *comb. nov.*

Cymbella latens G. Krasske 1937. Arch. Hydrobiol. 31: 43

Encyonema mesianum (Cholnoky) D.G. Mann, *comb. nov.*

Cymbella mesiana B.J. Cholnoky 1955.

Hydrobiologia 7: 160

Encyonema minutum (Hilse in Rabenhorst) D.G. Mann, *comb. nov.*

 Cymbella minuta W. Hilse 1862. In L. Rabenhorst, Algen Europ.: 1261

Encyonema muelleri (Hustedt) D.G. Mann, *comb. nov.*

 Cymbella muelleri F. Hustedt 1938. Arch. Hydrobiol. Suppl. 15: 425

Encyonema obscurum (Krasske) D.G. Mann, *comb. nov.*

 Cymbella obscura G. Krasske 1938. Arch. Hydrobiol. 33: 531

Encyonema paucistriatum (Cleve-Euler) D.G. Mann, *comb. nov.*

 Cymbella paucistriata A. Cleve-Euler 1934. Soc. Sci. Fenn. Comm. Biol. 4(14): 77

Encyonema perpusillum (A. Cleve) D.G. Mann, *comb. nov.*

 Cymbella perpusilla A. Cleve 1895. Bih. Kongl. Svenska Vetensk.-Akad. Handl. 21, 3(1): 19

Encyonema reichardtii (Krammer) D.G. Mann, *comb. nov.*

 Cymbella reichardtii K. Krammer 1985. Biblioth. Diat. 9: 32

Encyonema rugosum (Hustedt) D.G. Mann, *comb. nov.*

 Cymbella rugosa F. Hustedt 1955. Abh. naturw. Ver. Bremen 34: 67

Encyonema silesiacum (Bleisch in Rabenhorst) D.G. Mann, *comb. nov.*

 Cymbella silesiaca M. Bleisch 1864. In L. Rabenhorst, Algen Europ.: 1802

Encyonema spiculum (Hustedt) D.G. Mann, *comb. nov.*

 Cymbella spicula F. Hustedt 1938. Arch. Hydrobiol., Suppl. 15: 422

Encyonema subalpinum (Hustedt) D.G. Mann, *comb. nov.*

 Cymbella subalpina F. Hustedt 1942. Int. Rev. ges. Hydrobiol. 42: 98

Encyonema subturgidum (Hustedt) D.G. Mann, *comb. nov.*

 Cymbella subturgida F. Hustedt 1942. Int. Rev. ges. Hydrobiol. 42: 105

Encyonema tenuissimum (Hustedt) D.G. Mann, *comb. nov.*

 Cymbella tenuissima F. Hustedt 1942. Int. Rev. ges. Hydrobiol. 42: 100

Other species belonging to the genus include:

Encyonema caespitosum Kützing 1849

E. gracile Ehrenberg 1841

E. hebridicum Grunow ex Cleve 1891

E. lunatum (Wm. Smith in Greville) Van Heurck 1896

E. prostratum (Berkeley) Kützing 1844

E. triangulum (Ehrenberg) Kützing 1849

E. turgidum (Gregory) Grunow ex A. Schmidt 1875

FALLACIA A.J. Stickle & D.G. Mann, *gen. nov.*

Cellulae solitariae, aspectu valvae plerumque visae. Chromatophorum unicum ex laminis duabus ad cincturam appressis isthmo angusto connexis constans; isthmus ad epivalvam positus extensionibus lateralibus angustis interdum instructus; margines distales chromatophori extensionibus similaribus interdum instructi. Pyrenoides 1-2 invaginatae. Nucleus in statu vegetativo saepe, in statu dividenti semper eccentrice positus et tum latus idem ac in divisione proxima nunquam occupans; mores protoplasti ita oscillantes. Valvae lineares, lanceolatae vel ellipticae, polis plerumque obtuse rotundatis vel leviter apiculatis. Frons plana; limbus non profundus. Striae uniseriatae vel partim biseriatae, sternis lateralibus infra planum frontis impressis interruptae, extra conopeis subtiliter porosis ex parte vel omnino tectae. Areolae circulares vel ellipticae, hymenibus occlusae. Conopea libera vel cum valva ita consociata ut conopei super valvam cutis externa componant; margines conopeorum integrae vel extensionibus digitiformibus instructae. Canales inter conopea et valvam formati, secus margines omnes conopeorum vel solum ad polos vel solum per poros minutos conopeorum ipsorum extrinsecus aperti. Raphosternum plerumque angustum. Extrema centralia interna raphis latus primum versus flexa; extrema externa recta expansa. Fissurae terminales flexae, reflexae vel uncinatae. Cingulum ex taeniis apertis constans, quarum taenia prima latissima est.

Fallax est, quod valvae *Fallaciae* earum *Lyrellae* primo adspectu maxime similes sunt. Sed a *Lyrella Fallacia* forma chromatophori, moribus protoplasti, structura areolae, praesentia conopeorum porosorum - breviter fere omnino distinguenda. A *Sellaphora* praecipue differt porosis conopeis.

Holotypus: *Fallacia pygmaea* (Kützing) A.J. Stickle &
D.G. Mann, *comb. nov.*
　Navicula pygmaea F.T. Kützing 1849. Spec. Alg.:
　77

Fallacia aequorea (Hustedt) D.G. Mann, *comb. nov.*
　Navicula aequorea F. Hustedt 1939. Abh. naturw.
　Ver. Bremen 31: 621

Fallacia amphipleuroides (Hustedt) D.G. Mann, *comb.
nov.*
　Navicula amphipleuroides F. Hustedt 1955. Bull.
　Duke Univ. Mar. Stn., 6: 30

Fallacia auriculata (Hustedt) D.G. Mann, *comb. nov.*
　Navicula auriculata F. Hustedt 1944. Ber. dt. Bot.
　Ges. 61: 273

Fallacia bioculata (Grunow ex A. Schmidt) D.G.
Mann, *comb. nov.*
　Navicula bioculata A. Grunow 1881. Ex A. Schmidt,
　Atlas Diat.: taf. 70, fig. 9

Fallacia brachium (Hustedt) D.G. Mann, *comb. nov.*
　Navicula brachium F. Hustedt 1944. Ber. dt. Bot.
　Ges. 61: 272

Fallacia? clipeiformis (König) D.G. Mann, *comb. nov.*
　Navicula clipeiformis D. König 1959. Z. dt. geol.
　Ges. 111: 51

Fallacia cryptolyra (Brockmann) A.J. Stickle & D.G.
Mann, *comb. nov.*
　Navicula cryptolyra C. Brockmann 1950. Abh.
　Senckenb. naturf. Ges. 478: 19

Fallacia diploneoides (Hustedt) D.G. Mann, *comb.
nov.*
　Navicula diploneoides F. Hustedt 1955. Bull. Duke
　Univ. Mar. Stn., 6: 22

Fallacia? dithmarsica (König in Hustedt) D.G. Mann,
comb. nov.
　Navicula dithmarsica D. König 1964. In F. Hustedt,
　Rabenhorsts Krypt.-Fl. Deutschl. Österr. u.
　Schweiz 7, 3(3): 550

Fallacia egregia (Hustedt) D.G. Mann, *comb. nov.*
　Navicula egregia F. Hustedt 1942. Ber. dt. Bot. Ges.
　60: 64

Fallacia fenestrella (Hustedt) D.G. Mann, *comb. nov.*
　Navicula fenestrella F. Hustedt 1955. Bull. Duke
　Univ. Mar. Stn. 6: 30

Fallacia forcipata (Greville) A.J. Stickle & D.G.
Mann, *comb. nov.*
　Navicula forcipata R.K. Greville 1859. Quart. J.
　microsc. Sci. 7: 83

Fallacia fracta (Hustedt ex Simonsen) D.G. Mann,
comb. nov.

Navicula fracta F. Hustedt 1987. Ex R. Simonsen,
　Atlas Cat. Diat. Types Fr. Hustedt: 474

Fallacia frustulum (Hustedt) D.G. Mann, *comb. nov.*
　Navicula frustulum F. Hustedt 1945. Arch.
　Hydrobiol. 40: 921

Fallacia? gemmifera (Simonsen) D.G. Mann, *comb.
nov.*
　Navicula gemmifera R. Simonsen 1960. Kiel.
　Meeresf. 16: 128

Fallacia helensis (Schulz) D.G. Mann, *comb. nov.*
　Navicula helensis P. Schulz 1926. Bot. Arch. 13: 217

Fallacia hudsonis (Grunow ex Cleve) A.J. Stickle &
D.G. Mann, *comb. nov.*
　Navicula hudsonis A. Grunow 1891. Ex P.T. Cleve,
　Diatomiste 1: 77

Fallacia hummii (Hustedt) D.G. Mann, *comb. nov.*
　Navicula hummii F. Hustedt 1955. Bull. Duke
　Univ. Mar. Stn. 6: 23

Fallacia hyalinula (De Toni) A.J. Stickle & D.G.
Mann, *comb. nov.*
　Navicula hyalinula J.B. De Toni 1891. Syll. Bacill.:
　92

Fallacia? inattingens (Simonsen) D.G. Mann, *comb.
nov.*
　Navicula inattingens R. Simonsen 1959. Kiel.
　Meeresf. 15: 78

Fallacia indifferens (Hustedt) D.G. Mann, *comb. nov.*
　Navicula indifferens F. Hustedt 1942. Ber. dt. Bot.
　Ges. 60: 67

Fallacia insociabilis (Krasske) D.G. Mann, *comb. nov.*
　Navicula insociabilis G. Krasske 1932. Hedwigia 72:
　114

Fallacia litoricola (Hustedt) D.G. Mann, *comb. nov.*
　Navicula litoricola F. Hustedt 1955. Bull. Duke
　Univ. Mar. Stn. 6: 23

Fallacia lucens (Hustedt ex Salah) D.G. Mann, *comb.
nov.*
　Navicula lucens F. Hustedt 1953. Ex M.M. Salah
　1953. J. Roy. microsc. Soc. 72: 163

Fallacia lucinensis (Hustedt) D.G. Mann, *comb. nov.*
　Navicula lucinensis F. Hustedt 1950. Arch.
　Hydrobiol. 43: 350

Fallacia mitis (Hustedt) D.G. Mann, *comb. nov.*
　Navicula mitis F. Hustedt 1945. Arch. Hydrobiol.
　40: 919

Fallacia monoculata (Hustedt) D.G. Mann, *comb. nov.*
　Navicula monoculata F. Hustedt 1945. Arch.
　Hydrobiol. 40: 921

Fallacia naumannii (Hustedt) D.G. Mann, *comb. nov.*

Navicula naumannii F. Hustedt 1942. Arch.
Hydrobiol. 39: 115

Fallacia nummularia (Greville) D.G. Mann, *comb.
nov.*
Navicula nummularia R.K. Greville 1859. Trans.
Bot. Soc. Edinb. 6: 249

Fallacia ny (Cleve) D.G. Mann, *comb. nov.*
Navicula ny P.T. Cleve 1894. K. Svenska Vetensk.-
Akad Handl., Ny Foljd 26(2): 75

Fallacia nyella (Hustedt ex Simonsen) D.G. Mann,
comb. nov.
Navicula nyella F. Hustedt 1987. Ex R. Simonsen,
Atlas Cat. Diat. Types Fr. Hustedt: 490

Fallacia oculiformis (Hustedt) D.G. Mann, *comb. nov.*
Navicula oculiformis F. Hustedt 1955. Bull. Duke
Univ. Mar. Stn. 6: 22

Fallacia omissa (Hustedt) D.G. Mann, *comb. nov.*
Navicula omissa F. Hustedt 1945. Arch. Hydrobiol.
40: 918

Fallacia perlucida (Hustedt) D.G. Mann, *comb. nov.*
Navicula perlucida F. Hustedt 1937. Arch.
Hydrobiol., Suppl. 15: 250

Fallacia pseudoforcipata (Hustedt) D.G. Mann, *comb.
nov.*
Navicula pseudoforcipata F. Hustedt 1942. Abh.
naturw. Ver. Bremen 32: 199

Fallacia pseudomitis (Hustedt) D.G. Mann, *comb. nov.*
Navicula pseudomitis F. Hustedt 1950. Arch.
Hydrobiol., 43: 352

Fallacia pseudomuralis (Hustedt) D.G. Mann, *comb.
nov.*
Navicula pseudomuralis F. Hustedt 1937. Arch.
Hydrobiol., Suppl. 15: 245

Fallacia pseudony (Hustedt) D.G. Mann, *comb. nov.*
Navicula pseudony F. Hustedt 1955. Bull. Duke
Univ. Mar. Stn. 6: 23

Fallacia? pseudosemilyrata (Simonsen) D.G. Mann,
comb. nov.
Navicula pseudosemilyrata R. Simonsen 1959. Kiel.
Meeresf. 15: 78

Fallacia schaeferae (Hustedt) D.G. Mann, *comb. nov.*
Navicula schaeferae F. Hustedt 1964. Rabenhorsts
Krypt.-Fl. Deutschl. Österr. u. Schweiz, 7, 3(3):
545

Fallacia? semilyrata (Simonsen) D.G. Mann, *comb.
nov.*
Navicula semilyrata R. Simonsen 1959. Kiel.
Meeresf. 15: 78

Fallacia solutepunctata (Hustedt) D.G. Mann, *comb.
nov.*
Navicula solutepunctata F. Hustedt 1939. Abh,
Naturw. Ver. Bremen 31:638

Fallacia standeriella (Archibald) D.G. Mann, *comb.
nov.*
Navicula standeriella R.E.M. Archibald 1966. Nova
Hedwigia, Beih. 21: 262

Fallacia subforcipata (Hustedt) D.G. Mann, *comb.
nov.*
Navicula subforcipata F. Hustedt 1964. In
Rabenhorsts Krypt.-Fl. Deutschl., Österr. u.
Schweiz 7, 3(3): 533

Fallacia subhamulata (Grunow in Van Heurck) D.G.
Mann, *comb. nov.*
Navicula subhamulata A. Grunow 1880. In H. van
Heurck, Syn. Diat. Belg.: Pl.13

Fallacia sublucidula (Hustedt) D.G. Mann, *comb. nov.*
Navicula sublucidula F. Hustedt 1950. Arch.
Hydrobiol. 43: 354

Fallacia submitis (Hustedt) D.G. Mann, *comb. nov.*
Navicula submitis F. Hustedt 1945. Arch.
Hydrobiol. 40: 919

Fallacia subsulcatoides (Hustedt ex Simonsen) D.G.
Mann, *comb. nov.*
Navicula subsulcatoides F. Hustedt 1987. Ex R.
Simonsen, Atlas Cat. Diat. Types Fr. Hustedt:
460

Fallacia tenera (Hustedt) D.G. Mann, *comb. nov.*
Navicula tenera F. Hustedt 1937. Arch. Hydrobiol.,
Suppl.15: 259

Fallacia teneroides (Hustedt) D.G. Mann, *comb. nov.*
Navicula teneroides F. Hustedt 1956. Ergebn. dtsch.
limnol. Venezuela Exped. 1: 117

Fallacia tenerrima (Hustedt) D.G. Mann, *comb. nov.*
Navicula tenerrima F. Hustedt 1937. Arch.
Hydrobiol. Suppl. 15: 247

Fallacia umpatica (Cholnoky) D.G. Mann, *comb. nov.*
Navicula umpatica B.J. Cholnoky 1968. Bot. Mar.,
Suppl. 11: 65

Fallacia versicolor (Grunow) D.G. Mann, *comb. nov.*
Navicula versicolor A. Grunow 1874. Jahresb.
Komm. Unters. dtsch. Meere in Kiel 2: 89

Fallacia veterana (Hustedt ex Simonsen) D.G. Mann,
comb. nov.
Navicula veterana F. Hustedt 1987. Ex R.
Simonsen, Atlas Cat. Diat. Types Fr. Hustedt:
491

Fallacia vitrea (Østrup) D.G. Mann, *comb. nov.*
Frustulia vitrea E. Østrup 1901. Freshwat. Diat.

Faeroes Bot., Faeroes 1: 262

Fallacia vittata (Cleve) D.G. Mann, *comb. nov.*

Diploneis bioculata var. *vittata* P.T. Cleve 1894. K. Svenska Vetensk.-Akad. Handl., Ny Foljd 26(2): 80

Fallacia zonata (Hustedt) D.G. Mann, *comb. nov.*

Navicula zonata F. Hustedt 1956. Ergebn. dt. limnol. Venezuela-Exped. 1: 116

LUTICOLA D.G. Mann, *gen. nov.*

Cellulae solitariae, aspectu valvae plerumque visae. Chromatophorum unicum, centro eius pyrenoidem conspicuam continenti et ad cincturam appresso, lobis duobus sub valvas inde extensis; uterque lobus secus raphem indentatus. Valvae lineares, lanceolatae vel ellipticae, polis obtuse rotundatis vel capitatis. Frons plana a limbo non profundo plusminusve distincta; junctura frontis cum limbo raro spinifera. Margo in unoquoque quadrante valvae ad positionem medium inter centrum et polum incisus. Striae uniseriatae, unaquaque in fronte ex areolis pluribus circularibus et in limbo ex areola unica circulari constanti. Areolae frontis intus hymenibus intra strias confluentibus occlusae, hymenibus lamellas inter costas transapicales facientibus; areolae limborum hymenibus suis instructae. Ad juncturam frontis cum limbo est canalis longitudinalis in substantia valvae, canali *Neidii* aliquantum similis. In latere valvae stigmati opposito, stratum interius canalis structuram linguiformem vel laciniiformem centrum versus extensam interdum faciens. Raphosternum mediocriter angustum, ad centrum expansum et incrassatum staurum brevem formans stauro in latere uno stigmate unico penetrato. Apertura interna stigmatis curvata rimiformis labiata; apertura externa porus simplex. Extrema centralia interna raphis recta, simplicia vel leviter labiata; extrema centralia externa forma varia: obstipa, curvata vel a latere stigmatis abrupte flexa. Fissurae terminales ab extremibus centralibus plerumque ex adverso flexae. Cinctura ex taeniis apertis constans, unaquaque seriebus 1–2 pororum minutorum circularium instructa, parte interiori taeniae primae (valvocopulae) serie projecturarum grossarum.

Holotypus: *Luticola mutica* (Kützing) D.G. Mann,

comb. nov.

Navicula mutica F.T. Kützing 1844. Kies. Bacill.: 93

Luticola argutula (Hustedt) D.G. Mann, *comb. nov.*

Navicula argutula F. Hustedt 1955. Ber. dt. Bot. Ges. 68: 126

Luticola cohnii (Hilse) D.G. Mann, *comb. nov.*

Stauroneis cohnii W. Hilse 1860. Jahresb. schlesischen Ges. Vaterl. Kult. 38: 83

Luticola dapaliformis (Hustedt) D.G. Mann, *comb. nov.*

Navicula dapaliformis F. Hustedt 1966. Rabenhorsts Krypt.-Fl. Deutschl. Österr. u. Schweiz, 7, 3(4): 605

Luticola dapalis (Frenguelli) D.G. Mann, *comb. nov.*

Navicula dapalis J. Frenguelli 1941. Revta. Museo La Plata (N.S.) 3 Bot.: 248

Luticola dismutica (Hustedt) D.G. Mann, *comb. nov.*

Navicula dismutica F. Hustedt 1966. Rabenhorsts Krypt.-Fl. Deutschl. Österr. u. Schweiz, 7, 3(4): 595

Luticola gaussii (Heiden in Heiden & Kolbe) D.G. Mann, *comb. nov.*

Navicula muticopsis var. *gaussii* H. Heiden 1928. In H. Heiden & R.W. Kolbe, Deutsche Südpol.-Exped. 1901–1903, 8(5): 623

Luticola goeppertiana (Bleisch in Rabenhorst) D.G. Mann, *comb. nov.*

Navicula mutica var. *goeppertiana* M. Bleisch 1861. In L. Rabenhorst, Alg. Europ.: 1183

Luticola heufleriana (Grunow) D.G. Mann, *comb. nov.*

Stauroneis heufleriana A. Grunow 1863. Verh. zool.-bot. Ges. Wien 13: 155

Luticola incoacta (Hustedt ex Simonsen) D.G. Mann, *comb. nov.*

Navicula incoacta F. Hustedt 1987. Ex R. Simonsen, Atlas Cat. Diat. Types Fr. Hustedt: 499

Luticola inserata (Hustedt) D.G. Mann, *comb. nov.*

Navicula inserata F. Hustedt 1955. Ber. dt. Bot. Ges. 68: 125

Luticola lacertosa (Hustedt) D.G. Mann, *comb. nov.*

Navicula lacertosa F. Hustedt 1955. Ber. dt. Bot. Ges. 68: 123

Luticola lagerheimii (Cleve) D.G. Mann, *comb. nov.*

Navicula lagerheimii P.T. Cleve 1894. K. Svenska Vetensk.-Akad. Handl. 26(2): 131

Luticola mitigata (Hustedt) D.G. Mann, *comb. nov.*

Navicula mitigata F. Hustedt 1966. Rabenhorsts Krypt.-Fl. Deutschl. Österr. u. Schweiz, 7, 3(4):

591

Luticola monita (Hustedt) D.G. Mann, *comb. nov.*
  Navicula monita F. Hustedt 1966. Rabenhorsts
    Krypt.-Fl. Deutschl. Österr. u. Schweiz. 7. 3(4):
    590

Luticola murrayi (W. & G.S. West) D.G. Mann, *comb. nov.*
  Navicula murrayi W. & G.S. West 1911. Brit.
    Antarct. Exped. (1907-1909). Rep. Scient. Inv.1:
    285

Luticola muticoides (Hustedt) D.G. Mann, *comb. nov.*
  Navicula muticoides F. Hustedt 1949. Süssw.- Diat.
    Expl. Parc Nat. Alb., Miss. Damas: 82

Luticola muticopsis (Van Heurck) D.G. Mann, *comb. nov.*
  Navicula muticopsis H. van Heurck 1909. Expéd.
    Antarct. Belg., Diat.: 12

Luticola nivalis (Ehrenberg) D.G. Mann, *comb. nov.*
  Navicula nivalis C.G. Ehrenberg 1853. Ber. Akad.
    Wiss. Berlin 1853: 528

Luticola obligata (Hustedt) D.G. Mann, *comb. nov.*
  Navicula obligata F. Hustedt 1966. Rabenhorsts
    Krypt.-Fl. Deutschl. Österr. u. Schweiz 7. 3(4):
    592

Luticola palaearctica (Hustedt ex Simonsen) D.G.
    Mann, *comb. nov.*
  Navicula palaearctica F. Hustedt 1987. Ex R.
    Simonsen, Atlas Cat. Diat. Types Fr. Hustedt:
    499

Luticola paramutica (Bock) D.G. Mann, *comb. nov.*
  Navicula paramutica W. Bock 1963. Nova Hedwigia
    5: 237

Luticola peguana (Grunow in Cleve & Möller) D.G.
    Mann, *comb. nov.*
  Navicula mutica var. *peguana* A. Grunow 1879. In
    P.T. Cleve & J.D. Möller, Diat.: 188

Luticola plausibilis (Hustedt ex Simonsen) D.G.
    Mann, *comb. nov.*
  Navicula plausibilis F. Hustedt 1987. Ex R.
    Simonsen, Atlas Cat. Diat. Types Fr. Hustedt:
    498

Luticola pseudomutica (Hustedt) D.G. Mann, *comb. nov.*
  Navicula pseudomutica F. Hustedt 1955. Ber. dt.
    Bot. Ges. 68: 126

Luticola saxophila (Bock ex Hustedt) D.G. Mann,
    comb. nov.
  Navicula saxophila W. Bock 1966. In F. Hustedt,
    Rabenhorsts Krypt.-Fl. Deutschl. Österr. u.

Schweiz 7. 3(4): 599

Luticola seposita (Hustedt) D.G. Mann, *comb. nov.*
  Navicula seposita F. Hustedt 1942. Int. Rev. ges.
    Hydrobiol. **42**: 54

Luticola undulata (Hilse in Rabenhorst) D.G. Mann,
    comb. nov.
  Stauroneis undulata W. Hilse 1860. In L.
    Rabenhorst, Alg. Sachs.: 963

Luticola ventricosa (Kützing) D.G. Mann, *comb. nov.*
  Stauroneis ventricosa F.T. Kützing 1844. Kies.
    Bacill.: 105

Luticola voigtii (Meister) D.G. Mann, *comb. nov.*
  Navicula voigtii F. Meister 1932. Kiesel. Asien: 38

LYRELLA: new combinations

Lyrella abrupta (Gregory) D.G. Mann, *comb. nov.*
  Navicula lyra var. *abrupta* W. Gregory 1857. Trans.
    Roy. Soc. Edinb. 21: 486

Lyrella abruptoides (Hustedt) D.G. Mann, *comb. nov.*
  Navicula abruptoides F. Hustedt 1964. Rabenhorsts
    Krypt.-Fl. Deutschl. Österr. u. Schweiz 7. 3(3):
    515

Lyrella approximata (Greville) D.G. Mann, *comb. nov.*
  Navicula approximata R.K. Greville 1859. Trans.
    Bot. Soc. Edinb. 6: 247

Lyrella approximatella (Hustedt) D.G. Mann, *comb. nov.*
  Navicula approximatella F. Hustedt 1964.
    Rabenhorsts Krypt.-Fl. Deutschl. Österr. u.
    Schweiz 7. 3(3): 428

Lyrella approximatoides (Hustedt ex Simonsen) D.G.
    Mann, *comb. nov.*
  Navicula approximatoides F. Hustedt 1987. Ex R.
    Simonsen, Atlas Cat. Diat. Types Fr. Hustedt:
    484

Lyrella atlantica (A. Schmidt) D.G. Mann, *comb. nov.*
  Navicula lyra var. *atlantica* A. Schmidt 1874. Atlas
    Diat.: taf. 1, fig. 34

Lyrella baldjickiensis (Heiden in A. Schmidt) D.G.
    Mann, *comb. nov.*
  Navicula baldjickiensis H. Heiden 1905. In A.
    Schmidt, Atlas Diat.: taf. 258, figs 2-4

Lyrella barbara (Heiden in Heiden & Kolbe) D.G.
    Mann, *comb. nov.*
  Navicula barbara H. Heiden 1928. In H. Heiden &
    R.W. Kolbe, Deutsche Südpol.-Exped. 1901–
    1903, 8(5): 617

Lyrella barbitos (A. Schmidt) D.G. Mann, *comb. nov.*
  Navicula barbitos A. Schmidt 1888. Atlas Diat.:
    taf. 129, fig. 5

Lyrella californica (Greville) D.G. Mann, *comb. nov.*
 Navicula californica R.K. Greville 1859. Trans. Bot.
 Soc. Edinb. **6**: 248

Lyrella circumsecta (Grunow ex A.Schmidt) D.G.
 Mann, *comb. nov.*
 Navicula polysticta var. *circumsecta* A. Grunow
 1874. Ex A. Schmidt, Atlas Diat.: taf. 3, figs 27–
 28

Lyrella clavata (Gregory) D.G. Mann, *comb. nov.*
 Navicula clavata W. Gregory 1856. Trans. microsc.
 Soc. Lond. **4**: 46

Lyrella concilians (Cleve) D.G. Mann, *comb. nov.*
 Navicula concilians P.T. Cleve 1895. K. Svenska
 Vetensk.-Akad. Handl. **27**(3): 54

Lyrella copiosa (A. Schmidt) D.G. Mann, *comb. nov.*
 Navicula copiosa A. Schmidt 1888. Atlas Diat.:
 taf. 129, fig. 6

Lyrella crebra (Hustedt) D.G. Mann, *comb. nov.*
 Navicula crebra F. Hustedt 1964. Rabenhorsts
 Krypt.-Fl. Deutschl. Österr. u. Schweiz 7, 3(3):
 385

Lyrella diffluens (A. Schmidt) D.G. Mann, *comb. nov.*
 Navicula diffluens A. Schmidt 1874. Atlas Diat.: taf.
 2, fig. 15

Lyrella discreta (Hustedt) D.G. Mann, *comb. nov.*
 Navicula discreta F. Hustedt 1964. Rabenhorsts
 Krypt.-Fl. Deutschl. Österr. u. Schweiz 7, 3(3):
 496

Lyrella durandii (Kitton) D.G. Mann, *comb. nov.*
 Navicula durandii F. Kitton 1885. J. Roy. microsc.
 Soc., ser. 2, **5**: 1042

Lyrella excavata (Greville) D.G. Mann, *comb. nov.*
 Navicula excavata R.K. Greville 1866. Trans.
 microsc. Soc. Lond., N.S. **14**: 130

Lyrella exsul (A. Schmidt) D.G. Mann, *comb. nov.*
 Navicula exsul A. Schmidt 1874. Atlas Diat.: taf. 2,
 fig. 13

Lyrella genifera (A. Schmidt) D.G. Mann, *comb. nov.*
 Navicula genifera A. Schmidt 1874. Atlas Diat.: taf.
 2, fig. 8

Lyrella hennedyi (Wm. Smith) A.J. Stickle & D.G.
 Mann, *comb. nov.*
 Navicula hennedyi Wm. Smith 1856. Syn. Brit. Diat.
 II: 93

Lyrella hennedyella (Hustedt ex Simonsen) D.G.
 Mann, *comb. nov.*
 Navicula hennedyella F. Hustedt 1987. Ex R.
 Simonsen, Atlas Cat. Diat. Types Fr. Hustedt:
 486

Lyrella hennedyoides (Hustedt) D.G. Mann, *comb.
 nov.*
 Navicula hennedyoides F. Hustedt 1964.
 Rabenhorsts Krypt.-Fl. Deutschl. Österr. u.
 Schweiz 7, 3(3): 396

Lyrella illustris (Pantocsek) D.G. Mann, *comb. nov.*
 Navicula illustris J. Pantocsek 1892. Beitr. Kenntn.
 foss. Bacill. Ungarns 3: taf.2, fig.17

Lyrella illustrioides (Hustedt) D.G. Mann, *comb. nov.*
 Navicula illustrioides F. Hustedt 1964. Rabenhorsts
 Krypt.-Fl. Deutschl. Österr. u. Schweiz 7, 3(3):
 408

Lyrella imitans (A.Mann) D.G. Mann, *comb. nov.*
 Navicula imitans A. Mann 1925. Bull. U. S. Natl.
 Mus. 100(6): 104

Lyrella inaurata (Hendey) D.G. Mann, *comb. nov.*
 Navicula inaurata N.I. Hendey 1958. J. Roy.
 microsc. Soc. 77: 64

Lyrella inhalata (A. Schmidt) D.G. Mann, *comb. nov.*
 Navicula inhalata A. Schmidt 1874. Atlas Diat.:
 taf. 2, fig. 30

Lyrella investigata (Heiden in A. Schmidt) D.G.
 Mann, *comb. nov.*
 Navicula investigata H. Heiden 1905. In A. Schmidt,
 Atlas Diat.: taf. 258, fig. 6

Lyrella irrorata (Greville) D.G. Mann, *comb. nov.*
 Navicula irrorata R.K. Greville 1859. Trans. Bot.
 Soc. Edinb. **6**: 246

Lyrella irroratoides (Hustedt) D.G. Mann, *comb. nov.*
 Navicula irroratoides F. Hustedt 1964. Rabenhorsts
 Krypt.-Fl. Deutschl. Österr. u. Schweiz 7, 3(3):
 437

Lyrella lyroides (Hendey) D.G. Mann, *comb. nov.*
 Navicula lyroides N.I. Hendey 1958. J. Roy.
 microsc. Soc. 77: 60

Lyrella mediopartita (Grove ex A. Schmidt) D.G.
 Mann, *comb. nov.*
 Navicula mediopartita E. Grove 1896. Ex A.
 Schmidt, Atlas Diat.: taf. 204, fig. 16

Lyrella miranda (Hustedt) D.G. Mann, *comb. nov.*
 Navicula miranda F. Hustedt 1964. Rabenhorsts
 Krypt.-Fl. Deutschl. Österr. u. Schweiz 7, 3(3):
 397

Lyrella nebulosa (Gregory) D.G. Mann, *comb. nov.*
 Navicula nebulosa W. Gregory 1857. Trans. R. Soc.
 Edinb. **21**: 480

Lyrella novaeseelandiae (Hustedt) D.G. Mann, *comb.
 nov.*
 Navicula novaeseelandiae F. Hustedt 1964.

Rabenhorsts Krypt.-Fl. Deutschl. Österr. u.
Schweiz 7, 3(3): 406

Lyrella oamaruensis (Grunow ex A.Schmidt) D.G.
Mann, *comb. nov.*

Navicula oamaruensis A. Grunow 1888. Ex A.
Schmidt, Atlas Diat.: taf. 129, fig. 9

Lyrella perfecta (Pantoscek) D.G. Mann, *comb. nov.*
Navicula perfecta J. Pantoscek 1886. Beitr. Kenntn.
foss. Bacill. Ungarns 1: 28

Lyrella perplexoides (Hustedt) D.G. Mann, *comb. nov.*
Navicula perplexoides F. Hustedt 1964. Rabenhorsts
Krypt.-Fl. Deutschl. Österr. u. Schweiz 7, 3(3):
383

Lyrella praetexta (Ehrenberg) D.G. Mann, *comb. nov.*
Navicula praetexta C.G. Ehrenberg 1860. Ber. Akad.
Wiss. Berlin 1840: 214

Lyrella pseudoapproximata (Hendey) D.G. Mann,
comb. nov.

Navicula pseudoapproximata N.I. Hendey 1958. J.
Roy. microsc. Soc. 77: 64

Lyrella robertsiana (Greville) D.G. Mann, *comb. nov.*
Navicula robertsiana R.K. Greville 1865. Trans. Bot.
Soc. Edinb. 8: 235

Lyrella sandriana (Grunow) D.G. Mann, *comb. nov.*
Navicula sandriana A. Grunow 1863. Verh. zool.-
bot. Ges. Wien 13: 153

Lyrella schaarschmidtii (Pantocsek) D.G. Mann,
comb. nov.

Navicula schaarschmidtii J. Pantocsek 1886. Beitr.
Kenntn. foss. Bacill. Ungarns 2: 28

Lyrella semiapproximata (Hustedt ex Simonsen) D.G.
Mann, *comb. nov.*

Navicula semiapproximata F. Hustedt 1987. Ex R.
Simonsen, Atlas Cat. Diat. Types Fr. Hustedt:
483

Lyrella signatula (Hustedt) D.G. Mann, *comb. nov.*
Navicula signatula F. Hustedt 1964. Rabenhorsts
Krypt.-Fl. Deutschl. Österr. u. Schweiz 7, 3(3):
491

Lyrella spectabilis (Gregory) D.G. Mann, *comb. nov.*
Navicula spectabilis W. Gregory 1857. Trans. R.
Soc. Edinb. 21: 481

Lyrella subirroratoides (Hustedt) D.G. Mann, *comb.
nov.*

Navicula subirroratoides F. Hustedt 1964.
Rabenhorsts Krypt.-Fl. Deutschl. Österr. u.
Schweiz 7, 3(3): 442

Lyrella variolata (Cleve) D.G. Mann, *comb. nov.*
Navicula variolata P.T. Cleve 1892. Diatomiste 1: 76

Lyrella venusta (Janisch ex Cleve) D.G. Mann, *comb.
nov.*

Navicula venusta C. Janisch 1895. Ex P.T. Cleve, K.
Svenska Vetensk.-Akad. Handl. 27(3): 56

Lyrella zanzibarica (Greville) D.G. Mann, *comb. nov.*
Navicula zanzibarica R.K. Greville 1866. Trans.
microsc. Soc. Lond. N. S. 14: 129

Other species belonging here include:

Lyrella lyra (Ehrenberg) Karayeva 1978

MARTYANA F.E. Round, *gen. nov.*

Valvae ovato-ellipticae heteropolares, in cellulis
parvis plusminusve ovales, polo basali area
porellorum instructo; ad polum apicalem
(latiorem) frons minus elevata. Striae
uniseriatae, inter costas transapicales extra
prominentes positae. Areolae rimiformes ad
axem apicalem parallelae. Et spinae ligantes et
rimoportulae nullae. Copulae usque ad 5
apertae curvatae, apices versus angustiores.
Valvocopula lata; copulae ceterae angustae.
Cellulae ad grana arenae saepe affixae aquam
dulcem habitantes.

Holotypus: *Martyana martyi* (Héribaud) F.E. Round,
comb. nov.

Opephora martyi J. Héribaud 1902. Diat. Auvergne,
mém. 1: 43

MICROTABELLA F.E. Round, *gen. nov.*

Cellulae per polos vel frontes affixae, catenas
fractiflexas vel taeniformes facientes. Valvae
lanceolatae ad centrum interdum dilatatae ad
apices areis porellorum instructae. Frontis cum
limbo junctura inornata vel crista spinosa
praedita. Striae uniseriatae in limbo interdum
biseriatae, areolis circularibus vel subquadratis.
Sternum indistinctum vel nullum. Facies
interna frontis transapicaliter costata. Copulae
multae angustae porosae. Genus cosmopolitum
marinum; plantae epiphyticae.

Holotypus: *Microtabella interrupta* (Ehrenberg) F.E.
Round, *comb. nov.*

Tessella interrupta C.G. Ehrenberg 1838. Infus.: 202.

Microtabella delicatula (Kützing 1844) F.E. Round,
comb. nov.

Hyalosira delicatula F.T. Kützing 1844. Kies.
Bacill.: 125

OXYNEIS F.E. Round, *gen. nov.*

Cellulae inter se affixae catenas rectas vel
fractiflexas facientes. Valvae ellipticae vel
panduriformes. Frons plana ad margines in

limbum non profundum curvans, junctura frontis cum limbo spinis munita. Spinae furcatae, in catenis rectis cellulas inter se firme conjungentes. Striae uniseriatae e sterno angusto in limbum extensae. Rimoportula unica prope polum unum in stria posita. Porelli apicales in seriebus confertis radiantibus ordinati. Copulae apertae, unaquaeque septo brevi apicali et serie unica areolarum praedita. Habitat aquam dulcem magnopere acidam.

Holotypus: *Oxyneis binalis* (Ehrenberg) F. E. Round comb. nov.

Fragilaria binalis C.G. Ehrenberg 1854. Mikrogeol.: Pl. 14/52

PETRODICTYON D.G. Mann, *gen. nov.*

Cellulae solitariae, aspectu valvae vel cincturae visae. Et valva et cinctura heteropolaris. Chromatophorum unicum ad polum basalem (angustiorem) plicatum, supra in laminas duas lobatus ad valvas appressas expansum et explanatum. Valvae cuneati-ovatae. Systema raphis peripheriam totam valvae circumdans, supra frontem in carinam non profundam elevata. Frons laevis, in sectione transapicali leviter undulata, secus lineam mediam valvae et supra costas transapicales incrassatas leviter sulcata. Striae uniseriatae, inter se costis transapicalibus angustis separatae, areolis comparate grandibus subquadratis. Areolae simplices cribris voliferis extra occlusae. Costae transapicales secus lineam mediam valvae coalescentes sternum angustum longitudinale facientes; costae aliquae valde incrassatae e margine valvae ad lineam mediam frontis extensae (Nota bene! Respectu ontogenesis linea media frontis eadem est ac margines duae appressae et connatae), sub raphe pro fibulis fungentes, hic profundiores anteridum similes; fibulae aliae breviores, in costis transapicalibus item portatae. Inter fibulas sunt portulae circulares vel ellipticae, quae lumen cellulae cum canali sub raphe conjugunt. Extrema 'centralia' raphis (ad polum latiorem) tortuosa, extra nares intra structuras labiatas profundas facientes. Extrema 'polaria' (ad polum angustiorem) extra simplicia vel plusminusve dilatata, intra leviter labiata. Cinctura ex taeniis apertis paucis constans.

Forma frustuli et dispositione raphis *Surirellae*

simile, differt structura valvae ex costis transapicalibus angustis et striis uniseriatis constanti, areolis cribris occlusis, fibulis costiformibus longisque brevibusque, et extremis centralibus raphis.

Holotypus: *Petrodictyon gemma* (Ehrenberg) D.G. Mann, *comb. nov.*

Surirella gemma C.G. Ehrenberg 1839. Abh. Akad. Wiss. Berlin 1839: 156

Petrodictyon contiguum (A. Mann) D.G. Mann, *comb. nov.*

Surirella contigua A. Mann 1925. Bull. U. S. Natl. Mus. 100, 6: 152

Petrodictyon foliatum (A. Mann) D.G. Mann, *comb. nov.*

Surirella foliata A. Mann 1925. Bull. U. S. Natl. Mus. 100, 6: 152

Petrodictyon gemmoides Østrup) D.G. Mann, *comb. nov.*

Surirella gemmoides E. Østrup 1897. Meddel. Grønland 15: 277

PETRONEIS A.J. Stickle & D.G. Mann, *comb. nov.*

Cellulae solitariae aspectu valvae fere semper visae. Chromatophora 2 laminiformia transapicaliter et secus raphem indentata, uno eorum ad utramque valvam appresso. Valvae latae non profundae, lineares vel lanceolatae, polis rotundatis vel late apiculatis. A limbo non profundo frons non facile distinguenda. Striae uniseriatae. Areolae grandes circulares vel ellipticae (transapicaliter elongatae), intra cribris voliferis vel volis occlusae. Raphosternum centrale, ad centrum saepe valde dilatatum, aream centralem ellipticam vel transverse oblongam hic faciens. Fissurae externae raphis ad centrum in depressionem spatulatam vel cuneiformem aperientes; extrema centralia ipsa recta non expansa. Extrema interna centralia unciformia. Fissurae terminales ad latus secundum flexae, apicaliter dilatatae. Cinctura non profunda. Cingulum ex aliquot taeniis apertis constans, unaquaque seriebus transversalibus 1–2 pororum instructa.

Lyrella affinis, sed differt egestate sternorum lateralium in valva, id est areae lyriformis hyalini. A *Cosmioneide* differt velis areolarum, taeniis paucioribus, limbo prope polos non deminuto, etc.

Holotypus: *Petroneis humerosa* (Brébisson ex Wm.

Smith) A.J. Stickle & D.G. Mann, *comb. nov.*

Navicula humerosa A. de Brébisson 1856. Ex Wm.
Smith, Syn. Brit. Diat. II: 93

Petroneis arabica (Grunow ex A. Schmidt) D.G.
Mann, *comb. nov.*

Navicula arabica A. Grunow 1875. Ex A. Schmidt,
Atlas Diat.: taf. 6, figs 13–14.

Petroneis deltoides (Hustedt) D.G. Mann, *comb. nov.*

Navicula deltoides F. Hustedt 1966. Rabenhorsts
Krypt.-Fl. Deutschl. Österr. u. Schweiz 7, 3(4):
689

Petroneis granulata (Bailey) D.G. Mann, *comb. nov.*

Navicula granulata J.W. Bailey 1854. Smithson.
Contr. Knowl. 7: 10

Petroneis groveoides (Hustedt) D.G. Mann, *comb. nov.*

Navicula groveoides F. Hustedt 1966. Rabenhorsts
Krypt.-Fl. Deutschl. Österr. u. Schweiz 7, 3(4):
686

Petroneis japonica (Heiden in A. Schmidt) D.G.
Mann, *comb. nov.*

Navicula japonica H. Heiden 1903. In A. Schmidt,
Atlas Diat.: taf. 244, fig. 8

Petroneis latissima (Gregory) A.J. Stickle & D.G.
Mann, *comb. nov.*

Navicula latissima W. Gregory 1856. Trans.
microsc. Soc. Lond. 4: 40

Petroneis marina (Ralfs in Pritchard) D.G. Mann,
comb. nov.

Navicula marina J. Ralfs 1861. In A. Pritchard,
Hist. Infus., 4th Edn.: 903

Petroneis monilifera (Cleve) A.J. Stickle & D.G.
Mann, *comb. nov.*

Navicula monilifera P.T. Cleve 1895. K. Svenska
Vetensk.-Akad. Handl. 27(3): 43

Petroneis plagiostoma (Grunow in Cleve & Möller)
D.G. Mann, *comb. nov.*

Navicula plagiostoma A. Grunow 1879. In P.T.
Cleve & J.D. Möller, Diat.: 257

Petroneis punctigera (Hustedt) D.G. Mann, *comb. nov.*

Navicula punctigera F. Hustedt 1955. Bull. Duke
Univ. Mar. Stn. 6: 26

Petroneis subdiffusa (Hustedt) D.G. Mann, *comb. nov.*

Navicula subdiffusa F. Hustedt 1955. Bull. Duke
Univ. Mar. Stn. 6: 24

Petroneis transfuga (Grunow ex Cleve) D.G. Mann,
comb. nov.

Navicula transfuga A. Grunow 1883. In P.T. Cleve,
Vega-Exped. iakttag. 3: 511

PROSCHKINIA: new combinations

Proschkinia complanata (Grunow) D.G. Mann, *comb.
nov.*

Amphora complanata A. Grunow 1867. Hedwigia 6:
25

Proschkinia complanatoides (Hustedt ex Simonsen)
D.G. Mann, *comb. nov.*

Navicula complanatoides F. Hustedt 1987. Ex R.
Simonsen, Atlas Cat. Diat. Types Fr. Hustedt:
480

Proschkinia complanatula (Hustedt ex Simonsen)
D.G. Mann, *comb. nov.*

Navicula complanatula F. Hustedt 1987. Ex R.
Simonsen, Atlas Cat. Diat. Types Fr. Hustedt:
479

Proschkinia hyalosirella (Hustedt ex Simonsen) D.G.
Mann, *comb. nov.*

Navicula hyalosirella F. Hustedt 1987. Ex R.
Simonsen, Atlas Cat. Diat. Types Fr. Hustedt:
479

Proschkinia? longirostris (Hustedt) D.G. Mann, *comb.
nov.*

Navicula longirostris F. Hustedt 1930. Bacill.
(Pascher SüsswasserFl.): 285

Proschkinia poretzkajae (Koretkevich) D.G. Mann,
comb. nov.

Navicula poretzkajae O.S. Koretkevich 1959. Notul.
Syst. Sect. cryptog. Inst. bot. V.L. Komarov
Acad. Sci. U.R.S.S. 12: 96

PSAMMODICTYON D.G. Mann, *gen. nov.*

Cellulae solitariae aspectu valvae fere semper
visae. Chromatophora 2 laminiformia, uno
utrumque polum versus, lacuna angusta inter se
respectu planum transapicale obliqua. Valvae
latae non profundae, panduriformes vel late
lineares, polis rotundatis vel apiculatis. Frons in
sectione transapicali sigmoidea, a limbo distali
non distinguenda, sine crista marginali,
interdum sterna lateralia habens. Areolae
frontis ex parte vel ubique loculatae (rarissime
omnino simplices), loculis in striis tribus ad
angulum 120° sibi iacentibus ordinatis; loculus
extra per foramen unicum intra per poros 1–4
apertus, poris internis hymenibus occlusis.
Areolae limbi proximalis simpliciores non
loculatae. Systema raphis fibulatum carinatum
valde excentricum, inter frontem et limbum
proximalem non profundum positum.
Raphosternum angustum, interdum cristam
gracilem in utroque latere raphis extra ferens,

cristis ad centrum distantibus nodulum parvum includentibus, alibi approximatissimis. Extrema externa centralia raphis recta paulo dilatata, nodulo rotundato separata; extrema interna recta simplicia vel helictoglossam duplicem parvam facientia. Fissurae terminales apicaliter interdum dilatatae, unciformes vel flexae, aut latus ventrale aut latus dorsale versus curvantes. Fibulae breves costiformes vel tabulares. Frustula semper nitzschioidea, raphibus valvarum duarum inter se diagonaliter oppositis. Cinctura non profunda; cingulum ex taeniis apertis constans, prima seriebus pluribus porosum parvorum instructa, ceteris plerumque angustioribus non porosis.

Tryblionellae affinis, sed differt structura loculata frontis extremis, centralibus externis raphis, structura cingulari, etc.

Holotypus: *Psammodictyon panduriforme* (Gregory) D.G. Mann, *comb. nov.*

> *Nitzschia panduriformis* W. Gregory 1857. Trans. Roy. Soc. Edinb. **21**: 529

Psammodictyon areolatum (Hustedt) D.G. Mann, *comb. nov.*

> *Nitzschia areolata* F. Hustedt 1952. Ber. dt. Bot. Ges. **64**: 311

Psammodictyon bisculptum (A. Mann) D.G. Mann, *comb. nov.*

> *Nitzschia bisculpta* A. Mann 1925. Bull. U. S. Natl. Mus. **100**,6: 125

Psammodictyon bombiforme (Grunow in Cleve & Grunow) D.G. Mann, *comb. nov.*

> *Nitzschia constricta* (Gregory) Grunow, *nom. illeg.*, var. *bombiformis* A. Grunow 1880. In P.T. Cleve & A. Grunow, K. Svenska Vetensk.-Akad. Handl. **17**(2): 71

Psammodictyon constrictum (Gregory) D.G. Mann, *comb. nov.*

> *Tryblionella constricta* W. Gregory 1855. Quart. J. microsc. Sci. **3**: 40

Psammodictyon corpulentum (Hendey) D.G. Mann, *comb. nov.*

> *Nitzschia corpulenta* N.I. Hendey 1958. J. Roy. Microsc. Soc. **77**: 78

Psammodictyon ferox (Hustedt) D.G. Mann, *comb. nov.*

> *Nitzschia ferox* F. Hustedt 1952. Ber. dt. Bot. Ges. **64**: 311

Psammodictyon inductum (Hustedt) D.G. Mann, *comb. nov.*

> *Nitzschia inducta* F. Hustedt 1952. Ber. dt. Bot. Ges. **64**: 310

Psammodictyon mediterraneum (Hustedt in A. Schmidt), D.G. Mann, *comb. nov.*

> *Nitzschia mediterranea* F. Hustedt 1921. In A. Schmidt, Atlas Diat.: taf. 331, fig. 22.

Psammodictyon molle (Hustedt) D.G. Mann, *comb. nov.*

> *Nitzschia mollis* F. Hustedt 1952. Ber. dt. Bot. Ges. **64**: 312

Psammodictyon roridum (Giffen) D.G. Mann, *comb. nov.*

> *Nitzschia rorida* M.H. Giffen 1975. Bot. mar. **18**: 90

Psammodictyon rudum (Cholnoky) D.G. Mann, *comb. nov.*

> *Nitzschia ruda* B.J. Cholnoky 1968. Bot. mar., Suppl. **11**: 79

SCOLIONEIS D.G. Mann, *gen. nov.*

Cellulae solitariae, unaquaque circum axem apicalem torta, aspectu cincturae plerumque visae. Chromatophora 2, H-formia, ad valvas appressa, interdum in latere uno cellulae prope cincturam apparenter connexa, pyrenoidibus rotundatis pluribus continentia. Valvae lineares vel lanceolatae, polis acute vel obtuse rotundatis. Systema raphis propter torsionem valvae apparenter sigmoidea. Frons in sectione transapicali curvata, a limbo profundo non vere distinguenda. Striae uniseriatae simplices; valvae ubique unistratae. Areolae parvae circulares hymenibus occlusae, extra per rimas lineares vel furcatas aperientes. Raphosternum plerumque angustum, ad centrum expansum. Extrema centralia externa raphis simplicia; fissurae terminales rectae, ad marginem valvae extensae. Extrema centralia interna T-formia; helictoglossa elongata angusta. Cingulum ex taeniis apertis pluribus constans, taenia prima (valvocopula) serie una pororum, unaquaque taenia alia seriebus duabus instructa. Taenia prima secus marginem partis interioris dentata.

Scolioneis Scoliopleurae vel *Scoliotropidis* primo adspectu similis est, sed differt extremis raphis, structura simpliciore striarum et areolarum, et dispositione chromatophororum.

Holotypus: *Scolioneis tumida* (Brébisson ex Kützing) D.G. Mann, *comb. nov.*

> *Navicula tumida* A. de Brébisson 1849. Ex F.T.

Kützing, Spec. Alg.: 77

Scolioneis brunkseiensis (Hendey) D.G. Mann, *comb. nov.*

    Scoliopleura brunkseiensis N.I. Hendey 1964. Bacillariophyceae M.A.F.F. Fish. Investig. (ser. 4) 5: 235

SEMINAVIS D.G. Mann, *gen. nov.*

Cellulae solitariae dorsiventrales, aspectu cincturae plerumque visae. Chromatophora 2, ad latera cincturae appressa, dorsali majore quam ventrali. Valvae semilanceolatae, margine ventrali recto, margine dorsali curvato. Frons in limbum dorsalem profundum curvata; limbus ventralis plusminusve absens. Striae uniseriatae, inter costas transapicales prominentes intra apertae. Areolae apicaliter elongatae, intra hymenibus occlusae, extra per rimas angustas aperientes. Pars dorsalis raphosterni latior quam pars ventralis; pars ventralis (prima) latus dorsale versus ita producta, ut fissurae internae raphis laterales fiant, orificiis earum in apice cristae angustae positis (ut in *Navicula* sensu stricto). Extrema centralia interna raphis recta non expansa; extrema centralia externa parum expansa et latere ventrali obversa, vel alis e parte ventrali raphosterni productis obtecta. Fissurae terminales unciformes, latus dorsale versus curvantes. Pars dorsalis cincturae aliquantum latior quam pars ventralis. Cingulum ex elementis tribus non porosis apparenter constans, primo taenia aperta lata, secundo segmento lato hiatum inter extremis elementi primi occupanti, tertio angusto imperfecte cognito sed probabiliter taenia aperta.

Fabrica areolarum raphis et elementorum cincturae *Naviculae* et affinium similis, differt forma cellulae, quae valde dorsiventralis est. Ab *Amphora* fabrica frustulorum et dispositione chromatophororum differt.

Holotypus: *Seminavis gracilenta* (Grunow ex A. Schmidt) D.G. Mann, *comb. et stat. nov.*

    Amphora angusta var. *gracilenta* A. Grunow 1875. Ex A. Schmidt, Atlas Diat.: taf. 25, fig. 15

Seminavis? cymbelloides (Grunow) D.G. Mann, *comb. nov.*

    Amphora cymbelloides A. Grunow 1867. Hedwigia 6: 24

STAUROPHORA: new combinations

Staurophora amphioxys (Gregory) D.G. Mann, *comb. nov.*

    Stauroneis amphioxys W. Gregory 1856. Trans. microsc. Soc. Lond., N.S. 4: 48

Staurophora elata (Hustedt ex Simonsen) D.G. Mann, *comb. nov.*

    Stauroneis elata F. Hustedt 1987. Ex R. Simonsen, Atlas Cat. Diat. Types Fr. Hustedt: 462

Staurophora rossii (Hendey) D.G. Mann, *comb. nov.*

    Stauroneis rossii N.I. Hendey 1964. Bacillariophyceae M.A.F.F. Fish. Investig. (ser. 4) 5: 220

Staurophora? wislouchii (Poret. & Anisimova) D.G. Mann, *comb. nov.*

    Stauroneis wislouchii V.S. Poretzky & N.V. Anisimova 1933. Issled. Ozer S.S.S.R. 2: 51

Other species:

Staurophora salina (W.Smith) Mereschkowsky 1903

THALASSIONEIS F.E. Round, *gen. nov.*

Valvae lineares ad centrum dilatatae, polis rotundatis. Frons plana ad margines in limbum non profundum curvata. Striae uniseriatae aliquantum irregulares inter se late separatae. Areolae simplices parvae circulares, extra cribris indistinctis occlusae. Areae apicales porellorum aliquantum depressae, limbo ultra areas has aliquot areolis dispersis instructo. Rimoportula unica in dilatatione centrali polum unum versus posita, extra per porum parvum elongatum annulatum aperta, intra sessilis rima transapicali. Copulae non visae.

Holotypus: *T. signyensis* F.E. Round *sp. nov.*

Species characteribus ut in descriptione generis. Valvae 25 μm longae, 7-10 μm latae, striis 11-14 in 10 μm

Holotypus: BM 81503

TRYBLIONELLA: new combinations

Tryblionella acuta (Cleve) D.G. Mann, *comb. nov.*

    Nitzschia acuta P.T. Cleve 1878. Bih. K. Svenska Vetensk.-Akad. Handl. 5(8): 13

Tryblionella adducta (Hustedt) D.G. Mann, *comb. nov.*

    Nitzschia adducta F. Hustedt 1955. Bull. Duke Univ. Mar. Stn. 6: 43

Tryblionella aerophila (Hustedt) D.G. Mann, *comb. nov.*

    Nitzschia aerophila F. Hustedt 1942. Ber. dt. Bot. Ges. 60: 70

Tryblionella ardua (Cholnoky) D.G. Mann, *comb. nov.*

    Nitzschia ardua B.J. Cholnoky 1961. Hydrobiologia

17: 314

Tryblionella balatonis (Grunow in Cleve & Grunow) D.G. Mann, *comb. nov.*

Nitzschia balatonis A. Grunow 1880. In P.T. Cleve & A. Grunow, K. Svenska Vetensk.-Akad. Handl. **17**(2): 70

Tryblionella bathurstensis (Giffen) D.G. Mann, *comb. nov.*

Nitzschia bathurstensis M.H. Giffen 1970. Nova Hedwigia, Beih. **31**: 287

Tryblionella bicuneata (Grunow in Cleve & Grunow) D.G. Mann, *comb. nov.*

Nitzschia bicuneata A. Grunow 1880. In P.T. Cleve & A. Grunow, K. Svenska Vetensk.-Akad. Handl. **17**(2): 75

Tryblionella brightwellii (Kitton in Pritchard) D.G. Mann, *comb. nov.*

Nitzschia brightwellii F. Kitton 1861. In A. Pritchard, Hist. Infus., 4th edn.: 780

Tryblionella calida (Grunow in Cleve & Grunow) D.G. Mann, *comb. nov.*

Nitzschia calida A. Grunow 1880. In P.T. Cleve & A. Grunow, K. Svenska Vetensk.-Akad. Handl. **17**(2): 75

Tryblionella campechiana (Grunow) D.G. Mann, *comb. nov.*

Nitzschia campechiana A. Grunow 1880. J. Roy. microsc. Soc. **3**: 395

Tryblionella? chutteri (Archibald) D.G. Mann, *comb. nov.*

Nitzschia chutteri R.E.M. Archibald 1966. Nova Hedwigia, Beih. **21**: 265

Tryblionella coarctata (Grunow in Cleve & Grunow) D.G. Mann, *comb. nov.*

Nitzschia coarctata A. Grunow. In P.T. Cleve & A. Grunow, K. svenska VetenskAkad. Handl. **17**(2): 68

Tryblionella cocconeiformis (Grunow) D.G. Mann, *comb. nov.*

Nitzschia cocconeiformis A. Grunow 1880. J. Roy. microsc. Soc. **3**: 395

Tryblionella conformata (Hustedt) D.G. Mann, *comb. nov.*

Nitzschia conformata F. Hustedt 1952. Ber. dt. Bot. Ges. **64**: 313

Tryblionella davidsonii (Grunow & Dickie in Cleve & Grunow) D.G. Mann, *comb. nov.*

Nitzschia davidsonii A. Grunow & Dickie 1880. In P.T. Cleve & A. Grunow, K. Svenska

Vetensk.-Akad. Handl. **17**(2): 75

Tryblionella didyma (Hustedt) D.G. Mann, *comb. nov.*

Nitzschia didyma F. Hustedt 1952. Ber. dt. Bot. Ges. **64**: 309

Tryblionella? divergens (Hustedt) D.G. Mann, *comb. nov.*

Nitzschia divergens F. Hustedt 1952. Ber. dt. Bot. Ges. **64**: 312

Tryblionella graeffii (Grunow ex Cleve) D.G. Mann, *comb. nov.*

Nitzschia graeffii A. Grunow 1878. In P.T. Cleve, Bih. K. Svenska Vetensk.-Akad. Handl. **5**(8): 20

Tryblionella granulata (Grunow) D.G. Mann, *comb. nov.*

Nitzschia granulata A. Grunow 1880. J. Roy. microsc. Soc. **3**: 395

Tryblionella halteriformis (Hustedt) D.G. Mann, *comb. nov.*

Nitzschia halteriformis F. Hustedt 1952. Ber. dt. Bot. Ges. **64**: 311

Tryblionella hungarica (Grunow) D.G. Mann, *comb. nov.*

Nitzschia hungarica A. Grunow 1862. Verh. zool. bot. Ges. Wien **12**: 568

Tryblionella jelineckii (Grunow) D.G. Mann, *comb. nov.*

Nitzschia jelineckii A. Grunow 1863. Verh. zool. bot. Ges. Wien **13**: 144

Tryblionella limicola (Grunow) D.G. Mann, *comb. nov.*

Nitzschia limicola A. Grunow 1880. J. Roy. microsc. Soc. **3**: 395

Tryblionella littoralis (Grunow in Cleve & Grunow) D.G. Mann, *comb. nov.*

Nitzschia littoralis A. Grunow 1880. In P.T. Cleve & A. Grunow, K. Svenska Vetensk.-Akad. Handl. **17**(2): 75

Tryblionella magnacarina (Hohn & Hellerman) D.G. Mann, *comb. nov.*

Nitzschia magnacarina M.H. Hohn & J. Hellerman 1966. Trans. Am. microsc. Soc. **85**: 128

Tryblionella marginulata (Grunow in Cleve & Grunow) D.G. Mann, *comb. nov.*

Nitzschia marginulata A. Grunow 1880. In P.T. Cleve & A. Grunow, K. Svenska Vetensk.-Akad. Handl. **17**(2): 72

Tryblionella nicobarica (Grunow) D.G. Mann, *comb. nov.*

Nitzschia panduriformis var. *nicobarica* A. Grunow

1868. Reise 'Novara', Bot. Theil **1**: 97

Tryblionella obscurepunctata (Hustedt ex Simonsen) D.G. Mann, *comb. nov.*

Nitzschia obscurepunctata F. Hustedt 1987. Ex R. Simonsen, Atlas Cat. Diat. Types Fr. Hustedt: 451

Tryblionella partita (Hustedt) D.G. Mann, *comb. nov.*

Nitzschia partita F. Hustedt 1952. Ber. dt. Bot. Ges. **64**: 311

Tryblionella perversa (Grunow) D.G. Mann, *comb. nov.*

Nitzschia perversa A. Grunow 1880. J. Roy. microsc. Soc. **3**: 395

Tryblionella plicatula (Hustedt) D.G. Mann, *comb. nov.*

Nitzschia plicatula F. Hustedt 1953. Arch. Hydrobiol. **48**: 151

Tryblionella ponciensis (Hagelstein) D.G. Mann, *comb. nov.*

Nitzschia ponciensis R. Hagelstein 1938. N. Y. Acad. Sci., Sci. Survey Porto Rico & Virg. Isl. **8**(3): 401

Tryblionella princeps (Hanna & Grant) D.G. Mann, *comb. nov.*

Nitzschia princeps G.D. Hanna & W.M. Grant 1926. Proc. Calif. Acad. Sci., 4th ser. **15**: 159

Tryblionella pseudohungarica (Hustedt) D.G. Mann, *comb. nov.*

Nitzschia pseudohungarica F. Hustedt 1939. Arch. Hydrobiol., Suppl. **16**: 381

Tryblionella rabenhorstii (Grunow) D.G. Mann, *comb. nov.*

Nitzschia rabenhorstii A. Grunow 1880. J. Roy. microsc. Soc. **3**: 394

Tryblionella salvadoriana (Hustedt) D.G. Mann, *comb. nov.*

Nitzschia salvadoriana F. Hustedt 1952. Ber. dt. Bot. Ges. **64**: 308

Tryblionella sibula (Giffen) D.G. Mann, *comb. nov.*

Nitzschia sibula M.H. Giffen 1973. Bot. mar. **16**: 42

Tryblionella sulcata (Grunow in Cleve & Grunow) D.G. Mann, *comb. nov.*

Nitzschia sulcata A. Grunow 1880. In P.T. Cleve & A. Grunow. K. Svenska Vetensk.-Akad. Handl. **17**(2): 76

Tryblionella umbilicata (Hustedt.) D.G. Mann, *comb. nov.*

Nitzschia umbilicata F. Hustedt 1949. Expl. Parc Natl. Albert, Miss. H. Damas (1935-36) **8**: 129.

Tryblionella visurgis (Hustedt) D.G. Mann, *comb. nov.*

Nitzschia visurgis F. Hustedt 1957. Abh. naturw. Ver. Bremen **34**: 339

Tryblionella zebuana (A. Mann) D.G. Mann, *comb. nov.*

Nitzschia zebuana A. Mann 1925. Bull. U. S. Natl. Mus. **100**(6): 129

Other species belonging to the genus include:

Tryblionella acuminata Wm. Smith 1853

T. apiculata Gregory 1857

T. circumsuta (Bailey) Ralfs in Pritchard 1861

T. gracilis Wm. Smith 1853

T. lanceola Grunow ex Cleve 1878

T. levidensis Wm. Smith 1856

T. navicularis (Brébisson ex Kützing) Ralfs in Pritchard 1861

T. plana (Wm. Smith) Pelletan 1889

T. punctata Wm. Smith 1853

T. victoriae Grunow 1862

UROSOLENIA F.E. Round & R.M. Crawford *gen. nov.*

Cellulae delicatissimae facile praetermissae solitariae cylindricae. Chromatophora multa disciformia. Valvae conicae, extensione longo tenui capilliformi dentibus paucis parvis apicaliter munito. Areolae simplices irregulariter dispersae basin versus extensionis grandiores. Rimoportulae nullae. Cingulum ex taeniis dimidiis multis imbricatis irregulariter perforatis constans. Genus cosmopolitum aquam dulcem habitans.

Holotypus: *Urosolenia eriensis* (H.L. Smith) F.E. Round & R.M. Crawford *comb. nov.*

Rhizosolenia eriensis H.L. Smith 1872. The Lens **1**: 44

Appendix II:

List of genera that have been described or re-investigated recently but which we have not been able to include in the generic atlas. There are also many other rare genera or genera recorded only once which are listed in the Index Nominum Genericorum (Appendix III).

Centric and Araphid Pennate Genera

Adoneis Andrews & Rivera 1987. ref. G.W. Andrews & P. Rivera 1987. Diat. Res. 2: 1–14

Ailuretta Sims 1986. ref. P.A. Sims 1986. Diat. Res.1: 241–269

Alveoflexus Hendey & Sims 1984. ref. N.I. Hendey & P.A. Sims 1984. Bacillaria 7: 59–90

Ancoropsis Hendey & Sims 1984. ref. N.I. Hendey & P.A. Sims 1984. Bacillaria 7: 59–90

Bifibulatia Takano 1983. ref. H. Takano 1983. Bull. Tokai Reg. Fish. Res. Lab. 112: 13–26

Bonea Ross & Sims 1987. ref. R. Ross & P.A. Sims 1987. Bull. Br. Mus. nat. Hist. (Bot.) 16: 269–311

Dextradonator Ross & Sims 1980. ref. R. Ross & P.A. Sims 1980. Bacillaria 3: 115–127

Discodiscus Gombos 1980. ref. A.M. Gombos 1980. Bacillaria 3: 227–272

Hyperion Gombos 1983. ref. A.M. Gombos 1983. Bacillaria 6: 237– 244

Maluina Ross & Sims 1987. ref. R. Ross & P.A. Sims 1987. Bull. Br. Mus. nat. Hist. (Bot.) 16: 269–311

Miraltia Marino, Montresor & Zigone 1987. ref. D. Marino, M. Montresor & A. Zigone. Diat. Res. 2: 205–211

Monile Ross & Sims 1987. ref. R. Ross & P.A. Sims 1987. Bull. Br. Mus. nat. Hist. (Bot.) 16: 269–311

Pleurocyclos Casper & Scheffler 1986. ref. S.J. Casper & W. Scheffler 1986. Arch. Protistenk. 132: 295, 1–33

Protoraphis Simonsen 1970. ref. R. Simonsen 1970. Nova Hedwigia Beih. 31: 383–394

Pseudoguinardia von Stosch 1985. ref. H.A. von Stosch 1985. Brunonia 8: 293–348

Pseudoleyanella Takano 1985. ref. H. Takano 1985. Bull. Tokai Reg. Fish. Res. Lab. 115: 29–37

Stephanocostis Genkal & Kusmina 1985. ref. S.I. Genkal & G.V. Kusmina 1985. Inf. Bull. Biol. Inland Waters 67: 10

Strelnikovia Ross & Sims 1985. ref. R. Ross & P.A. Sims 1985. Bull. Br. Mus. nat. Hist. (Bot.) 13: 277–381

Tabulariopsis Williams 1988. ref. D.M. Williams 1988 Nova Hedwigia 47: 247–254

Thalassiosiropsis Hasle 1985. ref. G.R. Hasle 1985. Micropaleontology 31: 280–284

Trochosira Kitton 1871. ref. P.A. Sims 1988. Diat. Res. 3:245–57

Tumulopsis Hendey 1982. ref. N.I. Hendey 1982. Nova Hedwigia, Beih. 73: 275–280

Undatocystis Lupikina 1984. ref. Lupikina 1984. Trudy Geol. in.-ta. Akad. nauk SSSR. 385: 274

Raphid Pennate Genera

Banquisia T.B.B. Paddock, 1988. ref. T.B.B. Paddock 1988. Bibliotheca Diatomologia 16: 152 pp. Berlin: Cramer

Bennettella Holmes 1985. ref. R.W. Holmes 1985. Br. phycol. J. 20: 43–57

Caponea Podzorski 1984. ref. A.C. Podzorski 1984. Nova Hedwigia 40: 1–8

Denticulopsis Simonsen 1979. ref. F. Akiba & Y. Yanagisawa 1985. Init. Rep. Deep Sea Drill.Proj. 87: 483–554

Diadema Kemp & Paddock 1989. ref. K-D. Kemp & T.B.B. Paddock 1989. Diat. Res. 4:

Dimidiata Hajós 1974. ref. M. Hajós 1974. Nova Hedwigia, Beih. 45: 365–388

Ephemera T.B.B. Paddock 1988. ref. T.B.B. Paddock 1988. Bibliotheca Diatomologia 16: 152 pp. Berlin: Cramer.

Epipellis Holmes 1985. ref. R.W. Holmes 1985. Br. phycol. J. 20: 43–57

Gomphonitzschia Grunow 1868. ref. N.I. Hendey & P.A. Sims 1982. Bacillaria 5: 191–212

Krasskella Ross & Sims 1978. ref. R. Ross & P.A. Sims 1978. Bacillaria 1: 151–168

Manguinea T.B.B. Paddock 1988. ref. T.B.B. Paddock 1988. Bibliotheca Diatomologia 16: 152 pp. Berlin: Cramer.

Mannsia T.B.B. Paddock 1988. ref. T.B.B. Paddock 1988. Bibliotheca Diatomologia 16: 152 pp. Berlin: Cramer.

Membraneis T.B.B. Paddock, *gen. nov.* 1988. ref. T.B.B. Paddock 1988. Biblioteca Diatomologia 16: 152 pp. Berlin: Cramer.

Östrupia Heiden 1906. ref. H.-J. Schrader 1971. Nova Hedwigia **22**: 915–938

Perrya Kitton 1874. ref. T.B.B. Paddock & P.A. Sims 1981. Bacillaria **4**: 177–222

Plagiolemma T.B.B. Paddock 1988. ref. T.B.B. Paddock 1988. Bibliotheca Diatomologia **16**, 152 pp. Berlin: Cramer.

Protokeelia C.W. Reimer & J.J. Lee 1984. ref. C.W. Reimer & J.J. Lee 1984 Proc. Acad. Nat. Sci., Phila. **136**: 194

Rouxia Brun & Héribaud in Héribaud 1893. ref. F. Akiba & Y. Yanagisawa 1985. Init. Rep. Deep Sea Drill. Proj. **87**: 483–554

Simonsenia Lange-Bertalot 1979. ref. H. Lange-Bertalot 1979. Bacillaria **2**: 127–36

Staurotropis T.B.B. Paddock 1988. ref. T.B.B. Paddock 1988. Bibliotheca Diatomologia **16**: 152 pp. Berlin: Cramer.

Stilus T.B.B. Paddock 1988. ref. T.B.B. Paddock 1988. Bibliotheca Diatomologia **16**: 152 pp. Berlin: Cramer.

Appendix III:
Index nominum genericorum

Abas, Ross & Sims. Bacillaria, 3: 119, 1980. T: *A. wittii* (Grunow) Ross & Sims (= *Syringidium wittii* Grunow in Van Heurck (1880–1885).

Acanthoceras, Honigmann. Arch. Hydrobiol. Planktonk. 5: 76, 1909. T: *A. magdeburgense* Honigmann. Non Kützing, 1841.

Acanthodiscus, Pantocsek. Beitr. Kenntn. Foss. Bacill. Ungarns, 3: T. 11, F. 180, 1892. T: *A. rugosus* Pantocsek.

Achnantella, B. Gaillon. Tabl. Syn. Nemazoaires. <10 (Aperçu Hist. Nat. 34), 1833 + *Achnanthes* J. B. Bory de St.-Vincent, 1822. Superfluous substitute name.

Achnanthella, C. Lemaire. Dict. Universel Hist. Nat. (Orbigny) 1: 73, 1841 (*orth. var.*) + *Achnantella* B. Gaillon, 1833 + *Achnanthes* Bory de St. Vincent, 1822.

Achnanthepyla, M. Peragallo. 2me Expéd. Antarct. Franç. (1908–10), Bot. Diat. 1921. LT: *A. mesogongyla* (Grun. ex Cl. & Grun. 1880) Perag. (= *Achnanthes mesogongyla* Cl. ex Cl. & Grunow).

Achnanthes, Bory de St.-Vincent. Dict. Class. Hist. Nat. 1: 79, 27 Mai 1822. LT: *A. adnata* Bory de St. Vincent.

Achnanthidium, Kützing. Kies. Bacill. Diat. 75, pl. III, Sys. 13, 19, 7/9 Nov. 1844. LT: *A. microcephalum* Kützing (vide Rabenhorst Süssw. - Diat. Freunde Mikr. 25, 46, 1853).

Achnanthosigma, Reinhard. Bull. Soc. Imp. Naturalistes Moscou. 57(1): 298, figs 1, 2, 1882. T: *A. mereschkovskii* Reinhard.

Actinella, F. W. Lewis. Proc. Acad. Nat. Sci. Philadelphia, 1863: 343, pl. 1, f. 5. 1864. (*Nom. cons.*) T: *A. punctata* F. W. Lewis.

Actinoclava, O. Müller. Ber. Deutsch. Bot. Ges. 29: 661, pl. 26, f. 1. 1912. T: *A. frankei* O. Müller.

Actinocyclus, C. G. Ehrenberg. Ber. Bekanntm. Verh. Königl. Preuss. Akad. Wiss. Berlin, 2: 61, 1837. T: *A. octonarius* C. G. Ehrenberg.

Actinodictyon, Pantocsek. Beitr. Kenntn. Foss. Bacill. Ungarns, 2: 110, pl. 10, f. 176. 1889. T: *A. antiquorum* Pantocsek.

Actinodiscus, Greville. Trans. Microsc. Soc. London, Ser. 2. 11: 69, pl. 4, f. 11. 1863. T: *A. barbadensis* Greville.

Actinogonium, C. G. Ehrenberg. Ber. Bekanntm. Verh. Königl. Preuss. Akad. Wiss. Berlin, 1847: 54, 1847. T: *A. septenarium* C. G. Ehrenberg.

Actinogramma, C. G. Ehrenberg. Ber. Bekanntm. Verh. Königl. Preuss. Akad. Wiss. Berlin, 1872: 276, 1872.

Actinoneis, Cleve. Kongl. Svenska Vetenskapsakad. Handl. Ser. 2. 27(3): 185, 1896. The status of this name, whether genus or subgenus, is obscure. Vide Cleve. op. cit. p. 164. (For this date, see Ross in Bull. Brit. Mus. Nat. Hist. Bot. 3: 51 footnote (1963)).

Actinophaenia, Shadbolt. Trans. Microscop. Soc. London. Ser. 2. 2: 16, 1854. T: *A. splendens* Shadbolt.

Actinoptychus, C. G. Ehrenberg. Abh. Königl. Akad. Wiss. Berlin, 1841 (1): 400, 437, 1843. T: *A. senarius* (C. G. Ehrenberg) C. G. Ehrenberg, (= *Actinocyclus senarius* C. G. Ehrenberg).

Adoneis, Andrews & Rivera. Diatom Research, 2(2): 1–26, 1987. T: *A. pacifica* Andrews & Rivera.

Ailuretta, Sims. Diatom Research, 1: 254, 35–41, 74–75, 1986. T: *A. cerataulinoides* Sims.

Alloeoneis, Pantocsek. Beitr. Kenntn. Foss. Bacill. Ungarns, 2: 55, 1889. (*orth. var.*) + *Alloioneis* K. M. Schumann, 1878. (Schumann's genus).

Alloioneis, J. Schumann in Cleve. Bih. Kongl. Svenska Vetensk.-Akad. Handl. 5(8): 7, 1878. T: *Navicula alternans* K. M. Schumann. (Earlier ref. Verh. K. K. Zool. Bot. Ges. Wien., 1867: 73, 1867).

Allonitzschia, A. Mann. Bull. U. S. Natl. Mus. 100(6): 16, 1925. T: *A. munifica* A. Mann.

Alveoflexus, Hendey & Sims. Bacillaria. 7: 65, pl. 20, 9–15, 1984. T: *A. firthii* Hendey & Sims.

Ambistria, Lohman & Andrews. Geol. Survey Prof. Paper, Contrib. Paleont. 593-E: E 15, 1968. T: *A. hyalina* Lohman & Andrews.

Amphicampa, Rabenhorst. Fl. Eur. Algarum, 1: 257, 1864. (non (C. G. Ehrenberg) Ralfs, 1861). T: *A. alata* (C. G. Ehrenberg) Rabenhorst (= *Navicula* *alata* C. G. Ehrenberg) + *Entomoneis* C. G. Ehrenberg, 1845.

Amphicampa, (C. G. Ehrenberg) Ralfs in Pritchard. Hist. Infus. Ed. 4. 765, 1861. T: *A. mirabilis* (C. G. Ehrenberg) Ralfs (= *Eunotia mirabilis* C. G. Ehrenberg). *Eunotia* Sect. *Amphicampa* C. G. Ehrenberg, Mikrogeologie: 373 (*E. mirifica*: T. 33. F. VII (2)) (*E. mirabilis*), 1854.

Amphipentas, C. G. Ehrenberg. Ueber Thierarten Kreidebildung. 43, 1840. Abh. Königl. Akad. Wiss. Berlin, 1839: 123, 1841. T: *A. pentacrinus* C. G. Ehrenberg (Ber. Bekanntm. Verh. Königl. Preuss. Akad. Wiss. Berlin, 1840: 205, 1840).

Amphipleura, Kützing. Kies. Bacill. Diat. : 103, 7/9 Nov. 1844. LT: *A. pellucida* (Kützing) Kützing ((*Frustulia pellucida* Kützing) (Vide Van Heurck, Syn. Diat. Belgique, 179, 1885: Cleve. Kongl. Svenska Vetenskapsakad. Handl. 26(2): 106, 1894).

Amphiprora, C. G. Ehrenberg. Abh. Königl. Akad. Wiss. Berlin, 1841: 401, 1843. LT: *A. constricta* C. G. Ehrenberg (Vide Rabenhorst, Fl. Eur. Algarum, 1: 253, 258, 1864).

Amphiraphia, Chen & Zhu. Acta phytotax. sinica. 21: 450, 1–3, pl. 1, 1–7, 1983. T: *A. xizangensis* Chen & Zhu.

Amphitetras, C. G. Ehrenberg. Ueber noch jetzt lebende Thierarten. Kreidebildung, 42, 1840. Abh. Königl. Akad. Wiss. Berlin, 1839: 122, 1841. T: *A. antediluviana* C. G. Ehrenberg.

Amphitropis, Rabenhorst. Fl. Eur. Algarum, 3: 416, 1868. T: *Navicula alata* C. G. Ehrenberg + *Entomoneis* C. G. Ehrenberg , 1845. Superfluous substitute name for *Amphicampa* Rabenhorst, 1864. Non (C. G. Ehrenberg) Ralfs, 1861.

Amphora, C. G. Ehrenberg ex Kützing. Kies. Bacill. Diat. : 107, 7/9 Nov. 1844. LT: *A. ovalis* (Kützing) Kützing (= *Navicula amphora* C. G. Ehrenberg) (Vide C. S. Boyer, Proc. Acad. Nat. Sci. Philadelphia 79, Suppl. : 253, 1928). (Note the type of genus is *Navicula amphora* C. G. Ehrenberg. *Amphora ovalis* (Kützing) Kützing (*Frustulia ovalis* Kützing) is a taxonomic but not a nomenclatural synonym.

Anaulus, C. G. Ehrenberg. Ber. Bekanntm. Verh. Königl. Preuss. Akad. Wiss. Berlin, 1844: 197, 1844. T: *A. scalaris* C. G. Ehrenberg.

Ancoropsis, Hendey & Sims. Bacillaria 7: 62 pl. 1A, 1–8, 1984. T: *A. eocenica* Hendey & Sims.

Anisodiscus, Grunow ex Pantocsek. Beitr. Kenntn.

Foss. Bacill. Ungarns, 1: 67, 1886. T: *A. pantocsekii* Grunow ex Pantocsek.

Annellus, Tempère in Tempère & H. Peragallo. Diat. Monde Entier. Ed. 2. 60, 1908. T: *A. californicus* Tempère.

Anomoeoneis, Pfitzer. Bot. Abh. Morphol. Physiol. 2: 77, 1871. T: *A. sphaerophora* Pfitzer.

Anorthoneis, Grunow. Reise Fregatte Novara Bot. 1(1): 9, 1868. T: *A. excentrica* (Donkin) Grunow (= *Cocconeis excentrica* Donkin).

Antelminellia, Schütt. Pflanzenleben Hochsee, 17, 1893. T: *A. gigas* (Castracane) Schütt. (= *Ethmodiscus gigas* Castracane).

Anthemodiscus, J. W. Barker & Meakin. J. Quekett Microscop. Club, Ser. 4. 1: 252, 1944. T: *A. elegans* J. W. Barker & Meakin.

Anthodiscina, P. C. Silva. Taxon, 19: 941, 30 Dec. 1970. T: *A. floreata* (Grove et Sturt) P. C. Silva (= *Anthodiscus floreatus* Grove et Sturt). Substitute name for *Anthodiscus* Grove et Sturt, 1887. (Non G. F. W. Meyer, 1818).

Anthodiscus, E. Grove & G. Sturt. J. Quekett Microscop. Club, Ser. 2. 3: 65, 1887 (Non G. F. W. Meyer, 1818). T: *A. floreatus* E. Grove et G. Sturt).

Arachnodiscus, J. W. Bailey ex C. G. Ehrenberg. Ber. Bekanntm. Verh. Königl. Preuss. Akad. Wiss. Berlin, 1849: 63, 1849. (*Nom. rej.*). T: *A. ornatus* (C. G. Ehrenberg) C. G. Ehrenberg (= *Hemiptychus ornatus* C. G. Ehrenberg). *Nom. rej.* vs *Arachnoidiscus* Deane ex Pritchard, 1852. (*Nom. cons.*).

Arachnoidiscus Deane ex Pritchard. Hist. Infus. Ed. 3. 318, 1852. (*Nom. cons.*) T: *A. japonicus* Shadbolt ex Pritchard.

Arcocellulus, Hasle, von Stosch & Syvertsen. Bacillaria, 6: 54, 1983. T: *A. mammifer* Hasle, von Stosch & Syvertsen.

Arcus, Olshtynskaja. Paleontol. Sb. Lemburg, 15: 77, 1978. T: *A. kasjanicus* Olshtynskaja.

Ardissonea, De Notaris. Erbar. Crittog. Ital. Ser. 2. No. 334. 1870. T: *A. robusta* (Ralfs) De Notaris (= *Synedra robusta* Ralfs).

Ardissonia, Rabenhorst. Hedwigia, 10: 42, 1871 (*Orth. var.*) + *Ardissonea* De Notaris, 1870.

Argonauta, G. Karsten in Engler & Prantl. Nat. Pflanzenfam. Ed. 2, 2: 243, 1928. T: *A. grayii* G. Karsten (= *Grayia argonauta* E. Grove et Brun). Superfluous substitute name for 'Grayia' E. Grove et Brun ex A. Schmidt, 1892, non. W. J. Hooker et

Arnott, 1840 (= *Neograya* O. Kuntze, 1898).

Aristella , Kützing. Linnaea, 8: 563, 1834 (1833) (non (Trinius) Bertoloni, 1833). T: *A. minuta* Kützing.

Arthrogyra, C.G. Ehrenberg ex Ralfs in Pritchard. Hist. Infus. Ed. 4., 822, 1861.

Asterionella, Hassall. Microscop. Exam. Water. T. 2. (Lower Half) F. 5, 1850. T: *A. formosa* Hassall.

Asterodiscus, A. Johnson. Amer. J. Sci. Arts. Ser. 2. 13: 33, 1852.

Asterolampra, C. G. Ehrenberg. Ber. Bekanntm. Verh. Königl. Preuss. Akad. Wiss. Berlin, 1844: 73, 1844. T: *A. marylandica* C. G. Ehrenberg.

Asteromphalus, C. G. Ehrenberg. Ber. Bekanntm. Verh. Königl. Preuss. Akad. Wiss. Berlin, 1844: 198, 1844. LT: *A. darwinii* C. G. Ehrenberg (vide Boyer, Proc. Acad. Nat. Sci. Philadelphia 78, Suppl. 72, 1927).

Attheya, T. West. Trans. Microscop. Soc. London, Ser. 2, 8: 152, 1860. T: *A. decora* T. West.

Aulacocystis, A. H. Hassall. Hist. Brit. Freshwater Algae, 1: 437, 1845. + *Amphipleura* Kützing 1844 (by lectotypification). T: *A. pellucida* (Kützing) A. H. Hassall (*Frustulia pellucida* Kützing).

Aulacodiscus, C. G. Ehrenberg. Ber. Bekanntm. Verh. Königl. Preuss. Akad. Wiss. Berlin, 1844: 73, 1844 (*Nom. cons.*) T: *A. crux* C. G. Ehrenberg.

Aulacoseira, Thwaites. Ann. Mag. Nat. Hist., Ser. 2, 1: 167, 1848 . T: *A. crenulata* (C. G. Ehrenberg) Thwaites (= *Gallionella crenulata* C. G. Ehrenberg).

Auliscus, C. G. Ehrenberg. Ber. Bekanntm. Verh. Königl. Preuss. Akad. Wiss. Berlin, 1843: 270, 1843. T: *A. cylindricus* C. G. Ehrenberg.

Auricula, Castracane. Atti Acad. Pontif. Sci. Nuovi Lincei. 26: 407, 1873. (*Nom. cons.*). T: *A. amphitritis* Castracane.

Auriculopsis, Hendey. Fishery Invest., London, Ser. 4, 5: 158, 1964. T: *A. sparsipunctata* Hendey.

Azpeitia, M. Peragallo in Tempère & H. Peragallo. Diat. Monde Entier, Ed. 2, 326, 1912. T: *A. temperei* M. Peragallo.

Bacillaria, Gmelin. Syst. Nat. Ed. 13. 1(6): 3903, 1791. T: *B. paradoxa* Gmelin *Nom. illeg.* (= *Vibrio paxillifer* O. F. Müller).

Bacteriastrum, Shadbolt. Trans. Microscop. Soc. London, Ser. 2. 2: 14, 1854. LT: *B. furcatum* Shadbolt (vide C. S. Boyer, Proc. Acad. Nat. Sci. Philadelphia, 78 Suppl. 114, 1927).

Bacterosira, Gran. Nyt. Mag. Naturvidensk. 38:114,

1900. T: *B. fragilis* (Gran) Gran (= *Lauderia fragilis* Gran).

Bangia, Lyngbye. Tent. Hydrophytol. Dan: 82, 1819.

Baxteria, Van Heurck. Treat. Diat. (Trans. Baxter): 460, 1896. (non R. Brown ex W. J. Hooker, 1843 (*Nom. cons.*) T: *B. brunii* Van Heurck + *Baxteriopsis* G. Karsten, 1928.

Baxteriopsis, G. Karsten in Engler & Prantl. Nat. Pflanzenfam. Ed. 2, 2: 250, 1928. T: *B. brunii* (Van Heurck) G. Karsten (= *Baxteria brunii* Van Heurck). Substitute name for '*Baxteria*' Van Heurck, 1896, non R. Brown ex W. J. Hooker, 1843 (*Nom. cons.*).

Bellerochea, Van Heurck. Syn. Diat. Belgique, 203, 1885. T: *B. malleus* (Brightwell) Van Heurck (= *Triceratium malleus* Brightwell).

Benetorus, Hanna. Occas. Pap. Calif. Acad. Sci. 13: 15, 1927. T: *B. fantasmus* Hanna.

Bergonia, Tempère. Diatomiste, 1:70, 1891. T: *B. bardadensis* Tempère.

Berkeleya, Greville. Scott. Crypt. Fl. 5, T. 294, 1827. T: *B. fragilis* Greville.

Berkella, Ross & Sims. Bacillaria, 1: 156, 1978. T: *B. linearis* Ross & Sims.

Biblarium, C. G. Ehrenberg. Ber. Bekanntm. Verh. Königl. Preuss. Akad. Wiss. Berlin, 1843: 47, 1843. T: *B. glans* C. G. Ehrenberg.

Biddulphia, S. F. Gray. Nat. Arr. Brit. Pl. 1:294, 1 Nov., 1821. LT: *B. pulchella* S. F. Gray, (*Nom. illeg.*) (= *Conferva biddulphiana* J. E. Smith, *B. biddulphiana* (J. E. Smith) Boyer) (Vide Kützing, Linnaea 8: 567. 579, 1833).

Biddulphiopsis, von Stosch et Simonsen. Bacillaria, 7: 12, 1–35, 1984. T: *B. titiana* von Stosch et Simonsen (= *Cerataulus titianus* Grun.) Verh. zool. -bot. Ges. Wien 13, 159, pl. 4, 25a, b, 1863.

Bifibulatia, Takano. Bull. Tokai Reg. Fish. Res. Lab. 112: 15, 2, 11–18, 1983. T: *B. holsatica* (Helmcke & Krieger) Takano (= *Rhopalodia holsatica* Helmcke & Krieger).

Bogorovia, Jousé. Beih. Nova Hedw. 45: 350, 1974. T: *B. veniamini* Jousé.

Bonea Ross & Sims. Bull. Br. Mus. Nat. Hist. (Bot.), 16: 286, pl. 10; pl. 13, 99, 100, 1987 Nov. 26. T: *B. simulans* Ross & Sims.

Brachysira, Kützing. Algarum Aquae Dulcis German. 16: N. 153, 1836. T: *B. aponina* Kützing.

Brasiliella, C. Zimmermann. Broteria, Ser. Bot. 16:91, 1918. T: *B. helix* C. Zimmermann.

Brebissonia, Grunow. Verh. zool.-bot. Ges. Wien, 10: 512, 1860. (*Nom. cons.*) T: *B. boeckii* (Ehrenberg) O'Meara (Proc. Roy. Irish Acad. Ser. 2. 2: 338, 1875) (= *Cocconema boeckii* Ehrenberg).

Briggera, R. Ross & P. A. Sims. Bull. Br. Mus. Nat. Hist. (Bot.) 13: 291, 1985. T: *B. ornithocephala* (Greville) Ross & Sims (= *Hemiaulus ornithocephala*).

Brightwellia, Ralfs in Pritchard. Hist. Infus. Ed. 4. 940, 1861. T: *B. coronata* (Brightwell) Ralfs (= *Craspedodiscus coronatus* Brightwell).

Brockmanniella, Hasle, von Stosch & Syvertsen. Bacillaria, 6: 34, 1983. T: *B. brockmannii* (Hust.) Hasle, von Stosch & Syvertesen.

Brunia, J. Tempère ex G. B. De Toni. Syll. Algarum 2: 1193, 1894. (Non Linnaeus, 1753). T: *B. japonica* J. Tempère ex G. B. de Toni (= *Neobrunia* O. Kuntze, 1894).

Bruniopsis, G. Karsten in Engler & Prantl. Nat. Pflanzenfam. Ed. 2. 2; 218, 1928. = *Neobrunia* O. Kuntze, 1894. Superfluous substitute name for *Brunia* J. Tempère ex G. B. de Toni, 1894, Non Linnaeus, 1753.

Calodiscus, Rabenhorst. Süssw.-Diat. Freunde Mikr. 12, 1853. T: *C. superbus* Rabenhorst.

Caloneis, Cleve. Kongl. Svenska Vetenskapsakad. Handl. Ser. 2. 26(2): 46, 1894. LT: *C. amphisbaena* (Bory St.-Vincent) Cleve. (= *Navicula amphisbaena* Bory St. Vincent) (Vide C. S. Boyer, Proc. Acad. Nat. Sci. Philadelphia 79 Suppl. 306, 1928).

Campylodiscus, C. G. Ehrenberg ex Kützing. Kies. Bacill. Diat. 59, 7/9 Nov. 1844. LT: *C. clypeus* (C. G. Ehrenberg) C. G. Ehrenberg ex Kützing (*Cocconeis ? clypeus* C. G. Ehrenberg). (Vide C. S. Boyer, Proc. Acad. Nat. Sci. Philadelphia 79 Suppl. 548, 1928).

Campyloneis, Grunow. Verh. zool.-bot. Ges. Wien 12: 429, 1862. T: *C. argus* Grunow.

Campylopyxis, Medlin. Br. phycol. J. 20: 318, 1985. T: *C. garkeana* (Grun.) Medlin. (= *Rhoiconeis garkeana* Grun.).

Campylosira, Grunow ex Van Heurck. Syn. Diat. Belgique, 157, Mai 1885. T: *C. cymbelliformis* (A. Schmidt) Grunow ex Van Heurck (= *Synedra cymbelliformis* A. Schmidt).

Campylostylus, Shadbolt ex Boyer. Proc. Acad. Nat. Sci. Philadelphia, 78 Suppl. : 218, 1927. T: *C. striatus* Shadbolt ex Boyer *nom. superfl.*

(= *Synedra normanniana* Greville)
= *C. normannianus* (Greville) Gerloff, Natour et Rivera. Willdenowia 8: 271, 1978.

Campylostylus Shadbolt. (Ms.) 1849. Grev. Quart. Journ. Microsc. Sci. 10: 232, 1862. T: *C. striatus* Shadbolt. *Synedra normanniana* Grev.

Campylostylus Shadbolt ex van Heurck. Treat. Diat. (Trans. Baxter) 318, 1896. T: *C. striatus* Shadbolt ex van Heurck, *Nom. illeg.* (= *Synedra normaniana* Greville).

Capsula, Brun. Diatomiste, 2: 235, 1896. T: *C. barboi* J. Brun.

Carpartogramma, Kufferath. Expl. Hydrobiol. Lac. Tanganika (1946-47) Res. Sci. 4(3): 27, 1956. T: *C. jeanii* Kufferath. *C. karstenii* (Zanon) R. Ross (= *Schizostauron karstenii* Zanon).

Castracania, G. B. De Toni. Syll. Algarum, 2: 750, 1892. T: *C. boryana* (Pantocsek) G. B. De Toni (= *Salacia boryana* Pantocsek). Substitute name for *Salacia* Pantocsek, 1889. Non Linnaeus, 1771.

Catacombas, Williams et Round. Diat. Res. 1: 315, 1-9, 1986. T: *C. gaillonii* (Bory) Williams & Round (= *Synedra gaillonii* Bory).

Catenula, Mereschkowsky. Bot. Zap. 19: 97, 1903. LT: *C. pelagica* Mereschkowsky (Vide C. S. Boyer, Proc. Acad. Nat. Sci. Philadelphia, 79 Suppl., 492, 1928).

Catillus, Hendey. Beih. Nova Hedwig. 54: 375, 1977. T: *C. subimpletus* (Peragallo) Hendey (= *Cocconeis fluminensis* var? *subimpleta* Peragallo).

Centrodiscus, G. B. De Toni. Notarisia, 5: 922, 1890. (*Orth. var.*) + *Kentrodiscus* Pantocsek, 1889.

Centronella, Max Voigt. Forschungsber, Biol. Stat. Plön. 9: 41, 1902. T: *C. reicheltii* Max Voigt ('*Reichelti*').

Centroporus, Pantocsek. Beitr. Kenntn. Fos. Bacill. Ungarns. 2: 80, 1889. T: *C. crassus* Pantocsek.

Cerataulina, H. Peragallo. Diatomiste, 1: 103, 1892. T: *C. bergonii* (H. Peragallo) Schütt (In Engler & Prantl, Nat. Pflanzenfam. 1(1B): 95, 1896) (= *Cerataulus bergonii* H. Peragallo).

Cerataulus, C. G. Ehrenberg. Ber. Bekanntm. Verh. Königl. Preuss Akad. Wiss. Berlin, 1843: 270, 1843. T: *C. turgidus* (C. G. Ehrenberg) C. G. Ehrenberg (= *Denticella turgida* C. G. Ehrenberg).

Ceratodiscus, Meakin. J. Quekett Microscop. Club. Ser. 4. 1: 100, 1938. T: *C. barkeri* Meakin ('*Barkeri*').

Ceratoneis, C. G. Ehrenberg. Ber. Bekanntm. Verh. Königl. Preuss. Akad. Wiss. Berlin, 1839: 157, 1839. LT: *C. closterium*, C. G. Ehrenberg (Vide W. Smith, Ann. Mag. Nat. Hist. Ser. 2. 9: 9, 1852).

Ceratophora, G. B. De Toni. Notarisia, 5: 922, 1890 (*Orth. var.*) + *Keratophora* Pantocsek, 1889.

Cestodiscus, Greville. Trans. Microscop. Soc. London, Ser. 2. 13: 48, 1865. T: *C. johnsonianus* Greville.

Chaetoceros, C. G. Ehrenberg. Ber. Bekanntm. Verh. Königl. Preuss. Akad. Wiss. Berlin, 1844: 198, 1844. LT: *C. tetrachaeta* C. G. Ehrenberg (vide Boyer, Proc. Acad. Nat. Sci. Philadelphia, 78: Suppl. 104, 1927).

Charcotia, M. Peragallo. 2me Expéd. Antarct. Franç. (1908–10) Comm. J. Charcot, Bot. Diat. 76, 1921. (Non Hue, 1915). ('Comm.' is part of title, abbreviation for 'commande par').

Chasea, Hanna. J. Paleont. 8: 354, 1934. T: *C. bicornis* Hanna.

Cheloniodiscus, Pantocsek. Beitr. Kenntn. Foss. Bacill. Ungarns, 2: 285, 1889. T: *C. ananinensis* Pantocsek.

Chrysanthemodiscus, A. Mann. Bull. U.S. Natl. Mus. 100(6): 58, 1925. T: *C. floriatus* A. Mann.

Chuniella, G. Karsten. Wiss. Ergebn. Deutsch. Tiefsee-Exped. Valdivia, 2(2): 129, 1905.

Cistula, Cleve. Kongl. Svenska Vetenskapsakad. Handl. Ser. 2. 26(2): 124, 1894. T: *C. lorenziana* (Grunow) Cleve (= *Navicula lorenziana* Grunow).

Cladogramma, C. G. Ehrenberg. Mikrogeologie, T. 33, F. XIII., 1, 1854. T: *C. californicum* C. G. Ehrenberg.

Clavicula, Pantocsek. Beitr. Kenntn. Foss. Bacill. Ungarns. 1: 37, 1886. T: *C. polymorpha.*

Clavularia, Greville. Trans. Microscop. Soc. London, Ser. 2. 13: 24, 1865. T: *C. barbadensis* Greville.

Cleveamphora, Mereschowsky. Beih. Bot. Centralbl. 15: 27, 1903. T: *Amphora ovalis* C. G. Ehrenberg ex Kützing + *Amphora* C. G. Ehrenberg ex Kützing, 1844 (by lectotypification).

Cleveia, Pantocsek in Tempère. Diatomiste, 2: 162, 1895. T: *Alloeoneis castracanei* Pantocsek.

Clevia, Mereschkowsky. Beih. Bot. Centralbl. 15: 27, 1903. Non *Cleveia* Pantocsek, 1895.

Climacidium, C. G. Ehrenberg. Abh. Königl. Akad. Wiss. Berlin, 1869: 43, 1870.

Climacodium, Grunow. Reise Fregatte Novara Bot. 1(1): 102, 1868. T: *C. frauenfeldianum* Grunow.

Climaconeis, Grunow. Verh. zool.-bot. Ges. Wien, 12: 421, 1862. LT: *C. frauenfeldii* Grunow (Vide Van Heurck, Treat. Diat. (Trans. Baxter): 346, 1896) = *Climaconeis lorenzii* Grunow.

Climacosira, Grunow. Verh. zool.-bot. Ges. Wien, 12: 424, 1862. T: *C. mirifica* (W. Smith) Grunow (= *Rhabdonema mirificum* W. Smith).

Climacosphenia, C. G. Ehrenberg. Abh. Königl. Akad. Wiss. Berlin, 1841: 401 (1843). T: *C. moniligera* C. G. Ehrenberg.

Cocconeis, C. G. Ehrenberg. Abh. Königl. Akad. Wiss. Berlin, 1835: 173, 1837 (1836). LT: *C. scutellum* C. G. Ehrenberg (vide C. S. Boyer, Proc. Acad. Nat. Sci. Philadelphia 79 Suppl. 242, 1928).

Cocconema, C. G. Ehrenberg. Abh. Königl. Akad. Wiss. Berlin, 1830: 40, 62, 1832. T: *C. cistula* (C. G. Ehrenberg) C. G. Ehrenberg (= *Bacillaria cistula* C. G. Ehrenberg). *Cocconema* Ehrenberg in Hemprich et Ehrenberg, Symbolae Physicae. Pars Zoologica. Animalia evertebrata exclusis insectis. Phytozoa. Polygastrica et Rotatoria africana et asiatica: [7]. 1831.

Coenobiodiscus, A. R. Loeblich III, W. W. Wight & Darley. J. Phycol. 4: 27, 1968. T: *C. muriformis* A. R. Loeblich III, W. W. Wight & Darley. (= *Planktoniella muriformis* (Loeblich, Wight & Darley) Round).

Colletonema, Brébisson ex Kützing. Sp. Algarum, 105, 23–24 Jul. 1849. T: *C. viridulum* Brébisson ex Kützing.

Cometes, Edsbagge. Bot. Gothoburgensia 6: 56, 1968. T: *C. globosa*

Compositus, Vekschina Trudy Sib. nauchno-issled. Inst. Geol. Geofiz. miner. Syr'ya 8: 161, 1960. T: *C. ergenii* Vekschina.

Corethron, Castracane. Rep. Sci. Res. Voy. Challenger, Bot. 2: 85, 1886. LT: *C. criophilium* Castracane (Vide C. S. Boyer, Proc. Acad. Nat. Sci. Philadelphia 78 Suppl. 114, 1927).

Corinna, Heiberg. Consp. Crit. Diat. Dan. 53, 1863. T: *C. elegans* Heiberg.

Corona, Lefébure & Chenevière. Bull. Soc. Franc. Microscop. 7: 9, 1938. T: *C. magnifica* Lefebure et Chenevière.

Coscinodiscus, C. G. Ehrenberg. Abh. Königl. Akad. Wiss. Berlin, 1838: 128, 1840. LT: *C. argus*, C. G. Ehrenberg. (Vide R. Ross & P. A. Sims, Beih. Nova Hedwig. 45: 103, 1974).

Coscinophaena, C. G. Ehrenberg ex G. B. De Toni.

Syll. Algarum, 2: 1370, 1894.

Coscinosira, Gran. Nyt. Mag. Naturvidensk. 38: 115, 1900. T: *C. polychorda* (Gran) Gran (= *Coscinodiscus polychordus* Gran).

Cosmiodiscus, Greville. Trans. Microscop. Soc. London, Ser. 2. 14: 79, 1866

Cotyledon, Brun. Mém. Soc. Phys. Hist. Genève, 31(2, 1): 23, 1891 (Non Linnaeus, 1753). LT: *C. circularis* Brun. (Vide Van Heurck, Treat. Diat., Transl. Baxter: 519, 1896). + *Gutwinskiella* De Toni, 1894.

Craspedodiscus, C. G. Ehrenberg. Ber. Bekanntm. Verh. Königl. Preuss. Akad. Wiss. Berlin, 1844: 261, 1844. T: *C. elegans* C. G. Ehrenberg.

Craspedoporus, Greville. Trans. Microscop. Soc. London, Ser. 2. 11: 68, 1863. T: *C. ralfsianus* Grev.

Craticula, Grunow. Reise Fregatte Novara Bot. 1(1): 20, 1868. T: *C. perrotettii* Grunow.

Creswellia, Arnott ex Greville in W. Gregory. Trans. Roy. Soc. Edinburgh, 21: 538, 1857. T: *C. turris* Greville.

Crucidenticula, Akiba & Yangisawa. Init. Rep. Deep Sea Drilling Proj. 87, 486, pl. 1, 9; pl. 2, 1–7; pl. 5, 1–9, 1986 May. T: *C. nicobarica* (Grun.) Akiba & Yangisawa (= *Denticula nicobarica* Grun.).

Ctenodiscus G. B. De Toni. Syll. Algarum, 2: 1012, 1894. (*Orth. var.*) *Ktenodiscus* Pantocsek, 1889.

Ctenophora, Williams & Round. Diatom Research, 1, 330, 53–61, 1986 Nov. T: *C. pulchella* (Ralfs ex Kütz.) Williams & Round (= *Synedra pulchella* Ralfs ex Kutz.)

Cuneolus, Giffen. Bot. Mar. 13: 90, 1970. T: *C. minutus* Giffen.

Cussia, Schrader. Init. Rep. Deep Sea Drilling Project 24: 914, 1974. & Proc. Cal. Acad. Sci. Ser. 4: 39, 541, 1974. T: *C. lancettula*

Cyclophora, Castracane. Atti. Accad. Pontif. Sci. Nuovi Lincei, 31: 186, 1878. T: *C. tenuis* Castracane.

Cyclosira, H. Peragallo & M. Peragallo. Diat. Mar. France, 436, 1907. T: *C. bergonii* H. et M. Peragallo.

Cyclostephanos, Round, Arch. Protistenk. 125: 326, 1982. T: *C. dubius* (Fricke) Round.

Cyclotella (Kützing) Brébisson. Consid. Diatom. 19, 1838 *nom. cons.* T: *C. operculata* (C. A. Agardh) Brébisson (=*Frustulia operculata* C. A. Agardh) (*Frustulia* subgen. *Cyclotella* Kützing, Linnaea 8: 535, 1833). T: Tenstaedt, Erfurt, Germany, Kützing

139 (BM praep. coll. Diat. 17986), *typ. cons.*

Cylindropyxis, Hendey. Fishery Invest. London, Ser. 4. 5: 92, 1964. T: *C. tremulans* Hendey.

Cylindrotheca, Rabenhorst. Algen Sachsens, N. 801, 1859. T: *C. gerstenbergeri* Rabenhorst.

Cymatodiscus, Hendey. J. Roy. Microscop. Soc. London, Ser. 3. 77: 42, Oct. 1958. T: *C. planetophorus* (Meister) Hendey (= *Coscinodiscus* ? *Planetophorus* Meister).

Cymatoneis, Cleve. Kongl. Svenska Vetenskapsakad. Handl. Ser. 2. 26(2): 75, 1894. LT: *C. sulcata* (Greville) Cleve (= *Navicula sulcata* Greville) (Vide C. S. Boyer, Proc. Acad. Nat. Sci. Philadelphia, 79 Suppl. 306, 1928).

Cymatonitzschia, Simonsen. 'Meteor' Forsch. D. 19: 56, 1974. T: *C. marina* (Lewis) Simonsen.

Cymatopleura, W. Smith. Ann. Mag. Nat. Hist. Ser. 2. 7: 12, 1851 *nom. cons.* T: *C. solea* (Brébisson) W. Smith, *nom. illeg.* + *C. librile* (Ehrenberg) Pantocsek

Cymatosira, Grunow. Verh. zool.-bot. Ges. Wien, 12: 377, 1862. T: *C. lorenziana* Grunow.

Cymatotheca, Hendey. J. Roy. Microscop. Soc. London, Ser. 3. 77: 41, Oct. 1958. T: *C. weissflogii* (Grunow) Hendey (= *Euodia weissflogii* Grunow in Van Heurck).

Cymbella, C. A. Agardh. Consp. Crit. Diat. 1: 1830 *nom. cons.* T: *C. cymbiformis* C. Agardh, *typ. cons.*

Cymbellonitzschia, Hustedt in A. Schmidt, Atlas Diat. T. 352, F. 12, 13, 1924. T: *C. minima* Hustedt.

Cymbophora (Kützing) Brébisson. Consid. Diat. 13: 1838. T: *C. gastroides nob. Frustulia* Kütz.

Cymbosira, Kützing. Kies. Bacill. Diat. 77, 7/9 Nov. 1844. T: *C. agardhii* Kützing, *Nom illeg.* (= *Achnanthes seriata* C. A. Agardh.).

Cystopleura, Brébisson ex O. Kuntze. Rev. Gen. 2: 890, 5 Nov. 1891. T: *Frustulia picta* Kützing (= *Cystoplura turgida* (Ehrenberg) Kuntze) + *Epithema* Brébisson. Non *Epithema* Blume + *Epithemia* Brébisson ex Kützing.

Dactyliosolen, Castracane. Rep. Sci. Res. Voy. Challenger, Bot. 2: 75, 1886. T: *D. antarcticus* Castracane.

Debya, Pantocsek. Beitr. Kenntn. Foss. Bacill. Ungarns, 1: 65, 1886. T: *D. insignis* Pantocsek.

Debya, Rattray. J. Roy. Microscop. Soc. London, 1888: 909, 1888. (Non. Pantocsek, 1886). T: *D. oamaruensis* (Grunow) Rattray + *Rattrayella* De Toni, 1889.

Delphineis, Andrews. Beih. Nov. Hedwig. 54: 249, 1977. T: *D. angustata* (Pantoscek) Andrews (= *Rhaphoneis angustata* Pantocsek, 1886).

Dendrella, Bory de St.-Vincent. Dict. Class. Hist. Nat. 5: 393, 15 Mai, 1824. Encycl. Méth., Hist. Nat. Zoophytes 242, Jul. 1824 (*Nom. rej.*). Nom. rej. vs *Didymosphenia* M. Schmidt, 1899 (*Nom. cons.*).

Denticella, C. G. Ehrenberg. Infus. 210, 1838. T: *D. aurita* (Lyngbye) C. G. Ehrenberg (= *Diatoma aurita* Lyngbye) + *Odontella* C. A. Agardh, 1832.

Denticula, Kützing. Kies. Bacill. Diat. 43, 7/9 Nov. 1844. LT: *D. elegans* Kützing (Vide C. S. Boyer, Proc. Acad. Nat. Sci. Philadelphia, 79 Suppl. 530, 1928).

Denticulopsis, Simonsen. Bacillaria, 2: 63, 1979. T: *D. hustedtii* (Simonsen & Kanaya) Simonsen (= *Denticula hustedtii* Simonsen & Kanaya, Int. Rev. ges. Hydrobiol. 46: 501, 1961).

Desmogonium, C. G. Ehrenberg in Schomburgk. Reisen Brit.-Guiana 1840–44, 3 Fauna Fl: 539, 1848. T: *D. guianense* C. G. Ehrenberg.

Detonia, Frenguelli. Darwiniana, 9: 119, 1949. T: *D. superba* (Janisch) Frenguelli (= *Cocconeis superba* Janisch).

Detonula, Schütt ex G. B. De Toni. Syll. Algarum, 2: 1425, 1894 (? Type species, *Lauderiae* ?).

Dextradonator, Ross & Sims. Bacillaria, 3: 115, 1980. T: *D. eximius* (Grunow) Ross & Sims (= *Syringidium eximium* Grunow in Van Heurck (1880–85) pl. 106, figs 1, 3 (1883).

Diadema, K.-D. Kemp & T. B. B. Paddock. Diat. Res. 4: 39, 1989. T: *D. luxuriosa* (Greville) Kemp & Paddock (= *Navicula luxuriosa*).

Diadesmis, Kützing. Kies. Bacill. Diat. 109, 7/9 Nov. 1844. T: *D. confervacea*. F. Kützing.

Diatoma, A. P. de Candolle in Lamarck & A. P. de Candolle, Fl. Franc. Ed. 3. 2: 48, 17 Sept. 1805. (Non. Loureiro, 1790 (*Nom. rej.*) Nec Bory de-St. Vincent 1824 (*Nom. cons.*) + *Neodiatoma* Kanitz 1887.

Diatoma, Bory de St.-Vincent. Dict. Class. Hist. Nat. 5: 461, 15 Mai, 1824 (*Nom. cons.*) T: *D. vulgare* Bory de St.-Vincent '*Vulgaris*' (*Typ. cons.*)

Diatomella, Greville. Ann. Mag. Nat. Hist. Ser. 2. 15: 259, 1855. (*Nom. cons.*). T: *D. balfouriana* (W. Smith) Greville (= *Grammatophora* ? *balfouriana* W. Smith).

Dichomeneis, C. G. Ehrenberg. Ber. Bekanntm.

Verh. Königl. Preuss. Akad. Wiss. Berlin, 1861: 293, 1862. T: *D. subtilis* C. G. Ehrenberg.

Dickieia, Berkeley ex Kützing. Kies. Bacill. Diat. 119, 7/9 Nov. 1844. T: *D. ulvacea* Berkeley ex Kützing. *Dickieia* Berkeley et Ralfs, Ann. Mag. Nat. Hist. 14; 328, Nov. 1844.

Dicladia, C. G. Ehrenberg. Ber. Bekanntm. Verh. Königl. Preuss. Akad. Wiss. Berlin, 1844: 73, Feb. 1844. T: *D. capreolus* C. G. Ehrenberg (vide C. G. Ehrenberg, loc. cit. 271, 1844).

Dicladiopsis, G. B. De Toni. Syll. Algarum, 2: 1003, 1894. LT: *D. barbadensis* (Greville) De Toni (*Dicladia barbadensis* Greville). (cf. Ross et Sims, Bull. Brit. Mus. (Nat. Hist.) Bot. 13: 317, 1985).

Dictyocysta, C. G. Ehrenberg. Ber. Bekanntm. Verh. Königl. Preuss. Akad. Wiss. Berlin, 1854: 235, 1854.

Dictyolampra, C. G. Ehrenberg. Ber. Bekanntm. Verh. Königl. Preuss. Akad. Wiss. Berlin, 1847: 54, 1847. T: *D. stella* C. G. Ehrenberg.

Dictyoneis, P. T. Cleve. Diatomiste, 1: 14, 1890. LT: *D. marginata* (F. W. Lewis) P. T. Cleve (= *Navicula marginata* F. W. Lewis) (vide C. S. Boyer, Proc. Acad. Nat. Sci. Philadelphia, 79 Suppl. 343, 1928).

Dictyopyxis, C. G. Ehrenberg. Ber. Bekanntm. Verh. Königl. Preuss. Akad. Wiss. Berlin, 1844: 262, 1844.

Didymosphenia, M. Schmidt in A. Schmidt. Atlas Diat. T. 214, F. 1–12, 1899. (*Nom. cons.*). T: *D. geminata* (Lyngbye) M. Schmidt (= *Echinella geminata* Lyngbye) (*Typ. cons.*).

Dimeregramma, Ralfs in Pritchard. Hist. Infus. Ed. 4. 790, 1861. LT: *D. minor* (Gregory) Ralfs (= *Denticula minor* Gregory) (vide C. S. Boyer, Proc. Acad. Nat. Sci. Philadelphia, 78 Suppl. 192, 1927).

Dimeregrammopsis, M. Ricard. Cryptogamie-Algologie, 8: 1987. T: *D. furcigerum* (Grunow) Ricard.

Dimmerogramma, G. B. De Toni. Syll. Algarum, 2: 711, 1892. (*Orth. var.*) *Dimeregramma* Ralfs 1861.

Dimidiata, Hajós. Beih. Nova Hedwig. 45: 373, 1974. T: *D. saccula* Hajós.

Diomphala, C. G. Ehrenberg. Ber. Bekanntm. Verh. Königl. Preuss. Akad. Wiss. Berlin, 1842: 336, 1842. T: *D. clava-herculis* C. G. Ehrenberg ('*Clava herculis*') Nom. rej. vs. *Didymosphenia* M. Schmidt, 1899.

Diploneis, C. G. Ehrenberg ex Cleve. Kongl. Svenska Vetenskapsakad. Handl. Ser. 2. 26(2): 76, 1894. (Note: Ehrenberg 1840 sect. of *Pinnularia* ? Ber. Akad. Berlin 84: (1840? 1844?)).

Diplomenora, K. Blazé. Br. phycol. J. 19: 218, 1984. T: *D. cocconeiformis* (Schmidt) Blazé (= *Coscinodiscus cocconeiformis*).

Discodiscus, Gombos. Bacillaria, 3: 241, 1980. T: *D. tetraporus* (Brun) Gombos.

Disconeis, Cleve. Kongl. Svenska Vetenskapsakad. Handl. Ser. 2. 27(3): 180, 1896. Vide Cleve, op. cit. P. 164. (Note: Status of this name, whether genus or subgenus, is obscure). Vide Cleve, op. cit. p. 194.

Discoplea, Ehrenberg. Ber. Bekanntm. Verh. Königl. Preuss. Akad. Wiss. Berlin 1840: 208, 1840. T: *D. kuetzingii* Ehrenberg.

Discosira, Rabenhorst. Süssw. -Diat. Freunde Mikr. 12, 1853. T: *D. sulcata* Rabenhorst.

Discus, Stodder. Amer. J. Microscop. 4: 14, 1879. T: *D. porcelaineous* Stodder.

Disiphonia, C. G. Ehrenberg. Mikrogeologie 260, 1854. T: *D. australis* C. G. Ehrenberg *Nom. rej.* vs. *Diatomella* Greville 1855 (*Nom. cons.*)

Ditylium, G. B. De Toni. Syll. Algarum, 2: 1017, 1894, (*Orth. var.*) + *Ditylum* J. W. Bailey ex L. W. Bailey, 1862.

Ditylum, J. W. Bailey ex L. W. Bailey. Boston J. Nat. Hist. 7: 332, 1862. T: *D. trigonum* Bail.

Donkinia, Ralfs in Pritchard. Hist. Infus. Ed. 4. 920, 1861. T: *D. carinata* (Donkin) Ralfs. (= *Pleurosigma carinatum* Donkin).

Doryphora, Kützing. Kies. Bacill. Diat. 74, 7/9 Nov. 1844. T: *D. amphiceros* (C. G. Ehrenberg) Kützing (= *Rhaphoneis amphiceros* C. G. Ehrenberg).

Dossetia, Azpeitia. Diat. Espanola, 202, 1911. T: *D. temperei* Azpeitia.

Drepanotheca, Schrader. Beih. Nov. Hedwig. 28: 13, 1969. T: *D. macra* Schrader.

Druridgea, G. B. De Toni. Syll. Algarum, 2: 1365, 1894. (*Orth. var.*) + *Druridgia* Donkin, 1861

Druridgia, Donkin. Quart. J. Microscop. Sci. Ser. 2. 1. : 13, 1861. T: *D. geminata* Donkin, *Nom. superfl.* (= *D. compressa* (T. West) H. L. Smith).

Dycladia, Pantocsek. Beitr. Kenntn. Foss. Bacill. Ungarns, 3: T. 5, F. 70, 1892. (*Orth. var.*) + *Dicladia* C. G. Ehrenberg, 1844. T: *D. japonica* Pantocsek.

Echinodiscus, A. Mann. Bull. U. S. Natl. Mus. 100(6): 75, 1925. (Non. (A. P. de Candolle) Bentham 1837). T: *E. vermiculatus* A. Mann.

Ellerbeckia, R. M. Crawford. Algae and the Aquatic Environment (ed. F. E. Round) p. 421. Bristol, Biopress. 1988. T: *E. arenaria* (Moore ex Ralfs) Crawford.

Encyonema, Kützing. Linnaea, 8: 589, 1833. T: *E. paradoxum* Kützing.

Endictya, C. G. Ehrenberg. Ber. Bekanntm. Verh. Königl. Preuss. Akad. Wiss. Berlin, 1845. 71, 1845. T: *E. oceanica* C. G. Ehrenberg.

Entogonia, Greville. Quart. J. Microscop. Sci., Ser. 2. 3: 235, 1863.

Entomogaster, C. G. Ehrenberg. Abh. Königl. Akad. Wiss. Berlin, 1870: 52, 1871. T: *E. polygastricis bacillariis*.

Entomoneis, C. G. Ehrenberg. Ber. Bekanntm. Verh. Königl. Preuss. Akad. Wiss. Berlin, 1845. 71, 1845. T: *E. alata = Navicula alata*.

Entopyla, C. G. Ehrenberg. Ber. Bekanntm. Verh. Königl. Preuss. Akad. Wiss. Berlin, 1848: 7, 1848. T: *E. australis* (C. G. Ehrenberg) C. G. Ehrenberg (= *Surirella ? australis* C. G. Ehrenberg).

Epithelion, Pantocsek. Beitr. Kenntn. Foss. Bacill. Ungarns, 3: 48, T. 17, F. 253, 1892. T: *E. curvatum* Pantocsek.

Epithema, Brébisson. Consid. Diat. 16, 1838 (Non *Epithema* Blume 1826) + *Epithemia* Kützing, 1833. T: *E. pictum*.

Epithemia, Kützing. Kies. Bacill. Diat. 33, 7/9 Nov. 1844. LT: *E. turgida* (C. G. Ehrenberg) Kützing (=*Eunotia turgida* C. G. Ehrenberg) vide Boyer, Proc. Acad. Nat. Sci. Philadelphia, 79 Suppl. 488, 1928.

Ethmodiscus, Castracane. Rep. Sci. Res. Voy. Challenger Bot. 2. : 166, 1886. T: *E. punctiger* A. F. Castracane.

Eucampia, C. G. Ehrenberg. Ber. Bekanntm. Verh. Königl. Preuss. Akad. Wiss. Berlin, 1839: 156, 1839. T: *E. zodiacus* C. G. Ehrenberg.

Eucocconeis, Cleve. Kongl. Svenska Vetenskapsakad. Handl., Ser. 2. 27(3): 173, 1896. Status of this name, whether genus or subgenus, is obscure. Vide Cleve, op. cit. P. 164.

Eumeridion, Kützing. Kies. Bacill. Diat. 42, 7/9 Nov. 1844. T: *E. constrictum* (Ralfs) Kützing (= *Meridion constrictum* Ralfs).

Eunotia, C. G. Ehrenberg. Ber. Bekanntm. Verh.

Königl. Preuss. Akad. Wiss. Berlin, 2: 44, 1837. LT: *E. arcus* C. G. Ehrenberg (Vide C. S. Boyer, Proc. Acad. Nat. Sci. Philadelphia, 78, Suppl. 215, 1927).

Eunotogramma, Weisse. Mélanges Biol. Bull. Saint-Pétersbourg, 2: 243, 1855.

Euodia, J. W. Bailey ex Ralfs in Pritchard, Hist. Infus. Ed. 4. 852, 1861. (Non J. R. et J. G. A. Forster, 1776). T: *E. gibba* J. W. Bailey ex Ralfs.

Euphyllodium, Shadbolt. Trans. Microscop. Soc. London, Ser. 2. 2: 14, 1854. T: *E. spathulatum* Shadbolt. *Nom. rej.* vs *Podocystis* J. W. Bailey 1854 (*Nom. cons.*).

Eupleuria, Arnott. Quart. J. Microscop. Sci. 6: 89, 1858. T: *E. pulchella* G. A. W. Arnott.

Eupodiscus, C. G. Ehrenberg. Ber. Bekanntm. Verh. Königl. Preuss. Akad. Wiss. Berlin, 1844. 73, 1844 (*Nom. rej.*) T: *E. germanicus* (C. G. Ehrenberg) C. G. Ehrenberg. (*Tripodiscus germanicus* C. G. Ehrenberg). *Nom. rej.* vs *Eupodiscus* J. W. Bailey, 1851 (*Nom. cons.*).

Eupodiscus, Rattray. J. Roy. Microsc. Soc. 1888, 909, 1851. (*Nom. cons.*) T: *E. radiatus* J. W. Bailey (*Typ. cons.*).

Exilaria, Greville. Scot. Crypt. Fl. Vol. 5., 289, May 1827. T: *E. flabellata* Greville. *Nom. rej.* vs *Licmophora* C. A. Agardh, 1827.

Extubocellulus, Hasle, von Stosch & Syvertsen. Bacillaria, 6: 69, 1983. T: *E. spinifer* (Hargr. et Guill.) Hasle, von Stosch & Syvertsen.

Falcatella, Rabenhorst. Süssw.-Diat. Freunde Mikr. 46, 1853. T: *F. romana* Rabenh.

Falcula, Voigt. Rev. Algol. N. S. 5: 85, 1960.

Fenestrella, Greville. Trans. Microscop. Soc. London, Ser. 2. 11: 67, 1863. T: *F. barbadensis* Greville.

Flexibiddulphia, Simonsen. Atlas and Catalogue of the Diatom Types of Friedrich Hustedt. J. Cramer: Stuttgart, 1: 263, 1987 Aug. T: *F. semicircularis* (Brightwell) Simonsen (= *Triceratium semicirculare* Brightwell).

Florella, Navarro. Bot. Mar. 25: 248, 1982. T: *F. portoricensis* Navarro.

Fragilaria, Lyngbye. Tent. Hydrophytol. Dan. 182, 1819. LT: *F. pectinalis* (O. F. Müller) Lyngbye (= *Conferva pectinalis* O. F. Müller). (Vide C. S. Boyer, Proc. Acad. Nat. Sci. Philadelphia, 78, Suppl. 183, 1927).

Fragilariella, Hendey. J. Roy. Microscop. Soc. London, Ser. 3. 377: 54, Oct. 1958. T: *F. dusenii*

(Cleve) Hendey (= *Denticula? dusenii* Cleve).

Fragilariforma, D. M. Williams & F. E. Round. Diat. Res. 3, 265, 1988. T: *F. virescens* (Ralfs) Williams & Round.

Fragilariopsis, Hustedt in A. Schmidt. Atlas Diat. T. 299, F. 9-14, 1913. T: *F. antarctica* (Castracane) Hustedt (= *Fragilaria antarctica* Castracane, 1886. Non. A. Schwarz, 1887).

Frickea, Heiden ex A. Schmidt, Atlas Diat. T. 264, F. 1. April 1906. T: *F. lewisiana* (Greville) Heiden ex A. Schmidt (= *Navicula lewisiana* Greville).

Frustulia, C. A. Agardh. Syst. Algarum, XIII, 1, 1824 (*Nom. rej.* vs. *Frustulia*, Rabenhorst, 1853). (*Nom. cons.*). T: *F. obtusa*.

Frustulia, Rabenhorst. Süssw.-Diat. Freunde Mikr. 50, 1853 (*Nom. cons.*). T: *F. saxonica* Rabenhorst (*Typ. cons.*).

Fusotheca, Mereschkowsky. Trudy St. Petersburgsk. Obsc. Estestvoisp. 9: 446, 1878. T: *F. polaris* Mereschkowsky.

Gaillonella, Bory de St.-Vincent. Dict. Class. Hist. Nat. 7: 101, 5 Mar. 1825. T: *Conferva moniliformis* O. F. Müller + *Lysigonium* (Link 1820).

Gephyria, Arnott. Quart. J. Microscop. Sci. 6: 163, 1858. T: *G. telfairiae* Arnott.

Girodella, Gaillon ex Turpin. Mém. Mus. Hist. Nat. 15: 318, 1827. T: *G. comoides* (Dillwyn) Gaillon ex Turpin (= *Conferva comoides* Dillwyn).

Gladius, Forti & Schulz. Beih. Bot. Centralbl. 50(2): 242, 1932. T: *G. antiquus* Forti et Schulz.

Gleseria, Lupikina & Dolmatova, Bot. Zh. SSSR, 69, 1532, Tab. 1, 1984. T: *G. penzhica* Lupikina & Dolmatova

Gloionema, C. A. Agardh. Disp. Algarum Suec. 45, 1812. T: *G. paradoxum* C. A. Agardh.

Glorioptychus, Hanna. Occas. Pap. Calif. Acad. Sci. 13: 19, 1927. T: *G. callidus* Hanna.

Glyphodesmis, R. K. Greville. Quart. J. Microscop. Sci. Ser. 2. 2: 234, 1862. T: *G. eximia* R. K. Greville.

Glyphodiscus, Greville. Trans. Microscop. Soc. London, Ser. 2. 10: 91, 1862. T: *G. stellatus* Greville.

Gomphocaloneis, Meister. Kieselalg. Asien 41, 1932. T: *G. undulata* Meister.

Gomphocymbella, O. Müller. Bot. Jahrb. Syst. 36: 145, 1905. T: *G. vulgaris* (Kütz.) O. Müller.

Gomphogramma, A. Braun in Rabenhorst. Süssw.-Diat. Freunde Mikr. 33, 1853. T: *G. rupestre* A.

Braun.

Gomphoneis, Cleve. Kongl. Svenska Vetensk.-Akad. Handl., Ser. 2, 26(2): 73, 1894.

Gomphonella, Rabenhorst. Süssw.-Diat. Freunde Mikr. 61, 1853. T: *G. olivacea* Rabenh.

Gomphonema, C. A. Agardh. Syst. Algarum, XVI, 11, 1824 (*Nom. rej.*). T: *G. geminatum* (Lyngbye) C. A. Agardh. (= *Echinella geminata* Lyngbye) *Nom. rej.* vs. *Gomphonema* C. G. Ehrenberg, 1832. (*Nom. cons.*).

Gomphonema, C. G. Ehrenberg. Abh. Königl. Akad. Wiss. Berlin, 1831: 87, 1832 (*Nom. cons.*). T: *G. acuminatum* C. G. Ehrenberg (*Typ. cons.*).

Gomphonemopsis, L. Medlin in L. Medlin & F. E. Round. Diat. Res. 1: 207, 1986. T: *G. exigua* (Kütz.) Medlin (= *Gomphonema exiguum*).

Gomphonitzschia, Grunow. Reise Fregatte Novara Bot. 1(1): 7, 1868. T: *G. ungeriana* Grunow.

Gomphopleura, Reichelt ex Tempère. Diatomiste, 2: 80, 1894. T: *G. nobilis* Reichelt ex Tempère.

Gomphoseptatum, L. Medlin in L. Medlin & F. E. Round. Diat. Res. 1: 212, 1986. T: *G. aestuarii* (Cleve) Medlin (= *Gomphonema aestuarii*).

Gomphosphaeria, Kützing. Algarum Aquae Dulcis German. 16: 151, 1836. T: *G. aponina* Kützing. (*NOTE*: This belongs in the Cyanophyceae-Chroococcaceae).

Gomphotheca, Hendey & Sims. Bacill. 5: 199, 1982. T: *G. sinensis* (Skvortzow) Hendey & Sims.

Gonioceros, H. Peragallo & M. Peragallo. Diat. Mar. France, 468, 1907. T: *G. armatum* (T. West) H. et M. Peragallo (= *Chaetoceros armatum* T. West)

Goniothecium, C. G. Ehrenberg. Abh. Königl. Akad. Wiss. Berlin, 1, 1841: 401 (1843). T: *G. rogersii* C. G. Ehrenberg.

Gossleriella, Schütt. Pflanzenleben Hochsee, 20, 1893. T: *G. tropica* Schütt.

Grammatonema, Kützing. Phycol. German, 140, Jul-Aug. 1845. T: *G. striatulum* Kützing.

Grammatophora, C. G. Ehrenberg. Noch zahlreich jetzt lebende Thierarten Kreidebildung 46, 1840. Abh. Akad. Wiss. Berlin, 1839: 126, 152, 1841. LT: *G. angulosa* Ehrenberg (vide C. S. Boyer, Proc. Acad. Nat. Sci. Philadelphia 78 Suppl. 154, 1927). Ber. Akad. Wiss. Berlin, 1840: 161, Oct. 1840 (*NOTE: which may have been earlier than the above*).

Grammonema, C. A. Agardh. Consp. Crit. Diat. 63, 1832. T: *G. jurgensii* C. A. Agardh.

Grayia, E. Grove & Brun in A. Schmidt Atlas Diat., T. 172, Fig. 11, Mai 1892 (Non W. J. Hooker et Arnott, 1840). T: *G. argonauta* E. Grove et Brun = *Neograya* O. Kuntze, 1898.

Groentvedia, Hendey. Fishery Invest. London, Ser. 4. 5: 73, 1964. T: *G. elliptica* Hendey.

Grovea, A. Schmidt ex Van Heurck (= *Biddulphia pedalis* E. Grove et G. Sturt).

Grunoviella, Van Heurck. Treat. Diat. 332, 1896. T: *G. gemmata* (Grunow) Van Heurck (= *Sceptroneis ? gemmata* Grunow).

Grunowia, Rabenhorst. Fl. Eur. Algarum, 1: 16, 146, 1864. T: *G. sinuata* (Thwaites ex W. Smith) Rabenhorst.

Guinardia, H. Peragallo. Diatomiste, 1: 107, 1892. T: *G. flaccida* (Castracane) Peragallo.

Gutwinskiella, G. B. De Toni. Syll. Algarum, 2: 1323, 1894. T: *G. clypeolum* (Brun.) De Toni.

Gyrodiscus, Witt. Verh. Russ. -Kais. Mineral. Ges. St.-Petersburg. Ser. 2. 22: 161, 1886. T: *G. vortex* Witt.

Gyroptychus, A. Schmidt. Atlas Diat. T. 149, F. 19, Feb. 1890. T: *G. contabulatus* A. Schmidt.

Gyrosigma, Hassall. Hist. Brit. Freshw. Algae, 1: 435, 1845 (*Nom. cons.*). T: *G. hippocampus* (Ehrenberg) Hassall (*hippocampa*) (= *Navicula hippocampus* C. G. Ehrenberg).

Halionyx, C. G. Ehrenberg. Ber. Bekanntm. Verh. Königl. Preuss. Akad. Wiss. Berlin, 1844: 198, 1844.

Halurina, C. Zimmermann. Broteria, Ser. Bot. 16: 90, 1918. T: *H. itaparicana* C. Zimmermann.

Handmannia, M. Peragallo. Mitteil. Mikr. Ver. Linz. 1:14, 1913. T: *H. austriaca* M. Peragallo.

Hannaea, Patrick. Patrick & Reimer: Diatoms of U. S. 131: 1966. Monographs Acad. Nat. Sci. Philadelphia,13. T: *H. arcus* (Ehrenb.) Patr. (= *Navicula arcus* C. G. Ehrenberg).

Hantzschia, Grunow. Monthly Microscop. J. 18: 174, 1 Oct, 1877. (*Nom. cons.*). T: *H. amphioxys* (C. G. Ehrenberg) Grunow (= *Eunotia amphioxys* C. G. Ehrenberg) (*Typ. cons.*)

Haslea, Simonsen. 'Meteor' Forschungsergebnisse, Reihe D. 19:46, 1974. T: *H. ostrearia* (Gaillon).

Haynaldella, Pantocsek. Beitr. Kenntn. Foss. Bacill. Ungarns 3: 57, T. 32, F. 459, 1892. T: *H. antiqua* Pantocsek.

Haynaldia, Pantocsek. Beitr. Kenntn. Foss. Bacill. Ungarns, 2: 120, 1889. (Non Schur, 1866). T: *H.*

antiqua Pantocsek.

Heibergia, Greville. Trans. Micr. Soc. London New Ser. 13: 100, 1865. T: *H. barbadensis* Greville.

Heliopelta, C. G. Ehrenberg. Ber. Bekanntm. Verh. Königl. Preuss. Akad. Wiss. Berlin, 1844: 262, 1844. T: *H. leeuwenhoeckii.*

Helisella, Jurilj. Prir. Istraz, Hrvatske Slavonije, 24: 181, 1948. T: *H. glabra* Jurilj.

Helminthopsidella, Silva. Taxon, 19: 943, 30 Dec. 1970. T: *H. weissflogii* (Van Heurck) Silva (= *Helminthopsis weissfloggi* Van Heurck). Substitute name for *Helminthopsis* Van Heurck, 1896. Non Heer, 1877.

Helminthopsis, Van Heurck. Treat. Diat. 455, 1896 (Non O. Heer, 1877). T: *H. weissflogii* Van Heurck.

Hemiaulus, C. G. Ehrenberg. Ber. Bekanntm. Verh. Königl. Preuss. Akad. Wiss. Berlin, 1844: 199, 1844 (*Nom. rej.*). T: *H. antarcticus* C. G. Ehrenberg.

Hemiaulus, Heiberg. Krit. Overs. Danske Diat. 45, 1863 (*Nom. cons.*), non *Hemiaulus* Ehrenberg. T: *H. proteus* Heiberg.

Hemidiscus, G. C. Wallich. Trans. Microscop. Soc. London, Ser. 2. 8: 42, 1860. T: *H. cuneiformis* C. G. Wallich.

Hemiptychus, C. G. Ehrenberg. Ber. Bekanntm. Verh. Königl. Preuss. Akad. Wiss. Berlin, 1848: 7, 1848. T: *H. ornatus* C. G. Ehrenberg. *Nom. rej.* vs *Arachnoidiscus*, Deane ex Pritchard, 1825 (*Nom. cons.*).

Hemizoster, C. G. Ehrenberg. Ber. Bekanntm. Verh. König.. Preuss. Akad. Wiss. Berlin, 1844: 199, 1844. T: *H. tubulosus* C. G. Ehrenberg.

Hendeya, J. A. Long, D. P. Fuge & J. Smith. J. Paleontol. 20: 107, 1946. T: *H. dehiscens* J. A. Long, D. P. Fuge et J. Smith.

Henseniella, Schütt ex G. B. De Toni. Syll. Algarum, 2: 1425, 1894.

Henshawia, A. Mann. Bull. U. S. Natl. Mus. 100(6): 80, Pl. 17, F. 1, 2. 1925. T: *H. biddulphioides* A. Mann.

Hercotheca, C. G. Ehrenberg. Ber. Bekanntm. Verh. Königl. Preuss. Akad. Wiss. Berlin, 1844: 262, 1844. T: *H. mammillaris* C. G. Ehrenberg.

Heribaudia, M. Peragallo in Héribaud. Diat. Auvergne, 196, 1893. T: *H. ternaria* M. Peragallo.

Hesslandia, Cleve-Euler in Cleve-Euler & Hessland. Bull. Geol. Inst. Univ. Uppsala, 32: 170, 1947. T:

H. scanica Cleve-Euler.

Heterocampa, C. G. Ehrenberg. Abh. Königl. Akad. Wiss. Berlin, 1869: 44, 1870.

Heterodictyon, Greville. Trans. Microscop. Soc. London, Ser. 2. 11: 66, 1863. T: *H. rylandsianum* Grev.

Heteromphala, C. G. Ehrenberg. Ber. Bekanntm. Verh. Königl. Preuss. Akad. Wiss. Berlin, 1858: 10, 1859. T: *H. himantidium* C. G. Ehrenberg.

Heteroneis, Cleve. Kongl. Svenska Vetensk. -Akad. Handl. Ser. 2. 27(3): 182, 1891. Status of this name is obscure, whether genus or subgenus. Vide Cleve, op. cit. 164.

Heterostephania, C. G. Ehrenberg. Mikrogeologie, T. 35A, F. XIII. B. 4, 5, 1854. T: *H. rothii* C. G. Ehrenberg.

Himantidium, C. G. Ehrenberg. Ber. Bekanntm. Verh. Königl. Preuss. Akad. Wiss. Berlin, 1840: 212, 1841. T: *H. arcus* (C. G. Ehrenberg) C. G. Ehrenberg (= *Eunotia arcus* C. G. Ehrenberg).

Homoecladia, C. A. Agardh. Flora, 10:629, 28 Oct. 1827. T: *H. martiana* C. A. Agardh. *Nom. rej.* vs. *Nitzschia* Hassall, 1845 (*Nom. cons.*)

Hormophora, Jurilj. Acta Bot. Croat. 16: 93, 1957. (Non J. Agardh, 1892). T: *H. rogallii.*

Horodiscus, Hanna. Occas. Pap. Calif. Acad. Sci. 13: 21, 1927. T: *H. macroscriptus* Hanna.

Hustedtia, Meister. Kieselalg. Asien. 29, 1932. T: *H. mirabilis* Meister.

Hustedtiella, Simonsen. Kieler Meeresf. 16: 126, 1960. T: *H. baltica* Simonsen.

Huttonia, E. Grove & G. Sturt. J. Quekett Microscop. Club, Ser. 2. 3: 142, Aug. 1887. (Non Sternberg, 1837). LT: *H. alternans* E. Grove et G. Sturt (vide C. S. Boyer, Proc. Acad. Nat. Sci. Philadelphia, 78 Suppl. 144, 1927) = *Neohuttonia* O. Kuntze, 1898.

Huttoniella, G. Karsten in Engler & Prantl. Nat. Pflanzenfam. Ed. 2. 2: 243, 1928. T: *Neohuttonia* O. Kuntze, 1898. Superfluous substitute name for *Huttonia*, E. Grove et G. Sturt, 1887 . Non Sternberg 1837.

Hyalodictya, C. G. Ehrenberg. Abh. Königl. Akad. Wiss. Berlin, 1870: 52, 1871. T: *H. danae* C. G. Ehrenberg.

Hyalodiscus, C. G. Ehrenberg. Ber. Bekanntm. Verh. Königl. Preuss. Akad. Wiss. Berlin, 1845: 71, 1845. T: *H. laevis* C. G. Ehrenberg.

Hyalosira, Kützing. Kies. Bacill. Diat. 125, 7/9 Nov.

1844.

Hyalosynedra, Williams & Round. Diat. Res. 1, 318, 17–23, 1986 Nov. T: *H. laevigata* Williams & Round (= *Synedra laevigata* Grun.).

Hydrolinum, H. F. Link in C. G. D. Nees. Hor. Phys. Berol. 5, 16–20 Apr. 1820.

Hydrosera, G. C. Wallich. Quart. J. Microscop. Sci. 6: 251, 1858. T: *H. triquetra.*

Hydrosilicon, Brun. Mém. Soc. Phys. Genève, 31(2, 1): 31, 1891. T: *H. mitra* Brun.

Hydrosira, Mills. Index Gen. Spec. Diat. 866, 1934 (*Orth. var.*) + *Hydrosera* C. G. Wallich, 1858.

Hydrosirella, Hustedt. Ber. Deutsch. Bot. Ges. 64: 308, 1952. T: *H. robusta* Hustedt.

Hyperion, Gombos. Bacillaria, 6: 238, 1983. T: *H. titan* Gombos.

Iconella, Jurilj. Prir. Istraz. Hrvatske Slavonije, 24: 183, 1948. T: *I. variabilis* Jurilj.

Ikebea, Kormura. Trans. Proc. Paleont. Soc. Japan, 98: 134, 1975. T: *I. amphistriata* Komura.

Incisoria, Hajós & Stradner in Hajós. Int. Rep. Deep Sea Drilling Proj., 29: 937, Pl. 13, 15, 16; Pl. 36, 6, 1975. PS 993P T: *I. punctata* Hajós & Stradner.

Inoderma, Kützing. Algarum Aquae Dulcis German. N. 39–40, 1833. (non *Acherius*) S. F. Gray, 1821). T: *I. lamellosum* Kützing Lichen (Text not seen).

Insilella, C. G. Ehrenberg. Ber. Bekanntm. Verh. Königl. Preuss. Akad. Wiss. Berlin, 1845: 357, 1845. T: *I. africana* C. G. Ehrenberg.

Isodiscus, Rattray. J. Roy. Microscop. Soc. London, 1888: 919, 1888. T: *I. debyi.*

Isthmia, C. A. Agardh. Consp. Crit. Diat. 55, 1832. LT: *I. obliquata* (J. E. Smith) C. A. Agardh (= *Conferva obliquata* J. E. Smith) (Vide Kützing, Linnaea, 8: 579, 610, 1833).

Isthmiella, Cleve. Bih. Kongl. Svenska Vetensk. - Akad. Handl. 1(13): 10, 1873. T: *I. enervis* (C. G. Ehrenberg) Cleve (= *Isthmia enervis* C. G. Ehrenberg).

Janischia, Grunow in Van Heurck. Syn. Diat. Belg. T. 95, Bis. F. 10–11, 1883. T: *J. antiqua* Grunow.

Jousea, Gleser. Nov. Sist. Nizsh. Rast. 1967: 24, 7–9, 1967. T: *J. elliptica* (Jousé) Gleser (= *Fragilaria ? elliptica* Jousé)

Kannoa, Komura. Professor Saburo Kanno Memorial Volume 374, pl. 46, 1, 2, 1980 Mar. 25. T: *K. japonica* Komura.

Katahiraia, Komura. Trans. Proc. Paleont. Soc. Japan, 103: 385, Abb. 5, taf. 41, 1–5, 1976 Oct. 15.

T: *K. aspera* Komura.

Kentrodiscus, Pantocsek. Beitr. Kenntn. Foss. Bacill. Ungarns. 2: 75, 1889. T: *K. fossilis* Pantocsek.

Keratophora, Pantocsek. Beitr. Kenntn. Foss. Bacill. Ungarns, 2: 85, 1889. LT: *K. nitida* Pantocsek (cf. Ross et Sims in Bull. Brit. Mus. (Nat. Hist.) Bot. 13: 333, 1985).

Kidoa, Komura. Trans. Proc. Paleont. Soc. Japan, 116: 176, Taf. 24, 1–8; Abb. 2, 1–6 1979 Dec. 30. T: *K. graviarmata* Komura.

Kisseleviella, Sheshukova-Poretzkaya. Ucen. Zap. Leningradsk. Gosud. Univ., Ser. Biol. Nauk. 49: 206, 1962. T: *K. carina* Sheshukova-Poretzkeya

Kittonia, E. Grove & G. Sturt. J. Quekett Microscop. Club, Ser. 2(2): 74, 1887. T: *K. elaborata* Grove.

Klinodiscus, Jurilj. Prir. Istraz. Hrvatske Slavonije. 24: 193, 1948. T: *K. obliquus* Jurilj.

Kozloviella, Jousé. Beih. Nova Hedwig. 45: 351, 1974. T: *K. edita* Jousé

Krasskella, Ross & Sims. Bacillaria, 1: 154, 1978. T: *K. kriegerana* (Krasske) Ross & Sims.

Ktenodiscus, Pantocsek. Beitr. Kenntn. Foss. Bacill. Ungarns, 2:75, 1889. T: *K. hungaricus* Pantocsek.

Kuemmerlea, Krenner. Balkan-Kutat. Tud. Eredm. 3:92, 1926. T: *K. speciosa* Krenner.

Lamella, Brun. Diatomiste, 2: 78, 1894. T: *L. oculata* Brun.

Lampretodiscus, Pantocsek. Beitr. Kenntn. Foss. Bacill. Ungarns, 16:3, T. 16, F. 236, 58, 1892. T: *L. fasciculatus* Pantocsek.

Lampriscus, A. Schmidt. Atlas Diat. T. 80, F. 11, 1882. T: *L. kittonii* A. Schmidt.

Lauderia, Cleve. Bih. Kongl. Svenska Vetensk. - Akad. Handl. 1(11): 8, 1873. T: *L. annulata* Cleve.

Lauderiopsis, Ostenfeld in Ostenfeld & J. J. H. Schmidt. Vidensk. Meddel. Dansk. Naturhist. Forenh. Kjøbenhavn, 1901: 158, 1902. T: *L. costata* Ostenfeld.

Lepidodiscus, Witt. Verh. Russ.-Kais. Mineral. Ges. St. Petersburg, Ser. 2. 22: 163, 1886. T: *L. elegans* Witt.

Leptocylindrus, Cleve in C. G. J. Petersen. Vidensk. Udbytte Kanonb. 'Hauchs', 54: 1889. T: *L. danicus* Cleve.

Leptoscaphos, Schrader. Beih. Nova Hedwigia 28: 15, 1969. T: *L. punctatus* (E. Grove & G. Sturt) Schrader (= *Stoschia ? punctata* E. Grove et G. Sturt).

Leudugeria, Tempère ex Van Heurck. Treat. Diat.

539, 1896. T: *L. janischii* (Grunow) Tempère ex
Van Heurck (= *Euodia janischii* Grunow).

Leyanella, Hasle, von Stosch & Syvertsen. Bacillaria,
6: 50, 1983. T: *L. arenaria* Hasle, von Stosch &
Syvertsen.

Libellus, P. T. Cleve. Bih. Kongl. Svenska Vetensk. -
Akad. Handl. 1(13): 18, 1873. T: *L. grevilli* (C. A.
Agardh) P. T. Cleve ('*Grevillei*') (= *Schizonema
grevillii* C. A. Agardh).

Licmophora, C. A. Agardh. Flora, 10: 628, 28 Oct.
1827 (*Nom. cons.*) T: *L. argentescens* C. A.
Agardh. (*Typ. cons.*)

Licmosphenia, Mereschkowsky. Bot. Zap. 19: 89,
1903. T: *L. clevei*.

Lioneis, C. G. Ehrenberg. Ber. Bekanntm. Verh.
Königl. Preuss. Akad. Wiss. Berlin, 1861: 293,
1862. T: *L. paradoxa* C. G. Ehrenberg.

Liostephania, C. G. Ehrenberg. Ber. Bekanntm.
Verh. Königl. Preuss. Akad. Wiss. Berlin, 1847: 55,
1847. T: *L. rotula*.

Liparogyra, C. G. Ehrenberg. Ber. Bekanntm. Verh.
Königl. Preuss. Akad. Wiss. Berlin, 1848: 217,
1848.

Liradiscus, Greville. Trans. Microscop. Soc. London,
Ser. 2. 13: 4, 1865. T: *L. barbadensis*.

Liriogramma, Kolbe. Rep. Swedish Deep-Sea Exped.
1947/8 (Ed. H. Pettersson), 6(1): 39, 48, 1954, T: *L.
petterssonii* Kolbe (= *Asteromphalus petterssonii*).

Lisitzina, Jousé. Morskaya Mikropaleont. 1978. p. 47.
T: *L. ornata* Jousé.

Lithodesmioides, von Stosch. Brunonia, 9, 48, 60–84,
1986. T: *L. polymorphum*

Lithodesmium, C. G. Ehrenberg. Ber. Bekanntm.
Verh. Königl. Preuss. Akad. Wiss. Berlin, 1839:
156, 1839. T: *L. undulatum* C. G. Ehrenberg.

Longinata, Hajós. Init. Rep. Deep Sea Drilling Proj.
29: 935, Pl. 13, 12, 1975. T: *L. acuta* Hajós.

Lophotheca, Schrader. Beih. Nova Hedwig. 28: 68,
1969. T: *L. megala* Schrader.

Lyrella, Karajeva. Bot. Zh. Akad. Nauk. S. S. S. R.
63: 1595, 1978. T: *L. lyra* (C. G. Ehrenberg)
Karajeva (= *Navicula lyra* C. G. Ehrenberg).

Lysicyclia, C. G. Ehrenberg. Monatsber. Königl.
Preuss. Akad. Wiss. Berlin, 1856. T. Ad. P. 337, F.
29, 1856. T: *L. vogelii* C. G. Ehrenberg.

Lysigonium, Link in C. G. D. Nees. Horae Phys.
Berol. 4, 16–20 Apr. 1820. T: *Conferva
moniliformis* O. F. Müller. *Nom. rej.* vs '*Melosira*'
C. A. Agardh, 1824, Corr. Kützing 1833 (*Nom. et
Orth. cons.*)

Macrora, Hanna. Proc. Calif. Acad. Sci. Ser. 4. 20:
195, 1932. T: *M. stella* (Azpeitia) Hanna
(= *Pyxidicula stella* Azpeitia).

Maluina, R. Ross et P. A. Sims. Bull. Br. Mus. Nat.
Hist. (Bot.) 16: 285, pl. 9, 1987 Nov. 26. T: *M.
centralitenuis* (Ross et Sims) Ross et Sims (=
Hemiaulus centalitenuis Ross et Sims).

Mammidion, J. A. Long, D. P. Fuge & J. Smith. J.
Paleont. 20: 108, 1946. T: *M. elegans* J. A. Long,
D. P. Fuge et J. Smith.

Mammula, G. Karsten in Engler & Prantl. Nat.
Pflanzenfam. Ed. 2. 2: 228, 1928. T: *M. mammosa*
(E. Grove et G. Sturt) G. Karsten (= *Monopsia
mammosa* E. Grove et G. Sturt) + *Monopsia* E.
Grove et G. Sturt, 1887. Superfluous substitute
name.

Margaritum, H. Moreira Filho. Bol. Univ. Fed.
Parana 20 (Bot): 2, 1968. LT: *M. tenebro*
Leuduger-Fortmorel) H. Moreira Filho (=
Podosira tenebro Leuduger-Fortmorel).

Mastogloia, Thwaites ex W. Smith. Syn. Brit. Diat.
2:63, 1856. LT: *M. dansei* (Thwaites) Thwaites ex
W. Smith (*Danseii*) (= *Dickieia dansei* Thwaites)
(vide C. S. Boyer, Proc. Acad. Nat. Sci.
Philadelphia 79 Suppl. 327, 1928).

Mastogonia, C. G. Ehrenberg. Ber. Bekanntm. Verh.
Königl. Preuss. Akad. Wiss. Berlin, 1844: 263,
1844. LT: *M. crux* C. G. Ehrenberg (vide M.
Peragallo, Cat. Gen. Diat.: 483, 1897).

Mastoneis, Cleve. Kongl. Svenska Vetenskapsakad
Handl. Ser. 2. 26(2): 194, 1894. T: *M. biformis*
(Grunow) Cleve (= *Stauroneis biformis* Grunow).

Mediaria, Sheshukova-Poretzkaya. Ucen. Zap.
Leningradsk. Gosud Univ., Ser. Biol. Nauk. 49:
209, 1962. T: *M. splendida* Sheshukova.

Melonavicula, Christian. Amer. Monthly Microscop.
J. 7: 218, 1886. T: *M. marylandica* Christian.

Meloseira, C. A. Agardh. Syst. Algarum xiv, 8, 1824
+ *Melosira* C. A. Agardh, 1824, corr. Kützing,
1833 (*Nom. et Orth. cons.*) T: *M. nummuloides* C.
A. Agardh.

Melosira, C. A. Agardh. Syst. Algarum xiv, 8, 1824
(*Meloseira* corr. Kützing, Linnaea 8: 68, 1833).
(*Nom. et Orth. cons.*) T: *M. nummuloides* C. A.
Agardh (*Typ. cons.*).

NOTE: The basionym of *Fragilaria nummuloides*
(Dillwyn) Lyngbye is *Conferva nummuloides*
Dillwyn, Brit. Confervae. Agardh, whilst citing

Fragilaria nummuloides Lyngbye as a synonym of his *Meloseira nummuloides*, cites *Conferva nummuloides* Dillwyn as a synonym of *Meloseira discigera*. Accordingly, *Meloseira nummuloides* must be treated as a new name, not a new combination.

Meretrosulus, Hanna. Occas. Pap. Calif. Acad. Sci. 13: 24, 1927. T: *M. gracilis* Hanna.

Meridion, C. A. Agardh. Syst. Algarum xiv, 1824. T: *M. vernale* C. A. Agardh.

Meridium, E. M. Fries. Syst. Orbis Veg. 1: 355, 1825 (*Orth. var*) + *Meridion* C. A. Agardh 1824.

Mesasterias, C. G. Ehrenberg. Ber. Bekanntm. Verh. Königl. Preuss. Akad. Wiss. Berlin, 1872: 277, 1872. T: *M. abyssi* C. G. Ehrenberg.

Mesocena, C. G. Ehrenberg. Abh. Königl. Wiss. Berlin, Teil 1, p. 401, 1841 (1843). *NOTE:* This is probably not a diatom genus – VanLandingham, p. 2268, 1971.

Mesodictyon, Theriot & Bradbury in Theriot & Bradbury, Micropaleontology, 33, 357, Pls 1, 2, 1987. T: *M. magnum* Theriot, Bradbury & Krebs.

Micrampulla, Hanna. Occas. Pap. Calif. Acad. Sci. 13: 25, 1927. T: *M. parvula* Hanna.

Micromega, C. A. Agardh. Flora, 10: 628, 1827. T: *M. corniculatum* C. A. Agardh.

Microneis, Cleve. Kongl. Svenska Vetenskapsakad. Handl. Ser. 2. 27(3): 187, 1896. The status of this name, whether genus or subgenus, is obscure. Vide Cleve, op. cit. 164.

Microsiphonia, Weber. J. Phycol. 6: 151, 1970. T: *M. potamos* Weber.

Microtheca, C. G. Ehrenberg. Infus. Organ. 164, 1838. T: *M. octoceros* C. G. Ehrenberg.

Minidiscus, G. R. Hasle. Norweg. J. Bot. 20: 67, Mar. 1973. T: *M. trioculatus* (F. J. R. Taylor) G. R. Hasle (= *Coscinodiscus trioculatus* F. J. R. Taylor).

Minutocellus Hasle, von Stosch & Syvertsen. Bacillaria, 6: 38, 1983. T: *M. polymorphus* (Hargr. & Guill.) Hasle, von Stosch & Syvertsen (= *Bellerochea polymorpha* Hargraves & Guillard).

Moelleria, Cleve. Bih. Kongl. Svenska Vetensk.-Akad. Handl 1(11): 7, 1873. (Non Scopoli, 1777). T: *M. cornuta* Cleve.

Monema, Greville. Scott Crypt. Fl., 5: T. 286, Mai 1827. T: *M. quadripunctatum* (Lyngbye) Greville (= *Bangia quadripunctata* Lyngbye).

Monile, Ross et Sims. Bull. Brit. Mus. Nat. Hist. (Bot.), 16: 284, 1987. T: *M. laurentii* Ross et Sims.

Monnema, Meneghini. Atti Reale 1st. Veneto Sci. 1845/6: 138, 1846. (*Orth. var.*) + *Monema* Greville, 1827.

Monobrachia, Schrader in Schrader & Fenner. Init. Rep. D. S. D. P. 38: 989, Washington, 1976. T: *M. simplex* Schrader.

Monoceros, Goor. Recueil Trav. Bot. Neerl. 21: 303, 1924. T: *M. isthmiiforme* Goor.

Monogramma, C. G. Ehrenberg. Ber. Bekanntm. Verh. Königl. Preuss. Akad. Wiss. Berlin, 1843: 136, 1843.

Monopsia, E. Grove & G. Sturt. J. Quekett Microscop. Club, Ser. 2. 3: 141, Aug. 1887. T: *M. mammosa* E. Grove et G. Sturt.

Monopsis, G. B. De Toni et Levi-Morenos. Notarisia, 3: 469, Apr. 1888 (*Orth. var.*) + *Monopsia* E. Grove et G. Sturt, 1887. T: *M. mammosa* Grove et Sturt.

Muelleria (Frenguelli) Frenguelli. Revista Mus. La Plata, Secc. Paleontol. 3: 172, 1945. *Navicula* subg. *Muelleria* Frenguelli, Anales Soc. Cl. Argent. 97: 256, 1924. T: *M. linearis* (O. Müll) Freng. (= *Diploneis linearis* O. Müller).

Muelleriella, Van Heurck. Treat. Diat. 435, 1896. T: *M. limbata* (Ehrenberg.) Van Heurck.

Muelleriopsis, Hendey. Beih. Nova Hedwigia, 39: 87, 1972. T: *M. limbata* (Ehrenberg) Hendey (= *Pyxidicula limbata* C. G. Ehrenberg).

Myxobaktron, Schmidle. Hedwigia, 43: 415, 1904. T: *M. usterianum* Schmidle.

Nanoneis, Norris. Norweg. J. Bot. 20: 323, 1973. T: *N. hasleae* Norris.

Naunema, C. G. Ehrenberg. Infus. Organ, 233: 1838.

Navicula, Bory de St.-Vincent. Dict. Class. Hist. Nat. 2: 128, 31 Dec. 1822. T: *N. tripunctata* (O. F. Müller) Bory de St.-Vincent (Encycl. Méth. Hist. Nat., Zoophytes, 563, 1827). (= *Vibrio tripunctatus* O. F. Müller).

Naviculopsis, V. Nikolaev. Nov. Sist. Nizsh. Rast. (Bot. Inst. Akad. Nauk, SSSR), 1966: 21, 18 June 1966 (Non J. Frenguelli 1940) T: *N. septata* Nikolaev (= *Diatomella salena* var. *septata* (Nikolaev) Makarova).

Neidium, Pfitzer. Bot. Abh. Morphol. Physiol. 2: 39, 1871. LT: *N. affine* (C. G. Ehrenberg) Pfitzer (= *Navicula affine* C. G. Ehrenberg) (vide C. S. Boyer, Proc. Acad. Nat. Sci. Philadelphia, 79 Suppl. 320, 1928).

Nematoplata, Bory de St.-Vincent. Dict. Class. Hist. Nat. 1: 593, 27 Mai 1822. T: *N. bronchialis* (A. W. Roth) Bory de St.-Vincent (op. cit. 11: 499, 1827) (= *Conferva bronchialis* A. W. Roth).

Neobrunia, O. Kuntze. Bull. Herb. Boissier, 2: 477, July 1894. T: *N. japonica* (J. Tempère) O. Kuntze (Rev. Gen. 3(3): 417, 1898) (= *Brunia japonica* J. Tempère). Substitute name for *Brunia* Tempère ex De Toni 1894, non Linnaeus, 1753.

Neodelphineis, Takano. Bull. Tokai Reg. Fish. Res. Lab. 106: 45, 1982. T: *N. pelagica* Takano.

Neodenticula, Akiba & Yanagisawa. Init. Rep. D.S.D.P., 87, 490, pl. 21, 7-21; Pl. 22, 1-12, 1986 May. T: *N. kamtschatica* (Zabelina) Akiba & Yanagisawa (= *Denticula kamtschatica* Zabelina), Zabelina.

Neodiatoma, Kanitz. Syst. Veg. Janua, 5: 1887. Substitute name for *Diatoma* A. P. de Candolle, 1805, Non Loureiro 1790 (*Nom. rej.*). Nec. Bory de St.-Vincent 1824 (*Nom. cons.*).

Neofragilaria Williams & Round. Diat. Res. 2: 280, 1987. T: *N. virescens* (Ralfs) Williams & Round (= *Fragilaria virescens* Ralfs).

Neograya, O. Kuntze. Rev. Gen. Pl. 3(2): 74 Adnot: 3(3) 417, 28 Stp. 1898. T: *N. argonauta* (E. Grove et J. Brun) O. Kuntze (= *Grayia argonauta* E. Grove et J. Brun). Substitute name for *Grayia* E. Grove et J. Brun 1892, non W. J. Hooker et G. A. W. Arnott, 1840.

Neohuttonia, O. Kuntze. Rev. Gen. Pl. 3(3): 417, 28 Sept, 1898. LT: *N. alternans* (E. Grove et Sturt) O. Kuntze (= *Huttonia alternans* E. Grove et Sturt). (Vide C. S. Boyer, Proc. Acad. Nat. Sci. Philadelphia, 78 Suppl. 144, 1927). Substitute name for *Huttonia*, E. Grove et Sturt 1888 non K. Sternberg, 1837.

Neostreptotheca, von Stosch. Beih. Nova Hedwigia 54: 138, 1977. T: *N. subindica* von Stosch.

Neosynedra, Williams et Round. Diat. Res. 1, 332, 62-65, 1986 Nov. T: *N. provincialis* (Grun.) Williams & Round (= *Synedra provincialis* Grun.)

Nitzschia, Hassall. Hist. Brit. Freshwater Algae 1: 435, 1845 (*Nom. cons.*). T: *N. elongata* Hassall. *Nom. illeg.* (= *Bacillaria sigmoidea* Nitzsch. *N. sigmoidea* (Nitzsch) W. Smith).

Nitzschiella, Rabenhorst. Fl. Eur. Algarum, 1; 16, 163, 1864.

Noszkya, Lefébure & Chenevière. Bull. Soc. Franc. Microscop. 7: 10, 1938. T: *N. strix* Lefébure et Chenevière.

Nothoceratium, G. B. De Toni. Syll. Algarum, 2: 914, 1894.

Novilla, Heiberg. Consp. Crit. Diat. Dan. 100, 1863. T: *B. striatula* (Turpin) Heiberg (= *Surirella striatula* Turpin).

Odontella, C. A. Agardh. Consp. Crit. Diat. 56, 1832. T: *O. aurita* (Lyngbye) C. A. Agardh (= *Diatoma auritum* Lyngbye).

Odontidium, Kützing. Kies. Bacill. Diat. 44, 7/9 Nov. 1844.

Odontodiscus, C. G. Ehrenberg. Ber. Bekanntm. Verh. Königl. Preuss. Akad. Wiss. Berlin, 1845: 72, 1845.

Odontotropis, Grunow. Denkschr. Kaiserl. Akad. Wiss., Math. -Naturwiss. Kl. 48: 59, 1884. T: *O. cristata* Grunow.

Oestrupia, Heiden ex Hustedt. Ber. Deutsch. Bot. Ges. 53: 16, 1935. T: *O. powelli* (Lewis) Heiden. (= *Navicula powelli* Lewis).

Okedenia, Eulenstein ex G. B. De Toni. Syll. Algarum, 2: 229, 1891. T: *O. inflexa* (Brébisson) Eulenstein ex G. B. De Toni (= *Amphipleura inflexa* Brébisson ex Kützing).

Omphalopelta, C. G. Ehrenberg. Ber. Bekanntm. Verh. Königl. Preuss. Akad. Wiss. Berlin, 1844: 263, 1844.

Omphalopsis, Greville. Trans. Bot. Soc. Edinburgh, 7: 536. 1863. T: *O. australis* Greville.

Omphalotheca, C. G. Ehrenberg. Mikrogeologie, T. 35A, F. IX. 4, 1854. T: *O. hispida* C. G. Ehrenberg.

Oncosphenia, C. G. Ehrenberg. Ber. Bekanntm. Verh. Königl. Preuss. Akad. Wiss. Berlin, 1845: 72, 1845. T: *O. carpathica* C. G. Ehrenberg.

Opephora, P. Petit. Mission Sci. Cap Horn, 1882-1883, 5, Bot. 130, 1889. LT: *O. pacifica* (Grun.) P. Petit (= *Fragilaria pacifica* Grun.) Vide Patrick, R. & Reimer, C. The Diatoms of the United States, Monogr. Acad. Nat. Sci. Philadelphia, 13: 115, 1966.

Opephoropsis, Frenguelli. Revista Mus. La Plata, Secc. Paleont. 3: 205, 1945. T: *O. swartzii* (Grunow) Frenguelli (*Schwartzii*) (= *Fragilaria swartzii* Grunow). Erroneously spelt *Opheporopsis* (Vide Frenguelli, op. cit. 110, 207, 208, T. 13, F. 6.

Ophidocampa, C. G. Ehrenberg. Über Machtige Gebirgs-Schichten Mikrosk. Bacill. Mexico.

Reimpp. in Abh. Königl. Akad. Wiss. Berlin, 1869: 44, 1870.

Orthoneis, Grunow. Reise Fregatte Novara Bot. 1(1): 9, 1870.

Orthoseira, Thwaites. Ann. Mag. Nat. Hist. Ser. 2. 1: 167, 1848. T: *O. americana* (Kütz.) (= *Melosira americana* Kützing).

Östrupia, Heiden ex A. Schmidt. Atlas Diat. T. 264, F. 4, 5, 8, 9, April 1906. T: *Ö. quadriseriata* (Cl. et Grun.) = *Caloneis quadriseriata* Cl. et Grun.

Pachyneis, Simonsen. 'Meteor' Forschungsergbenisse, Reihe D, 19, p. 49, 1974. T: *P. gerlachii* Simonsen.

Palmeria, Greville. Ann. Mag. Nat. Hist. Ser. 3. 16: 1, 1865. T: *P. hardmaniana* Greville.

Pantocsekia, Grunow ex Pantocsek. Beitr. Kenntn. Foss. Bacill. Ungarns, 1: 47, 1886 (*Nom. cons.*) T: *P. clivosa* Grunow ex Pantocsek.

Papiliocellulus, Hasle, von Stosch & Syvertsen. Bacillaria, 6: 64, 1983. T: *P. elegans* Hasle, von Stosch & Syvertsen.

Pappia, Hajós. In Hajós & Rehakova, Chronostratigraphie und Neostratotypen. M. 5, Sarmatien (Miozän der zentralen Paratethys, Bd. IV): 555, 1974. T: *P. ocellata* Hajós.

Paralia, Heiberg. Consp. Crit. Diat. Dan. 33, 1863. T: *P. marina* (W. Smith) Heiberg (= *Orthoseira marina* W. Smith).

Parlibellus, E. J. Cox. Diat. Res. 3: 19, 1988. T: *P. delognei* (Van Heurck) Cox (= *Navicula delognei*).

Pentapodiscus, C. G. Ehrenberg. Ber. Bekanntm. Verh. Königl. Preuss. Akad. Wiss. Berlin, 1843: 165, 1843. T: *P. germanicus* C. G. Ehrenberg. *Nom. rej.* vs. *Aulacodiscus* C. G. Ehrenberg, 1844.

Peponia, Greville. Trans. Microscop. Soc. London, Ser. 2. 11: 75, pl. 5, fig. 25. T: *P. barbadensis* Greville.

Peragallia, Schütt. Ber. Deutsch. Bot. Ges. 13: 48, 1895. T: *P. meridiana* Schütt.

Periptera, C. G. Ehrenberg. Ber. Bekanntm. Verh. Königl. Preuss. Akad. Wiss. Berlin, 1844: 263, 1844.

Perissonoë, Andrews & Stoelzel. Proc. 7th Internat. Diat. Symp., 226: 1–8, 21, 23–28, 31–34, 1984. T: *Perissonoë parvula* (Greville) D. Williams 1988 (= *Amphitetras parvula* Greville).

Peristephania, C. G. Ehrenberg. Ber. Bekanntm. Verh. Königl. Preuss. Akad. Wiss. Berlin, 1854: 235, 1854. T: *P. eutycha* C. G. Ehrenberg.

Perithyra, C. G. Ehrenberg ex Van Heurck. Diat. Treat. 492, 1896. T: *P. denaria* (Ehrenberg).

Perizonium, Cohn & Janisch ex Rabenhorst. Fl. Eur. Algarum, 1: 19 'Perizonia' 228, 1864. T: *P. braunii* Cohn et Janisch ex Rabenhorst.

Peronia, Brébisson & Arnott ex Kitton. Quart. J. Microscop. Sci. Ser. 2. 8: 16, 1868 (*Nom. cons.*) T: *P. erinacea* Brébisson et Arnott ex Kitton, *Nom. illeg.* (= *Gomphonema fibula* Brébisson ex Kützing. 'P. fibula' (Brébisson ex Kützing) R. Ross).

Peroniopsis, Hustedt. Ber. Deutsch. Bot. Ges. 65: 275, 1952. T: *P. heribaudi* (Brun et Per.) Hustedt (= *Peronia Heribaudii* Brun et Per.)

Perrya, Kitton. Monthly Microscop. J. 12: 218, 1874. T: *P. pulcherrima* Kitton.

Petitia, M. Peragallo. In Tempère & H. Peragallo, Diat. Monde Entier, Ed. 2: 146, 1909. (Non N. J. Jacquin 1760). T: *P. temperei* M. Peragallo.

Phacodiscus, Meunier. Duc D'Orleans Camp. Arct. 1907, Bot. Microplankton, 281, 1910. T: *P. punctulatus* (W. Gregory) Meunier. (= *Coscinodiscus punctulatus* W. Gregory).

Phaeodactylum, Bohlin. Ofvers. Ventensk.-Akad. Forh., Stokh. 54: 507, 1897. T: *P. tricornutum* Bohlin.

Pharyngoglossa, Corda. Almanach Carlsbad, 5: 189, 1836. T: *P. sigmoidea* Corda.

Phlyctaenia, Kützing. Sp. Algarum, 96, 23/24 Jul. 1849.

Pinnularia, C. G. Ehrenberg. Ber. Bekanntm. Verh. Königl. Preuss. Akad. Wiss. Berlin, 1843: 45, 1843. (*Nom. cons.*). T: *P. viridis* (Nitzsch) C. G. Ehrenberg (= *Bacillaria viridis* Nitzsch) (*Typ. cons.*).

Pinnunavis, Okuno. Advances Phycol. Japan: 109, 1975. T: *P. elegans* (W. Smith) Okuno (= *Navicula elegans* W. Smith).

Placoneis, Mereschkowsky. Beih. Bot. Centralbl. 15:3, 1903. T: *P. exigua* (Greg.) Mereschkowsky.

Plagiodiscus, Grunow & Eulenstein. Hedwigia, 6: 8, 1867. T: *P. nervatus* Grunow.

Plagiodiscus, Jurilj. Prir. Istraz. Hrvatske Slavonije, 24: 187, 1948. (Non Grunow ex Eulenstein 1867). T: *P. glaber*.

Plagiogramma, Greville. Quart. J. Microscop. Sci. 7: 207, 1859. LT: *P. gregorianum* Greville *nom. illeg.* (= *Denticula staurophora* Gregory; *P. staurophorum* (Gregory) Heiberg) (vide C. S.

Boyer, Proc. Acad. Nat. Sci. Philadelphia, 78 Suppl., 177, 1927).

Plagiogrammopsis, Hasle, von Stosch & Syvertsen. Bacillaria, 6: 30, 1983. T: *P. vanheurckii* (Grun.) Hasle, von Stosch & Syvertsen (= *Plagiogramma vanheurckii* Grun. in Van Heurck).

Plagiotropis, Pfitzer. Bot. Abh. Morphol. Physiol. 2: 93, 1871. T: *P. baltica* Pfitzer.

Planktoniella, Schütt. Pflanzenleben Hochsee, 20, 1893. T: *P. sol* (G. C. Wallich) Schütt. (= *Coscinodiscus sol* G. C. Wallich).

Pleurocyclus, Casper & Scheffler. Arch. Protistenk. 132: 295, 1-33, 1986. T: *P. stechlinensis*.

Pleurodesmium, Kützing. Bot. Zeitung, 4: 248, 1846. T: *P. brebissonii* Kützing.

Pleurodiscina, Silva. Taxon, 19: 943, 30 Dec. 1970. T: *P. pantocsekii* (Barker et Meakin) Silva (= *Pleurodiscus pantocsekii* Barker et Meakin). Substitute name for *Pleurodiscus* Barker et Meakin 1944, non Lagerheim 1895 (Fossil).

Pleurodiscus, W. J. Barker & Meakin. J. Quekett Microscop. Club, Ser. 4. 1: 252, 1944. (Non Lagerheim 1895). T: *P. pantocsekii* W. J. Barker et Meakin. = *Pleurodiscina* Silva.

Pleuroneis, Cleve. Kongl. Svenska Vetenskapsakad. Handl. Ser. 2. 27(3): 181, 1896. The status of this name, whether genus or subgenus, is obscure: vide Cleve, op. cit. 164.

Pleurosigma, W. Smith. Ann. Mag. Nat. Hist. Ser. 2. 9: 2, 1852 (*Nom. cons.*). T: *P. angulatum* (Quekett) W. Smith (= *Navicula angulata* Quekett) (*Typ. cons.*).

Pleurosiphonia, C. G. Ehrenberg. Abh. Königl. Akad. Wiss. Berlin, 1870: 52, 1871.

Pleurosira, (Meneghini) Trevisan 1848. Saggio di una monografia delle alghe coccotalle, Padova: seminario, 112 pp. T: *Melosira (Pleurosira) thermalis* (Menegh.). Attir. Istit. Veneto Sc. Lett. Arti, 1845-46, pp. 43-231.

Pleurostauron, Grunow. Reise Fregatte Novara Bot. 1(1): 21, 1868 (*Orth. var.*) + *Pleurostaurum* Rabenhorst ex Janisch 1859. T: *P. javanicum* Grun.

Pleurostaurum, Rabenhorst ex Janisch. Hedwigia, 2: 25, 1859. T: *P. acutum* (W. Smith) Rabenhorst ex Janisch (= *Stauroneis acuta* W. Smith).

Ploiaria, Pantocsek ex Van Heurck. Beir. Kenntn. Foss. Bacill. Ungarns, Teil II Brackwasser Bacillarien, p. 83, 1889. T: *P. petasiformis* Pant.

Plumosigma, T. Nemoto. Sci. Rep. Whales Res. Inst. 'Tokyo', 11:111, Jun. 1956. (Bacillariophyceae/ Naviculaceae). T: *P. hustedti*.

Podiscus, J. W. Bailey. Amer. J. Sci. Arts. 46: 137, 1844. T: *Tripodiscus argus*, C. G. Ehrenberg + *Tripodiscus* C. G. Ehrenberg, 1841 *P. rogersi*.

Podocystis, J. W. Bailey. Smithsonian Contr. Knowl. 7(3): 11, 1854 (*Nom. cons.*). T: *P. americana* J. W. Bailey.

Pododiscus, Kützing. Kies. Bacill. Diat. 51, 7/9 Nov. 1844. T: *P. jamaicensis*.

Podosira, C. G. Ehrenberg. Ber. Bekanntm. Verh. Königl. Preuss. Akad. Wiss. Berlin, 1840: 161, 1840. T: *P. moniliformis* (Montagne) C. G. Ehrenberg (= *Trochiscia moniliformis* Montagne).

Podosphenia, C. G. Ehrenberg. Erkenntniss grosser organischer ausbildung kleinsten thierischen Organismen, 22, 1836: Abh. Königl. Akad. Wiss. Berlin, 1835: 173, 1837.

Polymyxus, L. W. Bailey. Boston J. Nat. Hist. 7: 341, 1862. T: *P. coronalis* L. W. Bailey.

Pomphodiscus, W. J. Barker & Meakin. J. Quekett Microscop. Club, Ser. 4. 2:144, 1946. T: *P. morenoenis* (J. A. Long, D. P. Fuge et J. Smith) W. J. Barker et Meakin (= *Craspedodiscus morenoensis* J. A. Long, D. P. Fuge et J. Smith).

Pontodiscus, Temniskova-Topalova & Sheshukova-Poretzkaya. Bot. Zh. 66: 1309, 1981. T: *P. gorbunovii* (Sheshuk.) Moiss. & Sheshuk. (= *Coscinodiscus gorbunovii* Sheshuk.).

Poretzkia, Jousé. Not. Syst. Sect. Crypt. Inst. Bot. 6(1-6): 73, 1949. T: *P. mirabilis*.

Porocyclia, C. G. Ehrenberg. Ber. Bekanntm. Verh. Königl. Preuss. Akad. Wiss. Berlin, 1848: 217, 1848. T: *P. dendrophila* C. G. Ehrenberg.

Porodiscus, Greville. Trans. Microscop. Soc. London, Ser. 2. 11: 63, 1863.

Porosira, E. Jørgensen. Bergens Mus. Skr. 7: 97, 1905. T: *P. glacialis* (Grunow) E. Jørgensen (= *Podosira hormoides* var. *glacialis* Grunow).

Porostauros, Habirshaw. Catal. Diat. 208, 1877 (*Orth. var.*) + *Prorostaurus* C. G. Ehrenberg, 1843. T: *P. splendens*.

Porosularia, Skvortzov. Quart. J. Taiwan Mus. 29: 407, 1976. T: *P. kolbei* Skvortzov.

Porpeia, J. W. Bailey ex Ralfs in Pritchard. Hist. Infus. Ed. 4. 850, 1861. T: *P. quadriceps* J. W. Bailey ex Ralfs.

Potamodiscus, Gerloff. Willdenowia 4: 359, 1968.

T: *P. kalbei* Gerloff. (*NOTE:* Not a diatom, see Gaarder, Fryxell & Hasle, Arch. Protistenk. 118, 346–351, 1976).

Praecymatosira Strelnikova. Issled. Fauni Morei, Istoriya Mikroplankt. Norvez. Marya 23(31): 63, 1979. T: *P. monomembranacea* (Schrader) Strelnikova (= *Pseudorutilaria monomembranaceae* Schrader).

Praeepithemia Jousé. Trudy Inst. Geogr. Acad. Nauk, SSSR 51(6): 243, 1952. T: *P. robusta* Jousé.

Pritchardia, Rabenhorst. Fl. Eur. Algarum, 1: 162, 1864. (Non Unger ex Endlicher, 1842) (*Nom. rej.*) Nec Seemann et H. Wendland 1862 (*Nom. cons.*).

Proboscia, Sündstrom. The marine genus *Rhizosolenia*. A new approach to the taxonomy, 99: 258–266, 1986 Jul. 15. T: *P. alata* (= *Rhizosolenia alata* Brightwell).

Proboscidea, Paddock & Sims. Bacillaria, 3: 175, 1980. T: *P. insecta* (Grunow in A. Schmidt) Paddock & Sims (= *Amphora insecta* Grunow in A. Schmidt).

Proboscineis, Butzin. Willdenowia, 10: 145, 1980. T: *P. insecta* (Grunow in A. Schmidt) Butzin (= *Amphora insecta* Grunow in A. Schmidt). Substitute name for *Proboscidea* nom. superfl. (Superfluous substitute name for *Thalassiophysa*).

Progonoia, Schrader. Beih. Nova Hedwigia, 28: 58, 1969. T: *P. didomatia* Schrader.

Prorostauros, Ralfs in Pritchard. Hist. Infus. Ed. 4, 915, 1861 (*Orth. var.*) + *Prorostaurus* C. G. Ehrenberg, 1843.

Prorostaurus, C. G. Ehrenberg. Ber. Bekanntm. Verh. Königl. Preuss. Akad. Wiss. Berlin, 1843: 136, 1843.

Proshkinia, Karayeva. Bot. Zh. 63: 1748, 1978. T: *P. bulnheimii* (Grunow in Van Heurck) Karayeva (= *Navicula bulnheimii* Grunow in Van Heurck).

Proteucylindrus Li & Chiang. Br. phycol. J. 14: 382, 1979. T: *P. taiwanensis* Li & Chiang.

Protoraphis, Simonsen. Beih. Nova Hedwigia, 31: 384, 1970. T: *P. hustedtiana* Simonsen.

Psammodiscus, Round & Mann. Ann. Bot. 46: 371, 1980. T: *P. nitidus* (Gregory) Round & Mann (= *Coscinodiscus nitidus* Gregory).

Pseudoamphiprora (Cleve) Cleve. Kongl. Svenska Vetensakad. Handl. Ser. 2. 26(2): 70, 1894. T: *Navicula arctica* Cleve (= *P. stauroptera* (J. W. Bailey) Cleve (*Amphora stauroptera* J. W. Bailey).

Navicula sect. *Pseudoamphiprora* Cleve, op. cit. 18(5): 13, 1881.

Pseudoaulacodiscus, Vekschina. Trudysib. nauchno-issled. Inst. Geol. Geofiz. miner. Syr'ya 15: 91, 1961. T: *P. jousae* Vekschina.

Pseudoauliscus, A. Schmidt. Atlas Diat. T. 32. F. 29. 15 Dec. 1875. T: *P. radiatus* (Aul.) Bailey.

Pseudocerataulus, Pantocsek. Beitr. Kenntn. Foss. Bacill. Ungarns, 2: 98, 1889. T: *P. kinkerii* Pantocsek.

Pseudodictyoneis, Cleve ex Pantocsek. Beitr. Kenntn. Foss. Bacill. Ungarns, 3. T. 1. F. 8, 1892. T: *P. hungarica* Cleve ex Pantocsek.

Pseudodimerogramma, Schrader. Init. Rep. D. S. D. P. 38: 993, 1976. T: *P. oligocenica* (Schrader & Fenner).

Pseudoeunotia, Grunow in Van Heurck. Syn. Diat. Belgique. T. 35. F. 22–23, Mai–Jun 1881. LT: *P. doliolus* (G. C. Wallich) Grunow (= *Synedra doliolus* G. C. Wallich) (Vide F. Hustedt in Rabenhorst, Krypt. Fl. Deutschl. Österr. Schweiz, 7, 2: 259, 1932).

Pseudogomphonema, Medlin. Diat. Res. 1: 214, 1986. T: *P. kamtschaticum* (Grunow) Medlin (= *Gomphonema kamtschaticum*).

Pseudoguinardia, von Stosch. Brunonia, 8: 307, 7–11, 1986. T: *P. recta* von Stosch.

Pseudohimantidium, Hustedt & Krasske, Arch. Hydrobiol. 38: 272, 1941. T: *P. pacificum* Hustedt et Krasske.

Pseudoleyanella, Takano. Bull. Tokai Reg. Fish. Res. Lab., 115: 30, 1–20, 1985 Feb. T: *P. lunata* Takano.

Pseudomastogloia, Pantocsek. Beitr. Kenntn. Foss. Ungarns 3: 89, 1905. T: *P. castracanei* (Pantocsek) Pantocsek (= *Alloeoneis castracanei* Pantocsek).

Pseudonitzschia, H. Peragallo in H. Peragallo & M. Peragallo, Diat. Mar. France, 263, 298 (*Pseudonitzschia*) 1900.

Pseudoperonia, Manguin. Mém. Mus. nat. Hist. n. s., Ser. b. T. 12 (Fasc. 2): 62, 1964. T: *P. andina* E. Manguin.

Pseudopodosira, Jousé in Proschkina-Lavrenko, 1949. Diat. Anal. T: *P. pileiformis* Jousé.

Pseudopyxilla, Forti. Nuova Notarisia, 20: 25, 1909. T: *P. tempereana*.

Pseudorutilaria, E. Grove & G. Sturt. J. Quekett Microscop. Club, Ser. 2. 2: 324, 1886. T: *P. monile* E. Grove & G. Sturt.

Pseudosolenia, Sündstrom. The marine genus *Rhizosolenia*. A new approach to the taxonomy, 95: 40–46, 247–257, 1986 Jul. 15. T: *P. calcar-avis* (Schultze) Sundström (= *Rhizosolenia calcar-avis* Schultze).

Pseudostaurosira, Williams & Round. Diat. Res. 2: 276, 1987. T: *P. brevistriata* (Grun. in Van Heurck) Williams & Round (= *Fragilaria brevistriata*)

Pseudostictodiscus, Grunow ex A. Schmidt. Atlas Diat. T. 74, F. 24–30, 28 Jan 1882. T: *P. angulatus* Grunow ex A. Schmidt.

Pseudotriceratium, Grunow. Denkschr. Kaiserl. Akad. Wiss., Math. - Naturwiss. Kl. 48: 83, 1884.

Pteroncola, Holmes & Croll. Proc. 7th Diat. Symp. 1984: 267. T: *P. marina* Holmes & Croll.

Pterotheca, Grunow ex Forti. Nuova Notarisia 20: 25, 1909 (Non Cassini 1816). T: *P. subuliformi* Grun.

Punctastriata, Williams & Round. Diat. Res. 2: 278, 1987. T: *P. linearis* Williams & Round.

Pyrgodiscus, Kitton ex Cleve. J. Quekett Microscop. Club, Ser. 2. 2: 173, 1885. T: *P. armatus* Kitton ex Cleve.

Pyrgupyxis, Hendey. Occas. Pap. Calif. Acad. Sci. 72: 3, 1969. T: *P. eocena* Hendey.

Pyxidicula, C. G. Ehrenberg. Org. Richt. kleinsten Raumes. Dritter Beitrag; 151, 1834, (*nom. rej.*) T: *P. operculata* (C. A. Agardh) Ehrenberg. : 165, 1838 (= *Frustulia operculata* C. A. Agardh, Flora 10: 627, 1827.

Pyxilla, Greville. Trans. Microscop. Soc. London, Ser. 2. 13: 1, 1865. T: *P. johnsoniana*.

Radiodiscus, Forti & Schulz. Beih. Bot. Centralbl. 50(2): 245, 1932. T: *R. cretaceus* Forti et Schulz. *NOTE:* The relative dates of this and *Radiodiscus* (Bale) Mills 1932 have not yet been established.

Radiodiscus (Bale) Mills. Trans. Microscop. Soc. London, Ser. 3. 52: 384, 1932. T: *R. hispidus* (Grunow) Manfred Voigt (J. Indian Bot. Soc. 30: 56, 1951) (= *Actinoptychus hispidus* Grunow) *Actinoptychus* subg. *Radiodiscus* Bale, J. Quekett Microscop. Club, Ser. 2. 12: 44, 1913. *NOTE:* The relative dates of this name and *Radiodiscus* Forti et Schulz 1932 have not yet been established.

Radiopalma, Brun. Mém. Soc. Phys. Genève, 31(2, 1): 42, 1891. T: *R. dichotoma* Brun.

Ralfsia, O'Meara. Proc. Roy. Irish Acad. Ser. 2. 2: 293, 1875. *NOTE:* non Berkeley 1831 (Phaeophyta).

Rancia, Mangin ex Chavaillon. Bull. Lab. Marit. Dinard, 20: 59, 1939. T: *R. triangularis* Mangin ex Chavaillon.

Raphidodiscus, H. L. Smith ex Christian. Microscope (Ann Arbor) 7: 68, 1887.

Rattrayella, G. B. De Toni. Notarisia, 4: 691, Jan. 1889. T: *R. oamaruensis* (Grunow) G. B. De Toni (= *Eupodiscus oamaruensis* Grunow). Substitute name for *Debya* Rattray 1888, non Pantocsek 1886.

Reicheltia, Van Heurck. Treat. Diat. 243, 1896. T: *R. nobilis* (Reichelt ex Tempère) Van Heurck (= *Gomphopleura nobilis* Reichelt ex Tempère) + *Gomphopleura* Reichelt ex Tempère 1894).

Reimeria, J P. Kociolek & E. R. Stoermer. Syst. Bot. 12: 457, 1987. T: *R. sinuata* (Gregory) Kociolek & Stoermer (= *Cymbella sinuata*).

Rhabdium, Wallroth. Fl. Cryptog. German, 2: 116, 1833: *R. obtusum* Wallroth.

Rhabdonema, Kützing. Kies. Bacill. Diat. 126, 7/9 Nov. 1844 (*Nom. cons.*) T: *R. minutum* Kützing (*Typ. cons.*).

Rhabdosira, C. G. Ehrenberg. Abh. Königl. Akad. Wiss. Berlin, 1869: 44, 1870. T: *R. moluccensis* C. G. Ehrenberg.

Rhaphidodiscus, H. L. Smith ex Christian. Microscope (Ann Arbor) 7: 68, 1887.

Rhaphidogloea, Kützing. Kies. Bacill. Diat. 110, 7/9 Nov. 1844.

Rhaphidophora, J. A. Long, D. P. Fuge & J. Smith. J. Paleontol. 20: 110, 1946 (Non Hasskarl 1842). T: *R. elegans* J. A. Long, D. P. Fuge et J. Smith.

Rhaphoneis, C. G. Ehrenberg. Ber. Bekanntm. Verh. Königl. Preuss. Akad. Wiss. Berlin, 1844: 74, 1844. LT: *R. amphiceros* (C. G. Ehrenberg) C. G. Ehrenberg (= *Cocconeis amphiceros* C. G. Ehrenberg) (vide C. S. Boyer, Proc. Acad. Nat. Sci. Philadelphia, 78 Suppl., 190, 1927)

Rhipidophora, Kützing. Kies. Bacill. Diat. 121, 7/9 Nov. 1844.

Rhizonotia, C. G. Ehrenberg. Ber. Bekanntm. Verh. Königl. Preuss. Akad. Wiss. Berlin, 1843: 139, 1843. T: *R. melo* C. G. Ehrenberg.

Rhizosolenia, C. G. Ehrenberg. Abh. Königl. Akad. Wiss. Berlin, 1841: 402, 1843 (*Nom. rej.*) T: *R. americana* C. G. Ehrenberg. *Nom. rej.* vs. *Rhizosolenia* Brightwell 1858 (*Nom. cons.*).

Rhoiconeis , G. B. De Toni. Syll. Algarum, 2:197,

1891. (*Orth. var.*) + *Rhoikoneis* Grunow 1863.
NOTE: This may have been introduced in a paper by G. B. De Toni in Notarisia 5: 1890 but this has not been seen by us.

Rhoicosigma, Grunow. Hedwigia, 6: 19, 1867. T: *R. reichardtianum* Grunow.

Rhoicosphenia, Grunow. Verh. zool. bot. Ges. Wien, 10: 511, 1860. T: *R. curvata* (Kützing) Grunow (= *Gomphonema curvata* Kützing).

Rhoikoneis, Grunow. Verh. zool. bot. Ges. Wien, 13: 147, 1863. T: *R. bolleana.*

Rhopalodia, O. Müller. Bot. Jahrb. 22: 57, 1895, (*nom. cons.*) T: *R. gibba* (C. G. Ehrenberg) O. Müller (= *Navicula gibba* Ehrenberg).

Riedelia, Jousé & Sheshukova-Poretzkaja. Nov. Sist. Nizsh. Rast. 8: 19, 1971. (Nom Chamisso 1832) (*Nom. rej.*), Nec D. Oliver 1883 (*Nom. cons.*). T: *R. mirabilis* Jousé.

Robinsonetta, Hanna & Brigger. Occas. Pap. Calif. Acad. Sci. 45: 20, 1964. T: *R. barboi* Hanna & Brigger.

Rocella, Hanna. J. Paleont. 4/5: 1930. T: *R. gemma* Hanna.

Roperia, Grunow ex Pelletan. Les Diatomées, 2: 158, 1889. T: *R. tessellata* (Roper) Grunow ex Pelletan (= *Eupodiscus tessellatus* Roper).

Rosaria, Carmichael ex W. H. Harvey in J. E. Smith. Engl. Fl. 5: 371, Aug. 1832 ('1833').

Rossia, Voigt. J. Microscop. Soc. 79: 95. T: *R. elliptica* Voigt.

Rossiella, Desikachary & Mahrshwari. J. Indian Bot. Soc. 37: 28, 1948. T: *R. paleacea* (Grunow) Desikachary & Mahrshwari (= *Stoschia paleacea* Grunow?).

Rouxia, Brun & Héribaud in Héribaud. Diat. Auvergne, 156, 1893. T: *R. peragalli* Brun et Héribaud.

Rutilaria, Greville. Quart. J. Microscop. Sci. Ser. 2. 3: 127, 1863. T: *R. epsilon* Kitton.

Rutilariopsis, Van Heurck. Treat. Diat. (Trans. Baxter), 459, 1896. T: *R. recens* (Cleve) Van Heurck (= *Rutilaria recens* Cleve).

Rylandsia, Greville & Ralfs ex Greville. Trans. Roy. Microscop. Soc. London, Ser. 2. 9: 67, 1861. T: *R. biradiata* Greville.

Salacia, Pantocsek. Beitr. Kenntn. Foss. Bacill. Ungarns, 2:68, 1889. (Non Linnaeus 1771). T: *S. boryana* Pantocsek + *Castracania* G. B. De Toni 1892.

Sameioneis, Russell & Norris. Pacific Sci., 25: 358, 1971. T: *S. rogallii* (Jurilj) Russell et Norris (= *Homophora rogallii* Jurilj).

Sawamuraia, Komura. Trans. Proc. Paleont. Soc. Japan, n. s. 103: 382, Abb. 2, taf. 40, 1, 1976 Oct. 15. T: *S. biseriata.*

Scalptrum, Corda. Almanach Carlsbad, 5, T. 5, F. 70, 1835. T: *S. striatula* Corda. *Nom. rej.* vs. *Gyrosigma* Hassall (*Nom. cons.*) et *Pleurosigma* W. Smith (*Nom. cons.*).

Sceletonema, G. B. De Toni. Syll. Algarum, 2: 157, 1894. (*Orth. var.*) + *Skeletonema* Greville, 1865.

Sceptroneis, C. G. Ehrenberg. Ber. Bekanntm. Verh. Königl. Preuss. Akad. Wiss. Berlin, 1844: 264, 1844. T: *S. caduceus* C. G. Ehrenberg.

Sceptronema, Takano. Bull. Tokai Reg. Fish. Res. Lab. 111: 26, 2, 15–20, 1983 Aug. T: *S. orientale* Takano.

Scheletonema, G. B. De Toni. Notarisia, 5: 915, 1890. (*Orth. var.*) + *Skeletonema* Greville 1865.

Schimperiella, G. Karsten. Wiss. Ergeb. Deutsch. Tiefsee-Exped. Valdivia 1898–1899, 2(2): 88, 1905.

Schizonema, C. A. Agardh. Syst. Algarum, XV, 1824.

Schizostauron, Grunow. Hedwigia, 6: 28, 1867. T: *S. lindigianum* Grunow.

Schmidtiella, Ostenfeld. Bot. Tidsskr. 25: 23, 1903. T: *S. pelagica* Ostenfeld.

Schroederella, Pavillard. Bull. Soc. Bot. France, 60: 126, 1913. T: *S. delicatula* (H. Peragallo) Pavillard (= *Lauderia delicatula* H. Peragallo).

Schuettia, G. B. De Toni. Syll. Algarum, 2: 1395, 1894.

Schulziella, Hanna & Forti. Atti R. Ist. Veneto, 92: 1280, 1933. T: *S. cretacea* (Forti et Schulz) Hanna et Forti (= *Radiodiscus cretaceus* Forti et Schulz) + *Radiodiscus* Forti et Schulz, 1932. *NOTE*: the question of whether this name is superfluous or legitimate depends on the relative dates of publication of *Radiodiscus* Forti et Schulz and *Radiodiscus* (Bale) Mills.

Scoliodiscus Jurilj Prirodoslovna Istrazivanja 26: 165, 1954. (Substitute name *Plagiodiscus* Jur.) non *Plagiodiscus* Grun. et Eul.

Scoliopleura, Grunow. Verh. zool.-bot. Ges. Wien, 10: 554, 1860. LT: *S. peisonis* Grunow (vide C. S. Boyer, Proc. Acad. Nat. Sci. Philadelphia 79 Suppl., 361, 1928).

Scoliotropis, Cleve. Kongl. Svenska Vetenskapsakad. Handl. Ser. 2. 26(2): 72, 1894. T: *S. latestriata*

(Brébisson) Cleve (= *Amphiprora latestriata* Brébisson).

Scoresbya, Hendey. Discovery Rep., 16: 346, 1937. T: *S. kempii* Hendey.

Secallia, Azpeitia. Diat. Española, 217: 1911. T: *S. cabelleroi* Azpeitia.

Sellaphora, Mereschkowsky. Ann. Mag. Nat. Hist. Ser. 7. 9: 186, 1902.

Semiorbis, Patrick. Patrick & Reimer: Diatoms of U.S. 163, 1966. Monographs Acad. Nat. Sci. Philadelphia. T: *S. hemicyclus* (Ehrenberg) Patrick (= *Synedra hemicyclus* C. G. Ehrenberg).

Semseyia, Pantocsek. Verh. Russ.-Kais. Mineral. Ges. St. Petersburg, Ser. 2. 39:644, 1902. T: *S. maeotica* Pantocsek.

Sheshukovia, Glezer. Bot. Zhur. 60(9): 1307, 1975. T: *S. kolbei* var. *uralensis* (Jousé) Glezer (= *Triceratium kolbei* var. *uralensis* Jousé).

Sigma, M. Peragallo in Héribaud. Ann. Biol. Lacustre, 10 (Diat. Travertins Auvergne): 100, 1920. T: *S. radiata* M. Peragallo.

Sigmatella, Kützing. Algarum Aquae Dulcis German. No. 2, 1833. T: *S. nitzschii* Kützing. *Nom. illeg.* (= *Bacillaria sigmoidea* Nitzsch). *Nom. rej.* vs. *Nitzschia* Hassall 1845 (*Nom. cons.*).

Simonsenia, Lange-Bertalot. Bacillaria, 2: 131, 1979. T: *S. delognei* (Grunow) Lange-Bertalot (= *Nitzschia delognei* Grunow).

Skeletonema, Greville. Trans. Microscop. Soc. London, Ser. 2. 13: 43, 1865. T: *S. barbadense* Greville.

Smithiella, H. Peragallo & M. Peragallo. Diat. mar. France, 343, 1901. T: *S. marina* (W. Smith) H. et M. Peragallo (= *Himantidium marinum* W. Smith).

Solium, Heiberg. Consp. Crit. Diat. Dan. 52, 1863. T: *S. exsculptum* Heiberg.

Spatangidium, Brébisson. Bull. Soc. Linn. Normandie, 2: 294, 1857.

Spermatogonia, Leuduger-Fortmorel. Ann. Jard. Bot. Buitenzorg, 11: 49, T. 4, F. 8, 1892. T: *S. antiqua* Leuduger-Fortmorel.

Sphenella, Kützing. Kies. Bacill. Diat. 83, 7/9 Nov. 1844.

Sphenosira, C. G. Ehrenberg. Abh. Königl. Akad. Wiss. Berlin, 1841: 402, 1843. T: *S. catena* C. G. Ehrenberg.

Sphinctocystis, Hassall. Hist. Brit. Freshw. Algae, 1: 436, 1845. T: *S. librilis* (C. G. Ehrenberg) Hassall

(= *Navicula librile* C. G. Ehrenberg) *Nom. rej.* vs. *Cymatopleura* W. Smith 1851.

Sphynctolethus, Hanna. Occas. Pap. Calif. Acad. Sci. 13: 31, 1927. T: *S. monstrosus* Hanna.

Spinigera, Heiden & Kolbe. Deutsch. Südpolar Exped. 1901–1903, 8, Bot. : 564, 1928. T: *S. bacillaris* Heiden & Kolbe

Spirodiscus, Jurilj. Prir. Istraz. Hrvatske Slavonije, 24: 185, 1948. (Non C. G. Ehrenberg 1832). T: *S. spiralis* (Kützing) Jurilj (= *Surirella spiralis* Kützing).

Staurogramma, Rabenhorst. Süssw.-Diat. Freunde Mikr., 50: 1853. T: *S. persicum* Rabenhorst.

Stauroneis, C. G. Ehrenberg. Ber. Bekanntm. Verh. Königl. Preuss. Akad. Wiss. Berlin, 1843: 45, 1843. LT: *S. phoenicenteron* (Nitzsch) C. G. Ehrenberg. (= *Bacillaria phoenicenteron* Nitzsch.) (vide C. S. Boyer, Proc. Acad. Nat. Sci. Philadelphia, 79 Suppl. 420, 1928).

Stauronella, Mereschkowsky. Ann. Mag. Nat. Hist. Ser. 7. 8: 430, 1901. T: *S. constricta* Mereschkowsky.

Staurophora, Mereschkowsky. Beih. Bot. Centralbl. 15: 20, 1903.

Stauropsis, Meunier. Duc D'Orléans Camp Arct. 1907, Bot., Microplankton, 318, 1910.

Stauroptera, C. G. Ehrenberg. Ber. Bekanntm. Verh. Königl. Preuss. Akad. Wiss. Berlin, 1843: 45, 1843. T: *S. semicruciata* C. G. Ehrenberg, *Nom. rej.* vs. *Pinnularia* C. G. Ehrenberg 1843 (*Nom. cons.*).

Staurosigma, Grunow ex Rabenhorst. Fl. Eur. Algarum, 1: 253, 1864. T: *S. ehrenbergii* Grunow ex Rabenhorst. *Nom. illeg.* (= *Stauroneis sigma* C. G. Ehrenberg).

Staurosira, C. G. Ehrenberg. Ber. Bekanntm. Verh. Königl. Preuss. Akad. Wiss. Berlin, 1843: 45, 1843.

Staurosira, P. Petit ex Pelletan. Diatomées, 2: 50, 1889. (Non C. G. Ehrenberg 1843).

Staurosirella, Williams & Round. Diat. Res. 2: 274, 1987. T: *S. lapponica* (Grun. in Van Heurck), Williams & Round (= *Fragilaria lapponica* Grunow in Van Heurck).

Stelladiscus, Rattray. Proc. Roy. Soc. Edinburgh, 16: 632, 1890. T: *S. stella* (Norman) Rattray (= *Asterolampra stella* Norman).

Stellarima, G. R. Hasle & P. A. Sims. Br. phycol. J. 21: 111, 1986. T: *S. microtrias* (Ehrenb.) Hasle & Sims (= *Symbolophora microtrias*).

Stenoneis, P. T. Cleve. Kongl. Svenska

Vetenskapsakad. Handl. Ser. 2. 26(2): 123, 1894.
T: *S. inconspicua* (Gregory) P. T. Cleve
(= *Navicula inconspicua* Gregory).

Stenopterobia, Brébisson ex Van Heurck. Treat.
Diat. 374, 1896.

Stephanocyclus, Skabitschevsky. Ukr. Bot. Zhr.
32(2): 205, 1975. T: *S. planum* (Fricke)
Skabitschevsky (= *Cyclotella meneghiana* var.
plana (?) Fricke).

Stephanodiscus, C. G. Ehrenberg. Ber. Bekanntm.
Verh. Königl. Preuss. Akad. Wiss. Berlin, 1845: 72,
1845. LT: *S. niagarae* C. G. Ehrenberg (vide C. S.
Boyer, Proc. Acad. Nat. Sci. Philadelphia 78,
Suppl. 60, 1927).

Stephanogonia, C. G. Ehrenberg. Ber. Bekanntm.
Verh. Königl. Preuss. Akad. Wiss. Berlin, 1844:
264, 1844.

Stephanopyxis (C. G. Ehrenberg) C. G. Ehrenberg.
Ber. Bekanntm. Verh. Königl. Preuss. Akad. Wiss.
Berlin, 1845: 80, 1845. T: *Pyxidicula aculeata*
C. G. Ehrenberg *Pyxidicula* subg. *Stephanopyxis*
C. G. Ehrenberg, op. cit., 1844: 264, 1844.

Stephanosira, C. G. Ehrenberg. Ber. Bekanntm.
Verh. Königl. Preuss. Akad. Wiss. Berlin, 1848:
217, 1848.

Stephanosira, G. Karsten. Wiss. Ergeb. Deutsch.
Tiefsee-Exped. Valdivia, 1898-1899, 2(2): 159, 1906.
(Non C. G. Ehrenberg 1848). T: *S. decussata*
G. Karsten.

Stictocyclus, A. Mann. Bull. U. S. Natl. Mus., 100(6):
146, 1925. T: *S. varicus* A. Mann, *Nom. illeg.*
(= *Actinocyclus stictodiscus* Grunow
= *S. stictodiscus* (Grunow) R. Ross).

Stictodesmis, Greville. Trans. Bot. Soc. Edinburgh,
7: 534, 1863. T: *S. australis* Greville.

Stictodiscus, Greville. Trans. Microscop. Soc.
London, Ser. 2. 9: 39, 1861. LT: *S. rota*
(C. G. Ehrenberg) Greville (= *Discoplea rota*
C. G. Ehrenberg) (vide C. S. Boyer, Proc. Acad.
Nat. Sci. Philadelphia, 78 Suppl., 69, 1927).

Stigmaphora G. C. Wallich. Trans. Microscop. Soc.
London, Ser. 2. 8: 43, 1860. T: *S. rostrata* Wallich.

Stoschia, Janisch ex Grunow in Van Heurck. Syn.
Diat. Belgique, T. 128, F. 6, 2 Mai 1883.
T: *S. mirabilis* Janisch ex Grunow.

Strangulonema, Greville. Trans. Microscop. Soc.
London, Ser. 2. 13: 43, 1865. T: *S. barbadense*
Greville.

Strelnikovia Ross et Sims. Bull. Brit. Mus. (Nat.

Hist.) Bot. 13: 324, 1985. T: *S. antiqua*
(Strelnikova) Ross et Sims (= *Rutilaria antiqua*
Strelnikova).

Streptotheca, W. H. Shrubsole. J. Quekett Microsc.
Club, Ser. 2., 4, 259-62, 1890. T: *S. thamesis*
Shrubsole.

Striatella, C. A. Agardh. Consp. Crit. Diat. 60, 1832.
LT: *S. unipunctata* (Lyngbye) C. A. Agardh
(= *Fragilaria unipunctata* Lyngbye) (vide
C. G. Ehrenberg, Infus. 202, 230, 1838).

Strombus, Schütt ex Castracane. Notarisia, 7: 1519,
1892. T: *S. pelagicus* Schütt ex Castracane.

Sturtgrovea, O. Kuntze. Rev. Gen. 3(2): 74 Adnot. 28
Sept. 1898 + *Monopsia* E. Grove et G. Sturt 1887.
Superfluous substitute name.

Sturtiella, Simonsen & Schrader. Beih. Nova
Hedwigia, 45: 152, 1974. T: *S. elegans* (Grove et
Sturt) Simonsen & Schrader (= *Craspedoporus
elegans* E. Grove et Sturt).

Styllaria, Draparnaud ex Bory de St.-Vincent. Dict.
Class. Hist. Nat. 2: 129, 31 Dec. 1822.
LT: *S. paradoxa* (Lyngbye) Bory de St.-Vincent
(= *Echinella paradoxa* Lyngbye) (vide ICBN 209,
1961). *Nom. rej.* vs *Licmophora* C. A. Agardh 1827
(*Nom. cons.*).

Stylobiblium, C. G. Ehrenberg. Ber. Bekanntm.
Verh. Königl. Preuss. Akad. Wiss. Berlin, 1845: 72,
1845. T: *S. clypeus.*

Subsilicea, von Stosch & Reimann. Beih. Nova
Hedwigia, 31: 12, 1970. T: *S. fragilarioides* von
Stosch & Reimann.

Suriraya, Pfitzer. Bot. Abh. Morphol. Physiol. 2: 107,
1871 + *Surirella* Turpin 1828.

Surirella, Turpin. Mém. Mus. Hist. Nat. 16: 363,
1828. T: *S. striatula* Turpin.

Symblepharis, C. G. Ehrenberg. Abh. Königl. Akad.
Wiss. Berlin, Phys. Kl. 1872: 390, T. 6(2), F. 9, 10,
1873. (Non Montagne 1837). T: *S. clara*
C. G. Ehrenberg.

Symbolophora, C. G. Ehrenberg. Ber. Bekanntm.
Verh. Königl. Preuss. Akad. Wiss. Berlin, 1844: 74,
1844. T: *S. trinitatis* C. G. Ehrenberg.

Syndendrium, C. G. Ehrenberg. Ber. Bekanntm.
Verh. Königl. Preuss. Akad. Wiss. Berlin, 1845: 73,
155, 1845. T: *S. diadema* C. G. Ehrenberg.

Syndetocystis, Ralfs ex Greville. Trans. Microscop.
Soc. London, 14: 125, 1866. T: *S. barbadensis*
Walker et Chase (Notes on some new and rare
diatoms 2-3: 6, 1887).

Syndetoneis, Grunow. Bot. Centralbl. 34: 26, 1888. T: *S. amplectans* (E. Grove et G. Sturt) Grunow (= *Hemiaulus amplectans* E. Grove et G. Sturt).

Synedra, C. G. Ehrenberg. Abh. Königl. Akad. Wiss. Berlin, 1830: 40, 1832, 1830. LT: *S. balthica* C. G. Ehrenberg. (op. cit. 1831) (vide Kützing, Kies. Bacill. Diat. 64, 7/9 Nov. 1844: Grunow, Verh. zool.-bot. Ges. Wien 12: 561, 1862 in Rabenhorst, Beitr. Kenntn. Algarum, 2: 7, 1865). *NOTE: Bacillaria ulna* is placed by Ehrenberg in *Navicula* in the paper in which he describes *Synedra*. He lists spp. of *Synedra* in 1831: *S. famelica* to *Nitzschia* by Grun. (1862) after synonymy in *Synedra* by Kütz. *S. lunans* and *S. bilumans* to *Ceratoneis* by Grun. (1865) *S. balthisa* and *S. ulna*.

Synedrosphenia, (H. Peragallo) Zaragosa. Azpeitia Moros, Asoc. Esp. Progr. Ci. Congr. 1908, 4(2): 220, 1911. *Synedra* subg. *Synedrosphenia* H. Peragallo in H. et M. Peragallo, Diat. Mar. France, 312, 1900. T: *S. giennensis* Azpeitia.

Syncyclia, C. G. Ehrenberg. Abh. Königl. Akad. Wiss. Berlin, 1835: 174, 1837. T: *S. salpa* C. G. Ehrenberg. (*NOTE*: Possibly title is Zusatze Erkenntn. Organ. 23, 1836 from Roy. Soc. Cat. Sci. Papers).

Syringidium, C. G. Ehrenberg. Ber. Bekanntm. Verh. Königl. Preuss. Akad. Wiss. Berlin, 1845: 357, 1845. *(nom. rej.)*. T: *S. bicorne* Ehrenberg.

Syrinx, Corda. Almanach Carlsbad, 5: 208, T. 4, F. 45–46, 1835. T: *S. annulatum* Corda.

Systephania, C. G. Ehrenberg. Ber. Bekanntm. Verh. Königl. Preuss. Akad. Wiss. Berlin, 1844: 264, 1844.

Szechenyia, Pantocsek. Beschr. Abbild. Foss. Bacill. Szliacs, 16, 1903. Verh. Vereins Natur-Heilk Presburg, 24: 14, 1904. T: *S. antiqua* Pantocsek.

Tabellaria, C. G. Ehrenberg ex Kützing. Kies. Bacill. Diat., 127, 1844. T: *T. flocculosa* (A. W. Roth) Kützing (= *Conferva flocculosa* Roth).

Tabularia, Williams & Round. Diat. Res. 1: 322, 24–32, 1986 Nov. T: *T. barbatula* (= *Synedra barbatula* Kütz.)

Tabulina, Brun. Mém. Soc. Phys. Genève, 30(9): 59, 1889. T: *T. testudo* Brun.

Temachium, Wallroth. Fl. Cryptog. German., 2: 116, 1833. T: *T. pectinale* Wallroth.

Temperea, Forti. Atti Ist. Veneto, 71: 718, 1912. Non M. Peragallo 1908. T: *T. miocenica* (Forti) Forti (op. cit. 72: 1591, 1913) (= *Aulacodiscus miocenicus* Forti) + *Temperella* Mills 1935.

Temperea, M. Peragallo in Tempère & H. Peragallo. Diat. Monde Entier, Ed. 2: 54, 1908. T: *T. mephistopheles* M. Peragallo.

Temperella, Mills. Index Gen. Sp. Diat., 1816–1932: 1596, 1935. T: *T. miocenica* (Forti) Mills (*Aulacodiscus miocenicus* Forti). Substitute name for *Temperea* Forti 1912, Non M. Peragallo 1908.

Terebraria, Greville. Trans. Microscop. Soc. London, Ser. 2. 12: 8 1864. T: *T. barbadensis* Greville.

Terpsinoë, C. G. Ehrenberg. Abh. Königl. Akad. Wiss. Berlin, 1841: 402, 1843. T: *T. musica* C. G. Ehrenberg.

Tessella, C. G. Ehrenberg. Abh. Königl. Akad. Wiss. Berlin, 1835: 173, 1837 ('1836'). T: *T. catena* C. G. Ehrenberg, *Nom. rej.* vs. *Rhabdonema* Kützing 1844 (*Nom. cons.*) (*NOTE*: No title or page no. of the 1836 reprint).

Tetrachaeta, C. G. Ehrenberg. Ber. Bekanntm. Verh. Königl. Preuss. Akad. Wiss. Berlin, 1844: 61, 1844.

Tetracyclus, Ralfs. Ann. Mag. Nat. Hist. 12: 105, 1843. T: *T. lacustris* Ralfs.

Tetragramma, C. G. Ehrenberg. Ber. Bekanntm. Verh. Königl. Preuss. Akad. Wiss. Berlin, 1843: 136, 1843.

Tetrapodiscus, C. G. Ehrenberg. Ber. Bekanntm. Verh. Königl. Preuss. Akad. Wiss. Berlin, 1843: 165, 1843. T: *T. germanicus* C. G. Ehrenberg, *Nom. rej.* vs. *Aulacodiscus* C. G. Ehrenberg 1844 (*Nom. cons.*) (see *Pintapodiscus*).

Thalassionema, Grunow ex Hustedt in Rabenhorst. Krypt.-Flora Deutschl. 7(2): 244, 1932. T: *T. nitzschioides* (Grunow) Grunow ex Hustedt (= *Synedra nitzschioides* Grunow).

Thalassiophysa, Conger. Smithsonian Misc. Collect. 122(14): 1, 1954. T: *T. rhipidis* Conger.

Thalassiosira, Cleve. Bih. Kongl. Svenska Vetenskakad. Handl. 1(13): 6, 1873. T: *T. nordenskioldii* Cleve.

Thalassiosiropsis, Hasle in Hasle & Syvertsen. Micropaleont. 31: 89, pls 1-5, 1985. T: *T. wittiana* (= *Coscinodiscus wittiana* Pant.).

Thalassiothrix, Cleve & Grunow. Kongl. Svenska Vetenskapsakad. Handl., Ser. 2. 17(2): 108, 1880. (*Orth. var.*) + *Thallasiothrix* Cleve & Grunow ex H. L. Smith 1879). T: *T. longissima* Cleve & Grunow.

Thaumatonema, Greville. Trans. Microscop. Soc.

London, Ser. 2. 11: 76, pl. V, Fig. 26, 1863.
T: *T. barbadense* Greville.

Thumia, Cleve ex Lefébure & Chenéviere. Bull. Soc. Franc. Microscop. 7: 11, 1938. T: *T. elegans* Cleve ex Lefébure et Chenéviere.

Tibiella, Bessey. Trans. Amer. Microscop. Soc. 21: 77, 1900. T: *T. punctata* (F. W. Lewis) Patrick (= *Actinella punctata* F. W. Lewis) + *Actinella* F. W. Lewis 1864 (*nom. cons.*).

Tortilaria, W. J. Barker & Meakin. J. Quekett Microscop. Club, Ser. 4. 2: 234, 1948. T: *T. briggerii* W. J. Barker et Meakin, *nom. superfl.* (+ *Triceratium swastika* J. A. Long, D. P. Fuge et J. Smith).

Toxarium, J. W. Bailey. Smithsonian Contr. Knowl. 7(3): 15, 1854. T: *T. undulatum* J. W. Bailey.

Toxonidea, Donkin. Trans. Microscop. Soc. London, Ser. 2. 6: 19, 1858. LT: *T. gregoriana* Donkin (vide C. S. Boyer, Proc. Acad. Nat. Sci. Philadelphia, 79 Suppl., 476, 1928).

Trachyneis, Cleve. Kongl. Svenska Vetenskapsakad. Handl, Ser. 2. 26(2): 190, 1894. T: *T. aspera* (C. G. Ehrenberg) Cleve (= *Navicula aspera* C. G. Ehrenberg) (vide C. S. Boyer, Proc. Acad. Nat. Sci. Philadelphia, 79 Suppl., 428, 1928).

Trachysphenia, Petit in Folin & Perier. Fonds de la Mer, 3: 190, 1877. T: *T. australis* Petit.

Tribrachia, A. Mann. Bull. U. S. Natl. Mus. 100(6): 160, 1925. T: *T. pellucida* A. Mann.

Triceratium, C. G. Ehrenberg. Ber. Bekanntm. Verh. Königl. Preuss. Akad. Wiss. Berlin, 1839: 156, 1839. LT: *T. favus* C. G. Ehrenberg (Vide A. Mann, Contr. U. S. Natl. Herb. 10: 295, 1907).

Trichotoxon, F. M. Reid & F. E. Round. Diat. Res. 2: 224, 1987. T: *T. reinboldi* (Van Heurck) Reid & Round (= *Synedra reinboldii*).

Trigonium, Cleve. Ofvers Kongl. Vetenskapsakad. Forhandl. 1867: 663, 1868. T: *T. arcticum* (Brightwell) Cleve (= *Triceratium arcticum* Brightwell).

Trinacria, Heiberg. Consp. Crit. Diat. Dan. 49, 1863. LT: *T. regina* Heiberg (vide C. S. Boyer, Proc. Acad. Nat. Sci. Philadelphia, 78 Suppl., 142, 1927).

Tripodiscus, C. G. Ehrenberg. Über noch zahlreich jetzt lebende Thierarten der Kreidebildung, 50, 1840. Abh. Königl. Akad. Wiss. Berlin, 1839: 130, 1841. T: *T. argus* C. G. Ehrenberg. *Nom. rej.* vs. *Aulacodiscus* C. G. Ehrenberg 1844 (*nom. cons.*).

Trochiscia, Montagne. Ann. Sci. Nat. Bot. Ser. 2. 8: 349, 1837. T: *T. moniliformis* Montagne.

Trochosira, Kitton. J. Quekett Microscop. Club, 2: 170, 1871.

Tropidoneis, Cleve. Diatomiste, 1: 53, 1891. T: *T. vitrea* (W. Smith) Cleve (Kongl. Svenska Vetensk.-Akad. Handl. Ser. 2. 26(2): 27, 1894. (= *Amphiprora vitrea* W. Smith) + *Plagiotropis* Pfitzer 1871.

Truania, Pantocsek. Beitr. Kenntn. Foss. Bacill. Ungarns, 1: 45, 1886. T: *T. archangelskiana* Pantocsek.

Tryblionella, W. Smith. Syn. Brit. Diat. 1: 35, 1853.

Tryblioptychus, Hendey. J. Roy. Microscop. Soc. Ser. 3. 77: 45, 1958. T: *T. cocconeiformis* (Cleve) Hendey.

Tschestnovia, Pantocsek. Beitr. Kenntn. Foss. Bacill. Ungarns, 2: 110. 1889. T: *T. mirabilis* Pantocsek.

Tubaformis, Gombos. Init. Rep. D.S.D.P. 71: 572, pl. 5, 1-6; pl. 6, 1, 2. 1983. T: *T. unicornis* Gombos.

Tubularia, Brun. Diatomiste, 2: 88, 1894. T: *T. pistillaris* Brun.

Tubulariella, Silva. Taxon, 19: 945, 30 Dec. 1970. T: *T. pistillaris* (Brun) Silva (= *Tubularia pistillaris* Brun) Substitute name for *Tubularia* Brun 1894. Non H. Roussell 1806.

Undatella, Paddock & Sims. Bacillaria 3: 169, 1980. T: *U. lineata* (Greville) Paddock & Sims (= *Amphiprora lineata* Greville).

Undatodiscus, Lupikina. Trudy Geol. In.-ta. Akad. nauk SSSR, 385: 274, pl. 77, 18-24; pl. 789, 1-4; pl. 82, 1984. T: *U. variabilis* (= *Aulacodiscus variabilis* Lupikina), Bot. Mat. 16: pl. 1, 1-6; pl. 2, 1, 1965.

Upothema, J. A. Long, D. P. Fuge & J. Smith. J. Paleont. 20: 116, 1946. T: *U. californica* J. A. Long, D. P. Fuge & J. Smith.

Valdiviella, A. F. W. Schimper in G. Karsten. Wiss. Ergebn. Deutsch. Tiefsee-Exped. Valdivia 1898–1899, 2(2): 369, 1907. T: *V. formosa* A. F. W. Schimper.

Vanheurckia, Brébisson. Ann. Soc. Phytol. Microgr. Belg. 1: 201, 1868. LT: *V. lewisiana* (Greville) Brébisson (= *Frustulia lewisiana* Greville) (vide ICBN 1954: 200, 1956).

Vanheurckiella, Pantocsek ex M. Peragallo. Cat. Gen. Diat. 968, 1897. T: *V. admirabilis* (Pantocsek ex M. Peragallo). (*NOTE:* Actually an animal (sponge spicule) but published as a diatom genus (Fossil).)

Vanhoeffenus, Heiden & Kolbe in Drygalski, Deutsch Südpol. Exped. 1901-1903. 8: 473. 1928. T: *V. antarcticus* Heiden et Kolbe.

Willemoesia, Castracane. Rep. Sci. Res. Voyage 'Challenger'. Bot. 2: 165. 1886. T: *W. humilis* (Rattray) Van Heurck (Treat. Diat. 537. 1896) (= *Coscinodiscus humilis* Rattray).

Wittia, Pantocsek. Beitr. Kenntn. Foss. Bacill. Ungarns. 2: 110. 1889. T: *W. insignis* Pantocsek.

Wrightia, O'Meara. Quart. J. Microscop. Sci. 7: 295. 1867.

Xanthiopyxis, (C. G. Ehrenberg) C. G. Ehrenberg. Ber. Bekanntm. Verh. Königl. Preuss. Akad. Wiss. Berlin. 1845: 56. 81. 1845. *Pyxidicula* subg. *Xanthiopyxis* C. G. Ehrenberg. op. cit. 1844: 264. 1844.

Xystotheca, Hanna. Proc. Calif. Acad. Sci. Ser. 4. 20: 226. 1932. T: *X. hustedtii* Hanna (*Hustedti*).

Yoshidaia, Komura. Trans. Proc. Paleont. Soc. Japan. 103: 389. abb. 9. taf. 41. 6–8. 1976 Oct 15. T: *Y. divergens* Komura.

Zotheca, Pantocsek. Res. Wiss. Erforsch. Balatonsees. 2(2. 1. Anhang): 84. 1902.

Zygoceros, C. G. Ehrenberg. Ber. Bekanntm. Verh. Königl. Preuss. Akad. Wiss. Berlin. 1839: 156. 1839. T: *Z. rhombus* C. G. Ehrenberg.

References

Admiraal, W. (1984). The ecology of estuarine sediment-inhabiting diatoms. In *Progress in Phycological Research* (ed. F. E. Round & D. J. Chapman) 3, 269–322. Bristol: Biopress.

Altman, P. L. & Dittmer, D. S. (ed.) (1962). *Growth (A Biological Handbook)*. Washington DC. Federation of American Societies for Experimental Biology.

Andersen, R. A., Medlin, L. K. & Crawford, R. M. (1986). An investigation of cell wall components of *Actinocyclus subtilis* (Bacillariophyceae). *J. Phycol.*, 22, 466–84.

Anderson, L. W. J. & Sweeney, B. M. (1978). Role of inorganic ions in controlling sedimentation rate of a marine centric diatom. *Ditylum brightwellii*. *J. Phycol.*, 14, 204–14.

Andrews, G. W. (1971). Some fallacies of quantitative diatom paleontology. *Nova Hedwigia. Beih.*, 39, 285–94.

Andrews, G. W. (1974). Systematic position and stratigraphic significance of the marine Miocene diatom *Raphidodiscus marylandicus*. In *Proceedings of the Second Symposium on Recent & Fossil Diatoms. Nova Hedwigia, Beih.*, 45, 231–50.

Andrews, G. W. (1975). Taxonomy and stratigraphic occurrence of the marine diatom genus *Raphoneis*. *Nova Hedwigia*, 53, 193–222.

Andrews, G. W. (1977). Morphology and stratigraphic significance of *Delphineis*, a new marine diatom genus. *Nova Hedwigia*, 54, 243–60.

Andrews, G. W. (1978). Marine diatom sequence in Miocene strata of the Chesapeake Bay region, Maryland. *Micropaleont.*, 24, 371–406.

Andrews, G. W. (1979). Morphologic variations in the Miocene diatom *Actinoptychus heliopelta* Grunow. *Nova Hedwigia*, 64, 79–98.

Andrews, G. W. (1980). Morphology and stratigraphic significance of *Delphineis*, a new marine diatom genus. *Nova Hedwigia, Beih.*, 54, 243–60.

Andrews, G. W. (1981a). Revision of the diatom genus *Delphineis* and morphology of *Delphineis surirella* (Ehrenberg) G. W. Andrews n. comb. In *Proceedings of the 6th Symposium on Recent and Fossil Diatoms, Budapest, Hungary.* (ed. R. Ross), pp. 81–92. Koenigstein: O. Koeltz.

Andrews, G. W. (1981b). *Achnanthes linkei* and the origin of monoraphid diatoms. *Bacillaria.*, 4, 29–40.

Andrews, G. W. (1986). Morphology and stratigraphic occurrence of the marine diatom genus *Brightwellia* Ralfs. In *Proceedings of the 8th International Diatom Symposium* (ed. M. Ricard), pp. 125–40. Koenigstein: O. Koeltz.

Andrews, G. W. (1988). Evolutionary trends in the marine diatom genus *Delphineis* G. W. Andrews. In *Proceedings of the 9th Diatom Symposium* (ed. F. E. Round), pp. 197–206. Bristol: Biopress & Koenigstein: Koeltz.

Andrews, G. W. & Rivera, P. (1987). Morphology and evolutionary significance of *Adoneis pacifica* gen. et sp. nov. (Fragilariaceae, Bacillariophyta), a marine araphid diatom from Chile. *Diat. Res.*, 2, 1–14.

Andrews, G. W. & Stoelzel, V. A. (1984). Morphology and evolutionary significance of *Perissonoë*, a new marine diatom genus. In *Proceedings of the 7th International Diatom Symposium* (ed. D. G. Mann), pp. 225–40. Koenigstein: O. Koeltz.

Anon. (1703). Two letters from a gentleman in the country, relating to Mr. Leeuwenhoek's letter in *Transaction*, no. 283. Communicated by Mr. C. *Phil. Trans. Roy. Soc. London*, 23 (288), 1494–1501.

Anon. (1975). Proposals for a standardization of diatom terminology and diagnoses. *Nova Hedwigia, Beih.*, 53, 323–54.

Baker, H. (1753). *Employment for the microscope*. pp. xiv + 442. London: R. Dodsley.

Barber, H. G. & Haworth, E. Y. (1981). *A Guide to the Morphology of the Diatom Frustule*. Freshwater Biological Association, scientific publication 42. 112pp.

Barron, J. A. (1985a). Miocene to Holocene planktic diatoms. In *Plankton Stratigraphy* (ed. H. M. Ball, J. B. Saunders & K. Perch-Beilson). Cambridge: Cambridge University Press.

Barron, J. A. (1985b). Late Eocene to Holocene diatom biostratigraphy of the equatorial Pacific Ocean, Deep Sea Drilling Project Leg 85. *Initial Rep. Deep Sea Drilling Project*, 85, 413–56. Washington.

Barron, J. A. (1987). Diatomite: Environmental and geological factors affecting its distrubition. In *Siliceous Sedimentary Rock – Hosted Ores and Petroleum* (ed. J. R. Hein), pp. 164–78. New York: Van Nostrand Reinhold Co. Ltd.

Barron, J. A., Keller, G. & Dunn, D. A. (1985). A multiple microfossil biochronology for the Miocene. *Geol. Soc. Am. Mem.* 163, 21–36.

Battarbee, R. W. (1984). Diatom analysis and the acidification of lakes. *Phil. Trans. Roy. Soc. Lond.*, B 305, 451–77.

Battarbee, R. W. (1986). Diatom analysis. In *Handbook of Holocene Palaeoecology and Palaeohydrology* (ed. B. E. Berglund), pp. 527–70. New York & London: J. Wiley.

Battarbee, R. W., Flower, R. J., Stevenson, A. C. & Rippey, B. (1985). Lake acidification in Galloway: a palaeoecological test of competing hypotheses. *Nature*, 314, 350–52.

Behre, K. (1956). Die Algenbesiedlung Seen um Bremen und Bremerhaven. *Veröff. Inst. Meeresforsch. Bremerhaven*, 4, 221–383.

Beklemishev, C. W., Petrikova, M. N. & Semina, G. I. (1961). On the cause of buoyancy of centric diatoms. *Trudy Inst. Okeanol.*, 51, 31–6.

Belcher, J. H., Swale, E. M. F. & Heron, J. (1966). Ecological and morphological observations on a population of *Cyclotella pseudostelligera* Hustedt. *J. Ecol.*, 54, 335–40.

Blackwell, J. (1969). Structure of β-chitin or parallel chain systems of poly-β(1-4)-N-acetyl-D-glucosamine. *Biopolymers*, **7**, 281–298.

Blackwell, J., Parker, D. D. & Rudall, K. M. (1967). Chitin fibres of the diatoms *Thalassiosira fluviatilis* and *Cyclotella cryptica*. *J. mol. Biol.*, **28**, 383–5.

Blank, G. S. & Sullivan, C. W. (1983). Diatom mineralization of silica acid. VII. Influence of microtubule drugs on symmetry and pattern formation in valves of *Navicula saprophila* during morphogenesis. *J. Phycol.*, **19**, 294–301.

Blank, G. S., Robinson, D. H. & Sullivan, C. W. (1986). Diatom mineralization of silicic acid. VIII. Metabolic requirements and the timing of protein synthesis. *J. Phycol.*, **22**, 382–9.

Blazé, K. (1984). Morphology and taxonomy of *Diplomenora* gen. nov. (Bacillariophyta). *Br. phycol. J.*, **19**, 217–25.

Boalch, G. T. (1974). The type material of the diatom genus *Bacteriastrum* Shadbolt. *Nova Hedwigia, Beih*, **45**, 159–63.

Boalch, G. T. (1975). The Lauder species of the diatom genus *Bacteriastrum* Shadbolt. *Nova Hedwigia, Beih.*, **53**, 185–9.

Bold, H. C. & Wynne, M. J. (1978). *Introduction to the Algae.* Englewood Cliffs: Prentice Hall.

Booth, B. & Harrison, P. J. (1979). Effect of silicate limitation on valve morphology in *Thalassiosira* and *Coscinodiscus* (Bacillariophyceae). *J. Phycol.*, **15**, 326–9.

Borowitzka, M. A., Chiappino, M. L. & Volcani, B. E. (1977). Ultrastructure of a chain-forming diatom *Phaeodactylum tricornutum*. *J. Phycol.*, **13**, 162–70.

Borowitzka, M. A. & Volcani, B. E. (1978). The polymorphic diatom *Phaeodactylum tricornutum*: ultrastructure of its morphotypes. *J. Phycol.*, **14**, 101–21.

Bory de Saint-Vincent, J. B. M. (1822). Dictionnaire classique d'Histoire Naturelle, 1. Paris.

Boyle, J. A., Pickett-Heaps, J. D. & Czarnecki, D. B. (1984). Valve morphogenesis in the pennate diatom *Achnanthes coarctata*. *J. Phycol.*, **20**, 563–73.

Brand, L. E., Murphy, L. S., Guillard, R. R. L. & Lee, H. (1981). Genetic variability and differentiation in the temperature niche component of the diatom *Thalassiosira pseudonana*. *Mar. Biol.*, **62**, 103–10.

Brightwell, T. (1856). Further observations on the genus *Triceratium*, with descriptions and figures of new species. *J. Microsc. Sci.*, **4**, 272–6.

Brogan, M. W. & Rosowski, J. R. (1988). Frustular morphology and taxonomic affinities of *Navicula complanatoides* (Bacillariophyceae). *J. Phycol.*, **24**, 262–73.

Brook, A. J. (1981). *The Biology of Desmids.* Botanical Monographs No. 16. Oxford: Blackwells.

Brooks, M. (1975a). Studies on the genus *Coscinodiscus* I. Light, transmission and scanning electron microscopy of *C. concinnus* Wm. Smith. *Bot. mar.*, **18**, 1–13.

Brooks, M. (1975b). Studies on the genus *Coscinodiscus* II. Light, transmission and scanning electron microscopy of *C. asteromphalus* Ehr. *Bot. mar.*, **18**, 15–27.

Brooks, M. (1975c). Studies on the genus *Coscinodiscus*. III. Light, transmission and scanning electron microscopy of *C. granii* Gough. *Bot. mar.*, **18**, 29–39.

Burke, J. F. & Woodward, J. B. (1963–74). A review of the genus *Aulacodiscus*. Staten Island Inst. Arts Sci., N. Y., 1–357.

Burckle, L. H. & Opdyke, N. D. (1977). Late Neogene diatom correlations in the Circum-Pacific. In *Proc. 1st Int. Congr. Pac. Neog. Strat., Tokyo* 1976. Kaiyo Shuppan, Tokyo.

Buzer, J. S. (1981). Diatom analyses of sediments from Lough Ine, Co. Cork, Southwest Ireland. *New Phytol.*, **89**, 511–33.

Cande, W. Z. & McDonald, K. L. (1985). *In vitro* reactivation of anaphase spindle elongation using isolated diatom spindles. *Nature*, **316**, 168–70.

Cande, W. Z. & McDonald, K. L. (1986). Physiological and ultrastructural analysis of elongating mitotic spindles reactivated *in vitro*. *J. Cell Biol.*, **103**, 593–604.

Canter, H. M. & Jaworski, G. H. M. (1982). Some observations on the alga *Fragilaria crotonensis* Kitton and its parasitism by two chytridiaceous fungi. *Ann. Bot.*, **49**, 429–46.

Canter, H. M. & Jaworski, G. H. M. (1983). A further study on parasitism of the diatom *Fragilaria crotonensis* Kitton by chytridiaceous fungi in culture. *Ann. Bot.*, **52**, 549–63.

Carpenter, E. J. & Guillard, R. R. L. (1971). Intraspecific differences in nitrate half-saturation constants for three species of marine phytoplankton. *Ecology*, **52**, 183–5.

Carter, H. J. (1865). Conjugations of *Navicula serians*, *N. rhomboides*, and *Pinnularia gibba*. *Ann. Mag. Nat. Hist.*, Ser. 3, **15**, 161–75.

Cavalier-Smith, T. (1986). The Kingdom Chromista: Origin and systematics. In *Progress in Phycological Research*. Vol. 4. (ed. F. E. Round & D. J. Chapman), pp. 309–347. Bristol: Biopress.

Chiappino, M. L. & Volcani, B. E. (1977). Studies on the biochemistry and fine structure of silica shell formation in diatoms VII. Sequential cell wall development in the pennate *Navicula pelliculosa*. *Protoplasma*, **93**, 205–21.

Cleve, P. T. (1894–5). Synopsis of the naviculoid diatoms. *Kongl. svenska VetenskAkad. Handl.*, **26**, 1–194; **27**, 1–219.

Cleve, P. T. & Grunow, A. (1980). Beiträge zur Kenntniss der arctischen Diatomeen. *Kongl. svenska VetenskAkad. Handl.*, **17**, 1–121.

Cleve-Euler, A. (1951–55). Die Diatomeen von Schweden und Finnland. I–V. *Kongl. svenska VetenskapsAkad. Handl.*, Ser. 4, 2(1), 1–163 (1951); Ser. 4, 3(3), 1–153 (1952); Ser. 4, 4(2), 1–158 (1953); Ser. 4, 4(5), 1–255 (1953); Ser. 4, 5(4), 1–232 (1955).

Coleman, A. W. (1985). Diversity of plastid DNA configuration among classes of eukaryote algae. *J. Phycol.*, **21**, 1–16.

Compère, P. (1982). Taxonomic revision of the diatom genus *Pleurosira* (Eupodiscaceae). *Bacillaria*, **5**, 165–90.

Cooksey, B. & Cooksey, K. E. (1980). Calcium is necessary for motility in the diatom *Amphora coffeaeformis*. *Pl. Physiol.*, **65**, 129–31.

Cooksey, B. & Cooksey, K. E. (1988). Chemical signal-response in diatoms of the genus *Amphora*. *J. Cell. Sci.*, **91**, 523–9.

Cooksey, K. E. (1981). Requirement for calcium in

adhesion of a fouling diatom to glass. *Appl. Environ. Microbiol.* **41**, 1378–82.

Cooksey, K. E. & Cooksey, B. (1986). Adhesion of fouling diatoms to surfaces: some biochemistry. In *Algal Fouling* (ed. L. V. Evans & K. D. Hoagland), pp. 41–53. Amsterdam: Elsevier.

Corliss, J. O. (1976). On lumpers and splitters of higher taxa in ciliate systematics. *Trans. Am. Microsc. Soc.* **95**, 430–43.

Cox, E. J. (1975a). A reappraisal of the diatom genus *Amphipleura* Kütz. using light and electron microscopy. *Br. phycol. J.*, **10**, 1–12.

Cox, E. J. (1975b). Further studies on the genus *Berkeleya* Grev. *Br. phycol. J.*, **10**, 205–17.

Cox, E. J. (1977a). The distribution of tube-dwelling diatom species in the Severn Estuary. *J. mar. biol. Ass. UK*, **57**, 19–27.

Cox, E. J. (1977b). Raphe structure in naviculoid diatoms as revealed by scanning electron microscopy. *Nova Hedwigia, Beih.*, **54**, 261–74.

Cox, E. J. (1978). Taxonomic studies on the diatom genus *Navicula* Bory. *Navicula grevillii* (C. A. Ag.) Heiberg and *N. comoides* (Dillwyn) H. & M. Peragallo. *Bot. J. Linn. Soc.*, **76**, 127–43.

Cox, E. J. (1979a). Studies on the diatom genus *Navicula* Bory. *Navicula scopulorum* Bréb. and a further comment on the genus *Berkeleya* Grev. *Br. phycol. J.*, **14**, 161–74.

Cox, E. J. (1979b). Studies on the diatom genus *Navicula* Bory. The typification of the genus. *Bacillaria*, **2**, 137–53.

Cox, E. J. (1981a). Mucilage tube morphology of three tube-dwelling diatoms and its diagnostic value. *J. Phycol.*, **17**, 72–80.

Cox, E. J. (1981b). The use of chloroplasts and other features of the living cell in the taxonomy of naviculoid diatoms. In *Proceedings of the 6th Symposium on Recent and Fossil Diatoms* (ed. R. Ross), pp. 115–33. Koenigstein: O. Koeltz.

Cox, E. J. (1981c). Observations on the morphology and vegetative cell division of the diatom *Donkinia recta*. *Helgol. wiss. Meeresunters.*, **34**, 497–506.

Cox, E. J. (1982). Taxonomic studies on the diatom genus *Navicula* Bory. IV. *Climaconeis* Grun., a genus including *Okedenia inflexa* (Bréb.) Eulenst. ex De Toni and members of *Navicula* sect. Johnsonieae *sensu* Hustedt. *Br. phycol. J.*, **17**, 147–68.

Cox, E. J. (1983a). Observations on the diatom genus *Donkinia* Ralfs in Pritchard. II. Frustular studies and intraspecific variation. *Bot. Mar.*, **26**, 553–66.

Cox, E. J. (1983b). Observations on the diatom genus *Donkinia* Ralfs in Pritchard. III. Taxonomy. *Bot. Mar.*, **26**, 567–80.

Cox, E. J. (1987). *Placoneis* Mereschkowsky: The re-evaluation of a diatom genus originally characterized by its chloroplast type. *Diat. Res.*, **2**, 145–57.

Cox, E. J. (1988). Taxonomic studies on the diatom genus *Navicula* Bory. V. The establishment of *Parlibellus* gen. nov. for some members of *Navicula* sect. *Microstigmaticae*. *Diat. Res.*, **3**, 9–38.

Cox, E. J. & Ross, R. (1980). The striae of pennate diatoms.

In *Proceedings of the 6th International Diatom Symposium* (ed. R. Ross), pp. 267–278. Koenigstein: O. Koeltz.

Crabtree, K. (1969). Post-glacial diatom zonation of limnic deposits in North Wales. *Mitt. Internat. Verein. Limnol.*, **17**, 165–71.

Craigie, J. S. (1974). Storage products. In *Algal physiology and Biochemistry*, Botanical Monographs No. 10, (ed. W. D. P. Stewart), pp. 206–35. Oxford: Blackwells.

Crawford, R. M. (1973a). The protoplasmic ultrastructure of the vegetative cell of *Melosira varians* C. A. Agardh. *J. Phycol.*, **9**, 50–61.

Crawford, R. M. (1973b). The organic component of the cell wall of the marine diatom *Melosira nummuloides* (Dillw.) C. Ag. *Br. phycol. J.*, **8**, 257–66.

Crawford, R. M. (1974a). The auxospore wall of the marine diatom *Melosira nummuloides* (Dillw.) C. Ag. and related species. *Br. phycol. J.*, **9**, 9–20.

Crawford, R. M. (1974b). The structure and formation of the siliceous wall of the diatom *Melosira nummuloides* (Dillw.) Ag. *Nova Hedwigia, Beih.*, **45**, 131–45.

Crawford, R. M. (1975). The taxonomy and classification of the diatom genus *Melosira* C. Ag. I. The type species *M. nummuloides* C. Ag. *Br. phycol. J.*, **10**, 323–38.

Crawford, R. M. (1978). The taxonomy and classification of the diatom genus *Melosira* C. A. Agardh. III. *Melosira lineata* (Dillw.) C. A. Ag. and *M. varians* C. A. Ag. *Phycologia*, **17**, 237–50.

Crawford, R. M. (1979a). Filament formation in the diatom genera *Melosira* C. A. Agardh and *Paralia* Heiberg. *Nova Hedwigia, Beih.*, **64**, 121–33.

Crawford, R. M. (1979b). Taxonomy and frustular structure of the marine centric diatom *Paralia sulcata*. *J. Phycol.* **15**, 200–210.

Crawford, R. M. (1981a). The siliceous components of the diatom cell wall and their morphological variation. In *Silicon and Siliceous Structures in Biological Systems* (ed. T. L. Simpson & B. E. Volcani), pp. 120–56. New York: Springer-Verlag.

Crawford, R. M. (1981b). The diatom genus *Aulacoseira* Thwaites: its structure and taxonomy. *Phycologia*, **20**, 174–92.

Crawford, R. M. (1981c). Some considerations of size reduction in diatom cell walls. In *Proceedings of the 6th Symposium on Recent and Fossil Diatoms* (ed. R. Ross), pp. 253–65. Koenigstein: O. Koeltz.

Crawford, R. M. (1988). A reconsideration of *Melosira arenaria* and *M. teres* resulting in a proposed new genus *Ellerbeckia*. In *Algae and the Aquatic Environment* (ed. F. E. Round), pp. 413–33. Bristol: Biopress.

Crawford, R. M. & Round, F. E. (1989). *Corethron* and *Mallomonas* – some striking morphological similarities. In *The Chromophyte Algae, problems and perspectives*. (ed. B. S. C. Leadbeater & W. A. Diver) pp. 203–303. Oxford: Oxford University Press.

Crawford, R. M. & Schmid, A. M. M. (1986). Ultrastructure of silica deposition in diatoms. In *Biomineralization in Lower Plants and Animals*. Systematics Association Special Volume No. 30, (ed. B. S. C. Leadbeater & R. Riding), pp. 291–314. Oxford: Oxford University Press.

Crawford, R. M., Canter, H. M. & Jaworski, G. H. M. (1985). A study of two morphological variants of the diatom *Fragilaria crotonensis* Kitton using electron microscopy. *Ann. Bot.*, **55**, 473–85.

Croll, D. A. & Holmes, R. W. (1982). A note on the occurrence of diatoms on the feathers of diving seabirds. *Auk*, **99**, 765–6.

Croome, R. L. & Tyler, P. P. (1983). *Mallomonas plumosa* (Chrysophyceae) a new species from Australia. *Br. phycol. J.*, **18**, 151–8.

Daniel, G. F., Chamberlain, A. H. L. & Jones, E. B. G. (1987). Cytochemical and electronmicroscopical observations on the adhesive materials of marine fouling diatoms. *Br. phycol. J.*, **22**, 101–18.

Darley, W. M. (1977). Biochemical composition. In *The Biology of Diatoms* (ed. D. Werner), pp. 198–223. Oxford: Blackwells.

Darley, W. M. & Volcani, B. E. (1969). Role of silicon in diatom metabolism. A silicon requirement for deoxyribonucleic acid synthesis in the diatom *Cylindrotheca fusiformis* Reimann & Lewin. *Exp. Cell. Res.*, **58**, 334–42.

Darley, W. M. & Volcani, B. E. (1971). Synchronized cultures: diatoms. *Methods Enzymol.*, **23A**, 85–96.

Davey, M. C. (1986). The relationship between size, density and sinking velocity through the life cycle of *Melosira granulata* (Bacillariophyta). *Diat. Res.*, **1**, 1–18.

Davey, M. C. (1987). Seasonal variation in the filament morphology of the freshwater diatom *Melosira granulata* (Ehrenb.) Ralfs. *Freshwater Biol.*, **18**, 5–16.

Davey, M. C. & Crawford, R. M. (1986). Filament formation in the diatom *Melosira granulata*. *J. Phycol.* **22**, 144–50.

Davis, P. H. & Heywood, V. H. (1963). *Principles of Angiosperm Taxonomy*. Edinburgh & London: Oliver & Boyd.

Dawson, P. A. (1973a.) The morphology of the siliceous components of *Didymosphaenia geminata* (Lyngb.) M. Schm. *Br. phycol. J.*, **8**, 65–78.

Dawson, P. A. (1973b). Further observations on the genus *Didymosphaenia* M. Schmidt – *D. sibirica* (Grun.) M. Schm. *Br. phycol. J.*, **8**, 197–201.

Dawson, P. A. (1974). Observations on diatom species transferred from *Gomphonema* C. A. Agardh to *Gomphoneis* Cleve. *Br. phycol. J.*, **9**, 75–82.

Denffer, D. von (1949). Die planktische Massenkultur pennaten Grunddiatomeen. *Arch. Mikrobiol.*, **14**, 159–202.

Dillwyn, L. W. (1809). *British Confervae*. London: W. Phillips. 87 pp. 110.

Dodge, J. D. (1983). A re-examination of the relationship between unicellular host and eucaryotic endosymbiont with special reference to *Glenodinium foliaceum* Dinophyceae. *Endocytobiology*, **2**, 1015–26.

Drebes, G. (1966). On the life history of the marine plankton diatom *Stephanopyxis palmeriana*. *Helgol. wiss. Meeresunters*, **13**, 101–14.

Drebes, G. (1969). Geschlechtliche Fortpflanzung der Kieselalge *Stephanopyxis turris* (Centrales). Begleitveröff. zum Film C 983. *Inst. Wiss. Film Göttingen*, 16 pp.

Drebes, G. (1972). The life history of the centric diatom *Bacteriastrum hyalinum* Lauder. *Nova Hedwigia, Beih.*, **39**, 95–110.

Drebes, G. (1974). *Marines Phytoplankton*. Stuttgart: G. Thieme Verlag. 186 pp.

Drebes, G. (1977a). Sexuality. In *The Biology of Diatoms*, Botanical Monographs No. 13, (ed. D. Werner), pp. 250–83. Oxford: Blackwells.

Drebes, G. (1977b). Cell structure, cell division, and sexual reproduction of *Attheya decora* West (Bacillariophyceae, Biddulphiineae). *Nova Hedwigia Beih.*, **54**, 167–78.

Drebes, G. & Schulz, D. (1981). *Anaulus creticus* sp. nov. a new centric diatom from the Mediterranean Sea. *Bacillaria*, **4**, 161–76.

Drew, G. & Nultsch, W. (1962). Spezielle Bewegungsmechanismen von Einzellern (Bakterien, Algen). In *Handbuch der Pflanzenphysiologie* (ed. W. Ruhland), **17**(2), 876–919. Berlin: Springer–Verlag.

Drum, R. W. & Hopkins, J. T. (1966). Diatom locomotion, an explanation. *Protoplasma*, **62**, 1–33.

Drum, R. W. & Pankratz, H. S. (1964). Pyrenoids, raphes and other fine structure in diatoms. *Amer. J. Bot.*, **51**, 405–18.

Drum, R. W. & Pankratz, H. S. (1965a). Fine structure of an unusual cytoplasmic inclusion in the diatom genus, *Rhopalodia*. *Protoplasma*, **60**, 141–49.

Drum, R. W. & Pankratz, H. S. (1965b). Locomotion and raphe structure of the diatom *Bacillaria*. *Nova Hedwigia*, **10**, 315–17.

Duke, E. L. & Reimann, B. E. F. (1977). The ultrastructure of the diatom cell. In *The Biology of Diatoms*, Botanical Monographs No. 13, (ed. D. Werner), pp. 65–109. Oxford: Blackwells.

Duke, E. L., Lewin, J. & Reimann, B. E. F. (1973). Light and electron microscope studies of diatom species belonging to the genus *Chaetoceros* Ehrenberg. I. *Chaetoceros septentrionale* Østrup. *Phycologia*, **12**, 1–9.

Dweltz, N. E., Colvin, J. R. & McInnes, A. G. (1968). Studies on the chitan (β-(1-4)-linked 2-acetamido-2-deoxy-D-glucan) fibres of the diatom *Thalassiosira fluviatilis* Hustedt. III. The structure of chitan from X-ray diffraction and electron microscope observations. *Can. J. Chem.*, **46**, 1513–21.

Eaton, J. W. & Moss, B. (1966). The estimation of numbers and pigment content in epipelic populations. *Limnol. Oceanogr.*, **11**, 584–95.

Edgar, L. A. (1979). Diatom locomotion: computer assisted analysis of cine film. *Br. phycol. J.*, **14**, 82–101.

Edgar, L. A. (1980). Fine structure of *Caloneis amphisbaena* (Bacillariophyceae). *J. Phycol.*, **16**, 621–72.

Edgar, L. A. (1982). Diatom locomotion: a consideration of movement in a highly viscous situation. *Br. phycol. J.*, **17**, 243–51.

Edgar, L. A. (1983). Mucilage secretions of moving diatoms. *Protoplasma*, **118**, 44–8.

Edgar, L. A. & Pickett-Heaps, J. D. (1982). Ultrastructural localisation of polysaccharides in the motile diatom *Navicula cuspidata*. *Protoplasma*, **113**, 10–22.

Edgar, L. A. & Pickett-Heaps, J. D. (1983). The mechanism

of diatom locomotion. I. An ultrastructural study of the motility apparatus. *Proc. Roy. Soc. London B*, **218**, 331–43.

Edgar, L. A. & Pickett-Heaps, J. D. (1984a). Valve morphogenesis in the pennate diatom *Navicula cuspidata*. *J. Phycol.*, **20**, 47–61.

Edgar, L. A. & Pickett-Heaps, J. D. (1984b). Diatom locomotion. In *Progress in Phycological Research*, Volume 3, (ed. F. E. Round & D. J. Chapman), pp. 47–88. Bristol: Biopress.

Edgar, L. A. & Zavortink, M. (1983). The mechanism of diatom locomotion. II. Identification of actin. *Proc. Roy. Soc. London B*, **218**, 345–8.

Ehrenberg, C. G. (1838). *Die Infusionsthierchen als vollkommene Organismen*. Leipzig: Leopold Voss. i–xviii + 548 pp.

Ehrlich, A., Crawford, R. M. & Round, F. E. (1982a). A study of the diatom *Cerataulus laevis* – the structure of the frustule. *Br. phycol. J.*, **17**, 195–203.

Ehrlich, A., Crawford, R. M. & Round, F. E. (1982b). A study of the diatom *Cerataulus laevis* – the structure of the auxospore and the initial cell. *Br. phycol. J.*, **17**, 205–14.

Eminson, D. & Moss, B. (1980). The composition and ecology of periphyton communities in freshwaters. I. The influence of host type and external environment on community composition. *Br. phycol. J.*, **15**, 429–46.

Engstrom, D. R., Swain, E. B. & Kingston, J. C. (1985). A palaeolimnological record of human disturbance from Harvey's Lake, Vermont: geochemistry, pigments and diatoms. *Freshwater Biol.*, **15**, 261–88.

Eppley, R. W. (1977). The growth and culture of diatoms. In *The Biology of Diatoms*, Botanical Monographs No. 13, (ed. D. Werner), pp. 24–64. Oxford: Blackwells.

Erben, K. (1959). Untersuchungen über Auxosporenentwicklung und Meioseauslösung an *Melosira nummuloides* (Dillw.) C. A. Agardh. *Arch. Protistenk.*, **104**, 165–210.

Ettl, H. (1978). Teilungsverhalten der Chromatophoren in bezug auf die Mitose während des Lebenszyklus von *Diatoma hiemale* var. *mesodon*. II. *Plant Syst. Evol.*, **129**, 315–22.

Ettl, H. & Brezina, V. (1975). Teilungsverhalten der Chromatophoren in bezug auf die Mitose während des Lebenszyklus von *Diatoma hiemale* var. *mesodon*. *Plant Syst. Evol.*, **124**, 187–203.

Evans, G. H. (1970). Pollen and diatom analyses of Late-Quaternary deposits in the Blelham basin, North Lancashire. *New Phytol.*, **69**, 821–74.

Evans, G. H. & Walker, R. (1977). The Late-Quaternary history of the diatom flora of Llyn Clyd and Llyn Glas, two small oligotrophic high mountain tarns in Snowdonia (Wales). *New Phytol.*, **78**, 221–36.

Evensen, D. L. & Hasle, G. R. (1975). The morphology of some *Chaetoceros* (Bacillariophyceae) species as seen in the electron microscopes. *Nova Hedwigia, Beih.*, **53**, 153–84.

Falk, M., Smith, D. G., McLachlan, J. & McInnes, A. G. (1966). Studies on chitan (β1-4-linked 2- acetamido-2-deoxy-D-glucan fibres of the diatom *Thalassiosira fluviatilis* Hustedt. II. Chemical, infrared, X-ray and proton magnetic resonance studies on chitan. *Can. J. Chem.*, **44**, 2269–81.

Fisher, N. S. (1977). On the differential sensitivity of estuarine and open-ocean diatoms to exotic chemical stress. *Am. Nat.*, **111**, 871–95.

Fisher, N. S., Graham, L. B., Carpenter, E. J. & Wurster, C. F. (1973). Geographic differences in phytoplankton sensitivity to PCBs. *Nature*, **241**, 548–9.

Flower, R. J. (1986). Two forms of *Tabellaria binalis* (Ehr.) Grun. in two acid lakes in Galloway, Scotland. In *Diatoms and Lake Acidity* (ed. J. P. Smol, R. W. Battarbee, R. B. Davis & J. Merilainen), pp. 45–54. The Hague: W. Junk.

Flower, R. J. & Battarbee, R. W. (1983). Diatom evidence for recent acidification of two Scottish lochs. *Nature*, **305**, 130–3.

Ford, C. W. & Percival, E. (1965). Carbohydrates of *Phaeodactylum tricornutum*. Part II. A sulphated glucuronomannan. *J. Chem. Soc.*, **1298**, 7035–41.

Francisco, A. de & Roth, L. E. (1977). The marine diatom, *Striatella unipunctata*. I. Cytoplasmic fine structure with emphasis on the Golgi apparatus. *Cytobiologie*, **14**, 191–206.

Frenguelli, J. (1945). Las Diatomeas del Platense. *Revta Mus. La Plata*, n. s. 3 (*secc. paleont.*), 77–221.

Fritsch, F. E. (1935). *The Structure and Reproduction of the Algae I*. Cambridge: Cambridge University Press, 791 pp.

Fryxell, G. A. (1978a). Chain-forming diatoms: three species of Chaetoceraceae. *J. Phycol.*, **14**, 62–71.

Fryxell, G. A. (1978b). Proposal for the conservation of the diatom *Coscinodiscus argus* Ehrenberg as the type of the genus. *Taxon*, **27**, 122–125.

Fryxell, G. A. & Hasle, G. R. (1971). *Corethron criophylum* Castracane: its distribution and structure. *Antarctic Res. Ser.*, **17**, 335–46.

Fryxell, G. A. & Hasle, G. R. (1972). *Thalassiosira eccentrica* sp. nov. and some related centric diatoms. *J. Phycol.*, **8**, 297–317.

Fryxell, G. A. & Hasle, G. R. (1974). Coscinodiscaceae: Some consistent patterns in diatom morphology. *Nova Hedwigia, Beih.*, **45**, 69–84.

Fryxell, G. A. & Medlin, L. K. (1981). Chain forming diatoms: evidence of parallel evolution in *Chaetoceros*. *Crypt. Algol.*, **2**, 3–29.

Fryxell, G. A. & Miller, W. I. III. (1978). Chain-forming diatoms: three araphid species. *Bacillaria*, **1**, 113–36.

Fryxell, G. A. & Semina, H. J. (1981). *Actinocyclus exiguus* sp. nov. from the southern parts of the Indian and Atlantic Oceans. *Br. phycol. J.*, **16**, 441–8.

Fryxell, G. A., Sims, P. A. & Watkins, T. P. (1986). *Azpeitia* (Bacillariophyceae): Related genera and promorphology. *Am. Soc. Plant Tax.*, **13**, 1–74.

Gallagher, J. C. (1980). Population genetics of *Skeletonema costatum* (Bacillariophyceae) in Narragansett Bay. *J. Phycol.*, **126**, 464–74.

Gallagher, J. C. (1982). Physiological variation and electrophoretic banding patterns of genetically different seasonal populations of *Skeletonema costatum*

(Bacillariophyceae). *J. Phycol.*, **18**, 148-62.

Gallagher, J. C. (1983). Cell enlargement in *Skeletonema costatum* (Bacillariophyceae). *J. Phycol.*, **19**, 539-42.

Geissler, U. (1970a). Die Variabilität der Schalenmerkmale bei den Diatomeen. *Nova Hedwigia*, **19**, 623-773.

Geissler, U. (1970b). Die Schalenmerkmale der Diatomeen – Ursachen ihrer Variabilität und Bedeutung für die Taxonomie. *Nova Hedwigia, Beih.* **31**, 511-35.

Geitler, L. (1927a). Die Reduktionsteilung und Copulation von *Cymbella lanceolata*. *Arch. Protistenk.*, **58**, 465-507.

Geitler, L. (1927b). Somatische Teilung, Reduktionsteilung, Copulation und Parthenogenese bei *Cocconeis placentula*. *Arch. Protistenk.*, **59**, 506-49.

Geitler, L. (1928). Copulation und Geschlechtsverteilung bei einer *Nitzschia*-Art. *Arch. Protistenk.*, **61**, 419-42.

Geitler, L. (1931). Der Kernphasenwechsel der Diatomeen. Mit einem Anhang: die Kernteilung von *Hydrosera*. *Beih. Bot. Centralbl.*, **48**, 1-14.

Geitler, L. (1932). Der Formwechsel der pennaten Diatomeen. *Arch. Protistenk.*, **78**, 1-226.

Geitler, L. (1937). Chromatophor, Chondriosomen, Plasmabewegung und Kernbau von *Pinnularia nobilis* und einigen anderen Diatomeen nach Lebendbeobachtungen. *Protoplasma*, **27**, 534-43.

Geitler, L. (1939a). Die Auxosporenbildung von *Synedra ulna*. *Ber. dt. bot. Ges.*, **57**, 432-6.

Geitler, L. (1939b). Gameten- und Auxosporenbildung von *Synedra ulna* im Vergleich mit anderen pennaten Diatomeen. *Planta*, **30**, 551-66.

Geitler, L. (1949). Die Differenzierung des Protoplasten der Diatomee *Synedra*. *Österr. Bot. Z.*, **95**, 345-61. 1-11.

Geitler, L. (1951a). Der Bau des Zellkerns von *Navicula radiosa* und verwandten Arten und die präanaphasische Trennung von Tochtercentromeren. *Österr. Bot. Z.*, **98**, 206-14.

Geitler, L. (1951b). Kopulation und Formwechsel von *Eunotia arcus*. *Österr. Bot. Z.*, **98**, 292-337.

Geitler, L. (1951c). Prägame Plasmadifferenzierung und Kopulation von *Eunotia flexuosa*. *Österr. Bot. Z.*, **98**, 395-402.

Geitler, L. (1952a). Untersuchungen über Kopulation und Auxosporenbildung pennater Diatomeen III. Gleichartigkeit der Gonenkerne und Verhalten des Heterochromatins bei *Navicula radiosa*. *Österr. Bot. Z.*, **99**, 469-82.

Geitler, L. (1952b). Oogamie, Mitose, Meiose und metagame Teilung bei der zentrischen Diatomee *Cyclotella*. *Österr. Bot. Z.*, **99**, 506-20.

Geitler, L. (1952c). Untersuchungen über Kopulation und Auxosporenbildung pennater Diatomeen. IV. V. *Österr. Bot. Z.*, **99**, 598-605.

Geitler, L. (1953). Allogamie und Autogamie bei der Diatomee *Denticula tenuis* und die Geschlechtsbestimmung der Diatomeen. *Österr. Bot. Z.*, **100**, 331-52.

Geitler, L. (1954a). Lebendbeobachtung der Gametenfusion bei *Cymbella*. *Österr. Bot. Z.*, **101**, 74-8.

Geitler, L. (1954b). Paarbildung und Gametenfusion bei *Anomoeoneis exilis* und einigen anderen pennaten Diatomeen. *Österr. Bot. Z.*, **101**, 441-52.

Geitler, L. (1958a). Fortpflanzungsbiologische Eigentümlichkeiten von *Cocconeis* und Vorarbeiten zu einer systematischen Gliederung von *Cocconeis placentula* nebst Beobachtungen an Bastarden. *Österr. Bot. Z.*, **105**, 350-79.

Geitler, L. (1958b). Notizen über Rassenbildung, Fortpflanzung, Formwechsel und morphologische Eigentümlichkeiten bei pennaten Diatomeen. *Österr. Bot. Z.*, **105**, 408-42.

Geitler, L. (1963a). Alle Schalenbildungen der Diatomeen treten als Folge von Zell- oder Kernteilungen auf. *Ber. dtsch. Bot. Ges.*, **75**, 393-6.

Geitler, L. (1963b). Rassenbildung bei *Gomphonema angustatum*. *Österr. Bot. Z.*, **110**, 481-3.

Geitler, L. (1968a). Kleinsippen bei Diatomeen. *Österr. Bot. Z.*, **115**, 354-62.

Geitler, L. (1968b). Auxosporenbildung bei einigen pennaten Diatomeen und *Nitzschia flexoides* n. sp. in der Gallerte von *Ophrydium versatile*. *Österr. Bot. Z.*, **115**, 482-90.

Geitler, L. (1969a). Notizen über die Auxosporenbildung einiger pennaten Diatomeen. *Österr. Bot. Z.*, **117**, 265-75.

Geitler, L. (1969b). Die Auxosporenbildung von *Nitzschia amphibia*. *Österr. Bot. Z.*, **117**, 404-10.

Geitler, L. (1970a). Pädogame Automixis und Auxosporenbildung bei *Nitzschia frustulum* var. *perpusilla*. *Österr. Bot. Z.*, **118**, 121-30.

Geitler, L. (1970b). Kleinsippen von *Gomphonema angustatum*. *Österr. Bot. Z.*, **118**, 197-200.

Geitler, L. (1970c). Die Entstehung der Innenschalen von *Amphiprora paludosa* unter acytokinetischer Mitose. *Österr. Bot. Z.*, **118**, 591-6.

Geitler, L. (1970d). Zwei morphologisch und fortpflanzungsbiologisch unterschiedene Kleinsippen pennater Diatomeen aus dem Neusiedler See. *Nova Hedwigia*, **19**, 391-6.

Geitler, L. (1972). Sippen von *Gomphonema parvulum*, Paarungsverhalten und Variabilität pennater Diatomeen. *Österr. Bot. Z*, **120**, 257-68.

Geitler, L. (1973). Auxosporenbildung und Systematik bei pennaten Diatomeen und die Cytologie von *Cocconeis*-Sippen. *Österr. Bot. Z.*, **122**, 299-321.

Geitler, L. (1975a). Lebendbeobachtung der Chromatophorenteilung der Diatomee *Nitzschia*. *Plant Syst. Evol.*, **123**, 145-52.

Geitler, L. (1975b). Formwechsel, sippenspezifischer Paarungsmodus und Systematik bei einigen pennaten Diatomeen. *Plant Syst. Evol.*, **124**, 7-30.

Geitler, L. (1977). Zur Entwicklungsgeschichte der Epithemiaceen *Epithemia*, *Rhopalodia* und *Denticula* (Diatomophyceae) und ihre vermütlich symbiotischen Sphäroidkörper. *Plant Syst. Evol.*, **128**, 259-75.

Geitler, L. (1979). On some pecularities in the life history of pennate diatoms hitherto overlooked. *Amer. J. Bot.*, **66**, 91-7.

Geitler, L. (1980). Zellteilung und Bildung von Innenschalen bei *Hantzschia amphioxys* und *Achnanthes coarctata*. *Plant Syst. Evol.*, **136**, 275-86.

Geitler, L. (1981). Die Lage des Chromatophors in Beziehung zur Systematik von *Cymbella*-Arten (Bacillariophyceae). *Pl. Syst. Evol.*, **138**, 153–6.

Geitler, L. (1982). Die infraspezifischen Sippen von *Cocconeis placentula* des Lunzer Seebachs. *Arch. Hydrobiol., Suppl.* **63**, 1–11.

Geitler, L. (1984). Ergänzungen zu älteren Listen der Typen Auxosporenbildung pennaten Diatomeen. *Arch. Hydrobiol.*, **101**, 101–4.

Geitler, L. (1985). Automixis bei pennaten Diatomeen. *Plant Syst. Evol.*, **150**, 303–6.

Gerloff, J. (1970). Elektronenmikroskopische Untersuchungen an Diatomeenschalen. VII. Der Bau der Schale von *Planktoniella sol* (Wallich) Schütt. *Nova Hedwigia, Beih.*, **31**, 203–34.

Gerloff, J. & Rivera, P. (1979). Der submikroskopische Bau der Schalen von *Cocconeis pediculus* (Bacillariophyceae). *Willdenowia*, **9**, 99–110.

Germain, H. (1981). *Flore des Diatomées*. Paris: Soc. Nouv. Edit. Boubée. 444 pp.

Gibson, R. A. (1978). *Pseudohimantidium pacificum*, an epizoic diatom new to the Florida Current (Western North Atlantic Ocean). *J. Phycol.*, **14**, 371–3.

Gibson, R. A. (1979). Observations of stalk production by *Pseudohimantidium pacificum* Hust. & Krasske (Bacillariophyceae, Protoraphidaceae). *Nova Hedwigia*, **31**, 899–915.

Gibson, R. A. & Navarro, J. N. (1981). *Chrysanthemodiscus floriatus* Mann (Bacillariophyceae) a new record for the Atlantic Ocean with comments on its structure. *Phycologia*, **20**: 338–41

Giffen, M. H. (1970). New and interesting marine and littoral diatoms from Sea Point, near Cape Town, South Africa. *Bot. Mar.*, **13**, 87–99.

Glezer, Z. I. (1975). K revizii roda *Triceratium* Ehr. *sensu* Hustedt, 1930 (Bacillariophyta) [Towards a revision of the genus *Triceratium* Ehr. *sensu* Hustedt, 1930]. *Bot. Zh. SSSR*, **60**, 1304–10.

Gmelin, J. F. (1791). In *Systema Naturae* Edition 13 1(6). Lipsiae.

Gombos, A. M., Jr (1980). The early history of the diatom family Asterolampraceae. *Bacillaria*, **3**, 227–72.

Gombos, A. M. (1982). Early and middle Eocene diatom evolutionary events. *Bacillaria*, **5**, 225–42.

Gombos, A. M. (1983). A new diatom genus from the early Paleocene. *Bacillaria*, **6**, 237–44.

Gombos, A. M., Jr & Ciesielski, P. F. (1983). Late Eocene to early Miocene diatoms from the Southwest Atlantic. *Initial Rep. Deep Sea Drilling Project*, **71**, 583–634.

Goodwin, T. W. (1974). Carotenoids and biliproteins. In *Algal Physiology and Biochemistry*, Botanical Monographs No. 10 (ed. W. D. P. Stewart). pp. 176–205. Oxford: Blackwells.

Gordon, R. & Drum, R. W. (1970). A capillarity mechanism for diatom gliding locomotion. *Proc. Nat. Acad. Sci. USA*, **67**, 338–44.

Granetti, B. (1968). Studio comparativo della struttura e del ciclo biologico di due diatomee di acqua dolce: *Navicula minima* Grun. e *Navicula seminulum* Grun. I. & II. *Giorn. Bot. Ital.*, **102**, 133–58, 167–85.

Granetti, B. (1977). Variazioni morfologiche, strutturali e biometriche dei frustuli di *Navicula gallica* (W. Smith) Van Heurck coltivata in vitro per alcuni anni. *Giorn. Bot. Ital.*, **111**, 227–61.

Granetti, B. (1978). Struttura di alcune valve teratologiche di *Navicula gallica* (W. Smith) Van Heurck. *Giorn. Bot. Ital.*, **112**, 1–12.

Gregory, W. (1857). On new forms of marine Diatomaceae found in the Firth of Clyde and in Loch Fine. *Trans. Roy. Soc. Edinb.* **21**, 473–542.

Greville, R. K. (1860). A monograph of the genus *Asterolampra*, including *Asteromphalus* and *Spatangidium*. *Trans. Microsc. Soc. Lond.*, **8** (N. S.), 102–24.

Griffith, J. W. (1855). On the conjugation of the Diatomaceae. *Ann. Mag. Nat. Hist.*, Ser. 2, **16**, 92–4.

Griffith, J. W. (1856). On the siliceous sporangial sheath of the Diatomaceae. *Ann. Mag. Nat. Hist.*, Ser. 2. **18**, 75–6.

Gross, F. & Zeuthen, E. (1948). The buoyancy of planktonic diatoms: a problem of cell physiology. *Proc. Roy. Soc. London B*, **135**, 382–9.

Gschöpf, O. (1952). Das Problem der Doppelplättchen und ihrer homologer Gebilde bei den Diatomeen. *Österr. Bot. Z.*, **99**, 1–36.

Guillard, R. R. L. (1968). B12 specificity of marine centric diatoms. *J. Phycol.*, **4**, 59–64.

Guillard, R. R. L. (1975). Culture of phytoplankton for feeding marine invertebrates. In *Culture of Marine Invertebrate Animals* (ed. W. L. Smith & M. H. Chanley), pp. 29–60. New York: Plenum.

Guillard, R. R. L. & Lorenzen, C. L. (1972). Yellow-green algae with cholorophyllide c. *J. Phycol.*, **8**, 10–14.

Guillard, R. R. L. & Ryther, J. H. (1962). Studies of marine planktonic diatoms. I. *Cyclotella nana* Hustedt, and *Detonula confervacea* (Cleve) Gran. *Can. J. Microbiol.*, **8**, 229–39.

Hajós, M. (1973). Diatomées du Pannonien Inférieur provenant du bassin néogène de Csákvár. *Acta Bot. Acad. Sci. Hung.*, **18**, 95–118.

Håkansson, H. (1986). A study of the *Discoplea* species (Bacillariophyceae) described by Ehrenberg. *Diat. Res.*, **1**, 33–56.

Håkansson, H. & Locker, S. (1981). *Stephanodiscus* Ehrenberg 1846, a revision of the species described by Ehrenberg. *Nova Hedwigia*, **35**, 117–50.

Hallegraeff, G. M. (1986). Taxonomy and morphology of the marine plankton diatoms *Thalassionema* and *Thalassiothrix*. *Diat. Res.*, **1**, 57–80.

Hanna, G. D. (1927). Cretaceous diatoms from California. *Occas. Pap. Calif. Acad. Sci.*, **13**, 48pp.

Happey-Wood, C. M. & Jones, P. (1988). Rhythms of vertical migration and motility in intertidal benthic diatoms with particular reference to *Pleurosigma angulatum*. *Diat. Res.*, **3**, 83–93.

Hargraves, P. E. (1976). Studies on marine plankton diatoms. III. Structure and classification of *Gossleriella tropica*. *J. Phycol.*, **12**, 285–91.

Hargraves, P. E. (1979). Studies on marine plankton diatoms. IV. Morphology of *Chaetoceros* resting spores.

Nova Hedwigia, Beih., **64**, 99–120.

Hargraves, P. E. (1984). Resting spore formation in the marine diatom *Ditylum brightwellii* (West) Grun. ex Van Heurck. In *Proceedings of the 7th International Diatom Symposium* (ed. D. G. Mann), pp. 33–46. Koenigstein: O. Koeltz.

Hargraves, P. E. (1986). The relationship of some fossil diatom genera to resting spores. In *Proceedings of the 8th International Diatom Symposium,* (ed. M. Ricard), pp. 67–80. Koenigstein: O. Koeltz.

Hargraves, P. E. & French, F. W. (1983). Diatom resting spores: significance and strategies. In *Survival Strategies of the Algae* (ed. G. A. Fryxell), pp. 49–68. Cambridge: Cambridge University Press.

Harper, M. A. (1977). Movements. In *The Biology of Diatoms,* Botanical Monographs No. 13, (ed. D. Werner). pp. 224–49. Oxford: Blackwells.

Harper, M. A. & Harper, J. T. (1967). Measurements of diatom adhesion and their relationship with movement. *Br. phycol. Bull.,* **3**, 195–207.

Hart, T. J. (1935). On the diatoms of the skin film of whales, and their possible bearing on problems of whale movements. *Discovery Reports,* **10**, 247–82.

Harwood, D. M. (1988). Upper Cretaceous and Lower Paleocene diatom and silicoflagellate biostratigraphy of Seymour Island, eastern Antarctic Peninsula. *Geol. Soc. Am. Mem.,* **169**, 55–129.

Hasle, G. R. (1972b). *Fragilariopsis* Hustedt as a section of the genus *Nitzschia* Hassall. *Nova Hedwigia, Beih.,* **39**, 111–19.

Hasle, G. R. (1972c). Two types of valve processes in centric diatoms. *Nova Hedwigia, Beih.,* **39**, 55–78.

Hasle, G. R. (1973a). Some Marine Plankton Genera of the Diatom Family Thalassiosiraceae. In *Proceedings of the 2nd International Symposium on Recent & Fossil Marine Diatoms, Nova Hedwigia, Beih.,* **45**, 1–68.

Hasle, G. R. (1973b). Thalassiosiraceae, a new diatom family. *Norw. J. Bot.,* **20**, 67–9.

Hasle, G. R. (1974). The "mucilage pore" of pennate diatoms. *Nova Hedwigia, Beih.,* **45**, 167–94.

Hasle G. R. (1975). Some living marine species of the diatom family Rhizosoleniaceae. *Nova Hedwigia, Beih.,* **53**, 99–140.

Hasle, G. R. (1977). Morphology and taxonomy of *Actinocyclus normanii* f. *subsalsa* (Bacillariophyceae). *Phycologia,* **16**, 321–8.

Hasle, G. R. (1978). Some freshwater and brackish water species of the diatom genus *Thalassiosira* Cleve. *Phycologia,* **17**, 263–92.

Hasle, G. R. & Mendiola, B. R. E. de (1967). The fine structure of some *Thassionema* and *Thalassiothrix* species. *Phycologia,* **6**, 107–25.

Hasle, G. R. & Semina, H. J. (1987). The marine planktonic diatoms *Thalassiothrix longissima* and *Thalassiothrix antarctica* with comments on *Thalassionema* spp. and *Synedra reinboldii. Diat. Res.,* **2**, 175–92.

Hasle, G. R. & Sims, P. A. (1986). The diatom genera *Stellarima* and *Symbolophora* with comment on the genus *Actinoptychus. Br. phycol. J.,* **21**, 97–114.

Hasle, G. R. & Syvertsen, E. E. (1980). The diatom genus *Cerataulina*: Morphology and taxonomy. *Bacillaria,* **3**, 79–113.

Hasle, G. R. & Syvertsen, E. E. (1981). The marine diatoms *Fragilaria striatula* and *F. hyalina. Striae,* **14**, 110–18.

Hasle, G. R., Heimdal, B. R. & Fryxell, G. A. (1971). Morphological variability in fasciculated diatoms as exemplified by *Thalassiosira tumida* (Janisch) Hasle. comb. nov. *Antarctic Res. Ser.,* **17**, 313–33.

Hasle, G. R., Stosch, H. A. von & Syvertsen, E. E. (1983). Cymatosiraceae, a new diatom family. *Bacillaria,* **6**, 9–156.

Haupt, W. (1983). Movement of chloroplasts under the control of light. *Prog. phycol. Res.,* **2**, 227–81.

Haworth, E. Y. (1969). The diatoms of a sediment core from Blea Tarn, Langdale. *J. Ecol.,* **57**, 429–39.

Haworth, E. Y. (1972). Diatom succession in a core from Pickerel Lake, northeastern South Dakota. *Bull. geol. Soc. Am.,* **83**, 157–72.

Haworth, E. Y. (1980). Comparison of continuous phytoplankton records with the diatom stratigraphy in the recent sediments of Blelham Tarn. *Limnol. Oceanogr.,* **25**, 1093–1103.

Haworth, E. Y. (1988). Distribution of diatom taxa of the old genus *Melosira* (now mainly *Aulacoseira*) in Cumbrian waters. In *Algae and the Aquatic Environment* (ed. F. E. Round), pp. 138–67. Bristol: Biopress.

Heath, I. B. & Darley, W. M. (1972). Observations on the ultrastructure of the male gametes of *Biddulphia levis* Ehr. *J. Phycol.,* **8**, 51–9.

Heath, M. R. & Spencer, C. P. (1985). A model of the cell cycle, and cell division phasing in a marine diatom. *J. Gen. Microbiol.,* **131**, 411–25.

Hecky, R. E., Mopper, K., Kilham, P. & Degens, E. T. (1973). The amino acid and sugar composition of diatom cell-walls. *Mar. Biol.,* **19**, 323–31.

Heinzerling, O. (1908). Der Bau der Diatomeenzelle. *Bibliotheca Bot.,* **69**, 1–88.

Helmcke, J.-G. & Krieger, W. (1953–77). Diatomeenschalen im elektronenmikroskopischen Bild. Parts 1–10 [1023 sets of electron micrographs].

Hendey, N. I. (1937). The plankton diatoms of the southern seas. *Discovery Rep.,* **16**, 151–364.

Hendey, N. I. (1964). An introductory account of the smaller algae of British coastal waters. Part V. Bacillariophyceae (Diatoms). *Ministry of Agriculture, Fisheries and Food, Fisheries Investigations, Series IV.* London: HMSO. 317 pp.

Hendey, N. I. (1969). *Pyrgupyxis,* a new genus of diatoms from a south Atlantic Eocene core. *Occas. Pap. Calif. Acad. Sci.* No. 72, 6 pp.

Hendey, N. I. & Crawford, R. M. (1977). Notes on the occurrence, distribution and fine structure of *Druridgea compressa* (West) Donkin. *Nova Hedwigia, Beih.,* **54**, 1–14.

Hendey, N. I. & Sims, P. A. (1982). A review of the genus *Gomphonitzschia* Grunow and the description of *Gomphotheca* gen. nov. an unusual marine diatom group from tropical waters. *Bacillaria,* **5**, 191–212.

Hendey, N. I. & Sims, P. A. (1987). Examination of some

fossil eupodiscoid diatoms with descriptions of two new species of *Craspedoporus* Greville. *Diat. Res.*, 2. 23–34.

Herth, W. (1978). A special chitin-fibril-synthesizing apparatus in the centric diatom *Cyclotella*. *Naturwissenschaften*, 65. 260.

Herth, W. (1979). The site of β-chitin fibril formation in centric diatoms II. The chitin-forming cytoplasmic structures. *J. Ultrastr. Res.*, 68. 16–27.

Herth, W. & Barthlott, W. (1979). The site of β-chitin fibril formation in centric diatoms I. Pores and fibril formation. *J. Ultrastr. Res.*, 68. 6–15.

Herth, W. & Zugenmaier, P. (1977). Ultrastructure of the chitin fibrils of the centric diatom *Cyclotella cryptica*. *J. Ultrastr. Res.*, 61. 230–9.

Hinz, I., Spurck, T. P. & Pickett-Heaps, J. D. (1986). Metabolic inhibitors and mitosis: III. Effects of dinitrophenol on spindle disassembly in *Pinnularia*. *Protoplasma*, 132. 85–9.

Hoban, M. A. (1983). Biddulphioid diatoms II: The morphology and systematics of the pseudocellate species. *Biddulphia biddulphiana* (Smith) Boyer. *B. alternans* (Bailey) Van Heurck & *Trigonium arcticum* (Brightwell) Cleve. *Bot. Mar.*, 26. 271–84.

Hoban, M. A., Fryxell, G. A. & Buck, K. R. (1980). Biddulphioid diatoms: resting spores in Antarctic *Eucampia* and *Odontella*. *J. Phycol.*, 16. 591–602.

Holdsworth, R. H. (1968). The presence of a crystalline matrix in pyrenoids of the diatom. *Achnanthes brevipes*. *J. Cell Biol.*, 37. 831–7.

Holmes, R. W. (1977). *Lauderia annulata* – a marine centric diatom with an elongate bilobed nucleus. *J. Phycol.*, 13. 180–3.

Holmes, R. W. (1985). The morphology of diatoms epizoic on cetaceans and their transfer from *Cocconeis* to two new genera, *Bennettella* and *Epipellis*. *Br. phycol. J.*, 20. 43–57.

Holmes, R. W. & Croll, D. A. (1984). Initial observations on the composition of dense diatom growths on the body feathers of three species of diving seabirds. In *Proceedings of the 7th International Diatom Symposium* (ed. D. G. Mann), pp. 265–77. Koenigstein: O. Koeltz.

Holmes, R. W. & Mahood, A. D. (1980). *Aulacodiscus kittonii* Arnott – distribution and morphology on the west coast of the United States. *Br. phycol. J.*, 15. 377–89.

Holmes, R. W. & Reimann, B. E. F. (1966). Variation in valve morphology during the life cycle of the marine diatom *Coscinodiscus concinnus*. *Phycologia*, 5. 233–44.

Holmes, R. W., Crawford, R. M. & Round, F. E. (1982). Variability in the structure of the genus *Cocconeis* Ehr. (Bacillariophyta) with special reference to the cingulum. *Phycologia*, 21. 370–81.

Honigmann, H. (1909). Beiträge zur Kenntnis des Süsswasser-planktons. *Arch. f. Hydrobiol. Planktonk.*, 5. 71–8.

Hoops, H. J. & Floyd, G. L. (1979). Ultrastructure of the centric diatom. *Cyclotella meneghiniana*: vegetative cell and auxospore development. *Phycologia*, 18. 424–35.

Hopkins, J. T. (1969). Diatom motility: its mechanism, and diatom behaviour patterns in estuarine mud. Ph.D.

thesis, University of London.

Hopkins, J. T. & Drum, R. W. (1966). Diatom motility: an explanation and a problem. *Br. phycol. Bull.*, 3. 63–7.

Hustedt, F. (1914). Die Bacillariaceen–Gattung *Tetracyclus* Ralfs. *Abh. Nat. Ver. Bremen*, 23. 90.

Hustedt, F. (1927–66). Die Kieselalgen Deutschlands. Österreichs und der Schweiz. In *Dr. L. Rabenhorst's Kryptogamen-Flora von Deutschland, Österreich und der Schweiz*. 7. Leipzig: Akademische Verlagsgesellschaft.

Hustedt, F. (1930). Bacillariophyta (Diatomaceae). In A. Pascher *Die Süsswasser-flora Mitteleuropas*. Heft 10. 466 pp. Jena: Gustav Fischer Verlag.

Hutchinson, G. E. (1967). *A Treatise on Limnology. Vol. 2. Introduction to Lake Biology and the Limnoplankton*. New York: J. Wiley.

Idei, M. & Kobayasi, H. (1989). The fine structure of *Diploneis finnica* with special reference to the marginal openings. *Diat. Res.* 4. 25–37.

Iyengar, M. O. P. & Subrahmanyan, R. (1944). On reduction division and auxospore formation in *Cyclotella meneghiniana*. *J. Indian bot. Soc.*, 23. 125–52.

Jarosch, R. (1962). Gliding. In *Physiology and Biochemistry of Algae* (ed. R. A. Lewin), pp. 573–81. New York & London: Academic Press.

Jaworski, G. H. M., Wiseman, S. W. & Reynolds, C. S. (1988). Variability in sinking rate of the freshwater diatom *Asterionella formosa*: the influence of colony morphology. *Br. phycol. J.* 23. 167–76.

John, J. (1986). Observations on the ultrastructure of *Gephyria media* W. Arnott. In *Proceedings of the 8th International Diatom Symposium* (ed. M. Ricard), pp. 155–162. Koenigstein: O. Koeltz.

Jolley, E. T. & Jones, A. K. (1977). The interaction between *Navicula muralis* Grunow and an associated species of *Flavobacterium*. *Br. phycol. J.*, 12. 315–28.

Jones, V. J., Stevenson, A. C. & Battarbee, R. W. (1986). Lake acidification and the land-use hypothesis: a mid-post-glacial analogue. *Nature*. 322. 157–8.

Karayeva, N. I. (1978a). New genus of the family Naviculaceae West. [In Russian]. *Bot. Zh.*, 63. 1593–6.

Karayeva, N. I. (1978b). A new suborder of diatoms. [In Russian]. *Bot. Zh.*, 63. 1747–50.

Karsten, G. (1896). Untersuchungen über Diatomeen. I. *Flora*, 82. 286–96.

Karsten, G. (1897). Untersuchungen über Diatomeen. II *Flora*, 83. 33–53.

Karsten, G. (1899). Die Diatomeen der Kieler Bucht. *Wiss. Meeresunters. N.F. (Kiel)*, 4. 17–205.

Karsten, G. (1928). Bacillariophyta (Diatomaceae). In *Die natürlichen Pflanzenfamilien, 2nd edn.* (ed. A. Engler & K. Prantl), 2: 105–303. Leipzig: W. Engelmann.

Kates, M. & Volcani, B. E. (1966). Lipid components of diatoms. *Biochem. Biophys. Acta*, 116. 264–78.

Kates, M. & Volcani, B. E. (1968). Studies on the biochemistry and fine structure of silicon shell formation in diatoms. Lipid components of the cell walls. *Z. Pflanzenphysiol.*, 60. 19–29.

Kim, W. H. & Barron, J. A. (1986). Diatom biostratigraphy of the upper Oligocene to lowermost Miocene San

715

Gregorio formation, Baja California Sur, Mexico. *Diat. Res.*, **1**, 169–87.

Knudson, B. M. (1952). The diatom genus *Tabellaria*. I. Taxonomy and morphology. *Ann. Bot. N.S.*, **16**, 421–40.

Knudson, B. M. (1953a). The diatom genus *Tabellaria*. II. Taxonomy and morphology of the plankton varieties. *Ann. Bot. N.S.*, **17**, 131–55.

Knudson, B. M. (1953b). The diatom genus *Tabellaria*. III. Problems of infra-specific taxonomy and evolution in *T. flocculosa. Ann. Bot. N.S.*, **17**, 597–609.

Kobayasi, H. & Yoshida, M. (1984). Diatoms found in small concrete pools as suitable material for observation in junior high school biology. *Bull. Tokyo Gakugei Univ.*, Sect. 4, **36**, 115–43.

Kociolek, J. P. & Rosen, B. H. (1984). Observations on North American *Gomphoneis* (Bacillariophyceae). I. Valve ultrastructure of *G. mammilla* with comment on the taxonomic status of the genus. *J. Phycol.*, **20**, 361–8.

Kociolek, J. P. & Stoermer, E. F. (1987). Ultrastructure of *Cymbella sinuata* and its allies (Bacillariophyceae), and their transfer to *Reimeria*, gen. nov. *Syst. Bot.*, **12**, 451–9.

Kociolek, J. P. & Stoermer, E. F. (1988a). A preliminary investigation of the phylogenetic relationships among the freshwater, apical pore field-bearning cymbelloid and gomphonemoid diatoms (Bacillariophyceae). *J. Phycol.*, **24**, 377–85.

Kociolek, J. P. & Stoermer, E. F. (1988b). Taxonomy and systematic position of the *Gomphoneis quadripunctata* species complex. *Diat. Res.*, **3**, 95–108.

Kociolek, J. P. & Stoermer, E. F. (1989). Chromosome numbers in diatoms: a review. *Diat. Res.*, **4**, 46–54.

Kolbe, R. W. (1955). Diatoms from Equatorial Pacific cores. *Rep. Swed. Deep Sea Exped.*, **6**, 1–49.

Kolkwitz, R. & Marsson, M. (1908). Ökologie der pflanzliche Saprobien. *Ber. Deutsch. Bot. Ges.*, **26**, 505–19.

Koppen, J. D. (1975). A morphological and taxonomic consideration of *Tabellaria* (Bacillariophyceae). *J. Phycol.*, **11**, 236–44.

Körner, H. (1970). Morphologie und Taxonomie der Diatomeengattung *Asterionella. Nova Hedwigia*, **20**, 557–724.

Krammer, K. (1980). Morphologic and taxonomic investigations of some freshwater species of the diatom genus *Amphora* Ehr. *Bacillaria*, **3**, 197–225.

Krammer, K. (1982). Valve morphology in the genus *Cymbella* C. A. Agardh. In *Micromorphology of Diatom Valves* (ed. J.-G. Helmcke & K. Krammer), XI. 298 pp.

Krammer, K. (1988a). The Gibberula-group in the genus *Rhopalodia* O. Müller (Bacillariophyceae) I. Observations on the valve moriophology. *Nova Hedwigia*, **46**, 277–303.

Krammer, K. (1988b). The Gibberula-group in the genus *Rhopalodia* O. Müller (Bacillariophyceae) II. Revision of the group and new taxa. *Nova Hedwigia*, **47**, 159–205.

Krammer, K. & Lange-Bertalot, H. (1986). Bacillariophyceae. I. Teil. Naviculaceae. In *Süsswasserflora von Mitteleuropa, Band 2/1*. 876 pp.

Krammer, K. & Lange-Bertalot, H. (1987). Morphology and taxonomy of *Surirella ovalis* and related taxa. *Diat. Res.*, **2**, 77–95.

Krammer, K. & Lange-Bertalot, H. (1988). Bacillariophyceae. 2. Teil. Bacillariaceae. Epithemiaceae. Surirellaceae. In *Süsswasserflora von Mitteleurope, Band 2/2*.

Krieger, W. (1927). Die Gattung *Centronella* Voigt. *Ber. Deutsch. Bot. Ges.*, **45**:

Kützing, F. T. (1844). *Die kieselschaligen Bacillarien oder Diatomeen*. Nordhausen. 152 pp.

Labeyrie, L. D. (1974). New approach to surface seawater paleotemperatures using $^{18}0/^{16}0$ ratios in silica of diatom frustules. *Nature*, **248**, 40–2.

Labeyrie, L. D. & Juillet, A. (1982). Oxygen isotopic exchangeability of diatom valve silica: interpretation and consequences for paleoclimatic studies. *Geochim. Cosmochim. Acta*, **46**, 967–75.

Labeyrie, L. D., Juillet, A. & Duplessy, J.-C. (1984). Oxygen isotopic stratigraphy: fossil diatoms vs foraminifera. In *Proceedings of the 7th International Diatom Symposium* (ed. D. G. Mann), pp. 477–91. Koenigstein: O. Koeltz.

Labeyrie, L. D., Pichon, J. J., Labracherie, M., Ippolito, P., Duprat, J. & Duplessy, J.-C. (1986). Melting history of Antarctica during the past 60,000 years. *Nature*, **322**, 701–6.

Lacalli, T. C. (1981). Dissipative structures and morphogenetic pattern in unicellular algae. *Phil. Trans. Roy. Soc. London*, B **294**, 547–88.

Lange-Bertalot, H. (1976). Eine Revision zur Taxonomie der *Nitzschiae Lanceolatae* Grunow. Die "klassischen" bis 1930 beschriebenen Süsswasserarten Europas. *Nova Hedwigia*, **28**, 253–307.

Lange-Bertalot, H. (1980a). Zur systematischen Bewertung der bandförmigen Kolonien bei *Navicula* und *Fragilaria. Nova Hedwigia*, **33**, 723–87.

Lange-Bertalot, H. (1980b). Ein Beitrag zur Revision der Gattungen *Rhoicosphenia* Grun., *Gomphonema* C. Ag., *Gomphoneis* Cl. *Bot. Notiser*, **133**, 585–94.

Lange-Bertalot, H. & Simonsen, R. (1978). A taxonomic revision of the *Nitzschiae lanceolatae* Grunow. 2. European and related extra-European fresh water and brackish water taxa. *Bacillaria*, **1**, 11–111.

Lauterborn, R. (1896). *Untersuchungen über Bau, Kernteilung und Bewegung der Diatomeen*. Leipzig: W. Engelmann. 165 pp.

Lebour, M. V. (1930). The planktonic diatoms of northern seas. *Roy. Soc. Publ.*, no. 116, 1–244.

Le Cohu, R. (1983). Observations sur deux espèces de diatomées du genre *Diatomella: Diatomella hustedtii* Manguin et *Diatomella ouenkoana* Maillard. *Crypt. Algol.*, **4**, 63–71.

Lee, J. J. & Reimer, C. W. (1984). Isolation and identification of endosymbiotic diatoms. In *Proceedings of the 7th International Diatom Symposium* (ed. D. G. Mann), pp. 327–43. Koenigstein: O. Koeltz.

Lee, J. J., McEnery, M. & Garrison, J. (1980a). Experimental studies of larger foraminifera and their symbionts from the Gulf of Elat on the Red Sea. *J. Foram. Res.*, **10**, 31–47.

Lee, J. J., Reimer, C. & McEnery, M. (1980b). The identification of diatoms isolated as endosymbionts from larger foraminifera from the Gulf of Elat (Red Sea) and the description of 2 new species. *Fragilaria shiloi*, sp. nov. and *Navicula reissii* sp. nov. *Bot. mar.*, **23**, 41–8.

Lee, J. J., McEnery, M., Röttger, R. & Reimer, C. (1980c). The isolation, culture, and identification of endosymbiotic diatoms from *Heterostegina depressa* d'Orbigny and *Amphistegina lessonii* d'Orbigny (larger foraminifera) from Hawaii. *Bot. mar.*, **23**, 297–302.

Lee, J. J., McEnery, M. E., Shilo, M. & Keiss, Z. (1979). Isolation and cultivation of diatom symbionts from larger foraminifera (Protozoa). *Nature*, **280**, 57–8.

Lee, J. J. & Xenophontes, X. (1989). The unusual life cycle of *Navicula muscatinei*. *Diat. Res.*, **4**, 69–77.

Lee, K. & Round, F. E. (1987). Studies on freshwater *Amphora* species. I. *Amphora ovalis*. *Diat. Res.*, **2**, 193–203.

Lee, K. & Round, F. E. (1988). Studies on freshwater *Amphora* species. II. *Amphora copulata* (Kütz.) Schoeman & Archibald. *Diat. Res.*, **3**, 217–25.

Lee, M.-J., Ellis, R. & Lee, J. J. (1982). A comparative study of photoadaptation in four diatoms isolated as endosymbionts from larger foraminifera. *Mar. Biol.*, **68**, 193–7.

Leeuwenhoek, A. van (1703). Concerning green weeds growing in water, and some animalcula found about them. *Phil. Trans. Roy. Soc. London*, **23** (283), 1304–11.

Leslie, R. J. & Pickett-Heaps, J. D. (1983). Ultraviolet microbeam irradiations of mitotic diatoms: investigation of spindle elongation. *J. Cell Biol.*, **96**, 548–61.

Leslie, R. J. & Pickett-Heaps, J. D. (1984). Spindle microtubule dynamics following ultraviolet-microbeam irradiations of mitotic diatoms. *Cell*, **36**, 717–27.

Lewin, J. C. (1958). The taxonomic position of *Phaeodactylum tricornutum*. *J. gen. Microbiol.* **18**, 427–32.

Lewin, J. C. (1961). The dissolution of silica from diatom walls. *Geochim. Cosmochim. Acta*, **21**, 182–9.

Lewin, J. C. (1974). Blooms of surf-zone diatoms along the coast of the Olympic Peninsula, Washington. III. Changes in the species composition of the blooms since 1925. *Nova Hedwigia, Beih.*, **45**, 251–7.

Lewin, J. C. & Guillard, R. R. L. (1963). Diatoms. *Ann. Rev. Microbiol.*, **17**, 373–141.

Lewin, J. C. & Hruby, T. (1973). Blooms of surf-zone diatoms along the coast of the Olympic Peninsula, Washington. II. A diel periodicity in buoyancy shown by the surf-zone diatom species, *Chaetoceros armatum* T. West. *Est. Coast. Mar. Sci.*, **1**, 101–5.

Lewin, J. C. & Mackas, D. (1972). Blooms of surf-zone diatoms along the coast of the Olympic Peninsula, Washington. I. Physiological investigations of *Chaetoceros armatum* and *Asterionella socialis* in laboratory cultures. *Mar. Biol.*, **16**, 171–81.

Lewin, J. C. & Norris, R. E. (1970). Surf-zone diatoms of the coasts of Washington and New Zealand (*Chaetoceros armatum* T. West and *Asterionella* spp.). *Phycologia*, **9**, 143–9.

Lewin, J. C. & Rao, V. N. R. (1975). Blooms of surf-zone diatoms along the coast of the Olympic Peninsula, Washington. VI. Daily periodicity phenomena associated with *Chaetoceros armatum* in its natural habitat. *J.*

Phycol., **11**, 330–8.

Lewin, J. C., Hruby, T. & Mackas, D. (1975). Blooms of surf-zone diatoms along the coast of the Olympic Peninsula, Washington. V. Environmental conditions associated with the blooms (1971 and 1972). *Est. Coast. Mar. Sci.*, **3**, 229–41.

Lewin, J. C., Lewin, R. A. & Philpott, D. E. (1958). Observations on *Phaeodactylum tricornutum*. *J. gen. Microbiol.*, **18**, 418–26.

Lewin, J. C., Reimann, B. E., Busby, W. F. & Volcani, B. E. (1966). Silica shell formation in synchronously dividing diatoms. In *Cell Synchrony – Studies in Biosynthetic Regulation* (eds I. L. Cameron & G. M. Padilla), pp. 169–88. New York: Academic Press.

Lewin, R. A. (1958). The mucilage tubes of *Amphipleura rutilans*. *Limnol. Oceanogr.*, **3**, 111–13.

Lewis, W. M. (1984). The diatom sex clock and its evolutionary significance. *Amer. Nat.*, **123**, 73–80.

Li, C.-W. & Chiang, Y.-M. (1977). The fine structure of the frustule of a centric diatom *Hydrosera triquetra* Wallich. *Br. phycol. J.*, **12**, 203–13.

Li, C.-W. & Volcani, B. E. (1984). Aspects of silicification in wall morphogenesis of diatoms. *Phil. Trans. Roy. Soc. Lond.*, B **304**, 519–28.

Li, C.-W. & Volcani, B. E. (1985a). Studies on the biochemistry and fine structure of silica shell formation in diatoms. VIII. Morphogenesis of the cell wall in a centric diatom, *Ditylum brightwellii*. *Protoplasma*, **124**, 10–29.

Li, C.-W. & Volcani, B. E. (1985b). Studies on the biochemistry and fine structure of silica shell formation in diatoms. IX. Sequential valve formation in a centric diatom, *Chaetoceros rostratum*. *Protoplasma*, **124**, 30–41.

Li, C.-W. & Volcani, B. E. (1985c). Studies on the biochemistry and fine structure of silica shell formation in diatoms. X. Morphogenesis of the labiate process in centric diatoms. *Protoplasma*, **124**, 147–56.

Li, C.-W. & Volcani, B.E. (1987). Four new apochlorotic diatoms. *Br. phycol. J.*, **22**, 375–82.

Liebisch, W. (1928). *Amphitetras antedeluviana* Ehr., sowie einige Beiträge zum Bau und zur Entwicklung der Diatomeenzelle. *Zeit. f. Botanik*, **20**, 225–71.

Liebisch, W. (1929). Experimentelle und kritische Untersuchungen über die Pektinmembran der Diatomeen unter besonderer Berücksichtigung der Auxosporenbildung und der Kratikularzustande. *Z. Bot.*, **22**, 1–65.

Lizitzin, A. P. (1971). Distribution of siliceous microfossils in suspension and in bottom deposits. In *The Micropaleontology of Oceans* (ed. B. M. Funnel & W. R. Riedel), pp. 173–95.

Locker, F. (1950). Beiträge zur Kenntnis des Formwechsels der Diatomeen an Hand von Kulturversuchen. *Österr. Bot. Z.*, **97**, 322–32.

Loeblich, A. R. III, Wight, W. W. & Darley, W. M. (1968). A unique colonial marine centric diatom *Coenobiodiscus muriformis* gen. et sp. nov. *J. Phycol.*, **4**, 23–9.

Lund, J. W. G. (1949). Studies on *Asterionella* I. The origin and nature of the cells producing seasonal maxima.

J. Ecol., **37**, 389–419.

Lund, J. W. G. (1950). Studies on *Asterionella formosa* Hass. II. Nutrient depletion and the spring maximum. *J. Ecol.*, **38**, 1–32.

Lund, J. W. G. (1954). The seasonal cycle of the plankton diatom *Melosira italica* (Ehr.) Kütz. subsp. *subarctica* O. Müll. *J. Ecol.*, **42**, 151–79.

Lund, J. W. G. (1955). Further observations on the seasonal cycle of *Melosira italica* (Ehr.) Kütz. subsp. *subarctica* O. Müll. *J. Ecol.*, **43**, 90–102.

Lund, J. W. G. (1965). The ecology of the freshwater phytoplankton. *Biol. Rev.*, **40**, 231–93.

Lund, J. W. G. (1971). An artificial alteration of the seasonal cycle of the plankton diatom *Melosira italica* subsp. *subarctica* in an English lake. *J. Ecol.*, **59**, 521–33.

Lund, J. W. G. & Reynolds, C. S. (1982). The development and operation of large limnetic enclosures in Blelham Tarn, English Lake District, and their contribution to phytoplankton ecology. *Progr. phycol. Res.*, **1**, 1–65.

Lund, J. W. G. Kipling, C. & LeCren, E. D. (1958). The inverted microscope method of estimating algal numbers and the statistical basis of estimations by counting. *Hydrobiologia*, **11**, 143–70.

McDonald, K. L., Edwards, M. K. & McIntosh, J. R. (1979). Cross-sectional structure of the central mitotic spindle of *Diatoma vulgare*. Evidence for specific interactions between antiparallel microtubules. *J. Cell Biol.*, **83**, 443–61.

McDonald, K. L., Pickett-Heaps, J. D., McIntosh, J. R. & Tippit, D. H. (1977). On the mechanism of anaphase spindle elongation in *Diatoma vulgare*. *J. Cell Biol.*, **74**, 377–88.

McIntosh, J. R., McDonald, K. L., Edwards, M. K. & Ross, B. M. (1979). Three-dimensional structure of the central mitotic spindle of *Diatoma vulgare*. *J. Cell Biol.*, **83**, 428–42.

Mackereth, F. J. H. (1965). Chemical investigation of lake sediments and their interpretation. *Proc. Roy. Soc. London, B* **161**, 295–309.

Mackereth, F. J. H. (1966). Some chemical observations on post-glacial lake sediments. *Phil. Trans. Roy. Soc. London, B* **250**, 165–213.

McLachlan, J., McInnes, A. G. & Falk, M. (1965). Studies on the chitan (chitin: poly-*N*-acetylglucosamine) fibers of the diatom *Thalassiosira fluviatilis* Hustedt. I. Production and isolation of chitan fibers. *Can. J. Bot.*, **43**, 707–13.

Mague, T. H., Mague, F. C. & Holm-Hansen, O. (1977). Physiology and chemical composition of nitrogen-fixing phytoplankton in the Central North Pacific Ocean. *Mar. Biol.*, **41**, 213–27.

Mahoney, R. K. & Reimer, C. W. (1986). Studies on the genus *Brebissonia* (Bacillariophyceae). I. Introduction and observations on *B. lanceolata* comb. nov. In *Proceedings of the 8th International Diatom Symposium* (ed. M. Ricard). pp. 183–90. Koenigstein: O. Koeltz.

Makarova, I. V. (1988). Diatomaceous algae of the seas of the U.S.S.R.: the genus *Thalassiosira* Cl. *Akad. Nauk C.C.C.P.*, 117 pp.

Mann, A. (1907). Report on the diatoms of the Albatross voyages in the Pacific Ocean, 1888-1904. *Contributions from the United States National Herbarium*, **10**(5), 221–419.

Mann, D. G. (1977). The diatom genus *Hantzschia* Grünow – an appraisal. *Nova Hedwigia, Beih.*, **54**, 323–54.

Mann, D. G. (1978). Studies in the family Nitzschiaceae (Bacillariophyta). Ph.D. Dissertation, University of Bristol.

Mann, D. G. (1980a). *Hantzschia fenestrata* Hust. (Bacillariophyta) – *Hantzschia* or *Nitzschia*. *Br. phycol. J.*, **15**, 249–60.

Mann, D. G. (1980b). Studies in the diatom genus *Hantzschia* II. *H. distinctepunctata*. *Nova Hedwigia*, **33**, 341–52.

Mann, D. G. (1981a). A note on valve formation and homology in the diatom genus *Cymbella*. *Ann. Bot.*, **47**, 267–9.

Mann, D. G. (1981b). Studies in the diatom genus *Hantzschia* 3. Infraspecific variation in *H. virgata*. *Ann. Bot.*, **47**, 377–95.

Mann, D. G. (1981c). Sieves and flaps: siliceous minutiae in the pores of raphid diatoms. In *Proceedings of the 6th Symposium on Recent and Fossil Diatoms*, (ed. R. Ross), pp. 279–300. Koenigstein: O. Koeltz.

Mann, D. G. (1982a). Structure, life history and systematics of *Rhoicosphenia* (Bacillariophyta) I. The vegetative cell of *Rh. curvata*. *J. Phycol.*, **18**, 162–76.

Mann, D. G. (1982b). Structure, life history and systematics of *Rhoicosphenia* (Bacillariophyta) II. Auxospore formation and perizonium structure of *Rh. curvata*. *J. Phycol.*, **18**, 264–74.

Mann, D. G. (1982c). Auxospore formation in *Licmophora* (Bacillariophyta). *Plant Syst. Evol.*, **139**, 289–94.

Mann, D. G. (1982d). The use of the central raphe endings as a taxonomic character (Notes for a monograph of the Bacillariaceae 1.) *Plant Syst. Evol.*, **141**, 143–52.

Mann, D. G. (1983). Symmetry and cell division in raphid diatoms. *Ann. Bot.*, **52**, 573–81.

Mann, D. G. (1984a). Protoplast rotation, cell division and frustule symmetry in the diatom *Navicula bacillum*. *Ann. Bot.*, **53**, 295–302.

Mann, D. G. (1984b). An ontogenetic approach to diatom systematics. In *Proceedings of the 7th International Diatom Symposium* (ed. D. G. Mann), pp. 113–41. Koenigstein: O. Koeltz.

Mann, D. G. (1984c). Auxospore formation and development in *Neidium* (Bacillariophyta). *Br. phycol. J.*, **19**, 319–31.

Mann, D. G. (1984d). Structure, life history and systematics of *Rhoicosphenia* (Bacillariophyta). V. Initial cell and size reduction in *Rh. curvata* and a description of the Rhoicospheniaceae fam. nov. *J. Phycol.*, **20**, 544–55.

Mann, D. G. (1984e). Observations on copulation in *Navicula pupula* and *Amphora ovalis* in relation to the nature of diatom species. *Ann. Bot.*, **54**, 429–38.

Mann, D. G. (1985). *In vivo* observations of plastid and cell division in raphid diatoms and their relevance to diatom systematics. *Ann. Bot.*, **55**, 95–108.

Mann, D. G. (1986a). Methods of sexual reproduction in *Nitzschia*: Systematic and evolutionary implications.

(Notes for a monograph of the Bacillariaceae 3). *Diat. Res.*, **1**, 193–203.

Mann, D. G. (1986b). *Nitzchia* subgenus *Nitzchia* (Notes for a monograph of the Bacillariaceae, 2). In *Proceedings of the 8th International Diatom Syposium* (ed. M. Ricard), pp. 215–26. Koenigstein: O. Koeltz.

Mann, D. G. (1987). Sexual reproduction in *Cymatopleura*. *Diat. Res.*, **2**, 97–112.

Mann, D. G. (1988a). The nature of diatom species: Analyses of sympatric populations. In *Proceedings of the 9th International Diatom Symposium* (ed. F. E. Round), pp. 293–304. Koenigstein: O. Koeltz and Bristol: Biopress.

Mann, D. G. (1988b). Why didn't Lund see sex in *Asterionella*? A discussion of the diatom life cycle in nature. In *Algae and the Aquatic Environment* (ed. F.E. Round), pp. 384–412. Bristol: Biopress.

Mann, D. G. (1989a). On auxospore formation in *Caloneis* and the nature of *Amphiraphia* (Bacillariophyta). *Plant Syst. Evol.*, **163**, 43–52.

Mann, D. G. (1989b). The diatom genus *Sellaphora*: separation from *Navicula*. *Br. phycol. J.*, **24**, 1–20.

Mann, D. G. (1989c). The species concept in diatoms: evidence for morphologically distinct, sympatric gamodemes in four epipelic species. *Plant Syst. Evol.* (in press).

Mann, D. G. & Marchant, H. J. (1989). The origins of the diatom and its life cycle. In *The Chromophyte Algae: problems and perspectives* (ed. B. S. C. Leadbeater & J. C. Green). Oxford: Oxford University Press.

Mann, D. G. & Stickle, A. J. (1988). Nuclear movements and frustule symmetry in raphid pennate diatoms. In *Proceedings of the 9th International Diatom Symposium* (ed. F. E. Round), pp. 281–9. Koenigstein: O. Koeltz and Bristol: Biopress.

Mann, D. G. & Stickle, A. J. (1989). Meioisis, nuclear cyclosis and auxospore formation in *Navicula sensu stricto* (Bacillariophyta). *Br. phycol. J.*, **24**, 167–81.

Manton, I. & von Stosch, H. A. (1966). Observations on the fine structure of the male gamete of the marine centric diatom *Lithodesmium undulatum*. *J. Roy. microsc. Soc.*, **85**, 119–134.

Manton, I., Kowallik, K. & von Stosch, H. A. (1969a). Observations on the fine structure and development of the spindle at mitosis and meiosis in a marine centric diatom (*Lithodesmium undulatum*) I. Preliminary survey of mitosis in spermatogonia. *J. Microsc.*, **89**, 295–320.

Manton, I., Kowallik, K. & von Stosch, H. A. (1969b). Observations on the fine structure and development of the spindle at mitosis and meiosis in a marine centric diatom (*Lithodesmium undulatum*). II. The early meiotic stages in male gametogenesis. *J. Cell Sci.*, **5**, 271–98.

Manton, I., Kowallik, K. & von Stosch, H. A. (1970a). Observations on the fine structure and development of the spindle at mitosis and meiosis in a marine centric diatom (*Lithodesmium undulatum*) III. The later stages of meiosis I in male gametogenesis. *J. Cell Sci.*, **6**, 131–57.

Manton, I., Kowallik, K. & von Stosch, H. A. (1970b). Observations on the fine structure and development of the spindle at mitosis and meiosis in a marine centric diatom (*Lithodesmium undulatum*) IV. The second meiotic division and conclusion. *J. Cell Sci.*, **7**, 407–43.

Marciniak, B. (1982). Late glacial and holocene new diatoms from a glacial lake Przedi Staw in the Piec Stawow Polskich Valley, Polish Tatra Mts. *Acta Geol. Scient. Hung.*, **25**, 161–71.

Mayama, S. & Kobayasi, H. (1986). Observation of *Navicula mobiliensis* var. *minor* Patr. and *N. goeppertiana* (Bleisch) H. L. Sm. In *Proceedings of the 8th International Diatom Symposium* (ed. M. Ricard), pp. 173–82. Koenigstein: O. Koeltz.

Mayama, S. & Kobayasi, H. (1989). Sequential valve development in the monoraphid diatom *Achnanthes minutissima* var. *saprophila*. *Diat. Res.*, **4**, 111–17.

Maynard Smith, J. (1978). *The Evolution of Sex*. Cambridge: Cambridge University Press.

Medlin, L. K. (1985). A reappraisal of the diatom genus *Rhoiconeis* and the description of *Campylopyxis*, gen. nov. *Br. phycol. J.*, **20**, 313–28.

Medlin, L. K. & Fryxell, G. A. (1984a). Structure, life history and systematics of *Rhoicosphenia* (Bacillariophyta). III. *Rhoicosphenia adolfi* and its relationship to *Rhoiconeis*. In *Proceedings of the 7th International Diatom Symposium* (ed. D. G. Mann), pp. 255–63. Koenigstein: O. Koeltz.

Medlin, L. K. & Fryxell, G. A. (1984a). Structure, life history and systematics of *Rhoicosphenia* (Bacillariophyta). IV. Correlation of size reduction with changes in valve morphology of *Rh. genuflexa*. *J. Phycol.*, **20**, 101–8.

Medlin, L. K. & Round, F. E. (1986). Taxonomic studies of marine gomphonemoid diatoms. *Diat. Res.*, **1**, 205–25.

Medlin, L. K., Crawford, R. M. & Andersen, R. A. (1986). Histochemical and ultrastructural evidence for the function of labiate process in the movement of centric diatoms. *Br. phycol. J.*, **21**, 297–301.

Medlin, L. K., Elwood, H. J., Stickel, S. & Sogin, M. L. (1988). The characterization of enzymically amplified eukaryotic 16s-like rRNA coding regions. *Gene*, **71**, 491–9.

Mereschkowsky, C. (1901a). On *Okedenia*. *Ann. Mag. Nat. Hist.*, Ser. 7, **8**, 415–23.

Mereschkowsky, C. (1901b). On *Stauronella*, a new genus of diatoms. *Ann. Mag. Nat. Hist.*, Ser. 7, **8**, 424–34.

Mereschkowsky, C. (1901c). Études sur l'endochrome des Diatomées. *Mem. Acad. Imp. Sci. St. Petersb.*, Ser. 8, **11**(6), 1–40.

Mereschkowsky, C. (1902). On *Sellaphora*, a new genus of diatoms. *Ann. Mag. Nat. Hist.*, Ser. 7, **9**, 185–195.

Mereschkowsky, C. (1902-3). Les types de l'endochrome. *Scripta Botanica Horti Univers. Imp. Petropol.*, **21**, 1–106 [Russian], 107–93 [French].

Mereschkowsky, C. (1903a). Nouvelles recherches sur la structure et la division des Diatomées. *Bull. Soc. Imp. nat. Moscou, N.S.*, **17**, 149–72.

Mereschkowsky, C. (1903b). Über *Placoneis*, ein neues Diatomeengenus. *Beih. bot. Centralbl.*, **15**, 1–30.

Meunier, A. (1910). *Microplankton des Mers de Barents et de Kara*. Duc d'Orleans Campagne Arctique de 1907,

Bulens, Brussels. 355 pp.

Migita, S. (1967). Sexual reproduction of centric diatoms *Skeletonema costatum*. *Bull. Jap. Soc. Scient. Fish.*, 33, 392-8.

Miller, W. I. III & Collier, A. (1978). Ultrastructure of the frustule of *Triceratium favus* (Bacillariophyceae). *J. Phycol.*, 14, 56-62.

Mills, F. W. (1933-5). *An index to the genera and species of the Diatomaceae and their synonyms 1816-1932.* London: Wheldon & Wesley, 1726 pp.

Mitchison, J. M. (1957). The growth of single cells. I. *Schizosaccharomyces pombe*. *Exp. Cell. Res.*, 13, 244-62.

Mitchison, J. M. (1971). *The Biology of the Cell Cycle.* Cambridge: Cambridge University Press. 313 pp.

Montgomery, R. T. & Miller, I. W. (1978). A taxonomic study of Florida Keys benthic diatoms based on scanning electron microscopy. In *Environmental and ecological studies of the diatom communities associated with the coral reefs of the Florida Keys* (ed. R. T. Montgomery). Ph.D. dissertation, Florida State University.

Morrow, A. C., Deason, T. R. & Clayton, D. (1981). A new species of the diatom genus *Eunotia*. *J. Phycol.*, 17, 265-70.

Moss, M. O., Gibbs, G., Gray, V. & Ross, R. (1978). The presence of a raphe in *Semiorbis hemicyclus* (Ehr.) Patr. *Bacillaria*, 1, 137-50.

Moyiseeva, A. I. & Genkal, S. I. (1987). On the freshwater species of the genus *Paralia* (Bacillariophyta). *Botanicheskii Zhurnal*, 72, 1500-4 (in Russian).

Müller, O. (1883). Das Gesetz der Zelltheilungsfolge von *Melosira* (*Orthosira*) *arenaria* Moore. *Ber. Dtsch. Bot. Ges.*, 1, 35-44.

Müller, O. (1884). Die Zellhaut und des Gesetz der Zelltheilungsfolge von *Melosira* (*Orthosira* Thwaites) *arenaria* Moore. *Jahrb. Wiss. Bot.*, 14, 232-90.

Müller, O. (1889). Auxosporen von *Terpsinöe musica* Ehr. *Ber. Dtsch. Bot. Ges.*, 7, 181-3.

Müller, O. (1895). Ueber Achsen, Orientierungs- und Symmetrieebenen bei den Bacillariaceen. *Ber. Dtsch. Bot. Ges.*, 13, 222-34.

Müller, O. (1903). Sprungweise Mutation bei Melosireen. *Ber. Dtsch. Bot. Ges.* 21: 326-33.

Müller, O. (1906). Pleomorphismus, Auxosporen und Dauersporen bei *Melosira* arten. *Jahrb. Wiss. Bot.* 43: 49-88.

Müller, O. F. (1783a). Strand- Parlebandet och Armbandet, tvanne microscopiska Strandvaxter. *Kongl. Vetenskaps Academiens, Nya Handlingar*, 4: 80-5 and Tab. III.

Müller, O. F. (1783b). Vaesen i Strandvandet. *Nye Samling af det Kongelige Danske Videnskabers Selskabs Skrifter*, 2, 277-86.

Müller, O. F. (1786). *Animalcula infusoria fluviatilia et marina*. Lvi + 367 pp. Havniae: N. Moller.

Murphy, L. S. & Bellastock, R. A. (1980). The effect of environmental origin on the response of marine diatoms to chemical stress. *Limnol. Oceanogr.*, 25, 160-5.

Nakajima, T. & Volcani, B. E. (1969). 3,4-Dihydroxyproline: a new amino acid in diatom cell walls. *Science*, 164, 1400-6.

Nakajima, T. & Volcani, B. E. (1970). *N*-Trimethyl-*L*-hydroxylysine phosphate and nonphosphorylated compound in diatom cell walls. *Biochem. Biophys. Res. Commun.*, 39, 28-33.

Navarro, J. N. (1981). A survey of the marine diatoms of Puerto Rico. I. Suborders Coscinodiscineae and Rhizosoleniineae. *Bot. Mar.*, 24, 427-39.

Nelson, D. M. & Brand, L. E. (1979). Cell division periodicity in 13 species of marine phytoplankton on a light-dark cycle. *J. Phycol.*, 15, 67-75.

Nelson, D. M., Goering, J. J., Kilham, S. S. & Guillard, R. R. L. (1976). Kinetics of silica acid uptake and rates of silica dissolution in the marine diatom *Thalassiosira pseudonana*. *J. Phycol.*, 12, 246-52.

Neville, A. C. (1986). The physics of helicoids. Multidirectional 'plywood' structure in biological systems. *Phys. Bull.*, 37, 74-6.

Neville, A. C. (1988). A pipe-cleaner molecular model for morphogenesis of helicoidal plant cell walls based on hemicellulose complexity. *J. theor. Biol.*, 131, 243-54.

Nikolaev, V. A. (1969). Species novae Bacillariophytorum epiphyticorum marninorum. *Nov. Syste. Plant. non Vasc.*, 6, 29-34.

Nikolaev, V. A. (1983). On the genus *Symbolophora* (Bacillariophyta). *Bot. Zh.*, 68, 1123-8.

Nikolaev, V. A. (1984). On the importance of the areola structure for the taxonomy of diatoms (Bacillariophyta). *Bot. Zh.*, 69, 1040-6. [In Russian, title and summary in English.]

Nipkow, F. (1927). Über das Verhaben der Skelette planktischer Kieselalgen im geschichteten Tiefschlam des Zurich- und Baldeggersees. *Z. Hydrol. Hydrogr. Hydrobiol.*, 4, 71-120.

Novarino, G. (1987). A note on the internal construction of the partectal ring of *Mastogloia lanceolata*. *Diat. Res.* 2, 213-7.

Okita, T. W. & Volcani, B. E. (1978). Role of silicon in diatoms. IX. Differential synthesis of DNA polymerases and DNA-binding proteins during silicate starvation and recovery in *Cylindrotheca fusiformis*. *Biochim. Biophys. Acta*, 519, 76-86.

Okita, T. W. & Volcani, B. E. (1980). Role of silica in diatom metabolism. X. Polypeptide labelling patterns during the cell cycle, silicate starvation and recovery in *Cylindrotheca fusiformis*. *Exp. Cell. Res.*, 125, 471-81.

Oliver, R. K., Kinnear, A. J. & Ganf, G. G. (1981). Measurements of cell density of three freshwater phytoplankters by density gradient centrifugation. *Limnol. Oceanogr.*, 26, 285-94.

Olson, R. J., Watras, C. & Chisholm, S. W. (1986a). Patterns of individual cell growth in marine centric diatoms. *J. gen. Microbiol.*, 132, 1197-1204.

Olson, R. J., Vaulot, D. & Chisholm, S. W. (1986b). Effects of environmental stresses on the cell cycles of two marine phytoplankton species. *Plant Physiol.*, 80, 918-25.

Opute, F. I. (1974). Studies on fat accumulation in *Nitzschia palea* Kütz. *Ann. Bot.*, 38, 889-902.

Ott, E. (1900). Untersuchungen über den Chromatophorenbau der Süsswasser-Diatomaceen und dessen Beziehung zur Systematik. *Sber. Akad. Wiss. Wien*, **109**, 769–801.

Paasche, E. (1980). Silicon. In *The Physiological Ecology of Phytoplankton*, Studies in Ecology 7, (ed. I. Morris), pp. 259–84. Oxford: Blackwells.

Paddock, T. B. B. (1978). Observations on the valve structures of diatoms of the genus *Plagiodiscus* and on some associated species of *Surirella*. *Bot. J. Linn. Soc.*, **76**, 1–25.

Paddock, T. B. B. (1986). Observations on the genus *Stauropsis* Meunier and related species. *Diat. Res.*, **1**, 89–98.

Paddock, T. B. B. (1988). *Plagiotropis* Pfitzer and *Tropidoneis* Cleve, a summary account. *Bibliotheca Diatomologica*, Vol. 16, 152pp, 37 pls.

Paddock, T. B. B. & Sims, P. A. (1977). A preliminary survey of the raphe structure of some advanced groups of diatoms (Epithemiaceae – Surirellaceae). *Nova Hedwigia, Beih.*, **54**, 291–322.

Paddock, T. B. B. & Sims, P. A. (1980). Observations on the marine diatom genus *Auricula* and two new genera *Undatella* and *Proboscidea*. *Bacillaria*, **3**, 161–96.

Paddock, T. B. B. & Sims, P. A. (1981). A morphological study of keels of various raphe–bearing diatoms. *Bacillaria*, **4**, 177–222.

Palmer, J. D. & Round, F. E. (1965). Persistent, vertical-migration rhythms in benthic microflora. I. The effect of light and temperature on the rhythmic behaviour of *Euglena obtusa. J. mar. biol. Ass. U.K.*, **45**, 567–82.

Palmer, J. D. & Round, F. E. (1967). Persistent, vertical-migration rhythms in benthic microflora. VI. The tidal and diurnal nature of the rhythm in the diatom *Hantzschia virgata. Biol. Bull. mar. biol. Lab., Woods Hole*, **132**, 44–55.

Pascher, A. (1921). Über die Übereinstimmungen zwischen den Diatomeen, Heterokonten und Chrysomonaden. *Ber. dt. Bot. Ges.*, **39**, 236–40.

Paterson, D. M. (1986). The migratory behaviour of diatom assemblages in a laboratory tidal micro-ecosystem examined by low temperature scanning electron microscopy. *Diat. Res.*, **1**, 227–39.

Paterson, D. M., Crawford, R. M. & Little, C. (1986). The structure of benthic diatom assemblages: A preliminary account of the use and evaluation of low-temperature scanning electron microscopy. *J. Exp. Mar. Biol. Ecol.*, **96**, 279–89.

Patrick, R. & Reimer, C. W. (1966). The diatoms of the United States I. *Acad. Nat. Sci. Philad., Monogr.* **13**, 688 pp.

Patrick, R. & Reimer, C. W. (1975). The diatoms of the United States II, part 1. *Acad. Nat. Sci. Philad., Monogr.* **13**, 213 pp.

Pavillard, T. (1924). Observations sur les diatomées, 4me série. Le genre *Bacteriastrum. Bull. Soc. bot. France*, **71**, 1084–90.

Pearsall, W. H. (1921). The development of vegetation in the English Lakes, considered in relation to the general evolution in glacial lakes and rock basins. *Proc. R. Soc. London B* **92**, 259–84.

Peragallo, H. (1891). Monographie du genre *Pleurosigma* et des genres alliés. *Le Diatomiste*, 1(4), 1–35.

Peragallo, H. (1892). Monographie du genre *Rhizosolenia* et de quelques genres voisins. *Le Diatomiste*, **1**, 79, 82, 99–117.

Peragallo, H. & Peragallo, M. (1897-1908). *Diatomées marines de France et des districts maritimes voisins.* Micrographie-Editeur, Grez-sur-Loing. 491 pp.

Pfitzer, E. (1871). Untersuchungen über Bau und Entwicklung der Bacillariaceen (Diatomaceen). In *Bot. Abhandl.* (ed. Hanstein), 1(2), 1–189.

Phipps, D. W., Jr & Rosowski, J. R. (1983). The morphology and integration of valves and bands in *Navicula mutica* var. *mutica* (Bacillariophyceae). *J. Phycol.*, **19**, 320–3.

Pickett-Heaps, J. D. (1983). Valve morphogenesis and the microtubule center in three species of the diatom *Nitzschia. J. Phycol.*, **19**, 269–81.

Pickett-Heaps, J.D. (1986). Mitotic mechanisms: an alternative view. *Trends in Biochemical Sciences*, **11**, 504–7.

Pickett-Heaps, J. D. & Kowalski, S. E. (1981). Valve morphogenesis and the microtubule center of the diatom *Hantzschia amphioxys. Europ. J. Cell Biol.*, **25**, 150–70.

Pickett-Heaps, J. D. & Tippit, D. H. (1978). The diatom spindle in perspective. *Cell*, **14**, 455–67.

Pickett-Heaps, J. D., Hill, D. R. A. & Wetherbee, R. (1986). Cellular movement in the centric diatom *Odontella sinensis. J. Phycol.* **22**, 334–39

Pickett-Heaps, J. D., McDonald, K. L. & Tippit, D. H. (1975). Cell division in the pennate diatom *Diatoma vulgare. Protoplasma*, **86**, 205–242.

Pickett-Heaps, J. D., Schmid, A.-M. M. & Tippit, D. H. (1984). Cell division in diatoms. *Protoplasma* **120**, 132–54.

Pickett-Heaps, J. D., Tippit, D. H. & Andreozzi, J. A. (1978a). Cell division in the pennate diatom *Pinnularia*. I. Early stages in mitosis. *Biol. Cellulaire*, **33**, 71–8.

Pickett-Heaps, J. D., Tippit, D. H. & Andreozzi, J. A. (1978b). Cell division in the pennate diatom *Pinnularia*. II. Later stages in mitosis. *Biol. Cellulaire*, **33**, 79–84.

Pickett-Heaps, J. D., Tippit, D. H. & Andreozzi, J. A. (1979a). Cell division in the pennate diatom *Pinnularia*. III. The valve and associated cytoplasmic organelles. *Biol. Cellulaire*, **35**, 195–8.

Pickett-Heaps, J. D., Tippit, D. H. & Andreozzi, J. A. (1979b). Cell division in the pennate diatom *Pinnularia*. IV. Valve morphogenesis. *Biol. Cellulaire*, **35**, 199–203.

Pickett-Heaps, J. D., Tippit, D. H. & Andreozzi, J. A. (1979c). Cell division in the pennate diatom *Pinnularia*. V. Observations on live cells. *Biol. Cellulaire*, **35**, 295–304.

Pickett-Heaps, J. D., Tippit, D. H. & Leslie, R. (1980a). Light and electron microscopic observations on cell division in two large pennate diatoms, *Hantzschia* and *Nitzschia*. I. Mitosis *in vivo. Europ. J. Cell Biol.*, **21**, 1–11.

Pickett-Heaps, J. D., Tippit, D. H. & Leslie, R. (1980b) Light and electron microscopic observations on cell division in two large pennate diatoms, *Hantzschia* and *Nitzschia*. II. Ultrastructure. *Europ. J. Cell Biol.*, **21**, 12–27.

Pickett-Heaps, J. D., Tippit, D. H. & Porter, K. R. (1982). Rethinking mitosis. *Cell*, **29**, 729–44.

Pickett-Heaps, J. D., Wetherbee, R. & Hill, D. R. A. (1988). Cell division and morphogenesis of the labiate process in the centric diatom *Ditylum brightwellii*. *Protoplasma*, **143**, 139–49.

Picket-Heaps, J. D., Cohn, S., Schmid, A.-M. M. & Tippit, D. H. (1988). Valve morphogenesis in *Surirella* Bacillariophyceae). *J. Phycol.*, **24**, 35–49.

Pocock, K. L. & Cox, E. J. (1982). Frustule structure in the diatom *Rhabdonema arcuatum* (Lyngb.) Kütz. *Nova Hedwigia*, **36**, 621–41.

Poulin, M., Bérard-Therriault, L. & Cardinal, A. (1986). *Fragilaria* and *Synedra* (Bacillariophyceae): A morphological and ultrastructural approach. *Diat. Res.*, **1**, 99–112.

Prézelin, B. B. & Boczar, B. A. (1986). Molecular bases of cell absorption and fluorescence in phytoplankton: potential applications to studies in optical oceanography. In *Progress in Phycological Research* (ed. F. E. Round & D. J. Chapman), **4**: 350–464. Bristol: Biopress.

Qi. Y.-Z., Reimer, C. W. & Mahoney, R. K. (1984). Taxonomic studies of the genus *Hydrosera*. I. Comparative morphology of *H. triquetra* Wallich and *H. whampoensis* (Schwarz) Deby with ecological remarks. In *Proceedings of the 7th International Diatom Symposium* (ed. D. G. Mann), pp. 213–24, Koenigstein: O. Koeltz.

Rattray, J. (1888a). A revision of the genus *Aulacodiscus* Ehrb. *J. Roy. Microsc. Soc.*, **8**, 337–82.

Rattray, J.(1888b). A revision of the genus *Auliscus* Ehrenberg and of some allied genera. *J. Roy. Microsc. Soc.*, **8**, 861– 920.

Rattray, J. (1890). A revision of the genus *Coscinodiscus* Ehr. and some allied genera. *Proc. Roy. Soc. Edinb.*, **16**, 449– 692.

Raven, J. A. (1980). Nutrient transport in micro-algae. *Adv. Microbiol. Physiol.*, **21**, 47–226.

Raven, J. A. (1983). The transport and function of silicon in plants. *Biol. Rev.*, **58**, 179–207.

Reid, F. M. & Round, F. E. (1987). The antarctic diatom *Synedra reinboldii*: Taxonomy, ecology and transference to a new genus, *Trichotoxon*. *Diat. Res.*, **2**, 219–27.

Reimann, B. (1960). Bildung, Bau, und Zusammenhang der Bacillariophyceenschalen. *Nova Hedwigia*, **2**, 349–73.

Reimann, B. E. F. & Lewin, J.C. (1964). The diatom genus *Cylindrotheca* Rabenhorst (with a reconsideration of *Nitzschia closterium*). *J. Roy. Microsc. Soc.*, **83**, 283–96.

Reimann, B. E. F. & Volcani, B. E. (1968). Studies on the biochemistry and fine structure of silica shell formation in diatoms. III. The structure of the cell wall of *Phaeodactylum tricornutum* Bohlin. *J. Ultrastruct. Res.*, **21**, 182–193.

Reynolds, C. S. (1984). *The Ecology of Freshwater Phytoplankton*. Cambridge: Cambridge University Press. 384 pp.

Ricard, M. (1970). Observations sur les diatomées marines du genre *Ethmodiscus* Castr. *Rev. Algol. N.S.*, **1**, 56–73.

Ricard, M. (1987). Diatomophycées. *Atlas du Phytoplankton Marin* (ed. A. Sournia), II, 297 pp. Paris: Editions du CNRS.

Ricard, M. & Gasse, F. (1972). *Ethmodiscus appendiculatus*

et *Ethmodiscus gazellae* en microscopie electronique a balayage. *Rev. Algol. N.S.*, **4**, 312–5.

Rieth, A. (1953). Zur Auxosporenbildung bei *Melosira nummuloides*. *Flora (Jena)*, **140**, 205–8.

Rines, J. E. B. & Hargraves, P. E. (1988). The *Chaetoceros* Ehrenberg (Bacillariophyceae) Flora of Narragansett Bay, Rhode Island, U.S.A. *Bibliotheca Phycologica*. **79**. 196pp.

Rivera, P. & Koch, P. (1984). Contributions to the diatom flora of Chile II. In *Proceedings of the 7th International Diatom Symposium* (ed. D.G. Mann), pp. 279–98. Koenigstein: O. Koeltz.

Rivera, P. S., Avaria, S. & Barrales, H. L. (1989). *Ethmodiscus rex* collected by net sampling off the coast of northern Chile. *Diat. Res.*, **4**, 131–42.

Robinson, D. H. & Sullivan, C. W. (1987). How do diatoms make silicon biominerals? *Trends Biochem. Sci.*, **12**, 151–4.

Robinson, M. (1982). Diatom analysis of early Flandrian lagoon sediments from East Lothian, Scotland. *J. Biogeogr.*, **9**, 207–21.

Roemer, S. C. & Rosowski, J. R. (1980). Valve and band morphology of some freshwater diatoms. III. Pre- and post-auxospore frustules and the initial cell of *Melosira roeseana*. *J. Phycol.*, **16**, 399–411.

Roessler, P. G. (1988). Effects of silicon deficiency on lipid composition and metabolism in the diatom *Cyclotella cryptica*. *J. Phycol.* **24**, 394–400.

Rosen, B. H. & Lowe, R. L. (1981). Valve ultrastructure of some confusing Fragilariaceae. *Micron*, **22**, 293–4.

Rosowski, J. R. (1980). Valve and band morphology of some freshwater diatoms. II. Integration of valves and bands in *Navicula confervacea* var. *confervacea*. *J. Phycol.*, **16**, 88–101.

Ross, R. (1976). Some Eocene diatoms from South Atlantic cores. II. *Rutilaria* Greville. *Occas. Pap. Calif. Acad. Sci.*, **123**, 21–7.

Ross, R. & Sims, P. A. (1970). Studies of *Aulacodiscus* with the scanning electron microscope. *Nova Hedwigia, Beih.*, **31**, 49–88.

Ross, R. & Sims, P. A. (1971). Generic limits in the Biddulphiaceae as indicated by the scanning electron microscope. In *Scanning Electron Microscopy: systematic and evolutionary applications* (ed. V.H. Heywood), pp. 155–77. London: Academic Press.

Ross, R. & Sims, P. A. (1972). The fine structure of the frustule in centric diatoms: a suggested terminology. *Br. phycol. J.*, **7**, 139–63.

Ross, R. & Sims, P. A. (1973). Observations on family and generic limits in the Centrales. *Nova Hedwigia, Beih.* **45**, 97–121.

Ross, R. & Sims, P. A. (1978). Notes on some diatoms from the Isle of Mull, and other Scottish localities. *Bacillaria*, **1**, 151–68.

Ross, R. & Sims, P. A. (1980). *Syringidium* Ehrenb. *Dextradonator* Ross & Sims, nov. gen. and *Abas* Ross & Sims, nov. gen. *Bacillaria*, **3**, 115–27.

Ross, R. & Sims, P. A. (1985). Some genera of the Biddulphiaceae (diatoms) with interlocking linking spines. *Bull. Br. Mus. Nat. Hist. (Bot.)*, **13**(3), 277–381.

Ross, R. & Sims, P.A. (1987). Further genera of the Biddulphiaceae (diatoms) with interlocking linking spines. *Bull. Br. Mus. Nat. Hist. (Bot.)*, 16, 269–311.

Ross, R., Sims, P. A. & Hasle, G. R. (1977). Observations on some species of the Hemiauloideae. *Nova Hedwigia, Beih.*, 54, 179–213.

Ross, R., Cox, E. J., Karayeva, N. I., Mann, D. G., Paddock, T. B. B., Simonsen, R. & Sims, P. A. (1979). An amended terminology for the siliceous components of the diatom cell. *Nova Hedwigia, Beih.*, 64, 513–33.

Roth, L. E. & de Francisco, A. (1977). The marine diatom, *Striatella unipunctata*. II. Siliceous structures and the formation of intercalary bands. *Cytobiologie*, 14, 207–21.

Round, F. E. (1953). An investigation of two benthic algal communities in Malham Tarn, Yorkshire. *J. Ecol.*, 41, 174–97.

Round, F. E. (1957). The late-glacial and post-glacial diatom succession in the Kentmere Valley deposit. *New Phytol.* 56, 98–126.

Round, F. E. (1961). The diatoms of a core from Esthwaite Water. *New Phytol.*, 60, 43–59.

Round, F. E. (1970). The genus *Hantzschia* with particular reference to *H. virgata* v. *intermedia* (Grün.) comb. nov. *Ann. Bot.*, 34, 75–91.

Round, F. E. (1972a). The problem of reduction of cell size during diatom cell division. *Nova Hedwigia*, 23, 291–303.

Round, F. E. (1972b). *Stephanodiscus binderanus* (Kütz.) Krieger or *Melosira binderana* Kütz. (Bacillariophyta, Centrales). *Phycologia*, 11, 109–17.

Round, F. E. (1972c). Some observations on colonies and ultrastructure of the frustule of *Coenobiodiscus muriformis* and its transfer to *Planktoniella*. *J. Phycol.*, 8, 221–31.

Round, F. E. (1973). On the diatom genera *Stephanopyxis* Ehr. and *Skeletonema* Grev. and their classification in a revised system of the Centrales. *Bot. mar.*, 16, 148–54.

Round, F. E. (1978a). *Stictocyclus strictodiscus* (Bacillariophyta): comments on its ecology, structure and classification. *J. Phycol.*, 14, 150–56.

Round, F. E. (1978b). The diatom genus *Chrysanthemodiscus* Mann. (Bacillariophyta). *Phycologia*, 17, 157–61.

Round, F. E. (1979a). The classification of the genus *Synedra*. *Nova Hedwigia, Beih.*, 64, 135–46.

Round, F. E. (1979b). A diatom assemblage living below the surface of intertidal sand flats. *Mar. Biol.* 54, 219–23.

Round, F. E. (1980). Forms of the giant diatom *Ethmodiscus* from the Pacific and Indian Oceans. *Phycologia*, 19, 307–16.

Round, F. E. (1981a). Morphology and phyletic relationships of the silicified algae and the archetypal diatom - monophyly or polyphyly? In *Silicon and Siliceous Structures in Biological Systems* (ed. T. L. Simpson & B. E. Volcani), pp. 97–128. New York: Springer-Verlag.

Round, F. E. (1981b). *The Ecology of Algae*. Cambridge: Cambridge University Press. 653 pp.

Round, F. E. (1981c). Some aspects of the origin of diatoms and their subsequent evolution. *BioSystems*, 14, 483–6.

Round, F. E. (1981d). *Cyclostephanos* - a new genus within the Sceletonemaceae. *Arch. Protistenk.*, 125, 323–9.

Round, F. E. (1981e). The diatom genus *Stephanodiscus*: an electron-microscopic view of the classical species. *Arch. Protistenk.*, 124, 455–70.

Round, F. E. (1982a). The diatom genus *Climacosphenia* Ehr. *Bot. mar.*, 25, 519–27.

Round, F. E. (1982b). Auxospore structure, initial valves and the development of populations of *Stephanodiscus* in Farmoor Reservoir. *Ann. Bot.*, 49, 447–9.

Round, F. E. (1984a). Structure of the cells, cell division and colony formation in the diatoms *Isthmia enervis* Ehr. and *I. nervosa* Kütz. *Ann. Bot.*, 53, 457–68.

Round, F. E. (1984b). The circumscription of *Synedra* and *Fragilaria* and their subgroupings. In *Proceedings of the 7th International Diatom Symposium* (ed. D. G. Mann), pp. 241–53. Koenigstein: O. Koeltz.

Round, F. E. (1988). A re-investigation of the diatom *Pinnularia cardinaliculus*. In *Algae and the Aquatic Environment* (ed. F. E. Round), pp. 434–45. Bristol: Biopress.

Round, F. E. & Brook, A. J. (1959). The phytoplankton of some Irish loughs and an assessment of their trophic status. *Proc. Roy. Irish Acad.*, 60(B), 167–91.

Round, F. E. & Crawford, R. M. (1981). The lines of evolution of the Bacillariophyta I. Origin. *Proc. Roy. Soc. London, B* 211, 237–60.

Round, F. E. & Crawford, R. M. (1989). Bacillariophyta. In *Handbook of Protoctists* (eds L. Margulis, D. J. Chapman, J. Corliss & M. Melkonian), (in press).

Round, F. E. & Hickman, M. (1971). Phytobenthos sampling and estimation of primary production. In *Methods for the Study of Marine Benthos* (ed. N. A. Holme & A. M. McIntyre), pp. 169–96, Oxford & Edinburgh: Blackwell.

Round, F. E. & Mann, D. G. (1980). *Psammodiscus* nov. gen Based on *Coscinodiscus nitidus*. *Ann. Bot.*, 46, 367–73.

Round, F. E. & Mann, D. G. (1981). The diatom genus *Brachysira*. I. Typification and separation from *Anomoeoneis*. *Arch Protistenk.*, 124, 221–31.

Round, F. E. & Palmer, J. D. (1966). Persistent, vertical-migration rhythms in benthic microflora II. Field and laboratory studies on diatoms from the banks of the River Avon. *J. mar. biol. Ass. UK*, 46, 191–214.

Salah, M. M. (1955). Some new diatoms from Blakeney Point (Norfolk). *Hydrobiologia*, 7, 88–102.

Schmid, A.-M. M. (1976). Morphologische und physiologische Untersuchungen an Diatomeen des Neusieder Sees: II. Licht- und rasterelektronenmikroskopische Schalenanalyse der umweltabhangigen Zyklomorphose von *Anomoeoneis sphaerophora* (Kg.) Pfitzer. *Nova Hedwigia*, 28, 309–51.

Schmid, A.-M. M. (1978). Zur Diatomeenflora Salzburg I. *Centronella reichelti* Voigt (Fragilariaceae). Neu für Österreich. *Floristische Mitteilungen aus Salzburg*, 5, 9–17.

Schmid, A.-M. M. (1979a). The development of structure in the shells of diatoms. *Nova Hedwigia, Beih.*, 64, 219–36.

Schmid, A.-M. M. (1979b). Influence of environmental factors on the development of the valve in diatoms. *Protoplasma*, 99, 99–115.

Schmid, A.-M. M. (1980). Valve morphogenesis in diatoms: a pattern-related filamentous system in pennates and the effect of APM, colchicine and osmotic pressure. *Nova Hedwigia*, 33, 811–47.

Schmid, A.-M. M. (1984a). Wall morphogenesis in *Thalassiosira eccentrica*: comparison of auxospore formation and the effect of MT-inhibitors. In *Proceedings of the 7th International Diatom Symposium* (ed. D. G. Mann), pp. 47–70. Koenigstein: O. Koeltz.

Schmid, A.-M. M. (1984b). Tricornate spines in *Thalassiosira eccentrica* as a result of valve–modelling. In *Proceedings of the 7th International Diatom Symposium* (ed. D. G. Mann), pp. 71–95. Koenigstein: O. Koeltz.

Schmid, A.-M. M. (1985). *Centronella reichelti* Voigt – A very unusual diatom in the surface sediments of the Grabensee. In *Contributions to the Paleolimnology of the Trumer Lakes (Salzburg) and the Lakes Mondsee, Attersee and Traunsee (Upper Austria)* (ed. D. Danielopol, R. Schmidt & E. Schultze), pp. 65–78. Limnologisches Institut, Österreichische Akademie der Wissenschaften.

Schmid, A.-M. M. (1986a). Organisation and function of cell structures in diatoms and their morphogenesis. In *Proceedings of the 8th International Diatom Symposium* (ed. M. Ricard), pp. 271–92. Koenigstein: O. Koeltz.

Schmid, A.-M. M. (1986b). Wall morphogenesis in *Coscinodiscus wailesii* Gran et Angst. II. Cytoplasmic events of valve morphogenesis. In *Proceedings of the 8th International Diatom Symposium* (ed. M. Ricard), pp. 293–314. Koenigstein: O. Koeltz.

Schmid, A.-M. M. (1987). Morphogenetic forces in diatom cell wall formation. In *Cytomechanics* (ed. J. Bereiter-Hahn, O. R. Anderson & W.-E. Reif), pp. 183–99. Berlin: Springer.

Schmid, A.-M. M. & Schulz, D. (1979). Wall morphogenesis in diatoms: deposition of silica by cytoplasmic vesicles. *Protoplasma*, 100, 267–88.

Schmid, A.-M. M. & Volcani, B. E. (1983). Wall morphogenesis in *Coscinodiscus wailesii* Gran and Angst. I. Valve morphology and development of its architecture. *J. Phycol.*, 19, 387–402.

Schmid, A.-M. M., Borowitzka, M. A. & Volcani, B. E. (1981). Morphogenesis and biochemistry of diatom cell walls. In *Cytomorphogenesis in Plants*, Cell Biology Monographs 8, (ed. O. Kiermayer), pp. 63–97. Vienna & New York: Springer-Verlag.

Schmidt, A, (*et al.*) (1874–1959). *Atlas der Diatomaceenkunde*. 472 plates. Leipzig: R. Reisland, Ascherleben.

Schnepf, E., Deichgräber, G. & Drebes, G. (1980). Morphogenetic process in *Attheya decora* (Bacillariophyceae, Biddulphiineae). *Plant Syst. Evol.*, 135, 265–77.

Schoeman, F. R. & Archibald, R. E. M. (1980). *The diatom flora of Southern Africa*, 6, 1–35. C.S.I.R. Special Rep. Wat. 50.

Schoeman, F. R. & Archibald, R. E. M. (1987). Observations on *Amphora* species (Bacillariophyceae) in the British Museum (Natural History). VI. Some species from the subgenus *Halamphora* Cleve. *Nova Hedwigia*, 44, 377–98.

Schoeman, F. R., Archibald, R. E. M. & Ashton, P. J. (1984). The diatom flora in the vicinity of the Pretoria Salt Pan, Transvaal, Republic of South Africa. Part III (final). *S. Afr. J. Bot.*, 3, 191–207.

Schrader, H.-J. (1971). Morphologische-systematische Untersuchungen an Diatomeen. I. Die Gattungen *Oestrupia* Heiden, *Progonoia* Schrader, *Caloneis* Cleve. *Nova Hedwigia*, 22, 915–38.

Schultz, M. E. (1971). Salinity related polymorphism in the brackish water diatom *Cyclotella cryptica*. *Can. J. Bot.*, 49, 1285–9.

Schultz, M. E. & Trainor, F. R. (1968). Production of male gametes and auxospores in the centric diatoms *Cyclotella meneghiniana* and *C. cryptica*. *J. Phycol.*, 4, 85–8.

Schultz, M. E. & Trainor, F. R. (1970). Production of male gametes and auxospores in a polymorphic clone of the centric diatom *Cyclotella*. *Can. J. Bot.*, 48, 947–51.

Schulz, D. & Jarosch, R. (1980). Rotating microtubules as a basis for anaphase spindle elongation in diatoms. *Europ. J. Cell Biol.*, 20, 249–53.

Schulz, D., Drebes, G., Lehmann, H. & Jank-Ladwig, R. (1984). Ultrastructure of *Anaulus creticus* Drebes & Schulz with special reference to its reduced ocelli. *Eur. J. Cell Biol.*, 33, 43–51.

Schütt, F. (1896). Bacillariales. In *Die natürlichen Pflanzenfamilien* (ed. Engler & Prantl), 1(1b), 31–153. Leipzig: W. Engelmann.

Semina, H. J. (1981). A morphological examination of the diatom *Thalassiothrix longissima* Cleve et Grunow. *Bacillaria*, 4, 147–60.

Semina, H. J. & Beklemishev, C. W. (1981). A promorphological approach to diatom cell structure. In *Proceedings of the 6th Symposium on Recent and Fossil Diatoms* (ed. R. Ross), pp. 211–30. Koenigstein: O. Koeltz.

Servant-Vildary, S. (1986). *Cyclotella* sp. fossiles de depôts d'eau douce d'age miocene en Espagne. In *Proceedings of the 8th International Diatom Symposium* (ed. M. Ricard), pp. 495–511. Koenigstein: O. Koeltz.

Sicko-Goad, L. (1986). Rejuvenation of *Melosira granulata* (Bacillariophyceae) resting cells from the anoxic sediments of Douglas Lake, Michigan. II. Electron microscopy. *J. Phycol.*, 22, 28–35.

Sicko-Goad, L., Stoermer, E. F. & Ladewski, B. G. (1977). A morphometric method for correcting phytoplankton cell volume estimates. *Protoplasma*, 93, 147–163.

Sicko-Goad, L., Stoermer, E. F. & Fahnenstiel, G. (1986). Rejuvenation of *Melosira granulata* (Bacillariophyceae) resting cells from the anoxic sediments of Douglas Lake, Michigan. I. Light microscopy and ^{14}C uptake. *J. Phycol.*, 22, 22–8.

Silva, P. C. (1980). Names of classes and families of living algae. *Regnum Vegetabile*, 102, 1–25.

Simola, H. (1977)). Diatom succession in the formation of annually laminated sediment in Lovojärvi, a small eutrophicated lake. *Ann. Bot. Fenn.*, 14, 143–8.

Simonsen, R. (1962). Untersuchungen zur Systematik und Ökologie der Bodendiatomeen der westlichen Ostsee. *Int. Rev. ges. Hydrobiol. Syst. Beih.*, 1, 144 pp.

Simonsen, R. (1970). Protoraphidaceae, eine neue Familie

der Diatomeen. *Nova Hedwigia*, 31, 377–94.

Simonsen, R. (1972). On the diatom genus *Hemidiscus* Wallich and other members of the so-called "Hemidiscaceae". *Veröff. Inst. Meeresforsch. Bremerhafen*, 13, 265–73.

Simonsen, R. (1974). The diatom plankton of the Indian Ocean Expedition of R/V *Meteor* 1964–5. *"Meteor" Forsch.-Ergebnisse Reihe D*, 19, 1–107.

Simonsen, R. (1975). On the pseudonodulus of the centric diatoms, or Hemidiscaceae reconsidered. *Nova Hedwigia*, *Beih.*, 53, 83–94.

Simonsen, R. (1979). The diatom system: ideas on phylogeny. *Bacillaria*, 2, 9–71.

Simonsen, R. (1987). *Atlas and Catalogue of the Diatom Types of Friedrich Hustedt*. 3 Vols. Berlin & Stuttgart: J. Cramer.

Sims, P. A. (1983). A taxonomic study of the genus *Epithemia* with special reference to the type species *E. turgida* (Ehrenb.) Kütz. *Bacillaria*, 6, 211–35.

Sims, P. A. (1986). *Sphynctolethus* Hanna, *Ailuretta*, gen. nov. and evolutionary trends within the Hemiauloideae. *Diat. Res.*, 1, 241–69.

Sims, P. A. (1988). The fossil genus *Trochosira*, its morphology taxonomy and systematics. *Diat. Res.*, 3, 245–57.

Sims, P. A. (1989). Some Cretaceous and Palaeocene species of *Coscinodiscus*: a micromorphological and systematic study. *Diat. Res.*, 4, (in press).

Sims, P. A. & Hasle, G. R. (1987). Two Cretaceous *Stellarima* species: *S. steinyi* and *S. distincta*; their morphology, palaeogeography and phylogeny. *Diat. Res.*, 2, 229–40.

Sims, P. A. & Holmes, R. W. (1983). Studies on the "kittonii" group of *Aulacodiscus* species. *Bacillaria*, 6, 267–92.

Sims, P. A. & Ross, R. (1988). Some Cretaceous and Palaeocene *Trinacria* (diatom) species. *Bull. Brit. Mus. (Nat. Hist.) Bot.*, 18, 275–322.

Singer, S. J. & Nicholson, G. L. (1972). The fluid mosaic model of the structure of cell membranes. *Science*, 175, 720–31.

Smith, J. E. & Sowerby, J. (1790–1814). *English Botany: or, coloured figures of British plants, with their essential characters, synonyms, and places of growth*. London: J. Sowerby.

Smith, W. (1853–6). *A synopsis of the British Diatomaceae*. Vols 1 and 2. London:

Sneath, P. H. A. & Sokal, R. R. (1973). *Numerical Taxonomy*. San Francisco: W. H. Freeman & Co.

Soranno, T. & Pickett-Heaps, J. D. (1982). Directionally controlled spindle disassembly after mitosis in the diatom *Pinnularia*. *Europ. J. Cell Biol.*, 26, 234–43.

Stager, J. C. (1984). The diatom record of Lake Victoria (East Africa): the last 17,000 years. In *Proceedings of the 7th International Diatom Symposium* (ed. D. G. Mann), pp. 455–76. Koenigstein: O. Koeltz.

Stager, J. C., Reinthal, P. N. & Livingstone, D. A. (1986). A 25,000-year history for Lake Victoria, East Africa, and some comments on its significance for the evolution of

cichlid fishes. *Freshwater Biol.*, 16, 15–19.

Starr, R. C. & Zeikus, I. A. (1987). UTEX – The culture collection of algae at the University of Texas at Austin. *J. Phycol.*, 23, Supplement. 47 pp.

Stauber, J. L. & Jeffrey, S. W. (1988). Photosynthetic pigments in fifty-one species of marine diatoms. *J. Phycol.*, 24, 158–72.

Stein, J. R. (ed.) (1973). *Handbook of phycological methods. Culture Methods and Growth Measurement*. Cambridge: Cambridge University Press. 448 pp.

Stephens, F. C. & Gibson, R. A. (1979a). Ultrastructural studies on some *Mastogloia* (Bacillariophyceae) species belonging to the group Ellipticae. *Bot. mar.*, 22, 499–509.

Stephens, F. C. & Gibson, R. A. (1979b). Observations of loculi and associated extracellular material in several *Mastogloia* (Bacillariophyceae) species. *Rev. Algol. N.S.*, 14, 211–32.

Stephens, F. C. & Gibson, R. A. (1980). Ultrastructural studies of some *Mastogloia* (Bacillariophyceae) species belonging to the group Sulcatae. *Nova Hedwigia*, 23, 219–248.

Steyaert, J. & Bailleux, E. M. (1975a). The structure of *Eucampia balaustium* Castr. as revealed by the stereo scanning electron microscope. *Nova Hedwigia*, 26, 195–204.

Steyaert, J. & Bailleux, E. M. (1975b). *Eucampia balaustium* n. sp. Castr. and *Molleria antarctica* n. sp. Castr. – Taxonomy and nomenclature. *Microscopy*, 32, 461–70.

Stickle, A. J. & Mann, D. G. (1988). Protoplast structure and behaviour in *Stauroneis* and its taxonomic implications. In *Proceedings of the 9th International Diatom Symposium*, (ed. F. E. Round), pp. 293–301. Bristol: Biopress & Koenigstein: Koeltz.

Stidolph, S. R. (1988). Observations and remarks on the morphology of the diatom genera *Gyrosigma* Hassall and *Pleurosigma* W. Smith. *Nova Hedwigia*, 47, 377–88.

Stockner, J. G. (1972). Paleolimnology as a means of assessing eutrophication. *Verh. Internat. Verein. Limnol.* 18, 1018–30.

Stockner, J. G. & Lund, J. W. G. (1970). Live algae in postglacial lake deposits. *Limnol. Oceanogr.*, 15, 41–58.

Stockwell, D. A. & Hargraves, P. E. (1986). Morphological variability within resting spores of the marine diatom genus *Chaetoceros* Ehrenberg. In *Proceedings of the 8th International Diatom Symposium* (ed. M. Ricard), pp. 81–95. Koenigstein: O. Koeltz.

Stoermer, E. F. (1967). Polymorphism in *Mastogloia*. *J. Phycol.* 3, 73–7.

Stoermer, E. F. & Håkansson, H. (1983). An investigation of the morphological structure and taxonomic relationships of *Stephanodiscus damasii*. *Bacillaria*, 6, 245–56.

Stoermer, E. F. & Ladewski, T. B. (1983). Quantitative analysis of shape variation in type and modern populations of *Gomphoneis herculeana*. *Nova Hedwigia, Beih.*, 73, 347–86.

Stoermer, E. F., Pankratz, H. S. & Drum, R. W. (1964). The fine structure of *Mastogloia grevillei* Wm. Smith. *Protoplasma*, 59, 1–13.

Stosch, H. A. von (1950). Oogamy in a centric diatom. *Nature*, 165, 531–2.

Stosch, H. A. von (1951a). Zur Entwicklungsgeschichte zentrischer Meeresdiatomeen. *Naturwissenschaften*, 38, 191–2.

Stosch, H. A. von (1951b). Entwicklungsgeschichtliche Untersuchungen an zentrischen Diatomeen. I. Die Auxosporenbildung von *Melosira varians*. *Arch. Mikrobiol.*, 16, 101–35.

Stosch, H. A. von (1954). Die Oogamie von *Biddulphia mobiliensis* und die bisher bekannten Auxosporenbildungen bei den Centrales. *VIIIe Congr. Intern. Bot., Rapp. Comm. Sect.*, 17, 58–68.

Stosch, H. A. von (1955). Zur Darstellung pflanzlicher Meristeme. *Z. wiss. Mikrosk.*, 62, 305–10.

Stosch, H. A. von (1956). Entwicklungsgeschichtliche Untersuchungen an zentrischen Diatomeen. II. Geschlechtszellenreifung, Befruchtung und Auxosporenbildung einiger grundbewohnender Biddulphiaceen der Nordsee. *Arch. Mikrobiol.*, 23, 327–65.

Stosch, H. A. von (1958a). Entwicklungsgeschichtliche Untersuchungen an zentrischen Diatomeen. III. Die Spermatogenese von *Melosira moniliformis* Agardh. *Arch. Mikrobiol.*, 31, 274–82.

Stosch, H. A. von (1958b). Kann die oogame Araphidee *Rhabdonema adriaticum* als Bindeglied zwischen den beiden grossen Diatomeengruppen angesehen werden? *Ber. dt. Bot. Ges.*, 71, 241–9.

Stosch, H. A. von (1962a). Über das Perizonium der Diatomeen. *Vortr. Gesamtgeb. Bot.*, 1, 43–52.

Stosch, H. A. von (1962b). Kulturexperiment und Ökologie bei Algen. *Kieler Meeresforsch.*, 18, Sonderheft, 13–27.

Stosch, H. A. von (1965). Manipulierung der Zellgrösse von Diatomeen im Experiment. *Phycologia*, 5, 21–44.

Stosch, H. A. von (1967). Diatomeen (in H. Ettl, D. G. Müller, K. Neumann, H. A. von Stosch & W. Weber: Vegetative Fortpflanzung, Parthenogenese und Apogamie bei Algen). In *Handbuch der Pflanzenphysiologie* (ed. W. Ruhland), 18: 657–81. Berlin: Springer-Verlag.

Stosch, H. A. von (1970). Methoden zur Präparation kleiner oder zarter Kieselelemente für die Elektronen- und Lichtmikroskopie, insbesondere von Diatomeen und bei geringen Materialmengen. *Z. wiss. Mikrosk.*, 70, 29–32.

Stosch, H. A. von (1974). Pleurax, seine Synthese und seine Verwendung zur Einbettung und Darstellung det Zellwände von Diatomeen, Peridineen und anderen Algen, sowie für eine neue Methode zur Electivfärbung von Dinoflagellaten-Penzern. *Arch. Protistenk.*, 116, 132–41.

Stosch, H. A. von (1975). An amended terminology of the diatom girdle. *Nova Hedwigia, Beih.*, 53, 1–35.

Stosch, H. A. von (1977). Observations on *Bellerochea* and *Streptotheca*, including descriptions of three new planktonic diatom species. *Nova Hedwigia, Beih.*, 54, 113–66.

Stosch, H. A. von (1980a). The two *Lithodesmium* species (Centrales) of European waters. *Bacillaria*, 3, 7–20.

Stosch, H. A. von (1980b). The 'endochiastic areola', a complex new type of siliceous structure in a diatom. *Bacillaria* 3, 21–40.

Stosch, H. A. von (1981). Structural and histochemical observations on the organic layers of the diatom cell wall. In *Proceedings of the 6th Symposium on Recent and Fossil Diatoms* (ed. R. Ross), pp. 231–52. Koenigstein: O. Koeltz.

Stosch, H. A. von (1982). On auxospore envelopes in diatoms. *Bacillaria*, 5, 127–56.

Stosch, H. A. von (1985). Some marine diatoms from the Australian region, especially from Port Phillip Bay and tropical north-eastern Australia. *Brunonia*, 8, 293–348.

Stosch, H. A. von & Drebes, G. (1964). Entwicklungsgeschichtliche Untersuchungen an zentrischen Diatomeen. IV. Die Planktondiatomee *Stephanopyxis turris* – ihre Behandlung und Entwicklungsgeschichte. *Helgol. wiss. Meeresunters.*, 11, 209–57.

Stosch, H. A. von & Fecher, K. (1979). "Internal thecae" of *Eunotia soleirolii* (Bacillariophyceae): development, structure and function as resting spores. *J. Phycol.*, 15, 233–43.

Stosch, H. A. von & Kowallik, K. V. (1969). Der von L. Geitler aufgestellte Satz über die Notwendigkeit einer Mitose für jede Schalenbildung von Diatomeen. Beobachtungen über die Reichweite und Überlegungen zu seiner zellmechanischen Bedeutung. *Österr. bot. Z.*, 116, 454–74.

Stosch, H. A. von & Reimann, B. E. F. (1970). *Subsilicea fragilarioides* gen. et spec. nov., eine Diatomee (Fragilariaceae) mit vorwiegend organischer Membran. *Nova Hedwigia, Beih.*, 31, 1–36.

Stosch, H. A. von & Simonsen, R. (1984). *Biddulphiopsis*, a new genus of the Biddulphiaceae. *Bacillaria*, 7, 9–36.

Stosch, H. A. von, Theil, G. & Kowallik, K. V. (1973). Entwicklungsgeschichtliche Untersuchungen an zentrischen Diatomeen. V. Bau und Lebenszyklus von *Chaetoceros didymum*, mit Beobachtungen über einige andere Arten der Gattung. *Helgol. wiss. Meeresunters.*, 25, 384–445.

Subrahmanyan, R. (1947). On somatic division, reduction division, auxospore-formation and sex-differentiation in *Navicula halophila* (Grün.) Cl. *J. Indian Bot. Soc.*, Iyengar Commem. Vol., 239–266.

Sullivan, C. W. (1986). Silicification by diatoms. In *Silicon Biochemistry*, Ciba Foundation Symp. 121 (ed. D. Evered & M. O'Connor), pp. 59–89. Wiley.

Sullivan, C. W. & Volcani, B. E. (1973). Role of silicon in diatom metabolism. III. The effects of silicic acid on DNA polymerase, TMP kinase and DNA synthesis in *Cylindrotheca fusiformis*. *Biochim. Biophys. Acta*, 308, 212–9.

Sullivan, C. W. & Volcani, B. E. (1976). Role of silicon in diatom metabolism. VII. Silicic acid-stimulated DNA synthesis in toluene-permeabilized cells of *Cylindrotheca fusiformis*. *Exp. Cell Res.*, 98, 23–30.

Sullivan, C. W. & Volcani, B. E. (1981). Silicon in the cellular metabolism of diatoms. In *Silicon and siliceous structures in biological systems* (ed. T. L. Simpson & B. E.

Volcani), pp. 15–42. New York: Springer-Verlag.

Sullivan, M. J. (1979). Taxonomic notes on epiphytic diatoms of Mississippi Sound, USA. *Nova Hedwigia. Beih.,* **64**, 241–9.

Sullivan, M. J. (1986). A light and scanning microscopical study of *Eupodiscus radiatus* Bailey (Eupodiscaceae). In *Proceedings of the 8th International Diatom Symposium* (ed. M. Ricard), pp. 113–123. Koenigstein: O. Koeltz.

Sullivan, M. J. (1988). A morphological study of the marine diatom *Glyphodesmis eximia*: The type of the genus. In *Proceedings of the 9th International Diatom Symposium* (ed. F.E. Round), pp. 361–70. Bristol: Biopress & Koenigstein: Koeltz.

Sundback, K. (1987). The epipsammic marine diatom *Opephora olsenii* Moller. *Diat. Res.,* **2**, 241–9.

Sundback, K. & Medlin, L. K. (1986). A light and electron microscopic study of the epipsammic diatom *Catenula adhaerens* Mereschkowsky. *Diat. Res.,* **2**, 283–90.

Sundstrom, B. G. (1984). Observations on *Rhizosolenia clevei* (Bacillariophyceae) and *Richelia intracellularis* Schmidt (Cyanophyceae). *Bot. mar.,* **27**, 345–53.

Sundstrom, B. G. (1986). The marine diatom genus *Rhizosolenia*. A new approach to the taxonomy. Doctoral Diss. (LUNDBS/NBB–1008), 1–196.

Syvertsen, E. E. (1979). Resting spore formation in clonal cultures of *Thalassiosira antarctica* Comber, *T. nordenskioeldii* Cleve and *Detonula confervacea* (Cleve) Gran. *Nova Hedwigia, Beih.,* **64**, 41–63.

Syvertsen, E. E. (1985). Resting spore formation in the Antarctic diatoms *Coscinodiscus furcatus* Karsten and *Thalassiosira australis* Peragallo. *Polar Biol.,* **4**, 113–9.

Syvertsen, E. E. & Hasle, G. R. (1983). The diatom genus *Eucampia*: morphology and taxonomy. *Bacillaria*, **6**, 169–210.

Taguchi, S., Hirata, J. A. & Laws, E. A. (1987). Silicate deficiency and lipid synthesis of marine diatoms. *J. Phycol.,* **23**, 260–7.

Takano, H. (1981). New and rare diatoms from Japanese marine waters. VII. *Bull. Tokai Reg. Fish. Res. Lab.,* **105**, 45–57.

Takano, H. (1983). New and rare diatoms from Japanese marine waters. XI. Three new species epizoic on copepods. *Bull. Tokai Reg. Fish. Res. Lab.,* **111**, 23–35.

Tappan, H. (1980). *The Paleobiology of Plant Protists.* San Francisco: W. H. Freeman & Co. 1028 pp.

Taylor, D. L. (1972). Ultrastructure of *Cocconeis diminuta* Pantocsek. *Arch. Mikrobiol.,* **81**, 136–45.

Taylor, F. J. R. (1967). Phytoplankton of the south western Indian Ocean. *Nova Hedwigia,* **12**, 433–76.

Teixeira, C. (1958). A new genus and a new species of diatom from Brazilian marine waters. *Bol. Inst. Oceanogr. São Paulo.*

Thaler, F. (1972). Beitrag zur Entwicklungsgeschichte und zum Zellbau einiger Diatomeen. *Österr. Bot. Z.,* **120**, 313–347.

Theriot, E. & Stoermer, E. F. (1981). Some aspects of morphological variation in *Stephanodiscus niagarae* (Bacillariophyceae). *J. Phycol.,* **17**, 64–72.

Theriot, E., Stoermer, E. F. & Håkansson, H. (1987).

Taxonomic interpretation of the rimoportula of freshwater genera in the centric diatom family Thalassiosiraceae. *Diat. Res.,* **2**, 251–65.

Theriot, E., Håkansson, H., Kociolek, J. P., Round, F. E. & Stoermer, E. F. (1987). Validation of the centric diatom genus name *Cyclostephanos. Br. phycol. J.,* **22**, 345–7.

Thompson, R. (1984). A global review of palaeomagnetic results from wet lake sediments. In *Lake Sediments and Environmental History* (ed. J. W. G. Lund & E. Y. Haworth), pp. 145–64. Leicester: Leicester University Press.

Tippit, D. H. & Pickett-Heaps, J. D. (1977). Mitosis in the pennate diatom *Surirella ovalis. J. Cell Biol.,* **73**, 705–27.

Tippit, D. H., McDonald, K. L. & Pickett-Heaps, J. D. (1975). Cell division in the centric diatom *Melosira. Cytobiologie,* **12**, 52–73.

Tippit, D. H., Pickett-Heaps, J. D. & Leslie, R. (1980). Cell division in two large pennate diatoms *Hantzschia* and *Nitzschia* III. A new proposal for kinetochore function during prometaphase. *J. Cell Biol.,* **86**, 402–16.

Tippit, D. H., Schulz, D. & Pickett-Heaps, J. D. (1978). Analysis of the distribution of spindle microtubules in the diatom *Fragilaria. J. Cell Biol.,* **79**, 737–63.

Tschermak-Woess, E. (1953). Über auffallende Strukturen in den Pyrenoiden einiger Naviculoideen. *Österr. Bot. Z.,* **100**, 160–78.

Tschermak-Woess, E. (1973). Über die bisher vergeblich gesuchte Auxosporenbildung von *Diatoma. Österr. Bot. Z.,* **121**, 23–7.

Utermöhl, H. (1958). Zur Vervollkommung der quantitätiven Phytoplankton-Methodik. *Mitt. Internat. Verein. Limnol.,* **9**, 1–38.

VanLandingham, S. L. (1967–79). *Catalogue of the fossil and recent genera and species of diatoms and their synonyms.* 8 vols. Vaduz: J. Cramer. 4654 pp.

Venkateswarlu, N. & Round, F. E. (1973). Observations on *Aulacodiscus amherstia* sp. nov. from the Bay of Bengal. *Br. phycol. J.* **8**, 163–73.

Voigt, M. (1943). Sur certaines irregularitées dans la structure des diatomées. *Notes Bot. Chim.,* **4**.

Voigt, M. (1960a). *Falcula,* un nouveau genre de diatomées de la Méditerranée. *Rev. algol.,* **5**, 85–8.

Voigt, M. (1960b). A new diatom genus from East Asia. *J. Roy. Microsc. Soc.,* ser. 3, **79**, 95–6.

Volcani, B. E. (1978). Role of silicon in diatom metabolism and silicification. In *Biochemistry of Silicon and Related Problems* (ed. G. Bendz & I. Lindqvist), pp. 177–204. New York: Plenum Press.

Volcani, B. E. (1981). Cell wall formation in diatoms: morphogenesis and biochemistry. In *Silicon and Siliceous Structures in Biological Systems* (ed. T. L. Simpson & B. E. Volcani), pp. 157–200. New York: Springer-Verlag.

Wahrer, R. J., Fryxell, G. A. & Cox, E. R. (1985). Studies in pennate diatoms; valve morphologies of *Licmophora* and *Campylostylus. J. Phycol.,* **21**, 206–17.

Walker, R. (1978). Diatom and pollen studies of a sediment profile from Melynllyn, a mountain tarn in Snowdonia, North Wales. *New Phytol.,* **81**, 791–804.

Walsby, A. E. (1988). Buoyancy in relation to the ecology

of the freshwater phytoplankton. In *Algae and the Aquatic Environment* (ed. F. E. Round), pp. 125–37. Bristol: Biopress.

Walsby, A. E. & Reynolds, C. S. (1980). Sinking and floating. In *The Physiological Ecology of Phytoplankton* (ed. I. Morris), pp. 371–412. Oxford: Blackwells.

Walsby, A. E. & Xypolyta, A. (1977). The form resistance of chitan fibres attached to the cells of *Thalassiosira fluviatilis* Hustedt. *Br. phycol. J.*, **12**, 215–23.

Waterkeyn, L. & Bienfait, A. (1987). Localisation et rôle des β-1,3-glucanes (callose et chrysolaminarine) dans le genre *Pinnularia* (Diatomées). *La Cellule*, **74**, 198–226.

Watkins, T. P. & Fryxell, G. A. (1986). Generic characterization of *Actinocyclus*: Consideration in light of three new species. *Diat. Res.*, **1**, 291–312.

Werner, D. (1971a). Der Entwicklungscyclus mit Sexualphase bei der marinen Diatomee *Coscinodiscus asteromphalus* I. Kultur und Synchronisation von Entwicklungstadien. *Arch. Mikrobiol.*, **80**, 43–9.

Werner, D. (1971b). Der Entwicklungscyclus mit Sexualphase bei der marinen Diatomee *Coscinodiscus asteromphalus* II. Oberflächenabhängige Differenzierung während der vegetativen Zellverkleinerung. *Arch. Mikrobiol.*, **80**, 115–33.

Werner, D. (1971c). Der Entwicklungscyclus mit Sexualphase bei der marinen Diatomee *Coscinodiscus asteromphalus* III. Differenzierung und Spermatogenese. *Arch. Mikrobiol.*, **80**, 134–46.

Werner, D. (1978). Regulation of metabolism by silicate in diatoms. In *Biochemistry of Silicon and Related Problems* (ed. G. Bendz & I. Lindqvist), pp. 149–76. New York: Plenum Press.

Wesenberg-Lund, C. (1908). *Plankton investigations of the Danish lakes. General Part: the Baltic freshwater plankton, its origin and variation.* Copenhagen: Gyldendalske Boghandel. 389 pp.

Whatley, J. M. & Whatley, F. R. (1981). Chloroplast evolution. *New Phytol.*, **87**, 233–47.

Wiedling, S. (1948). Beiträge zur Kenntnis der vegetativen Vermehrung der Diatomeen. *Bot. Notiser*, **1948**, 322–54..

Williams, D. M. (1985). Morphology, taxonomy and inter-relationships of the ribbed araphid diatoms from the genera *Diatoma* and *Meridion* (Diatomaceae: Bacillariophyta). *Bibliotheca Diatomologica*, **8**, 5–228.

Williams, D. M. (1986). Comparative morphology of some species of *Synedra* with a new definition of the genus. *Diat. Res.*, **1**, 131–52.

Williams, D. M. (1987). Observations on the genus *Tetracyclus* (Bacillariophyta). I. Valve and girdle structure of extant species. *Br. phycol. J.*, **22**, 383–99.

Williams, D. M. (1988). *Tabulariopsis*, a new genus of marine araphid diatom, with notes on the taxonomy of *Tabularia* (Kütz.) Williams et Round. *Nova Hedwigia*, **47**, 247–54.

Williams, D. M. & Round, F. E. (1986). Revision of the genus *Synedra* Ehrenb. *Diat. Res.*, **1**, 313–39.

Williams, D. M. & Round, F. E. (1987). Revision of the genus *Fragilaria*. *Diat. Res.*, **2**, 267–88.

Williams, D. M. & Round, F. E. (1988). Phylogenetic systematics of *Synedra*. In *Proceedings of the 9th International Diatom Symposium* (ed. F. E. Round), pp. 303–15. Bristol: Biopress & Koenigstein: Koeltz.

Wood, A. M., Lande, R. & Fryxell, G. A. (1987). Quantitative genetic analysis of morphological variation in an Antarctic diatom grown at two light intensities. *J. Phycol.*, **23**, 42–54.

Wood, E. J. F. (1959). An unusual diatom from the Antarctic. *Nature*, **184**, 1962–3.

Taxonomic index

Full accounts of the genera in the atlas are indicated by **bold** numbers.
The classification of the genera is given on pp. 125–9.

Subject index

Printed in the United Kingdom
by Lightning Source UK Ltd.
125361UK00001B/45-48/A